Hans-Joachim Hübschmann

Handbook of GC/MS

Related Titles

M. McMaster

GC/MS

A Practical User's Guide

2008

ISBN 978-0-470-10163-6

J. T. Watson, O. D. Sparkman

Introduction to Mass Spectrometry

Instrumentation, Applications, and Strategies for Data Interpretation

2007

ISBN: 978-0-470-51634-8

W. R. Külpmann

Clinical Toxicological Analysis

Procedures, Results, Interpretation

2 Volumes

2009

ISBN: 978-3-527-31890-2

I. Eidhammer, K. Flikka, L. Martens, S.-O. Mikalsen

Computational Methods for Mass Spectrometry Proteomics

2007

ISBN: 978-0-470-51297-5

H.H. Maurer, K. Pfleger, A. Weber

Mass Spectral and GC Data of Drugs, Poisons, Pesticides, Pollutants and Their Metabolites

2 Volumes

2007

ISBN 978-3-527-31538-3

P. Rösner, T. Junge, F. Westphal, G. Fritschi

Mass Spectra of Designer Drugs

Including Drugs, Chemical Warfare Agents, and Precursors

2 Volumes

2007

ISBN: 978-3-527-30798-2

Hans-Joachim Hübschmann

Handbook of GC/MS

Fundamentals and Applications

Second, Completely Revised and Updated Edition

WILEY-VCH Verlag GmbH & Co. KGaA

The Autor

Dr. Hans-Joachim Hübschmann
Thermo Fisher Scientific
Advanced Mass Spectometry
Hanna-Kunath-Strasse 11
28199 Bremen
Germany

Library of Congress Card No.: applied for

British Library Cataloguing-in-Publication Data:
A catalogue record for this book is available from the British Library

Bibliographic information published by the Deutsche Nationalbibliothek
The Deutsche Nationalbibliothek lists this publication in the Deutsche Nationalbibliografie; detailed bibliographic data are available in the Internet at http://dnb.d-nb.de.

Printed in the Federal Republic of Germany
Printed on acid-free paper

Composition ProSatz Unger, Weinheim
Printing Strauss GmbH, Mörlenbach
Bookbinding Litges & Dopf GmbH, Heppenheim

ISBN: 978-3-527-31427-0

Dedicated to my wife Gudrun
and my children Maren, Colja, Jessica and Sebastian

Foreword

It is an excellent move that you look into this book!

Analytical chemists want to be efficient and rapid: we are interested in a given task and the results should be available the next morning. This suggests taking the simplest route: "inject and see", there is no time to fiddle about technology! The vendor of the possibly expensive instrumentation might have highlighted the simplicity of his apparatus.

This is a fundamental error. Efficient analysis presupposes a significant amount of time being devoted to understanding the method and the instrumentation. Not doing this in the beginning all too often exacts a high price at a later stage, e.g. in terms of a laborious and awkward method, endless troubleshooting and poor results.

Knowledge of the technology is a prerequisite to make the best choices for a straight and simple method – from sample preparation to injection, chromatographic resolution and detection. If we are honest, we know that a staggering amount of our time is lost to troubleshooting, and unless we have a deep insight into the technology, this troubleshooting is likely to be frustrating and ineffective (problems tend to recur). Hence investing time into understanding the technology is a wise investment for rapid (and reliable) analysis.

Additionally, efficient analysts devote a substantial part of their time to keeping up with technology in order to keep their horizons open: we cannot always anticipate what might be useful tomorrow, and a brilliant alternative may not come to mind if one were not acquainted with the possibility beforehand. To investigate technology only in the context of a given, possibly urgent task is shortsighted. Admittedly, it takes discipline to absorb technical information when the current necessity may not be immediately apparent. However, it pays back many times. It may also be difficult to convince a boss that the investment into reading basic texts and experimenting with puzzling phenomena is essential to be an efficient analyst – unless he was an analyst himself and knows firsthand the demanding nature of analytical chemistry!

It is great that an old hand in the field like Hans-Joachim Hübschmann took his time to bring the present knowledge into such a concise and readable form.

Continue reading!

Fehraltorf, Switzerland
May 2008

Koni Grob

Handbook of GC/MS: Fundamentals and Applications, Second Edition. Hans-Joachim Hübschmann

ISBN: 978-3-527-31427-0

Preface to the Second Edition

Mass spectrometers identify and quantify molecules by the direct detection of the ionized species. This is in contrast to many other analytical methods that measure the interaction with a molecule e.g., magnetic resonance or UV extinction. The unbiased, highly selective detection of either an accurate mass, or structural fragmentation reactions, makes MS today, more than ever, an indispensable analytical tool to achieve highest accuracy and ultimate compound confirmation. Mass spectrometry in hyphenation with gas or liquid chromatography has become the success story in analytical instrumentation, covering a never expected wealth of applications, from daily routine quality control, to confirmatory analysis with legal impact.

Chromatography, in this context, is often not at the top of the list when discussing GC/MS technologies, but has received increased attention through its role as the technology driver towards new and further extended GC/MS applications. Emerging and newly developed sampling technologies have found increased use in routine applications such as instrumental online cleanup strategies, large volume injection techniques, and the strong bias to increased speed of chromatographic separations. The common endeavour of many new trends is speed of analysis, especially in the quest for a reduced sample cleanup to allow higher throughput at a lower cost of analysis. Clear evidence of the current vitality index in chromatography is the increased participation and high number of contributions at international and local analytical conferences with presentations on well-prepared solutions covering a large diversity of application areas.

Obviously, the pendulum is swinging back from an “everything is possible” LC/MS approach towards GC/MS for proven solutions. This is not for sensitivity reasons but because of the practical approach providing a very general electron ionisation technique compared to the often experienced ion suppression effects known from electrospray LC/MS ion generation. The increased requirement for target compound analysis in trace analysis with legal implications further consolidates the vital role of GC/MS for the analysis of volatile and semi-volatile compounds, as this is the typical situation, e.g., in food safety and doping applications.

Selectivity is key. Sufficient sensitivity for standard and clean samples is a technical minimum requirement and is not the critical issue for employing GC/MS instrumentation any more. Reliable quantitation in complex matrix samples at the lowest limits, and certainly the compliance to international regulations, is driving methodologies forward. Due to the increased requirement for multi-component trace determinations in critical matrices, and the high cost for manual sample preparation, the high target compound selectivity of the mass

Handbook of GC/MS: Fundamentals and Applications, Second Edition. Hans-Joachim Hübschmann

ISBN: 978-3-527-31427-0

spectrometer is increasingly required. In this context, instrumental off-line and even more on-line sample preparation using pressurizes liquid extractions, and online LC-GC pre-separations or solid phase extractions, have become a major trend that is expected to grow further. Highly efficient ionization and selective analyzer technologies, including MS/MS and accurate mass capabilities will advance GC/MS into even higher integrated sample preparation solutions.

GC/MS has expanded rapidly into new areas of application, not leaving development in the known traditional use aside. Environmental analysis has become important as never before, partly due to the implementation of the UN Stockholm Convention Program on persistent organic pollutants. Forensic and toxicological analysis covering drug screening, tracing of drugs and explosives and general unknown analysis, petrochemical applications with the task of crude oil maturity analysis for new exploration sites, and the pharmaceutical applications for quality control, counterfeit and the investigation of natural products, metabolism and kinetics are still challenging applications.

Fairly new challenges arise from the widespread tasks in homeland security to quickly identify chemicals hazardous to human health and the environment, e.g., with the large number of pesticides or toxins as ricin. For food safety assurance GC/MS and LC/MS became the most widely applied analytical techniques for trace and residue analysis. The global trade of food and feed together with the increased public awareness of food safety issues combined with a global brand recognition, generated a primary focus on regulatory compliance testing and law enforcement as a global analytical challenge, not only for GC/MS.

The second English edition of the Handbook of GC/MS accommodates the new trends in GC/MS with a significant revision and extension covering emerging new techniques and referencing recent leading applications. With regard to sample preparation, new pressurized fluid extraction and online solid phase solutions have been added. New separation strategies with fast GC, multidimensional gas chromatography and column switching are covered both in the fundamental section as well as featuring important applications. The section mass spectrometry has been expanded with a focus on increased and high resolution and accurate mass analyser techniques, including time-of-flight and accurate mass quantifications using isotope dilution and lock mass techniques.

The applications section of the Handbook received a major revision. A number of new leading applications with a special focus on widely employed environmental, forensic and food safety examples including isotope ratio mass spectrometry monitoring are discussed. Special focus was put on multi-component analysis methods for pesticides using fast GC and highly selective MS/MS methods. A fast GC application using high resolution GC/MS for the European priority polyaromatic hydrocabons is referenced.

The strengths of automated and on-line SPE-GC/MS method are featured for contaminants from water using multidimensional GC. Other new SPME applications are demonstrated with the determination of polar aromatic amines and PBBs. Another focal point with the presentations of new key applications is the analysis of dioxins, PCBs and brominated flame retardants PBDEs with examples of the congener specific analysis of technical mixtures, the application of fast GC methods and the isotope dilution quantitation for confirmatory analysis.

The identification and quantitation of toxins with the analysis of trichothecenes and other mycotoxins is covering as well such poisoning cases with the highly poisonous toxin ricin, that became of highest public interest due to several recently reported incidents. An exciting extension of GC/MS to high boiling and polymer substances by analytical pyrolysis is described by the analysis of glycol and derivatives, the characterization of natural waxes and the quantitative pyrolysis polymers.

This expanded and even more comprehensive compilation of up-to-date technical GC/MS fundamentals, operational know-how and shaping practical application work could not have been accomplished without the great support of many specialists and practising experts in this field. Sincere thanks for valuable discussion and provision of data and recent publications for review go to Jan Blomberg (Shell International Chemicals B.V., Amsterdam, The Netherlands), William Christie (The Scottish Crop Research Institute SCRI, Invergowrie, Dundee, Scotland), Inge de Dobeleer (Interscience B.V., Breda, Netherlands), Werner Engewald (Leipzig University, Institute of Analytical Chemistry, Leipzig, Germany), Konrad Grob (Kantonales Labor Zürich, Switzerland), Thomas Läubli (Brechbühler AG, Schlieren, Switzerland), Hans-Ulrich Melchert (Robert Koch Institute, Berlin, Germany), Frank Theobald (Environmental Consulting, Cologne, Germany), Nobuyoshi Yamashita (National Institute of Advanced Industrial Science and Technology AIST, Tsukuba, Japan). For the generous support with the permission to use current application material I also would like to thank Peter Dawes (SGE, Victoria, Australia) and Wolfgang John (Dionex GmbH, Idstein, Germany).

The helpful criticism and valuable contributions of many of my associates at Thermo Fisher Scientific in Austin, Bremen, Milan and San Jose notably Andrea Cadoppi, Daniela Cavagnino, Meredith Conoley, Dipankar Ghosh, Brody Guggenberger, Joachim Gummersbach, Andreas Hilkert, Dieter Juchelka, Dirk Krumwiede, Fausto Munari, Scott T. Quarmby, Reinhold Pesch, Harry Richie, Trisa C. Robarge and Giacinto Zilioli is gratefully acknowledged. Their experience in well-versed applications and critical technical discussions always provided a stimulating impact on this project.

It is my pleasure to thank the many colleagues and careful readers of the first issues whose kind comments and encouragement have aided me greatly in compiling this new revised 2nd edition of the Handbook of GC/MS.

Sprockhövel, July 2008 *Hans-Joachim Hübschmann*

Despite all efforts, errors or misleading formulations may still exist. The author appreciates comments and reports on inaccuracies to allow corrections in future editions to the correspondence email address: Hans-Joachim.Huebschmann@Thermofisher.com

Preface to the First Edition

More than three years have elapsed since the original German publication of the Handbook of GC/MS. GC/MS instrument performance has significantly improved in these recent years. GC/MS methodology has found its sound place in many "classical" areas of application of which many application notes are reported as examples in this handbook. Today the use of mostly automated GC/MS instrumentation is standard. Furthermore GC/MS as a mature analytical technology with a broad range of robust instruments increasingly enters additional analytical areas and displaces the "classical" instrumentation.

The very positive reception of the original German print and the wide distribution of the handbook into different fields of application has shown that comprehensive information about functional basics as well as the discussion about the practical use for different applications is important for many users for efficient method development and optimization.

Without the support from interested users and the GC/MS community concerned, the advancement and actualisation of this handbook would not be possible. My special thanks go to the active readers for their contribution to valuable discussions and details. Many of the applications notes have been updated or replaced by the latest methodology.

I would like to express my personal thanks to Dr. Brody Guggenberger (ThermoQuest Corp., Austin, Texas), Joachim Gummersbach (ThermoQuest GmbH, Egelsbach), Gert-Peter Jahnke (ThermoQuest APG GmbH, Bremen), Prof. Dr. Ulrich Melchert (Robert-Koch-Institut, Berlin), Dr. Jens P. Weller (Institut für Rechtsmedizin der Medizinischen Hochschule, Direktor Prof. Dr. med. H. D. Tröger, Hannover), and Dr. John Ragsdale jr. (ThermoQuest Corp., Austin, Texas) for their valuable discussions and contributions with application documentation and data.

My sincere thanks to Dr. Elisabeth Grayson for the careful text translation.

I wish all users of this handbook an interesting and informative read. Comments and suggestions concerning further improvement of the handbook are very much appreciated.

Sprockhövel, August 2000 *Hans-Joachim Hübschmann*

Handbook of GC/MS: Fundamentals and Applications, Second Edition. Hans-Joachim Hübschmann

ISBN: 978-3-527-31427-0

Contents

Handbook of GC/MS: Fundamentals and Applications, Second Edition. Hans-Joachim Hübschmann

ISBN: 978-3-527-31427-0

1
Introduction

Detailed knowledge of the chemical processes in plants and animals and in our environment has only been made possible through the power of modern instrumental analysis. In an increasingly short time span more and more data are being collected. The absolute detection limits for organic substances lie in the attomole region and counting individual molecules per unit time has already become a reality. We are making measurements at the level of background contamination. Most samples subjected to chemical analysis are now mixtures, as are even blank samples. With the demand for decreasing detection limits, in the future effective sample preparation and separation procedures in association with highly selective detection techniques will be of critical importance for analysis. In addition the number of substances requiring detection is increasing and with the broadening possibilities for analysis, so is the number of samples. The increase in analytical sensitivity is exemplified in the case of 2,3,7,8-TCDD.

Year	Instrumental technique	Limit of detection [pg]
1967	GC/FID (packed column)	500
1973	GC/MS (quadrupole, packed column)	300
1976	GC/MS-SIM (magnetic instrument, capillary column)	200
1977	GC/MS (magnetic sector instrument)	5
1983	GC/HRMS (double focusing magnetic sector instrument)	0.15
1984	GC/MSD-SIM (quadrupole benchtop instrument)	2
1986	GC/HRMS (double focusing magnetic sector instrument)	0.025
1989	GC/HRMS (double focusing magnetic sector instrument)	0.010
1992	GC/HRMS (double focusing magnetic sector instrument)	0.005
2006	GCxGC/HRMS (using comprehensive GC)	0.0003

Capillary gas chromatography is today the most important analytical method in organic chemical analysis for the determination of individual substances in complex mixtures. Mass spectrometry as the detection method gives the most meaningful data, arising from the direct determination of the substance molecule or of fragments. The results of mass spectrometry are therefore used as a reference for other indirect processes and finally for confirmation of the facts. The complete integration of mass spectrometry and gas chromatography

Handbook of GC/MS: Fundamentals and Applications, Second Edition. Hans-Joachim Hübschmann

ISBN: 978-3-527-31427-0

into a single GC/MS system has shown itself to be synergistic in every respect. While at the beginning of the 1980s mass spectrometry was considered to be expensive, complicated and time-consuming or personnel-intensive, there is now hardly a GC laboratory which is not equipped with a GC/MS system. At the beginning of the 1990s mass spectrometry became more widely recognised and furthermore an indispensable detection procedure for gas chromatography. The simple construction, clear function and an operating procedure, which has become easy because of modern computer systems, have resulted in the fact that GC/MS is widely used alongside traditional spectroscopic methods. The universal detection technique together with high selectivity and very high sensitivity have made GC/MS important for a broad spectrum of applications. Benchtop GC/MS systems have completely replaced in many applications the stand-alone GC with selective detectors today. Out of a promising process for the expensive explanation of spectacular individual cases, a universally used analytical routine method has developed within a few years. The serious reservations of experienced spectroscopists wanting to keep mass spectrometry within the spectroscopic domain, have been found to be without substance because of the broad success of the coupling procedure. The control of the chromatographic procedure still contributes significantly to the exploitation of the analytical performance of the GC/MS system (or according to Konrad Grob: chromatography takes place in the column!). The analytical prediction capabilities of a GC/MS system are, however, dependent upon mastering the spectrometry. The evaluation and assessment of the data is leading to increasingly greater challenges with decreasing detection limits and the increasing number of compounds sought or found. At this point the circle goes back to the earlier reservations of renowned spectroscopists.

The high performance of gas chromatography lies in separation of the substance mixtures. With the introduction of fused silica columns GC has become the most important and powerful method of analysing complex mixtures of products. GC/MS accommodates the current trend towards multimethods or multicomponent analyses (e.g. of pesticides, solvents etc) in an ideal way. Even isomeric compounds, which are present, for example in terpene mixtures, in PCBs and in dioxins, are separated by GC, while in many cases their mass spectra are almost indistinguishable. The high efficiency as a routine process is achieved through the high speed of analysis and the short turn-round time and thus guarantees a high productivity with a high sample throughput. Adaptation and optimisation for different tasks only requires a quick change of column. In many cases, however, and here one is relying on the explanatory power of the mass spectrometer, one type of column can be used for different applications by adapting the sample injection technique and modifying the method parameters.

The area of application of GC and GC/MS is limited to substances which are volatile enough to be analysed by gas chromatography. The further development of column technology in recent years has been very important for application to the analysis of high-boiling compounds. Temperature-stable phases now allow elution temperatures of up to 500 °C. A pyrolyser in the form of a stand-alone sample injection system extends the area of application to involatile substances by separation and detection of thermal decomposition products. A typical example of current interest for GC/MS analysis of high-boiling compounds is the determination of polyaromatic hydrocarbons, which has become a routine process using the most modern column material. It is incomprehensible that, in spite of an obvious detection problem, HPLC is still frequently used in parallel to GC/MS to determine polyaromatic hydrocarbons in the same sample.

The coupling of gas chromatography with mass spectrometry using fused silica capillaries has played an important role in achieving a high level of chemical analysis. In particular in the areas of environmental analysis, analysis of residues and forensic science the high information content of GC/MS analyses has brought chemical analysis into focus through sometimes sensational results. For example, it has been used for the determination of anabolic steroids in cough mixture and the accumulation of pesticides in the food chain. With the current state of knowledge GC/MS is an important method for monitoring the introduction, the location and fate of man-made substances in the environment, foodstuffs, chemical processes and biochemical processes in the human body. GC/MS has also made its contribution in areas such as the ozone problem, the safeguarding of quality standards in foodstuffs production, in the study of the metabolism of pharmaceuticals or plant protection agents or in the investigation of polychlorinated dioxins and furans produced in certain chemical processes, to name but a few.

The technical realisation of GC/MS coupling occupies a very special position in instrumental analysis. Fused silica columns are easy to handle, can be changed rapidly and are available in many high quality forms. The optimised carrier gas streams show good compatibility with mass spectrometers. Coupling can therefore take place easily by directly connecting the GC column to the ion source of the mass spectrometer. The operation of the GC/MS instrument can be realised because of the low carrier gas flow in the widely used benchtop instruments even with a low pumping capacity. Only small instruments are therefore necessary, and these also accommodate a low pumping capacity. A general knowledge of the construction and stable operating conditions forms the basis of smooth and easily learned service and maintenance. Compared with GC/MS coupling, LC/MS coupling, for example, is still much more difficult to control, not to mention the possible ion surpression by matrix effects.

The obvious challenges of GC and GC/MS lie where actual samples contain involatile components (matrix). In this case the sample must be processed before the analysis appropriately. The clean-up is generally associated with enrichment of trace components. In many methods there is a trend towards integrating sample preparation and enrichment in a single instrument. Even today the headspace and purge and trap techniques, thermodesorption, SPME (solid phase microextraction) or SFE (supercritical fluid extraction) are coupled on-line with GC/MS and got further miniaturized and integrated stepwise into the data system for smooth control. Development will continue in this area in future, and as a result will move the focus from the previously expensive mass spectrometer to the highest possible sample throughput and will convert positive substance detection in the mass spectrometer into an automatically performed evaluation.

The high information content of GC/MS analyses requires powerful computers with intelligent programs to evaluate them. The evaluation of GC/MS analyses based on data systems is therefore a necessary integral component of modern GC/MS systems. Only when the evaluation of mass spectrometric and chromatographic data can be processed together can the performance of the coupling process be exploited to a maximum by the data systems. In spite of the state of the art computer systems, the performance level of many GC/MS data systems has remained at the state it was 20 years ago and only offers the user a coloured data print-out. The possibilities for information processing have remained neglected on the part of the manufacturers and often still require the use of external programs (e.g. the characterisation of specimen samples, analysis of mixtures, suppressing noise etc).

Nonetheless development of software systems has had a considerable effect on the expansion of GC/MS systems. The manual evaluation of GC/MS analyses has become practically impossible because of the enormous quantity of data. A 60-minute analysis with two spectra per second over a mass range of 500 mass units gives 3.65 million pairs of numbers! The use of good value but powerful PCs allows the systems to be controlled but gives rapid processing of the relevant data and thus makes the use of GC/MS systems economically viable.

The Historical Development of the GC/MS Technique

The GC/MS technique is a recent process. The foundation work in both GC and MS which led to the current realisation was only published between the middle and the end of the 1950s. At the end of the 1970s and the beginning of the 1980s a rapid increase in the use of GC/MS in all areas of organic analysis began. The instrumental technique has now achieved the required level for the once specialised process to become an indispensable routine procedure.

1910 The physicist J.J. Thompson developed the first mass spectrometer and proved for the first time the existence of isotopes (^{20}Ne and ^{22}Ne). He wrote in his book 'Rays of Positive Electricity and their Application to Chemical Analysis': '*I have described at some length the application of positive rays to chemical analysis: one of the main reasons for writing this book was the hope that it might induce others, and especially chemists, to try this method of analysis. I feel sure that there are many problems in chemistry which could be solved with far greater ease by this than any other method*'. Cambridge 1913. In fact, Thompson developed the first isotope ratio mass spectrometer (IRMS).

1910 In the same year M.S. Tswett published his book in Warsaw on 'Chromophores in the Plant and Animal World'. With this he may be considered to be the discoverer of chromatography.

1918 Dempster used electron impact ionisation for the first time.

1920 Aston continued the work of Thompson with his own mass spectrometer equipped with a photoplate as detector. The results verified the existence of isotopes of stable elements (e.g. ^{35}Cl and ^{37}Cl) and confirmed the results of Thompson.

1929 Bartky and Dempster developed the theory for a double-focusing mass spectrometer with electrostat and magnetic sector.

1934 Mattauch and Herzog published the calculations for an ion optics system with perfect focusing over the whole length of a photoplate.

1935 Dempster published the latest elements to be measured by MS, Pt and Ir. Aston thus regarded MS to have come to the end of its development.

1936 Bainbridge and Jordan determined the mass of nuclides to six significant figures, the first accurate mass application.

1937 Smith determined the ionisation potential of methane (as the first organic molecule).

1938 Hustrulid published the first spectrum of benzene.

1941 Martin and Synge published a paper on the principle of gas liquid chromatography, GLC.

1946 Stephens proposed a time of flight (TOF) mass spectrometer: velocitron.

1947 The US National Bureau of standards (NBS) began the collection of mass spectra as a result of the use of MS in the petroleum industry.

1948 Hipple described the ion cyclotron principle, known as the 'Omegatron' which now forms the basis of the current ICR instruments.

1950 Gohlke published for the first time the coupling of a gas chromatograph (packed column) with a mass spectrometer (Bendix TOF, time of flight).

1950 The Nobel Prize for chemistry was awarded to Martin and Synge for their work on gas liquid chromatography (1941).

1950 From McLafferty, Biemann and Beynon applied MS to organic substances (natural products) and transferred the principles of organic chemical reactions to the formation of mass spectra.

1952 Cremer and coworkers presented an experimental gas chromatograph to the ACHEMA in Frankfurt; parallel work was carried out by Janák in Czechoslovakia.

1952 Martin and James published the first applications of gas liquid chromatography.

1953 Johnson and Nier published an ion optic with a 90° electric and 60° magnetic sector, which, because of the outstanding focusing properties, was to become the basis for many high resolution organic mass spectrometers (Nier/Johnson analyser).

1954 Paul published his fundamental work on the quadrupole analyser.

1955 Wiley and McLaren developed a prototype of the present time of flight (TOF) mass spectrometer.

1955 Desty presented the first GC of the present construction type with a syringe injector and thermal conductivity detector. The first commercial instruments were supplied by Burrell Corp., Perkin Elmer, and Podbielniak Corp.

1956 A German patent was granted for the QUISTOR (quadrupole ion storage device) together with the quadrupole mass spectrometer.

1958 Paul published information on the quadrupole mass filter as
- a filter for individual ions,
- a scanning device for the production of mass spectra,
- a filter for the exclusion of individual ions.

1958 Ken Shoulders manufactured the first 12 quadrupole mass spectrometers at Stanford Research Institute, California.

1958 Golay reported for the first time on the use of open tubular columns for gas chromatography.

1958 Lovelock developed the argon ionisation detector as a forerunner of the electron capture detector (ECD, Lovelock and Lipsky).

1962 U. von Zahn designed the first hyperbolic quadrupole mass filter.

1964 The first commercial quadrupole mass spectrometers were developed as residual gas analysers (Quad 200 RGA) by Bob Finnigan and P.M. Uthe at EAI (Electronic Associates Inc., Paolo Alto, California).

1966 Munson and Field published the principle of chemical ionisation.

1968 The first commercial quadrupole GC/MS system for organic analysis was supplied by Finnigan Instruments Corporation to the Stanford Medical School Genetics Department.

1978 Dandenau and Zerenner introduced the technique of fused silica capillary columns.

1978 Yost and Enke introduced the triple-quadrupole technique.

1982 Finnigan obtained the first patents on ion trap technology for the mode of selective mass instability and presented the ion trap detector as the first universal MS detector with a PC data system (IBM XT).

1989 Prof. Wolfgang Paul, Bonn University received the Nobel Prize for physics for work on ion traps, together with Prof. Hans G. Dehmelt, University of Washington in Seattle, and Prof. Norman F. Ramsay, Harvard University.

2000 A. Makarov published a completely new mass analyzer concept called "Orbitrap" suitable for accurate mass measurements of low ion beams.

2005 Introduction of a new type of hybrid Orbitrap mass spectrometer by Thermo Electron Corporation, Bremen, Germany, for MS/MS and very high resolution and accurate mass measurement on the chromatographic time scale.

2
Fundamentals

2.1
Sample Preparation

The preparation of analysis samples is today already an integral part of practical GC/MS analysis. The current trend is clearly directed to automated instrumental techniques and limits manual work to the essential. The concentration processes in this development are of particular importance for coupling with capillary GC/MS, as in trace analysis the limited sample capacity of capillary columns must be compensated for. It is therefore necessary both that overloading of the stationary phase by the matrix is avoided and that the limits of mass spectrometric detection are taken into consideration. To optimise separation on a capillary column, strongly interfering components of the matrix must be removed before applying an extract. The primarily universal character of the mass spectrometer poses conditions on the preparation of a sample which are to some extent more demanding than those of an element-specific detector, such as ECD or NPD unless highly selective techniques as MS/MS or high resolution accurate measurements are applied. The clean-up and analyte concentration, which forms part of sample preparation, must therefore in principle always be regarded as a necessary preparative step for GC/MS analysis. The differences in the concentration ranges between various samples, differences between the volatility of the analytes and that of the matrix and the varying chemical nature of the substances are important for the choice of a suitable sample preparation procedure.

Off-line techniques (as opposed to on-line coupling or hyphenated techniques) have the particular advantage that samples can be processed in parallel and the extracts can be subjected to other analytical processes besides GC/MS. On-line techniques have the special advantage of sequential processing of the samples without intermediate manual steps. The on-line clean-up allows an optimal time overlap which gives the sample preparation the same amount of time as the analysis of the preceding sample. This permits maximum use of the instrument and automatic operation.

On-line processes generally offer potential for higher analytical quality through lower contamination from the laboratory environment and, for smaller sample sizes, lower detection limits with lower material losses. Frequently total sample transfer is possible without taking aliquots or diluting. Volatility differences between the sample and the matrix allow, for example, the use of extraction techniques such as the static or dynamic (purge and trap) headspace techniques as typical GC/MS coupling techniques. These are already used as on-line techniques in many laboratories. Where the volatility of the analytes is insufficient, other

Handbook of GC/MS: Fundamentals and Applications, Second Edition. Hans-Joachim Hübschmann

ISBN: 978-3-527-31427-0

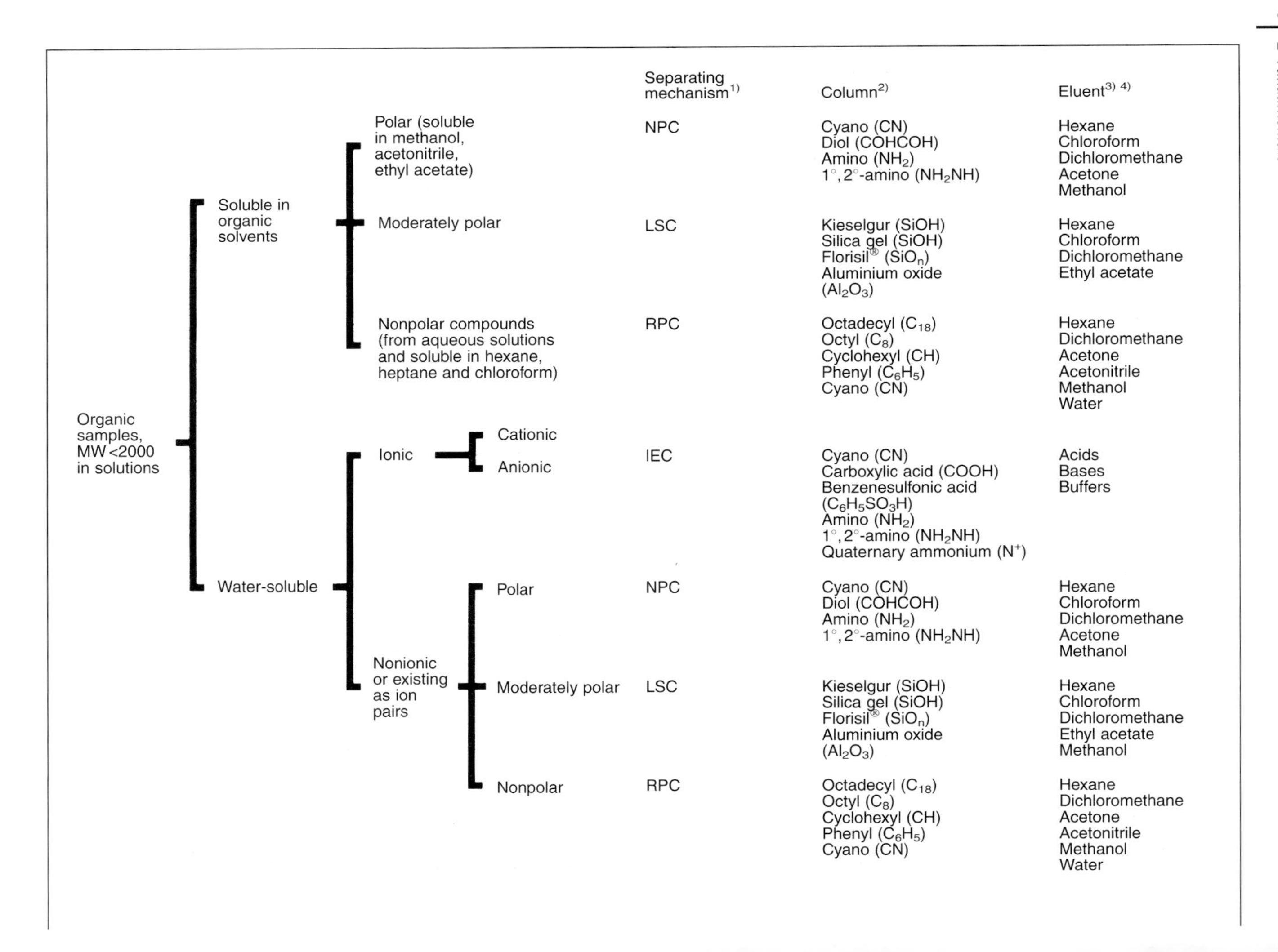
Separating mechanism[1)]
Column[2)]
Eluent[3) 4)]
Organic samples, MW <2000 in solutions
Soluble in organic solvents
Polar (soluble in methanol, acetonitrile, ethyl acetate)
NPC
Cyano (CN) Diol (COHCOH) Amino (NH_2) 1°, 2°-amino (NH_2NH)
Hexane Chloroform Dichloromethane Acetone Methanol
Moderately polar
LSC
Kieselgur (SiOH) Silica gel (SiOH) Florisil® (SiO_n) Aluminium oxide (Al_2O_3)
Hexane Chloroform Dichloromethane Ethyl acetate
Nonpolar compounds (from aqueous solutions and soluble in hexane, heptane and chloroform)
RPC
Octadecyl (C_{18}) Octyl (C_8) Cyclohexyl (CH) Phenyl (C_6H_5) Cyano (CN)
Hexane Dichloromethane Acetone Acetonitrile Methanol Water
Water-soluble
Ionic
Cationic
Anionic
IEC
Cyano (CN) Carboxylic acid (COOH) Benzenesulfonic acid ($C_6H_5SO_3H$) Amino (NH_2) 1°, 2°-amino (NH_2NH) Quaternary ammonium (N^+)
Acids Bases Buffers
Nonionic or existing as ion pairs
Polar
NPC
Cyano (CN) Diol (COHCOH) Amino (NH_2) 1°, 2°-amino (NH_2NH)
Hexane Chloroform Dichloromethane Acetone Methanol
Moderately polar
LSC
Kieselgur (SiOH) Silica gel (SiOH) Florisil® (SiO_n) Aluminium oxide (Al_2O_3)
Hexane Chloroform Dichloromethane Ethyl acetate Methanol
Nonpolar
RPC
Octadecyl (C_{18}) Octyl (C_8) Cyclohexyl (CH) Phenyl (C_6H_5) Cyano (CN)
Hexane Dichloromethane Acetone Acetonitrile Methanol Water

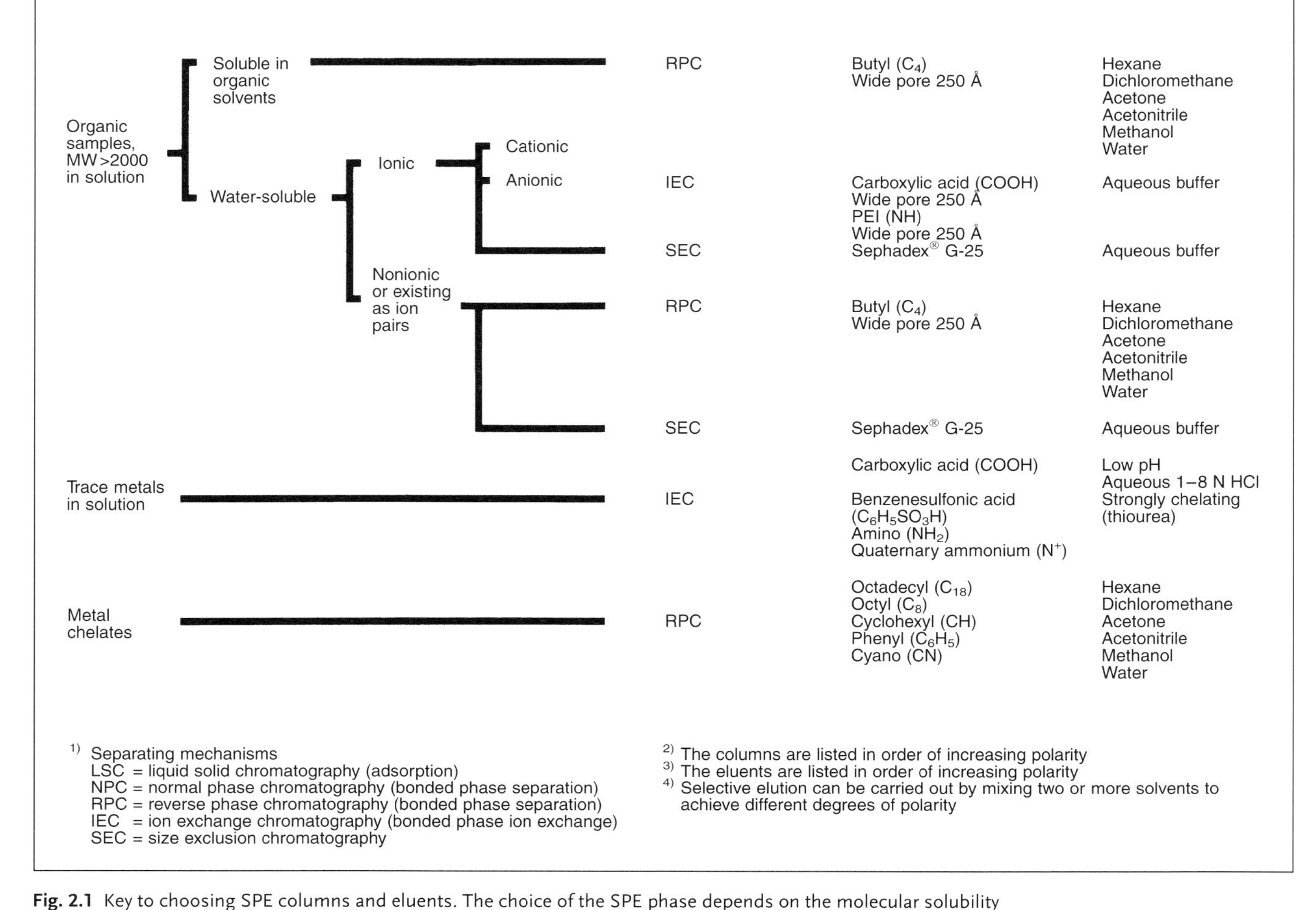

Fig. 2.1 Key to choosing SPE columns and eluents. The choice of the SPE phase depends on the molecular solubility of the sample in a particular medium and on its polarity. The sample matrix is not considered (J. T. Baker).

extraction procedures e.g. thermal extraction, pyrolysis or online SPE techniques are being increasingly used on-line. Solid phase extraction in the form of microextraction, LC/GC coupling, or extraction with supercritical fluids show high analytical potential here.

2.1.1
Solid Phase Extraction

From the middle of the 1980s solid phase extraction (SPE) began to revolutionise the enrichment, extraction and clean-up of analytical samples. Following the motto 'The separating funnel is a museum piece', the time-consuming and arduous liquid/liquid extraction has increasingly been displaced from the analytical laboratory. Today the euphoria of the rapid and simple preparation with disposable columns has lessened as a result of a realistic consideration of their performance levels and limitations. A particular advantage over the classical liquid/liquid partition is the low consumption of expensive and sometimes harmful solvents. The amount of apparatus and space required is low for SPE. Parallel processing of several samples is therefore quite possible. Besides an efficient clean-up, the necessary concentration of the analyte frequently required for GC/MS is achieved by solid phase extraction.

In solid phase extraction strong retention of the analyte is required, which prevents migration through the carrier bed during sample application and washing. Specific interactions between the substances being analysed and the chosen adsorption material are exploited to achieve retention of the analytes and removal of the matrix. An extract which is ready for analysis is obtained by changing the eluents. The extract can then be used directly for GC and GC/MS in most cases. The choice of column materials permits the exploitation of the separating mechanisms of adsorption chromatography, normal-phase and reversed-phase chromatography, and also ion exchange and size exclusion chromatography (Fig. 2.1).

The physical extraction process, which takes place between the liquid phase (the liquid sample containing the dissolved analytes) and the solid phase (the adsorption material) is common to all solid phase extractions. The analytes are usually extracted successfully because the interactions between them and the solid phase are stronger than those with the solvent or the matrix components. After the sample solution has been applied to the solid phase bed, the analytes become enriched on the surface of the SPE material. All other sample components pass unhindered through the bed and can be washed out. The maximum sample volume that can be applied is limited by the breakthrough volume of the analyte. Elution is achieved by changing the solvent. For this there must be a stronger interaction between the elution solvent and the analyte than between the latter and the solid phase. The elution volume should be as small as possible to prevent subsequent solvent evaporation.

In analytical practice two solid phase extraction processes have become established. Cartridges are mostly preferred for liquid samples (Figs. 2.2 and 2.3). If the GC/MS analysis reveals high contents of plasticisers, the plastic material of the packed columns must first be considered and in special cases a change to glass columns must be made. For sample preparation using slurries or turbid water, which rapidly lead to deposits on the packed columns, SPE disks should be used. Their use is similar to that of cartridges. Additional contamination, e.g. by plasticisers, can be ruled out for residue analysis in this case (Fig. 2.4).

A large number of different interactions are exploited for solid phase extraction (Fig. 2.2). Selective extractions can be achieved by a suitable choice of adsorption materials. If the eluate is used for GC/MS the detection characteristics of the mass spectrometer in particular

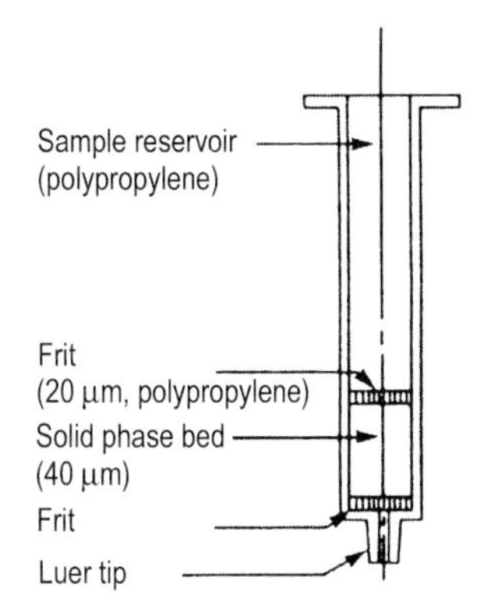

Fig. 2.2 Construction of a packed column for solid phase extraction (J. T. Baker).

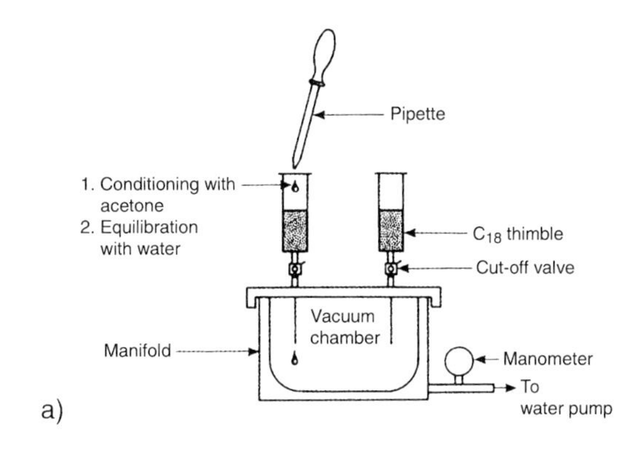

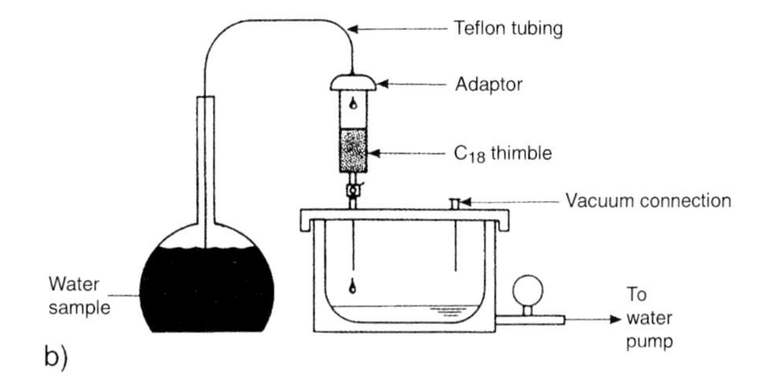

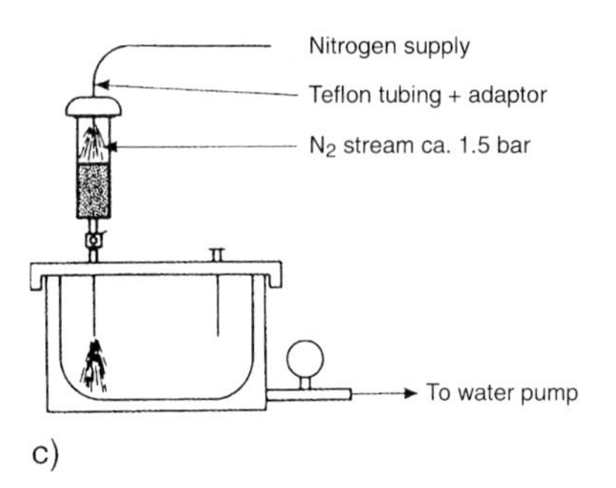

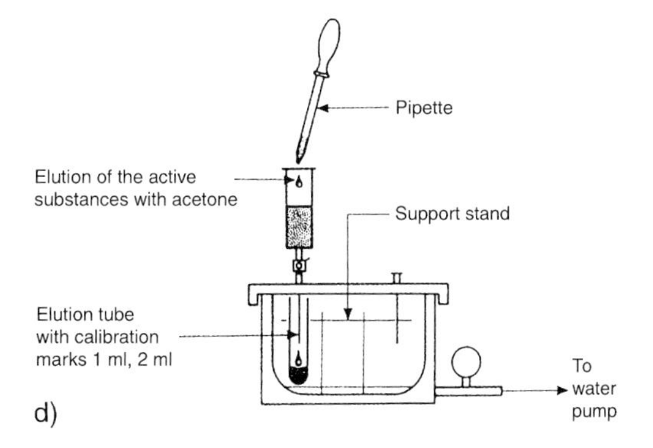

Fig. 2.3 Enrichment of a water sample on solid phase cartridges with C_{18}-material (Hein/Kunze 1994):
(a) conditioning,
(b) loading (extraction)/washing,
(c) drying,
(d) elution of active substances.

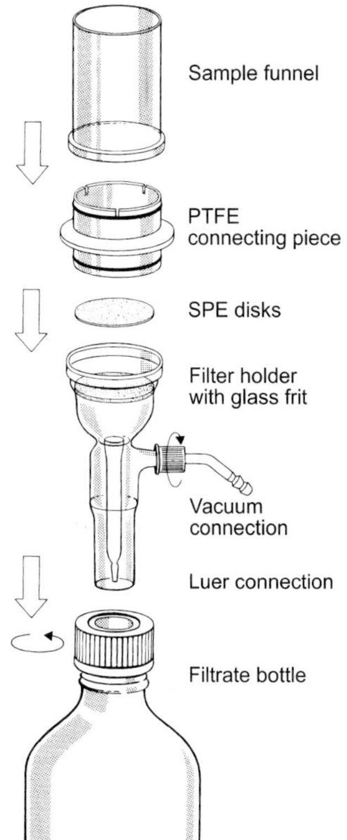

Fig. 2.4 Apparatus for solid phase extraction with SPE disks (J. T. Baker).

must be taken into account. Unlike an electron capture detector (ECD), which is still widely used as a selective detector for substances with a high halogen content in environmental analysis, a GC/MS system can be used to detect a wide range of nonhalogenated substances. The use of a selected ion monitoring technique (SIM, MID) therefore requires better purification of the extracts obtained by SPE. In the case of the processing of dioxins and PCBs from waste oil, for example, a silica gel column charged with sulfuric acid is also necessary. Extensive oxidation of the nonspecific hydrocarbon matrix is thus achieved. The quality and reproducibility of SPE depends on criteria comparable to those which apply to column materials used in HPLC.

2.1.1.1 **Solid Phase Microextraction**

The solvent-free extraction technique *solid phase microextraction* (SPME) recently developed by Prof. J. Pawliszyn (University of Waterloo, Ontario, Canada) is an important step towards the instrumentation and automation of the solid phase extraction technique for on-line sample preparation and introduction to GC/MS. It involves exposing a fused silica fibre coated with a liquid polymeric material to a sample containing the analyte. The typical dimensions of a fibre are 1 cm × 100 μm. The analyte diffuses from gaseous or liquid samples into the fibre surface and partitions into the coating according to the first partition coefficient. The

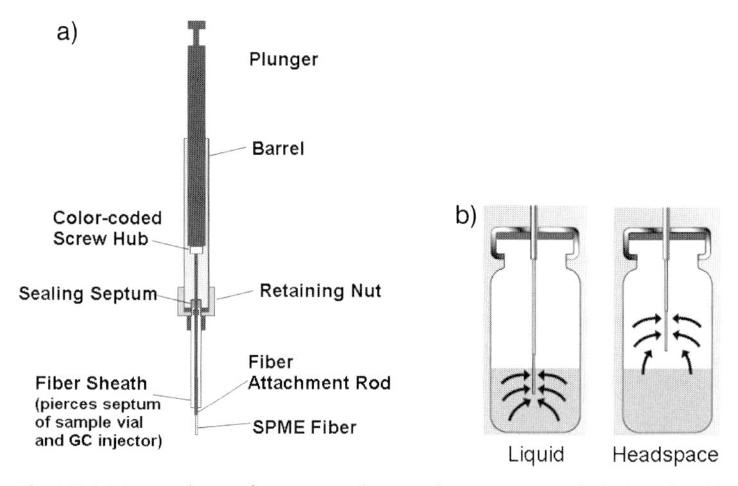

Fig. 2.5 (a) SPME plunger for autosampler use; (b) SPME principle for liquid and headspace application.

agitated sample can be heated during the sampling process if desired to achieve maximum recovery and precision for quantitative assays. After equilibrium is established, the fibre with the analyte is withdrawn from the sample and transferred into a GC injector system manually or via an autosampler. The analyte is desorbed thermally from the coating.

SPME offers several advantages for sample preparation including reduced time per sample, less manual sample manipulation resulting in an increased sample throughput and, in addition, the elimination of organic solvents and reduced analyte loss.

An SPME unit consists of a length of fused silica fibre coated with a phase similar to those used in chromatography columns. The phase can be mixed with solid adsorbents, e. g. divinylbenzene polymers, templated resins or porous carbons. The fibre is attached to a stainless steel plunger in a protective holder used manually or in a specially prepared autosampler. The plunger on the syringe retracts the fibre for storage and piercing septa, and exposes the fibre for extraction and desorption of the sample. A spring in the assembly keeps the fibre retracted, reducing the chance of it being damaged (Fig. 2.5 a and b).

The design of a portable, disposable SPME holder with a sealing mechanism gives flexibility and ease of use for on-site sampling. The sample can be extracted and stored by placing the tip of the fibre needle in a septum. The lightweight disposable holders can be used for the lifetime of the fibre. SPME could also be used as an indoor air sampling device for GC and GC/MS analysis. For this technique to be successful, the sample must be stable to storage.

A major shortcoming of SPME is the lack of fibres that are polar enough to extract very polar or ionic species from aqueous solutions without first changing the nature of the species through prior derivatisation. Ionic, polar and involatile species have to be derivatised to GC-amenable species before SPME extraction.

Since the introduction of SPME in 1993, a variety of applications have been established, including the analysis of volatile analytes and gases consisting of small molecules. However, SPME was limited in its ability to retain and concentrate these small molecules in the fibre coating. Equilibria were obtained rapidly and distribution constants were low, which resulted in high minimum detection limits. The thickness of the phase is important in capturing small molecules, but a stronger adsorbing mechanism is also needed. The ability to coat porous carbons on a fibre has enabled SPME to be used for the analysis of small molecules at trace levels.

The amount of analyte extracted is dependent upon the distribution constant. The higher the distribution constant, the higher is the quantity of analyte extracted. Generally a thicker film is required to retain small molecules and a thinner film is used for larger molecules with high distribution constants. The polarity of the fibre and the type of coating can also increase the distribution constant (Table 2.1).

Table 2.1 Types of SPME fibres currently available.

Nonpolar fibres

- Cellulose acetate/polyvinylchloride (PVC), alkane selective (Farajzadeh 2003)
- Powdered activated carbon (PAC), for BTEX and halocarbons (Shutao 2006)
- Polydimethylsiloxane (PDMS), 7, 30, 100 μm coating

Bipolar fibres

- Carboxen/Polydimethylsiloxane(CAR/PDMS), 75, 85 μm coating
- Divinylbenzene/Carboxen/Polydimethylsiloxane (DVB/CAR/PDMS), 30, 50 μm coating
- Divinylbenzene/Carboxen/Polydimethylsiloxane (DVB/CAR/PDMS), 30, 50 μm coating
- Polydimethylsiloxane/Divinylbenzene (PDMS/DVB), 65 μm coating

Polar fibres

- Carbowax/Divinylbenzene (CW/DVB), 70 μm coating
- Carbowax/Templated Resin (CW/TPR), 50 μm coating
- Polyacrylate (PA), 85 μm coating

Blended fibre coatings contain porous material, such as divinylbenzene (DVB) and Carboxen, blended in either PDMS or Carbowax.

Table 2.2 Summary of method performance for several analysis examples

Parameter	Formaldehyde	Triton-X-100	Phenylurea pesticides	Amphetamines
LOD	2–40 ppb	1.57 μg/L	< 5 μg/L	1.5 ng/mL
Precision	2–7%	2–15%	1.6–5.6%	9–15%
Linear range	50–3250 ppb	0.1–100 mg/L	10–10 000 μg/L	5–2000 ng/mL
Linearity (r_2)	0.9995	0.990	0.990	0.998
Sampling time	10–300 s	50 min	8 min	15 min

The principle of solid phase micro extraction recently found its extension in the stir bar sorptive extraction (SBSE). The extraction is performed with a glass coated magnetic stir bar which is coated with polydiemthylsiloxane (PDMS). The difference to SPME is the capacity of the sorptive PDMS phase. While in the SPME fibre a volume of about 15 μL is used, with SBSE a significantly enlarged volume of up to 125 μL is available. The larger volume of sorption phase provides a better phase ratio with increased recovery resulting in up to 250 fold lowered detection limits. Many samples can be extracted at the same time, typically extraction times run up to 60 min. Transfer to the GC is achieved by thermal desorption of the stir bar in a suitable injection system. Thermolabile compounds can be alternatively dissolved by liquids (SBSE/LD). The excellent sensitivity and reproducibility has been demonstrated in the multi-residue detection of endocrine disrupting chemicals in drinking water (Serodio 2004).

2.1.2 Supercritical Fluid Extraction

Supercritical fluid extraction (SFE) has the potential of replacing conventional methods of sample preparation involving liquid/liquid and liquid/solid extractions (the current soxhlet extraction steps) in many areas of application. The soxhlet extraction (Fig. 2.6) has, up to now, been the process of choice for extracting involatile organic compounds from solid matrices, such as soil, sewage sludge or other materials.

The prerequisite for the soxhlet extraction with organic solvents is a sample which is as anhydrous as possible. Time-consuming drying of the sample (freeze drying) is therefore necessary. The time required for a soxhlet extraction is, however, ca. 15–30 h for many samples in environmental analysis. This leads to the running of many extraction columns in parallel. The requirement for pure solvents (several hundred ml) is therefore high.

While the development of chromatographic separation methods has made enormous advances in recent years, equivalent developments in sample preparation, which keep up with

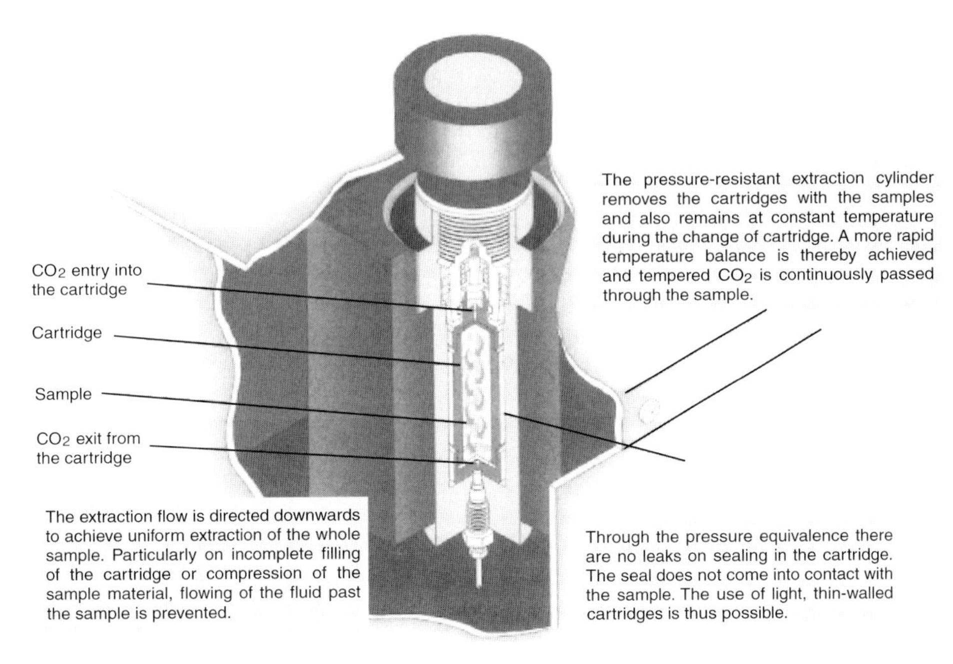

Fig. 2.6 SFE extraction system employing the principle of pressure equivalence (the cartridge with the sample is in a pressure-resistant cylinder, the internal and external pressures on the cartridge are kept equal, ISCO).

the possibilities of modern GC and GC/MS systems, have not been made. The technical process for extraction of natural products (e. g. plants/perfume oils, coffee/caffeine) with supercritical carbon dioxide has already been used for a long time and is particularly well known in the area of perfumery for its pure and high-value extracts.

Only in recent the years has the process achieved increased importance for instrumental analysis. SFE is a rapid and economical extraction process with high percentage recovery which gives sample extracts which can be used for residue analysis usually without further concentration or clean-up. SFE works within the cycle time of typical GC and GC/MS analyses, can be automated, and avoids the production of waste solvent in the laboratory. The EPA (US Environmental Protection Agency) is working on the replacement of extraction processes with dichloromethane and freons used up till now, with particular emphasis on environmental analysis. The first EPA method published was the determination of the total hydrocarbon content of soil using SFE sample preparation (EPA # 3560). The EPA processes for the determination of polyaromatic hydrocarbons (EPA # 3561) and of organochlorine pesticides and PCBs (EPA # 3562) using SFE extraction and detection with GC/MS systems soon followed. In the fundamental work of Lehotay and Eller on SFE pesticide extraction, carbamates, triazines, phosphoric acid esters and pyrethroids were all analysed using a GC/MS multimethod.

Supercritical fluids are extraction agents which are above their critical pressures and temperatures during the extraction phase. The fluids used for SFE (usually carbon dioxide) provide particularly favourable conditions for extraction (Fig. 2.7). The particular properties of a supercritical fluid arise from a combination of gaseous sample penetration, liquid-like solubilising capacity and substance transport (Table 2.3). The solubilising capacity of supercritical fluids reaches that of the liquid solvents when their density is raised. The maximum solubility of an organic compound can, nevertheless, be higher in a liquid solvent than in a supercritical phase. The solubility plays an important role in the efficiency of a technical process. However, in residue analysis this parameter is of no practical importance because of the extremely low concentration of the analytes.

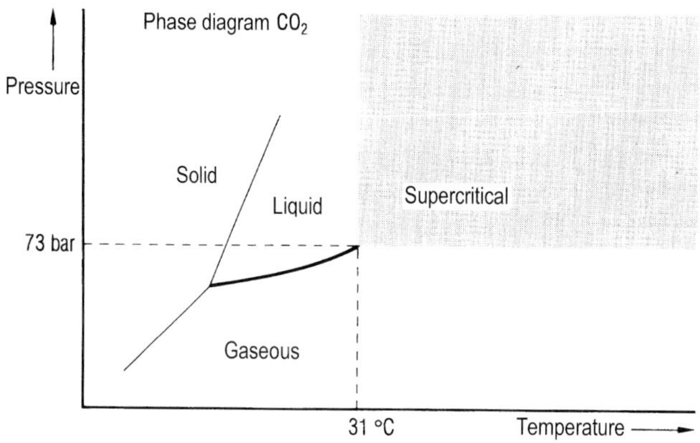

Fig. 2.7 Phase diagram.

Physical data for CO_2:		Critical data:	
Mp	−78.5 °C	Pressure	73.8 bar
Bp	−56.6 °C	Temperature	31.1 °C
Vapour pressure	57.3 bar at 20 °C	Density	0.464 g/L

Table 2.3 Comparison of the physical properties of different aggregate states (in orders of magnitude).

Phase	Density $[g \cdot cm^3]$	Diffusion $[cm^2 \cdot s^{-1}]$	Viscosity $[g \cdot cm^{-1} \cdot s^{-1}]$
Gas	10^{-3}	10^{-1}	10^{-4}
Supercritical fluid	0.1–1.0	10^{-3}–10^{-4}	10^{-3}–10^{-4}
Liquid	1	$<10^{-5}$	10^{-2}

The rate of extraction with a supercritical fluid is determined by the substance transport limits. Since supercritical fluids have diffusion coefficients an order of magnitude higher and viscosities an order of magnitude lower than those of liquid extraction agents, SFE extractions can be carried out in a much shorter time than, for example, the classical soxhlet extraction. Quantitative SFE extractions are typically finished in ca. 10–60 min, while normal solvent extractions for the same quantity of sample take several hours, often even overnight.

The solubilising capacity in an SFE step can easily be controlled via the density of the supercritical fluid, whereas the solubilising capacity of a liquid extraction agent is essentially constant. The density of the supercritical fluid is determined by the choice of pressure and temperature during the extraction. Raising the extraction temperature to ca. 60–80 °C also has a favourable effect on the swelling properties of the sample matrix. High-boiling polyaromatic hydrocarbons are even extracted from real life samples at temperatures above 150 °C. This makes it necessary to use pressures of above 500 bar to achieve maximum extraction.

At constant temperature low pressures favour the extraction of less polar analytes. The continuation of the extraction at higher pressures then elutes more polar and higher molecular weight substances. This allows the optimisation of the extraction for a certain class of compound by programming the change in pressure. In this way extracts are produced by SFE which are ready for analysis and generally require no further clean-up or concentration. The SFE extracts are often cleaner than those obtained by classical solvent extraction. Selective extractions and programmed optimisation are possible through the collection of extract fractions.

Carbon dioxide, which is currently used as the standard extraction agent in SFE, is gaseous under normal conditions and evaporates from the extracts after depressurising the supercritical fluid. Various techniques are used to obtain the extract. An alternative to simply passing the extract into open, unheated vessels involves freezing out the extracts in a cold trap. Here the restrictor is heated to ca. 150 °C and the extract is frozen out in an empty tube or one filled with adsorbent. The necessary cooling is achieved using liquid CO_2 on the outside of the trap. For many volatile substances this type of receiver leads to abnormally low values for quantitative analyses because of aerosol formation. The direct collection of the extract by adsorption on solid material (e.g. C_{18}) or passing the extract by means of a heated restrictor into a cooled, pressurised solvent has proved successful in many areas of application. The extraction of samples containing fats can also be carried out reliably using this method. Where adsorption is used, elution with a suitable solvent is carried out after the extraction as with solid phase extraction (SPE). If the extract is passed directly into a cooled solvent, the contents of the receiver can be further processed immediately. The choice of the appropriate collection technique is critical when modifiers are used. Even the addition of 10% of the modifier results in the capacity of the cold trap being exceeded. Where the extract is collected on C_{18} cartridges increasing quantities of modifier can lead to premature desorption, because, in general, good solvents for the analyte are used as modifiers.

The extracts thus obtained can be analysed directly using GC, GC/MS or HPLC. Only in the processing of samples containing fats is the removal of the fats necessary before the subsequent analysis. This can be readily achieved using SPE. On the other hand, extracts from liquid extractions must be subjected to a clean-up before analysis and also concentrated because of their high dilution. The steps require additional time and are often the cause of low percentage recoveries.

Carbon dioxide has further advantages as a supercritical fluid. It is nontoxic, inert, can be obtained in high purity, and is clearly more economical with regard to the costs of obtaining it and disposing of it than liquid halogenated extraction agents. Because of the critical temperature of CO_2 (31 °C) SFE can be carried out at low temperatures so that thermolabile compounds can be extracted.

The mode of function of the analytical SFE unit (Fig. 2.8) has clear parallels with that of an HPLC unit. An analytical SFE unit consists of a pump, the thermostat (oven) in which the sample is situated in an extraction thimble, and a collector for the extract.

Carbon dioxide of the highest purity ('for SFE/SFC', 'SFE grade', total purity <99.9998 vol. %) is used as the extraction agent and is available from stock from all major gas suppliers. The pumping unit consists of a syringe or piston pump. The liquid CO_2 is pumped out of the storage bottle, via a riser and compressed. Piston pumps require CO_2 bottles with an excess helium pressure (helium headspace ca. 120 bar when full). On the high pressure side the liquid modifier is passed into the compressed CO_2 with a suitable pump.

The extraction vessels for SFE are either situated in pressure-resistant cells (pressure equivalent extraction) or are the pressurised cells themselves. Nonextractable plastics or aluminium with a low thermal mass are used as the construction materials. Pressure-resistant vessels are made exclusively from stainless steel and are specified to resist up to the range of 500 bar or higher. The extraction vessels should be filled as completely as possible with the sample to avoid empty spaces. They should be extracted in a vertical position to prevent the CO_2 from flowing past the sample. Carbon dioxide flow from above to below has been shown to be most favourable, as even with compression homogeneous penetration of the sample on the base of the extraction chamber is guaranteed. For analytical purposes extraction thimbles with volumes of 0.5 mL to 50 mL are usual. The cap of the extraction vessels consists of a stainless steel frit with a pore size of 2 µm. In the case of pressure-resistant stainless steel vessels frits made of PEEK material are used as seals for the screw-on caps.

The CO_2 converted into a critical state through compression and variation of the temperature of the oven is passed continuously through the sample. In a static extraction step the extraction thimble is first filled with the supercritical fluid. At the selected oven temperature and through the effects of a selected modifier, in this step of ca. 15–30 min the sample matrix is digested and partition of the analytes between the matrix and the fluid is achieved. The swelling of the matrix has been shown to be of critical importance for real life samples as it leads to higher and more stable the yields of extract.

The dynamic extraction step is introduced by using a 6-way valve (Fig. 2.8). In this step supercritical CO_2 is passed continuously through the equilibrated sample and under the chosen conditions extractable analytes are transported out of the extraction chamber.

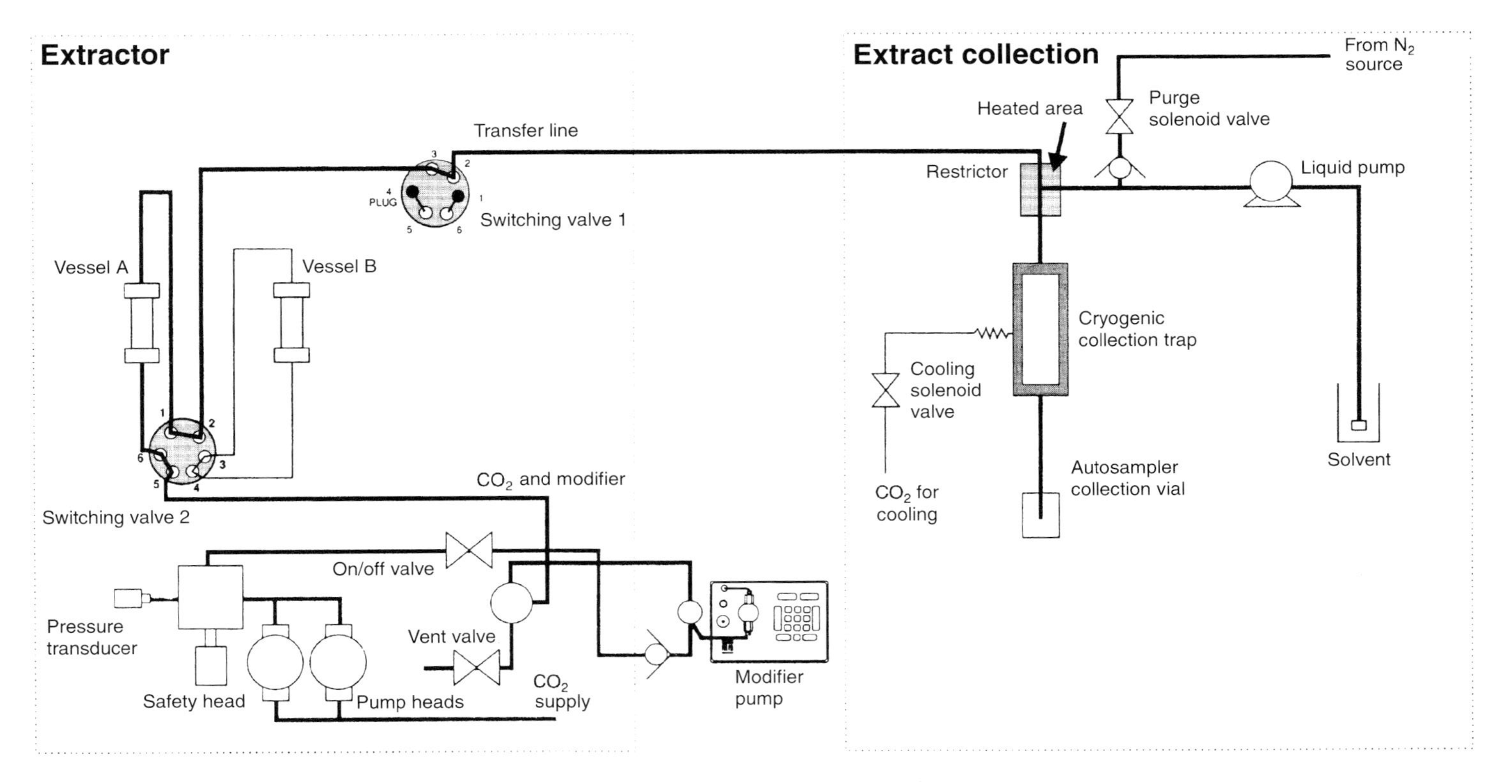

Fig. 2.8 Flow diagram of an SFE unit with cold trap and fraction collection, off-line coupling to GC (after Suprex).

The pressure ratios in the extraction chamber and in the tube carrying the extract are maintained by a restrictor at the end of the carrier tube. The dynamic extraction is quantitative and thus purifies the system for the next sample. The time required depends on the sample volume chosen.

At the restrictor the supercritical CO_2 is depressurised and the sample extract collected. A wide variety of restrictor constructions have been used with varying degrees of success. The choice of restrictor significantly affects the sample throughput of the SFE unit, the reproducibility of the process, and the breadth of application of SFE with regard to volatility of the analytes and the use of predried or moist samples. The earlier constructions with steel or fused silica restrictors with a defined opening size (e.g. integral restrictors) have not proved successful. The depressurising of the CO_2 usually took place in small glass vials. These were filled with solvent into which the restrictor was dipped. A portion of the volatile substances was lost and the necessary heat of evaporation for depressurising was not supplied adequately. All samples which were not carefully freeze-dried led here to a sudden freezing out of the extracted water and thus to frequent blocking of the restrictor. Consequently at this stage of the development the results of the process had low reproducibility.

A significant advance in restrictor technology came with the introduction of variable restrictors (Fig. 2.9). The width of the opening of the variable restrictor is controlled by the flow rate and has a fine adjustment mechanism in order to achieve a constant flow of the supercritical fluid. Diminutions in flow with simultaneous deposition of the matrix or particles at the restriction are thus counteracted. In this construction the restriction is situated

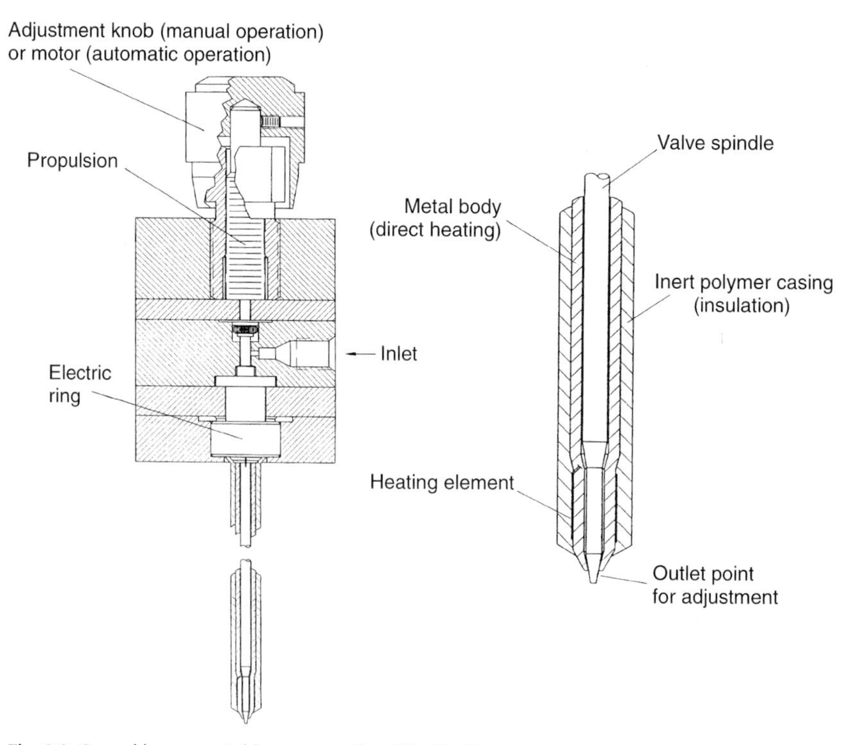

Fig. 2.9 Coaxial heater, variable restrictor for SFE (ISCO).

in a zone which can be heated to 200 °C, which completely prevents water from freezing out and provides the necessary energy for the depressurising of the CO_2.

Polar solvents, such as water, methanol, acetone, ethyl acetate or even toluene are predominantly used as modifiers in SFE (Table 2.4). The modifiers are added to swell the matrix and improve its surface activity and to increase the polarity of the CO_2. In many cases the use of modifiers leads to an increase in the yield of the extraction and even ionic compounds can be quantitatively extracted. The choice of modifier and the adjustment of its concentration are usually carried out empirically. Experience shows that modifiers which are themselves good solvents for the analytes can also be used successfully in SFE (Tables 2.5 and 2.6). Changes in the critical parameters of the fluid on addition of the modifier should be taken into account. Suitable software programs for calculating the quantities have been available (e.g. SFE-Solver, ISCO Corp., Lincoln, NE, USA). While methanol is the most widely used modifier, ethyl acetate, for example, has been shown to be particularly effective in the extraction of illegal drugs from hair (see Section 4.30).

The modifier can be added in two different ways. The simplest involves the direct addition of the modifier to the sample in the extraction vessel. As many samples, in particular most foodstuffs, already have a high water content, in such cases lyophilising the sample is not necessary. Here a suitable restrictor is necessary. The water contained in the sample can be used as the modifier for the extraction and the addition of other modifiers is unnecessary. In other cases a few mL of the modifier are added to the sample. The direct addition procedure is suitable for all extractions which employ the static extraction step.

By using an additional modifier pump (HPLC pump, syringe pump) a preselected quantity of modifier can be mixed continuously with the supercritical carbon dioxide on the high pressure side. This elegant procedure which is generally controlled by the system's software in modern instruments allows both the static step and also the continuous dynamic extraction in the presence of the modifier.

In SFE the sample quantity used has a pronounced effect on the concentrations of the analytes in the extract and on the length of the extraction. For residue analysis using SFE, sample volumes of up to 10 mL are generally used. Larger quantities of sample increase the time required for the quantitative extraction and also increase the CO_2 consumption, which is ca. three to ten times the empty volume of the extraction thimble. Further shortening of the extraction time and lowering of the sample volume is possible using on-line extraction.

On coupling with GC and GC/MS the possible water content of the extracts must be taken into consideration. The GC column must be chosen so that water can be chromatographed as a peak and does not affect the determination of the analytes. All stationary phases of medium and high polarity are suitable for this purpose. Generally it must be assumed that if there is a constant water background the response of the analytes in the mass spectrometer will be strongly impaired. An accumulation with decreasing response factors can be observed in the course of a day with ion sources with unheated lens systems.

On-line coupling with GC/MS can be easily realised with a cold injection system (e.g. PTV, programmable temperature vaporiser. Here a fused silica column as a restriction and the transfer line from the 6-way valve of the SFE unit are connected to the injector (see Fig. 2.8). The injector can be filled with a small quantity of an adsorption material, e.g. Tenax, depending on the task required. During the extraction the injector is kept at a low temperature with the split open, to ensure the expansion of the supercritical CO_2 and the trapping of the analyte. At the end of the extraction the 6-way valve of the SFE unit switches the carrier gas into

Table 2.4 Development of an SFE method for the extraction of PCDDs from fly ash (after Onuschka).

Experiment no.	1	2	3	4	5	6	7	8	9	10	11	12	13	14
Pressure	350	400	400	350	400	400	400	400	350	350	400	350	350	400
Replicate number	2	1	1	2	1	2	1	3	2	2	3	2	2	2
Supercritical fluid	N_2O	CO_2	N_2O	N_2O	CO_2	N_2O	N_2O	N_2O	CO_2	N_2O	N_2O	N_2O	N_2O	N_2O
Modifier identity	M	–	T	M	T	T	T	M	M	M	M	M	M	M
Pretreatment	–	F	F	–	F	F	F	HCl	T	T	T	E	–	–
Time-static [min]	–	64	–	120	64	64	32	64	40	40	60	50	40	64
Time-dynamic [min]	60	16	60	30	16	16	8	16	10	10	12	16	10	16
Program no.	B	A	B	C	A	A	A	A	C	C	C	D	A	A

Remarks:

Codes:

M	methanol 2%
T	toluene 5%
F	formic acid
E	pre-extraction with ethylene
HCl	hydrochloric acid treatment

Program A: 2 min static equilibration repeated 8 times followed by 30 seconds of leaching and recharging with a fresh supercritical fluid. This program, which takes 20 min to complete the 15 step run, was usually repeated 2 to 4 times.

Program B: Dynamic leaching only.

Program C: 20 steps static extraction; 20 min static equilibration time + 3 min purging, usually repeated 2 to 6 times. 20 step program consisting of 14 min static extraction followed by 30 seconds of leaching supplied 13 times. The program requires a total of 20 min and was repeated 2 to 6 times.

Program D: 12 steps static extraction; 15 min equilibration time + 3 min purging. 12 steps static extraction; 6 static steps, the first one lasting 5 min, the rest are set for 2 min duration. Leaching steps are set for 30 seconds. This program was repeated up to 7 times.

Table 2.5 Extraction yields from the SFE of PCBs for selected modifiers (after Langenfeld, Hawthorne et al.).

PCB isomer no.	1	2	3	4
18–2,2′,5	3.46 (2)	71 (12)	61 (1)	56 (4)
28–2,4,4′	2.21 (5)	63 (3)	72 (1)	65 (5)
52–2,2′,5,5′	4.48 (1)	71 (4)	74 (1)	70 (8)
44–2,2′,3,5′	1.07 (11)	94 (14)	101 (1)	88 (10)
66–2,3′,4,4′	0.93 (1)	68 (4)	73 (1)	66 (1)
101–2,2′,4,5,5′	0.82 (1)	84 (3)	87 (1)	81 (4)
118–2,3′,4,4′,5	0.51 (2)	87 (4)	92 (2)	84 (5)
138–2,2′,3,4,4′,5′	0.57 (2)	93 (2)	95 (4)	79 (7)
180–2,2′,3,4,4′,5,5′	0.16 (6)	106 (1)	109 (2)	99 (8)

(1) Certified contents in μg/g of a sediment sample according to NIST, determined by two sequential Soxhlet extractions of 16 h.

% recovery (% RSD) for the addition of the modifiers
(2) 10% methanol
(3) 10% methanol/toluene 1:1
(4) 1% acetic acid

SFE conditions: ISCO model 260D and SFX 2–10, 400 bar, 80 °C, 0.5 g sediment, 2.5 mL extraction cell, extraction agent CO_2, addition of modifier to the extraction cell (10% addition corresponds to 2.5 μL modifier), 5 min static and 10 min dynamic extraction.

Table 2.6 Extraction yields from the SFE of polyaromatic hydrocarbons for selected modifiers (after Langenfeld, Hawthorne et al.).

Polyaromatic hydrocarbon	1	2	3	4	5
Phenanthrene	4.5 (7)	99 (5)	100 (2)	238 (7)	265 (3)
Fluoranthene	7.1 (7)	84 (7)	96 (2)	175 (5)	195 (3)
Pyrene	7.2 (7)	69 (7)	81 (2)	138 (3)	151 (2)
Benzo[a]anthracene	2.6 (12)	63 (3)	94 (4)	97 (6)	89 (1)
Chrysene + triphenylene	5.3 (5)	67 (2)	90 (4)	56 (5)	67 (2)
Benzo[b,k]fluoranthene	8.2 (5)	55 (3)	108 (8)	66 (1)	67 (1)
Benzo[a]pyrene	2.9 (17)	20 (8)	60 (8)	32 (1)	34 (1)
Benzo[ghi]perylene	4.5 (24)	7 (5)	26 (11)	16 (4)	13 (11)
Indeno[1,2,3-cd]pyrene	3.3 (15)	14 (7)	57 (13)	35 (10)	28 (9)

(1) Certified contents in μg/g of a dust sample (urban air particulate matter) according to NIST, determined by soxhlet extraction over 48 h.

% recovery (% RSD) for the addition of modifiers
(2) 10% methanol
(3) 10% toluene
(4) 10% methanol/toluene 1:1
(5) 10% acetic acid

SFE conditions: ISCO model 260D and SFX 2-10, 400 bar, 80 °C, 0.3 g dust, 2.5 mL extraction cell, extraction agent CO_2, addition of modifier to the extraction cell (10% addition corresponds to 2.5 μL modifier), 5 min static and 10 min dynamic extraction.

Note: The high recoveries for some polyaromatic hydrocarbons are real and were verified using blank values and carry-over experiments.

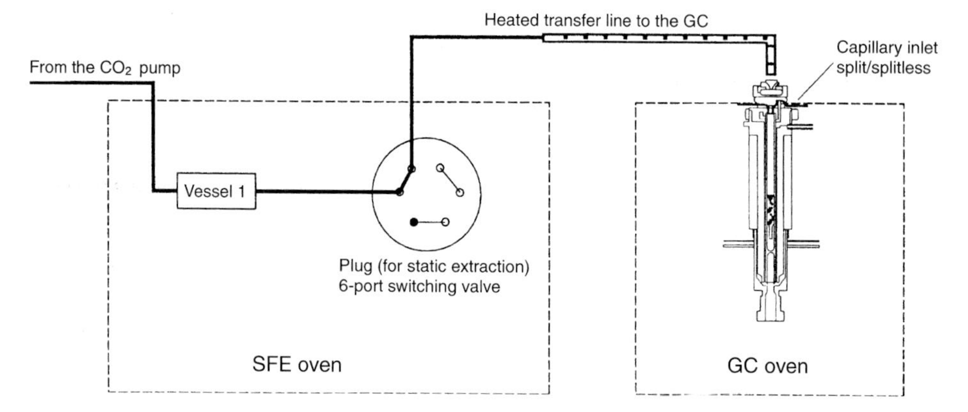

Fig. 2.10 On-line coupling of SFE with GC and GC/MS via a heated transferline (connection directly at the 6-port switching valve) and PTV cold injection system.

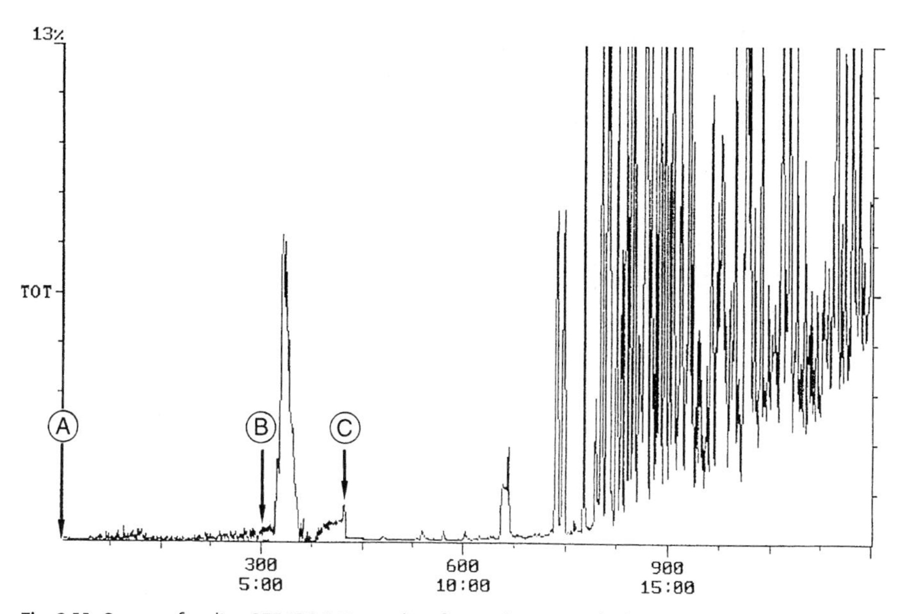

Fig. 2.11 Course of on-line SFE/GC/MS coupling for a polyaromatic hydrocarbon analysis
(A) Start of the dynamic extraction and data recording. Injector cold, split open
(B) End of the dynamic extraction and start of the injection Injector heats up, split shut
(C) Start of the GC temperature program Injector hot, split open

the transfer line to prevent further passage of CO_2 on to the GC column. Injection requires controlled heating of the injector with the split closed and the start-up of the analysis program (Fig. 2.11).

The principal advantages of on-line SFE are that the manual steps between extraction and analysis are omitted and high sensitivity is achieved. All the extract is transferred quantitatively to the chromatography column. The quantity of sample with a particular expected analyte content applied must be modified according to the capacity of the GC column. As far as the mass spectrometer is concerned, no special measures need to be taken on account of the on-line coupling. The quality of the analysis which can be achieved corresponds without limitations to the requirements of residue analysis.

The example of the extraction of polychlorinated dibenzodioxins (PCDDs) from fly ash clearly shows the comparison between the conventional soxhlet extraction and the way in which the current method is progressing. For the extraction 25 mg portions of homogenised fly ash were weighed out. The determination of the PCDDs was carried out with labelled internal standards using high resolution GC/MS. In the experiments presented here the variables sample pretreatment, fluid, modifier, pressure, and extraction time and program were systematically varied (Figs. 2.12 and 2.13).

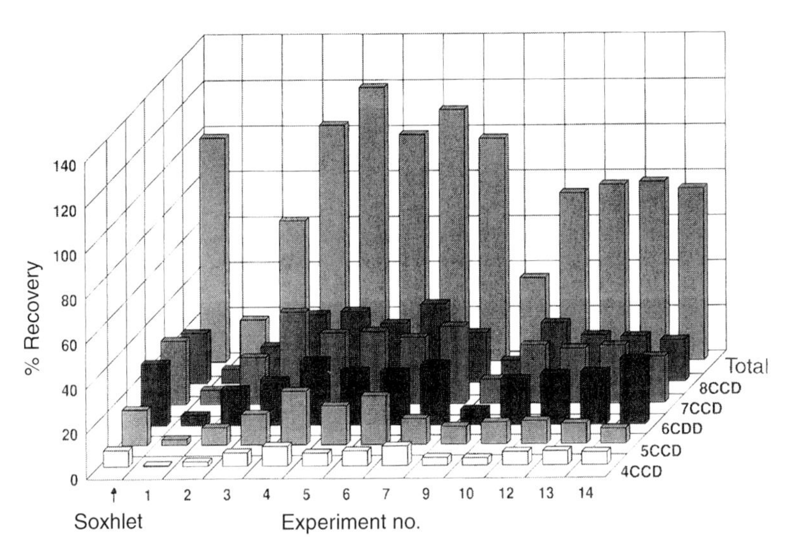

Fig. 2.12 Recoveries in SFE for PCDD extraction from fly ash compared with Soxhlet extraction (after Onuschka).

Besides the analytical aspects, the ecological and economic aspects of SFE are also important. An environmental analysis technique which produces large quantities of potentially environmentally hazardous waste is paradoxical and not acceptable in the long term. At the same time the laboratory personnel have to handle smaller quantities of harmful solvents. The critical analysis of the costs during one year shows that the SFE procedure for sample preparation in the routine laboratory reduces the cost to two-thirds that of the conventional soxhlet extraction.

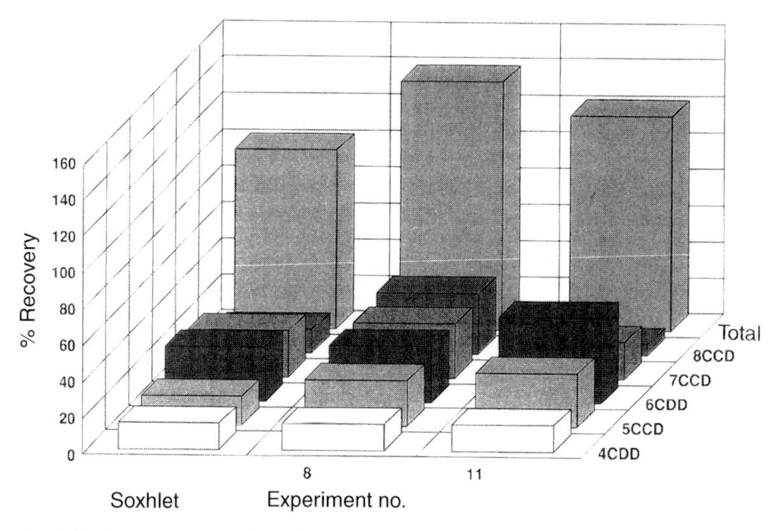

Fig. 2.13 Experiments 8 and 11 of the PCBD extraction from Table 2.6 compared with the Soxhlet extraction (after Onuschka).

2.1.3 Pressurized Fluid Extraction

The trend to automatized instrumental extraction methods has also reached the time consuming Soxhlet extraction process. A comparison of modern extraction methods like microwave assisted extraction (MAE), supercritical fluid extraction (SFE), solid phase micro extraction (SPME) and pressurized fluid extraction (PFE) with the traditional Soxhlet method, finds that PFE has a widespread application due to a number of positive features, including the most practical ones: ease-of-use, versatility and reproducibility. There are both strong analytical and economical advantages with PFE; the high extraction efficiency, savings in solvent consumption (and waste generation) and the short extraction times led to a steadily growing area of applications also for biological samples (Wenzel 1998, Li 2004). PFE also limits glassware handling, and facilitates a safe laboratory working environment.

Pressurized Fluid Extraction (PFE) is the general term for an automated technique for extracting solid and semisolid samples with liquid, organic or aqueous solvents. The PFE method is also marketed under the term "Accelerated Solvent Extraction" (ASE) by Dionex Corp., USA, covered by a number of international patents (Dionex 2006). PFE takes advantage of a high energy regime for the extraction process provided by high pressures at high temperatures. Special care has to be taken concerning thermolabile components for which a cold extraction method as the microwave assisted extraction might be the appropriate alternative. Automated PFE instrumentation allows the independent control of temperature and pressure conditions for each sample cell. This control is critical for a high analyte recovery and reproducibility.

At the typical PFE conditions the solvent characteristics change compared to the traditional Soxhlet technique and allow an improved separation of target analytes from the matrix (Hubert 2001). For instance the extraction of relatively polar phenols from soil sam-

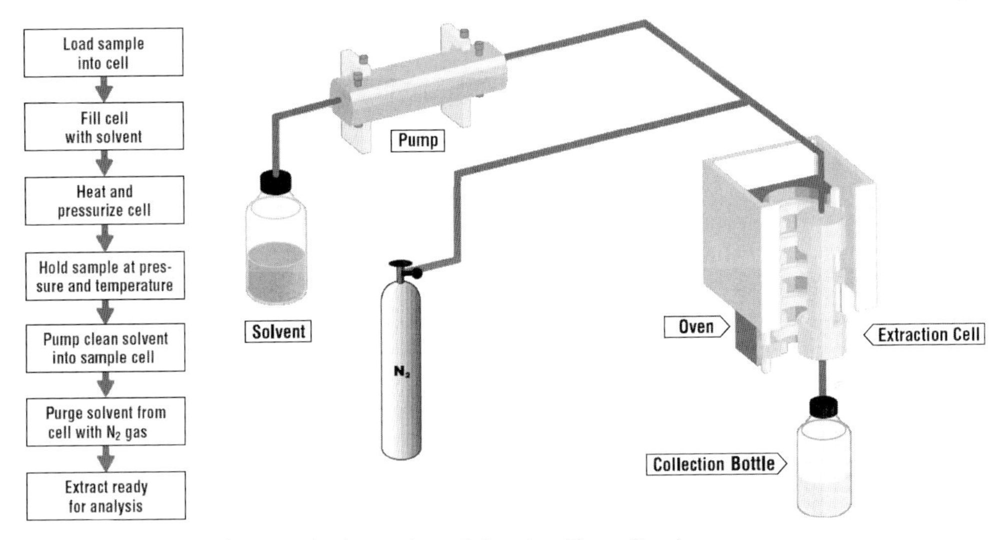

Fig. 2.14 Schematics of a pressurized extraction unit (courtesy Dionex Corp.).

ples is possible using the non polar solvent n-hexane. PFE extractions typically use the known organic solvents n-hexane, cyclo-hexane, toluene, dichloromethane, methanol, acetone as applied in Soxhlet methods. For biological samples and the extraction of water soluble components also water based media in mixtures with organic solvents are applied (Curren 2002). This variation gives access to polar compounds for subsequent LC/MS analysis. Due to a mostly sharp matrix/analyte separation by PFE, the collected extracts require less time consuming clean-up typically employing SPE or GPC steps.

Traditional techniques, like Soxhlet, can take 4–48 hours. With PFE, analyte recoveries equivalent to those obtained using traditional extraction methods can be achieved in only 15 to 30 minutes. Although PFE uses the same aqueous and organic solvents as traditional extraction methods, it uses them more efficiently. A typical PFE extraction is done using 50–150 mL of solvent, see Table 2.7 for comparison.

Table 2.7 Comparison of standard laboratory extraction techniques.

Techniques	Solvent volume per sample [mL]	Extraction time per sample [h]	Cost per sample [a] (in US $)
Manual Soxhlet	200–500	4–48	23.50
Automated Soxhlet	50–100	1–4	16.25
Sonication	100–300	0.5–1	20.75
SFE	8–50	0.5–2	17.60
Microwave	50–100	0.5–1	15.25
ASE	15–40	0.2–0.3	11.20

a) Cost per sample in US $ based on 2000 samples per year, average values, based on US comparison (Dionex 2006)

Approximately 75% of all PFE extractions are completed in less than 20 minutes using the standard PSE extraction conditions (100 °C, 1500 psi). But, PFE extractions are matrix dependent and require further optimization when applied to different matrices. Method development starts with standard conditions and evaluates both the recovery in the first step and the result of a second extraction of the given new sample. Also the initial application of a standard matrix e.g. sand is a helpful step. If these initial parameters don't provide the recoveries desired, the temperature is increased to improve the efficiency of the extraction. Adding static cycles, increasing static time, and selecting a different solvent are additional variables that can be used to optimize a method.

Recommended Extraction Conditions (US EPA 1998)

For semivolatiles, organophosphorus pesticides, organochlorine pesticides, herbicides, and PCBs:

Oven temperature:	100 °C
Pressure:	1500–2000 psi
Static time:	5 min (after 5 min pre-heat equilibration)
Flush volume:	60% of the cell volume
Nitrogen purge:	60 sec at 150 psi (purge time may be extended for larger cells)
Static cycles:	1

For PCDDs/PCDFs:

Oven temperature:	150–175 °C
Pressure:	1500–2000 psi
Static time:	5–10 min (after 5 min pre-heat equilibration)
Flush volume:	60–75% of the cell volume
Nitrogen purge:	60 sec at 150 psi (purge time may be extended for larger cells)
Static cycles:	2 or 3

PFE extraction is accepted for use in U.S. EPA SW-846 Method 3545A, which can be used in place of Methods 3540, 3541, 3550, and 8151. Method 3545A can be applied to the extraction of base/neutrals and acids (BNAs), chlorinated pesticides and herbicides, polychlorinated biphenyls (PCBs), organophosphorus pesticides, dioxins and furans and total petroleum hydrocarbons (TPH).

Scope and Application of EPA Method 3545A

1.1 Method 3545A is a procedure for extracting water insoluble or slightly water soluble organic compounds from soils, clays, sediments, sludges, and waste solids. The method uses elevated temperature (100–180 °C) and pressure (1500–2000 psi) to achieve analyte recoveries equivalent to those from Soxhlet extraction, using less solvent and taking significantly less time than the Soxhlet procedure. This procedure was developed and validated on a commercially available, automated extraction system.

1.2 This method is applicable to the extraction of semivolatile organic compounds, organophosphorus pesticides, organochlorine pesticides, chlorinated herbicides, PCBs, and PCDDs/PCDFs, which may then be analyzed by a variety of chromatographic procedures.

(US EPA Method 3545A, Jan 1998)

2.1.4 Online Liquid Chromatography Clean-up

The necessary effort of sample preparation in trace analysis is driving cost and limiting productivity not only in the GC/MS laboratory today. Especially a series of manual steps for extraction, evaporation and transfer involved are responsible for long analysis times, high cost and also the reason for contaminations and analyte losses. Coupling LC clean-up on-line with the GC separation opens a great potential for a significant reduction in cost per sample, full automation and high productivity for standardized analyses. An on-line coupling of GPC or HPLC as clean-up procedure is a mature technique first described by Majors 1980. Commercial solutions are available since more than 20 years and established especially for routine analysis in many laboratories covering a wide range of applications e.g. for pesticides, PCBs, dioxins or plasticizers (Grob 1990, De Paolo 1992, Jongenotter 1999). A specific advantage for trace analyses is the concentrated eluate allowing lower detection limits and reproducibility. High extract dilutions introducing additional contaminations with a subsequent need for evaporation as known from Soxhlett or SPE extractions are avoided.

The instrumental set-up is built by an additional LC system with autosampler, HPLC separation column and switching valve in front of the GC (see Fig. 2.15). An on-line LC detector is necessary during method development with spiked samples. The GC is equipped with an injector suitable for large volume injections as a regular PTV equipped with a packed insert liner and LVI vent valve or an on-column injector with retention gap and solvent vapor exit. Conventional liquid extraction techniques can be replaced with this setup completely and run fully automated for large sample series.

The significantly higher separation efficiency in GPC or regular HPLC columns (number of available theoretical plates) can be exploited for group type separations. Eluents applied for the on-line clean-up need to be compatible with gas chromatography as typical for normal phase LC (reversed phase separations need fraction collection and solvent exchange). The analyte containing fractions of the LC eluate are cut via a sample loop, or more flexible for

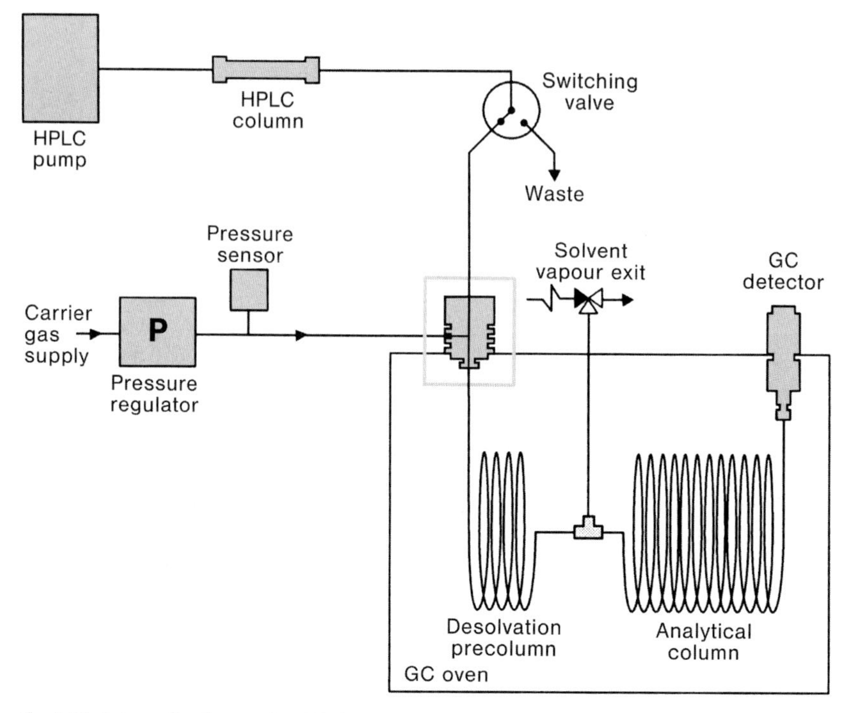

Fig. 2.15 Schematic of an on-line LC clean-up GC system (Trestianu 1996)

optimization, using a time controlled transfer to the GC injector. Typically, injection volumes of 100 μL to 1000 μL are employed using concurrent solvent evaporation or retention gap large volume injection (LVI) techniques as described in Section 2.2.3 GC/MS Inlet Systems.

2.1.5 Headspace Techniques

One of the most elegant possibilities for instrumental sample preparation and sample transfer for GC/MS systems is the use of the headspace technique (Fig. 2.16). Here all the frequently expensive steps, such as extraction of the sample, clean-up and concentration are dispensed with. Using the headspace technique the volatile substances in the sample are separated from the matrix. The latter is not volatile under the conditions of the analysis. The tightly closed sample vessels, which, for example, are used for the static headspace procedure, can frequently even be filled at the sampling location. The danger of false results (loss of analyte) as a result of transportation and further processing is thus reduced.

The extraction of the analytes is based on the partition of the very and moderately volatile substances between the matrix and the gas phase above the sample. After the partition equilibrium has been set up the gas phase contains a qualitatively and quantitatively representative cross-section of the sample and is therefore used for analysing the components to be determined. All involatile components remain in the headspace vial and are not analysed. For this reason the coupling of headspace instruments and GC/MS systems is particularly fa-

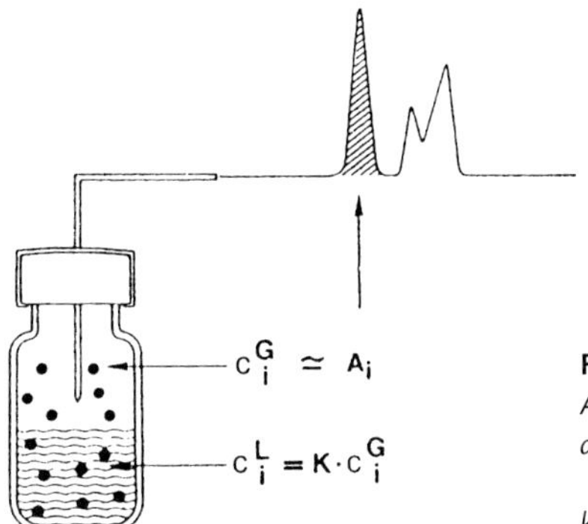

Fig. 2.16 Principle of headspace analysis (after Kolb)
A_i area of the GC signal of the ith component.
c_{Gi}, c_{Li} concentrations in the gas and liquid phases respectively
k' partition coefficient

vourable. Since the interfering organic matrix is not involved, a longer duty cycle of the instrument and outstanding sensitivity are achieved. Furthermore headspace analyses are easily and reliably automated for this reason and achieve a higher sample throughput in a 24-hour operating period.

There are limitations to the coupling of headspace analysis with GC/MS systems if moisture has to be driven out of the sample. In certain cases water can impair the focusing of volatile components at the beginning of the GC column. Impairment of the GC resolution can be counteracted by choosing suitable polar and thick film GC columns, by removing the water before injection of the sample, or by using a simple column connection.

It is also known that water affects the stability of the ion source of the mass spectrometer detector, which nowadays is becoming ever smaller. In the case of repeated analyses the effects are manifested by a marked response loss and poor reproducibility. In such cases special precautions must be taken, in particular in the choice of ion source parameters.

The headspace technique is very flexible and can be applied to the most widely differing sample qualities. Liquid or solid sample matrices are generally used, but gaseous samples can also be analysed readily and precisely using this method. Both qualitative and, in particular, quantitative determinations are carried out coupled to GC/MS systems.

There are two methods of analysing the volatiles of a sample which have very different requirements concerning the instrumentation: the static and dynamic (purge and trap) headspace techniques. The areas of use overlap partially but the strengths of the two methods are demonstrated in the different types of applications.

2.1.5.1 Static Headspace Technique

The term headspace analysis was coined in the early 1960s when the analysis of substances with odours and aromas in the headspace of tins of food was developed. The equilibrium of volatile substances between a sample matrix and the gas phase above it in a closed static system is the basis for static headspace chromatography (HSGC). The term headspace is often used without the word static, e. g. headspace analyses, headspace sampler etc.

An equilibrium is set up in the distribution of the substances being analysed between the sample and the gas phase. The concentration of the substances in the gas phase then remains constant. An aliquot is taken from the gas phase and is fed into the GC/MS system via a transfer line (Fig. 2.17).

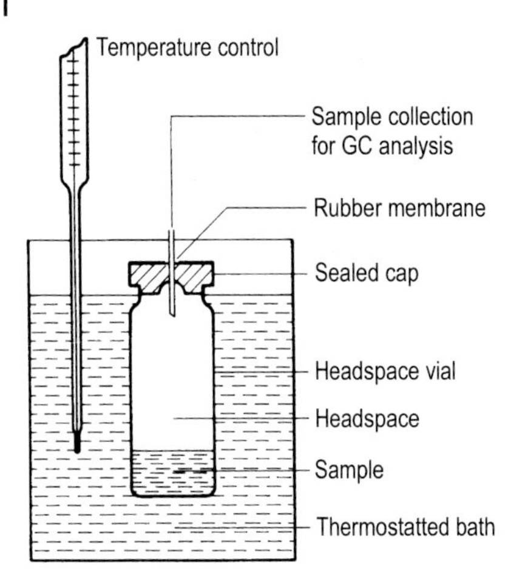

Fig. 2.17 Sample handling in the static headspace method (after Hachenberg).

The static headspace method is therefore an indirect analysis procedure, requiring special care in performing quantitative determinations. The position of the equilibrium depends on the analysis parameters (e. g. temperature) and also on the sample matrix itself. The matrix-dependence of the procedure can be counteracted in various ways. The matrix can be standardised, e. g. by addition of Na_2SO_4 or Na_2CO_3. Other possibilities include the addition method, internal standardisation, or the multiple headspace extraction procedure (MHE) published by Kolb in 1981 (Fig. 2.18).

> **Static Headspace Analysis**
>
> In static headspace analysis the samples are taken from a closed static system (closed headspace bottle) after the thermodynamic equilibrium (partition) between the liquid/solid matrix and the headspace above it has been established.

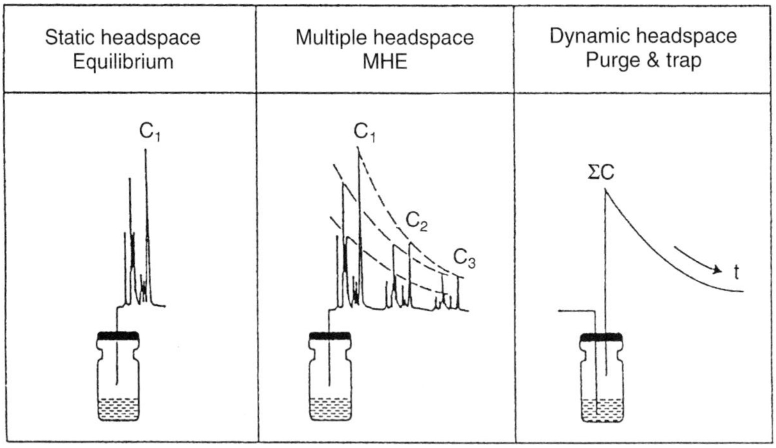

Fig. 2.18 Gas extraction techniques (after Kolb).

For coupling of HSGC with mass spectrometry the internal standard procedure has proved particularly successful for quantitative analyses. Besides the headspace-specific effects, possible variations in the MS detection are also compensated for. The best possible precision is thus achieved for the whole procedure. The MHE procedure can be used in the same way.

In static headspace analysis the partition coefficient of the analytes is used to assess and plan the method. For the partition coefficient k of a volatile compound the following equation is valid:

$$k = \frac{c_S}{c_G} = \frac{\text{concentration in the sample}}{\text{concentration in the gas phase}} \qquad (1)$$

The partition coefficient depends on the temperature at equilibrium. This must therefore be kept constant for all measurements.

Rearranging the equation gives:

$$c_S = k \cdot c_G \qquad (2)$$

As the peak area A determined in the GC/MS analysis is proportional to the concentration of the substance in the gas phase c_G, the following is valid:

$$A \approx c_G = \frac{1}{k} \cdot c_S \qquad (3)$$

To be able to calculate back to the concentration of a substance c_0 in the original sample, the mass equilibrium must be referred to:

$$M_0 = M_S + M_G \qquad (4)$$

The quantity M_0 of the volatile substance in the original sample has been divided at equilibrium into a portion in the gas phase M_G and another in the sample matrix M_S. Replacing M by the product of the concentration c and volume V gives:

$$c_0 \cdot V_0 = c_S \cdot V_S + c_G \cdot V_G \qquad (5)$$

By using the definition of the partition coefficient given in equation (2), the unknown parameter c_S can be replaced by $k \cdot c_G$:

$$c_0 \cdot V_0 = k \cdot c_G \cdot V_S + c_G \cdot V_G \qquad (6a)$$

$$c_0 = c_G \cdot \frac{V_S}{V_0} \cdot \left(k + \frac{V_G}{V_S}\right) \qquad (6b)$$

The starting concentration c_0 of the sample, assuming $V_0 = V_S$, is given by:

$$c_0 = c_G \cdot \left(k + \frac{V_G}{V_S}\right) \qquad (7)$$

As the peak area determined is proportional to the concentration of the volatile substance in the gas phase c_G, the following equation is valid, corresponding to the proportionality between the gas phase and the peak area:

$$c_0 \approx A \cdot (k + \beta) \tag{8}$$

whereby β is the phase ratio V_G/V_S (see equation (7)) and therefore describes the degree of filling of the headspace vessel (Ettre, Kolb 1991).

The effects on the sensitivity of a static headspace analysis can easily be derived from the ratio given in equation (8):

$$A \approx c_0 \cdot \frac{1}{k + \beta} \quad \text{with } \beta = V_G/V_S \tag{9}$$

The peak area determined from a given concentration c_0 of a component depends on the partition coefficient k and the sample volume V_S (through the phase ratio β).

For substances with high partition coefficients (e.g. ethanol, isopropanol, dioxan in water) the sample volume has no effect on the peak area determined. However, for substances with small partition coefficients (e.g. cyclohexane, trichloroethylene, xylene in water), the phase ratio β determines the headspace sensitivity. Doubling the quantity of sample leads in this case to a doubling of the peak area (Fig. 2.19). For all quantitative determinations of compounds with small partition coefficients, an exact filling volume must be maintained.

The sensitivity of detection in static headspace analysis can also be increased through the lowering of the capacity factor k by setting up the equilibrium at a higher temperature. Substances with high partition coefficients profit particularly from higher equilibration tempera-

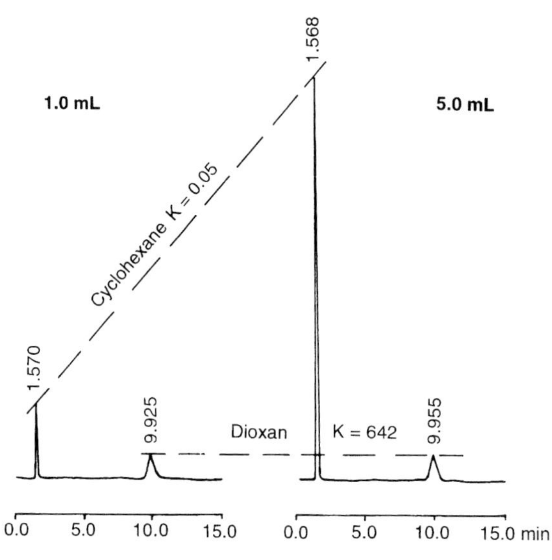

Fig. 2.19 Sample volume and sensitivity in the static headspace for cyclohexane ($K = 0.05$) and dioxan ($K = 642$) with sample volumes of 1 mL and 5 mL respectively (after Kolb).

tures (Table 2.8, e.g. alcohols in water). Raising the temperature is, however, limited by an increased matrix evaporation (water) and the danger of bursting the sample vessel or the seal.

Changes in the sample matrix can also affect the partition coefficient. For samples with high water contents an electrolyte can be added to effect a salting out process (Fig. 2.20). For example, for the determination of ethanol in water the addition of NH_4Cl gives a twofold and addition of K_2CO_3 an eightfold increase in the sensitivity of detection. The headspace sensitivity can also be raised by the addition of nonelectrolytes. In the determination of residual monomers in polystyrene dissolved in DMF, adding increasing quantities of water, e.g.

Table 2.8 Partition coefficients of selected compounds in water (after Kolb).

Substance	40 °C	60 °C	80 °C
Tetrachloroethane	1.5	1.3	0.9
1,1,1-Trichloroethane	1.6	1.5	1.2
Toluene	2.8	1.6	1.3
o-Xylene	2.4	1.3	1.0
Cyclohexane	0.07	0.05	0.02
n-Hexane	0.14	0.04	<0.01
Ethyl acetate	62.4	29.3	17.5
n-Butyl acetate	31.4	13.6	7.6
Isopropanol	825	286	117
Methyl isobutyl ketone	54.3	22.8	11.8
Dioxan	1618	642	288
n-Butanol	647	238	99

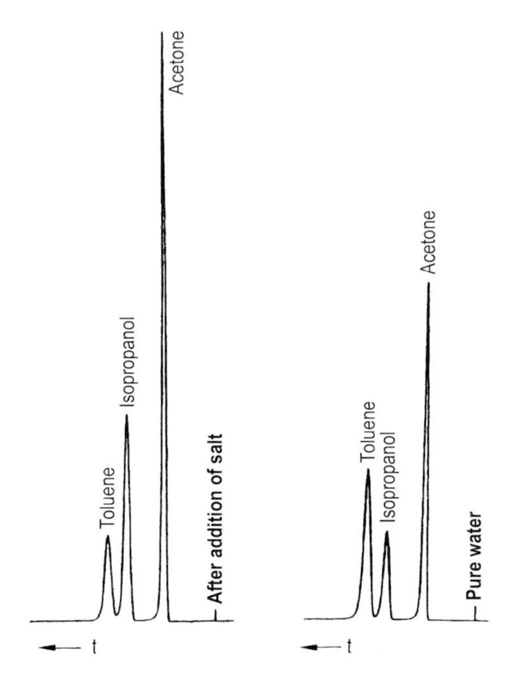

Fig. 2.20 Matrix effects in static headspace. The effect of salting out on polar substances (after Kolb).

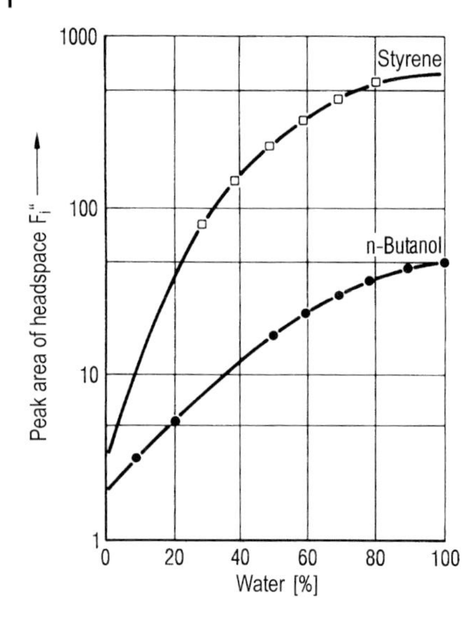

Fig. 2.21 Increasing water concentrations in the determination of styrene as a residual monomer lead to a sharp increase in response (after Hachenberg).

in the determination of styrene and butanol, leads to an increase in peak area for styrene by a factor of 160 and for n-butanol by a factor of only 25 (Fig. 2.21).

To shorten the equilibration time the most up-to-date headspace samplers have devices available for mixing the samples. For liquid samples vigorous mixing strongly increases the phase boundary area and guarantees rapid delivery of analyte-rich sample material to the phase boundary.

Teflon-coated stirring rods have not proved successful because of losses through adsorption and cross-contamination. However, shaking devices which mix the sample in the headspace bottle through vertical or rotational movement have proved effective (Table 2.9). A particularly refined shaker uses changing excitation frequencies of 2–10 Hz to achieve optimal mixing of the contents at each degree of filling (Fig. 2.22).

The resulting equilibration times are in the region of 10 min on using shaking devices compared with 45–60 min without mixing.

In addition quantitative measurements are more reliable. The use of shaking devices in static headspace techniques lowers the relative standard deviations to less than 2%.

Table 2.9 Comparison of the analyses with/without shaking of the headspace sample (volatile halogenated hydrocarbons), average values in ppb, standard deviations (after Tekmar).

Substance	Without shaking			With shaking		
Ethylbenzene	353	18	(5.2%)	472	8	(1.7%)
Toluene	336	20	(5.9%)	411	4	(1.0%)
o-Xylene	324	13	(4.1%)	400	7	(1.8%)
Benzene	326	18	(5.4%)	372	5	(1.3%)
1,3-Dichlorobenzene	225	13	(5.6%)	255	5	(2.1%)
1,2,4-Trichlorobenzene	207	9	(4.2%)	225	6	(2.5%)
Bromobenzene	213	11	(5.2%)	220	5	(2.1%)

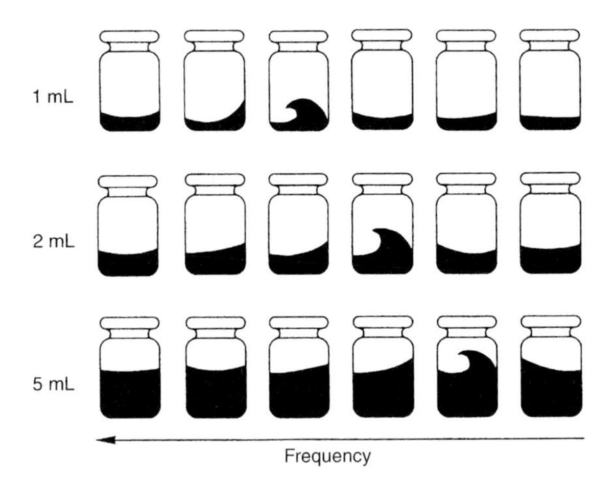

Fig. 2.22 Effect of variable shaking frequencies (2–10 Hz) on different depths of filling in headspace vessels (Perkin Elmer).

Static Headspace Injection Techniques

- *Pressure Balanced Injection* (Fig. 2.23):
  Variable quantity injected: can be controlled (programmed) by the length of the injection process, injection volume = injection time x flow rate of the GC column, no drop in pressure to atmospheric pressure, depressurising to initial pressure of column.
  Change in pressure in the headspace bottle: reproducible pressure build-up with carrier gas, mixing, initial pressure in column maintained during sample injection.
  Sample losses: none known.

- *Syringe Injection* (Fig. 2.24):
  Variable injection volume through pump action: easily controlled, drop in pressure to atmospheric on transferring the syringe to the injector.
  Pressure change in the headspace bottle: varies with the action of the syringe plunger (volume removed) on sample removal from the sealed bottle, compensation by injection of carrier gas necessary.
  Sample losses: possible through condensation on depressurising to atmospheric pressure, losses through evaporation from the syringe after it has been removed from the septum cap of the bottle, in comparison the largest surface contact with the sample.

- *Sample Loop* (Fig. 2.25):
  Quantity injected on the instrument side: changes in the volume injected by disconnection of the sample loop, drop in pressure to atmospheric on filling the sample loop.
  Pressure change in the headspace bottle: complete depressurising of the pressure built up in the bottle through tempering, the expansion volume must be a multiple of the sample loop to be able to fill it reproducibly.
  Sample losses: for larger sample loops attention must be paid to dilution with carrier gas during the pressure build-up phase, possible condensation on depressurising to atmospheric pressure.

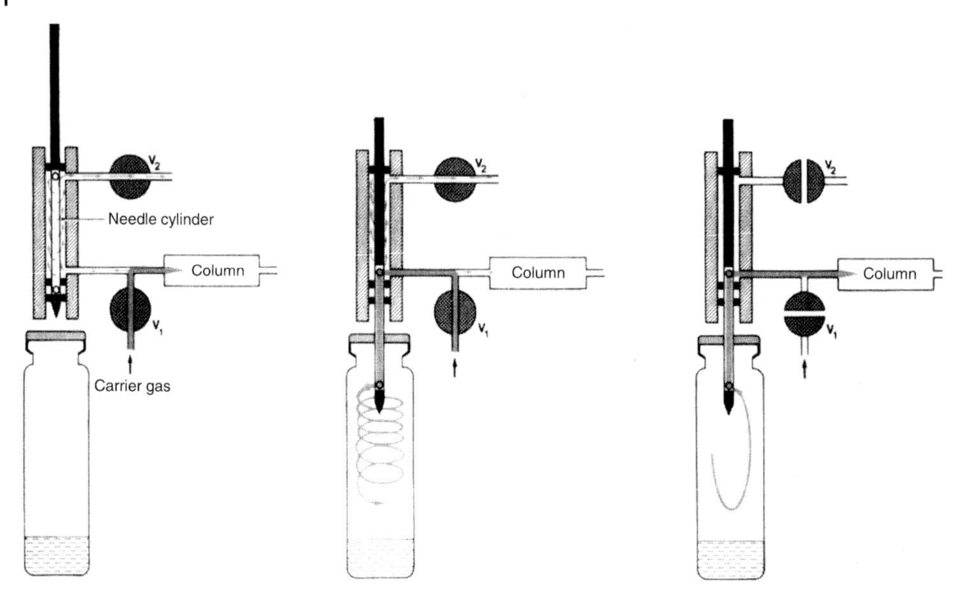

Fig. 2.23 Injection techniques for static headspace: the principle of pressure balanced injection (Perkin Elmer).

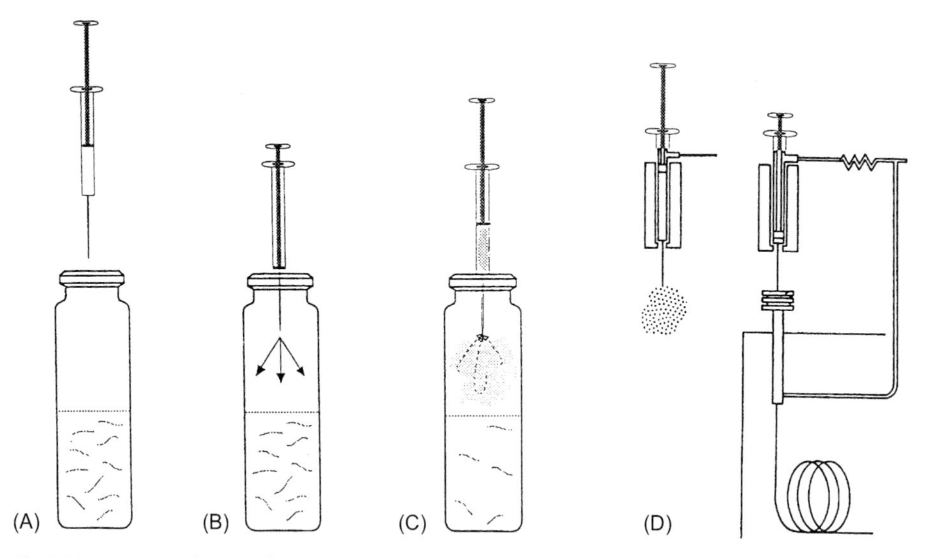

Fig. 2.24 Injection techniques for static headspace: the principle of transfer with a gastight syringe. (A) Sample heating, (B) Pressure, (C) Sampling, (D) Inject.

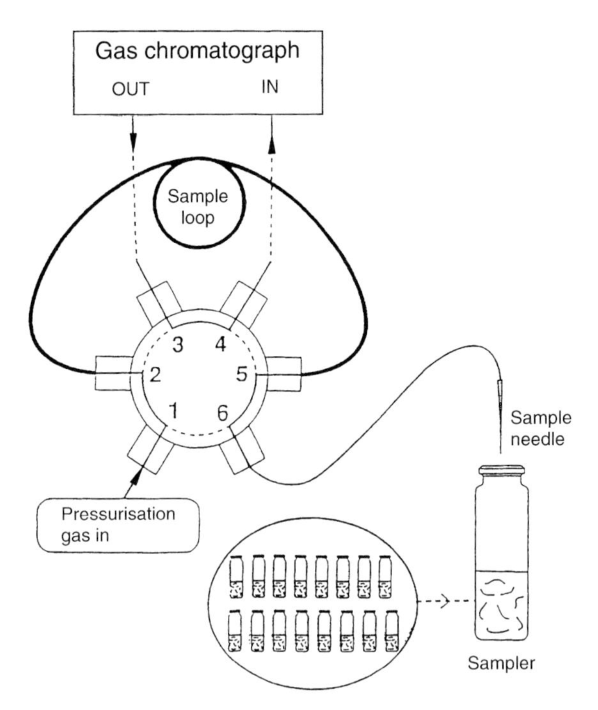

Fig. 2.25 Injection techniques for static headspace: the principle of application with a sample loop (Tekmar).
Sequence:
1 Heat sample at a precise temperature over a set period of time
2 Pressurise with carrier gas
3 Fill the sample loop
4 Inject by switching the 6-port switching valve

2.1.5.2 **Dynamic Headspace Technique (Purge and Trap)**

The trace analysis of volatile organic compounds (VOCs, volatile organic carbons) is of continual interest, e.g. in environmental monitoring, in outgasing studies on packaging material or in the analysis of flavours and fragrances. The dynamic headspace technique, known as purge and trap (PAT), is a process in which highly and moderately volatile organic compounds are continuously extracted from a matrix and concentrated in an adsorption trap. The substances driven out of the trap by thermal desorption reach the gas chromatograph as a concentrated sample plug where they are finally separated and detected.

Early purge and trap applications in the 1960s already involved the analysis of body fluids. In the 1970s purge and trap became known because of the increasing requirement for the testing of drinking water for volatile halogenated hydrocarbons. The analysis of drinking water for the determination of a large number of these compounds in ppq quantities (concentrations of less than 1 μg/L) only became possible through the concentration process which was part of purge and trap gas chromatography (PAT-GC). This technique is now frequently used to detect residues of volatile organic compounds in the environment and in the analysis of liquid and solid foodstuffs. In particular, the coupling with GC/MS systems allows its use as a multicomponent process for the automatic analysis of large series of samples.

An analytically interesting variant of the procedure is the fine dispersion of liquid samples in a carrier gas stream. This so-called spray and trap process uses the high surface area of the sample droplets for an effective gas extraction. The procedure also works very well with foaming samples. A detergent content of up to 0.1% does not affect the extraction result. The spray and trap process is therefore particularly suitable for use with mobile GC/MS analyses.

Modes of Operation of Purge and Trap Systems

A purge and trap analysis procedure consists of three main steps:

1. Purge phase with simultaneous concentration
2. Desorption phase
3. Baking out phase

1. The Purge Phase

During the purge phase the volatile organic components are driven out of the matrix. The purge gas (He or N_2) is finely divided in the case of liquid samples (drinking water, waste water) by passing through a special frit in the base of a U-tube (frit sparger, Fig. 2.26A). The surface of the liquid can be greatly increased by the presence of very small gas bubbles and the contact between the liquid sample and the purge gas maximised.

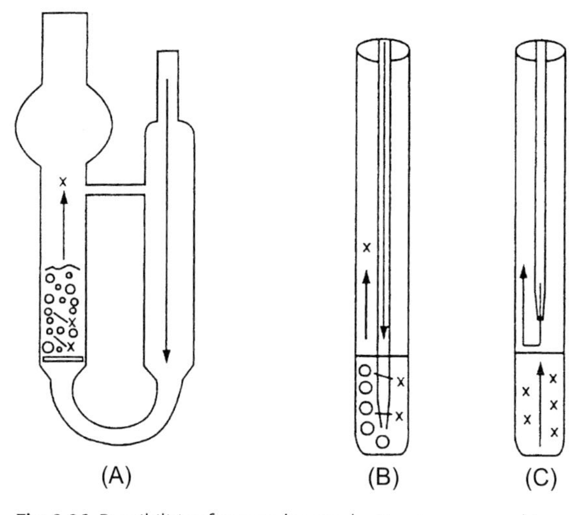

Fig. 2.26 Possibilities for sample introduction in purge and trap.
(A) U tube with/without frits (*fritless/frit sparger*) for water samples.
(B) Sample vessel (*needle sparger*) for water and soil samples (solids).
(C) Sample vessel (*needle sparger*) for foaming samples for determination of the headspace sweep.

The purge gas extracts the analytes from the sample and transports them to the trap. The analytes are retained in this trap and concentrated while the purge gas passes out through the vent. The desorption gas, which comes from the carrier gas provision of the GC, enters this phase via the 6-port switching valve in the gas chromatograph and maintains the constant gas flow for the column (Fig. 2.27).

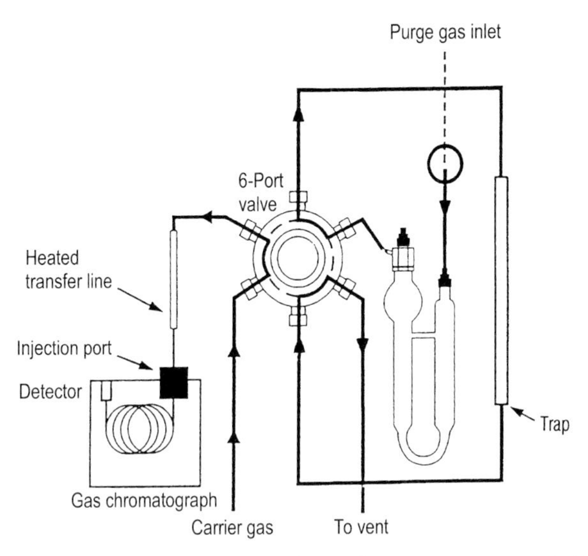

Fig. 2.27 Gas flow schematics of purge & trap-GC coupling, switching of phases at the 6-port switching valve. Bold line: Purge and baking out phase; dotted line: Desorption phase.

Solid samples are analysed in special vessels (needle sparger, see Fig. 2.26 B and C) into which a needle with side openings is dipped. For foaming samples the headspace sweep technique can be used.

The total quantity of volatile organic compounds removed from the sample depends on the purge volume. The purge volume is the product of the purge flow rate and the purge time. Many environmental samples are analysed at a purge volume of 440 mL. This value is achieved using a flow rate of 40 mL/min and a purge time of 11 min. A purge flow rate of 40 mL/min gives optimal purge efficiency. Changes in the purge volume should consequently only be made after adjusting the purge time. Although a purge volume of 440 mL is optimal in most cases, some samples may require larger purge volumes for adequate sensitivity to be reached (Fig. 2.28).

The purge efficiency is defined as that quantity of the analytes which is purged from a sample with a defined quantity of gas. It depends upon various factors. Among them are: purge volume, sample temperature, and nature of the sparger (needle or frit), the nature of the substances to be analysed and that of the matrix. The purge efficiency has a direct effect on the percentage recovery (the quantity of analyte reaching the detector).

Control of Purge Gas Pressure During the Purge Phase

The adsorption and chromatographic separation of volatile halogenated hydrocarbons is improved significantly by regulating the pressure of the purge gas during the purge phase. By additional back pressure control during this phase, a very sharp adsorption band is formed in the trap, from which, in particular, the highly volatile components profit, since a broader distribution does not occur. The danger of the analytes passing through the trap is almost completely excluded under the given conditions.

During the desorption phase the narrow adsorption band determines the quality of the sample transfer to the capillary column. The result is clearly improved peak symmetry,

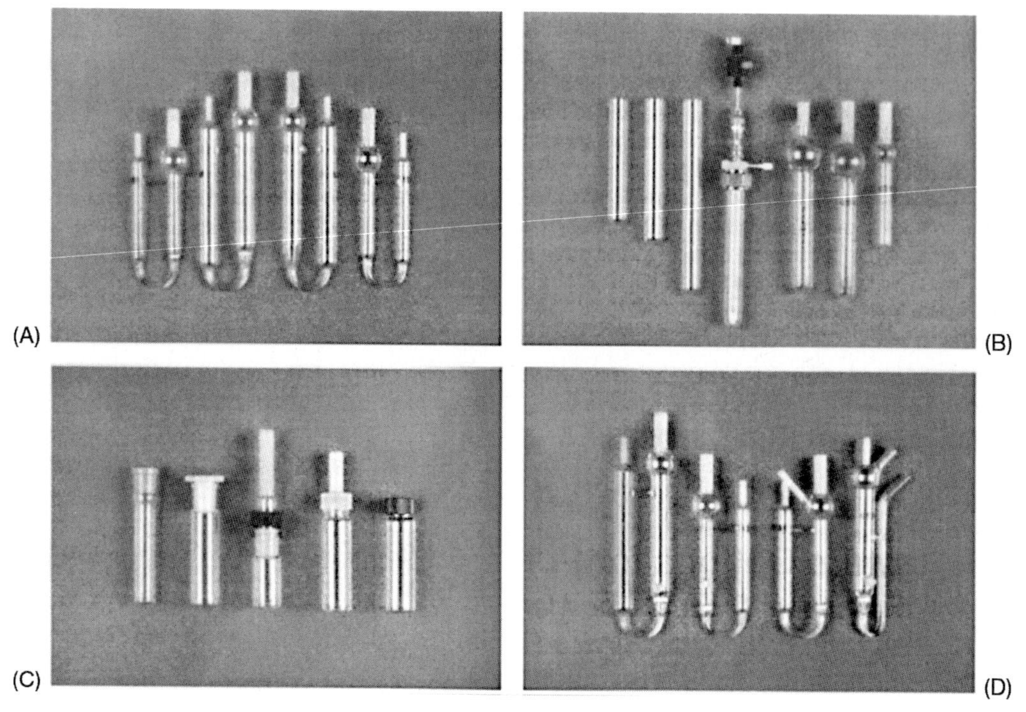

Fig. 2.28 Glass apparatus for the purge and trap technique (Tekmar).
(A) U-tube with/without frit (5 mL and 25 mL sizes).
(B) Needle sparger: left: single use vessels, middle: glass needle with frits, right: vessels with foam retention, 5 mL, 20 mL and 25 mL volumes.
(C) Special glass vessels: 25 mL flat-bottomed flask, 40 mL flask with seal, 20 mL glass (two parts) with connector, 40 mL glass with flange, 40 mL screw top glass.
(D) U-tube for connection to automatic sample dispensers: left: 25 mL and 5 mL vessel with side inlet, right: 5 mL vessel with upper inlet for sample heating, 25 mL special model with side inlet.

and thus better GC resolution and an improvement in the sensitivity of the whole procedure.

The Dry Purge Phase
To remove water from a hydrophobic adsorption trap (e.g. with a Tenax filling) a dry purge phase is introduced. During this step most of the water condensed in the trap is blown out by dry carrier gas. Purge times of ca. 6 min are typical.

2. The Desorption Phase

During the desorption phase the trap is heated and subjected to a backflush with carrier gas. The reversal of the direction of the gas flow is important in order to desorb the analytes in the opposite direction to the concentration by the trap. In this way narrow peak bands are obtained.

The time and temperature of the desorption phase affect the chromatography of the substances to be analysed. The desorption time should be as short as possible but sufficient to transfer the components quantitatively on to the GC column. Most of the analytes are transferred to the GC column during the first minute of the desorption step. The desorption time is generally 4 min.

The temperature of the desorption step depends on the type of adsorbent in the trap (Table 2.10). The most widely used adsorbent, Tenax, desorbs very efficiently at 180 °C without forming decomposition products. The peak shape of compounds eluting early can be improved by inserting a desorb-preheat step. Here the trap is preheated to a temperature near the desorption temperature before the valve is switched for desorption and before the gas flows freely through the trap. Gas is not passed through the trap during the preheating step, but the analytes are nevertheless desorbed from the carrier material. When the gas stream is passed through the trap after switching the valve, it purges the substances from the trap in a concentrated carrier gas cloud. Highly volatile compounds which are not focused at the beginning of the column thus give rise to a narrower peak shape. A preheating temperature of 5 °C below the desorption temperature has been found to be favourable.

VOCARB material, which is used in all current applications involving volatile halogenated hydrocarbons, can be desorbed at higher temperatures than Tenax (up to 290 °C). At higher desorption temperatures, however, the possibility of catalytic decomposition of some substances must be taken into account (see also Section 2.1.5).

Moisture Removal

Water is driven out of aqueous or moist samples, most of which is disposed of by the dry purge step, particularly in Tenax adsorption traps. Residual moisture would be transferred to the GC column during the desorption step. As the resolution of highly volatile substances on capillary columns would be impaired and the detection by the mass spectrometer would be affected, additional devices are used to remove the water. A moisture control system (MCS) found as a dedicated device in earlier purge or trap systems guaranteed maximum efficiency and retention of the residual moisture during the desorption process (Fig. 2.29, see p. 47).

If the dew point for water is reached in the MCS, a stationary water phase is formed which, for example, for BTEX and volatile halogenated hydrocarbons, the analytes can pass through unaffected. Polar components can also pass through the MCS system at moderate temperatures with a short retardation. When the desorption phase is finished, the tubing of the MCS is dried by baking out in the countercurrent (Fig. 2.30, see p. 48).

In particular, where the purge and trap technique is used in capillary gas chromatography and when using ECD or mass spectrometers as detectors, reliable removal of the moisture driven out is necessary.

3. Baking Out Phase

When required, the trap can be subjected to a baking out phase (trap-bake mode) after the desorption of the analytes. During desorption involatile organic compounds are released from the trap at elevated temperatures and driven off. Baking out conditions the trap material for the next analysis.

Table 2.10 List of trap materials used in the purge and trap procedure with details of applications and recommended analysis parameters.

Trap No.	Adsorbent	Application	Drying possible?	Drying time [min]	Desorb pre-heat [°C]	Desorb temp. [°C]	Bake-out temp. [°C]	Bake-out time [min]	Conditioning temp. [°C] time [min]	Remarks
1	Tenax	All substances down to CH_2Cl_2	Yes	2–6	220	225	230	7–10	230 10	Low response with brominated substances, high back pressure, background with benzene, toluene, ethylbenzene
2	Tenax Silica gel	All substances except freons	No	–	220	225	230	10–12	230 10	Low response with brominated substances, high back pressure, background with benzene, toluene, ethylbenzene
3	Tenax Silica gel Activated charcoal	All substances including freons	No	–	220	225	230	10–12	230 10	Low response with brominated substances, high back pressure, background with benzene, toluene, ethylbenzene
4	Tenax Activated charcoal	All substances down to CH_2Cl_2 and gases	No	–	220	225	230	7–10	230 10	Low response with brominated substances, high back pressure, background with benzene, toluene, ethylbenzene
5	OV-1 Tenax Silica gel Activated charcoal	All substances including freons	No	–	220	225	230	10–12	230 10	Low response with brominated substances, high back pressure, background with benzene, toluene, ethylbenzene
6	OV-1 Tenax Silica gel	All substances except freons	No	–	220	225	230	10–12	230 10	Low response with brominated substances, high back pressure, background with benzene, toluene, ethylbenzene

Table 2.10 (continued)

Trap No.	Adsorbent	Application	Drying possible?	Drying time [min]	Desorb pre-heat [°C]	Desorb temp. [°C]	Bake-out temp. [°C]	Bake-out time [min]	Conditioning temp. [°C] time [min]	Remarks
7	OV-1 Tenax	All substances down to CH_2Cl_2	Yes	2–6	220	225	230	7–10	230 10	Low response with brominated substances, high back pressure, background with benzene, toluene, ethylbenzene
8	Carbopak B Carbosieve SIII	All substances including freons	Yes	11	245	250	260	4–10	260 20–30	Losses of CCl_4
9	VOCARB 3000	All substances including freons	Yes	1–3	245	250	260	4	290 4 h	Response factors see Table 2.11
10	VOCARB 4000	All substances including freons	Yes	1–3	245	250	260	4	270 4 h	Low response with chlorinated compounds, high back pressure, quantitative losses of chloroethyl vinyl ethers
11	BTEXTRAP Carbopak B Carbopak C	All substances including freons	Yes	1–3	245	250	260	4	270	–
12	Tenax GR Graphpac-D	All substances including freons	Yes	1–4	245	250	260	12	–	–

Trap nos. 1–8: Tekmar; trap nos. 9–11: SUPELCO; trap no. 12: Alltech

Table 2.11 Response factors and standard deviations for components with EPA method 624 using a VOCARB 3000 trap and 5 ml of sample (250 °C desorption temperature, reference bromochloromethane/difluorobenzene, concentrations in ppb, Supelco).

Compound	20	50	100	150	200	Average	Standard deviation	% Relative standard deviation
Methyl chloride	0.933	0.962	0.984	0.906	1.287	1.014	0.139	13.7
Vinyl chloride	1.037	1.167	1.466	1.536	1.333	1.308	0.185	14.1
Methyl bromide	1.018	1.257	1.079	1.077	1.290	1.144	0.108	9.5
Ethyl chloride	0.419	0.582	0.503	0.557	0.536	0.519	0.056	10.9
Trichlorofluoromethane	0.972	1.533	1.266	1.491	1.507	1.354	0.213	15.8
1,1-Dichloroethylene	0.873	1.207	1.203	1.264	1.180	1.145	0.139	12.1
Dichloromethane	1.550	1.134	1.231	1.325	1.11	1.270	0.159	12.5
1,2-Dichloroethylene	0.973	0.991	0.944	1.124	1.007	1.008	0.062	6.1
1,1-Dichloroethane	2.064	2.093	2.013	2.186	2.175	2.106	0.066	3.1
Chloroform	2.513	2.368	2.193	2.419	2.373	2.373	0.104	4.4
Tetrachloroethane	1.205	1.472	1.302	1.325	1.476	1.356	0.105	7.7
Carbon tetrachloride	1.067	1.426	1.350	1.240	1.335	1.284	0.124	9.6
Benzene	0.947	0.949	0.906	0.973	1.024	0.960	0.039	4.0
1,2-Dichloroethane	0.046	0.045	0.047	0.050	0.049	0.047	0.002	4.2
Trichloroethylene	0.412	0.495	0.485	0.504	0.540	0.487	0.042	8.6
1,2-Dichloropropane	0.535	0.530	0.513	0.537	0.549	0.533	0.012	2.2
Bromodichloromethane	0.082	0.080	0.084	0.088	0.087	0.084	0.003	3.4
2-Chloroethyl vinyl ether	0.043	0.039	0.043	0.047	0.041	0.043	0.003	6.2
cis-1,3-Dichloropropene	0.898	0.977	1.001	1.086	1.082	1.009	0.070	6.9
Toluene	0.861	1.208	1.141	1.250	1.283	1.149	0.151	13.2
trans-1,3-Dichloropropene	0.312	0.300	0.280	0.299	0.289	0.296	0.011	3.7
1,1,2-Trichloroethane	0.593	0.486	0.488	0.526	0.507	0.520	0.039	7.5
Tetrachloroethylene	0.364	0.488	0.492	0.497	0.558	0.480	0.063	13.2
Dibromochloromethane	0.818	0.712	0.731	0.787	0.785	0.767	0.039	5.1
Chlorobenzene	1.135	1.116	1.117	1.183	1.242	1.159	0.048	4.2
Ethylbenzene	0.561	0.508	0.476	0.504	0.543	0.518	0.030	5.8
Bromoform	1.070	0.882	0.920	1.013	0.918	0.961	0.070	7.3
1,1,2,2-Tetrachloroethane	0.082	0.069	0.070	0.078	0.065	0.073	0.006	8.4
1,3-Dichlorobenzene	1.277	1.215	1.237	1.338	1.451	1.304	0.085	6.5
1,4-Dichlorobenzene	1.397	1.279	1.291	1.391	1.509	1.374	0.084	6.1
1,2-Dichlorobenzene	1.351	1.233	1.188	1.284	1.350	1.281	0.064	5.0

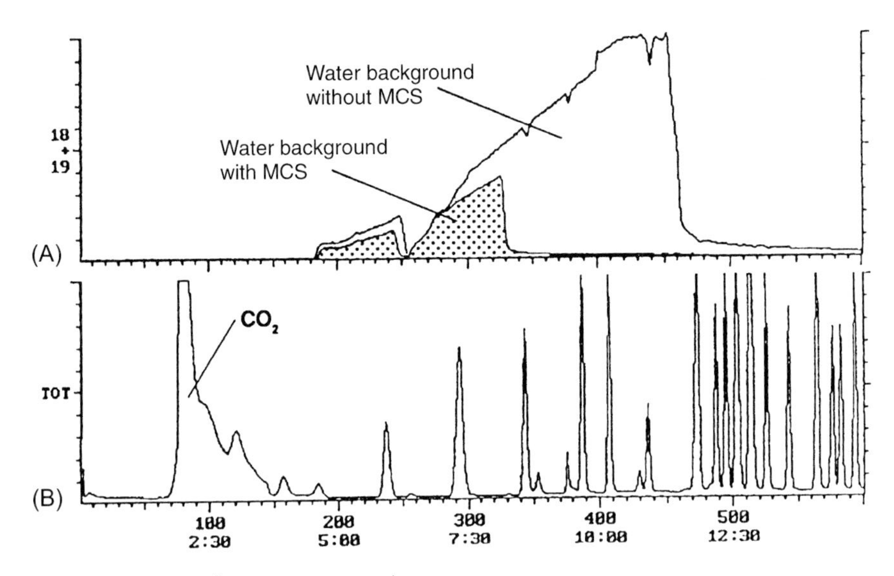

Fig. 2.29 GC/MS analysis using purge and trap.
(A) Mass chromatogram for water (m/z 18+19) with and without removal (without MCS) of the moisture driven out.
(B) Total ion chromatogram with volatile substances (volatile halogenated hydrocarbons) after removal of the water (with MCS).

Coupling of Purge and Trap with GC/MS Systems

There are two fundamentally different possibilities for the installation of a purge and trap instrument coupled to a GC/MS system, which depend on the areas of use and the number of purge and trap samples to be processed per day.

In many laboratories the most flexible arrangement involves connecting a purge and trap system to the gas supply of the GC injector. The purge and trap concentrator is connected in such a way that either manual or automatic syringe injection can still be carried out. The carrier gas is passed from the GC carrier gas regulation to the central 6-port valve of the purge and trap system (see Fig. 2.27). From here the carrier gas flows back through the continuously heated transfer line to the injector of the gas chromatograph. A particular advantage of this type of installation lies in the fact that the injector can still be used for liquid samples. This allows the manual injection of control samples or the operation of liquid autosamplers. In addition it guarantees that the whole additional tubing system remains free from contamination even when the purge and trap instrument is on standby.

When the sample is transferred to the GC column it should be ensured that adequate focusing of the analytes is achieved. For this purpose GC columns are available with film thicknesses of ca. 1.8 μm or more, which have already been used for the analysis of volatile halogenated hydrocarbons using GC/MS systems (types 502, 624 or volatiles). The injector system is operated using a split ratio which can be selected depending on the detection limit required.

A complete splitless injection of the analytes on to the GC column can only be achieved with cryofocusing (see also Section 2.2.3.7). Here the column is connected to the transfer

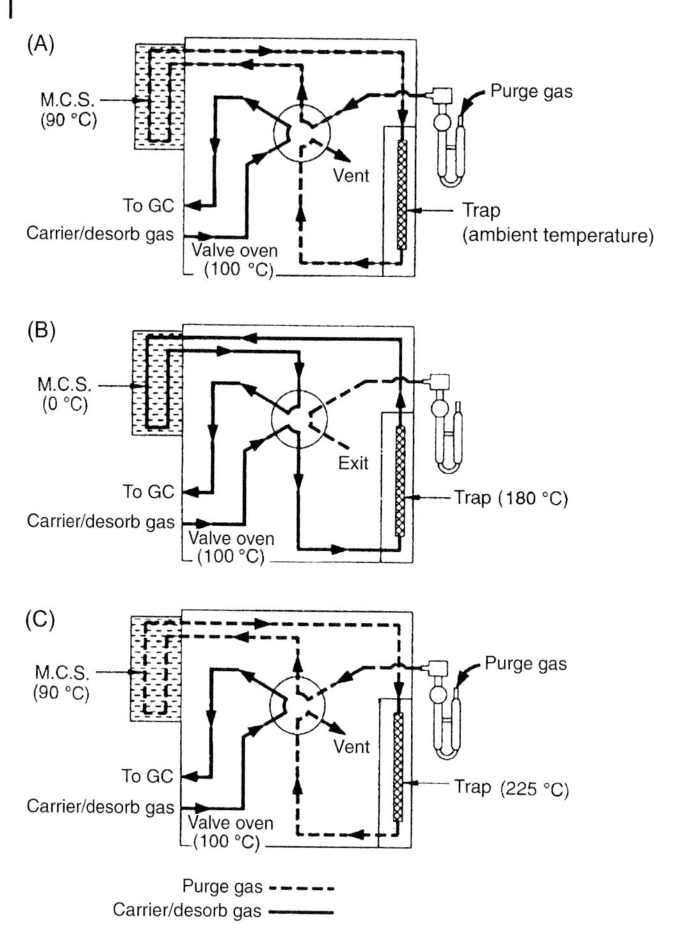

Fig. 2.30 Scheme showing the three phases of the purge and trap cycle with water removal (MCS, Tekmar). (A) Purge phase, (B) Desorption phase, (C) Baking out phase

line or the central 6-port valve of the purge and trap apparatus in such a way that the beginning of the column passes through a zone ca. 10 cm long which can be cooled by an external cooling agent, such as liquid CO_2 or N_2, to –30 °C to –150 °C. All components which reach the GC column after desorption from the trap can be frozen out in a narrow band at the beginning of the analytical column. This enables the highest sensitivities to be achieved, in particular with mass spectrometers. As large quantities of water are also concentrated together with the analytes in the case of moist or aqueous samples, removal of moisture in the desorption phase is particularly important with this type of coupling. This can, for example, be effected by means of a moisture control system (MCS) (Fig. 2.30). Insufficient removal of water leads to the deposition of ice and blockage of the capillaries. The consequences are poor focusing or complete failure of the instrument.

2.1.5.3 **Headspace versus Purge and Trap**

Both instrumental extraction techniques have specific advantages and disadvantages when coupled to GC and GC/MS. This should be taken into consideration when choosing an analysis procedure. In particular, the nature of the sample material, the concentration range for the measurement and the work required to automate the analyses for large numbers of samples play a significant role. The recovery and the partition coefficient, and thus the sensitivity which can be achieved, are relevant to the analytical assessment of the procedure. For both procedures it must be possible to vaporise the substances being analysed below 150 °C and then to partition them in the gas phase. The vapour pressure and solubility of the analytes, as well as the extraction temperature, affect both procedures (Fig. 2.31).

How then do the techniques differ? For this the terms recovery and sensitivity must be defined. For both methods, the recovery depends on the vapour pressure, the solubility and the temperature. The effects of temperature can be dealt with because it is easy to increase the vapour pressure of a compound by raising the temperature during the vaporisation step. With the purge and trap technique the term *percentage recovery* is used. This is the amount of a compound which reaches the gas chromatograph for analysis relative to the amount which was originally present in the sample. If a sample contains 100 pg benzene and 90 pg reach the GC column, the *percentage recovery* is 90%. In the static headspace technique, a simple expression like this cannot be used because it is possible to use a large number of types of vial, injection techniques and injection volumes, which always apply aliquots to the analysis.

The commonly used term connected with the static headspace method is the *partition coefficient*, as mentioned above. The partition coefficient is defined as the quantity of a compound in the sample divided by the quantity in the vapour phase. Therefore, the smaller the partition coefficient is, the higher the sensitivity. It should be noted that a partition coefficient is only valid for the analysis parameters at which it has been determined. These include the temperature, the size of the sample vial, the quantity of sample (weight and

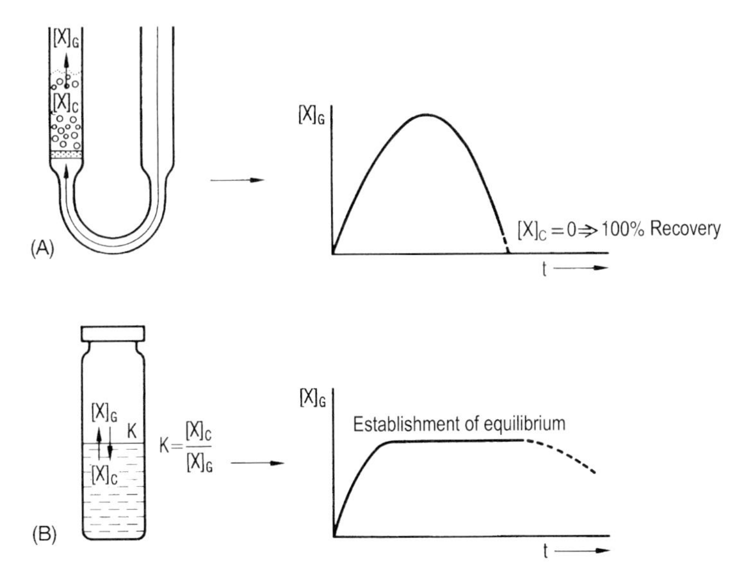

Fig. 2.31 Comparison of the purge and trap and static headspace techniques. (A) Purge and trap, (B) Static headspace

volume), the nature of the matrix and the size of the headspace. After the partition coefficient, the quantity injected is the next parameter affecting headspace sensitivity. The quantity injected is limited by a range of factors. For example, only a limited quantity can be removed from the headspace of a closed vessel. Attempts to remove a larger quantity of sample vapour would lead to a partial vacuum in the sample vessel. This is extraordinarily difficult to reproduce. Furthermore, only a limited quantity can be injected on to a GC column without causing peak broadening. For larger quantities of sample cryofocusing is necessary. Capillary columns require cold trapping at injection quantities of more than ca. 200 μL.

An alternative injection system involves pressure balanced injection. Here a needle is passed through the septum into the headspace in order to create pressure in the headspace vial which is at least as high as the column pre-pressure. After equilibrium has been established this pressure is released during injection on to the GC column over a short programmable time interval. This allows larger quantities to be injected. It is impossible to measure the exact quantity injected; however, the reproducibility of this method is extremely high.

First Example: Volatile Halogenated Hydrocarbons

A comparison with actual concentration values makes the differences between the static headspace and purge and trap techniques very clear. The percentage recovery for the purge and trap technique is, for example, for an environmental sample of volatile halogenated hydrocarbons: 95% for chloroform, 92% for bromodichloromethane, 87% for chlorodibromomethane and 71% for bromoform. For a sample of 5 mL, which contains 1 ppb of each substance (i. e. a total quantity of 5 ng of each compound), 4.75 ng, 4.6 ng, 4.35 ng and 3.55 ng are recovered. In a typical static headspace system with a sample vessel of 21 mL containing 15 mL of sample at 70 °C the partition coefficients for the corresponding volatile halogenated hydrocarbons are: 0.3, 0.9, 1.5 and 3.0. This means that the quantities in the 5 ml of headspace are: 11.5 ng, 7.9 ng, 6.0 ng and 3.8 ng. On injection of 20 μL of the headspace gas mixture on to a standard capillary column, the quantities injected are: 0.05 ng, 0.03 ng, 0.02 ng and 0.015 ng. For a larger injection (0.5 mL) using cryofocusing, the quantities injected are: 1.2 ng, 0.8 ng, 0.6 ng, and 0.4 ng. The purge and trap technique is therefore more sensitive than the static headspace procedure for these volatile halogenated hydrocarbons by factors of 4.1, 5.8, 7.2 and 9.3 (Table 2.12).

Static Headspace	**Dynamic Headspace**
Headspace	Purge and trap
Extraction: waiting for establishment of equilibrium	Continuous disturbance of the equilibrium by the purge gas
Intermediate steps: closed vessel at constant temperature	Enrichment of the substances driven off in a trap
Sample injection: Removal of a preselected volume from the headspace	Thermal desorption from the trap

Table 2.12 Lower application limits [μg/L] for headspace and purge and trap techniques in the analysis of water from the river Rhine (after Willemsen, Gerke, Krabbe).

Substance	Lower application limits [μg/L] for headspace	Lower application limits [μg/L] for purge and trap
Dichloromethane	1	0.5
Chloroform	0.1	0.05
1,1,1-Trichloroethane	0.1	0.02
1,2-Dichloroethane	5	0.5
Carbon tetrachloride	0.1	0.02
Trichloroethylene	0.1	0.05
Bromodichloromethane	0.1	0.02
1,1,2-Trichloroethane	0.5	0.5
Dibromochloromethane	0.1	0.05
Tetrachloroethylene	0.1	0.02
1,1,2,2-Tetrachloroethane	0.1	0.02
Bromoform	0.1	0.2
1,1,2,2-Tetrachloroethane	0.1	0.02
Benzene	5	0.1
Toluene	5	0.05
Chlorobenzene	5	0.02
Ethylbenzene	5	0.1
m-Xylene	5	0.05
p-Xylene	5	0.05
o-Xylene	5	0.1
Triethylamine	–	0.5
Tetrahydrofuran	–	0.5
1,3,5-Trioxan	–	0.5

Second Example: Cooking Oils

For compounds with poorer recoveries, e.g. in the analysis of free aldehydes in cooking oils, the difference is even clearer. At 150 °C the purge and trap analysis gives recoveries of 47%, 59% and 55% for butanal, 2-hexenal and nonan. For a sample of 0.5 mL, containing 100 ppb of the compounds, 24 ng, 30 ng and 28 ng are recovered. In the static headspace analysis, the partition coefficients for these compounds at 200 °C are all higher than 200. Assuming the value of 200, the quantity of each of these compounds in 5 mL of headspace is 0.7 ng. For an injection of 0.5 mL the quantity injected is therefore only 0.07 ng. The differences in sensitivity favouring purge and trap analysis are therefore 343, 428 and 400!

Third Example: Residual Solvents in Plastic Sheeting

Comparable ratios are obtained in the analysis of a solid sample, for example, the analysis of residual solvents in a technical product. A run using the purge and trap technique and 10 mL of sample at 150 °C gave a recovery of 63% for toluene. The sample contained 1.6 ppm, which corresponds to a quantity of 101 ng. The partition coefficient in the static headspace technique at 150 °C (for a sample of 1 g) is 95. The quantity of residual solvent in 19 mL of headspace is therefore 17 ng. For an injection of 0.5 mL, 0.4 ng are injected. The quantity injected is therefore smaller by a factor of 250 than that in the purge and trap analysis. Furthermore, the reproducibility of this analysis was 7% for the purge and trap technique and 32% for the static headspace analysis (relative standard deviation).

The theoretically achievable or effectively necessary sensitivity is not the only factor deciding the choice of procedure. The specific interactions between the analytes and the matrix, the performance of the detectors available, and the legally required detection limits play a more important role.

Besides the sensitivity, there are other aspects which must be taken into account when comparing the purge and trap and static headspace techniques.

The static headspace technique is very simple and relatively quick. The procedure is well documented in the literature, and for many applications the sensitivity is more than adequate, so that its use is usually favoured over that of the purge and trap technique. There are areas of application where good results are obtained with the static headspace technique which cannot be improved upon by the purge and trap method. These include: the determination of alcohol in blood, of free fatty acids in cell cultures, of ethanol in fermentation units or drinks, and residual water in polymers. This also applies to studies on the determination of ionisation constants of acids and bases and the investigation of gas phase equilibria.

However, for many other samples specific problems besides sensitivity arise on use of the static headspace technique, which can be overcome using the purge and trap procedure. In every static headspace system all the compounds present in the headspace are injected, not only the organic analytes. This means that an air peak is also obtained. All air contaminants (a widely occurring problem in many laboratories) are visible and oxygen can impair the service life of the capillary column at high temperatures. In addition a larger quantity of water is injected than in the purge and trap method. For drinks containing carbonic acid CO_2 can lead to the build-up of excess pressure. Even at room temperature an undesirably high pressure can build up in the sample vials, which leads to flooding of the GC column during the analysis. In addition, a very large quantity of CO_2 is injected. Heating the sample enhances these effects. For safety reasons sample vials with safety caps to guard against excess pressure have to be used in these cases. Dust particles lead to further problems in the case of powder samples. To achieve the necessary sensitivity for an analysis, the sample often has to be heated to a temperature which is higher than that necessary in the purge and trap technique. This can lead to thermal decomposition, which is frequently observed with foodstuffs. In addition, oxygen cannot be eliminated from the sample before heating in a static headspace system. It is therefore impossible to prevent oxidation of the sample contents. During the thermostatting phase additional problems arise as a result of the septum used. Substances emitted from the septum (e.g. CS_2 from butyl rubber septums) falsify the chromatogram. The permeability of the septum to oxygen presents a further hazard.

Advantages of the Static Headspace Technique

- The static headspace can be easily automated. All commercial headspace samplers operate automatically for 20, 40 or 100 samples. For manual qualitative preliminary samples a gas-tight syringe is satisfactory.
- Samples, which tend to foam or contain unexpectedly high concentrations of analyte, do not generally lead to faults or cross-contamination.
- All sample matrices (solid, liquid or gaseous) can be used directly usually without expensive sample preparation.

- Headspace samplers are often readily portable and, when required, can be rapidly connected to different types of GC instruments.
- The sample vessels (special headspace vials with caps) are only intended for single use. There is no additional workload of cleaning glass equipment and hence cross-contamination does not occur.
- Headspace vials can be filled and sealed at the sampling point outside the laboratory in certain cases. This dispenses with transfer of the sample material and eliminates the possibility of loss of sample. Moreover, the danger of inclusion of contamination from the laboratory environment is reduced.
- Because of the high degree of automation, the cost of analysing an individual sample is kept low.

Disadvantages of the Static Headspace Technique

- On filling headspace vials a corresponding quantity of air is enclosed in the vial unless filling is carried out under an inert gas atmosphere. However, this is very expensive. During thermostatting undesirable side reactions can occur as a result of atmospheric oxygen in the sample.
- During injection from headspace vials air regularly gets into the GC system and can affect sensitive column materials.
- In the case of moist or aqueous samples, a considerable quantity of water vapour gets on to the column. This requires special measures to ensure the integrity of the early eluting peaks.
- On use of mass spectrometers special attention must be paid to the stability of ion sources. On insufficient heating the surface of the source and the lens system become increasingly coated with moisture in the course of the work (caused by the injection of water vapour). As a result the focusing of the mass spectrometer can change and this can impair quantitative work.
- Quantitation is matrix-dependent. Standardisation measures are necessary, such as addition of carbonate, internal standards and MHE procedures.
- The ability of substances to be analysed is limited by the maximum possible filling of the headspace vials and by the partition coefficients. For small partition coefficients larger quantities of sample do not lead to an increase in sensitivity. If the maximum possible equilibration temperature is being used, it is almost impossible to increase the sensitivity further.
- The quantity injected is limited and where the sensitivity is insufficient, multiple extraction with cryofocusing is necessary.
- Undesired blank values can be obtained through contaminated air (laboratory air) which gets into the headspace vial or through bleeding of the septum caps.
- Excess pressure can cause headspace vials to burst (e.g. with drinks containing carbonic acid or high equilibration temperatures). This always puts instruments out of operation for long periods and results in considerable clean-up costs. Only the use of special vial caps (with spring rings) together with special sealed vials which can release excess pressure into the atmosphere prevents bursting.

Advantages of the Purge and Trap Technique

- The sample quantity (maximum 25–50 ml) can easily be adapted to give the required sensitivity.
- A pre-purge step can remove atmospheric oxygen from the sample, even at room temperature if required.
- No septum or other permeable material is placed in the way of the sample.
- Substances with high partition coefficients in the static headspace can be determined with good yields.
- There are no excess pressure problems. The entire gas stream is passed pressure-free through the trap to the vent.
- Water vapour can be kept completely out of the GC/MS system by means of a dry purge step and a moisture control system (MCS).
- In automatic operations mixing in an internal standard without manual measures is quite straightforward.

Disadvantages of the Purge and Trap Technique

- Foaming samples require special treatment, the headspace sweep technique.
- The purge vessels (made of glass) must be cleaned carefully. Economical single-use vessels are currently only available in polymer materials.
- Larger quantities of sample require longer purge times.
- For highly contaminated samples the breakthrough volume of the trap must be taken into consideration. If the baking out step is inadequate, the danger of a carry-over exists.
- Coupling with capillary GC necessitates the use of small split ratios or cryofocusing.

2.1.6
Adsorptive Enrichment and Thermodesorption

To determine volatile organic compounds, GC/MS coupling is the method of choice. In the analysis of air or gas samples an extraordinarily large number of components of the most widely differing classes of compounds and over a very wide concentration range have to be considered. Usually, the concentration of the substances of interest is too low for the direct measurement of an air sample, and therefore enrichment on suitable adsorbents is necessary. The concentration on solid adsorption material allows the accumulation of organic components from large volumes of gas. Typical areas of use include soil air, workplace monitoring, gases from landfill sites, air pollution, and air inside buildings.

Besides a high storage capacity to achieve high breakthrough volumes (BTV), the adsorption material is expected to have a low affinity for water (air moisture). This is not only important for GC/MS coupling. Neither water nor CO_2 should have a negative effect on the breakthrough volume of the organic components. The surrounding air generally has a high moisture content. However, particular precautions must be taken in the case of combustion gases. The desorption of the enriched components from the carrier materials used should be complete and without thermal changes.

The adsorption material used must have a high thermal stability and must not contribute to the background of the analysis. The expected adsorption and stability properties should not change for a large number of analyses, even on repeated use.

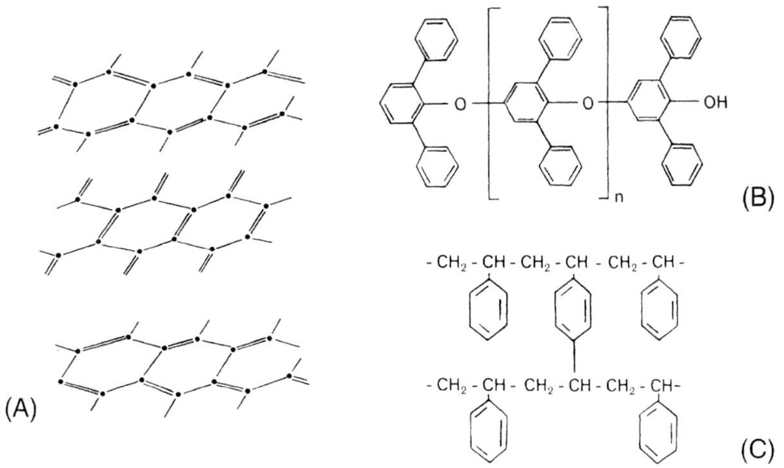

Fig. 2.32 Surface model for common adsorbents (Supelco).
(A) Carbotrap, surface area ca. 100 m^2/g, uniform charge distribution over all carbon atom centres.
(B) Tenax (2,6-diphenyl-p-phenylene oxide), surface area ca. 24 m^2/g, nonuniform charge distribution, the charge is essentially localised on the oxygen atoms. (Tenax-TA has replaced Tenax-GC as a new material of higher purity; Tenax-GR is a graphitised modification).
(C) Amberlite XAD-2, surface area ca. 300 m^2/g, nonuniform charge distribution, less polar than Tenax (XAD-4, ca. 800 m^2/g).

In air analysis adsorption materials, such as Tenax, Carbotrap and XAD resins, are generally used (see Fig. 2.32 and Table 2.13). Pure activated charcoal is indeed an outstanding adsorbent for all organic compounds; however, these can only be sufficiently desorbed when displaced using liquid solvents. Complete thermal desorption requires extremely high temperatures (>600 °C). This can lead to pyrolytic decomposition of the organic compounds which are then no longer detected in the residue analysis.

A recent development for retaining the high adsorptivity of activated charcoal with simultaneously favourable desorption properties involves the use of graphitised carbon black (Carbotrap, Carboxen). Its surface consists of graphite crystals. The associated hydrophobic properties and the exploitation of nonspecific interactions have led to wide use of these carrier materials.

Like Tenax, graphitised carbon blacks also have a low affinity for water. These adsorption materials can be dried with a dry gas stream, e.g. the GC carrier gas (in the direction of adsorption!), without significant loss of material (dry purge).

VOCARB traps are special combinations of the adsorption materials Carbopack (graphitised carbon black, GCB) and Carboxen (carbon molecular sieve). These combinations have been optimised in the form of VOCARB 3000 and VOCARB 4000 for volatile and involatile compounds respectively, corresponding to the EPA methods 624/1624 and 542.2. VOCARB 4000 exhibits higher adsorptivity for less volatile components, such as naphthalenes and trichlorobenzenes. However, it shows catalytic activity towards 2-chloroethyl vinyl ether (complete degradation!), 2,2-dichloropropane, bromoform and methyl bromide.

Volatile polar components are enriched on highly polar carrier materials. The combination of Tenax TA and silica gel has proved particularly successful for the enrichment of polar compounds.

Table 2.13 Adsorption materials and frequently described areas of use.

XAD-4	C_1/C_2-chlorinated hydrocarbons R11 halogenated narcotics vinyl chloride ethylene oxide styrene
Tenax TA 35-60 mesh	for boiling points 80–200 °C C_1/C_2-chlorinated hydrocarbons general solvents, volatile halogenated hydrocarbons BTEX phenols
Porapak	high-boiling, nonpolar substances halogenated narcotics ethylene oxide
Carbotrap/VOCARB	for boiling points –15 to –120 °C C_1/C_2-chlorinated hydrocarbons volatile halogenated hydrocarbons BTEX styrene
Molecular sieve 5 Å	N_2O

To cover a wider range of molecular sizes various adsorption materials are combined with one another. For example, the Carbotrap multibed adsorption tube (Fig. 2.33) consists of three materials: Carbotrap C with its low surface area of 12 m^2/g is used to enrich high molecular weight components, such as alkylbenzenes, polyaromatic hydrocarbons or PCBs, directly at the inlet. All the more volatile substances pass through to the subsequent layers. A layer of Carbotrap particles (Carbotrap B) separated by glass wool is characterised by the higher particle size of 20/40 mesh. Volatile organic substances, in particular, are excellently adsorbed on Carbotrap and thermally desorbed with high recoveries. To adsorb C_2-hydrocarbons Carbosieve S-III material with a particularly high surface area of 800 m^2/g and a pore

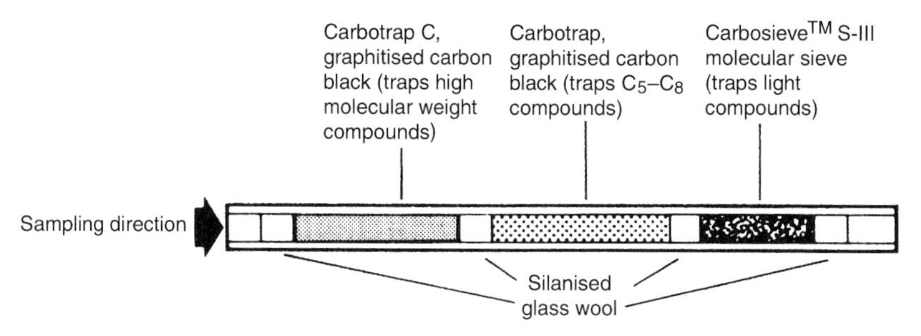

Fig. 2.33 Multibed adsorption/desorption tube Carbotrap 300 (Supelco).
Carbotrap, Carbotrap C: graphitised carbon black, GCB, surface area ca. 12 m^2/g.
Carbosieve S-III: carbon molecular sieve, surface area ca. 800 m^2/g, pore size 15–40 Å

Table 2.14 Breakthrough volumes [L] for Carbotrap 300 adsorption/desorption tubes (Supelco).

Substance	Carbosieve S-III	Carbotrap	Carbotrap C
Vinyl chloride	158		
Chloroform		1.1 a)	
1,2.Dichloroethane		0.4	
1,1,1-Trichloroethane		2.7 a)	
Carbon tetrachloride		4.7 a)	
1,2.Dichloropropane		6.8	
Trichloroethylene		2.5	
Bromoform		1.7	
Tetrachloroethylene		2.2	
Chlorobenzene		316	
n-Heptane		262	
1-Heptene		284 a)	
Benzene		2.3	
Toluene		130	
Ethylbenzene			12.9
p-Xylene			11.2
m-Xylene			11.0 a)
o-Xylene			11.0 a)
Cumene			27.8 a)

a) Theoretical value.

size of 15–40 Å is placed at the end of the adsorption tube (patent, BASF AG, Ludwigshafen) (Table 2.14). The hydrophobic properties of the Carbosieve material allow its use in atmospheres with high moisture contents.

2.1.6.1 Sample Collection

Sample collection can be either passive or active (Fig. 2.34). For passive collection diffusion tubes with special dimensions are used. The content of substances in the surrounding air is integrated, taking the collection time into account.

Active collection devices require a calibrated pump with which a predetermined volume is drawn through the adsorption tube. Having estimated the expected concentrations, for example for indoor air 100 mL/min and for outdoor air 1000 mL/min, the air is drawn through the prepared adsorption tube over a period of 4 h. After sample collection the adsorption tube must be closed tightly to exclude additional uncontrolled contamination.

Brown and Purnell carried out thorough investigations on the determination of breakthrough volumes. The latter generally vary widely with the collection rate. On use of Tenax as the adsorbent the ideal collection rate is 50 mL/min, in any case, however, between 5 and 600 mL/min. Moisture does not affect the breakthrough volumes with Tenax (unlike other porous materials). Furthermore, sample collection is greatly affected by temperature. An increase in temperature increases the breakthrough volume (ca. every 10 K doubles the volume).

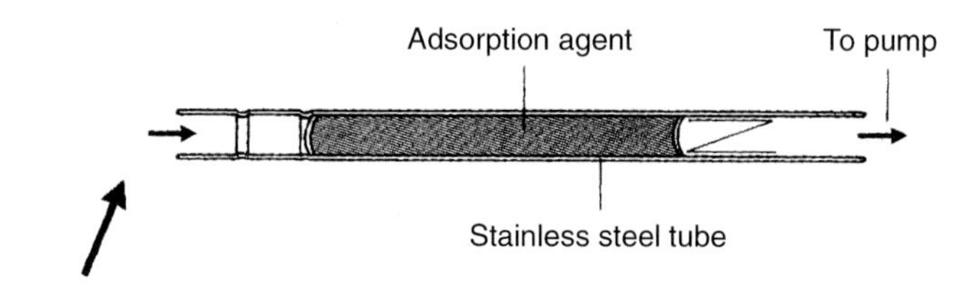

(A) **10–200 mL/min**

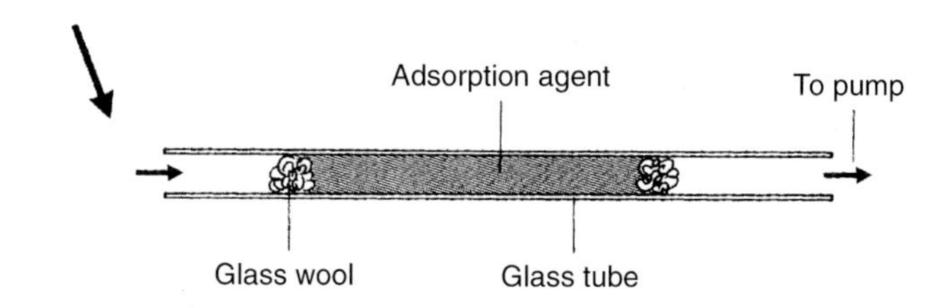

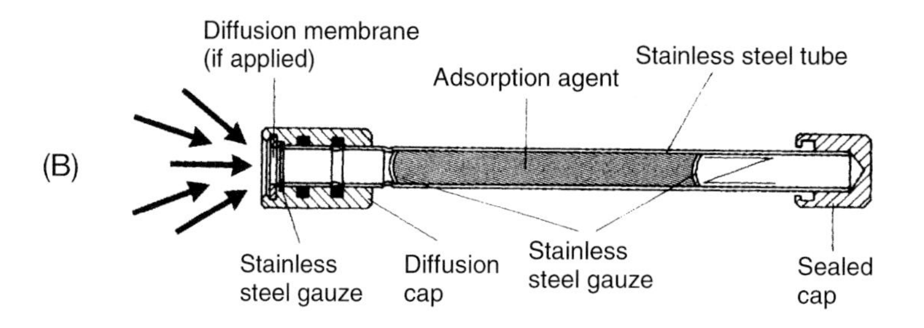

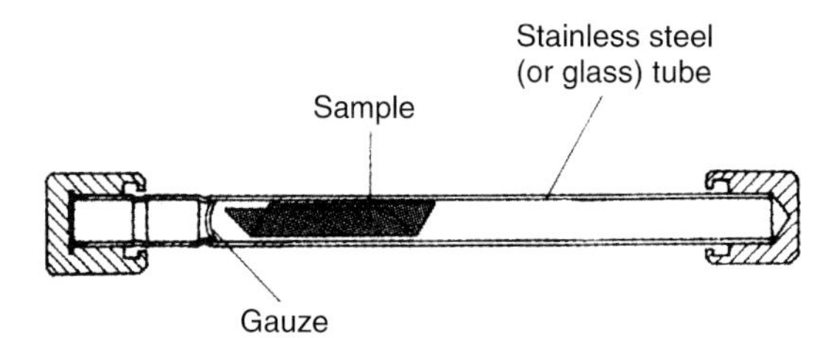

(C)

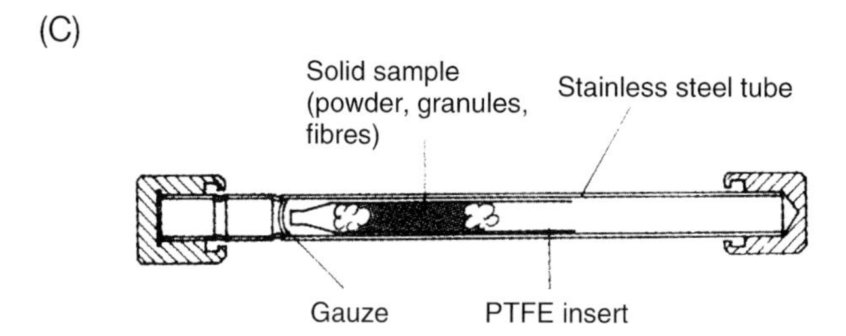

Fig. 2.34 Sample collection with thermodesorption tubes (Perkin-Elmer).
(A) Active sample collection with pump (e.g. personal air sampler),
(B) Passive sample collection by diffusion,
(C) Direct introduction of solid samples

2.1.6.2 **Calibration**

In the calibration of thermodesorption tubes the same conditions should predominate as in sample collection. Methods such as the liquid application of a calibration solution to the adsorption materials or the comparison with direct injections have been shown to be unsatisfactory. Alternatively Certified Reference Standards (CRS) are available (e.g. Markes Int., UK). CRS tubes are recommended in many key standard methods (e.g. US EPA Method TO-17) for auditing purposes and as a means of establishing analytical quality control. CRS tubes are often certified traceable to primary standards, have a minimum shelf life of typically 6 month. They are available ready for use with concentrations varying from 10 ng to 100 µg per component. Chromatograms from a shipping blank, and an example analysis of a CRS tube should be supplied along with the CRS Certificate.

CRS tubes are available loaded with benzene, toluene and xylene at levels of 25 ng or 1 çg per component; TO-17 standards at 25 ng per component of benzene, toluene, xylene, dichloromethane, 1,1,1-trichloroethane, 1,2,4-trimethylbenzene, methyl-t-butyl ether, butanol, ethyl acetate, methylethyl ketone. Custom CRS tubes are also available on a variety of sorbents for a wide range of compounds from different vendors.

The process for the preparation of standard atmospheres by continuous injection into a regulated air stream has been described in the VDI guidelines (Fig. 2.35). In this process an individual component or a mixture of the substances to be determined is continuously charged to an injector through which air is passed (complementary gas), using a thermostatted syringe burette. The air quantity (up to 500 ml/min) is adjusted using a mass flow

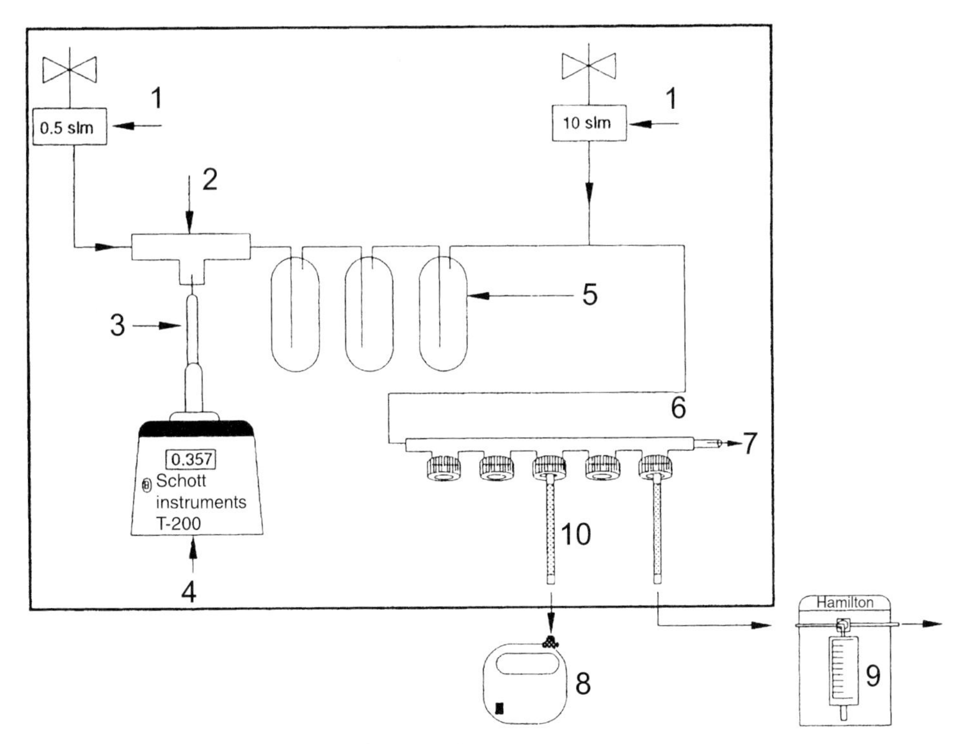

Fig. 2.35 Principle of a calibration unit for thermodesorption.

meter. The complementary gas can be diluted by mixing with a second air stream (dilution gas up to 10 l/min). Moistening the gas can be carried out inside or outside the apparatus. In this way concentrations in the ppm range can be generated. For further dilution, e. g. for calibration of pollution measurements, a separate dilution stage is necessary. The gas samples to be tested are drawn out of the calibration station into a glass tube with several outlets. In this case active or passive sample collection is possible. Continuous injection has the advantage that the preparation of mixtures is very flexible (Tables 2.15 and 2.16).

Table 2.15 Evaluation of test tubes which were prepared using the calibration unit by continuous injection ($n = 10$).

Component	Mean value	Standard deviation	Relative standard deviation [%]
1,1,1-Trichlorethane	562.6	4.95	0.88
Dichloromethane	538.6	7.93	1.47
Benzene	753.3	9.48	1.25
Trichloroethylene	627.3	16.7	2.67
Chloroform	626.6	12.1	1.94
Tetrachloroethylene	698.2	6.11	0.88
Toluene	1074	6.88	0.64
Ethylbenzene	358.1	2.91	0.81
p-Xylene	736.3	4.85	0.66
m-Xylene	731.6	4.77	0.65
Styrene	755.8	4.60	0.61
o-Xylene	389.5	3.49	0.89

Table 2.16 Analysis results of BTEX determination of certified samples after calibration with the calibration unit.

Mass [µg]	Benzene	Toluene	m-Xylene
Measured value	1.071	1.136	1.042
Required value (certified)	1.053	1.125	1.043
Standard deviation (certified)	0.014	0.015	0.015

2.1.6.3 Desorption

The elution of the organic compounds collected involves extraction by a solvent (displacement) or thermal desorption. Pentane, CS_2 or benzyl alcohol are generally used as extraction solvents. CS_2 is very suitable for activated charcoal, but cannot be used with polymeric materials, such as Tenax or Amberlite XAD, because decomposition occurs. As a result of displacement with solvents the sample is extensively diluted, which can lead to problems with the detection limits on mass spectrometric detection. With solvents additional contamination can occur. The extracts are usually applied as solutions. The readily automated static headspace technique can also be used for sample injection. This procedure has also proved to be effective for desorption using polar solvents, such as benzyl alcohol or ethylene glycol monophenyl ether (1% solution in water after Krebs).

In thermal desorption, the concentrated volatile components are released by rapid heating of the adsorption tube and after preliminary focusing, usually within the instrument, are injected into the GC/MS system for analysis (Table 2.17). Automated thermodesorption gives better sensitivity, precision and accuracy in the analysis. The number of manual steps in sample processing is reduced. Through frequent re-use of the adsorption tube and complete elimination of solvents from the analysis procedure, a significant lowering of cost per sample is achieved.

Table 2.17 Desorption temperatures for common adsorption materials and possible interfering components which can be detected by GC/MS.

Adsorbent	Desorption	Maximum temperature	Interfering components
Carbotrap	Up to 330 °C	> 400 °C	Not determined
Tenax	150–250 °C	375 °C	Benzene, toluene, trichloroethylene
Molecular sieve	250 °C	350 °C	Not determined
Porapak	200 °C	250 °C	Benzene, xylene, styrene
VOCARB 3000	250 °C	> 400 °C	Not determined
VOCARB 4000	250 °C	> 400 °C	Not determined
XAD-2/4	150 °C	230 °C	Benzene, xylene, styrene

Thermodesorption has now become a routine procedure because of program-controlled samplers. The individual steps are prescribed by the user in the control program and monitored internally by the instrument. For the sequential processing of a large number of samples, autosamplers with capacities of up to 100 adsorption tubes are commercially available.

The adsorption tubes are fitted with temporary Teflon caps, septums or sealed by use of SafeLok technique, a patented check valve design diffusion locking technology at either end of the tube (Markes Inc., UK). The sealing of the tubes protects the sample from ingress or loss of volatiles at all stages of the monitoring process. Tubes using check valves are not suitable for diffusive monitoring. The Teflon caps are removed inside the instrument before measurement and the adsorption tube is inserted into the desorption oven. Before the measurement is carried out the tightness of the seal is tested by monitoring an appropriate carrier gas pressure for a short time. Carrier gas is passed through the adsorption tubes in the desorption oven at temperatures of up to 400 °C (in the reverse direction to the adsorption!). The components released are stored in a cold trap inside the apparatus. The sample is transferred to the GC column by rapidly heating the cold trap. This two-stage desorption and the use of the multiple split technique enable the measuring range to be adapted to a wide range of substance concentrations. The sample quantity can be adapted to the capacity of the capillary column used through suitable split ratios both before and after the internal cold trap (Fig. 2.36).

Thermodesorption is now usually carried out automatically. The high performance achieved does not mean that it has no disadvantages. A gas sample in an adsorption tube for-

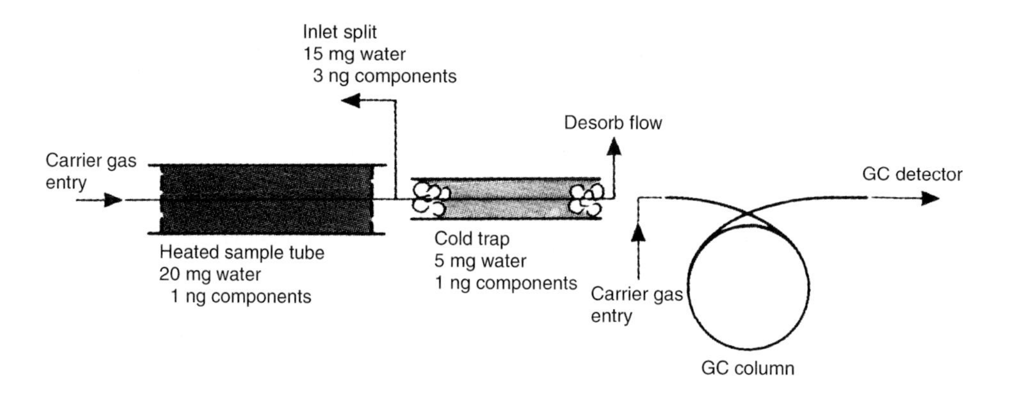

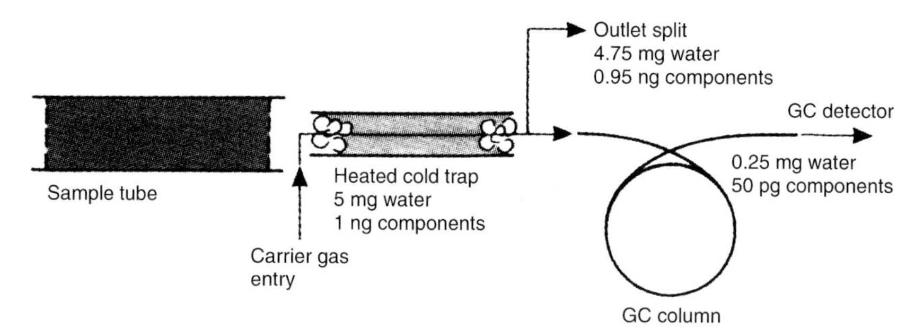

Fig. 2.36 Multiple split technique for thermodesorption (Perkin-Elmer).

merly always was unique. With a recent development by Markes International SecureTD-Q with a quantitative re-collection of a split flow for repeat analysis it overcomes the historical one-shot limitation of thermal desorption methods and simplifies method/data validation.

For high throughput applications an electronic tube tagging using RFID tube tags is available for industry standard sorbent tubes and 4.5-inch (DAAMS) tubes.

Possible and, for certain carrier materials, already known decomposition reactions have already been mentioned. For this reason another method is favoured by the EPA for air analysis. Passivated, nickel-coated canisters (ca. 2 L, maximum up to 15 L) are evacuated for sample collection. The samples collected on opening the canister can be measured several times in the laboratory. Suitable samplers are used, which are connected on-line with GC/MS. Cryofocusing is used to concentrate the analytes from the volumes collected. If required, the sample can be dried with a semipermeable membrane (Nafion drier) or by condensation of the water (moisture control system, see also Section 2.1.4.2). Adsorption materials are not used in these processes.

2.1.7
Pyrolysis and Thermal Extraction

The use of pyrolysis extends the area of use of GC/MS coupling to samples which cannot be separated by GC because they cannot be desorbed from a matrix or evaporated without decomposition. In analytical pyrolysis a large quantity of energy is passed into a sample so that fragments which can be gas chromatographed are formed reproducibly. The pyrolysis reaction initially involves thermal cleavage of C-C bonds, e. g. in the case of polymers. Thermally induced chemical reactions within the pyrolysis product are undesired side reactions and can be prevented by reaching the pyrolysis temperature as rapidly as possible. The reactions initiated by the pyrolysis are temperature-dependent. To produce a reproducible and quantifiable mixture of pyrolysis products, the heating rates and pyrolysis temperatures, in particular, should be kept constant. The sample and its contents can be characterised using the chromatographic sample trace (pyrogram) or by the mass spectroscopic identification of individual pyrolysis products.

The use of pyrolysis apparatus with GC/MS systems imposes particular requirements on them. The sample quantity applied must correspond to the capacity of commercial fused silica capillary columns. It is usually in the μg range or less. By selecting a suitable split ratio peak equivalents for the pyrolysis products can be achieved in the middle ng range. These lie completely within the range of modern GC/MS systems. The small sample quantities are also favourable for analytical pyrolysis in another aspect. The reproducibility of the procedure increases as the sample quantity is lowered as more rapid heat transport through the sample is possible (Fig. 2.37). Side reactions in the sample itself and as a result of reactive pyrolysis products are increasingly eliminated.

Because of the small sample quantities and the high reproducibility of the results, analytical pyrolysis has experienced a renaissance in recent years. Both in the analysis of polymers

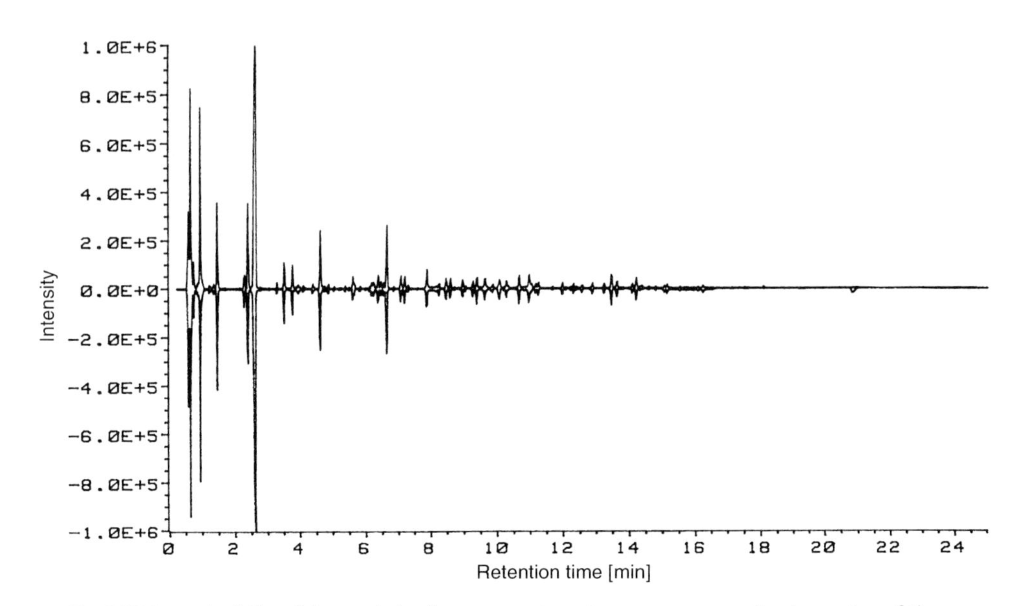

Fig. 2.37 Reproducibility of the pyrolysis of an automotive paint on two consecutive days using a foil pyrolyser coupled to GC/MS, mirrored representation (Steger, Audi AG).

with regard to quality, composition and stability, and in the areas of environmental analysis, foodstuffs analysis and forensic science, pyrolysis has become an important analytical tool, the significance of which has been increased immensely by coupling with GC/MS.

Analytical pyrolysis is currently dominated by two different processes: high frequency pyrolysis (Curie point pyrolysis) and foil pyrolysis. The processes differ principally through the different means of energy input and the different temperature rise times (TRT). Both pyrolysis processes can easily be connected to GC and GC/MS systems. The reactors currently used are constructed so they can be placed on top of GC injectors (split operation) and can be installed for short term use only if required.

Analytical Pyrolysis Procedures

Procedure	Foil	High frequency
Carrier	Pt foil	Fe/Ni alloys
Pyrolysis temperature	Can be freely selected up to 1400 °C	Fixed Curie temperatures
TRT	<8 ms	ca. 30 to 100 ms

2.1.7.1 Foil Pyrolysis

In foil pyrolysis the sample is applied to a thin platinum foil (Fig. 2.38). The thermal mass of this device is extremely low. After application of a heating current any desired temperature up to ca. 1400 °C can be achieved within milliseconds. The extremely high heating rate results in high reproducibility. The temperature of the Pt foil can be controlled by its resistance. However, the temperature can be measured and controlled more precisely and rapidly from the radiation emitted by the Pt foil. An exact calibration of the pyrolysis temperature can be carried out and the course of the pyrolysis is recorded by this feedback alone. Besides endothermic pyrolyses, exothermic processes can also be detected and recorded.

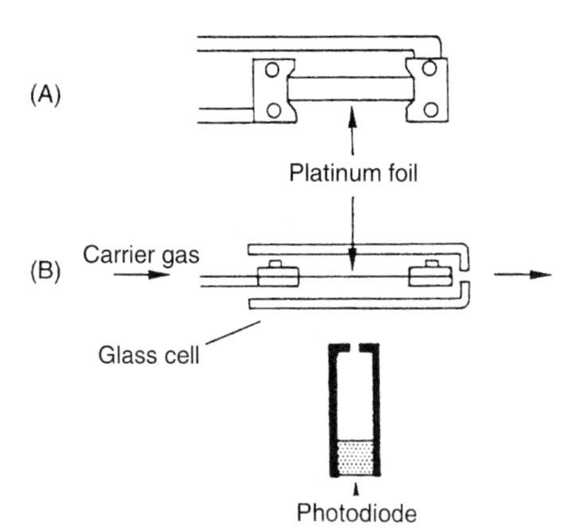

Fig. 2.38 Scheme of a Pt foil pyrolyser with temperature control by means of fibre optic cable (Pyrola).
(A) View of the Pt foil with carrier gas and current inlets.
(B) Side view of the pyrolysis cell with the glass cell (ca. 2 mL volume, pyrex) and photodiode under the Pt foil for calibration and monitoring of the pyrolysis temperature.

Pyrolysis Nomenclature

- Pyrolysis:
  A chemical degradation reaction initiated by thermal energy alone.
- Oxidative pyrolysis:
  A pyrolysis which is carried out in an oxidative atmosphere (e.g. O_2).
- Pyrolysate:
  The total products of a pyrolysis.
- Analytical pyrolysis:
  The characterisation of materials or of a chemical process by instrumental analysis of the pyrolysate.
- Applied pyrolysis:
  The production of commercially usable materials by pyrolysis.
- Temperature/time profile (TTP):
  The graph of temperature against time for an individual pyrolysis experiment.
- Temperature rise time (TRT):
  The time required by a pyrolyser to reach the pyrolysis temperature from the start time.
- Flash pyrolysis:
  A pyrolysis which is carried out with a short temperature rise time to achieve a constant final temperature.
- Continuous pyrolyser:
  A pyrolyser where the sample is placed in a preheated reactor.
- Pulse pyrolyser:
  A pyrolyser where the sample is placed in a cold reactor and then rapidly heated.
- Foil pyrolyser:
  A pyrolyser where the sample is applied to metal foil or band which is directly heated as a result of its resistance.
- Curie point pyrolyser:
  A pyrolyser with a ferromagnetic sample carrier which is heated inductively to its Curie point.
- Temperature-programmed pyrolysis:
  A pyrolysis where the sample is heated at a controlled rate over a range of temperatures at which pyrolysis occurs.
- Sequential pyrolysis:
  Pyrolysis where the sample is repeatedly pyrolysed for a short time under identical conditions (kinetic studies).
- Fractionated pyrolysis:
  A pyrolysis where the sample is pyrolysed under different conditions in order to investigate different sample fractions.
- Pyrogram:
  The chromatogram (GC, GC/MS) or spectrum (MS) of a pyrolysate.

2.1.7.2 Curie Point Pyrolysis

High frequency pyrolysis uses the known property of ferromagnetic alloys of losing their magnetism spontaneously above the Curie temperature (Curie point). At this temperature a large number of properties change, such as the electrical resistance or the specific heat. Above the Curie temperature ferromagnetic substances exhibit paramagnetic properties. The possibility of reaching a defined and constant temperature using the Curie point was first realised by W. Simon in 1965. In a high frequency field a ferromagnetic alloy does not absorb any more energy above its Curie point and remains at this temperature. As the Curie temperature is substance-dependent, another Curie temperature can be used by changing the material. If a sample is applied to a ferromagnetic material and is heated in an energy-rich high frequency field (Fig. 2.39), the pyrolysis takes place at the temperature determined by the choice of alloy and thus its Curie point. The temperature rise curves for various metals and alloys are shown in Fig. 2.40.

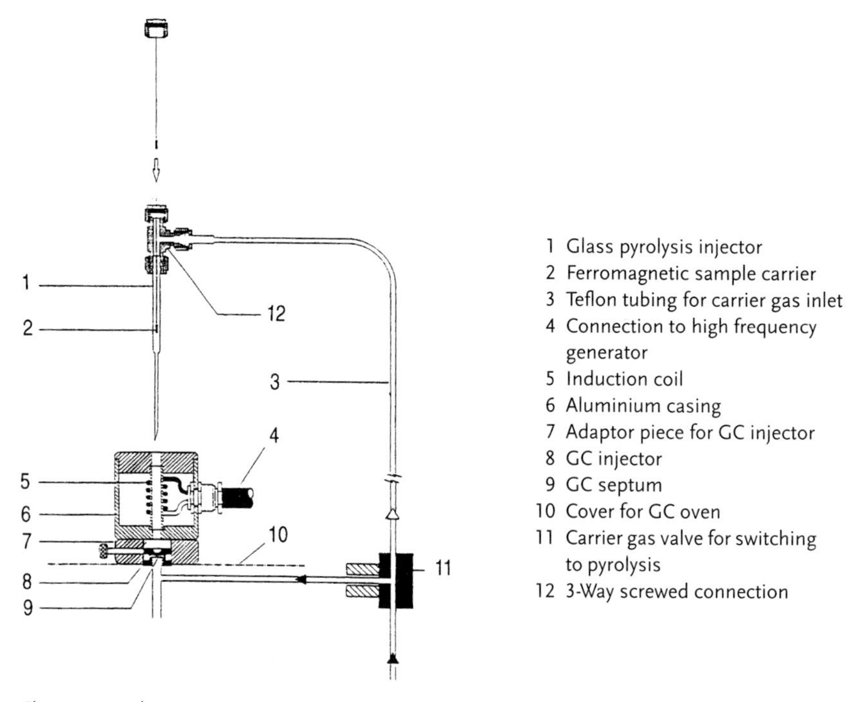

Fig. 2.39 Pyrolysis injector (Curie point hydrolysis) (after Fischer).

In practice sample carriers in the form of loops, coils or simple wires made of different alloys at fixed temperature intervals are used. A disadvantage of Curie point pyrolysis is the longer temperature rise time of ca. 30–100 ms required to reach the Curie point compared with foil pyrolysis. There are also effects due to the not completely inert surface of the sample carrier. They manifest themselves in the inadequate reproducibility of the pyrolysis process for analytical purposes. Copolymers with thermally reactive functional groups (e. g. free OH, NH_2 or COOH groups) cannot be analysed by Curie point pyrolysis.

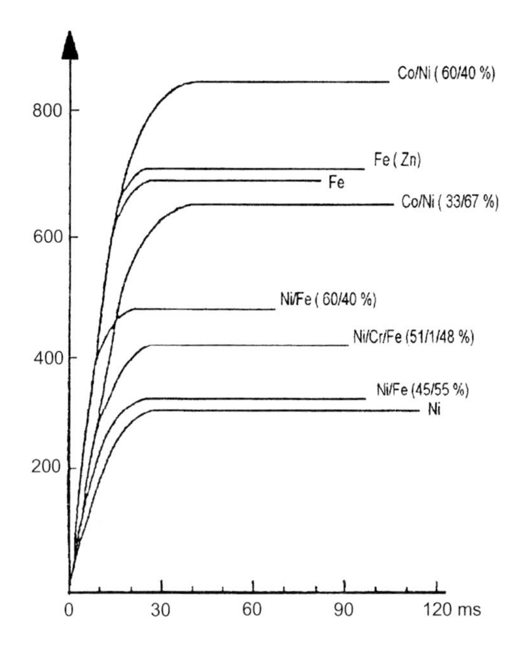

Fig. 2.40 Curie temperatures and temperature/time profiles of various ferromagnetic materials (after Simon).

For the analytical assessment of the coupling of pyrolysers with GC and GC/MS systems, a high-boiling mixture of cholesterol with n-alkanes has been proposed (Gassiot-Matas). Figure 2.41 shows the evaporation of cholesterol (500 ng) with the C_{34}- and C_{36}-n-alkanes (50 ng of each). Before the intact cholesterol is detected, its dehydration product appears. The intensity of this peak increases with small sample quantities and increasing temperatures in the region of substance transfer in the GC injector. The high peak symmetries of the signals of the alkanes, the cholesterol and its dehydration product indicate that the coupling is functioning well.

There are various possibilities for the evaluation of pyrograms obtained with a GC/MS system. With classical FID detection the pyrogram pattern is only compared with known standards. However, with GC/MS systems the mass spectra of the individual pyrolysis products can be evaluated. By using libraries of spectra, substances and substance groups can be identified. GC/MS pyrograms can be selectively investigated for trace components even in complex separating situations by using the characteristic mass fragments of minor components. The comparison of sample patterns becomes meaningful through the choice of mass chromatograms of substance-specific fragment ions. GC/MS pyrograms with its full pattern information can be stored and compared for identity or similarity. Suitable software systems are commercially available providing a numerical measure on similarity instead a visual inspection only (Chromsearch, Axel Semrau, Germany; Xaminer, Thermo Fisher, USA).

In the library search for the spectra of pyrolysis products special care must be taken. Commercial spectral libraries consist of spectra of particular substances which have been fully characterized, which is not the case for the majority of pyrolysis products. Classical mass spectrometry fragmentation rules apply for interpretation of search results. Depending on the sample material, however, an extremely large number of reaction products are formed on pyrolysis, which cannot be completely separated even under the best GC conditions. If the detection is sensitive enough this situation is shown clearly by the mass chromatogram. Also, for polymers there is a typical appearance of homologous fragments, which must also be taken

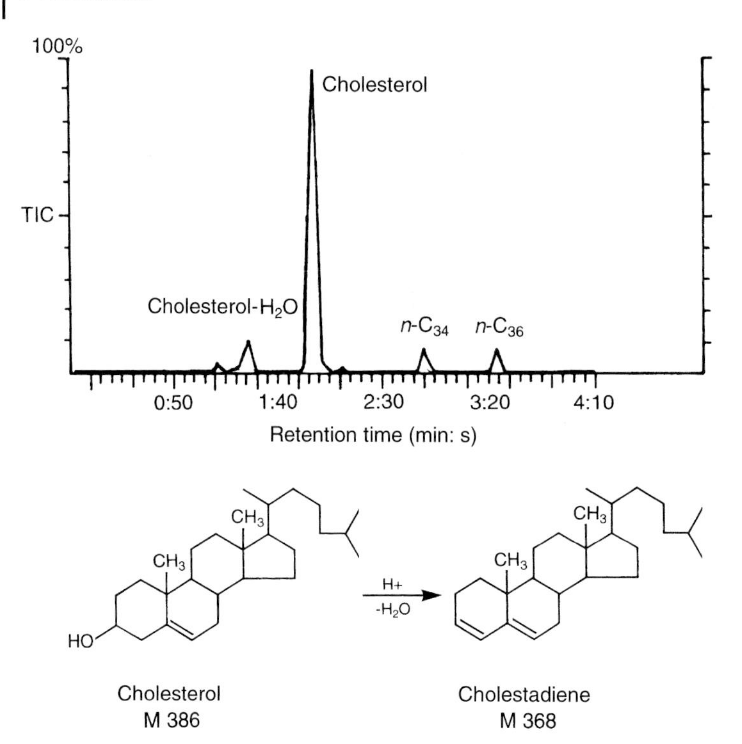

Fig. 2.41 Analysis of cholesterol and n-alkanes (C_{34}, C_{36}) for testing the pyrolysis coupling to the GC injector (after Richards 1988).
Conditions: GC column 4 m × 0.22 mm ID × 0.25 μm CP-Sil5, 200–320 °C, 30 °C/min, GC-ITD

into account. These series can also be shown easily using mass chromatograms through the choice of suitable fragment ions.

Quantitative determinations using pyrolysis benefit particularly from the selectivity of GC/MS detection. The precision is comparable to that of liquid injections.

2.1.7.3 **Thermal Extraction**

Besides the processes of foil and Curie point pyrolysis described, thermal extraction (Fig. 2.42) covers the area between pyrolysis and thermodesorption. In a thermal extraction device the sample (manual or automatic) is placed in a quartz oven, which can be operated at a constant temperature of up to 750 °C. This allows both the thermal extraction of volatile components from solid samples as well as their pyrolysis. Other technical solutions allow the temperature control of the pyrolysis chamber for a thermal desorption step with collection of the evaporated material either on a dedicated trap (e.g. Markes Inc., UK; Frontier Laboratories, Japan) or in the GC injector (PTV with sorbent, PYROL AB, Sweden).

The difference between thermal extraction and the analytical pyrolysis systems described above lies in the consideration of the sample quantity. The typical capacity of a thermal extractor is between 0.1 mg and ca. 500 mg sample material. This allows weighing out and reweighing. Even inhomogeneous materials can be investigated in this way.

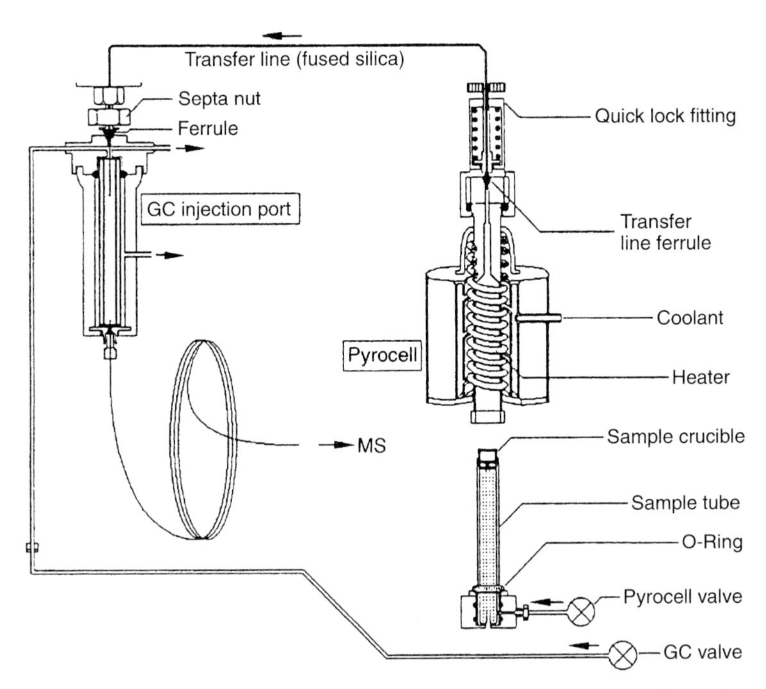

Fig. 2.42 Thermal extractor, cross section and coupling to a GC injector (after Ruska).

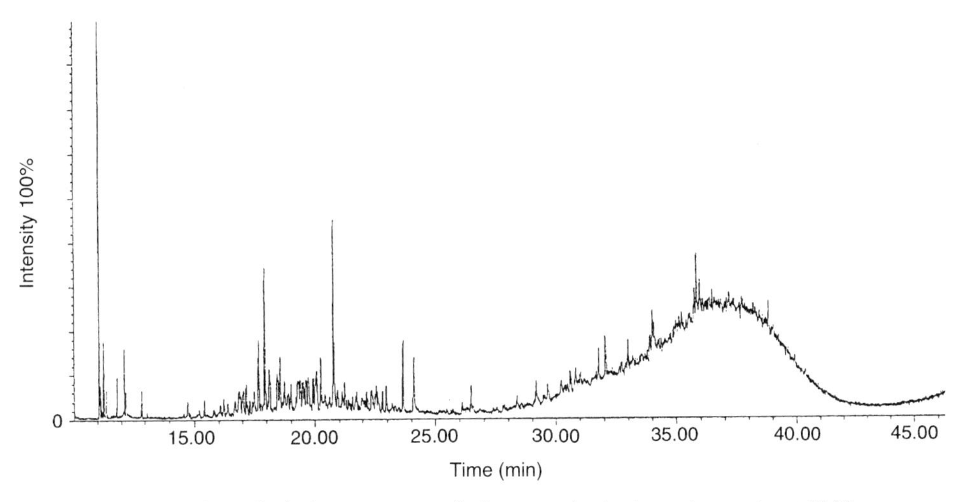

Fig. 2.43 GC/MS analysis of volatile components of a floor covering by thermal extraction at 80 °C (after Ruska).

The wide temperature range allows a broad spectrum of applications. The stripping of volatile compounds, e.g. solvents, volatile halogenated hydrocarbons and BTEX, is carried out typically below 80 °C (Fig. 2.43). At temperatures of up to 350 °C involatile compounds are released from the sample, for example from plastics or composite materials. Pyrolytic cleavage reactions begin at temperatures above 350 °C. Because of the large sample quantities and the indirect heating of the sample in an oven, the temperature in the sample increases relatively slowly. The pyrograms obtained can be compared with pyrolysis patterns which take precedence.

In the field of environmental analysis thermal extraction is proposed by an EPA method for the quantitative analysis of semivolatile compounds from solid sample materials. The method EPA 8275 is a thermal extraction capillary GC/MS procedure for the rapid and quantitative determination of targeted PCBs and PAHs in soils, sludges and solid wastes (EPA 1996).

2.2 Gas Chromatography

2.2.1 Fast Gas Chromatography

The pressure of workload in modern analytical laboratories necessitates an increased throughput of samples and hence method development is largely focused on productivity. However, speed of chromatographic analysis and enhancement of peak separation generally require analytical method optimization in opposite directions. Basically, two adjacent approaches are increasingly used to combine advancements in speed and peak separation. In the existing conventional GC ovens, the use of narrow bore columns below 0.1 mm i.d. offers a viable practical "Fast GC" solution for almost every GC and GC/MS with electronic pressure regulation. Alternatively the installation of cartridges for direct column heating are today commonly used even for „Ultra Fast“ GC separations.

A recent publication has studied the performance of different columns featuring different lengths and internal diameters at different heating rates and with different carrier gas types and flow rates (Facchetti 2002).

2.2.1.1 Fast Chromatography

The term "Fast GC" compares to the conventionally used fused silica columns of typical dimensions with lengths of 30 m or longer and inner diameters of 0.25 mm or higher. These column types became standard since the introduction of the fused silica capillaries used for most applications in GC/MS. With improved and partially automated sample preparation methods (see e.g. Section 2.1.3), GC analysis time is becoming the rate determining step for productivity in many laboratories. A reduction of GC/MS analysis time is possible with the utilization of alternatively available techniques using a reduced column length and inner diameter with appropriate film thicknesses and oven temperature programming.

A very good example describing the progress in GC separation technology and advancement in GC productivity is given with an optimisation strategy starting from a packed column separation (Facchetti 2005). The sample used for this comparison is the 16 component standard mixture used in EPA 610 analysis of polynuclear aromatic hydrocarbons (PAH). The EPA method is antiquated and describes a 45 minute analysis using a packed column containing OV17. All columns used in this example were 5% phenyl polysiphenylene-siloxane coated Trace TR-5MS columns (Thermo Fisher Scientific, Bellefonte, PA, USA) with di-

Table 2.18 Compounds in the EPA 610 standard on packed column.
Note: Four pairs of peaks remain unresolved using the EPA 610 packed column technique, marked with *.

Component	Retention time [min]
1. Naphthalene	4.50
2. Acenaphthylene	10.40
3. Acenaphthene	10.80
4. Fluorene	12.60
5. Phenanthrene	15.90*
6. Anthracene	15.90*
7. Fluoranthene	19.80
8. Pyrene	20.60
9. Benzo(a)anthracene	24.70*
10. Chrysene	24.70*
11. Benzo(b)fluoranthene	28.00*
12. Benzo(k)fluoranthene	28.00*
13. Benzo (a) pyrene	29.40
14. Dibenzo(a,h)anthracene	36.20*
15. Indeno(1,2,3-cd)pyrene	36.20*
16. Benzo(ghi)perylene	38.60

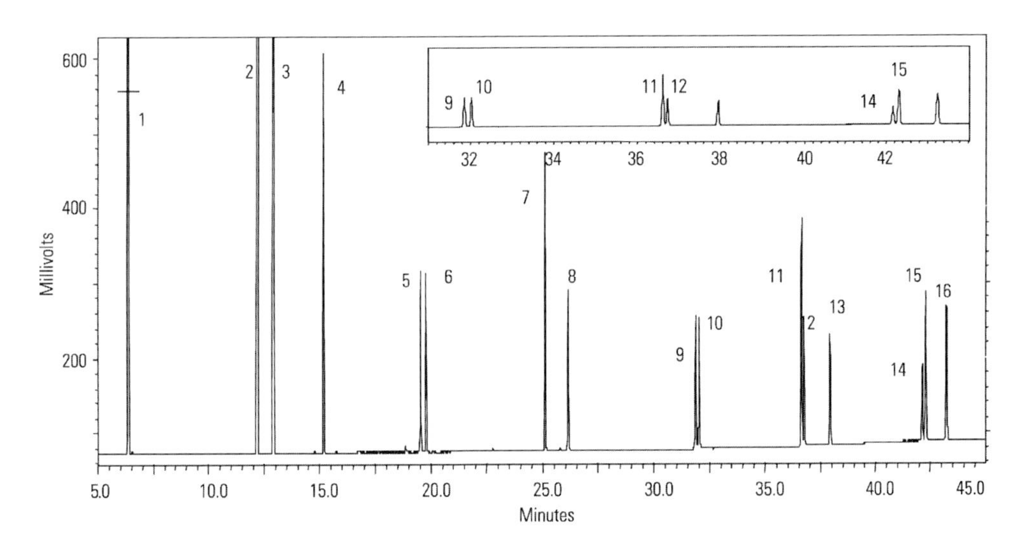

Fig. 2.44 Simulation of a packed column separation performed on a 30 m capillary column. Carrier gas has been changed from nitrogen to helium and compared with a 3% OV17 packed column, the 5% phenyl-methylpolysiloxane capillary column separates 4 previously unresolved peak pairs in a similar run time. Column: 30 m Trace TR-5MS; ID 0.25 mm, Film 0.25 μm; Initial Temp: 100 °C; Rate 1: 5 °C/min to 300 °C; Carrier Gas: Helium; Gas Flow: 1.5 mL/min. (constant flow); Detector: FID; Split Ratio: 50:1; Injection vol.: 1.0 μL by TriPlus auto sampler; **Sample run rate is 1 per hour** (includes oven cooling)

Table 2.19 Variables affecting GC run time.

Parameter	Run time effects
Temperature ramp rate	Higher starting temperature and steeper heating rate → peaks elute faster
Column length	Shorter column → shorter run time
Carrier gas flow rate	Each gas type has an optimum linear flow range within which speed can be adjusted
Film thickness	Thicker film → more interaction of solutes → longer run time Thinner film → Fast Chromatography
Column internal diameter	Reducing internal diameter increases column efficiency → shorter run-times using "Fast Chromatography"
Ultra fast technology	Combination of short column, small internal diameter, thin film, special ultra fast heating coil for ultra fast separations

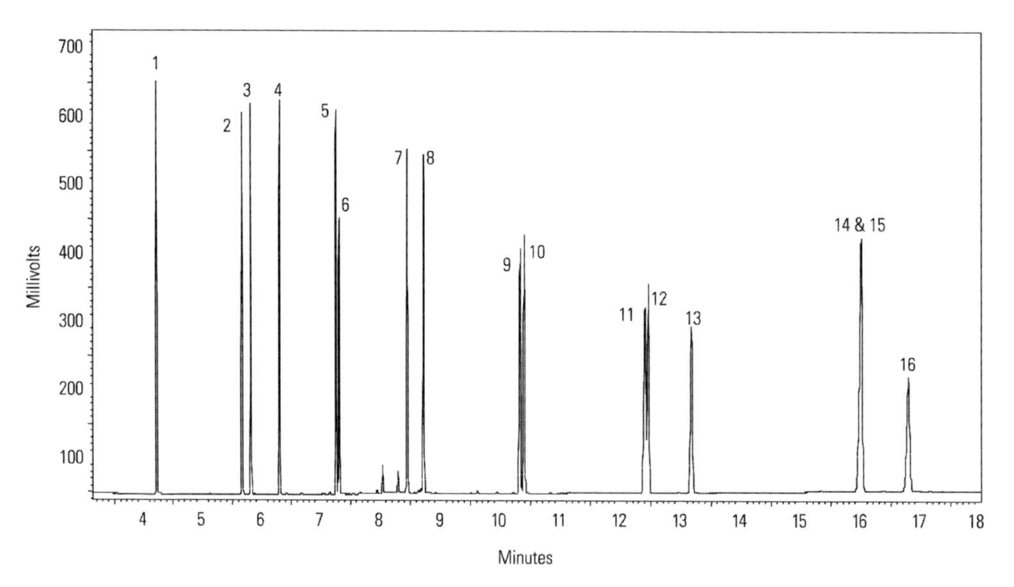

Fig. 2.45 Optimisation of the oven program has cut the run time by more than half without changing any other parameters. There is some loss of resolution between peaks 14 and 15 compared with the 45 min run, however the extra specificity of a MS detector would resolve these peaks.
Column: 30 m Trace TR-5MS; ID 0.25 mm, Film 0.25 μm; Initial Temp: 90 °C hold 1 min; Rate 1: 25 °C/min to 290 °C; Rate 2: 4 °C/min to 320 °C hold 5 min; Carrier Gas: Helium; Gas Flow: 1.5 mL/min. (constant flow); Detector: FID; Split Ratio: 50:1; Injection vol.: 0.5 μL manual injection; **Sample run rate is 2 per hour** (includes oven cooling).

mensions as specified with each chromatogram. Helium is the carrier gas in each case. Flow rates and oven programs were optimized to the column dimensions. The chromatograms were run in either constant flow or constant pressure mode as indicated using a Trace Ultra GC with a TriPlus autosampler (Thermo Fisher Scientific, Milan, Italy) fitted with a FID detector.

Initially, capillary GC methods attempted to simulate the packed column method with a run time of 45 minutes. The example below in Fig. 2.44 shows a capillary separation where the oven program is slowed down to keep the run time similar to the packed column method, and helium has been substituted for the original nitrogen carrier gas. However, with the increasing pressures on analysis time, and the flexibility within the EPA 610 method, it has been possible to greatly reduce run time by altering simple GC variables.

A number of variables within the GC setup can be manipulated to alter the run time, bearing in mind the need to maintain resolution, elution order and sensitivity as outlined in Table 2.19.

The first strategy for reducing run time should be to modify the oven ramp by increasing the ramp where the peaks are well separated with large retention time spaces and applying a slow ramp where extra separation is required (see Fig. 2.45).

Reducing column length is the second strategy for reducing run time. Note the need to increase the oven ramp rate in order to elute the high boiling materials, whilst retaining peak shape (see Fig. 2.46 with the switch to a 15 m column).

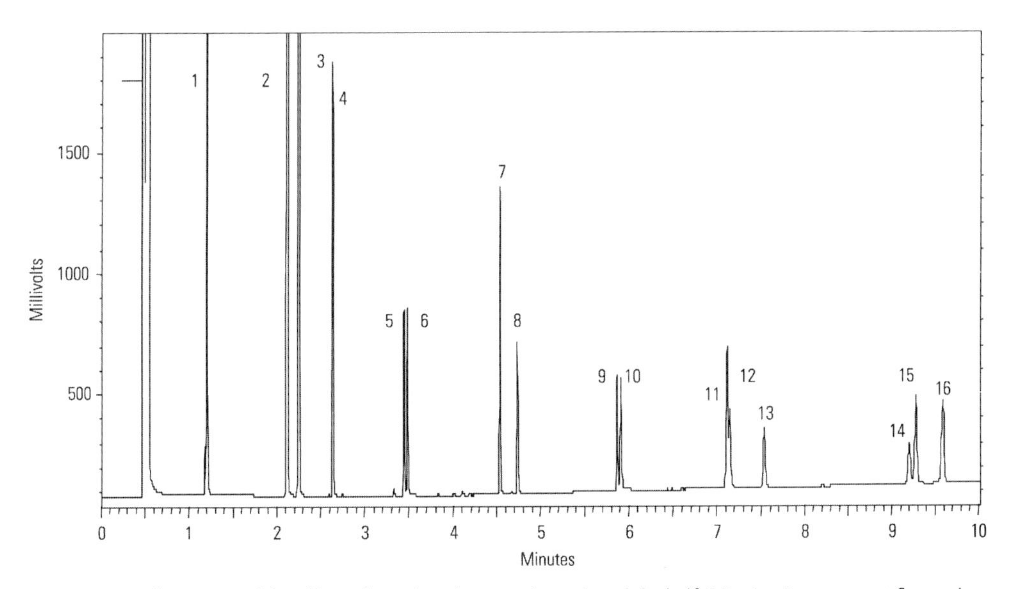

Fig. 2.46 Illustration of the effect of simply reducing column length by half (15 m) using constant flow reducing the run time to 10 min.
Column: 15m Trace TR-5MS; ID 0.25 mm, Film 0.25 μm; Initial Temp: 120 °C hold 0.2 min; Rate 1: 25 °C/min to 260 °C; Rate 2: 7 °C/min to 300 °C hold 3 min; Carrier Gas: Helium; Gas Flow: 1.5 mL/min constant flow; Detector: FID; Split Ratio: 50:1; Injection vol.: 1.0 μL by TriPlus auto sampler; **Sample run rate is 4 per hour** (includes oven cooling).

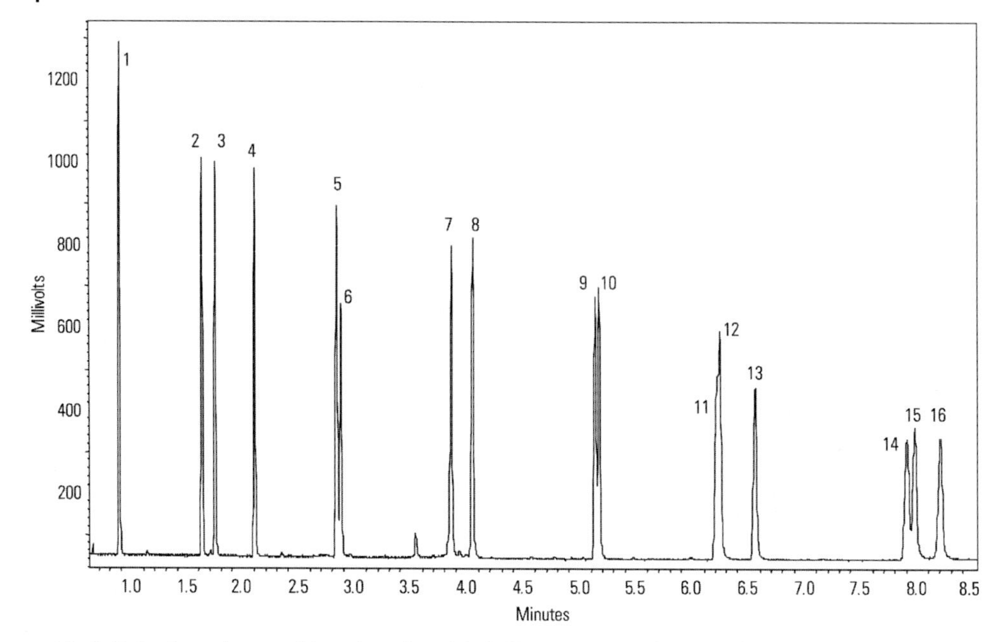

Fig. 2.47 Further reduction of the column length by half to 7 m but with larger ID, illustrates the effect of film thickness on run time. The larger ID makes the column less efficient. Therefore to maintain the resolution a similar oven program is needed, making the analysis time only slightly shorter than on the 15 m column. Column: 7 m Trace TR-5MS; ID 0.32 mm, Film 0.25 μm; Initial Temp: 120 °C hold 1 min; Rate 1: 25 °C/min to 250 °C; Rate 2: 10 °C/min to 300 °C hold 5 min; Gas Flow: 1.5 mL/min constant flow mode; Carrier Gas: Helium; Detector: FID; Split Ratio: 50:1; Injection vol.: 0.5 μL manual injection; **Sample run rate is 5 per hour** (includes oven cooling).

2.2.1.2 **Ultra Fast Chromatography**

The term Ultra Fast Gas Chromatography describes an advanced column heating technique that is taking advantage of short, narrow bore capillary columns, and very high temperature programming rates. This technique offers to shorten the analysis time by a factor up to 30 compared to conventional capillary gas chromatography. Ultra fast GC conditions typically apply the use of short 0.1 mm i.d. narrow-bore capillary columns of 2.5 to 5 meters in length with elevated heating rates of 100 to 1200 °C/min providing peak widths of 100 ms or less, see Fig. 2.49. At this point the limiting factor for sample throughput becomes the ability of the column to cool quickly enough between runs. In addition to a fast heating/cooling regime, a fast conventional or MS detection with an acquisition rate better than 25 Hz for reliable peak identification is required in, e.g. a restricted scan, SIM/MID or TOF mode, to achieve the necessary sampling rate across individual peaks. Currently, TOF and the fast scanning quadrupole technologies are the only MS technologies that are able to provide the required speed of detection for a ultra fast GC/MS at data rates of 50 Hz or better with a partial or full mass spectral information.

The special instrumentation required for ultra fast chromatography comprises a dedicated ultra fast column module (UFM), see Fig. 2.51, comprising a specially assembled fused silica column for direct resistive heating, wrapped with a heating element and a temperature

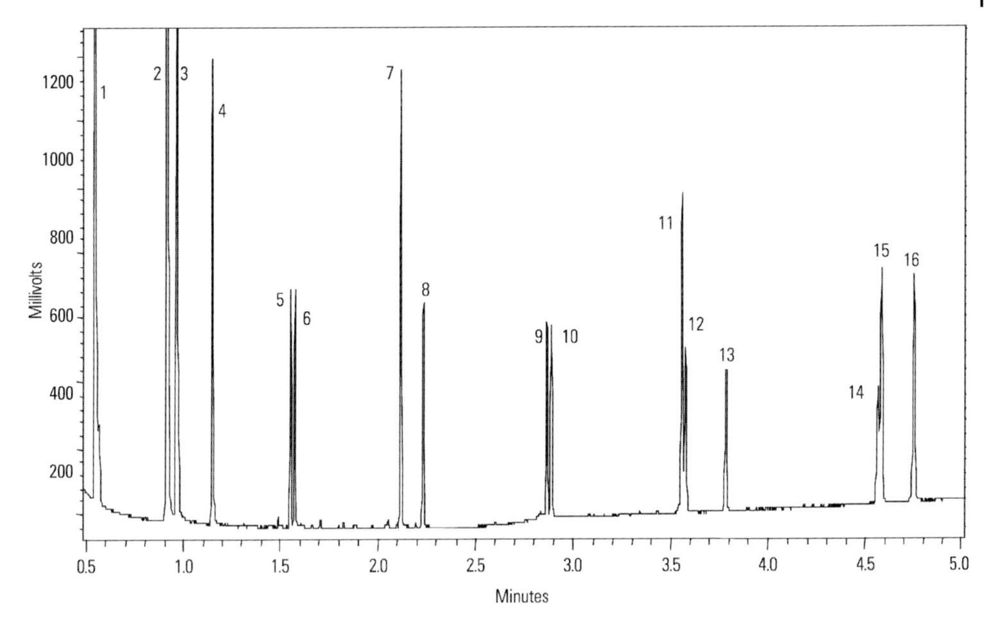

Fig. 2.48 The combination of using a narrow 0.1 mm column coated with a thinner film is the strategy for fast analysis. In comparison with the 7 m 0.25 μm film column, the longer 10 m x 0.1 μm film is faster. In addition, using constant pressure has enhanced the analysis speed down to under 5 minutes. The effect of running at constant pressure is to greatly speed up the linear velocity at the beginning of the run where the peaks are well separated. Later in the run, when the temperature is high, the velocity is lower allowing more dwell time on the column for the later peaks which elute close together.

sensor, see Fig. 2.52 (Facchetti 2002). The assembly is held in a compact metal cage to be installed inside the regular GC oven (which is not active during ultra fast operation). Temperature programming rates can be achieved as high as 1200 °C/min. UFM modules are commercially available providing different column lengths and diameters.

Another compelling example in Fig. 2.53 shows the ultra fast analysis of a mineral oil performed in only 2 minutes still maintaining enough resolution to separate pristane and phytane from n-C17 and n-C18.

2.2.2 Two Dimensional Gas Chromatography

The search for innovative solutions to increase the limiting peak capacities in gas chromatography came up many years ago with two dimensional separations. The development of instrumental techniques focused in the beginning on the transfer of specific peak regions of interest, the unresolved "humps", to a second separation column. This technique, widely known as "heart cutting", found many successful applications in solving critical research oriented projects but its use was not widespread in routine laboratories

The objective to increase the peak capacity for the resolution of entire complex chromatograms and not only of few discrete congested areas could be satisfied first with the introduction of a comprehensive two-dimensional gas chromatography (GCxGC), the most powerful

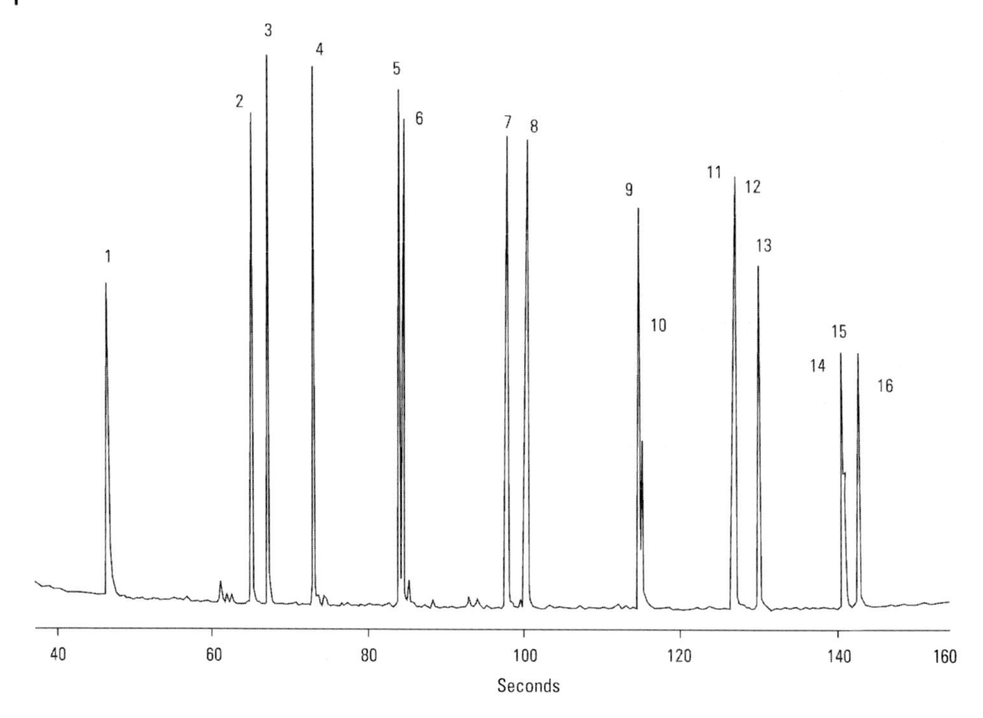

Fig. 2.49 Ultra Fast column and detector can run sample in just 160 seconds improving sample throughput to approx 12 per hour.; Column: 5 m Ultra Fast TR5-MS; ID 0.1 mm, Film 0.1 μm; Flow Rate: 1.0 mL/min; Carrier Gas: Hydrogen; Oven: 40 °C hold 0.3 min; Rate 1: 2 °C/s to 330 °C; **Sample run rate is 12 per hour** (includes column cooling).

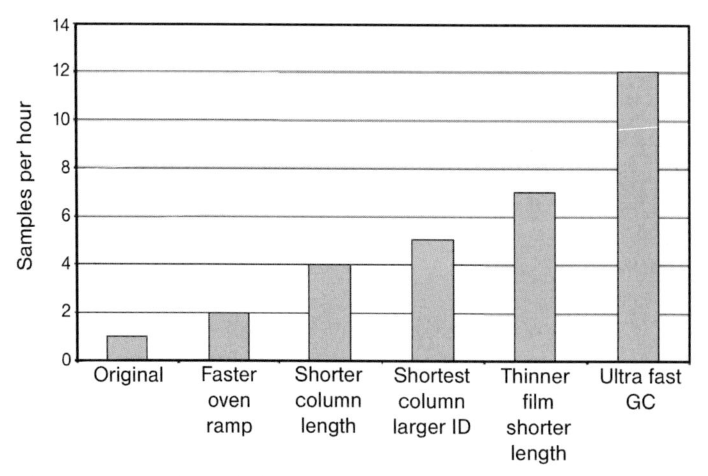

Fig. 2.50 Increased sample throughput by progress in column technology.

Fig. 2.51 UFM column module installed in a regular GC oven (Mega, Legnano, Italy; TRACE GC Ultra, Thermo Fisher Scientific, Milan, Italy).

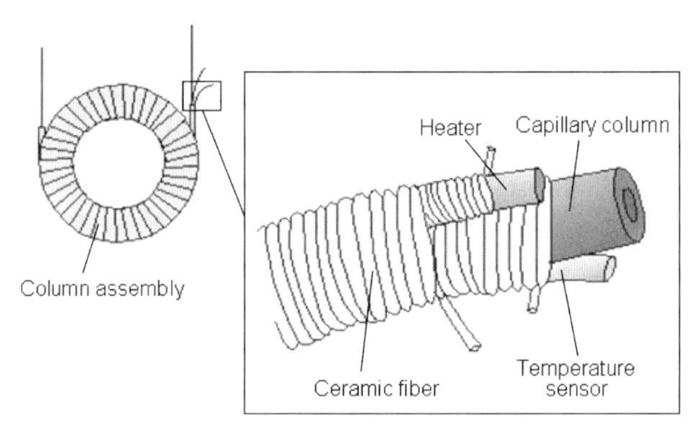

Fig. 2.52 Column wrapping detail with heating element and temperature sensor for ultra fast capillary chromatography.

two-dimensional solution today (Liu 1991). While the initial ideas of multidimensional GC relied on the use of two regular capillary columns (MDGC, selectivity tuning), current comprehensive two-dimensional chromatography benefits from the alignment with fast GC on the second separation dimension (Beens 2000, Bertsch 2000). The result is a significantly improved separation power. With the commercial availability of GC×GC instrumentation this technique found avid utilization for a widespread range of applications within the first few years (Marriott 2003).

Particularly for the separation of complex mixtures, GC×GC showed far greater resolution and a significant boost in signal to noise, with no increase in analysis times. Further, the hyphenation of GC×GC with MS detection providing three independent analytical dimensions made this technique ideal for the measurement of organic components within complex samples such as those from environmental, petrochemical and biological analysis. The significantly increased chromatographic resolution in GC×GC allows separation of

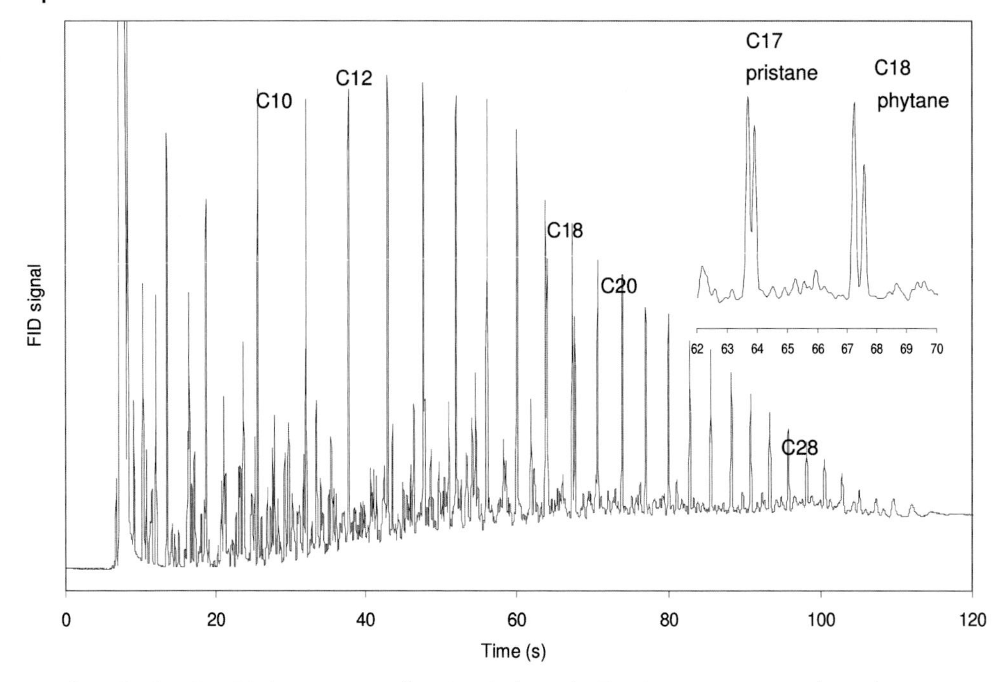

Fig. 2.53 Ultra Fast GC chromatogram of a mineral oil sample. The chromatogram was obtained using a 5 m, 0.1 mm ID, 0.1 μm film thickness SE54 column with a temperature program from 40 (12 s) to 350 °C (6 s) at 180 °C/min.

many previously undetectable components. Fast scanning mass spectrometers are required to maintain the excellent chromatographic resolution, provided today with full scan data by time-of-flight MS, or in selected ion monitoring mode also by quadrupole or high resolution MS systems. More than 15,000 peaks could be detected in an ambient air sample from the city of Augsburg by thermal desorption comprehensive two-dimensional GC with time-of-flight MS detection (Welthagen 2003).

The great advantage of the combination of multiple chromatographic steps is the increase in peak capacity. Peak capacity is the maximum number of peaks that can be resolved in a given time frame. The more peaks a combination of techniques is able to resolve, the more complex samples can be analysed. When a sample is separated using two dissimilar columns, the maximum peak capacity Φ_{max} will be the product of the individual column's peak capacity Φ_n.

$$\Phi_{max} = \Phi_1 \times \Phi_2$$

For example, if each separation mode generates peak capacities of 1100 in the first dimension and 30 in the second, the theoretical peak capacity of the 2D experiment will be 33,000, a huge gain in separation space, which would theoretically compare to the separation power of a 12,000 m column in the normal single dimension analysis. To achieve this gain, however, the two techniques should be totally orthogonal, that is, based upon completely different separation mechanisms.

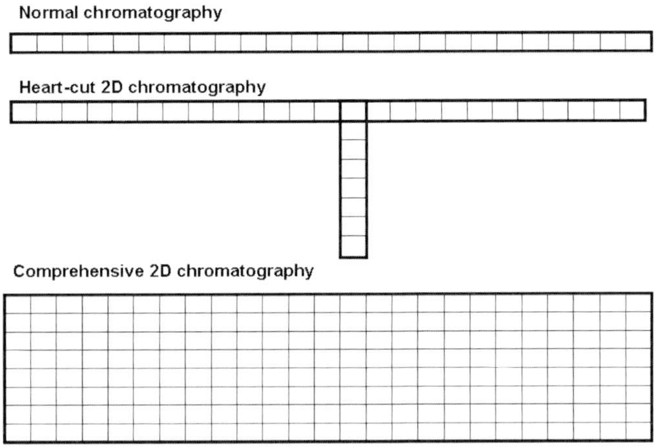

Normal chromatography

Heart-cut 2D chromatography

Comprehensive 2D chromatography

Comparison of peak capacities in normal, heart-cut and comprehensive two-dimensional chromatography

2.2.2.1 Heart Cutting

The goal of heart cutting is the increase of peak capacity for target substances in congested regions of unresolved compounds. Two capillary columns are connected in series, typically by means a valveless flow switching system. One or more short retention time slices are sampled onto a second separation column. A valveless switching device is ideally suited for high speed flow switching. It is based on the principle of pressure balancing and was introduced by Deans already in 1968. This technique found a wider recognition first with state-of-the-art EPC units, which allowed the integration of column switching as part of regular GC methods for routine applications (Deans 1968, MacNamara 2003, Majors 2005).

Heart cutting requires a sample specific individual setup of the analysis strategy. The sections of interest have to be previously identified and selected by retention time from the first chromatogram to be "cut" into the second column. The choice of column length is application specific and does not interfere with the "cutting" process. The second dimension column can be of the same but recommended of a different film polarity. A monitor detector is required, typically a universal FID, which continuously observes the separation on the first dimension. A second detector, often a MS detector, acquires data from the 2nd dimension in now "higher" chromatographic resolution.

2.2.2.2 Comprehensive GC × GC

True multidimensional chromatography requires two independent (orthogonal) separations mechanisms and the conservation of the first separation into the 2nd dimension. Comprehensive GC×GC today is the most developed and most powerful multi-dimensional chromatographic technique. The technique has been widely accepted and applied to the analysis of complex mixtures. Commercial instrumentation is available at a mature technological standard for routine application (see Fig. 2.54).

In contrast to heart cutting, the complete first dimension effluent is separated in slices (which can be interpreted as a continuous sequence of cuts) and forms a three dimensional data space on a second dimension with the retention times in both dimensions and the peak intensities by classical or MS detection. No monitor detector for the first dimension, or prior identification of an individual retention time window is required, and no increase in total analysis time is involved.

The timing regime of GC×GC should be relatively slow with a regular capillary column for the first dimension and a short fast GC column for the second dimension separation. The second dimension column should be of a different polarity than the first one to provide different kind of interactions in substance class separations. Typically, non polar columns

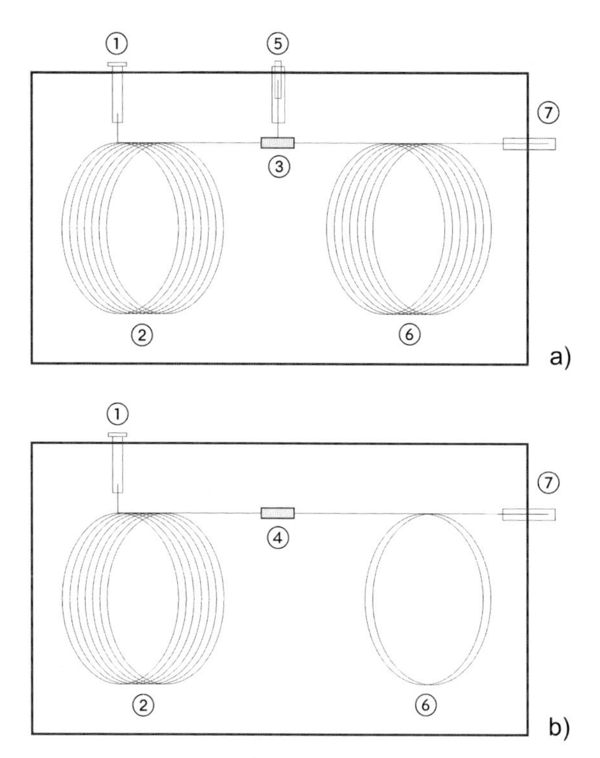

Fig. 2.54 Two-dimensional GC Principles.
a) Heartcutting with monitor detector. b) Comprehensive GC×GC.
1 Injector
2 Primary separation column, 1st dimension
3 Flow switching device
4 Modulator device
5 Monitor detector
6 Secondary separation column, 2nd dimension
7 MS transfer line or conventional detector, 2nd dimension

are used in the first dimension for a separation along the substance boiling points. Polar film interactions characterize the further enhanced separation on the 2nd dimension, which takes place at an only low temperature gradient, due to the speed of elution. Column lengths in the 2nd dimension are as short as 1–2 m for fast GC conditions (see also Section 2.2.1). In total, the duration of analysis in GC×GC is not increased compared to conventional techniques and the same sample throughput can be achieved.

The operation of the interface between the both columns, called the modulator, is key to GC×GC. The modulator operates in an alternating trap and inject mode with a modulation frequency related to the fast second dimension separation. Short elution sequences are integrated from the first column and injected for the second dimension separation. Each injection pulse generates a high speed chromatogram.

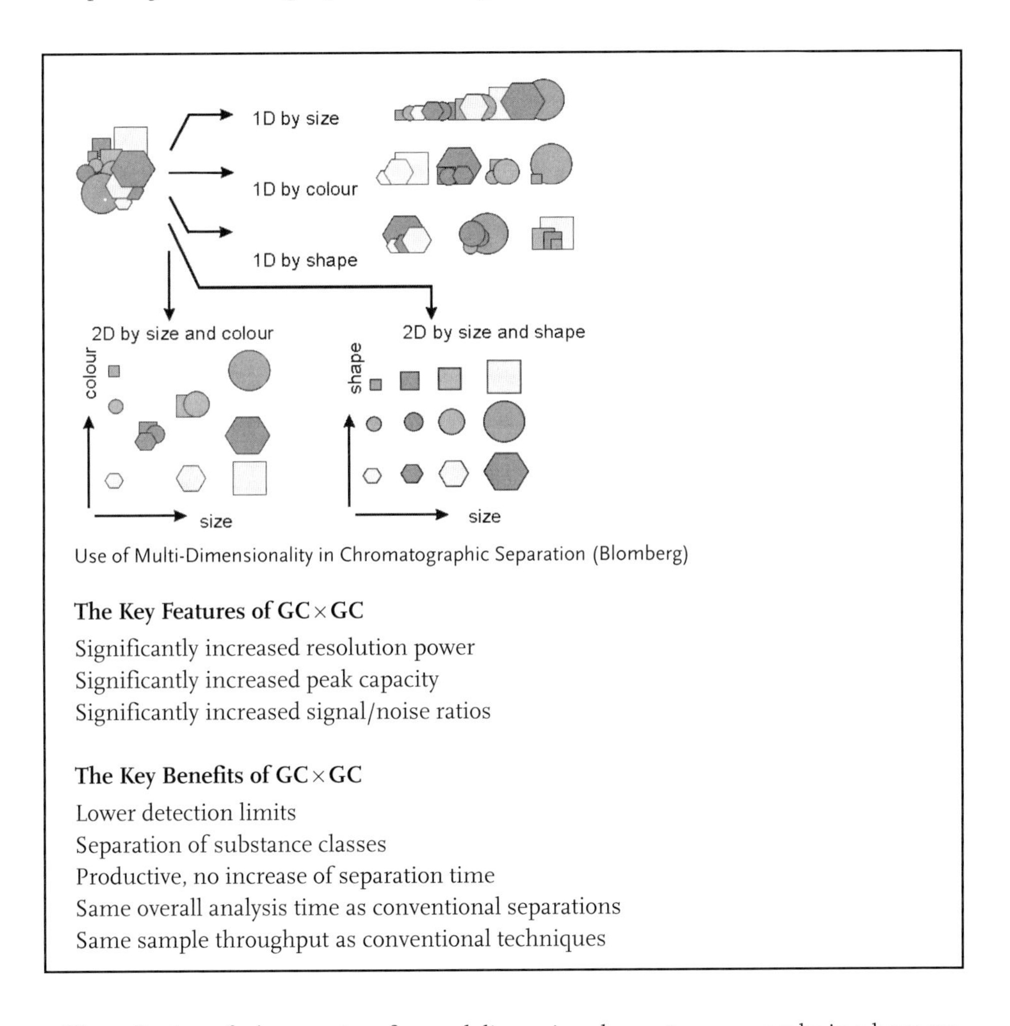

Use of Multi-Dimensionality in Chromatographic Separation (Blomberg)

The Key Features of GC×GC

Significantly increased resolution power
Significantly increased peak capacity
Significantly increased signal/noise ratios

The Key Benefits of GC×GC

Lower detection limits
Separation of substance classes
Productive, no increase of separation time
Same overall analysis time as conventional separations
Same sample throughput as conventional techniques

The collection of a large series of second dimension chromatograms results in a large amplification of the column peak capacities as the first dimension peaks are distributed on multiple fast second dimension chromatograms. The effect of increased separation can be visua-

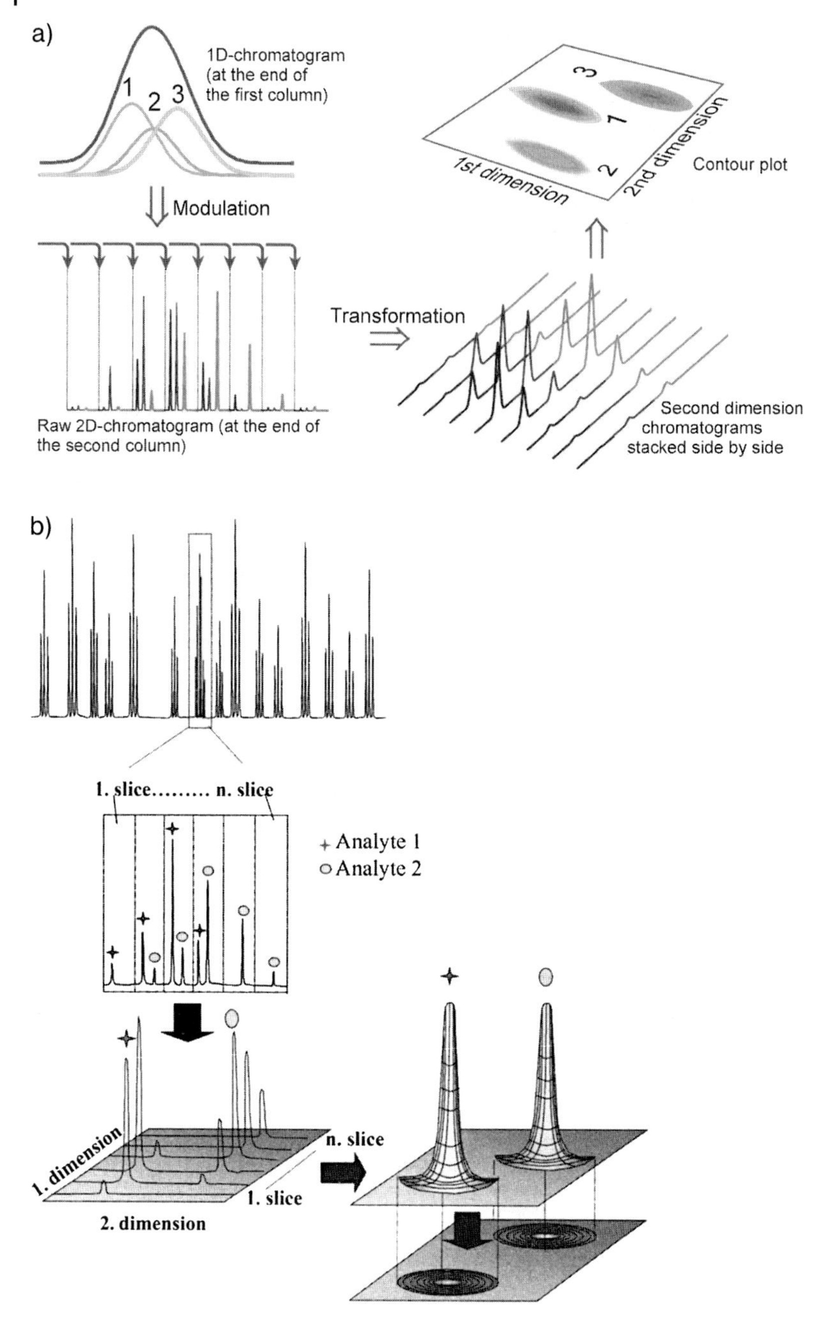

Fig. 2.55 (a) Visualisation of the principle of multidimensional GC separation with the final generation of the contour plot of an unresolved coelution of 3 substances in the 1st dimension (Cavagnino 2003); (b) Construction of a 3-dimensional contour plot for analytes 1 and 2 from a modulated chromatogram (Kellner 2004).

lized best with 3D contour or peak apex plots, illustrating best the commonly used term "comprehensive" chromatography, see Fig. 2.55 a. (Cavagnino 2003). The very high separation power offered by the comprehensive two-dimensional gas chromatography allows, even in the case of complex matrices, a good separation of target compounds from the interferences with a significant increase in signal to noise. This emerging technology has substantially enhanced the chromatographic resolution of complex samples and proven its great potential to further expand the capabilities of modern GC/MS systems.

2.2.2.3 **Modulation**

The key to successful application of GC×GC is the ability to trap and modulate efficiently the first column effluent into the second dimension. It is important to modulate faster than the peak width of first dimension peaks, so that multiple second dimension peaks (slices) are obtained (see Fig. 2.55 a). The moderator unit has to provide minimized bandwidths in sample transfer to retain the first dimension resolution and also allow the rapid remobilization. Practically a high peak compression is achieved by trapping the effluent from the first column on a short band inside of the modulator, being responsible for the significant increase of S/N values (Patterson 2006). This requires a high frequency handling of the column effluent with a reproducible trapping and controlled substance release. Practically applied modulation frequencies range from 4 to 6 s duration depending on the 2nd dimension column parameters.

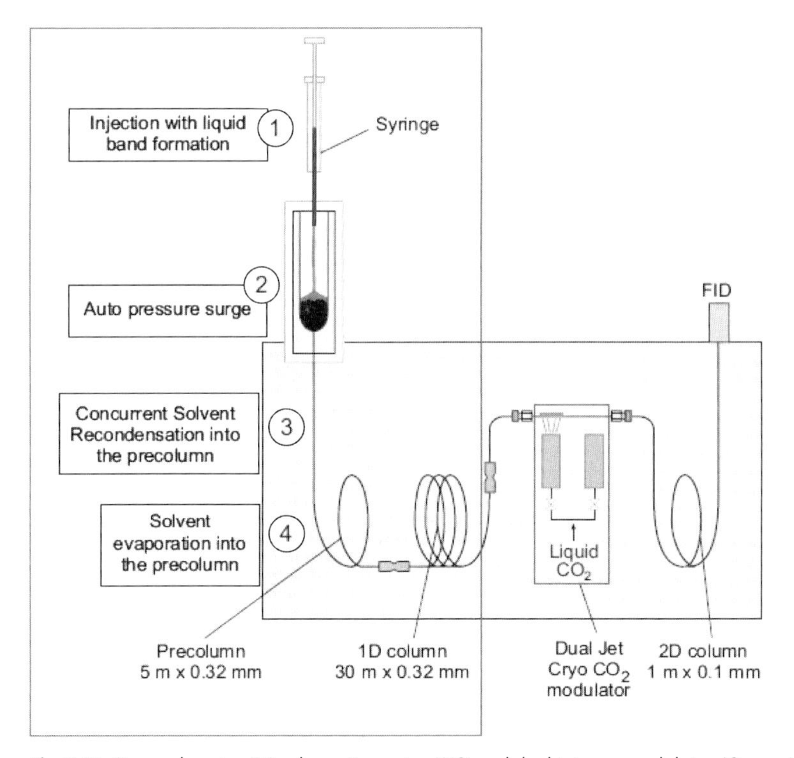

Fig. 2.56 Comprehensive GC schematics using LVSI and dual jet cryo modulator (Cavagnino 2003).

Thermal modulator solutions for substance trapping are:

- Trapping by thick stationary film and heat pulse to release substances.
- Cryogenic zone placed over a modulator capillary. Substances are released when removed and resume movement by oven temperature. Also dual jet cryo modulators have been introduced and commercially available to decouple the processes of collection from the 1st column and sampling into the 2nd dimension, see Figs. 2.57 and 2.58 a (Beens 1998). The operation principle of a dual cryo jet modulator is illustrated in Fig. 2.58 b (Kellner 2004).
- Jet-pulsed modulators, synchronizing hot and cold jet pulses.
- Flow switching valve solutions, used mainly for heart cutting purposes.

Fig. 2.57 Dual jet cryo modulator, using a stretched column region (Cavagnino 2003).

The availability of commercial instrumentation providing method controlled modulators, able to cool via cryojets, chilled air, liqu. CO_2 or liqu. N_2 and heat the trapping areas for injection rapidly, has made GC × GC straightforwardly applicable today to perform routinely.

2.2.2.4 **Detection**

The high speed conditions of the secondary column deliver very sharp peaks of typically 200 ms base width or below and hence require fast detectors with an appropriate high detection rate. Due to the strong peak band compression obtained during the modulation step, low detection limits are reached. Fast scanning quadrupole MS (scan speed >10,000 Da/s) have been applied by scanning a restricted mass range with an acquisition rate of 25 Hz as a cost effective alternative to TOF-MS for compound identification and quantitation (Mondello 2007). Even sector mass spectrometers have been used successfully in the fast SIM/MID technique, with excellent sensitivity in environmental trace analysis (Patterson 2006). TOF detection and fast scanning quadrupole instruments in a restricted mass range offer full scan analysis capabilities allowing a detailed peak deconvolution for the generation of pure mass spectra also for unresolved trace components, see Fig. 2.59 (also see Section on "Deconvolution").

With regard to sensitivity and specificity GC×GC/TOF-MS has proven to compare well with the classical HRGC/HRMS methods. Isotope ratio measurements of the most intense ions for both natives and isotopically labelled internal standards ensured the required spe-a)

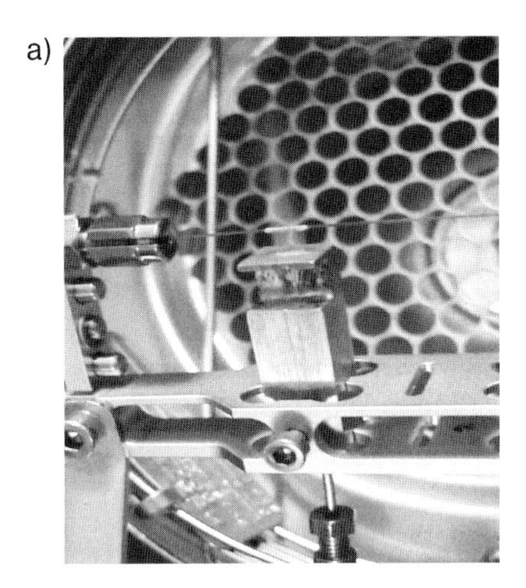

b)

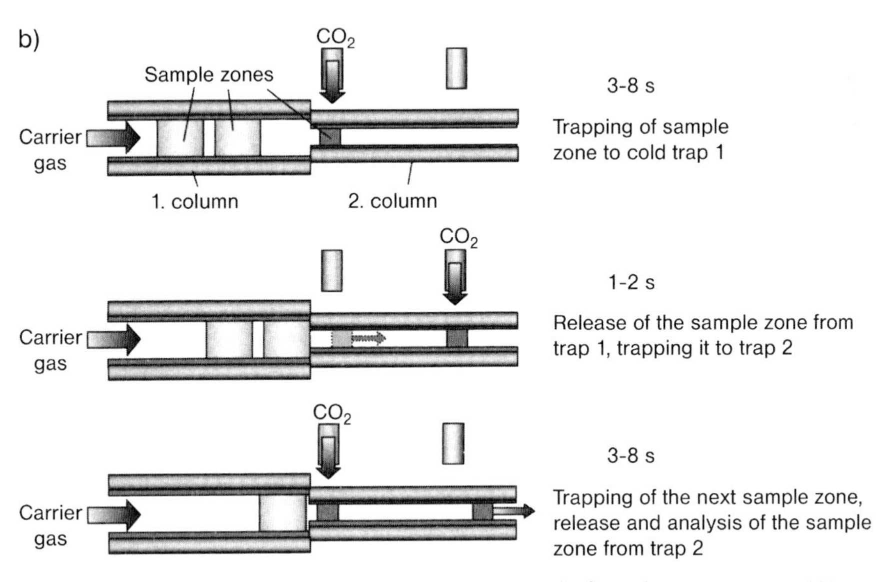

Fig. 2.58 (a) Dual jet cryo modulator in trapping operation on the first column (Cavagnino 2003). (b) Principle of dual jet cryo modulation (Kellner 2004).

cificity. Potentially interfering matrix compounds are kept away from the compounds to be measured in the two-dimensional chromatographic space (Focant 2004).

2.2.2.5 **Data Handling**

Data systems for comprehensive GC require special tools extending the available features of most current GC/MS software suites. Because of the 3D matrix of multiple chromatograms and conventional or MS detection, special tools are required to display and evaluate comprehensive GC separations. The time/response data streams are converted to a matrix format for the two-dimensional contour plot providing a colour code of peak in-

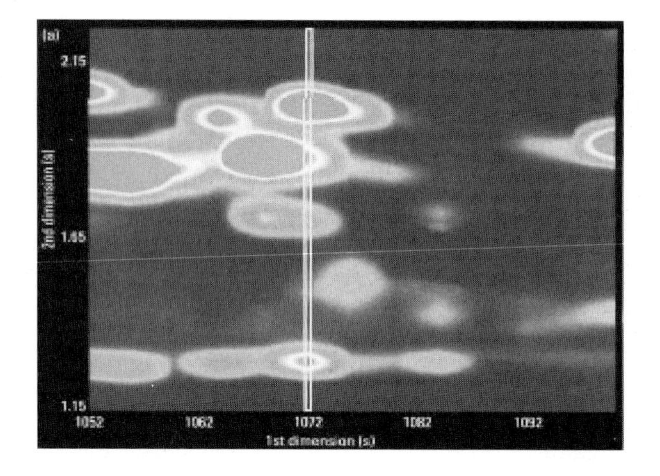

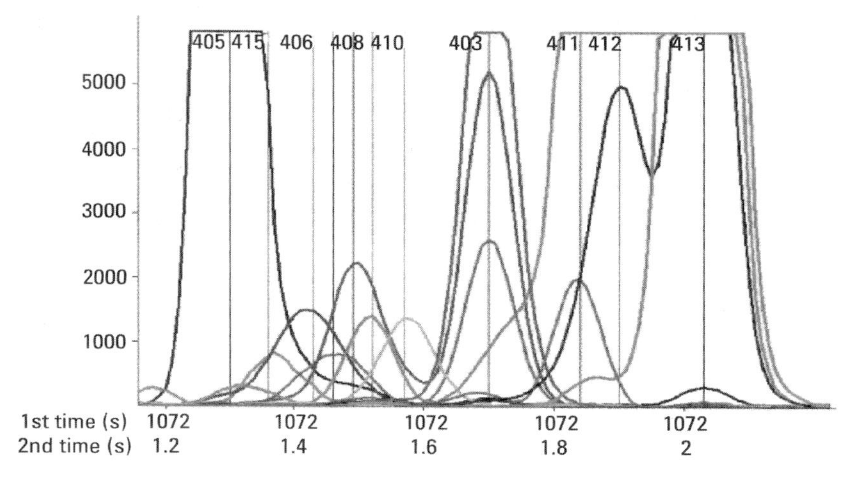

Fig. 2.59 Peak deconvolution in full scan GCxGC/TOF-MS (Dimandja 2004, reprinted with permission from Analytical Chemistry, Copyright American Chemical Society).
Top: 3D contour plot, scan 1072 highlighted.
Bottom: scan 1072, each vertical indicates a deconvoluted compound.

tensities (Fig. 2.59), or the three-dimensional surface plot generation (Fig. 2.60), for visualization.

Integrated software programs are commercially available providing total peak areas also for quantification purposes (Cavagnino 2003, Reichenbach 2004). Dedicated to the comprehensive qualitative sample characterization with GC × GC/TOF-MS analyses, as well as the quantitative analysis of specific mixture analytes, a GC/MS software package is commercially available (Leco 2005).

Advanced chemometric processing options are an active area of current research and will advance the GC × GC technology in the near future. In comprehensive GC using TOF detection (GC × GC/MS), data also can be visualized and processed as an image. Each resolved compound produces a two-dimensional peak and can be visually distinguished through pseudo-

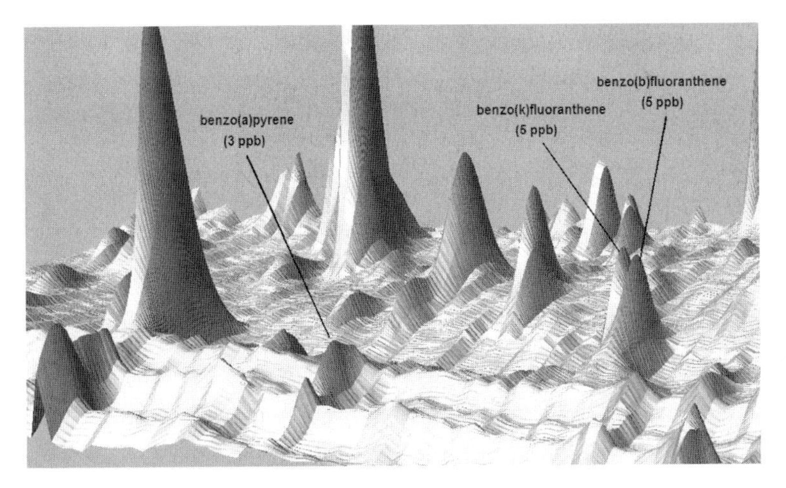

Fig. 2.60 3D Peak view of a GCxGC analysis of a cigarette smoke extract (Cavagnino 2003).
Columns set:

Precolumn:	5 m deactivated retention gap, 0.32 mm i.d.
1st dimension column:	RTX-5 30 m, 0.32 mm i.d., 0.25 μm df
2nd dimension column:	BPX-50 1 m, 0.1 mm i.d., 0.1 μm df
Carrier:	Helium
Operative conditions:	
Oven:	90 °C (8 min) to 310 °C @ 3 °C/min
Prog. Flow:	2.5 mL/min (4min) to 0.8 mL/min @ 5 mL/min/min
Inj. Vol.:	30 μL splitless
Splitless time:	0.8 min
Moderator:	CO_2 Dual-Jet modulator
Modulation time:	6 s
Data system:	HyperChrom for acquisition and data processing

colour mapping, generating a three-dimensional surface. Digital image processing methods, such as creating a difference image (by subtraction of chromatograms) or addition image (by addition of chromatograms in different colours) will provide insights previously not available to analytical samples. For GC×GC/MS, mass spectral searching can be used in conjunction with pattern matching (Hollingsworth 2006). Those features demonstrate the great potential of GC×GC/TOFMS with its large peak capacity at a largely improved speed of analysis.

2.2.2.6 Moving Capillary Stream Switching

Another often used valveless flow switching device for heart cutting purposes or switching the flow direction to different detectors is the Moving Capillary Stream Switching system (MCSS). The MCSS is a miniaturized and very effective flow switching system based on the position of capillary column ends in a small transfer tube. The column effluent is not directed by pressure variations but "delivered" to the required column by moving the end of the delivering column close to the column outlet of choice into an auxiliary carrier gas stream, see Fig. 2.61 (Sulzbach 1990).

Up to five capillary columns can be installed to a small (a few cm long) hollow tip of glass, the "glass dome" (see Fig. 2.61), which is located inside the GC oven. The column ends are

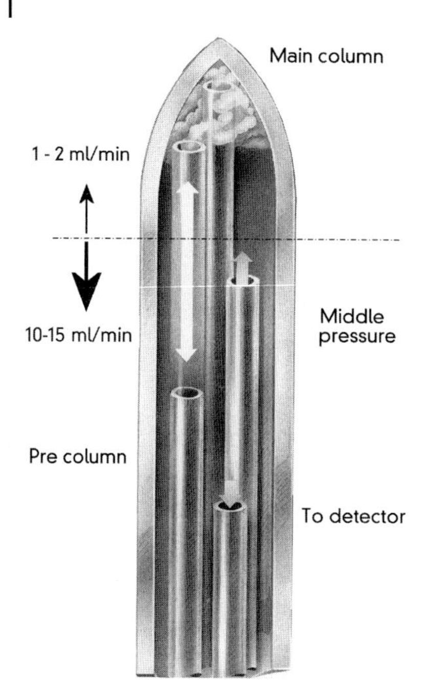

Fig. 2.61 Visualisation of MCSS glass dome operation – the flow of the movable pre-column is directed to the main column (reprinted with permission of Brechbuehler AG, Switzerland).

positioned at different locations inside the glass dome. The delivering pre-column inlet is movable in position over a length of about 1 cm. One of the installed columns is used for a variable make up flow of carrier gas from a second independent regulation. This make-up gas facilitates the parallel coupling of MS detectors in parallel to classical detectors or IRMS to prevent the access of the ions source vacuum to the split region. Instead of an additional detector the additional line can also be used for monitoring the middle pressure between the two columns. Commercial products are available for multidimensional GC covering a variable split range.

The analyte carrier stream fed into the dome by the pre-column can be split into a variable stream upwards, serving a second analytical column, and another stream downwards depending on the position of the pre-column end, e. g., to another column outlet, which can lead to a monitor detector. If the delivering column is pushed fully up, the eluate enters the second column and quantitatively transfers the analytes eluted in that position. There is no change of the pressure conditions in the whole switching system during the transfer period. Therefore, the carrier gas flow rates through both of the columns are always kept stable. Hence there is no influence of the number or duration of switching processes on the retention times on either column. Analytes can only contact inert glass surfaces, fused silica and the column coating; likewise when the sample is injected in any heated injector. Furthermore, the system is free of diffusion and maintains the chromatographic resolution of the pre-column. No influences from ambient atmosphere could be observed using ECD, MS or IRMS detection.

The movement of the pre-column is electrically actuated from outside of the GC oven, completely automated and sequence controlled through the time functions of the GC (see

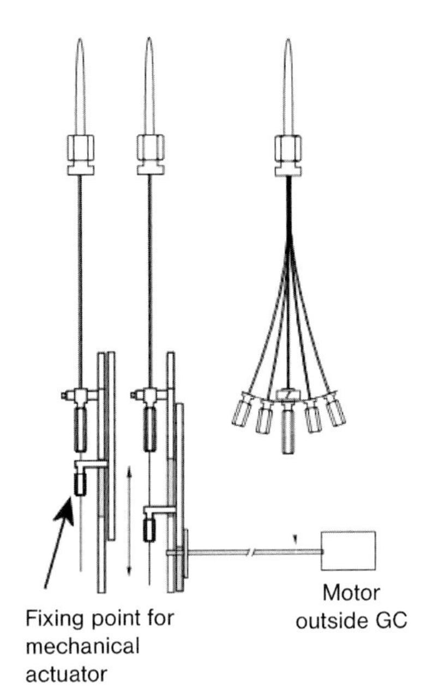

Fig. 2.62 MCSS actuation by external stepper motor.

Fig. 2.62). The MCSS can be used for a number of tasks affording the increased power of multidimensional heart-cutting GC separation (Fig. 2.63). The range of applications covers the classical MDGC in a single or double oven concept, as well as detector switching, back-flush, preparative GC and more.

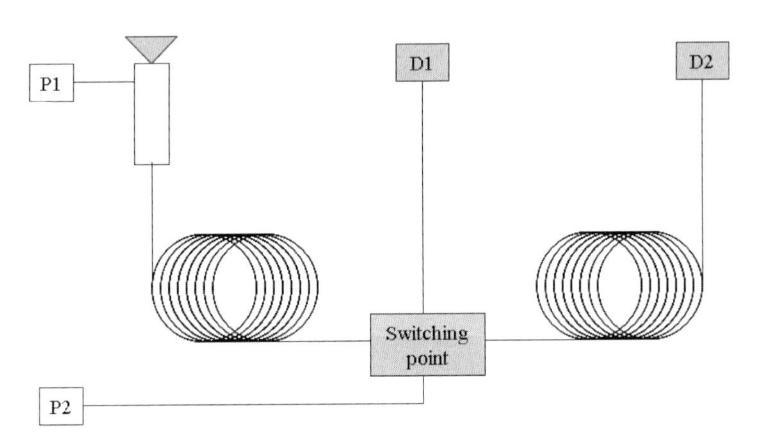

Fig. 2.63 Column Switching Schematics with pre-column, MCSS flow switch at the switch point and main separation column (P1, P2 carrier gas regulations, D1 monitor detector, D2 main detector).

MCSS Applications

Chiral separations of essential oils (FID/MS)
Classification of fuels (FID)
Matrix exclusion (FID/PND/ECD/MS)
Sniffing devices in parallel to detectors (MS)
Coplanar PCBs (double ECD/MS/IRMS)
Pesticides (ECD)
Preparative sampling (Microprep Trap)
All kinds of GC-IRMS applications
Parallel GC-MS-IRMS detection
and many others

2.2.3 GC/MS Sample Inlet Systems

Much less attention was paid to this area at the time when packed columns were used. On-column injection was the state of the art and was in no way a limiting factor for the quality of the chromatographic separation. With the introduction of capillary techniques in the form of glass or fused silica capillaries, high resolution gas chromatography (HRGC) maintained its presence in laboratories and GC and GC/MS made a great technological advance. Many well known names in chromatography are associated with important contributions to sample injection: Desty, Ettre, Grob, Halasz, Poy, Schomburg and Vogt among others. The exploitation of the high separating capacity of capillary columns now requires perfect control of a problem-orientated sample injection technique.

According to Schomburg, sample injection should satisfy the following requirements:
- Achieving the optimal efficiency of the column.
- Achieving a high signal/noise ratio through peaks which are as steep as possible in order to be certain of the detection and quantitative determination of trace components at sufficient resolution (no band broadening).
- Avoidance of any change in the quantitative composition of the original sample (systematic errors, accuracy).
- Avoidance of statistical errors which are too high for the absolute and relative peak areas (precision).
- Avoidance of thermal and/or catalytic decomposition or chemical reaction of sample components.
- Sample components which cannot be evaporated must not reach the column or must be removed easily (guard column). Involatile sample components lead to decreases in separating capacity through peak broadening, and shortening of the service life of the capillary column.
- In the area of trace analysis it is necessary to transfer the substances to be analysed to the separating system with as little loss as possible. Here the injection of larger sample volumes (up to >100 μL) is desirable.

- Simple handling, service and maintenance of the sample injection system play an important role in routine applications.
- The possibility of automation of the injection is important, not only for large numbers of samples, but also as automatic injection is superior to manual injection for achieving a low standard deviation.

Careless injection of a sample extract frequently overlooks the outstanding possibilities of the capillary technique. In all modern GC and GC/MS systems, sample injection is of fundamental importance for the quality of the chromatographic analysis. Poor injection cannot be compensated for even by the choice of the best column material. This also applies to the choice of detectors. The use of a mass spectrometer as a detector can be much more powerful if the chromatography is of the best quality. In fact the use of GC/MS systems shows that as soon as the GC is coupled to an MS system, the chromatography is rapidly degraded to an inlet route. Effort is put too quickly into the optimisation of the parameters of the MS detector without exploiting the much wider possibilities of the GC. Each GC/MS system is only as good as the chromatography allows!

The starting point for the discussion of sample inlet systems is the target of creating a sample zone at the top of the capillary column for the start of the chromatography which is as narrow as possible. This narrow sample band principally determines the quality of the chromatography as the peak shape at the end of the separation cannot be better (narrower, more symmetrical etc.) than at the beginning. As an explanatory model, chromatography can be described as a chain of distillation plates. The number of separation steps (number of plates) of a column is used by many column producers as a measure of the separating capacity of a column. In this sense sample injection means application to the first plate of the column. In capillary GC the volume of such a plate is less than 0.01 µl. The sample extracts used in trace analysis are generally very dilute, making larger quantities of solvent (>1 µl) necessary. This shows the importance of sample injection techniques. In this connection Pretorius and Bertsch should be cited: "When the column is described as the heart of chromatography, sample injection can be identified as the Achilles heel. It is the least understood and the most confusing aspect of modern gas chromatography."

In practice, different types of injectors are used (see Table 2.17). The sample injection systems are classified as hot or cold according to their function. A separate section is dedicated to direct on-column injection.

2.2.3.1 Carrier Gas Regulation

The carrier gas pressure regulation of GC injectors found in commercial instrumentation follows two different principles which influence injection modes and the adaptation of automated sample preparation devices. As most of the modern GC instruments are equipped with electronic flow control units (EPC) the sensor control loop must be taken special care of when modifying the carrier gas routes through external devices such as headspace, purge & trap or thermal desorption units. In these installations the carrier gas flow is often directed from the GC regulation to the external device and again returned to the GC injector.

Also, when applying large volume injection techniques, knowledge of the individual pressure regulation scheme is required for successful operation.

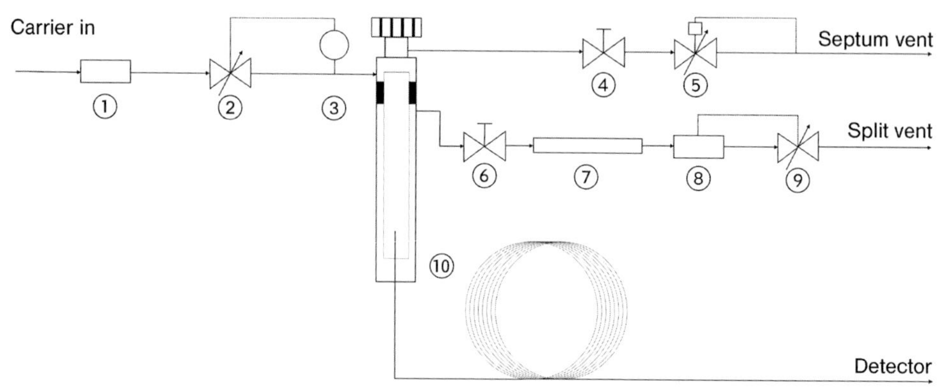

Fig. 2.64 Injector forward pressure regulation.
1 Carrier gas inlet filter
2 Proportional control valve
3 Electronic pressure sensor
4 Septum purge on/off valve
5 Septum purge regulator
6 Split on/off valve
7 Split outlet filter cartridge
8 Electronic flow sensor
9 Proportional control valve
10 Injector with column installed

Forward Pressure Regulation

The classical control of column flow rates is achieved by a "forward pressure" regulation. The inlet carrier gas flow is regulated by the electronic pressure control valve in front of the injector. The pressure sensor is installed close to the injector body and hence to the column head for fast feedback on the regulation either in the carrier gas supply line to the injector, the septum purge or split exit line. The split exit usually is regulated further down the line by a separate flow regulating unit (see Fig. 2.64). An in-line filter cartridge is typically used to prevent the lines from accumulating trace contaminations in the carrier gas, which needs to be exchanged frequently.

The forward injector pressure regulation is the most simple and versatile regulation design used in all splitless injection modes as well as the cold on-column injector. It uniquely allows the large volume injection in the split/splitless mode using the concurrent solvent recondensation (CSR) and the closure of the septum vent during the splitless phase. Also split and splitless injections for narrow bore columns, as used in Fast GC , benefit from the forward pressure regulation with a stable and very reproducible regulation at low carrier gas flow that is not usually possible with mechanical or electronic mass flow controllers. Also, the switch of injection techniques in one injector e. g. the PTV body between regular split/splitless and on-column injection with a retention gap is possible without any modification of the injector gas flow regulation.

The adaptation of an external device is straightforward, with the pressure sensor installed close to the injector head which is generally the case with commercial forward regulated GC instruments. If the pressure sensor is installed in the split line of the injector, it has to be moved close to the inlet to provide a short carrier gas regulation loop. Carrier gas flow and pressure in the external device is controlled by the EPC module of the GC.

Large volume injections are also straightforward using forward pressure regulated injectors. Typically, the regular split/splitless device can be used for large volume injections of up to 10 or 50 µL without any hardware modification, exploiting the concurrent solvent recondensation effect. For injections up to 10 µL, only the injection volume programming of the autosampler and the oven program has to be adjusted accordingly.

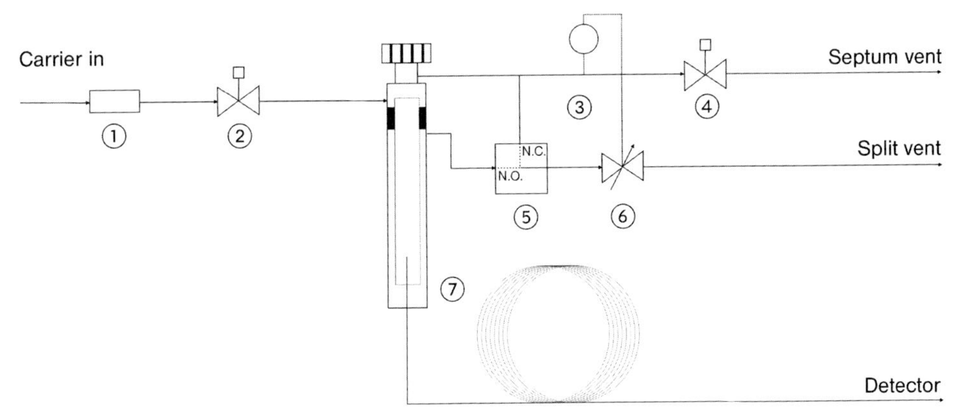

Fig. 2.65 Injector back pressure regulation.
1 Carrier gas inlet filter
2 Mass flow regulator
3 Electronic pressure sensor
4 Septum purge regulator
5 Solenoid valve
6 Electronic pressure control valve
7 Injector with column installed

Back Pressure Regulation

The term "back pressure" describes the position of the electronic pressure control valve behind the injector in the split exit line, in combination with a mass flow controller in front of the injector (see Fig. 2.65). In this widely used carrier gas regulation scheme for split/splitless injectors, the pressure sensor is typically found in the septum purge line close to the injector body to insure a pressure measurement close to the column head. A filter cartridge may be used to protect the regulation unit from any carrier gas contamination.

The total carrier gas flow of the back pressure regulated injector is set by the mass flow controller and is typically held constant independently of the column head pressure. The column flow is then adjusted and kept constant by varying the split exit flow according to the required column flow conditions. This allows the exact and independent setting of split flows from column flows. However, when the backpressure regulated inlet is used in split mode, a large amount of sample passes through the regulation in the split exit line. This mode of operation requires special care using a filter cartridge with frequent exchange to prevent the system from flow restrictions, chemical attack and, finally, clogging.

When adapting an external inlet device, the outlet of the mass flow controller is directed on the shortest route to the external device. A sample gas line from the external device is returned to the injector either using the regular carrier inlet, or by means of a needle through the septum. The pressure sensor must be relocated from the split line close to the electronic pressure control valve to achieve correct flows and sufficient pressure control stability with a short feedback loop due to the additional flow restrictions in the external device. The carrier gas flow through the external device is now regulated by the mass flow controller. The GC inlet pressure at the injector and the effective column flow is regulated independently by the EPC module through the electronic pressure control valve.

Alternatively, an external auxiliary EPC module for feeding the external device may be recommended.

2.2.3.2 The Microseal Septum

The Merlin Microseal Septa is a unique long-life replacement for the conventional silicon septa on split/splitless and PTV injectors. Functionally, the Microseal provides a two step sealing mechanism (see Fig. 2.66): A double O–ring type top wiper rib around the syringe needle improves resistance to particulate contamination, and a spring assisted duckbill seals the injection port. The Microseal can be easily installed on many injector types by removing the existing septa nut without any modifications. It can be used at injection port temperatures up to 400 °C. It is commercially available for operation up to 30 psi or up to 100 psi with the high pressure series.

The seal is made from Viton material, a high temperature resistant fluorocarbon elastomer also providing high resistance to wear. It greatly reduces shedding of septum particles into the injection port liner. Because the syringe needle does not pierce a septum layer, it is eliminating a major source of silicon septum bleed and ghost peaks. Due to this sealing mechanism it is especially suited for trace analysis applications requiring particular low background conditions.

The Microseal is usable in either manual or autosampler applications with 0.63 mm diameter (23 ga) blunt tip syringes. The low syringe insertion force makes it ideal for manual injections. The longer lifetime for many thousands of injections reduces the chances for septum leaks especially during extended autosampler runs.

Microseal is a trademark of Merlin Instrument Company.

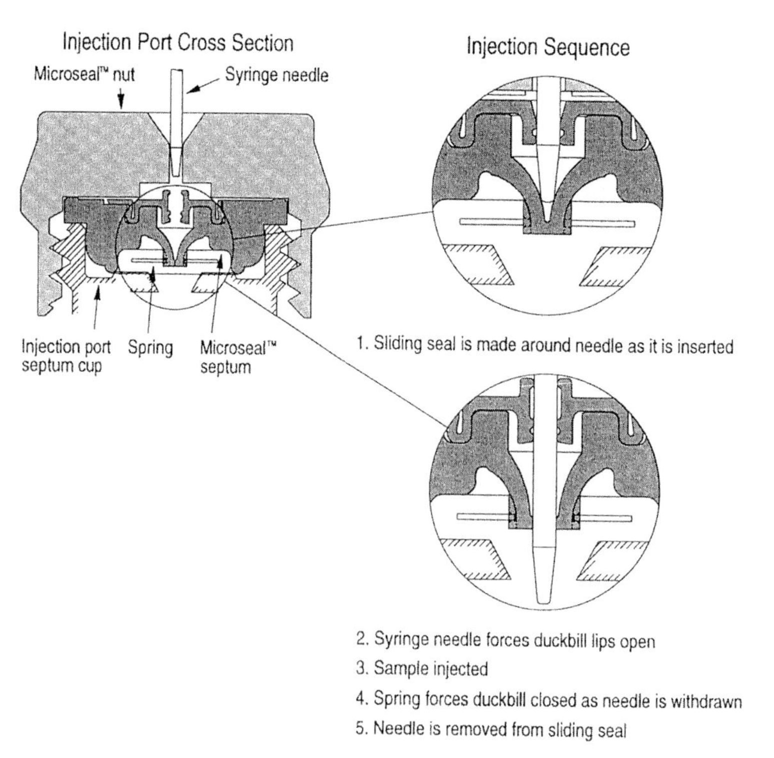

Fig. 2.66 2-Step injection process with the Merlin Microseal. Top: the needle is first penetrating the dust wiper rib; bottom: the needle is penetrating the spring loaded duck bill seal into the injector liner (Merlin Instr. Comp.).

2.2.3.3 Hot Sample Injection

Classical injection techniques involve applying the sample solution in constantly heated injectors. Both the solvent and the dissolved sample evaporate in an evaporation tube specially fitted for the purpose (insert) and mix with the carrier gas. Temperatures of ca. 200 °C to above 300 °C are used for evaporation. The operating procedures of split injection and total sample transfer (splitless) differ according to whether there is partial or complete transfer of the solvent/sample on to the column.

The problem of discrimination on injection into hot injectors arises with the question of what is the best injection technique. Figure 2.67 shows the effects of various injection techniques on the discrimination between various alkanes. While in the chain length range of up to ca. C_{16} hardly any differences are observed, discrimination can be avoided for higher-boiling, long-chain compounds by a suitable choice of injection technique. For example, filling the injection syringe with a solvent/derivatising agent plug before drawing up the sample and observing the needle temperature during a short period after insertion into the injector.

- *Solvent Flush/Hot Needle:* This is the injection procedure of choice for hot injectors and gives favourable discrimination properties. Circa 0.5–1 µl solvent or derivatising agent are first taken up in the syringe, then 0.5 µl air and finally the sample extract (sandwich technique). Before the injection the liquid plug is drawn up into the body of the syringe. The volumes can thus be read better on the scale. The injection involves first inserting the needle into the injector, waiting for the needle to warm up (ca. 2 s) and then rapidly injecting the sample. The solvent flush/hot needle procedure can be carried out manually or using a programmable autosampler.
- *Hot Needle:* This technique is definitely the most frequently used variant. Only the sample extract is drawn up into the syringe; the plug is thus drawn up into the body of the syringe

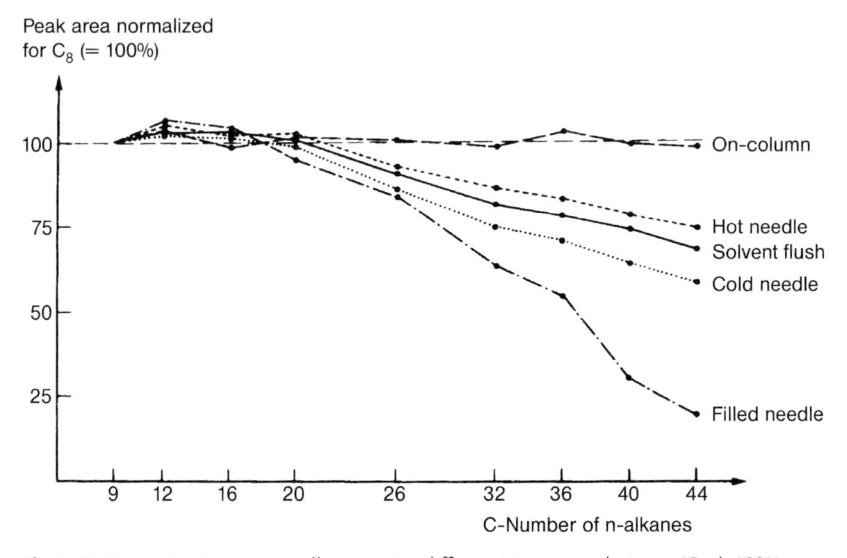

Fig. 2.67 Discrimination among alkanes using different injection techniques (Grob 2001). The peak areas are normalised to C_8 = 100%.

so that the volume can be read off on the scale. After inserting the needle, there is a pause for warming up and then the sample is rapidly injected.

- *Solvent Flush/Cold Needle:* Here the sandwich technique described above is used to fill the syringe. The injection is, however, carried out very quickly (usually using an autosampler) without waiting for the injection needle to warm up. This technique is most used with filled liners (e.g. glass wool, absorbent) allowing the formation of a liquid band inside the insert liner for the deposit of the liquid sample in the adsorbent at the bottom part of the liner. Evaporating liquid from the adsorbent leads to a local temperature decrease keeping the analytes focused in a small region for the transfer to the column.
- *Cold Needle:* Only the sample extract is drawn up into the syringe; the plug is held in the body of the syringe in such a way that the volume on the scale can be read. The injection is carried out very rapidly (usually with an autosampler) without waiting for the needle to warm up allowing liquid band formation with packed liners as well.
- *Filled Needle:* This injection procedure is no longer up-to-date and should be avoided with hot vaporizing injectors. It is associated with certain types of syringe which, on measuring out the sample extracts, can only allow the liquid plug into the injection needle. Warning: certain automatic liquid sample injectors use this procedure.

Split Injection

After evaporation of the liquid sample in the insert, with the split technique the sample/carrier gas stream is divided. The larger, variable portion leaves the injector via the split exit and the smaller portion passes on to the column. The split ratios can be adapted within wide limits to the sample concentration, the sensitivity of the detector and the capacity of the capillary column used. Typical split ratios mostly lie in the range 1:10 to 1:100 or more. The start values for the temperature program of a GC oven are independent of the injection procedure using the split technique. If the oven temperature is kept below the boiling point of the solvent at a given pressure, the re-concentration of the solvent into the column needs to be considered when calculating the split ratio. In this case more sample enters into the column and consequently the split ratio is not as calculated by the measured flow ratios. For that reason it is a good practice to keep the column temperature sufficiently high to avoid any recondensation during the split injection, if the split ratio is important.

For concentrated samples the variation of the split ratio and the volume applied represents the simplest method of matching the quantity of substance to the column load and to the linearity of the detector. Even in residue analysis the split technique is not unimportant. By increasing the split stream the carrier gas velocity in the injector is increased and this allows the highly accelerated transport of the sample cloud past the orifice of the column. This permits a very narrow sample zone to be applied to the column. To optimise the process the possibility of a split injection at a split ratio of less than 1:10 should also be considered. In particular, on coupling with static headspace or purge and trap techniques better peak profiles and shorter analysis times are achieved. The smallest split ratio which can be used depends on the internal volume of the insert. For liquid injection in hot split mode in general a large diameter liner for a reduced carrier gas speed is required to allow the sample to vaporize and mix with carrier gas before reaching the split point. Small diameter liners can induce a partial splitting as liquid that produce non repeatable data.

A disadvantage of the split injection technique is the uncontrollable discrimination with regard to the sample composition. This applies particularly to samples with a wide boiling point range. Quantitation with an external standard is particularly badly affected by this. Because of the deviation of the effective split ratio from that set up, this value should not be used in the calculation. Quantitation with an internal standard or alternatively the standard addition procedure must be used.

Total Sample Transfer (Splitless Injection)

With the total sample transfer technique the sample is injected into the hot injector with the split valve closed (Fig. 2.68). The volume of the injector insert must be able to hold the solvent/sample vapour cloud completely. Because of this special insert liners (vaporiser) are recommended for splitless operations. Depending on the insert used and the solvent, there is a maximum injection volume which allows the vapour to be held in the insert. Inserts which are too small lead, on explosive evaporation of the solvent, to expansion of the sample vapour beyond the inserts into the cold regions of the injector, and are causing a probable loss by the septum purge. Pressure waves of subsequent injections are bringing back deposited material as carry over from the split line into the next analysis. Inserts which are too

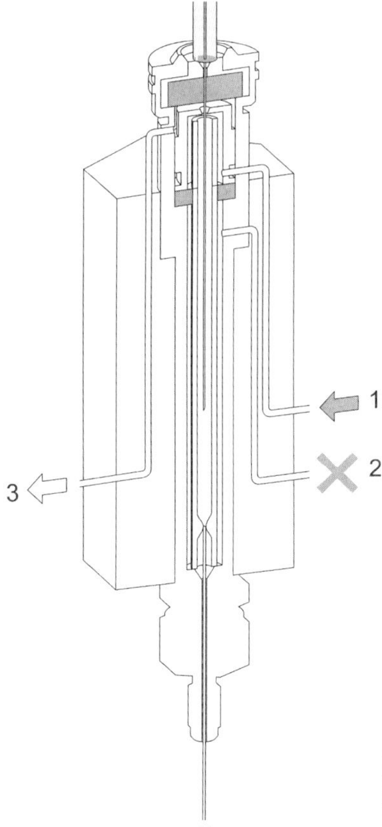

Fig. 2.68 Hot split/splitless injector (Thermo Fisher Scientific).
1 Carrier gas inlet
2 Split exit line
3 Septum purge line

wide lead to significant dilution of the sample cloud and thus to prolonged transfer times and losses through diffusion. A recent compromise made by many manufacturers involves insert volumes of 1 mL for injection volumes of about 1–2 μL. Splitless injections require an optimized needle length as stated by the manufacturer that defines the point of evaporation inside the liner. The septum flush should not be completely closed even with splitless injection. With many commercial SSL injectors the closure of the septum purge is optional and can be used to favor the auto-pressure surge. For application of surge pressures during injection the septum purge will be closed. On correct choice of insert the sample cloud does not reach the septum so that the low purge flow does not have any effect on the injection itself.

The carrier gas flushes the sample cloud continuously from the injector on to the column. This process generally takes ca. 30–90 s for complete transfer depending on the insert volume, sample volume and carrier gas flow. There is an exponential decrease in concentration caused by mixing and dilution with the carrier gas. Longer transfer times are generally not advisable because the sample band becomes broadened on the column. Ideal transfer times allow three to five times the volume of the insert to be transferred to the column. This process is favoured by high carrier gas flow rates and an increased head pressure during the transfer process (surge pressure). Hydrogen is preferred to helium as the carrier gas with respect to sample injection. For the same reason a column diameter of 0.32 mm is preferred for the splitless technique compared to narrower diameters, in order to be able to use optimal flow rates in the injector. In the same way a pressure surge step is available in modern GC instrumentation using electronic pressure regulation to improve the sample transfer.

As the transfer times are long compared with the split injection and lead to a considerable distribution of the sample cloud on the column, the resulting band broadening must be counteracted by a suitable temperature of the column oven. Only at start temperatures below the boiling point of the solvent sufficient refocusing of the sample is achieved because of the solvent effect. A rule of thumb is that the solvent effect operates best at an oven temperature of 10–15 °C below the boiling point of the solvent. The solvent condensing on the column walls acts temporarily as an auxiliary phase, accelerates the transfer of the sample cloud on to the column, holds the sample components and focuses them at the beginning of the column with increasing evaporation of solvent into the carrier gas stream (solvent peak). Here the use of “GC compatible solvents” that are miscible with the stationary phase becomes important. “Incompatible” solvents do not generate a suitable solvent effect and need to be avoided or applied in split mode. Only after the sample transfer is complete the split valve is opened until the end of the analysis to prevent the further entry of sample material or contaminants on to the column. The splitless technique therefore requires working with temperature programs. Because of the almost complete transfer of the sample on to the column, total sample transfer is the method of choice for residue analysis.

Because of the longer residence times in the injector, with the splitless technique there is an increased risk of thermal or catalytic decomposition of labile components. There are losses through adsorption on the surface of the insert, which can usually be counteracted by suitable deactivation. Much more frequently there is an (often intended) deposit of involatile sample residues in the insert or septum particles collect. This makes it necessary to regularly check and clean the insert.

Concurrent Solvent Recondensation

The large volume injection possible on a regular splitless injector did not garner much attention until the transfer mechanism of the sample vapour into the column involved was further investigated and fully understood. The concurrent solvent recondensation technique (CSR) permits the injection of increased sample amounts up to 50 μL by using a conventional split/splitless injector with forward pressure regulation. This can be achieved in a very simple and straightforward way, since all processes are self-regulating. Moreover, the technique is robust towards contaminants and therefore it is suitable when complex samples have to be injected in larger amounts (Magni 2003).

Due to the recondensation of the solvent inside of the capillary column causing a pressure drop at the beginning of the column, a strong pressure difference occurs between column and injector liner that significantly speeds up the sample transfer. The sample and solvent vapour is "sucked down" from the insert by the lower pressure inside of the column and continuously condensed to form the liquid band at the beginning of the column (see Fig. 2.69). For proper operation of the CSR technique the SSL injector has to be operated in the forward pressure regulation. A back pressure regulation would compensate for the occurring inlet pressures causing severe sample loss.

Sample volumes of up to 10 μL can be injected by using the regular 0.25 mm capillary columns while sample volumes of up to 50 μL require the wider diameter of an empty retention gap with 0.53 mm i.d. to accept the complete amount of liquid injected and some silanized glass wool in the insert liner (focus liner).

The practical benefit of CSR is the flexibility for the injection of a wider range of diluted sample volumes reducing significantly the time for extract pre-concentration. As the CSR injection is performed with the regular SSL injector hardware without any modifications, only the autosampler injection volume, when using regular 10 μL syringes, has to be programmed in the methods. For increasing volumes the initial isothermal time needs to be extended accordingly up to the end of the solvent evaporation from the pre-column. The increased solvent vapor cloud needs to leave the condensation region (band formation) before starting the oven program. The remaining oven temperature program remains unchanged.

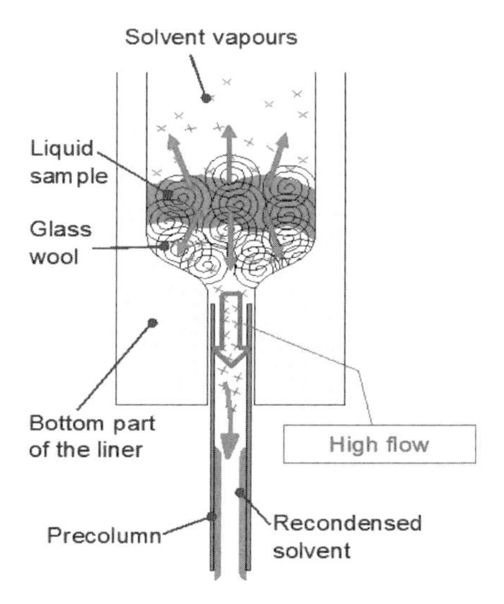

Fig. 2.69 Principle of concurrent solvent recondensation in splitless injection (Magni 2003).

Concurrent Solvent Recondensation
Key steps (see Fig. 2.69)

Restricted evaporation rate:
Injection with liquid band formation: The collection of the liquid on the glass wool allows a slow evaporation from a "single" droplet.

Increased transfer rate:
Auto-pressure surge due to the large amount of solvent: A temporary increase of the inlet pressure rapidly drives the vapours into the column where the recondensation process starts. The concurrent solvent recondensation in the pre-column is generating a strong suction effect.

Solvent evaporation in the pre-column:
The oven temperature is kept below the boiling point up to the end of solvent evaporation from the pre-column.

2.2.3.4 Cold Injection Systems

The cold sample injection technique involves injecting the liquid 'cold' sample directly on to the column (on-column, see Section 2.2.1.4) or into a specially constructed vaporiser. The sample extract is ejected from the syringe needle in liquid form into the insert at temperatures which are well below the boiling point of the solvent. Heating only begins after the syringe needle has been removed from the injection zone. In the area of residue analysis programmed temperature vaporiser (PTV) cold injection systems for split and splitless injection are becoming more widely used, as are on-column systems for exclusively splitless injection.

The cold injection of a liquid sample eliminates the selective evaporation from the syringe needle in all systems, which, in the case of hot injection procedures, leads to discrimination against high-boiling components. Discrimination is also avoided as a result of explosive evaporation of the solvent in hot injectors. Colder regions of the injection system, such as the septum area or the tubing leading to the split valve, can serve as expansion areas if the permitted injection volume is exceeded and can retain individual sample components through adsorption. The individual concepts of cold injection differ in the transfer of the sample to the column and in the possibility for using the split exit. There are definite advantages and limitations for the operation of injectors in practice and for the areas of use envisaged. These exist both between different cold injection techniques and in comparison with hot injection techniques.

The PTV Cold Injection System

The temperature-programmed evaporation with split or splitless operations using the currently available injectors is based on the systematic work of Poy (1981) and Schomburg (1981) (Figs. 2.70 and 2.71). In particular, emphasis was placed on the precise and accurate execution of quantitative analyses of complex mixtures with a wide boiling point range. Particularly at the beginning of the experiments, absence of discrimination for substances up to above C_{60} was documented and its suitability for involatile substances, such as polyaromatic

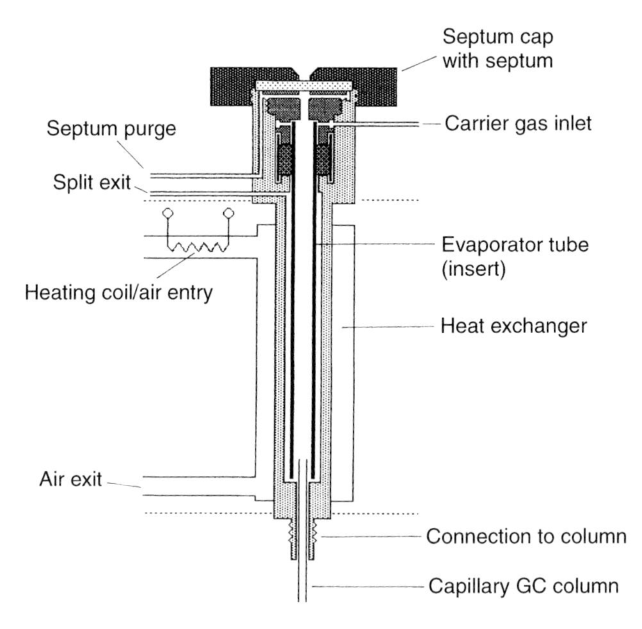

Fig. 2.70 PTV with air as the heating medium from the design by Poy.

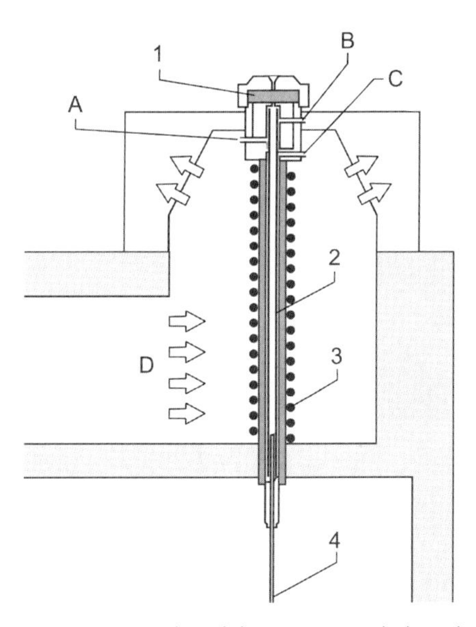

Fig. 2.71 PTV split/splitless injector with direct heating (Thermo Fisher Scientific).

A Carrier gas inlet	1 Septum
B Septum purge line	2 Injector body
C Split exit line	3 Direct heating
D Active cooling	4 Analytical column

hydrocarbons, was stressed. Examples of the analysis of triglycerides demonstrate the injection of samples up to above C_{50} and the analysis of crude oil fractions up to C_{90}! The PTV process combines the advantages of the hot and on-column injection techniques.

There are many advantages of cold split and splitless sample injection:

- Discrimination as a result of fractionated evaporation effects from the syringe needle does not occur.
- A defined volume of liquid can be injected reproducibly.
- The sample components evaporate as a result of controlled heating of the injection area in the order of their boiling points. The solvent evaporates first and leaves the sample components in the injection area without causing a distribution of the analytes in the injector as a result of explosive evaporation.
- Aerosol and droplet formation is avoided through fractional evaporation.
- As the evaporator does not have to take up the complete expansion volume of the injected sample solution, smaller inserts with smaller internal volumes can be used. The consequently more rapid transfer to the column lowers band broadening of the peaks and thus improves the signal/noise ratio.
- If the boiling points of the sample components and the solvent differ by more than 100 °C larger sample volumes (up to more than 100 μl) can be injected (solvent split).
- Impurities and residues which cannot be evaporated do not get on to the column.
- With concentrated samples the possibility can be used of adapting the injection to the capacity of the column and the dynamic range of the detector by selecting the split ratio.

Currently the PTV cold injection technique is mainly used for residue analysis in the areas of plant protection agents, pharmaceuticals, polyaromatic hydrocarbons, brominated flame retardants, dioxins and PCBs. However, the enrichment of volatile halogenated hydrocarbons, the direct analysis of water and the formation of derivatives in the injector also demonstrate the broad versatility of the PTV type injector.

The PTV Injection Procedure

PTV Total Sample Transfer

The splitless injection for the injection of a maximum sample equivalent on to the column is the standard requirement for residue analysis. The sample is taken up in a suitable solvent and injected at low temperature. The split valve is closed. The PTV temperature at injection should correspond to the boiling point of the solvent at 1013 mbar. During the injection phase the oven temperature is kept below the PTV temperature. It should be below the boiling point of the solvent in order to exploit the necessary solvent effect according to Grob for focusing the substances at the beginning of the column (Fig. 2.72). If the focusing of the substances is unsuccessful or insufficient, the peaks of the components eluting early are broad and are detected with a low signal/noise ratio.

The injector is heated a few seconds after the injection when the solvent has already evaporated and has reached the column (Fig. 2.73). Typically this time interval is between 5 and 30 s. For high-boiling substances in particular longer residence times have been found to be favourable. The heating rate should be moderate in order to achieve a smooth evaporation of the sample components required for transfer to the column. Heating rates of ca. 200–

300 °C/min have proved to be suitable. The optimal heating rate depends on the dimensions of the insert liner and the flow rate of the carrier gas in the insert.

PTV Total Sample Transfer

- Split valve closed
- PTV at the boiling point of the solvent
- Oven temperature below the boiling point of the solvent
- Start of PTV heating ca. 10 s after injection
- Start of the GC temperature program ca. 30–120 s after injection
- PTV remains hot until the end of the analysis

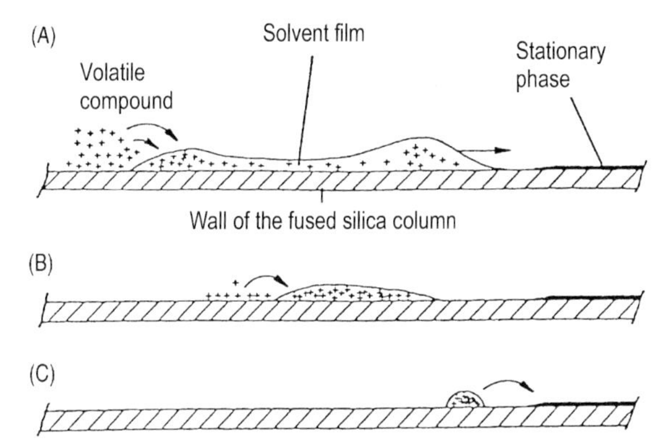

Fig. 2.72 Refocusing by means of the solvent effect in splitless injection (after Grob).
(A) Condensation of solvent at the beginning of the capillary column by lowering the oven temperature below the boiling point. The condensed solvent acts as a stationary phase and dissolves the analytes. At the same time this front migrates and evaporates in the carrier gas flow.
(B) The continuous evaporation of the solvent film concentrates the analytes on to a narrow ring (band) in the column. For this process no stationary phase at the beginning of the column is necessary.
(C) The substances concentrated in a narrow band meet the stationary phase. The separation begins with a sharp band.

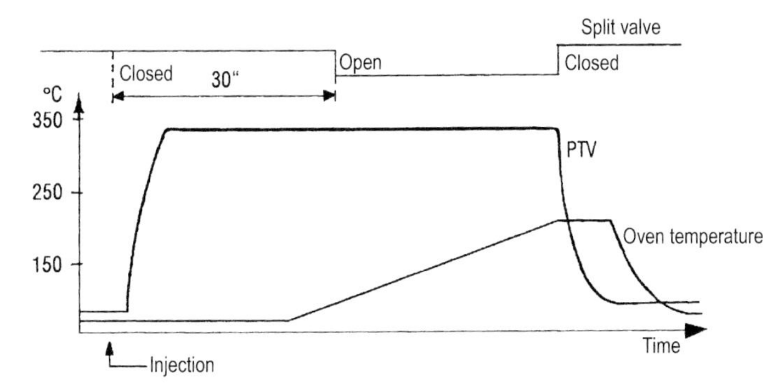

Fig. 2.73 PTV total sample transfer (splitless injection).

Filling the insert with silanised glass wool has proved effective for the absorption of the sample liquid and for rapid heat exchange. However, the glass wool clearly contributes to enlarging the active surface area of the injector and should therefore only be used in the analysis of noncritical compounds (alkanes, chlorinated hydrocarbons etc.). For the injection of polar or basic components an empty deactivated insert is recommended.

After the transfer of the substances to be analysed to the column, which is complete after ca. 30–120 s, the temperature program of the GC oven is started and the injector purged of any remaining residues by opening the split valve. If required, a second PTV heating ramp can follow to bake out the insert at elevated temperatures. The high injection temperature of the PTV is retained until the end of the analysis to keep the injector free from possible adsorptions and the accumulation of impurities from the carrier gas inlet tubing. The cooling of the PTV is adjusted so that both the PTV and the oven are ready for the next analysis at the same time.

A special form of the PTV cold injection system consists of the column and insert connected via a press-fit attachment. Total sample transfer is possible in the same way as the classical PTV injection, but no split is present to allow flushing of the injection area after evaporation. Injectors of this type can only be used for total sample transfer (e. g. SPI injector, Fig. 2.74).

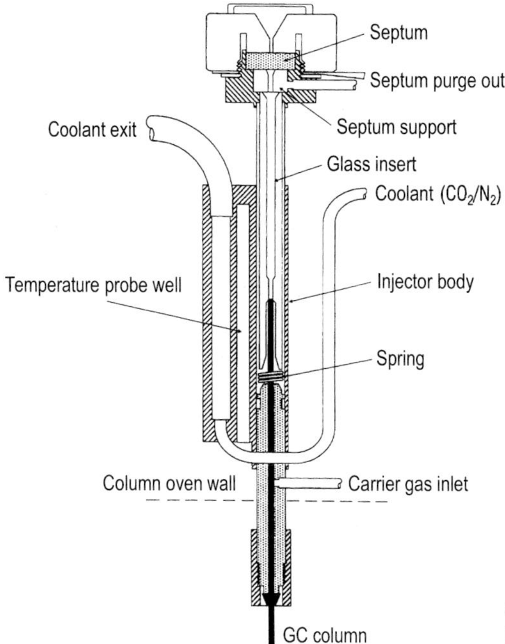

Fig. 2.74 Septum-equipped programmable injector, SPI (Finnigan/Varian).

PTV Split Injection

In this mode of operation the split valve is open throughout (Fig. 2.75). In classical sample application this mode of injection is suitable for concentrated solutions, whereby the column loading can be adapted to its capacity and the nature of the detector by regulating the split flow. Cold injection systems are characterised by the particular dimensions of the inserts. This allows a high carrier gas flow rate at the split point. Compared to hot split injectors,

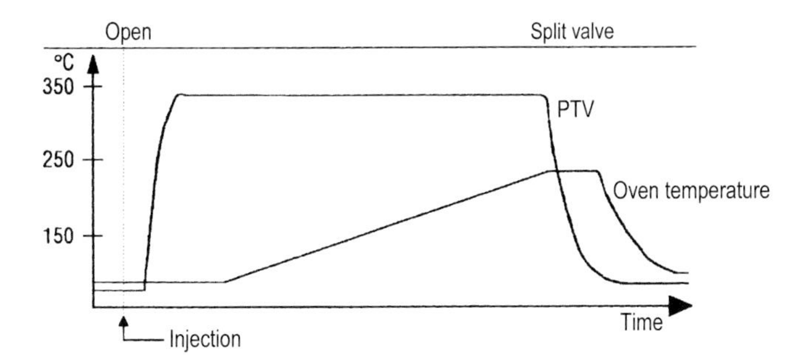

Fig. 2.75 PTV split injection.

smaller split ratios of ca. 1:5 are possible. In an individual case it is possible to work in the split mode and thus achieve better signal/noise ratios and shorter analysis times, compared with total sample transfer

PTV Split Injection

- Split valve open
- PTV below the boiling point of the solvent
- Start of PTV heating ca. 10 s after injection
- Oven temperature to be chosen freely
- PTV remains hot until the end of the analysis

All other PTV adjustments of time and temperature are completely unchanged compared with total sample transfer! Of particular importance for the split injection using the PTV cold injection system is the fact that the sample is injected into the cold injector (see also Section 2.2.3.5). The choice of start temperature of the GC oven is now no longer coupled to the boiling point of the solvent because of the solvent effect and is chosen according to the retention conditions.

PTV Solvent Split Injection

This technique is particularly suitable for residue analysis for the injection of solutions of diluted extracts and with low concentrations, which would undergo loss of analyte on further concentration (Fig. 2.76). The solvent split mode allows most efficient sample throughput shortening time-consuming pre-concentration steps. With suitable parameter settings very large solvent quantities, also for a fraction collection from an online sample preparation and being limited only by practical considerations, can be applied.

To inject larger sample volumes (from ca. 2 µl to well over 100 µl) it is advantageous for the insert to be filled at least with silanised glass wool or packing of Tenax, Chromosorb Supelcoport or another inert carrier material ca. 0.5–1 cm wide. The split valve is open during the injection phase. The PTV is kept at a temperature which corresponds to the boiling point of the solvent. The oven temperature is below the PTV temperature and thus below the boiling point of the solvent. The maximum injection rate depends upon how much solvent per unit time can be evaporated in the insert and carried out through the split tubing by the carrier gas. The at-once injection of large volumes can best be tested easily by apply-

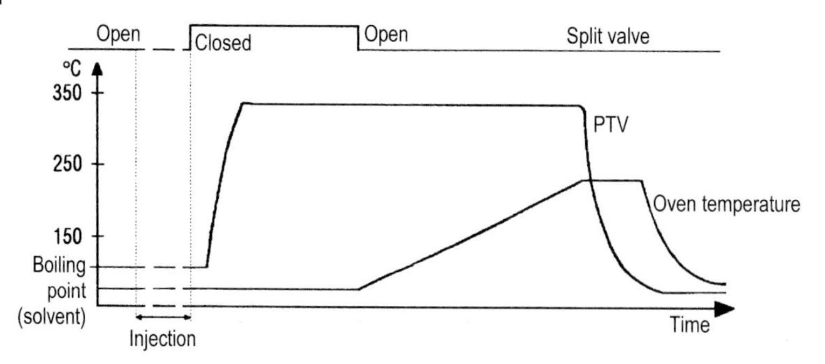

Fig. 2.76 PTV solvent split injection.

ing the desired amount of solvent with the liner outside of the injector. During injection a high split flow of >100 mL/min is recommended to focus the analytes on the packing material by local consumption of evaporation heat. For ease of method development of the injection the data recording of the GC or GC/MS system is started so that the course of the solvent peak can be followed (Fig. 2.77).

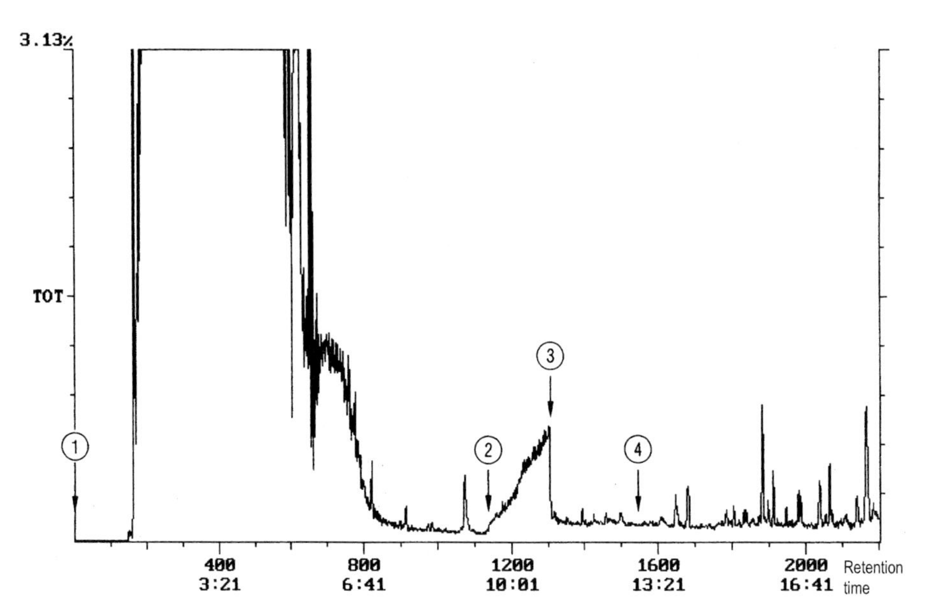

Fig. 2.77 The course of an analysis with large quantities of solvent for the PTV cold injection system during method development: 100 μL PCB solution (200 pg/μL). After the dead time a broad solvent peak is registered. After the solvent peak is eluted, the split valve is closed and the PTV heated up. A smaller roughly triangular solvent peak is produced from adsorbed material. After injection the split valve is opened again and the temperature program started. The peaks eluting show good resolution and are free from tailing.

1 Start of the injection: split open
2 Start of PTV heating: split closed
3 Baking out the PTV: split open
4 Start of the GC temperature program: split open and typically the start of the data acquisition

When quantities of more than 10 μL are applied, the split valve is only closed when the end of the solvent peak begins to show on the display, meaning that the injector is free of most of the solvent. After the split valve has been closed, the PTV injector can be heated up as usual. After ca. 30–120 s the transfer of the enriched sample from the insert to the column is complete and the GC temperature program can be started. The split valve then opens and the PTV is held at the injection temperature until the end of the analysis.

The PTV solvent split mode also allows the derivatisation of substances in the injector. This procedure simplifies sample preparation considerably. The sample extract is treated with the derivatising agent (e.g. TMSH for methylation) and injected into the PTV. The excess solvent is blown out during the solvent split phase. The derivatisation reaction takes place during the heating phase and the derivatised substances pass on to the column.

PTV Large Volume Injection

The PTV large volume injection mode (LVI-PTV) allows the repeated automated injection of sample volumes in the range up to 100 μL and even more. The large volume mode requires that the sample components are less volatile than the used solvent. The LVI-PTV operation needs the additional installation of a heated solvent split valve to prevent solvent vapour to condensate and plug the split exhaust line. The operation is further facilitated by a backflush valve below of the injector system which is recommended to prevent large amounts of solvent vapour entering the analytical column and for cleaning the injector during the heat-off step after injection.

As described in the previous section for PTV solvent split injections, a wide liner needs to be filled either by silanised quartz wool or other suitable packing material to retain the sample and prevent solvent from rinsing through the liner. The solvent capacity of the liner can easily be tested outside of the injector by adding the intended amount of solvent. No rinsing of the solvent may be observed when holding the insert liner upright on a sheet of paper.

Table 2.20 Effective solvent boiling points for different inlet pressures in °C.

Solvent	BP standard	100 kPa	200 kPa	300 kPa	400 kPa	500 kPa
iso-Pentane	28	49	65	77	87	98
Diethylether	35	54	72	84	93	101
n-Pentane	35	57	72	84	93	102
Dichloromethane	40	60	75	87	96	105
MTBE	55	72	91	104	118	124
Methylacetate	57	72	91	104	115	124
Chloroform	61	81	97	109	121	130
Methanol	65	82	97	107	116	124
n-Hexane	69	95	111	124	136	145
Ethylacetate	77	97	114	126	136	145
Cyclohexane	81	106	122	137	149	159

Comparison of large volume injection methods

	LV-On-column with solvent vapor exit	LV-PTV with solvent split	LV-Split/Splitless with concurrent solvent recondensation
Typical injection volume	150 μL	100 μL	30 μL
Robust versus complex matrix	No	Yes	Yes
Volatiles analysis	Yes	Need optimization	Yes
Suitability for thermolabile and actives compounds	High	Medium	Medium
Solvent vented	Yes	Yes	No
Requires uncoated pre-column	Yes	No*	Yes
Number of parameters	Medium	Large	Small**
Software assisted set-up	Yes	No	Yes

* It can be necessary if large amount of solvent is retained for improving volatiles recovery.
** Up to 10 μL with regular 0.25 ID columns.

PTV Cryo-enrichment

This injection technique is suitable for trace analysis and for concentrating volatile compounds, which are present in large volumes of gas, for example in gas sampling, thermodesorption or the headspace technique. Sample transfer into the injector is carried out while the PTV is in trapping mode. For this the insert is filled with a small quantity (ca. 1–2 mg, ca. 1–3 cm wide) of Tenax, Carbosieve or another thermally stable adsorbent. The PTV injector is cooled with liquid CO_2 or nitrogen until injection. During the injection the split valve is open as in the solvent split mode. Like the total sample transfer method, the oven temperature is kept correspondingly low in order to focus the components at the beginning of the column.

The gaseous sample is passed slowly through the injector and the organic substances retained on the adsorbent. Before injection the PTV can be heated to a low temperature with the split still open to dry the Tenax material if required. For transfer of the sample to the GC column, the split valve is closed and the PTV is heated to effect the total sample transfer of the concentrated components. After ca. 30–120 s the temperature program of the oven is started and the split valve is opened (Fig. 2.79). The PTV remains at the same temperature until the end of the analysis. This last step is particularly important in cryo-enrichment because the cooled adsorbent could become enriched with residual impurities from the carrier gas, which would lead to ghost peaks in the subsequent analysis.

2.2.3.5 **Injection Volumes**

For the discussion of the optimal injection volumes for hot and cold injection systems, the use of a 25 m long capillary column with an internal diameter of 0.32 mm is assumed, which operates with a linear carrier gas velocity of 30 cm/s. Under these conditions a carrier gas flow of ca. 2 mL/min through the column is expected. For the total sample transfer

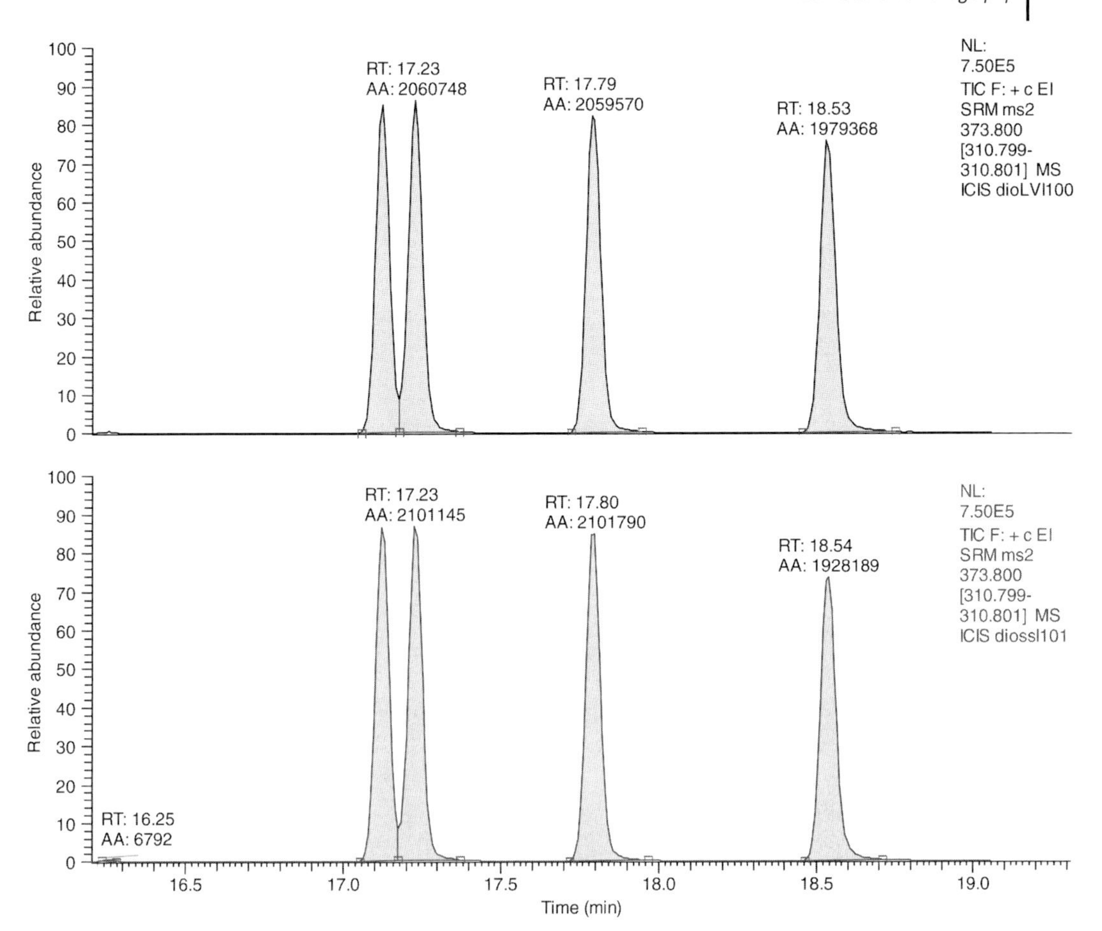

Fig. 2.78 Comparison of regular and large volume injection.
EPA 1613 CS3 standard, mass chromatograms of the hexachlorofuran congener peaks with retention time and peak area.
Thermo Scientific TSQ Quantum GC, column: TR-5MS, 30 m × 0.25 mm ID × 0.1 μm film
Trace GC with PTV: packed liner with deactivated glass wool
Top: 80 μL PTV-LVI injection of the 1 : 80 diluted standard solution
Bottom: 1 μL PTV splitless injection of the undiluted standard solution

mode the carrier gas velocity for both types of injector can be determined by considering the insert volumes in each case. The expansion volumes of the solvents used must also be taken into account. These are shown in Table 2.21.

To determine the time required to flush the contents of the injector completely on to the column at least twice the insert volume is applied, corresponding to a yield of 90–95%. Table 2.22 gives details for hot and cold injectors.

These considerations make it clear that the choice of capillary column and the appropriate injection parameters and volumes only allow the maximum performance of both hot and cold injection systems to be exploited. In particular, cold injection systems function less well if they are improperly used as hot injectors with solvent volumes which are normally used

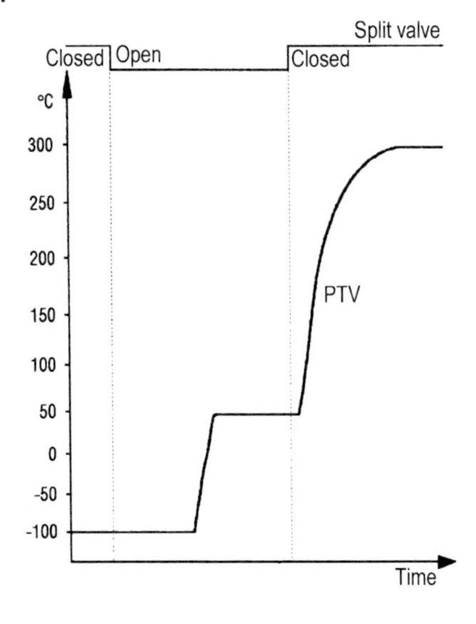

Fig. 2.79 PTV cryo-enrichment at −100 °C (with an optional fractionation step here at 50 °C) and injection with total sample transfer.

Table 2.21 Typical solvent expansion volumes.

Injection volume	H_2O	CS_2	CH_2Cl_2	Hexane	iso-Octane
0.1 µL	142 µL	42 µL	40 µL	20 µL	16 µL
0.5 µL	710 µL	212 µL	200 µL	98 µL	78 µL
1.0 µL	1420 µL	423 µL	401 µL	195 µL	155 µL
2.0 µL	2840 µL	846 µL	802 µL	390 µL	310 µL
3.0 µL	4260 µL	1270 µL	1200 µL	585 µL	465 µL
4.0 µL	5680 µL	1690 µL	1600 µL	780 µL	620 µL
5.0 µL	7100 µL	2120 µL	2000 µL	975 µL	775 µL

The expansion volumes given here refer to an injection temperature of 250 °C and a column pressure of 0.7 bar (10 psi). The values for other temperatures and pressures can be calculated according to $V_{Exp.} = 1/P \cdot n \cdot R \cdot T$.

with all commercial hot split or splitless injectors. The probability that the sample cloud will expand beyond the insert increases with the volatility of the solvent. This then leads to losses and tailing as a result of interactions with active surfaces. This would not occur if the PTV were used according to the instructions.

For hot injectors capillary columns with an internal diameter of 0.32 mm are more suitable, as here the transfer of the sample from the injector to the column takes place without diffusion and a more rapid injection is permitted. However, on coupling with MS detectors, the flow rate of ca. 2 ml/min can exceed the maximum compatible flow rate in some quadrupole systems. Columns with an internal diameter of 0.25 mm can be used as an alternative and then should be used in hot injectors with a narrow 2.0 mm internal diameter insert and an injection volume of 0.5 µl, to obtain optimal results. If inserts with a wide internal diameter (4 mm) are used, the solvent effect must be particularly exploited for focusing. It leads to accelerated emptying of the insert through solvent recondensation and a slip-stream

Table 2.22 Injection times for hot and cold injection (GC column 25 m × 0.32 mm).

Injector type	Hot injector	Cold injection system
Split valve	Closed	Closed
Internal diameter of insert	4 mm	2 mm
Interior volume	1 mL	300 μL
Flow rate	0.3 cm/s [a)]	1 cm/s
Minimum time with split closed	60 s [b)]	18 s
Maximum injection volume (250 °C)		
CH_2Cl_2	2 μL	(0.5 μL) [c)]
Hexane	5 μL	(1 μL)

a) This low flow rate already causes diffusion of the analytes injected against the carrier gas stream. The peaks are broader and the signal/noise ratio is poorer.
b) On use of a 25 m × 0.25 mm internal diameter column the carrier gas rate in the insert of the hot injector is only ca. 0.1 cm/s (4 mm insert) and would require a period with the split closed of at least 3 min. On the other hand the cold injection system only requires the split to be closed for at least 54 s.
c) The given injection volumes are valid for the case where a sample is injected into a heated cold injection system which contradicts the use envisaged for the construction. The injection volumes for cold injection are much higher.

in the direction of the injector. Furthermore, the start temperature of the program must be well below the boiling point of the solvent and there should be an isotherm of 1–3 min at the beginning of the oven program.

The use of capillary columns of 25 m in length and 0.25 mm internal diameter can be recommended without limitations for cold injection systems. The coupling to a mass spectrometer is particularly favourable because of the generally low flow rates. Also, the injection of larger sample volumes is straightforward and does not impair the operation of the mass spectrometer. For these reasons the use of cold injection systems for GC/MS is particularly recommended. This applies all the more to the throughput of large numbers of samples, as here automation of the system is indispensable. Unlike on-column injectors, cold injection systems can be used with autosamplers without special modifications.

In spite of the many advantages of the cold injection system, in practice there are also limits to its use (Table 2.23). These include the analysis of particularly thermally labile substances. Because of the low injection temperature cold injection systems should be particularly suitable for labile substances. However, during the heating phase, the residence times of the substances in the insert are long enough to initiate thermal decomposition. In this case, only on-column injection can be used because it completely avoids external evaporation of the sample for transfer to the column (see Section 2.2.1.4). To test for thermal decomposition, Donike suggested the injection of a mixture of the same quantities of fatty acid TMS esters (C_{10} to C_{22} thermolabile) and n-alkanes (C_{12} to C_{32} thermally stable). If no thermal decomposition takes place, all the substances should appear with the same intensity.

Table 2.23 Choice of a suitable injector system.

Characteristics of the sample	(1) Hot split	(2) Hot splitless	(3) PTV split	(4) PTV splitless	(5) PTV solvent split	(6) On-column
Concentrated samples	+	–	+	–	–	–
Trace analysis	–	+	–	+	+	+
Extreme dilution	–	–	–	–	+	+
Narrow boiling range	+	+	+	+	+	+
Wide boiling range	–	≈	+	+	–	+
Volatile substances	+	+	+	+	–	–
Involatile substances	–	≈	+	+	+	+
With involatile matrix	+	+	+	+	+	–
Thermally labile substances	≈	–	≈	≈	≈	+
Can be automated	+	+	+	+	+	+

+ recommended, ≈ can be used, – not recommended

2.2.3.6 On-column Injection

In the era of packed columns on-column injection (although not known as such at the time) was the state of the art. The difference between that and the present procedure lies in coping with the small diameters of capillary columns which have caused the term “filigree” to be applied to the on-column technique. Schomburg introduced the first on-column injector for capillary columns in 1977 under the designation of direct injection and described the process of sample injection very precisely. A year later a variant using syringe injection was developed by Grob. Today’s commercial on-column injection involves injecting the sample directly on to the column (internal diameter 0.32 mm) in liquid form using a standard syringe with a 75 mm long steel canula of 0.23 mm external diameter. More favourable dimensioning is possible on use of a retention gap with an internal diameter of 0.53 mm, which can even be used with standard canulas. The use of retention gaps allows autosamplers to be employed for on-column sample injection.

The injector itself does not have a complicated construction and can be serviced easily and safely. The carrier gas feed is situated and the capillary fixed centrally in the lower section. The middle section carries a rotating valve as a seal and in the upper section there is the needle entry point which allows central introduction of the syringe needle. In all types of operation the whole injector block remains cold and warming by the oven is also prevented by a surrounding air stream.

Small sample volumes of less than 10 μL can be injected directly on to a capillary column. Larger volumes require a deactivated retention gap of appropriate dimensions (Fig. 2.80). The retention gap is connected to the column with connectors with no dead volume (e.g. press fit connectors). For injection, the column should be operated at a high flow rate (ca. 2–3 mL/min for He). During the injection the oven temperature must be below the effective boiling point of the solvent. This corresponds to a temperature of ca. 10–15 °C above the boiling point of the solvent under normal conditions because of the pressure ratios in the

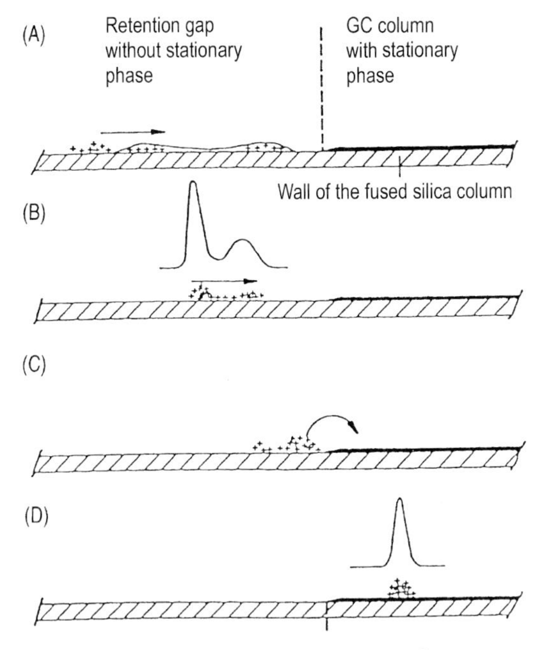

Fig. 2.80 Formation of the start band on use of a retention gap (after Grob).
(A) Starting situation after injection: the retention gap is wetted by the sample extract.
(B) The solvent evaporates and leaves an undefined substance distribution in the retention gap.
(C) The substance reaches the stationary phase of the GC column. Considerable retention corresponding to the capacity factor of the column begins.
(D) The analytes injected with the sample have been concentrated in a narrow band at the beginning of the column; this is the starting point for chromatographic separation.

on-column injector. As a rule of thumb, the boiling point increases by 1 °C for every 0.1 bar of pre-pressure. The injection is carried out by pressing the syringe plunger down rapidly (ca. 15–20 μL/s) to avoid the liquid being sucked up between the needle and the wall of the column through capillary effects, otherwise the subsequent removal of the syringe would result in loss of sample. For larger sample volumes, from ca. 20–30 μL, the injection rate must, however, be lowered (to ca. 5 μL/s) as the liquid plug causes increasing resistance and reversed movement of the liquid must be avoided (Fig. 2.81). The temperature program with the heating ramp can then begin when the solvent peak is clearly decreasing.

On-column Injection

- Flow rate ca. 2–3 mL/min
- Oven temperature at the effective boiling point of the solvent (normal boiling point plus 1 °C per 0.1 bar pre-pressure)
- Rapid injection of small volumes
- Oven temperature to be kept at the evaporation temperature
- Only start the heating program after elution of the solvent peak

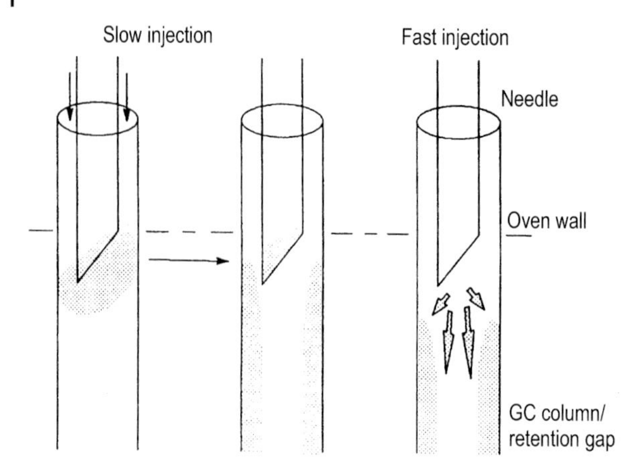

Fig. 2.81 Slow and fast on-column injections. For the slow injection the exterior of the needle is contaminated and part of the sample is lost.

The on-column technique has the following advantages over the cold injection system:

- The danger of thermal decomposition during the injection is practically eliminated. A substance evaporates at the lowest possible temperature and is heated to its elution temperature at the maximum.
- Only on-column injection allows the quantitative and reproducible sample injection on to the column without losses. Involatile substances, in particular, are transferred totally without discrimination.
- Defined volumes can be injected with high reproducibility. Standard deviations of ca. 1% can be achieved.
- The injection volumes can be varied within wide ranges without additional optimisation. At the beginning of the heating program, only the duration of elution of the solvent peak needs to be taken into account.

For use in GC/MS systems attention should be paid to the injection at the necessary high flow rates in connection with the maximum permitted carrier gas loading of the mass spectrometer. As the injection system is always cold, moisture can accumulate from samples or insufficiently purified carrier gas (Fig. 2.82). Because of this, the sensitivity of the mass spectrometer could be reduced. The effect is easy to detect because the mass spectrometer continually registers the water background and because the on-column injector can easily be opened at a particular time so that the carrier gas can pass out via the splitter.

If the water background decreases after the dead time, this effect must be taken into consideration during planning of the analysis, as the response behaviour of substances can change during the working day.

Other system-related limitations to the on-column technique for certain types of sample are:

- Concentrated samples may not be injected on-column. In these cases preliminary dilution is necessary or the use of the split injector is better (minimum on-column injection volume 0.2 μL).
- Samples containing a matrix rapidly lead to quality depletion. Here a retention gap is absolutely necessary. However, impurities in the retention gap quickly give rise to

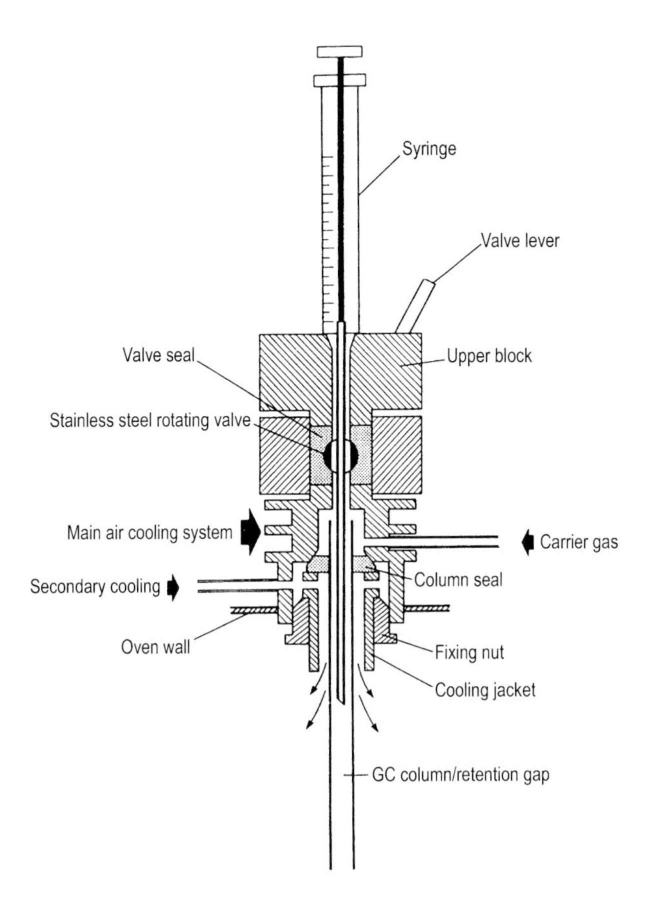

Fig. 2.82 On-column injector (Carlo Erba).

adsorption and peak broadening because of the small capacity, so that changing it regularly is necessary. For dirty samples hot or PTV sample injection is preferred.

- Samples containing volatile components are not easy to control, as focusing by the solvent effect is inadequate under certain circumstances. If changing the solvent does not help, changing to a hot or cold split or splitless system is recommended.

PTV On-column Injection

A regular PTV injector can also successfully be used for on-column injections. This method allows the straightforward automation of on-column injections using regular autosamplers with standard syringes. The insert liner used needs to be equipped with a restrictor at the top. This restrictor functions as a needle guide into the column. Syringes with the regular 0.47 o.d. needles can be used allowing the direct injection into wide-bore columns or retention gaps (pre-columns) of equivalent dimensions e. g. 0.53 mm i.d. For this purpose the column inlet is pushed up until it gets positioned at the bottom side of the restrictor site of the insert liner.

In PTV on-column mode, the injector body and the column oven is set for injection to a temperature below the solvent boiling point (see Table 2.20). After a short injection time of up to 20 s the injector is heated up for sample transfer. The oven program is started right after the completed sample transfer into the analytical column. The split valve remains opened with a low flow rate of only a few mL/min. The PTV temperature is maintained high throughout the chromatographic run as usual.

2.2.3.7 **Cryofocusing**

Cryofocusing should not be regarded as an independent injection system, but nevertheless should be treated individually in the list of injectors because the static headspace and purge and trap systems can be coupled directly to a cryofocusing unit as a GC injector. Furthermore, some thermodesorption systems already contain a cryofocusing unit. Many simpler instruments, however, do not, and require external focusing to ensure their proper function. Generally, for the direct analysis of air or gases from indoor rooms or at the workplace, or of emissions, a cryofocusing unit is required for concentration and injection on to the GC column.

Packed columns can effectively concentrate the substances contained in gaseous samples at the beginning of the column because of their high sample capacity. Capillary columns under normal working conditions cannot form sufficiently narrow initial bands for the substances to be detected from the gas volumes being handled in direct air analysis or on heating up traps (purge and trap, thermodesorption) because of their comparatively low sample capacity. Additional effective cooling is necessary. The entire oven space can be cooled, but this has a major disadvantage: the requirement in terms of time and cooling agents is immense.

A special cryofocusing unit cools down the beginning of the GC column and allows on-column focusing of the analytes without requiring a retention gap. The column film present improves the efficiency at the same time by acting as an adsorbent. For this purpose the capillary column is inserted into a 1/16 inch stainless steel tube via an opening in the

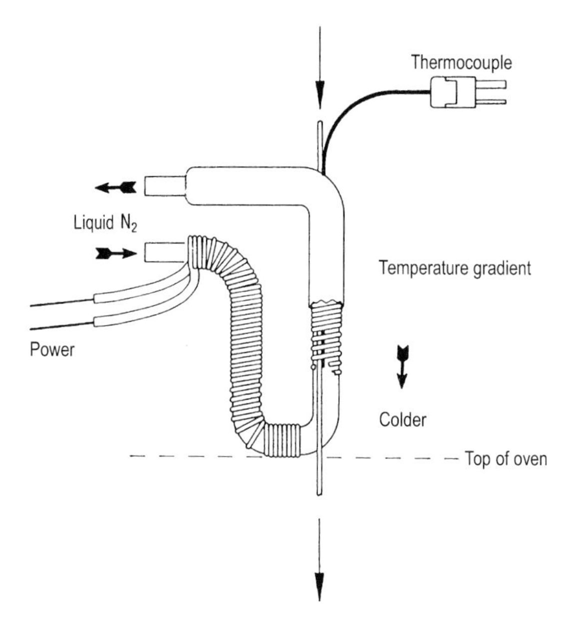

Fig. 2.83 Cut away diagram of a cryofocusing unit for use with fused silica capillary columns (Tekmar).

oven lid of the chromatograph and is connected via a connecting piece to the transfer line of the sampler (e.g. headspace, purge and trap, canister). This stainless steel tube is firmly welded to a cold finger and a thermoelement (Fig. 2.83). The tube is also surrounded by a differential heating coil which heats the inlet side more rapidly than the exit to the GC oven. In this way a uniform temperature gradient is guaranteed in all the phases of the operation. Because of the gradient, on cooling, e.g. with liquid nitrogen, all the analytes are focused into a narrow band at the beginning of the capillary column, as the substances migrate more slowly at the start of the band than at the end. After concentration and focusing, the chromatographic separation starts with the heating up of this region. On careful control of the gradient the heating rate does not affect the efficiency of the column (heating rates between 100 and 2000 °C/min).

Because of on-column focusing the cryotrap has the same sample capacity as the column has for these analytes. A breakthrough of the cryotrap as a result of too high an analyte concentration can be prevented by changing the column. Overloading the column would, in any case, result in poor separation. The diagram in Fig. 2.84 gives the temperatures at which a breakthrough of the analytes must be reckoned with in cryofocusing. Larger film thicknesses and internal diameters permit higher focusing temperatures which favour the mobilisation of the analytes from the column film.

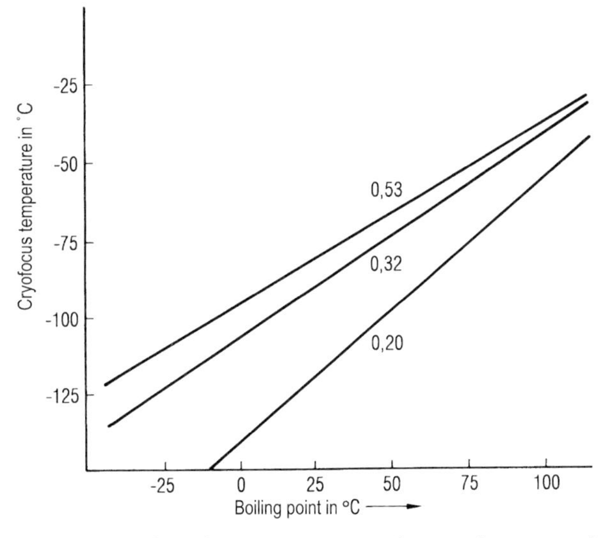

Fig. 2.84 Breakthrough temperatures in cryofocusing for various column internal diameters as a function of the boiling point of the analyte (Tekmar).
Film thicknesses:
0.25 μm at 0.20 mm internal diameter
1.0 μm at 0.32 mm internal diameter
3.0 μm at 0.53 mm internal diameter

2.2.4
Capillary Columns

There are no hard and fast rules for the choice of column for GC/MS coupling. The choice of the correct phase is made on the usual criteria: "like dissolves like". If substances exhibit no interaction with the stationary phase, there will be no retention and the substances leave the column in the dead time. The polarity of the stationary phase should correspond to the polarity of the substances being separated (see Table 2.24, pp. 120–127). Less polar substances are better separated on nonpolar phases and vice versa

Change to More Strongly Polar Stationary Phases

- Weaker retention of nonpolar compounds
- Stronger retention of polar compounds
- Shift of compounds with specific interactions

For coupling with mass spectrometry (GC/MS), the carrier gas flow and the specific noise of the column (column bleed) are included in the criteria governing the choice of column. When considering the optimal carrier gas flow, the maximum loading of the mass spectrometer must be taken into account. The limit for small benchtop mass spectrometers with quadrupole analysers is frequently ca. 1–2 mL/min. Larger instruments can generally tolerate higher loads because of their more powerful pumping systems. Ion trap mass spectrometers can be operated with a carrier gas flow of up to 3 mL/min and even higher with an external ion source. These conditions limit the column diameter which can be used. There can be no compromises concerning column bleed in GC/MS. Column bleed generally contributes to chemical noise where MS is used as the mass-dependent detector, and curtails the detection limits. The optimisation of a particular signal/noise ratio can also be effected in GC/MS by selecting particularly thermally stable stationary phases with a low tendency to bleed. For use in residue analysis, stationary phases for high-temperature applications have proved particularly useful (Figs. 2.85 and 2.86). Besides the phase itself, the film thickness also plays an important role. Thinner films and shorter columns have a lower tendency to cause column bleed.

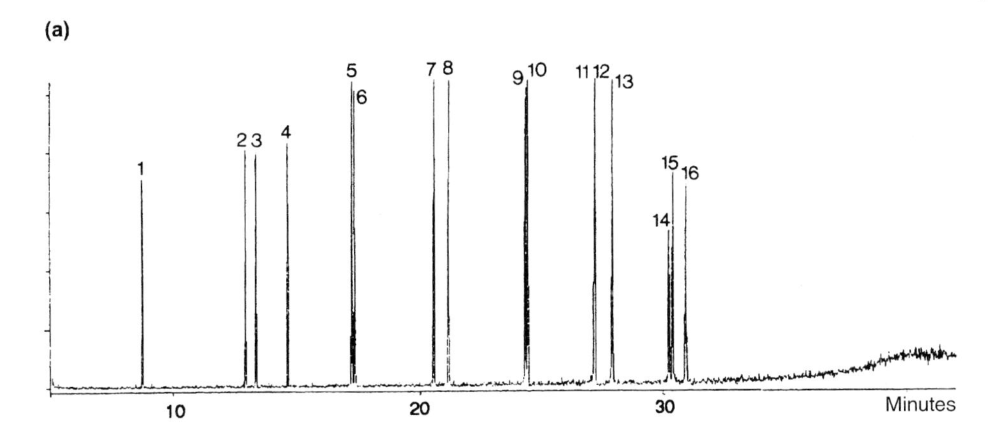

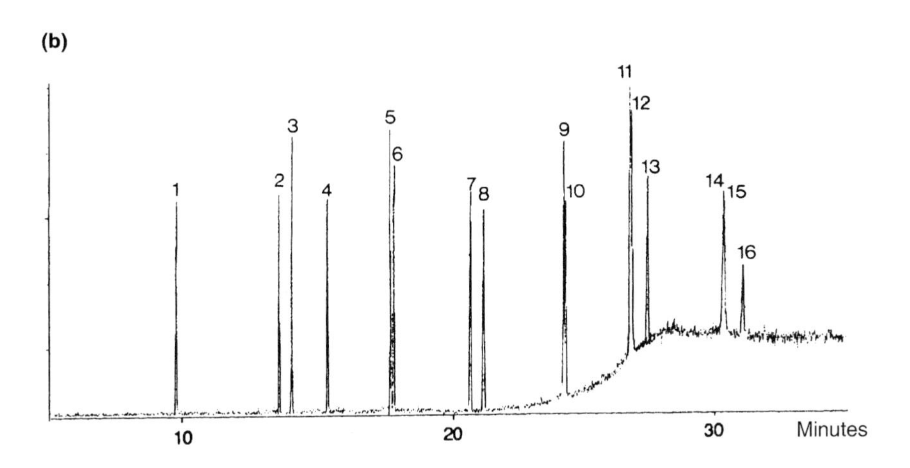

Polynuclear aromatic hydrocarbons

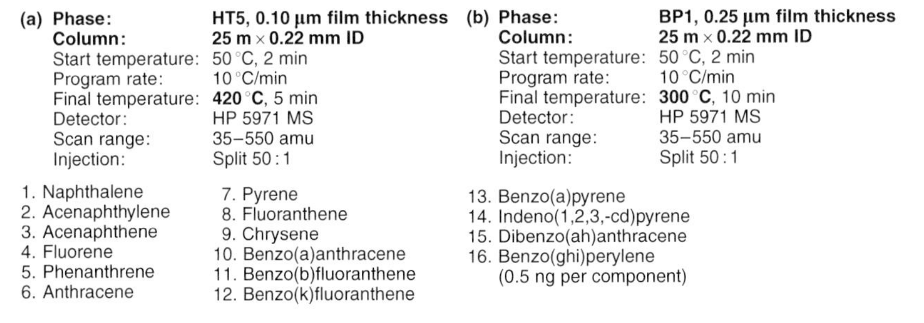

(a)		(b)	
Phase:	**HT5, 0.10 μm film thickness**	**Phase:**	**BP1, 0.25 μm film thickness**
Column:	**25 m × 0.22 mm ID**	**Column:**	**25 m × 0.22 mm ID**
Start temperature:	50 °C, 2 min	Start temperature:	50 °C, 2 min
Program rate:	10 °C/min	Program rate:	10 °C/min
Final temperature:	**420 °C**, 5 min	Final temperature:	**300 °C**, 10 min
Detector:	HP 5971 MS	Detector:	HP 5971 MS
Scan range:	35–550 amu	Scan range:	35–550 amu
Injection:	Split 50 : 1	Injection:	Split 50 : 1

1. Naphthalene
2. Acenaphthylene
3. Acenaphthene
4. Fluorene
5. Phenanthrene
6. Anthracene
7. Pyrene
8. Fluoranthene
9. Chrysene
10. Benzo(a)anthracene
11. Benzo(b)fluoranthene
12. Benzo(k)fluoranthene
13. Benzo(a)pyrene
14. Indeno(1,2,3,-cd)pyrene
15. Dibenzo(ah)anthracene
16. Benzo(ghi)perylene
    (0.5 ng per component)

Fig. 2.85 Comparison of a conventional silicone phase (lower trace: SGE BP1, dimethylsiloxane) with a high temperature phase (upper trace: SGE HT5, siloxane-carborane) using a polyaromatic hydrocarbon standard (SGE). The long temperature program for the high temperature phase up to 420 °C allows the elution of components 14, 15 and 16 during the heating-up phase at an elution temperature of ca. 350 °C. Sharp narrow peaks with low column bleed give better detection conditions for high-boiling substances (GC/MS system HP-MSD 5971).

Table 2.24 Composition of stationary phases for fused silica capillary columns with column designations for comparable phases from different manufacturers, arranged in order of increasing polarity, and columns with special phases (selection).

Phase composition	Agilent/J&W
100 % dimethyl polysiloxane	HP-1, HP-101, HP-1MS, Ultra-1, DB-1, DB-1MS, DB-1ht, SE-30
95 % dimethyl/5 % diphenyl polysiloxane	HP-5, HP-5MS, PAS-5, DB-5, DB-5.625, DB-5ht, SE-54
95 % dimethyl/5 % diphenyl polysilarylene	HP-5TA, DB-5ms
6 % cyanopropylphenyl/94 % dimethyl polysiloxane	HP-1301, HP-624, DB-1301, DB-624
80 % dimethyl/20 % diphenyl polysiloxane	–
65 % dimethyl/35 % diphenyl polysiloxane	HP-35, HP-35MS, DB-35
14 % cyanopropylphenyl/86 % dimethyl polysiloxane	HP-1701, PAS-1701, DB-1701
trifluoropropylmethyl polysiloxane	DB-210, DB-200
50 % dimethyl/50 % diphenyl polysiloxane	DB-17, DB-17HT, DB-608
100 % methylphenyl polysiloxane	HP-17, HP-50+
35 % dimethyl/65 % diphenyl polysiloxane	–
50 % cyanopropylmethyl/50 % phenylmethyl polysiloxane	HP-225, DB-225
polyethylene glycol (PEG)	HP-20M, Inno Wax, DB-Wax, Carbowax 20M, HP-Wax, DB-Wasetr
PEG for amines and basic compounds	CAM
PEG for acidic compounds	HP-FFAP, DB-FFAP, OV-351
90 % biscyanopropyl/10 % cyanopropylphenyl polysiloxane	–
70 % biscyanopropyl/30 % cyanopropylphenyl polysiloxane	–
biscyanopropyl polysiloxane	–
permethylated β-cyclodextrin	Cyclodex-β
1,2,3-tris(cyanoethoxy)propane	–
proprietary phase	–
phenyl polycarborane-siloxane	–
proprietary phase	DB-XLB
Column features	
built in guard column	DuraGuard
Silcosteel-treated stainless steel	ProSteel

Alltech	Macherey-Nagel	PerkinElmer	Phenomenex	Quadrex
AT-1, SE-30				
AT-1MS, EC-1	Optima 1, Optima-1ms	Elite-1	ZB-1	007-1
AT-5, SE-54				
AT-5	Optima 5, Optima-5ms	PE-2	ZB-5	007-2
–	–	–	ZB-5MS	–
AT-624	Optima 1301, Optima 624	–	ZB-624	007-1301
AT-20	–	PE-7	–	007-7
AT-35, AT-35MS, EC-35	–	Elite-35, Elite-35ms	ZB-35	007-11
AT-1701	Optima 1701	PE-1701	ZB-1701	007-1707
AT-210	Optima 210	–	–	007-210
–			–	
AT-50, AT-50MS	Optima 17	PE-17	ZB-50	007-17
–	–	–	–	400-65HT, 007-65HT
AT-225	Optima 225	PE-225	–	007-225
AT-Wax, Carbowax,				
AT-WAXMS	Permabond CW 20M, Optima WAX	PE-CW	ZB-WAX	007-CW
–	–	–	–	–
AT-1000, FFAP	Permabond FFAP, Optima FFAP	PE-FFAP	–	007-FFAP
AT-Silar	–	–	–	007-23
–	–	–	–	–
–	–	–	–	–
Chiraldex-β	–	–	–	–
–	–	–	–	–
–	–	–	–	–
–	–	–	–	–
–	–	–	–	–
–	–	–	Guardian	–
–	–	–	–	Ultra-Alloy

Table 2.24 (continued)

Phase composition	Restek	SGE
100% dimethyl polysiloxane	Rxi-1ms, Rtx-1, Rtx-1MS, Stx-1HT	–
95% dimethyl/5% diphenyl polysiloxane	Rxi-5ms, Rtx-5, Rtx-5MA, XTI-5, Stx-5HT	BP-5
95% dimethyl/5% diphenyl polysilarylene	Rtx-5Sil MS	BPX-5
6% cyanopropylphenyl/94% dimethyl polysiloxane	Rtx-1301, Rtx-624	BP-624
80% dimethyl/20% diphenyl polysiloxane	Rtx-20	–
65% dimethyl/35% diphenyl polysiloxane	Rtx-35, Rtx-35MS	BPX-35, BPX-608
14% cyanopropylphenyl/86% dimethyl polysiloxane	Rtx-1701	BP-1701
trifluoropropylmethyl polysiloxane	Rtx-200, Rtx-200MS	–
50% dimethyl/50% diphenyl polysiloxane	Rtx-17	–
100% methylphenyl polysiloxane	Rtx-50	BPX-50
35% dimethyl/65% diphenyl polysiloxane	Rtx-65, Rtx-65TG	–
50% cyanopropylmethyl/50% phenylmethyl polysiloxane	Rtx-225	BP-225
polyethylene glycol (PEG)	Stabilwax, Rtx-WAX	BP-20, SolGelWAX
PEG for amines and basic compounds	Stabilwax-DB	–
PEG for acidic compounds	Stabilwax-DA	BP-21
90% biscyanopropyl/10% cyanopropylphenyl polysiloxane	Rtx-2330	–
70% biscyanopropyl/30% cyanopropylphenyl polysiloxane		BPX-70
biscyanopropyl polysiloxane	Rt-2560	–
permethylated β-cyclodextrin	Rt-βDEXm	Cydex-β
1,2,3-tris(cyanoethoxy)propane	Rt-TCEP	–
proprietary phase	Rtx-440	HT-5
phenyl polycarborane-siloxane	Stx-500	–
proprietary phase	Rtx-XLB	–
Column features		
built in guard column	Integra-Guard	–
Silcosteel-treated stainless steel	MXT	AlumaClad

Supelco	Thermo Scientific	USP Nomenclature	Varian/Chrompack
Equity-1, SPB-1, SP-2100			VF-1MS, CP Sil 5 CB
SPB-1 Sulfur, SE-30, MDN-1	TR-1, TR-1MS	G1, G2, G38	CP Sil 5 CB MS
Equity-5, SPB-5, PTE-5, SE-54, SAC-5			VF-5MS, CP Sil 8 CB,
PTE-5 QTM, MDN-5,	TR-5	G27, G36, G41	CP Sil 8 CB MS
MDN-5S	TR-5MS	–	VF-5MS
SPB-1301	TR-V1	G43	–
SPB-20, VOCOL	–	G28, G32	–
SPB-35, SPB-608, MDN-35	TR-35MS	G42	VF-35MS
SPB-1701	TR-1701	G46	CP Sil 19 CB
–	–	G6	VF-200MS
–	–		
SP-2250, SPB-50	TR-50MS	G3	CP Sil 24 CB
–	–	G17	TAP-CB
–	–	G7, G19	CP Sil 43 CB
Supelcowax-10, Carbowax PEG 20M	TR-WAX, TR-WaxMS	G14, G15, G16, G20, G39	CP Wax 52 CB
Carbowax-Amine	–	–	CP Wax 51
Nukol, SP-1000	TR-FFAP	G25, G35	CP Wax 58 CB
SP-2330, SP-2331, SP-2380	–	G48	CP Sil 84
–	TR-FAME		–
SP-2560	–		–
β-DEX	–	–	CP-Cyclodextrin β
TCEP	–	–	CP-TCEP
–	–	–	–
–	–	–	–
–	–	–	–
–	–	–	EZ-Guard
Metallon	–	–	Ultimetal

Table 2.24 (continued)

Application	Agilent/J&W
Organic analysis-EPA methods 502.2, 524.2, 601, 602, 624, 8010, 8020,	–
Organic analysis-EPA methods 502.2, 524.2, 601, 602, 624, 8010, 8020,	HP-624, HP-VOC, DB-624, DB-502.2, DB-VRX
Organochlorine pesticides-EPA methods 8081, 608, and CLP pesticide	–
Organophosphorus pesticides-EPA method 8141A	–
Organochlorine pesticides-EPA methods 8081, 608, and CLP pesticide	HP-5, PAS-5, DB-5, DB-35, DB-608, HP-608,
	PAS-1701, DB-1701, DB-17, HP 50, HP-35
ASTM test method D2887	DB-2887
PONA analysis	HP-PONA, DB-Petro
Simulated distillation	–
Amines and basic compounds	–
Fatty Acid Methyl Esters (FAMES) (70% cyanopropyl polysilphenyl-siloxane)	–
Blood alcohol analysis	DB-ALC1, DB-ALC2
Residual solvents in pharmaceuticals	–
Residual solvents in pharmaceuticals	–
Fragrances and flavors	HP-20M, Carbowax 20M
Explosives (8% phenyl polycarbonane-siloxane)	–
Dioxins and furans	–
PCB congeners	–

Alltech	Macherey-Nagel	PerkinElmer	Phenomenex	Quadrex
–	–	–	–	–
AT-624	–	PE-502	OV-624	007-624, 007-502
–	–	–	–	–
–	–	–	–	–
AT-5, AT-35, AT 50, AT-Pesticides	–	PE-2, PE-608, PE-1701	ZB-5, ZB-35 ZB-1701, ZB-50	007-2, 007-608, 007-17, 007-1701
–	–	–	–	007-1-10V-1.0F
AT-Petro	–	–	–	007-1-10V-0.5F
–	–	–	–	–
–	–	–	–	–
–	–	–	–	–
–	–	–	–	–
–	–	–	–	–
–	–	–	–	–
–	–	–	–	007-CW
–	–	–	–	–
–	–	–	–	–
–	–	–	–	–

Table 2.24 (continued)

Application	Restek
Organic analysis-EPA methods 502.2, 524.2, 601, 602, 624, 8010, 8020,	Rtx-VMS, Trx-VGC
Organic analysis-EPA methods 502.2, 524.2, 601, 602, 624, 8010, 8020,	Rtx-VRX, Rtx-502.2, Rtx-624, Rtx-Volatiles
	Rtx-CLPesticides, Rtx-CLPesticides2,
Organochlorine pesticides-EPA methods 8081, 608, and CLP pesticide	Stx-CLPesticides, Stx-CLPesticides2
Organophosphorus pesticides-EPA method 8141A	Rtx-OPPesticides, Rtx-OPPesticides2
Organochlorine pesticides-EPA methods 8081, 608, and CLP pesticide	Rtx-5, Rtx-35, Rtx-50, Rtx-1701
ASTM test method D2887	Rtx-2887
PONA analysis	Rtx-1PONA
Simulated distillation	MXT-500 Sim Dist
Amines and basic compounds	Rtx-5 Amine
Fatty Acid Methyl Esters (FAMES) (70% cyanopropyl polysilphenyl-siloxane)	FAMEWAX
Blood alcohol analysis	Rtx-BAC1, Rtx-BAC2
Residual solvents in pharmaceuticals	Rtx-G27
Residual solvents in pharmaceuticals	Rtx-G43
Fragrances and flavors	Rt-CW20M F&F
Explosives (8% phenyl polycarbonane-siloxane)	Rtx-TNT, Rtx-TNT2
Dioxins and furans	Rtx-Dioxin, Rtx-Dioxin2
PCB congeners	Rtx-PCB

SGE	Supelco	Thermo Scientific	USP	Varian/Chrompack
–	–	TR-524	–	–
–	VOCOL, SPB-624	TR-524	–	CP Sil 13 CB
–	–	TR-5MS	–	–
–	–	TR-5MS	–	–
BP-5, BP-10, BP-608	SPB-5, SPB-608, SPB-1701	TR-5, TR-35MS, TR-50MS, TR-1701	G3	CP Sil 8 CB, CP Sil 19 CB
–	Petrocol 2887, Petrocol EX2887	TR-SimDist	–	CP-SimDist-CB
BP1-PONA	Petrocol DH	–	–	CP Sil PONA CB
HT5	–	–	–	–
–	PTA-5	–	G50	–
–	Omegawax	TR-FAME	–	–
–	–	–	–	–
–	–	–	G27	–
–	OVI-G43	TR-V1	G43	–
BP-20M	–	TR-WaxMS	–	–
–	–	TR-8095	–	–
–	–	–	–	–
–	–	–	–	–

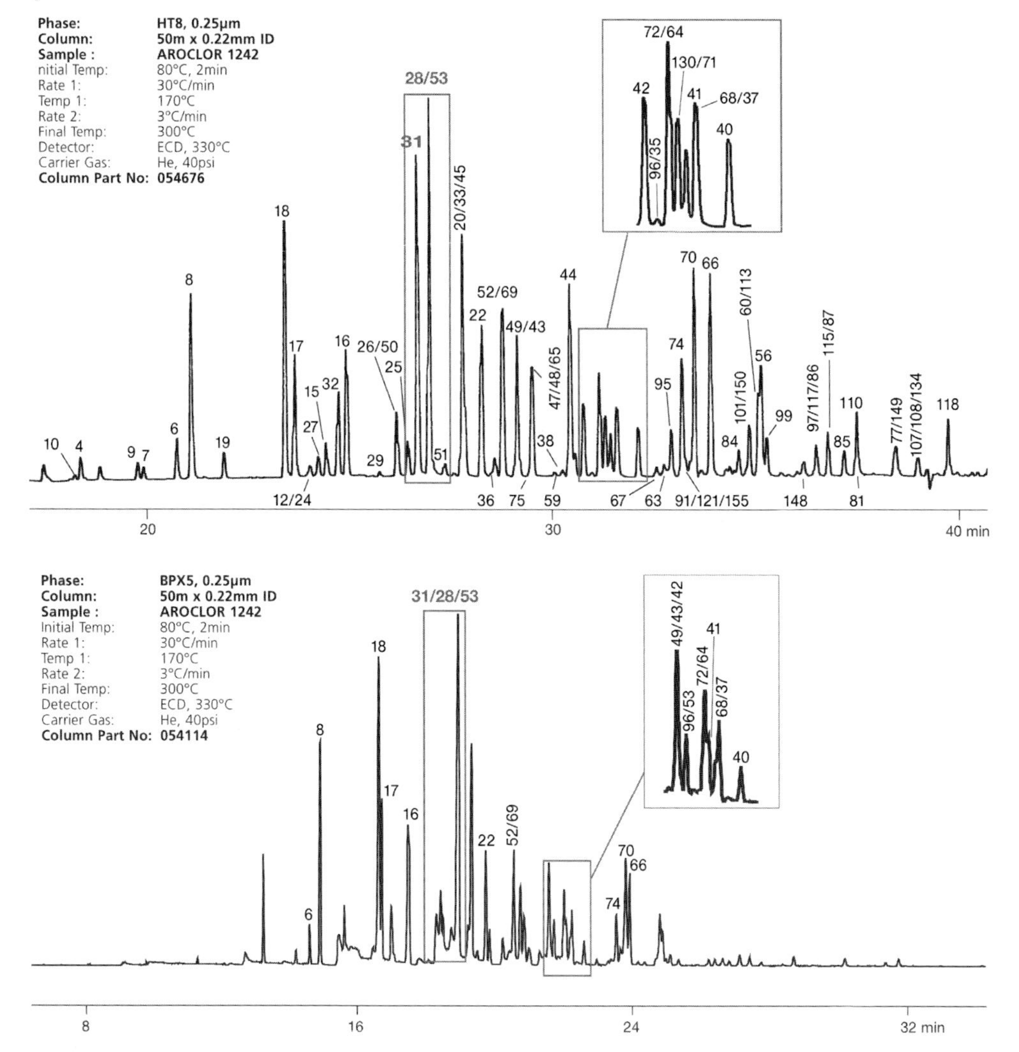

Fig. 2.86 An example of the different selectivities of stationary phases for the Aroclor mixture 1242 (see in particular the separation of the critical congeners 31, 28 and 53 (Courtesy SGE))
Top: Carborane phase – HT8 50 m × 0.22 mm × 0.25 µm
Bottom: 5%-Phenyl phase – BPX5 50 m × 0.22 mm × 0.25 µm

2.2.4.1 **Sample Capacity**

The sample capacity is the maximum quantity of an analyte with which the phase can be loaded (Table 2.25). An overloaded column exhibits peak fronting, i.e. an asymmetrical peak which has a gentle gradient on the front and a sharp slope on the back side. This effect can increase until a triangular peak shape is obtained, a so-called “shark fin”. Overloading occurs rapidly if a column of the wrong polarity is chosen. The capacity of a column depends on the internal diameter, the film thickness and the solubility of a substance in the phase.

Table 2.25 Sample capacities for common column diameters.

Internal diameter	0.18 mm	0.25 mm	0.32 mm	0.53 mm
Film thickness	0.20 µm	0.25 µm	0.25 µm	1.00 µm
⇒ **Sample capacity**	<50 ng	50–100 ng	400–500 ng	1000–2000 ng
Theoretical plates per metre of column	5300	3300	2700	1600
Optimal flow rate at				
20 cm/s helium	0.3 mL/min	0.7 mL/min	1.2 mL/min	2.6 mL/min
40 cm/s hydrogen	0.6 mL/min	1.4 mL/min	2.4 mL/min	5.2 mL/min

Sample Capacity

- Increases with internal diameter
- Increases with film thickness
- Increases with solubility

2.2.4.2 **Internal Diameter**

The internal diameter of a column used in capillary gas chromatography varies from 0.1 mm (microbore capillary) via 0.18 mm and 0.25 mm (narrow bore) and the standard columns with 0.32 mm to 0.53 mm (megabore, halfmil). For direct coupling with mass spectrometers, in practice only the columns up to 0.32 mm internal diameter are used. Megabore columns are mainly used for retention gaps (deactivated, no stationary phase) or to replace packed columns in specially designed GC instruments.

The internal diameter affects the resolving power and the analysis time (Fig. 2.87). Basically, at constant film thicknesses lower internal diameters are preferred in order to achieve higher chromatographic resolution. As the flow per unit time decreases at a particular carrier gas velocity, the analysis time increases. In the case of complex mixtures, changing to a column with a smaller internal diameter gives better separation of critical pairs of compounds (Table 2.26). In practice it has been shown that even changing from 0.25 mm internal diameter to 0.20 mm allows an improvement in the separation of, for example, PCBs. For sample application to narrow bore columns certain conditions must be adhered to, depending on the type of injector (see Section 2.2.3).

Table 2.26 Effect of column diameter and linear carrier gas velocity on the flow rate.

Internal diameter	**Linear carrier gas velocity**		**Flow rate**	
	He	H_2	He	H_2
0.18 mm	30–45 cm/s	45–60 cm/s	0.5–0.7 mL/min	0.7–0.9 mL/min
0.25 mm	**30–45 cm/s**	**45–60 cm/s**	**0.9–1.3 mL/min**	**1.3–1.8 mL/min**
0.32 mm	30–45 cm/s	45–60 cm/s	1.4–2.2 mL/min	2.2–2.8 mL/min
0.54 mm	30–45 cm/s	45–60 cm/s	4.0–6.0 mL/min	6.0–7.9 mL/min

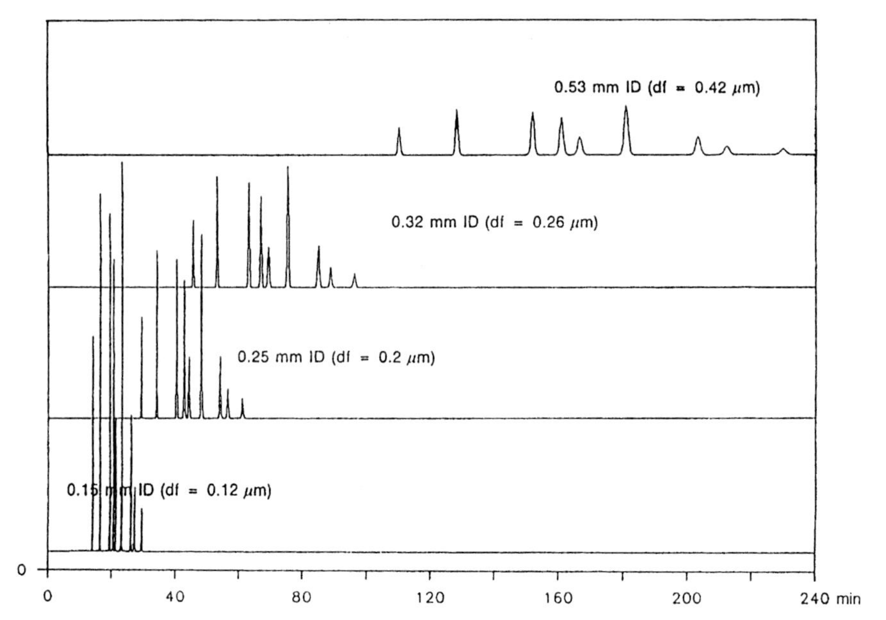

Fig. 2.87 Effect of increasing column internal diameter together with increasing film thickness on peak height and retention time. (The columns have the same phase ratio!) (Chrompack).

Smaller Internal Diameter (at Identical Film Thickness)

- Increases the resolution
- Increases the analysis times

2.2.4.3 Film Thickness

The variation in the film thickness at a given internal diameter and column length gives the user the possibility of carrying out special separation tasks. As a rule, thick films are used for volatile compounds and thin films for high-boilers and trace analysis. Thick film columns with coatings of more than 1.0 μm can separate extremely low boiling compounds well, e.g. volatile halogenated hydrocarbons. Through the large increase in capacity with thicker films, it is even possible to dispense with additional oven cooling during injection in the analysis of volatile halogenated hydrocarbons using headspace or purge and trap. Start temperatures above room temperature are usual. However, thick film columns exhibit severe column bleed at elevated temperatures.

For the residue analysis of all other substances, thin film columns with coating thicknesses of ca. 0.1 μm have proved to be very effective in GC/MS. Thin film columns give narrow rapid peaks and can be used in higher temperature ranges without significant column bleed. The elution temperatures of the compounds decrease with thin films and, at the same program duration, the analysis can be extended to compounds with higher molecular weights and thermolabile compounds e.g. the decabromodiphenylether (PBDE 209). The

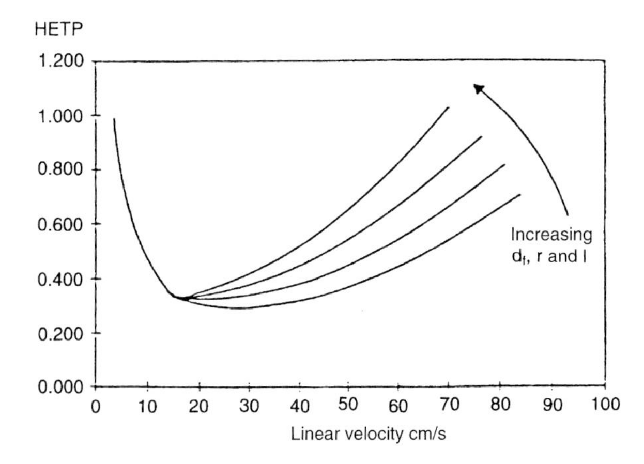

Fig. 2.88 Effect of film thickness (d_f), internal diameter (r) and length (l) on the van Deemter curves for helium as the carrier gas.

duration of the analysis for a given compound becomes shorter, but the capacity of the column decreases also limiting the load for matrix samples (Fig. 2.88).

Increasing Film Thicknesses

- Improve the resolution of volatile compounds
- Increase the analysis time
- Increase the elution temperatures

The Relationship Between Film Thickness and Internal Diameter

The phase ratio of a capillary column is determined by the ratio of the volume of the gaseous mobile phase (internal volume) to the volume of the stationary phase (coating). From Table 2.27 the phase ratio can be read off for each combination of film thickness and internal diameter (assuming the same film and the same column length). High values mean good separation; the same values show combinations with the same separating capacity. For GC/MS optimal separations can be planned and also other conditions, such as carrier gas flow and column bleed, can be taken into consideration. To achieve better separation it is possible to change to a smaller film thickness at the same internal diameter or to keep the film thickness and choose a higher internal diameter. For example, a Fast GC column of 0.18 mm ID and 0.10 μm film has almost the double phase ratio than the commonly used 0.25 mm ID column with 0.25 μm film. The phase ratio is tripled when switching to a 0.25 mm ID column with 0.1 μm film which is typical for trace analysis applications. Using Table 2.27 the separation efficiency can easily be optimized to the required conditions.

2.2.4.4 Column Length

The analytical column should be as short as possible. The most common lengths for standard columns are 30 or 60 m. Greater lengths are not necessary in residue analysis with GC/MS systems even for separating complex PCB or volatile halogenated hydrocarbon mixtures.

Table 2.27 Effect of column diameter and film thickness on the phase ratio.

Internal diameter	Film thickness **0.10** μm	0.25 μm	0.50 μm	1.0 μm	1.50 μm	2.0 μm	3.0 μm	5.0 μm
0.18 mm	450	180	90	45	30	23	15	9
0.25 mm	**625**	**250**	**125**	**63**	**42**	**31**	**21**	**13**
0.32 mm	800	320	160	80	53	40	27	16
0.53 mm	1325	530	265	128	88	66	43	27

Shorter columns would be desirable for simpler separations, but they are with the same diameter at the limit of the maximum flow for the mass spectrometer used. Here the switch to fast GC applications using smaller diameters should be considered. Doubling the column length only results in an improvement in the separation by a factor of 1.4 ($\sqrt{2}$) while the analysis time is doubled (and the cost of the column also!). For isothermal chromatography, the retention time is directly proportional to the column length while with programmed operations the retention time is essentially determined by the elution temperature of a compound. On changing to a longer column, the temperature program should always be optimised again to achieve optimal retention times.

Doubling the Column Length

- Resolution only increases by a factor of 1.4
- Costs are doubled
- Retention times are doubled
- Sample throughput (productivity) is cut by half
- The temperature program must be optimised again

2.2.4.5 Adjusting the Carrier Gas Flow

The maximum separating capacity of a capillary column can only be exploited with an optimised carrier gas flow. With direct GC/MS coupling the carrier gas flow is affected slightly by the vacuum on the detector side. In practice the adjustment is no different from that in classical GC systems. Only with instruments with electronic pressure control (EPC) is adjustment necessary for direct coupling with MS. The separating efficiency of a capillary column is given as theoretical plates in chromatographic terminology (see Section 2.2.5). A high analytical separating capacity is always accompanied by a large number of plates (number of separation steps) so the height equivalent to a theoretical plate (HETP) decreases. In van Deemter curves, the HETP is plotted against the carrier gas velocity (not flow!). The minimum of one of these curves gives the optimal adjustment for a particular carrier gas for isothermal operations.

For helium the optimal carrier gas velocity for standard columns is ca. 24 cm/s. As the viscosity of the carrier gas increases in the course of a temperature program, in practice at

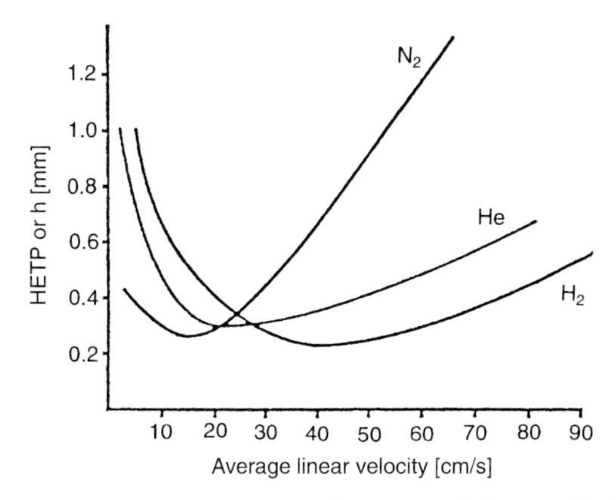

Fig. 2.89 van Deemter curves for nitrogen, helium and hydrogen as the carrier gas. The values refer to a standard column of 30 m length, 0.25 mm internal diameter and 0.25 μm film thickness (Restek).

the lower start temperature a higher velocity is used (ca. 30 cm/s). Velocities which are too high lower the efficiency.

With hydrogen as the carrier gas a much higher velocity (>40 cm/s) can be used, which leads to significant shortening of the analysis time. Furthermore, with hydrogen the right hand branch of the van Deemter curve is very flat, so that a further increase in the gas velocity is possible without impairing the efficiency (Figs. 2.89 and 2.90). The separating efficiency is retained and time is gained. Because of the often very limited pumping capacity in commercial GC/MS systems (see Section 2.4.1), little use can be made of these advantages with hydrogen. Caused by the expected shortage of the natural helium supply and the already increasing prices hydrogen becomes an economical alternative. Hydrogen generators are available for on-site premium quality supply.

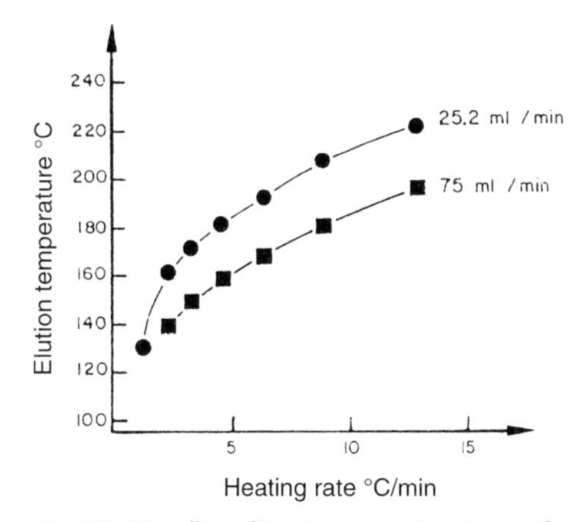

Fig. 2.90 The effect of heating rate and carrier gas flow on the elution temperature (retention temperature) (after Karasek).

2.2.4.6 Properties of Stationary Phases

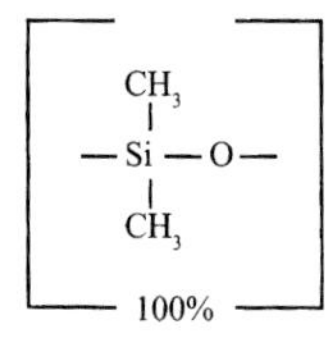

Fig. 2.91 100 % Dimethyl-polysiloxane

Polarity:	least polar bonded phase
Use:	boiling point separations for solvents, petroleum products, pharmaceuticals
Properties:	minimum temperature –60 °C maximum temperature 340–430 °C helix structure

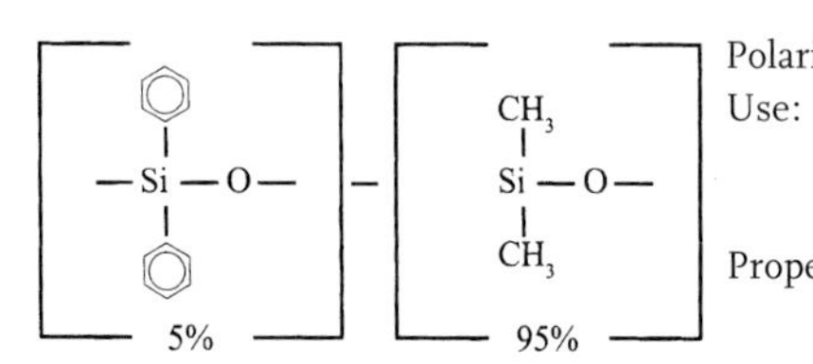

Fig. 2.92 5 % Diphenyl-95 % dimethyl-polysiloxane

Polarity:	nonpolar, bonded phase
Use:	boiling point, point separations for aromatic compounds, environmental samples, flavours, aromatic hydrocarbons
Properties:	minimum temperature –60 °C maximum temperature 340 °C

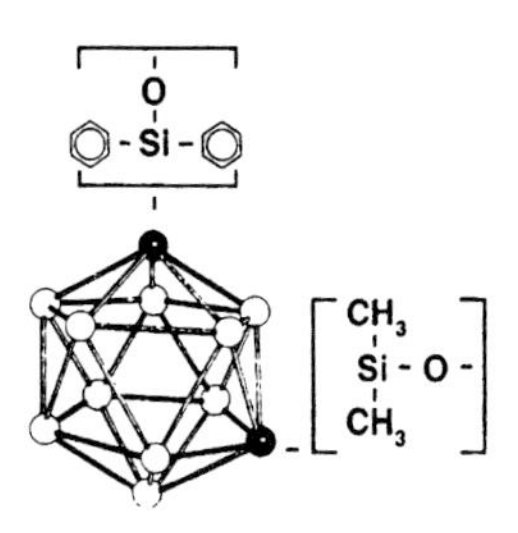

Fig. 2.93 Siloxane-carborane, comparable to 5 % phenyl

Use:

- ideal for GC/MS coupling because of very low bleeding
- all environmental samples
- all medium and high molecular weight substances
- polyaromatic hydrocarbons, PCBs, waxes, triglycerides

Properties: minimum temperature –10 °C
maximum temperature 480 °C (highest operating temperature of all stationary phases, aluminium coated, 370 °C polyimide coated), high temperature phase

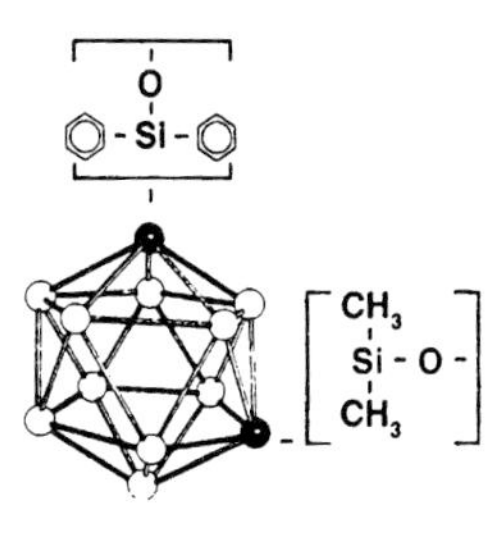

Fig. 2.94 Siloxane-carborane, comparable to 8 % phenyl

Polarity: weakly polar, similar to 8 % phenylsiloxane

Use:

- ideal for GC/MS coupling because of very low bleeding
- all environmental samples, can be used universally
- volatile halogenated hydrocarbons, solvents – polyaromatic hydrocarbons, pesticides, only column that separates all PCB congeners

Properties: minimum temperature –20 °C
maximum temperature 370 °C

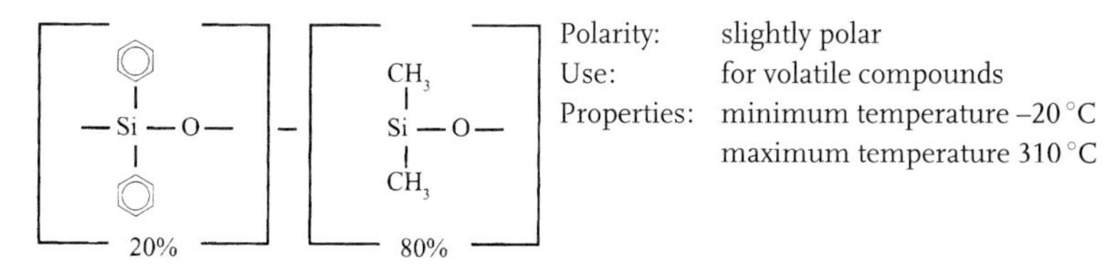

Fig. 2.95 20% Diphenyl-80% dimethyl-polysiloxane

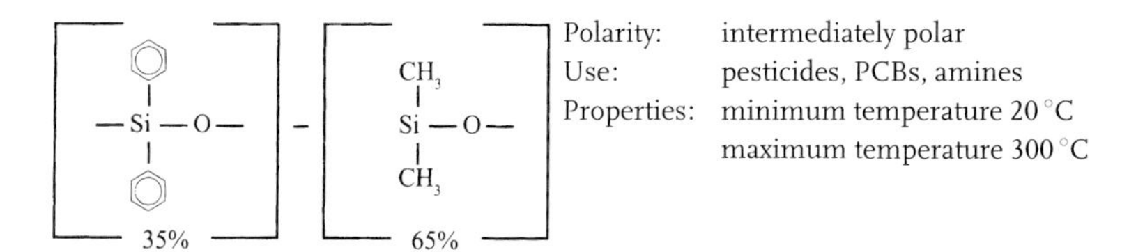

Fig. 2.96 35% Diphenyl-65% dimethyl-polysiloxane

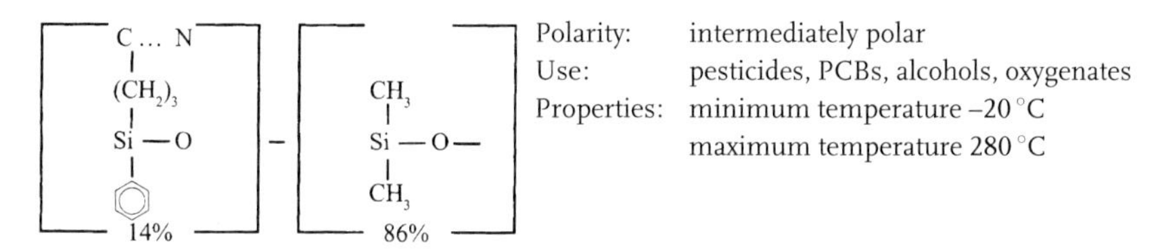

Fig. 2.97 14% Cyanopropylphenyl-86% dimethyl-polysiloxane

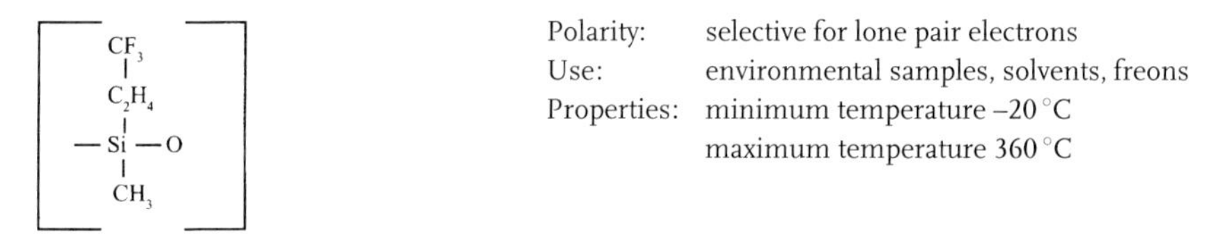

Fig. 2.98 100% Trifluoropropylmethyl-polysiloxane

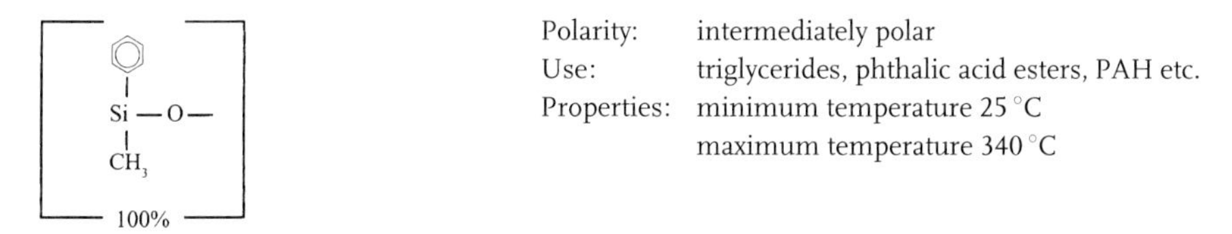

Fig. 2.99 50% Diphenyl-50% dimethyl-polysiloxane

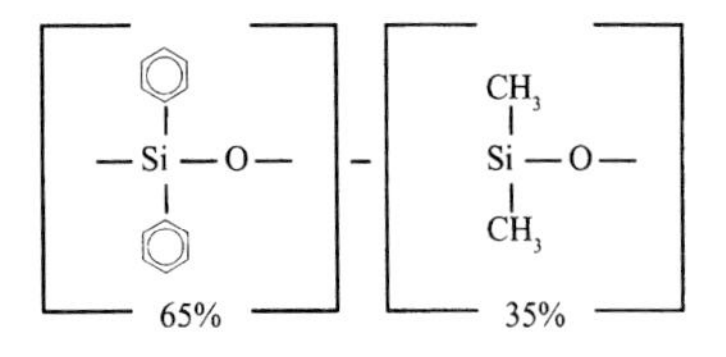

Polarity: medium polarity
Use: triglycerides, free fatty acids, terpenes
Properties: minimum temperature 50 °C
maximum temperature 340 °C

Fig. 2.100 65 % Diphenyl-35 % dimethyl-polysiloxane

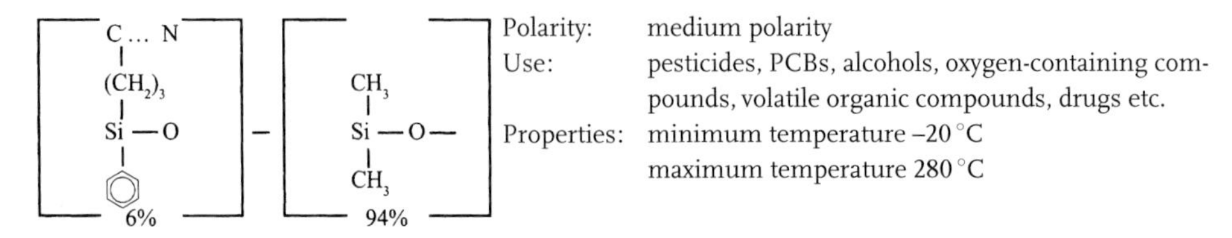

Polarity: medium polarity
Use: pesticides, PCBs, alcohols, oxygen-containing compounds, volatile organic compounds, drugs etc.
Properties: minimum temperature –20 °C
maximum temperature 280 °C

Fig. 2.101 6 % Cyanoprophylphenyl-94 % dimethyl-polysiloxane

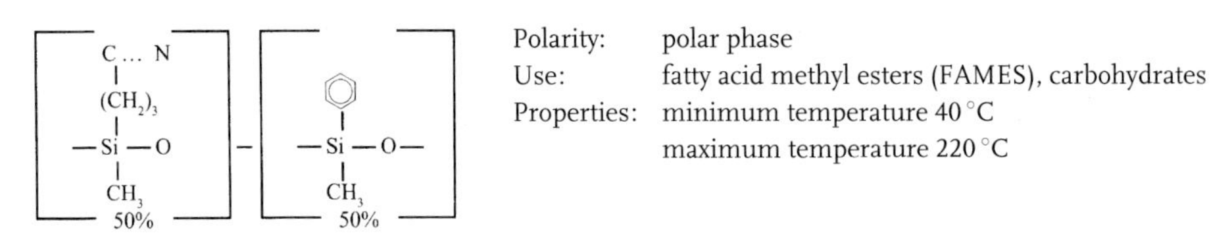

Polarity: polar phase
Use: fatty acid methyl esters (FAMES), carbohydrates
Properties: minimum temperature 40 °C
maximum temperature 220 °C

Fig. 2.102 50 % Cyanopropylmethyl-50 % phenylmethyl-polysiloxane

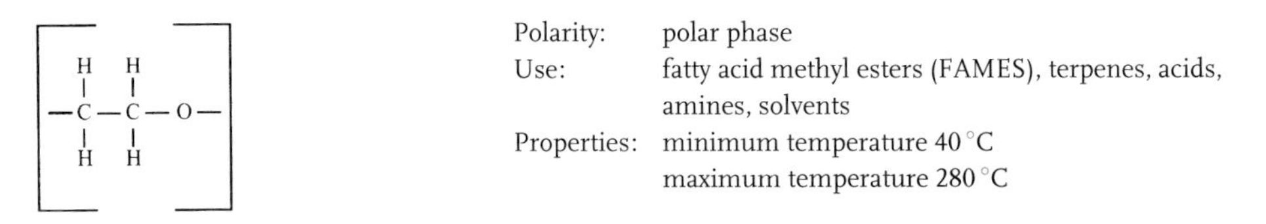

Polarity: polar phase
Use: fatty acid methyl esters (FAMES), terpenes, acids, amines, solvents
Properties: minimum temperature 40 °C
maximum temperature 280 °C

Fig. 2.103 100 % Carbowax polyethyleneglycol 20M

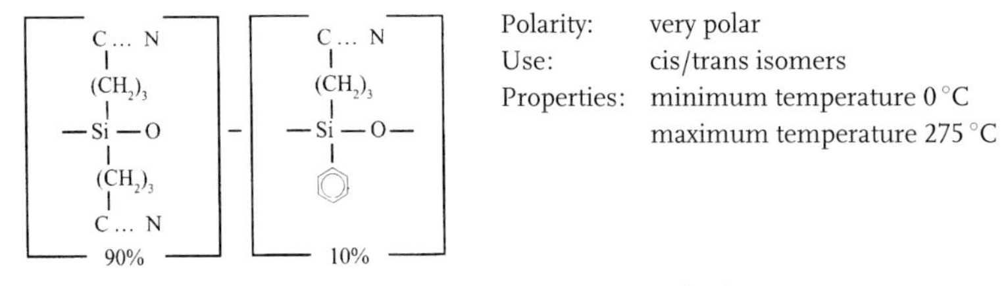

Polarity: very polar
Use: cis/trans isomers
Properties: minimum temperature 0 °C
maximum temperature 275 °C

Fig. 2.104 90 % Biscyanopropyl-10 % phenylcyanopropyl-polysiloxane

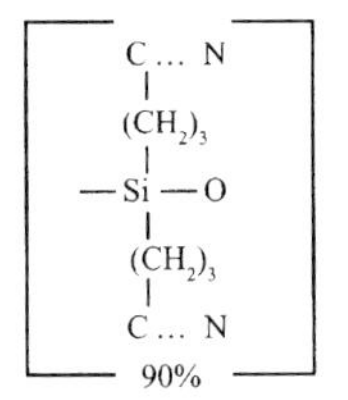

Polarity: very polar
Use: fatty acid methyl esters (FAMES)
Properties: minimum temperature 20 °C
maximum temperature 250 °C

Fig. 2.105 100% Biscyanopropyl-polysiloxane

2.2.5
Chromatography Parameters

All chromatography processes are based on the multiple repetition of a separation process, such as the continuous dynamic partition of the components between two phases.

In a model chromatography can be regarded as a continuous repetition of partition steps. The starting point is the partition of a substance between two phases in a separating funnel. Suppose a series of separating funnels is set up which all contain the same quantity of phase 1. As this phase remains in the separating funnels, it is known as the stationary phase.

The sample is placed in the first separating funnel dissolved in a second phase (the auxiliary phase). After establishing equilibrium through shaking, phase 2 is transferred to separating funnel 2. The auxiliary phase thereby becomes the mobile phase. Fresh mobile phase is placed in the first separating funnel etc. (Fig. 2.106).

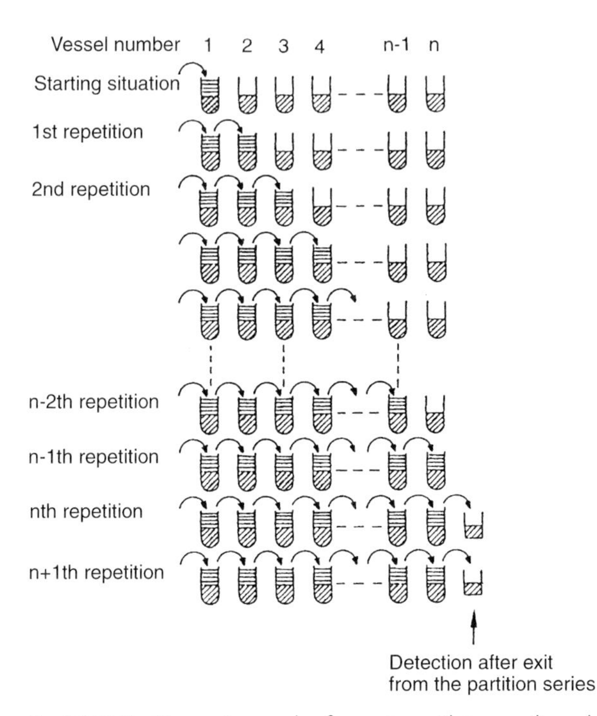

Fig. 2.106 Partition series: mode of operation with two auxiliary phases.

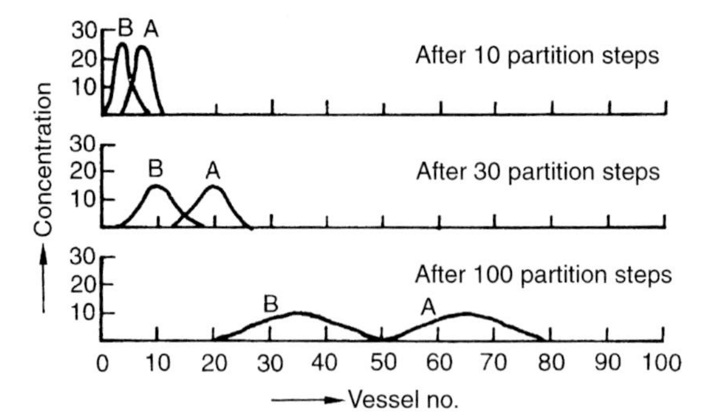

Fig. 2.107 Partition of substances A and B after 10, 30 and 100 separating steps.
After 10 steps A and B are hardly separated, after 30 steps quite well and after 100 steps practically completely. The two substances are partitioned among an ever increasing number of vessels and the concentrations decrease more and more ($\alpha_A = 2 : \alpha_B = 0.5$).

The results of this type of partition with 100 vessels and two substances A and B are shown in Fig. 2.107. The prerequisite for this is the validity of the Nernst equation. For detection the concentrations of A and B in the vessels are determined.

With the model described, so many separating steps are carried out that the mobile phase leaves the system of 100 vessels and the individual components A and B are removed, one after the other, from the series of vessels. This process is known as elution.

2.2.5.1 The Chromatogram and its Meaning

The substances eluted are transported by the mobile phase to the detector and are registered as Gaussian curves (peaks). The peaks give qualitative and quantitative information on the mixture investigated.

Qualitative: The retention time is the time elapsing between injection of the sample and the appearance of the maximum of the signal. The retention time of a component is always constant under the same chromatographic conditions. A peak can therefore be identified by a comparison of the retention time with a standard (pure substance).

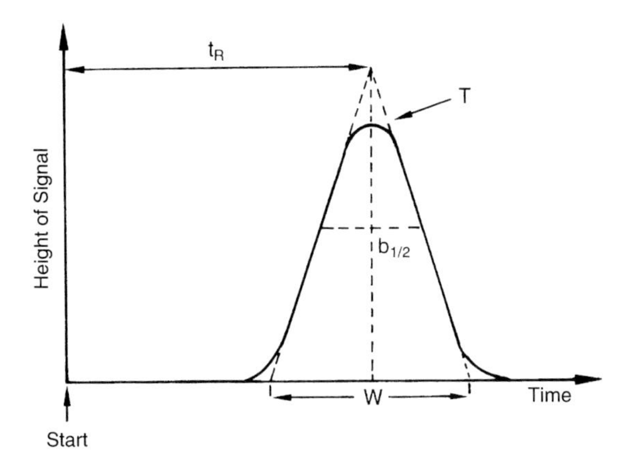

Fig. 2.108 Parameters determined for an elution peak.
t_R retention time
$b_{1/2}$ half width
W base width

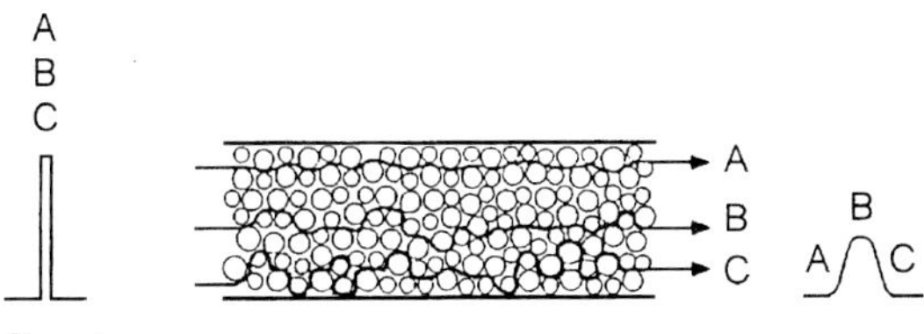

Fig. 2.109 Eddy diffusion in packed columns (multipath effect).

Quantitative: The height and area of a peak is proportional to the quantity of substance injected. Unknown quantities of substance can be determined by a comparison of the peak areas (or heights) with known concentrations.

In the ideal case the peaks eluting are in the shape of a Gaussian distribution (bell-shaped curve, Fig. 2.108). A very simple explanation of this shape is the different paths taken by the molecules through the separating system (multipath effect), which is caused by diffusion processes (Eddy diffusion) (Fig. 2.109).

Under defined conditions the time required for elution of a substance A or B at the end of the separating system, the **retention time** t_R, is characteristic of the substance. It is measured from the start (sample injection) to the peak maximum (Fig. 2.110).

At a constant flow rate t_R is directly proportional to the retention volume V_R.

$$V_R = t_R \cdot F \tag{10}$$

where F = flow rate in mL/min

The retention volume shows how much mobile phase has passed through the separating system until half of the substance has eluted (peak maximum!).

2.2.5.2 Capacity Factor k'

The retention time t_R depends on the flow rate of the mobile phase and the length of the column. If the mobile phase moves slowly or the column is long, t_0 is large and so is t_R. Thus t_R is not suitable for the comparative characterisation of a substance, e. g. between two laboratories.

It is better to use the capacity factor, also known as the k' value, which relates the net retention time t'_R to the dead time:

$$k' = \frac{t'_R}{t_0} = \frac{t_R - t_0}{t_0} \tag{11}$$

Thus, the k' value is independent of the column length and the flow rate of the mobile phase and represents the molar ratio of a particular component in the stationary and mobile phases. Large k' values mean long analysis times.

The k' value is related to the partition coefficient K as follows:

$$k' = K \cdot \frac{V_1}{V_g} \tag{12}$$

where V_1 = volume of the stationary phase
V_g = volume of the mobile phase

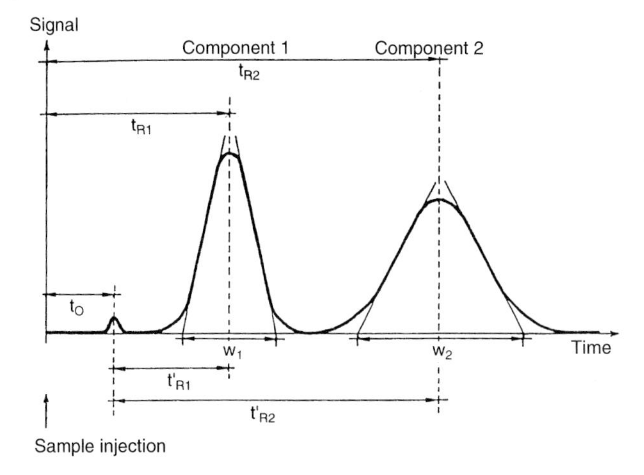

Fig. 2.110 The chromatogram and its parameters.

W *Peak width* of a peak. $W = 4\sigma$ with σ = the standard deviatiion of the Gaussian peak.

t_0 *Dead time* of the column; the time which the mobile phase requires to pass through the column. The linear velocity u of the solvent is calculated from

$$u = \frac{L}{t_0} \quad \text{with } L = \text{length of the column}$$

A substance which is not retarded, i.e. a substance which is not held by the stationary phase, appears at t_0 at the detector.

t_R *Retention time*: the time between the injection of a substance and the recording of a peak maximum.

t'_R *Net retention time.* From the diagram it can be seen that $t_R = t_0 + t'_R$.
t_0 is the same for all eluted substances and is therefore the residence time in the mobile phase. The substances separated differ in their residence times in the stationary phase t'_R. The longer a substance stays in the stationary phase, the later it is eluted.

The capacity factor is therefore directly proportional to the volume of the stationary phase (or for adsorbents, their specific surface area in m^2/g).

α is a measure of the relative retention and is given by:

$$\alpha = \frac{k'_2}{k'_1} = \frac{K_2}{K_1} \qquad (k'_2 > k'_1) \tag{13}$$

In the case where $\alpha = 1$, the two components 1 and 2 are not separated because they have the same k' values.

The relative retention α is thus a measure of the selectivity of a column and can be manipulated by choice of a suitable stationary phase. (In principle this is also true for the choice of the mobile phase, but in GC/MS helium or hydrogen are, in fact, always used.)

2.2.5.3 **Chromatographic Resolution**

A second model, the theory of plates, was developed by Martin and Synge in 1941. This is based on the functioning of a fractionating column, then as now a widely used separation technique. It is assumed that the equilibrium between two phases on each plate of the column has been fully established. Using the plate theory, mathematical relationships can be derived from the chromatogram, which are a practical measure of the sharpness of the separation and the resolving power.

The chromatography column is divided up into theoretical plates, i.e. into column sections in the flow direction, the separating capacity of each one corresponding to a theoretical plate. The length of each section of column is called the height equivalent to a theoretical plate (HETP). The HETP value is calculated from the length of the column L divided by the number of theoretical plates N:

$$\text{HETP} = \frac{L}{N} \qquad \text{in mm} \tag{14}$$

The number of theoretical plates is calculated from the shape of the eluted peak. In the separating funnel model it is shown that with an increasing number of partition steps the substance partitions itself between a larger number of vessels. A separation system giving sharp separation concentrates the substance band into a few vessels or plates. The more plates there are in a separation system, the sharper the eluted peaks.

The number of theoretical plates N is calculated from the peak profile. The retention time t_R at the peak maximum and the width at the base of the peak measured as the distance between the cutting points of the tangents to the inflection points with the base line are determined from the chromatogram (see Fig. 2.108).

$$N = 16 \cdot \left(\frac{t_g}{W}\right)^2 \tag{15}$$

where t_R = retention time
W = peak width

For asymmetric peaks the half width (the peak width at half height) is used:

$$N = 8\ \ln 2 \cdot \left(\frac{t_g}{W_n}\right)^2 \tag{16}$$

where t_R = retention time
W_n = peak width at half height

Consequence: A column is more effective, the more theoretical plates it has (Fig. 2.111).

The width of a peak in the chromatogram determines the resolution of two components at a given distance between the peak maxima (Fig. 2.112). The resolution R is used to assess the quality of the separation:

$$R \approx \frac{\text{retention difference}}{\text{peak width}} \tag{17}$$

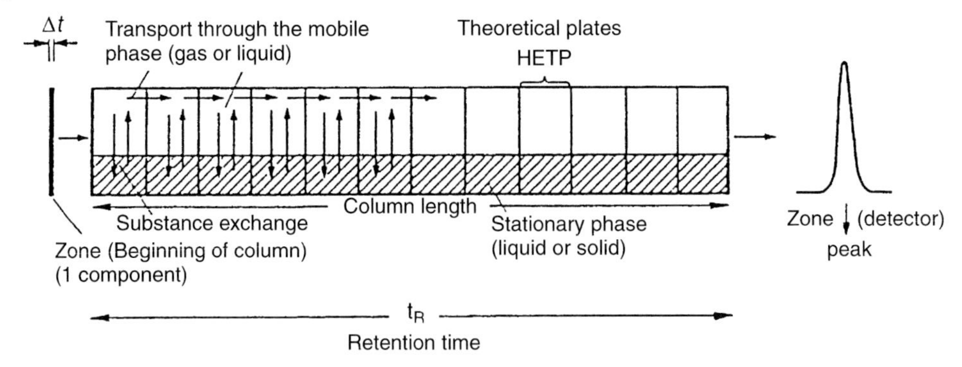

Fig. 2.111 Substance exchange and transport in a chromatography column are optimal when there are as many phase transfers as possible with the smallest possible expansion of the given zones (after Schomburg).

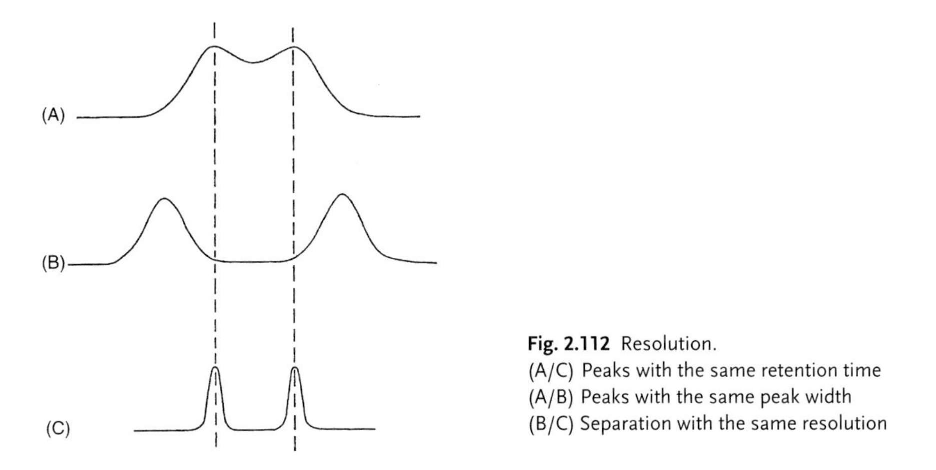

Fig. 2.112 Resolution.
(A/C) Peaks with the same retention time
(A/B) Peaks with the same peak width
(B/C) Separation with the same resolution

The resolution R of two neighbouring peaks is defined as the quotient of the distance between the two peak maxima, i.e. the difference between the two retention times t_R and the arithmetic mean of the two peak widths:

$$R = 2 \cdot \frac{t_{R2} - t_{R1}}{W_1 + W_2} = 1.198 \cdot \frac{t_{R2} - t_{R1}}{W_{h1} + W_{h2}}$$

where W_h = peak width at half height

Figure 2.113 shows what one can expect optically from a value for R calculated in this way. At a resolution of 1,0 the peaks are not completely separated, but it can definitely be seen that there are two components. The tangents to the inflection points just touch each other and the peak areas only overlap by 2%.

For the precise determination of the peak width the tangents to the inflection points can be drawn in manually (Fig. 2.114). For a critical pair, e.g. stearic acid (C_{18-0}) and oleic acid (C_{18-1}) the construction of the tangents is shown in Fig. 2.115.

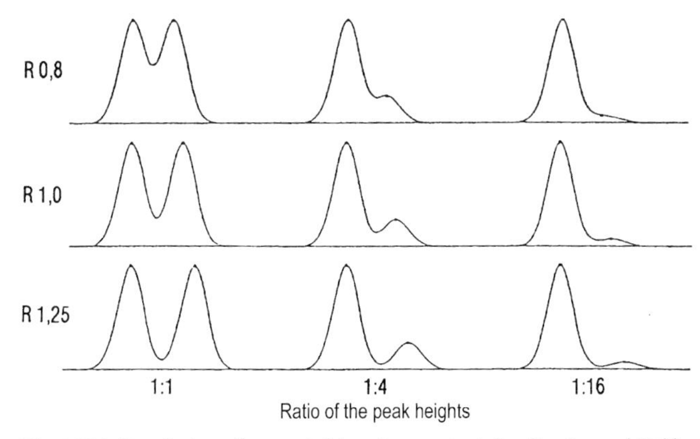

Fig. 2.113 Resolution of two neighbouring peaks (after Snyder and Kirkland).

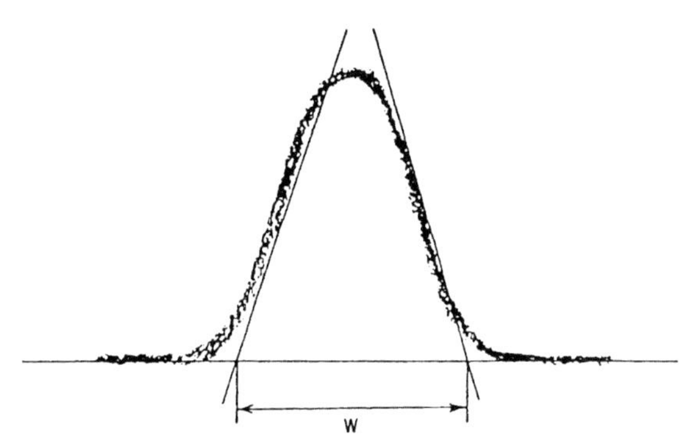

Fig. 2.114 Manual determination of the peak width using tangents to the inflection points.

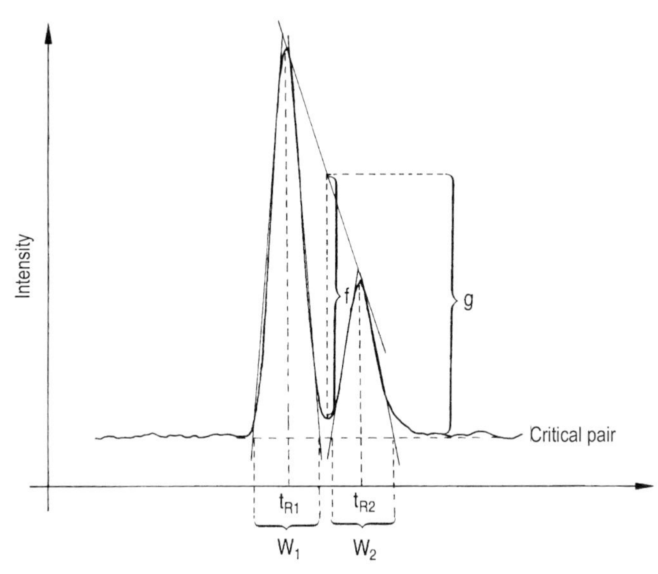

Fig. 2.115 Determination of the resolution and peak widths for a critical pair.

2.2.5.4 Factors Affecting the Resolution

Rearranging the resolution equation and putting in the capacity factor $k' = (t_R - t_0)/t_0$, the selectivity factor $\alpha = k'_2/k'_1$ and the number of theoretical plates N gives an important basic equation for all chromatographic elution processes.

The resolution R is related to the selectivity α (relative retention), the number of theoretical plates N and the capacity factor k' by:

$$R = \underbrace{\frac{1}{4}(\alpha - 1)}_{\text{I.}} \cdot \underbrace{\frac{k'}{1 + k'}}_{\text{II.}} \cdot \underbrace{\sqrt{N}}_{\text{III.}} \tag{18}$$

The Selectivity Term

R is directly proportional to $(\alpha - 1)$. An increase in the ratio of the partition coefficients leads to a sharp improvement in the resolution, which can be achieved, for example, by changing the polarity of the stationary phase for substances of different polarities.

As the selectivity generally decreases with increasing temperature, difficult separations must be carried out at as low a temperature as possible.

The change in the selectivity is the most effective of the possible measures for improving the resolution. As shown in Table 2.28, more plates are required to achieve the desired resolution when α is small.

Table 2.28 Relationship between relative retention α and the chromatographic resolution R.

Relative retention α	$R = 1.0$	$R = 1.5$
1.005	650 000 plates	1 450 000 plates
1.01	163 000	367 000
1.05	7 100	16 000
1.10	3 700	8 400
1.25	400	900
1.50	140	320
2.0	65	145

Figure 2.116 shows the effect of relative retention and number of plates on the separation of two neighbouring peaks:

- At high relative retention the number of theoretical plates in the column does not need to be large to achieve satisfactory resolution (a). The column is poor but the system is selective.
- A high relative retention and large number of theoretical plates give a resolution which is higher than the optimum. The analysis is unnecessarily long (b).
- At the same (small) number of theoretical plates as in (a), but at a smaller relative retention, the resolution is strongly reduced (c).
- If the relative retention is small, a large number of theoretical plates are required to give a satisfactory resolution (d).

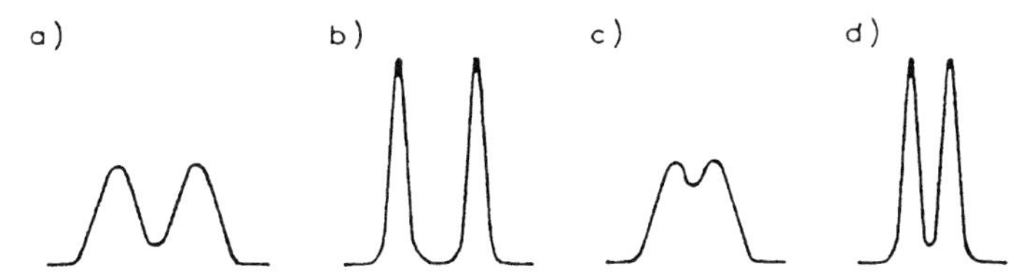

Fig. 2.116 Relative retention, number of plates and resolution.

The Retardation Term

Here the resolution is directly proportional to the residence time of a component in the stationary phase based on the total retention time. If the components only stayed in the mobile phase ($k' = 0$!) there would be no separation.

For very volatile or low molecular weight nonpolar substances there are only weak interactions with the stationary phase. Thus at a low k' value the denominator $(1 + k')$ of the term is large compared with k' and R is therefore small.

This also applies to columns with a small quantity of stationary phase and column temperatures which are too high. To improve the resolution a larger content of stationary phase can be chosen (greater film thickness).

The Dispersion Term

The number of plates N characterises the performance of a column (Fig. 2.117). However the resolution R only increases with the square root of N. As N is directly proportional to the column length L, the performance is only proportional to the square root of the column length.

Doubling the column length therefore only increases the resolution by a factor of 1.4. Since the retention time t_R is proportional to the column length, for an improvement in the resolution by a factor of 1.4, the analysis time is doubled.

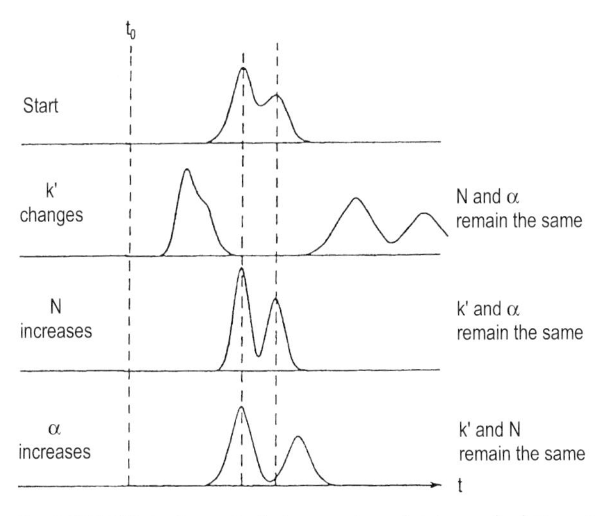

Fig. 2.117 Effect of capacity factor, number of plates and relative retention on the chromatogram (after L. R. Snyder and J. J. Kirkland).

2.2.5.5 Maximum Sample Capacity

The maximum sample capacity can be derived from the equations concerning the resolution. Under ideal conditions (see Fig. 2.115):

$$\frac{f}{g} = 100\% \tag{19}$$

where f = the area under the line connecting the peak maxima
g = the height of the connecting line above the base line, measured in the valley between the peaks

The maximum sample capacity of a column is reached if f/g falls below 90% for a critical pair. If too much sample material is applied to a column, the k' value and the peak width are no longer independent of the size of the sample, which ultimately affects the identification and the quantitation of the results (Fig. 2.118).

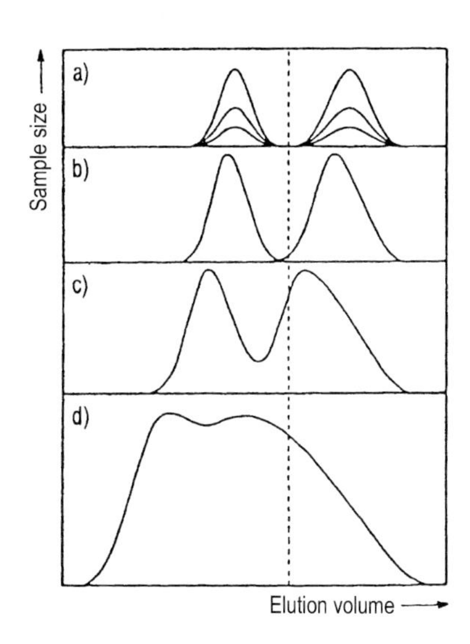

Fig. 2.118 Change in the chromatogram with increasing sample size (after L. R. Snyder and J. J. Kirkland).
(a) Constant k' values
(b–d) Increasing changes to the retention behaviour through overloading

2.2.5.6 Peak Symmetry

In exact quantitative work (integration of the peak areas) a maximum asymmetry must not be exceeded, otherwise there will be errors in determining the cut-off point of the peak with the base line.

For practical reasons the peak symmetry T is determined at a height of 10% of the total peak height (Fig. 2.119):

$$T = \frac{b_{0.1}}{a_{0.1}} \tag{20}$$

where $a_{0.1}$ = distance from the peak front to the maximum measured at 0.1 h
$b_{0.1}$ = distance from the maximum to the end of the peak measured at 0.1 h

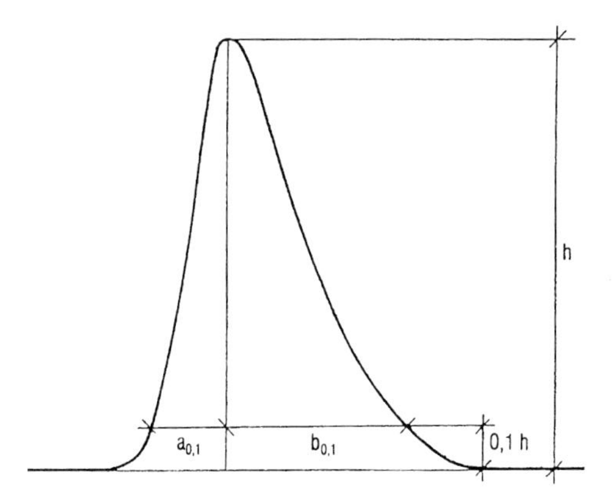

Fig. 2.119 Asymmetric peak.

T should ideally be 1.0 for a symmetrical peak, but for a practical quality measure it should not be greater than 2.5. If the tailing exceeds higher values, there will be errors in the quantitative area measurement because the point where the peak reaches the base line is very difficult to determine.

2.2.5.7 **Optimisation of Flow**

The flow of the mobile phase affects the rate of substance transport through the stationary phase. High flow rates allow rapid separation. However, the efficiency is reduced because of the slower exchange of substances between the stationary and mobile phases and leads to peak broadening. On the other hand, peak broadening caused by diffusion of the components within the mobile phase is only hindered by increasing the flow rate.

The aim of flow optimisation for given column properties at a given temperature and with a given carrier gas is to find the flow rate which gives either the maximum number of separation steps or at adequate efficiency the shortest possible analysis time.

The height equivalent to a theoretical plate (HETP) and the number of theoretical plates N depends on the flow rate of the mobile phase u according to van Deemter. The linear flow rate u of the mobile phase is calculated from the chromatogram:

$$u = \frac{L}{t_0} \tag{21}$$

with L = length of the column in cm
t_0 = dead time in s

The following affect the optimum flow rate:

1. The *Eddy diffusion* (Fig. 2.120) on the peak broadening. This effect is independent of flow and, naturally, for packed columns only, dependent on the nature of the packing material and the density of packing. For open capillary (tabular) columns, Eddy diffusion does not occur.

2. The *axial diffusion* on peak broadening. This diffusion occurs in and against the direction of flow and decreases with increasing flow rate.

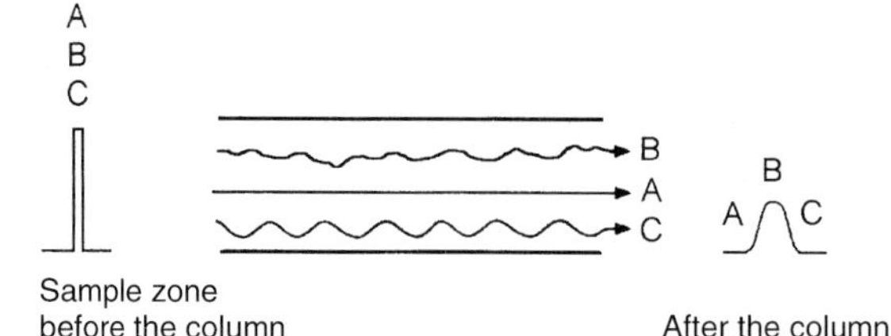

Fig. 2.120 Path differences for lamina flow in capillary columns (multipath effect, caused by turbulence at high flow rates).

3. Incomplete partition equilibrium. The transfer of analyte between the stationary and mobile phases only has a finite rate relative to that of the mobile phase, corresponding to the diffusion rates. The contribution to peak broadening increases with increasing flow rate of the mobile phase.

For the maximum efficiency of the separation (Fig. 2.121) a flow rate u_{min} must be chosen as a compromise between these opposing effects. The position of the minimum is affected by:

- the quantity of the stationary phase (e. g. film thickness),
- the particle size of the packing material (for packed columns),
- the diameter of the column,
- the nature of the mobile phase (diffusion coefficient, viscosity).

For practical reasons the effective flow rates in GC/MS analyses are set above the optimum flow rates for increased speed and sample throughput. In this case the right branch of the van Deempter graph needs to be considered where the slope of the curve is low (Fig. 2.121). The small loss in column resolution by increased flow rates is more than compensated by the advantage of a short analysis time which holds true for helium and especially for hydrogen. Constant flow rate conditions maintain chromatographic peak resolution in ramped temperature programs.

In particular in GC × GC the flow rate of the second short column is accepted to be significantly above the ideal flow rate using the direct connection to the modulator. A split device

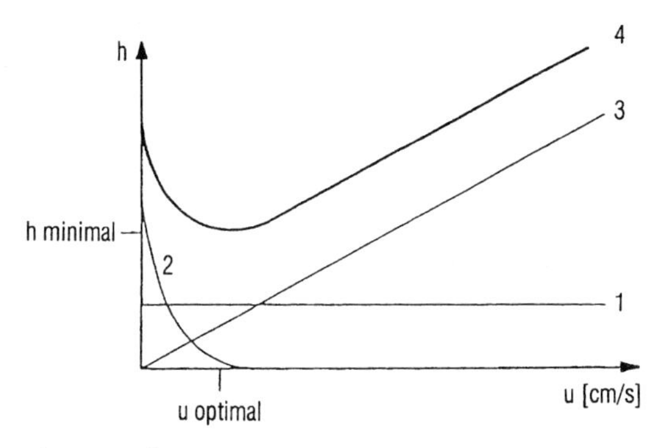

Fig. 2.121 The van Deemter curve.
1 Proportion of Eddy diffusion and flow distribution at band broadening
2 Proportion of longitudinal diffusion
3 Proportion of substance exchange phenomena
4 Resulting curve H(u), called the van Deemter curve

would be needed between modulator and second column for the independent adjustment of the flow rate. Most GC×GC application do not use a flow adjustment for the 2nd column which is due to the cut in substance concentration and hence sensitivity of the method, and, this is most probably the most important factor, the overwhelming increase in peak separation and S/N by using a high flow compromise.

Definition of Chromatographic Parameters		
Carrier gas velocity	$v = L/t_0$	The average linear carrier gas velocity has an optimal value for each column with the lowest possible height equivalent to a theoretical plate (see van Deemter); v is independent of temperature. L = length of column t_0 = dead time
Partition coefficient	$K = c_l/c_g$	Concentration of the substance in the stationary phase (liquid) divided by the concentration in the mobile phase (gas). K is constant for a particular substance in a given chromatographic system.
	$K = k' \cdot \beta$	K is also expressed as the product of the capacity ratio (k') and the phase ratio (β).
Capacity ratio (partition ratio)	$k' = K \cdot V_l/V_g$	Determines the retention time of a compound. V_l = volume of the stationary phase V_g = volume of the gaseous mobile phase
	$k' = t_R - t_0/t_0$	t_R = retention time of the substance t_0 = dead time
Phase ratio	$\beta = r/2d_f$	r = internal column radius d_f = film thickness
Number of theoretical plates	$N = 5.54\,(t_R/W_h)^2$	The number of theoretical plates is a measure of the efficiency of a column. The value depends on the nature of the substance and is valid for isothermal work. N = number of theoretical plates t_R = retention time of the substance W_h = peak width at half height

Term	Formula	Description
Height equivalent to a theoretical plate (HETP)	$h = L/N$	Is a measure of the efficiency of a column independent of its length. L = length of the column N = number of theoretical plates
Resolution	$R = 2\ (t_j - t_i)/(W_j + W_i)$	Gives the resolving power of a column with regard to the separation of components i and j (isothermally). t_i = retention time of substance i t_j = retention time of substance j $W_{i,j}$ = peak width at half height of substances i, j
Separation factor	$\alpha = k'_j / k'_i$	Measure of the separation of the substances i, j
Trennzahl number	$TZ = \dfrac{t_{R(x+1)} - t_{R(x)}}{W_{h(x+1)} + W_{h(x)}} - 1$	The trennzahl number is, like the resolution, a means of assessing the efficiency of a column and is also used for temperature-programmed work. TZ gives the number of components which can be resolved between two homologous n-alkanes.
Effective plates	$N_{eff.} = 5.54\ ((t_{R(i)} - t_0)\ /\ W_{h(i)})$	The effective number of theoretical plates takes the dead volume of the column into account.
Retention volume	$V_R = t_R \cdot F$	Gives the carrier gas volume required for elution of a given component. F = carrier gas flow
Kovats index	$KI = 100 \cdot c + 100\ \dfrac{\log{(t'_R)_x} - \log{(t'_R)_c}}{\log{(t'_R)_{c+1}} - \log{(t'_R)_c}}$	The Kovats index is used for isothermal work. t'_R = corrected retention times for standards and substances $t'_R = t_R - t_0$
Modified Kovats index	$RI = 100 \cdot c + 100\ \dfrac{(t'_R)_x - (t'_R)_c}{(t'_R)_{c+1} - (t'_R)_c}$	The modified Kovats index according to van den Dool and Kratz is used with temperature programming.

2.2.6
Classical Detectors for GC/MS Systems

Classical detectors are important for the consideration of GC/MS coupling if an additional specific means of detection is to be introduced parallel to mass spectrometry. The parallel coupling of a flame ionisation detector (FID) does not lead to results which are complementary to those of mass spectrometry, as both detection processes give practically identical chromatograms as the response factors for most of the organic substances are comparable. Parallel detection with a thermal conductivity detector is not used in practice as the mass spectrometric analysis of gases is generally carried out with special instruments (RGA, residual gas analyser, mass range $< 100\ \mu$).

Additional information can, however, be obtained with element-specific detectors. The detection limits which can be achieved with an electron capture detector (ECD) or a nitrogen/phosphorus detector (NPD) are usually comparable to those attainable using a mass spectrometer. On dividing up the carrier gas flow, the ratio of the two parts must be considered when planning such a setup. Normally a larger proportion is passed into the mass spectrometer so that in residue analysis low concentrations of substances do not fall below the detection limit. The use of such flow dividers for quantitative determinations must be checked in an individual case, as a constant division cannot be expected for all boiling point ranges.

Applications can cover the rapid screening e.g. on the intensity of halogenated compounds using an ECD with an intelligent decision for a subsequent MS analysis of the same positively screened sample for a mass selective quantitation.

2.2.6.1 FID

With the flame ionisation detector (FID) the substances to be detected are burned in a hydrogen flame and are thus partially ionised (Table 2.29). As the jet is at a negative potential, positive ions are neutralised. The corresponding electrons are captured at the ring-shaped collector electrode to give a signal current (Fig. 2.122). The electrode is at a potential which is ca. 200 V more positive than the jet.

Table 2.29 Reactions in the FID.

Pyrolysis:	CH_3^o, CH_2^o, CH^o, C^o
Excited radicals:	O_2^*, OH^*
Ionisation:	$CH_2^o + OH^* \rightarrow CH_3O^+ + e^-$
	$CH^o + OH^* \rightarrow CH_2O^+ + e^-$
	$CH^o + O_2^* \rightarrow CHO_2^+ + e^-$
	$C^o + OH^* \rightarrow CHO^+ + e^-$

Provided that only hydrogen burns in the flame, only radical reactions occur. No ions are formed. If organic substances with C-H and C-C bonds get into the flame, they are first pyrolysed. The carbon-containing radicals are oxidised by oxygen and the OH radicals formed in

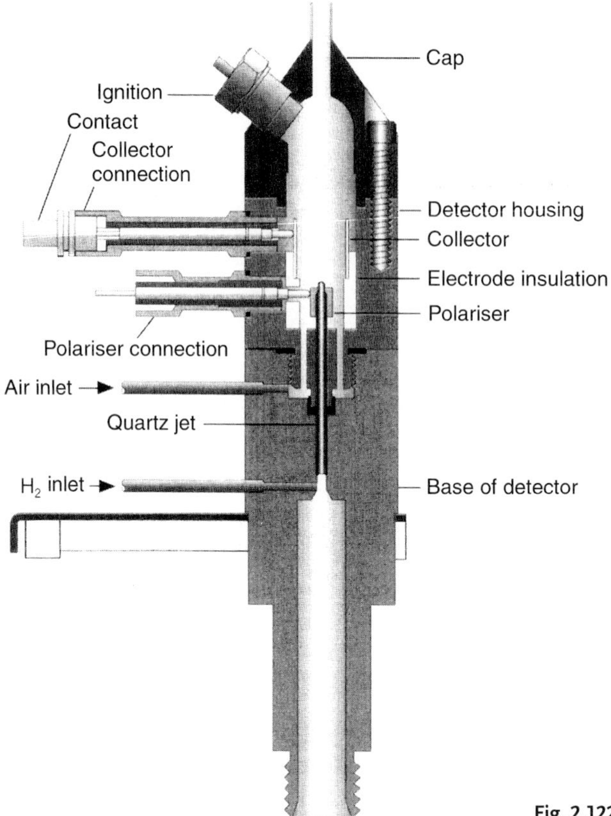

Fig. 2.122 Construction of an FID (Finnigan).

the flame. The excitation energy leads to ionisation of the oxidation products. Only substances with at least one C-H or C-C bond are detected, but not permanent gases, carbon tetrachloride or water.

If a reactor (hydrogenator, methaniser) is connected before the FID, the latter can be converted into an extremely sensitive detector for permanent gases such as CO and CO_2. The oxygen specific detector (O-FID) uses two reactors. In the first hydrocarbons are decomposed into carbon, hydrogen and carbon monoxide at above 1300 °C. CO is then converted into methane in the hydrogenation reactor and detected with the FID. With the O-FID, for example, oxygen-containing components in fuels can be detected.

FID Flame **I**onisation **D**etector

Universal detector

Advantages: High dynamics
High sensitivity
Robust

Use:	Hydrocarbons, e.g. fuels, odorous substances, BTX, polyaromatic hydrocarbons etc Comparison of diesel with petrol Important all round detector
Limits:	As it is universal, its performance is poor for trace analysis in complex matrices Low response for highly chlorinated or brominated substances.

2.2.6.2 NPD

A nitrogen-phosphorous detector (NPD) is a modified FID which contains a source of alkali situated between the jet and the collector electrode on a Pt wire, for the specific detection of nitrogen or phosphorus (Fig. 2.123).

The alkali beads are heated to red heat both electrically and in the flame and are excited to alkali emission. They are always at a negative potential compared with the collector electrode. For the detection of phosphorus the jet is earthed. The electrons emitted by the hydrocarbon parts of the molecule cannot exceed the negative potential of the beads and do not

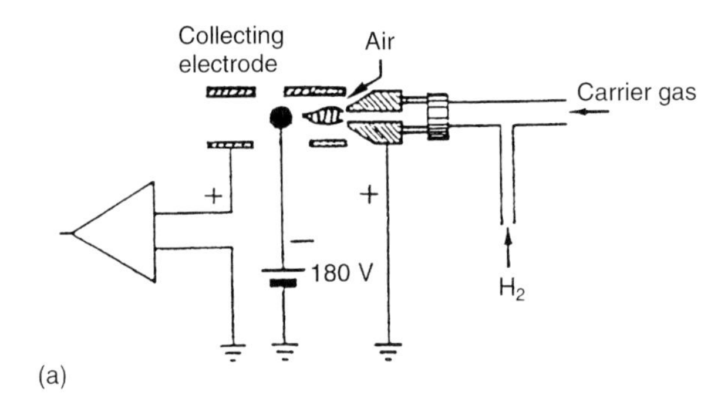

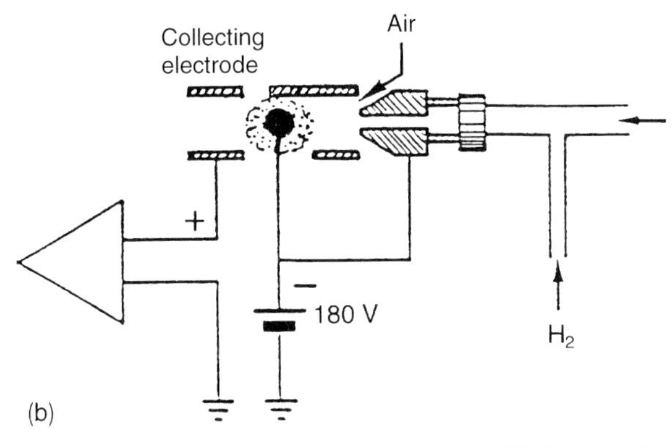

Fig. 2.123 (a) Diagram of an NPD (P operation); (b) diagram of an NPD (N operation).

Table 2.30 Reactions with P compounds in the NPD.

$\bar{O} = \dot{P}$	+	A*	→	$[\bar{O} = \bar{P}]^-$	+	A^+
$\bar{O} = \dot{P} = \bar{O}$	+	A*	→	$[\bar{O} = P = \bar{O}]^-$	+	A^+
$[\bar{O} = \bar{P}]^-$	+	OH°	→	HPO_2	+	e^-
$[\bar{O} = P = \bar{O}]^-$	+	OH°	→	HPO_3	+	e^-
HPO_3	+	H_2O	→	H_3PO_4		

A = alkali

reach the collector electrode, but are earthed. The electrons from the specific alkali reaction reach the collector electrode unhindered (Table 2.30).

Phosphorus-containing substances are first converted in the flame into phosphorus oxides with an uneven number of electrons. Anions formed in the alkali reaction by the addition of an electron are now oxidised by OH radicals. The electrons added are now released and produce a signal current.

Like phosphorus, nitrogen has an uneven number of electrons. Under the reducing conditions of the flame cyanide and cyanate radicals are formed, which can undergo the alkali reaction (Table 2.31). For this the input of hydrogen and air are reduced. Instead of the flame the hydrogen burns in the form of a cold plasma around the electrically heated alkali beads.

Table 2.31 Reactions with N compounds in the NPD.

Pyrolysis of CNC compounds			→	C ≡ N|°		
CN°	+	A*	→	CN^-	+	A*
CN^-	+	H°	→	HCN	+	e^-
CN^-	+	OH°	→	HCNO	+	e^-

A = alkali

In order to form the required cyanide and cyanate radicals the C-N structure must already be present in the molecule. Nitro compounds are detected, but not nitrate esters, ammonia or nitrogen oxides. By taking part in the alkali reaction the cyanide radical receives an electron. Cyanide ions are formed, which react with other radicals to give neutral species. The electron released provides the detector signal.

NPD Nitrogen/**P**hosphorus **D**etector

Specific detector

Advantages:	High selectivity and sensitivity Ideal for trace analyses
Use:	Only for N- and P-containing compounds Plant protection agents Chemical warfare gases, explosives Pharmaceuticals
Limits:	Additional detector for ECD or MS Quantitative measurements with an internal standard are recommended To some extent time-consuming optimisation of the Rb beads

2.2.6.3 **ECD**

The electron capture detector (ECD) consists of an ionisation chamber, which contains a nickel plate, on the surface of which a thin layer of the radioactive isotope ^{63}Ni has been applied (ca. 10–15 mC, Fig. 2.124). The carrier gas (N_2 or Ar/10% methane) is first ionised by the β radiation. The free electrons migrate towards the collector electrode and provide the background current of the detector. Substances with electronegative groups reduce the background current by capturing electrons and forming negative molecular ions. The main reactions in the ECD are dissociative electron capture (Table 2.32) and electron capture (Table 2.33, see also Section 2.3.4.2). Negative molecular ions can recombine with positive carrier gas ions.

Electron capture is more effective, the slower the electrons move. For this reason a more sensitive ECD is now operated using pulsed DC voltage. By changing the pulse frequency the background current generated by the electrons is kept constant. The pulse frequency thus becomes the actual detector signal.

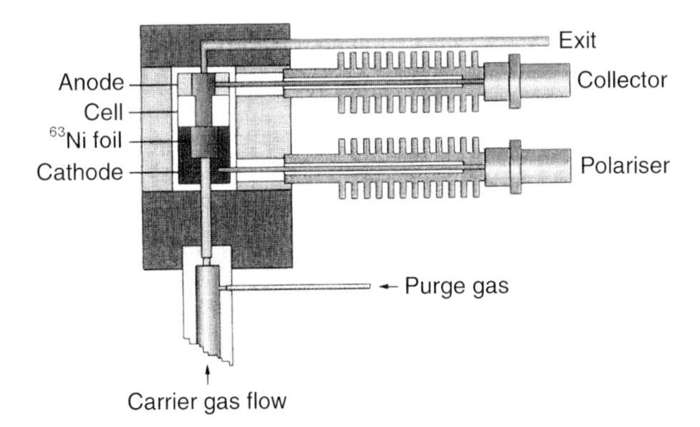

Fig. 2.124 Construction of an ECD (Finnigan).

Table 2.32 Substance reactions in the ECD.

1. Dissociative electron capture

AB	+	e^-	→	A^0	+	B^-

2. Addition of an electron

AB	+	e^-	→	$(AB^-)^*$

Table 2.33 Basic reactions in the ECD.

CG			$\xrightarrow{\beta}$	CG^+	+	e^-
Electron capture						
M	+	e^-	→	M^-		
Recombination						
M^-	+	CG^+	→	M^o	+	CG

CG = Carrier gas

ECD Electron Capture Detector

Specific detector

The ECD reacts with all electronegative elements and functional groups, such as -F, -Cl, -Br, $-OCH_3$, $-NO_2$ with a high response. All hydrocarbons (generally the matrix) remain transparent, although present.

Advantages:	Selectivity for Cl, Br, methoxy, nitro groups Transparency of all hydrocarbons (= matrix) High sensitivity
Disadvantages:	Radioactive radiator, therefore handling authorisation necessary Sensitive to misuse Mobile only under limited conditions
Use:	Typical detector for environmental analysis Ideal for trace analysis Plant protection agents PCBs, dioxins Volatile halogenated hydrocarbons, freons
Limits:	Substances with low halogen contents (Cl_1, Cl_2) only have a low response (Table 2.34) Volatile halogenated hydrocarbons with low chlorine levels are better detected with FID or MS Limited dynamics Multipoint calibration necessary

Table 2.34 Conversion rates in the ECD for molecules with different degrees of chlorination.

Molecule	Conversion rate [$cm^3 \cdot mol^{-1} \cdot s^{-1}$]	Main product
CH_2Cl_2	1×10^{-11}	Cl^-
$CHCl_3$	4×10^{-9}	Cl^-
CCl_4	4×10^{-7}	Cl^-
CH_3CCl_3	1×10^{-8}	Cl^-
$CH_2ClCHCl_2$	1×10^{-10}	Cl^-
CF_4	7×10^{-13}	M^-
CF_3Cl	4×10^{-10}	Cl^-
CF_3Br	1×10^{-8}	Br^-
CF_2Br_2	2×10^{-7}	Br^-
C_6F_6	9×10^{-8}	M^-
$C_6F_5CF_3$	2×10^{-7}	M^-
$C_6F_{11}CF_3$	2×10^{-7}	M^-
SF_6	4×10^{-7}	M^-
Azulene	3×10^{-8}	M^-
Nitrobenzene	1×10^{-9}	M^-
1,4-Naphthoquinone	7×10^{-9}	M^-

2.2.6.4 PID

The photo ionization detector (PID) operates on the principle of energy absorption of photons emitted by a UV lamp in the detector housing. The absorbed energy leads to ionization of the molecule by release of an electron from the excited molecule M* according to Eq. (1).

$$M + h \cdot \nu \rightarrow M^* \rightarrow M^+ + e^- \quad (1)$$

The high selectivity of the PID is based on the different energy levels of the emitted emission lines by the used UV lamps. The different types of lamps are filled either with argon, hydrogen, krypton or other gases to cover an ionisation energy range from 8.4 to 11.8 eV. Ionisation occurs only if the ionization potential of the molecules M is lower than the energy level of the emitted UV bands.

The number of ions produced is proportional to the absorption coefficient of the molecule and the intensity of the lamp.

Lamp type	Typical applications
8.4 eV	Amines, PAH
9.6 eV	Volatile aromatics, BTEX
10.2 eV	General application
11.8 eV	Aldehydes, ketones

The UV lamp is easily removable for exchange to applications requiring different selectivity. While the lamp housing is kept relatively cool (<100 °C) the quartz ionisation chamber containing the two measurement electrodes is the heated part of the PID (>300 °C). It is closed by a quartz or alkalifluoride window for entry of the UV light. Due to the non-destructive nature of the PID a second detector can be put in series or is mounted on top, see Fig. 2.125 with a non-selective FID.

The PID is mainly used for the analysis of aromatic pollutants, e.g. BTEX or PAH, or halogenated compounds in environmental applications. Many EPA methods e.g. 602, 502 or 503.1 cover priority pollutants in surface or drinking water. Typical for the PID is the detection of impurities air, also with mobile detectors. This is due to the fact that the energy of the UV-lamp is sufficient to ionize the majority of organic air contaminants, but insufficient to ionize the air components oxygen, nitrogen, water, argon and carbon dioxide.

PID Photo**i**onisation **D**etector

Universal and selective detector

The energy-rich radiation of a UV lamp (Fig. 2.125) ionises the substances to be analysed more or less selectively, depending on the energy content (Table 2.35); measurement of the overall ion flow.

Advantages: The selectivity can be chosen, see Fig. 2.126
High sensitivity
No gas supply is required
Robust, no maintenance required
Ideal for mobile and field analyses

Use: Field analysis of volatile halogenated hydrocarbons, BTEX, polyaromatic hydrocarbons etc.

Limits: Only for substances with low ionisation potentials (≤ 10 eV)
Consult manufacturers' data

Table 2.35 Ionisation potentials of selected analytes.

Substance	Ionisation potential [eV]	Substance	Ionisation potential [eV]
Helium	24.59	Isobutyraldehyde	9.74
Argon	15.76	Propene	9.73
Nitrogen	15.58	Acetone	9.69
Hydrogen	15.43	Benzene	9.25
Methane	12.98	Methyl isothiocyanate	9.25
Ethane	11.65	N, N-Dimethylformamide	9.12
Ethylene	11.41	2-Iodobutane	9.09
2-Chlorobutane	10.65	Toluene	8.82
Acetylene	10.52	n-Butylamine	8.72
n-Hexane	10.18	o-Xylene	8.56
2-Bromobutane	9.98	Phenol	8.50
n-Butyl acetate	9.97		

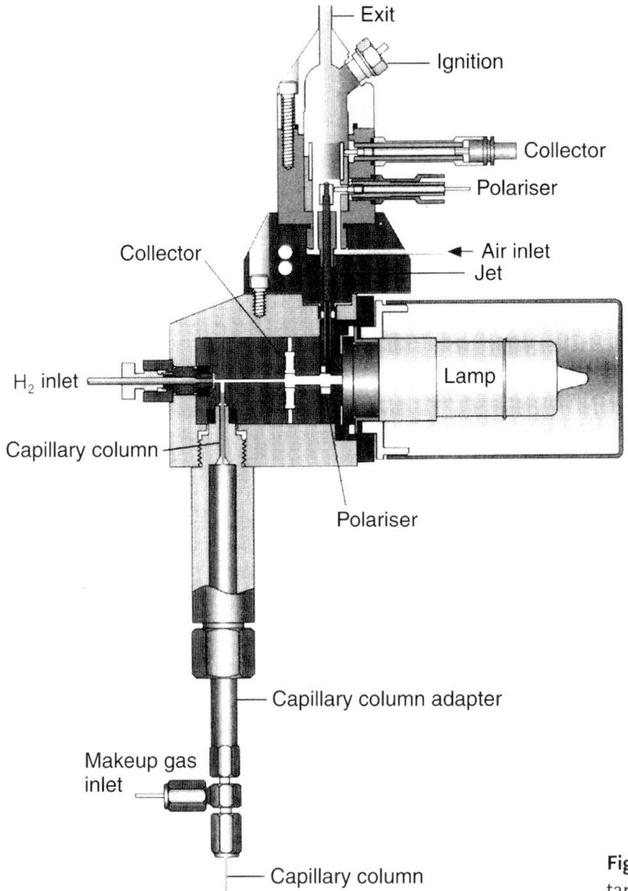

Fig. 2.125 Construction of a PID/FID tandem detector (Finnigan).

2.2.6.5 **ELCD**

In the electrolytical conductivity detector (ELCD) the eluate from the GC column passes into a Ni reactor in which all substances are completely oxidised or reduced at an elevated temperature (ca. 1000 °C).

The products dissociate in circulating water which flows through a cell where the conductivity is measured between two Pt spirals (Fig 2.127). The change in conductivity is the measurement signal. Ionic compounds can be measured without using the reactor.

The Hall detector has a comparable function to the ELCD and is a typical detector for packed or halfmil columns (0.53 mm internal diameter) because of the large volume of its measuring cell. Because of the latter significant peak tailing occurs with capillary columns. Special constructions for use with normal bore columns (<0.32 mm internal diameter) are available with special instructions.

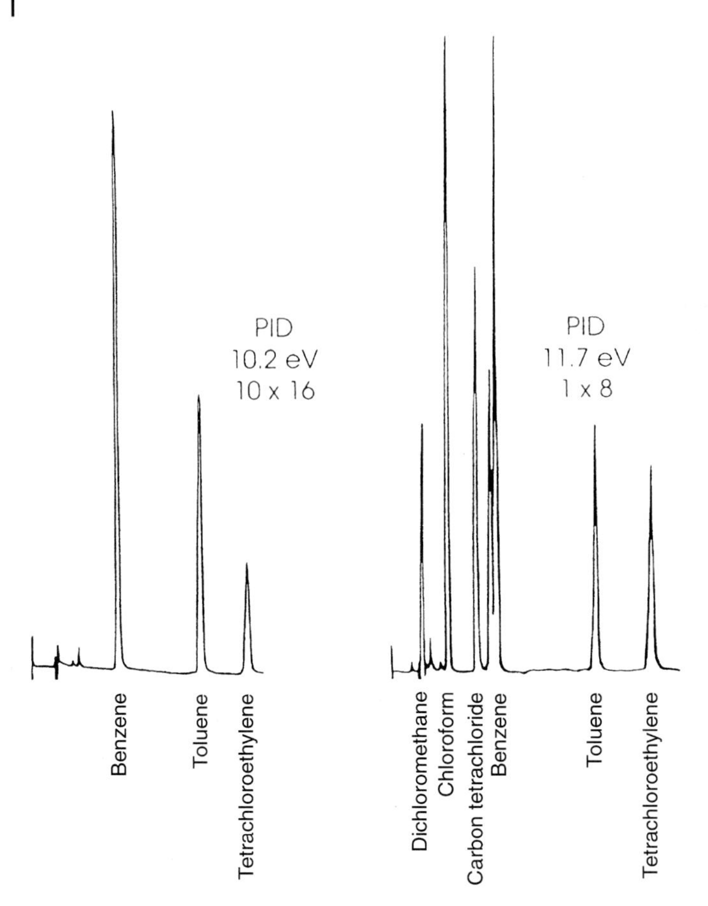

Fig. 2.126 Selectivity of the PID at 10.2 eV and 11.7 eV for a mixture of aromatic and chlorinated hydrocarbons (HNU).

ELCD Electrolytical **C**onductivity **D**etector

Selective detector

Advantages:
- Can be used for capillary chromatography.
- Selectivity can be chosen for halogens, amines, nitrogen, sulfur.
- High sensitivity at high selectivity.
- Can be used for halogen detection without a source of radioactivity, therefore no authorisation necessary.
- Simple calibration, as the response is directly proportional to the number of heteroatoms in the analyte, e.g. the proportion of Cl in the molecule.

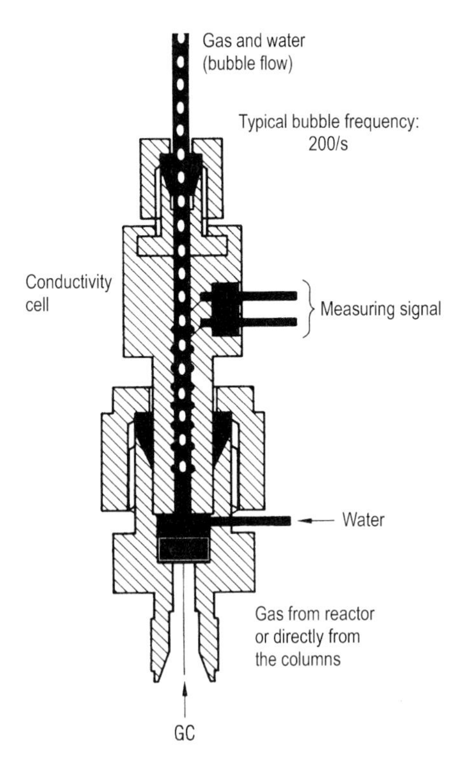

Fig. 2.127 Construction of an ELCD (patent Fraunhofer Gesellschaft, Munich).

Use:	• Environmental analysis, e.g. volatile halogenated hydrocarbons, PCBs. • Selective detection of amines, e.g. in packagings or foodstuffs. • Determination of sulfur-containing components.
Limits:	• The sensitivity in the halogen mode just reaches that of ECD, so that its use is particularly favourable in association with concentration procedures, such as purge and trap or thermodesorption.

2.2.6.6 **FPD**

Flamephotometric detectors (FPD) are used as one- or two-flame detectors. In a hydrogen-rich flame P- and S-containing radicals are in an excited transition state. On passing to the ground state a characteristic band spectrum is emitted (S: 394 nm, P: 526 nm). The flame emissions initiated by the eluting analytes (chemiluminescence) are determined using an optical filter and amplified by a photomultiplier (Fig. 2.128).

The analytical capability of the FPD can be expanded by connecting a second photomultiplier tube with a different optical filter on the same detector base, e.g. to monitor P and S containing substances in parallel.

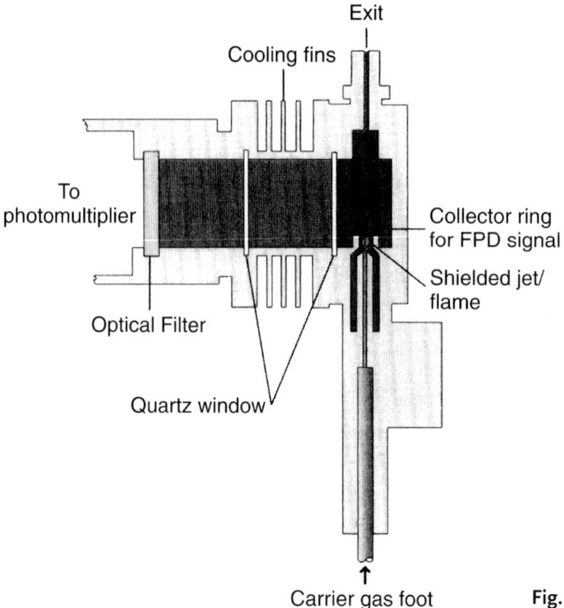

Fig. 2.128 Construction of an FPD (Finnigan).

FPD Flame**p**hotometric **D**etector

Selective detector

Advantages:
- In phosphorus mode, comparable selectivity to FID with high dynamics.
- High selectivity in sulfur mode.

Use:
- Mostly selective for sulfur and phosphorus compounds (e. g. plant protection agents).
- Detection of sulfur compounds in a complex matrix.

Limits:
- Adjustment of the combustion gases important for reproducibility and selectivity.
- In sulfur mode quenching effect possible because of too high a hydrocarbon matrix (double flame necessary).
- Sensitivity in sulfur mode not always sufficient for trace analysis.

2.2.6.7 PDD

The pulsed discharge detector (PDD) is a universal and highly sensitive non-radioactive and non-destructive detector, also known as a helium photoionisation detector. It is based on the principle of the photoionisation by radiation arising from the transition of diatomic helium to the dissociative ground state (Fig. 2.129).

The response to organic compounds is linear over five orders of magnitude with minimum detectable quantities in the low picogram range. The response to fixed gases is positive with minimum detectable quantities in the low ppb range. The performance of the detector is negatively affected by the presence of any impurities in the gas flows (carrier, discharge), therefore, the use of a high quality grade of helium (99.999% pure or better) as carrier and discharge gases is strongly recommended. Because even the highest quality carrier gas may contain some water vapour and fixed gas impurities, a helium purifier is typically included as part of the detector system.

The PDD detector consists of a quartz cell supplied from the top with ultra pure helium as discharge gas that reaches the discharge zone consisting of two of electrodes connected to a high voltage pulses generator (Pulsed Discharge Module). The eluates from the column, flowing counter to the flow of helium from the discharge zone, are ionized by photons at high energy arising from metastable Helium generated into the discharge zone. The resulting electrons are accelerated and measured as electrical signal by the collector electrode. The discharge and carrier gas flows are opposite. For this reason, it is necessary that the discharge gas flow is greater than the carrier gas flow to avoid that the eluates from the column reach the discharge zone with consequent discharge electrodes contamination.

The discharge and carrier gas flow out together from the bottom of the cell where it is possible to measure the sum of both at the outlet on the back of the instrument.

The PDD chromatograms show a great similarity to the classical FID detector and offers comparable performance without the use of a flame or combustible gases. The PDD in helium photoionisation mode is an excellent replacement for flame ionization detectors in petrochemical or refinery environments, where the flame and use of hydrogen can be problematic. In addition, when the helium discharge gas is doped with a suitable noble gas, such as argon, krypton, or xenon (depending on the desired cutoff point), the PDD can function as a specific photoionisation detector for selective determination of aliphatics, aromatics, amines, as well as other species.

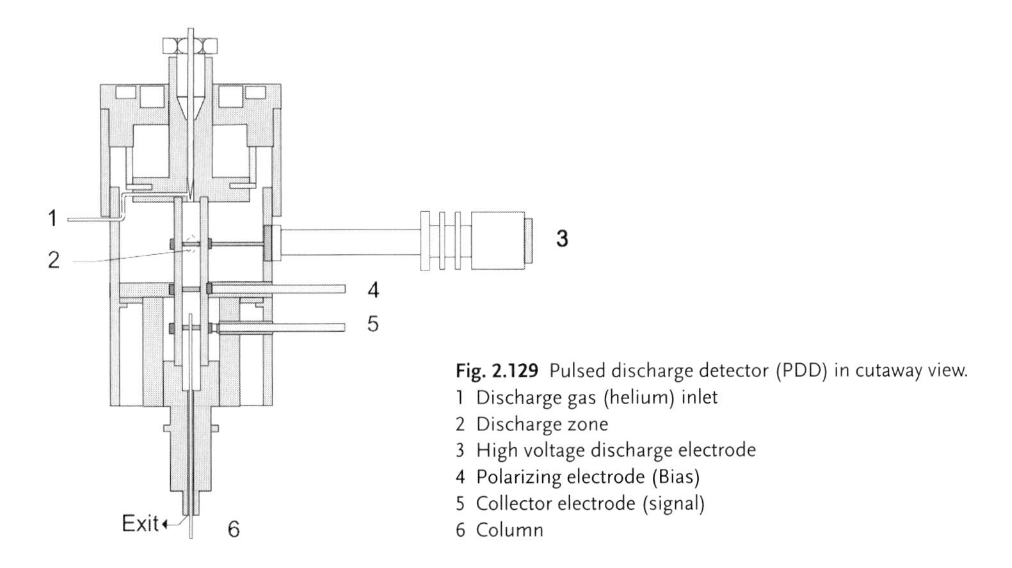

Fig. 2.129 Pulsed discharge detector (PDD) in cutaway view.
1 Discharge gas (helium) inlet
2 Discharge zone
3 High voltage discharge electrode
4 Polarizing electrode (Bias)
5 Collector electrode (signal)
6 Column

Some PDD detectors also offer an electron capture mode being selective for monitoring high electron affinity compounds such as freons, chlorinated pesticides, and other halogen compounds. For this type of compound, the minimum detectable quantity (MDQ) is at the femtogram (10–15) or picogram (10–12) level. The PDD is similar in sensitivity and response characteristics to a conventional radioactive ECD, and can be operated at temperatures up to 400 °C. For operation in this mode, He and CH_4 are introduced just upstream from the column exit.

2.2.6.8 **Connection of Classical Detectors Parallel to the Mass Spectrometer**

In principle there are two possibilities for operating another detector parallel to the mass spectrometer. The sample can be already divided in the injector and passed through two identical columns. The second more difficult but controllable solution is the split of the eluate at the end of the column.

The division of the sample on to two identical columns can be realised easily and carried out very reliably, but different retention times need to be considered due to the vacuum impact on the MS side. Since the capacities of the two columns are additive, the quantity of sample injected can be adjusted in order to make use of the operating range of the mass spectrometer. Two standard columns can be installed for most injectors without further adaptation being necessary. In the simplest case the connection can be made using ferrules with two holes in them. It must be ensured that there is a good seal. A better connection involves an adaptor piece with a separate screw-in joint for each column. With this construction reliable positioning of the columns in the injector is also possible. Suitable adaptors can be obtained for all common injectors.

In GC/MS the division of the flow at the end of the column requires a considerable higher effort because the direct coupling of the branch to the mass spectrometer causes reduced pressure in the split. For this reason the use of a simple Y piece is seldom possible with standard columns. The consequence would be a reversed flow through the detector connected in parallel caused by the wide vacuum impact of the MS into the column. The effect is equivalent to a leakage of air into the mass spectrometer. A split at the end of the column must therefore be carried out with high flow rates or using a makeup gas. Precise split of the eluent is possible, for example, using the glass cap cross divider (Fig. 2.130). Here the column, the transfer capillaries to the mass spectrometer and the parallel detector and a makeup gas inlet all meet. By choosing the internal diameter and the position of the end of the column in the glass cap (also known as the glass dome) the ratio of the split can easily be adjusted. The advantage of this solution lies in the free choice of column so that small internal diameters with comparatively low flow rates can also be used.

The calculation of gas flow rates through outlet splitters with fixed restrictors is desirable to have a means of estimating the rate of gas flow through a length of fixed restrictor tubing of specified dimensions. Conversely, for other applications it is necessary to estimate the length and ID of restriction tubing required to yield a desired flow rate at some specified head pressure. Within limits, the following calculation can be used for this purpose.

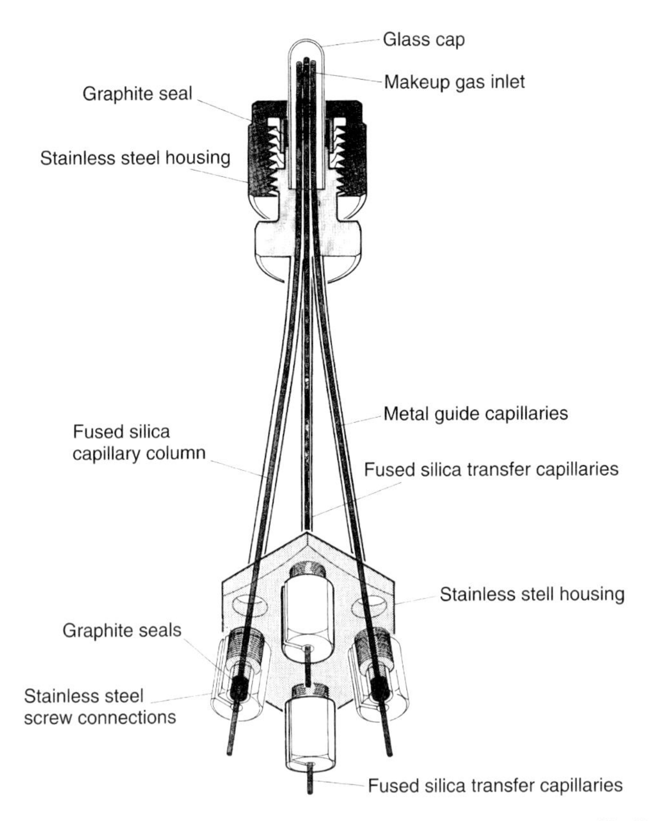

Fig. 2.130 Flow divider after Kaiser, Rieder (Seekamp Company, Werkhoff Splitter, Glass Cap Cross).

$$V = \frac{\pi \cdot P \cdot r^4 \cdot (7.5)}{l \cdot \eta}$$

with:

V = volumetric gas flow rate [cm^3/min]
P = pressure differential across the tube [dynes/cm]
= (PSI) · (68947.6)
r = tube radius [cm]
l = tube length [cm]
η = gas viscosity [poise (dyne-seconds/cm^2)]

One of the obvious limitations to this calculation is the critical nature of the radius measurement. Since this is a fourth power term, small errors in the ID measurement can result in relatively large errors in the flow rate. This becomes more critical with smaller diameter tubes (i.e. less than 100 micron). A 100× microscope (if available) is a convenient tool for determining the exact ID of a particular length of restriction tubing.

The gas viscosity values can be determined from the graph of gas viscosity against temperature for hydrogen, helium and nitrogen (Fig. 2.131).

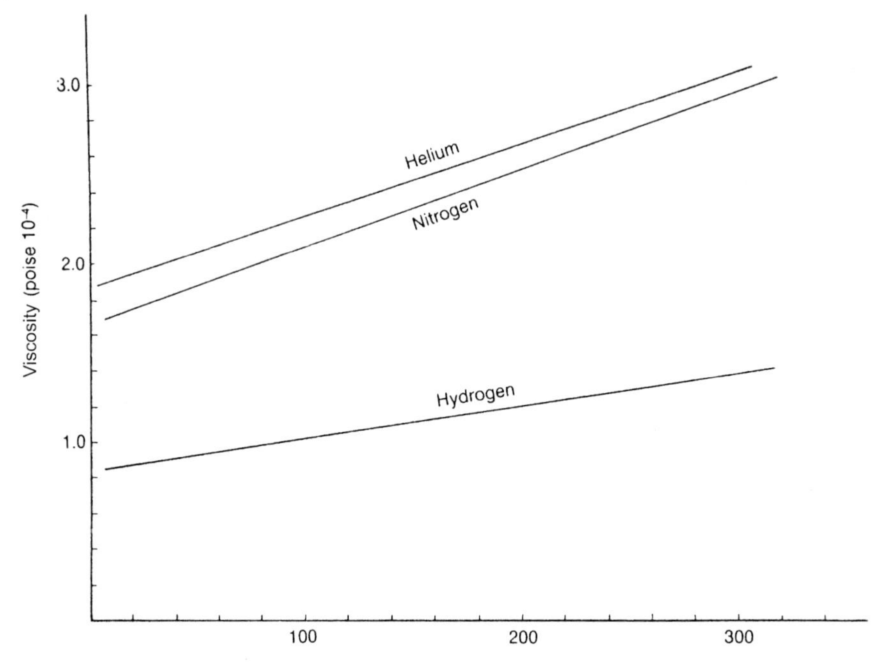

Fig. 2.131 Graphs to determine viscosity for hydrogen, helium and nitrogen.

2.3 Mass Spectrometry

> *"Looking back on my work in MS over the last 40 years, I believe that my major contribution has been to help convince myself, as well as other mass spectrometrists and chemists in general, that the things that happen to a molecule in the mass spectrometer are in fact chemistry, not voodoo; and that mass spectrometrists are, in fact, chemists and not shamans".*
>
> Seymour Meyerson
> Research Dept., Amoco Corp.

Mass spectrometers are instruments for producing and analysing mixtures of ions for components of differing mass and for exact mass determination. The substances to be analysed are fed to an ion source and ionised. In GC/MS systems there is continuous transport of substances by the carrier gas into the ion source. Mass spectrometers basically differ in the construction of the analyser as a beam or ion storage instrument. The performance of a mass spectrometric analyser is determined by the resolving power for differentiation between masses with small differences, the mass range and the transmission required to achieve high detection sensitivity.

2.3.1 Resolving Power and Resolution in Mass Spectrometry

The resolving power of a mass spectrometer describes the smallest mass differences which can be separated by the mass analyser. Resolution and resolution power in mass spectrometry today are defined differently depending on the analyser or instrument type, without the indication of the definition employed. The new IUPAC definitions of terms used in mass spectrometry provide a precise definition.

Mass Resolution

Smallest mass difference Δm between two equal magnitude peaks so that the valley between them is a specified fraction of the peak height.

Mass Resolving Power

In a mass spectrum, the observed mass divided by the difference between two masses that can be separated: $m/\Delta m$. The procedure by which Δm was obtained and the mass at which the measurement was made shall be reported as full width at half maximum (FWHM) or 10% valley.

The mass resolving power R can be calculated by the comparison of two adjacent mass peaks m_1 and m_2 of about equal height. R is defined as

$$m_1/\Delta m$$

with m_1 the lower value of two adjacent mass peaks. Δm of the two peaks is taken at an overlay (valley) of 10% or 50%.

In the 10% valley definition, the height from the baseline to the junction point of the two peaks is 10% of the full height of the two peaks. Each peak at this point is contributing 5% to the height of the valley.

Because it is difficult to get two mass spectral peaks of equal height adjacent to one another the practical method of calculating Ám as typically done by the instruments software is using a single mass peak of a reference compound. The peak width is measured in 5% height for the 10% valley definition and at 50% peak height for the FWHM definition. The resolving power calculated using the FWHM method gives values for R that are about twice that determined by the 10% valley method. This is due to the intercept theorems calculation in a triangle in which the ratios of height to width are equal:

$$h_1/w_1 = h_2/w_2$$

$$w_2 = (h_2/h_1) \times w_1$$

and as the peak height above 5% intensity compares to the half peak height very close by a factor of 2:

$$h_2 \sim 2 \times h_1$$

following

$$w_2 \approx (2 \times h_1/h_1) \text{ x } w_1$$
$$\approx 2 \times w_1$$

for
h1 = half peak height
w1 = peak width at 50 % height
h2 = peak height above 5 % to top
w2 = peak width at 5 %

following for the mass peak width that

$$\Delta m \text{ (5\% peak height)} \approx 2 \times \Delta m \text{ (50\% peak height)}$$

consequently

$$R \text{ (10\% valley)} \times 2 \approx R \text{ (FWHM)}$$

In practice a resolution power R 60,000 at 10 % valley compares directly to a specification of R 120,000 at FWHM.

The mass resolving power for magnetic sector instruments is historically given with a 10 % valley definition. A mass peak in high resolution magnetic sector MS is typically triangular with this analyser type. The peak width in 5 % height is of valuable diagnostic use for non optimal analyser conditions. Therefore this method provides an excellent measure for the quality of the peak shape together with the resolution power which would not be available at the half maximum condition. The typical broad peak base was initially observed in time of flight mass spectrometers and caused the 10 % valley definition would only calculate poor resolution values for this type of analyser. The more practical approach to describe the resolution power of TOF analysers is a measurement at half peak height, or as it is usually stated, at the FWHM. Both resolution power conditions compare by a factor of two as outlined above.

Constant Resolution Power Over the Mass Range

Double focusing mass spectrometers, using both electric and magnetic fields to separate ions, and time-of-flight analyzers (TOF), operate at constant mass resolving power. At a resolving power of 1000, these instruments separate ions of m/z 1000 and m/z 1001. In this example Δm is 1 and M is 1000, therefore R = 1000/1 or 1000. For practical use, this property of constant resolving power over the entire mass range means, that with a resolving power of 1000, values of 0.1 Da can be separated at m/z 100, i. e. $R = 1000 = m/\Delta m = 100/0.1$. This implies that the visible distance of mass peaks on the m/z scale decreases with increasing m/z, see Fig. 2.132.

High resolution mass spectrometers are using data system controlled split systems at the entrance and exit of the ion beam to and from the analyser for resolution adjustment. Practically, this can be done either manually or under software control. The adjustment is performed first by closing the entrance slit to get a flat top peak. Next the exit slit is closed accordingly until the intensity starts to decrease and a triangular peak is formed (see Fig. 2.133). No other adjustments in the focussing conditions are necessary to achieve the resolution setting.

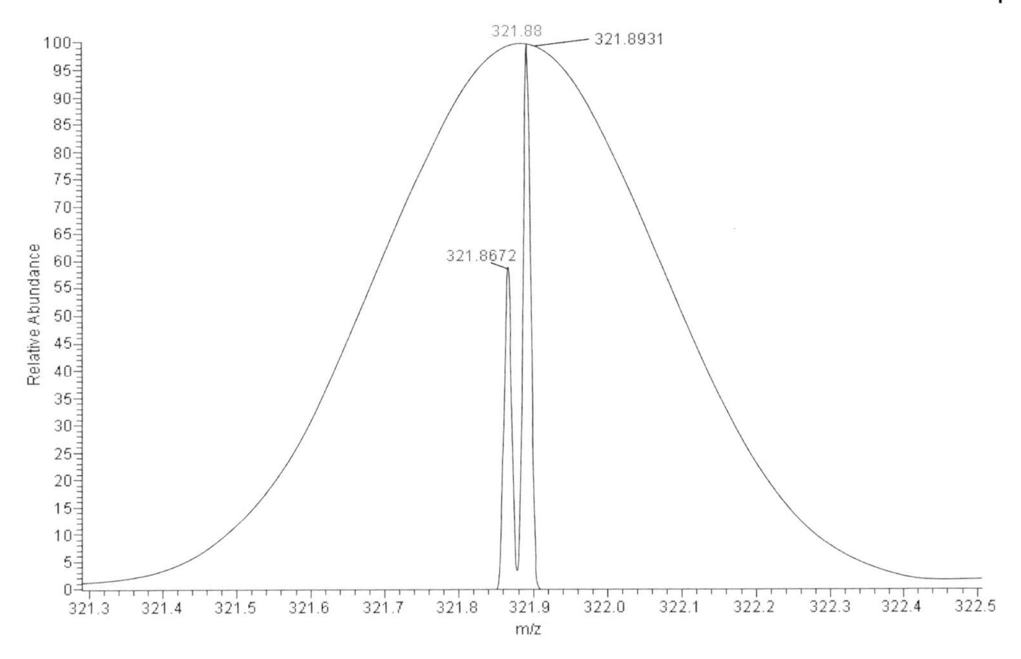

Fig. 2.132 Low and high resolved 2,3,7,8-Tetrachlorodibenzodioxin mass peak, m/z 321.8931 at R = 10,000 (10% valley). The background interference of m/z 321.8672 cannot be resolved at low resolution.

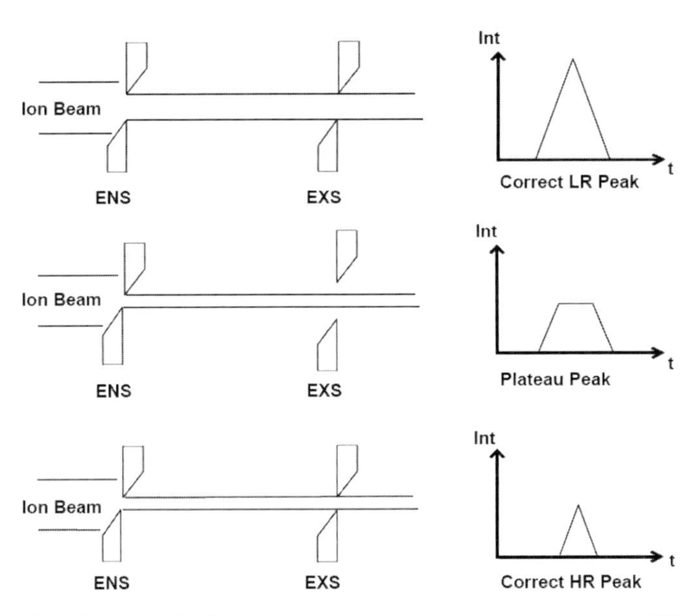

Fig. 2.133 Principle of the high resolution adjustment (ENS entrance slit, EXS exit slit).

Constant Resolution Over the Mass Range

Quadrupole and ion trap mass analysers show constant peak widths and mass differences over the entire mass range, hence both analyser types operate at constant mass resolution with increasing resolving power (see Fig. 2.145). This means, that the ability to separate ions at m/z 100 and m/z 1000 is the same. If Δm is 1 mass unit at m/z 100, the resolution at m/z 100 is 1 and the resolving power R is 100/1 or 100. If Δm is 1 at m/z 1000, the resolution at m/z 1000 is also 1, but the resolving power R at m/z 1000 is 1000/1 or 1000. Consequently, these types of analysers operate at increasing resolving power with increasing m/z value. Accordingly, the maximum resolving power, which is usually specified with a commercial quadrupole system, is dependent on the maximum specified mass range of the employed quadrupole hardware. The maximum resolution obtained with commercial quadrupole systems is typically at about 0.7 to 1.0 (FWHM) constant Da throughout the entire mass range, only hyperbolic shaped quadrupole rods of a certain length allow for narrower mass peak widths down to 0.1 Da, see Fig. 2.135, without losing ion transmission.

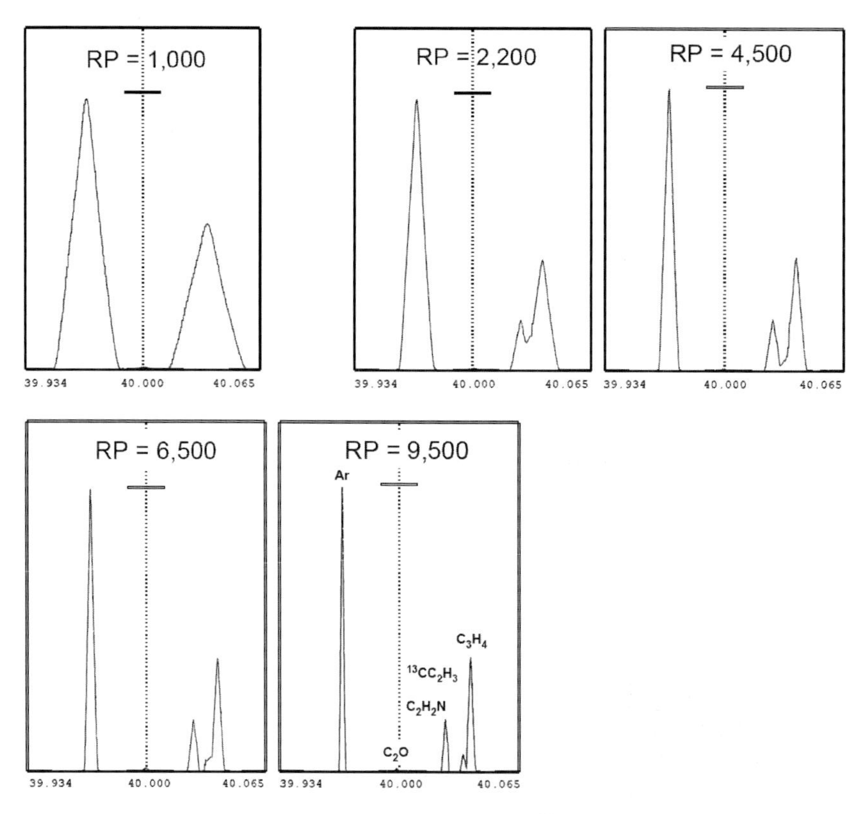

Fig. 2.134 Example of the separation of the two compounds $^{13}C_1C_2H_3$ (M = 40.0268) and C_3H_4 (M = 40.0313): R = 9500 for m = 40.0 Da; Δm = 0.0045 Da.

Ar^+	39.9624 Da
C_2O^+	39.9949 Da
$C_2H_2N^+$	40.0187 Da
$^{13}C_1C_2H_3^+$	40.0268 Da
$C_3H_4^+$	40.0313 Da

"High Resolution" with Quadrupole Analysers

Already in 1968 a publication by Dawson and Whetten dealt with the resolution capabilities of quadrupoles (Dawson 1968). The theoretical investigation covered round rods as well as hyperbolic rods and indicated the higher resolution potential of quadrupole analysers using hyperbolic rods (see Fig. 2.134). Especially with precisely machined hyperbolic rods sufficient ion transmission is achieved even at higher resolution power, making this technology especially useful for target compound analysis.

The maximum resolution power of quadrupole analysers depends on the number of cycles an ion is exposed to the electromagnetic fields inside of the quadrupole assembly, hence of the length of the rods.

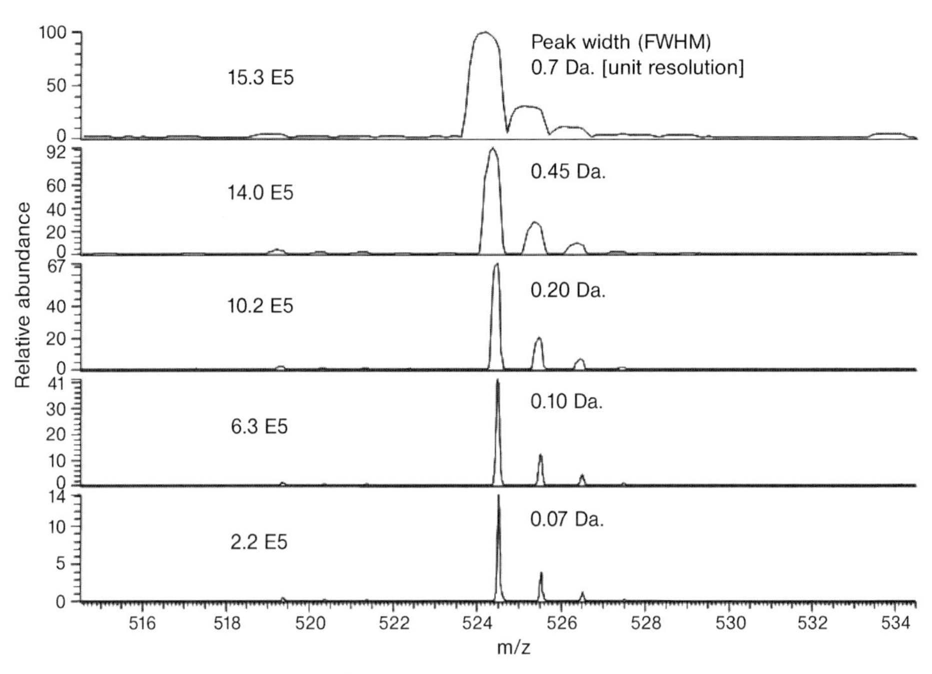

Fig. 2.135 Increased peak resolution of hyperbolic quadrupoles see Fig 2.136. Note the decrease in peak hight of factor 2 at 0.1 Da peak width compared to unit resolution (courtesy Thermo Fisher Scientific).

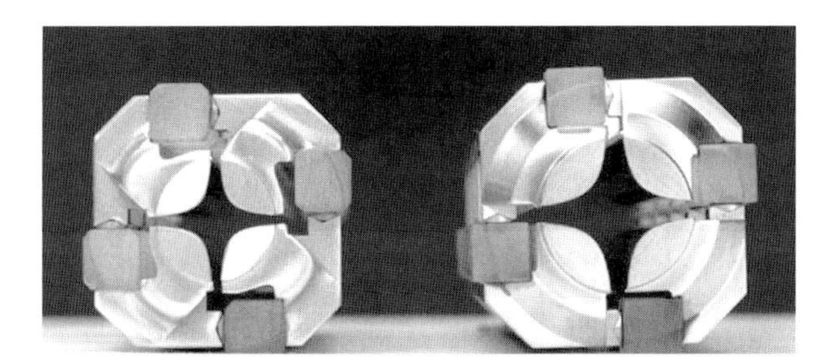

Fig. 2.136 Low resolution (left, TSQ 7000) and higher resolution (right, TSQ Quantum) quadrupole analysers (courtesy Thermo Fisher Scientific).

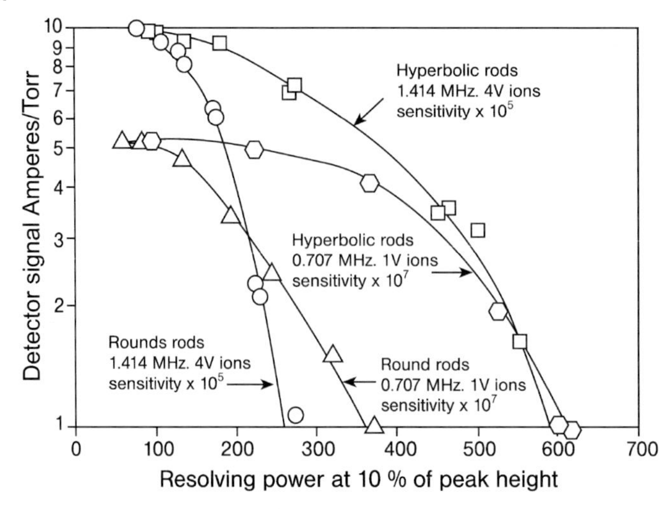

Fig. 2.137 Ion transmission of round and hyperbolic quadrupole systems (resolving power ag. ion transmission), first documented investigation by Brubaker 1968 (two curves left: round rods; two curves right: hyperbolic rods).

A recent development for "higher resolution" quadrupole analysers has been published and commercialized for a selected ion monitoring technique in MS/MS named „H-SRM", high resolution selected reaction monitoring. Hyperbolic quadrupole rods of a true hyperbolic pole face, high-precision 4-section design with a rod length of approx. 25 cm length are employed in triple quadrupole GC and LC/MS/MS systems for the selective quantification of target compounds. The increased field radius of 6 mm provides a significantly increased ion transmission for trace analysis (HyperQuad technology, patented Thermo Fisher Scientific).

By operating the first quadrupole in higher resolution mode the selectivity for MS/MS analysis of a parent target ion within a complex matrix is significantly increased. Higher S/N values for the product ion peak lead to significantly lowered method detection limits (MDL). Using the hyperbolic quadrupole rods the reduction from 0.7 Da peak width to 0.1 Da reduces overall signal intensity only by 30% but S/N increases by a factor better than 10. In many cases at low levels it even first allows to detect the target peak well above the background. By using round rods the drop in signal intensity would be close to 100%, and would lose all sample signal. Hence round rods cannot be applied for a higher resolved and higher selective detection mode.

To eliminate the uncertainty associated with possible mass drifts in H-SRM acquisitions, algorithms that correct the calibration table for mass position and peak width have been developed (Jemal 2003, Liu 2006). Data are obtained from H-SRM quantitation at a peak width setting of 0.1 Da at FWHM for the precursor ion. Deviations detected for mass position or peak width in internal standards are used to adjust the calibration. The data is then used to determine the deviations if any for all target calibrant ions. The calibration correction method is submitted and executed within the sequence of data acquisition. A scan-by-scan calibration correction method (CCM) has been applied to accommodate a variety of factors that influence precursor ion mass drift and ensure H-SRM performance eliminating signal

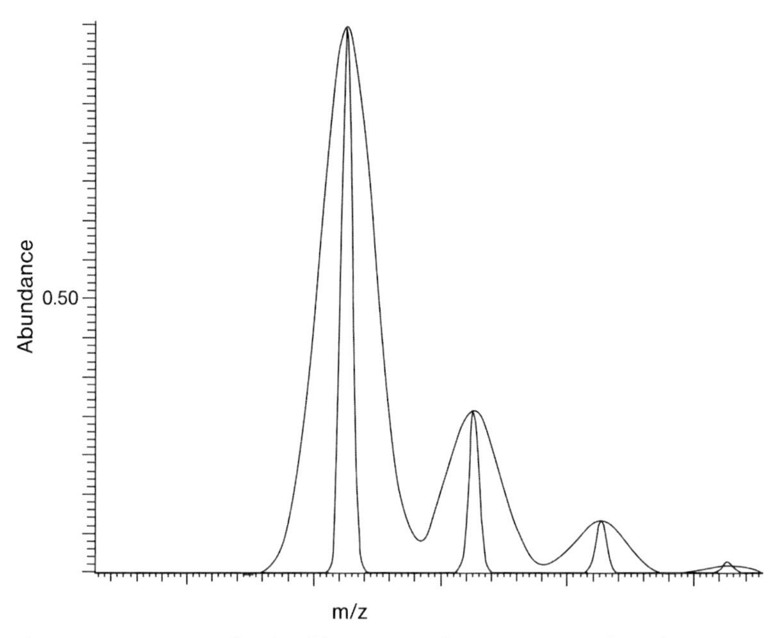

Fig. 2.138 Comparison of peak widths at 0.7 u and 0.1 u at FWHP with resolution power of R 10 000 at m/z 1000.

roll-off. With real life samples exhibiting strong matrix backgrounds a significant increase in S/N is achieved by using the H-SRM technique.

Analytical benefits of highly resolving quadrupole systems in GC/MS/MS analysis are the separation of ions with the same nominal m/z value for inceased selectivity, in particular the high resolution precursor ion selection for MS/MS mode.

Table 2.36 Increase in ion transmission for low and higher resolution quadrupoles at 0.2 da mass peak width FWHM, 100% equals 0.7 FWHM (Jemal 2003).

Quadrupole type	SIM	SRM
Higher resolution	70%	53%
Low resolution	11%	0.15%

Mass Analysers for Accurate Mass Measurement

Mass analyser	Maximum resolution power (FWHM)	Mass accuracy	Dynamic range	Inlet methods	MS/MS capability
Time-of-flight	5000–15 000	5 ppm	10^2–10^3	GC/MS, LC/MS	in hybrid systems
Hyperbolic triple quadrupole	10 000	5 ppm	10^4–10^6	GC/MS, LC/MS	typical application
Magnetic sector	120 000–150 000	2 ppm	10^5–10^6	GC/MS, LC/MS, Direct Inlets	limited
FT orbitrap	500 000	<1 ppm	10^3–10^4	LC/MS	extensive use
FT ion cyclotron resonance	2 000 000	<0,2 ppm	10^3–10^4	LC/MS	extensive use

2.3.1.1 High Resolution

The flight paths of ions with different m/z values follow a different course as a result of the magnetic and electric field in a magnetic sector instrument (Fig. 2.139). Splitting systems can mask ion beams. Spectra can be recorded by continually changing the parameters of the instrument, e. g. the acceleration voltage. The width of the ion beam is determined by the source slit at the ion source. The beams must not overlap, or only to a very small extent so that ions of different masses can be registered consecutively (Fig. 2.140).

The resolution A of neighbouring signals (Fig. 2.141) for magnetic sector instruments is calculated according to

$$A = \frac{m}{\Delta m} \tag{22}$$

with
m = mass and
Δm = distance between neighbouring masses

According to this formula the resolution is dimensionless.

By determination of the exact mass the empirical formula of a molecular ion (and of fragment ions) can be determined if the precision of the measurement is high enough (Fig. 2.142). The more the mass increases, the more interference can arise.

A selection of the large number of realistic chemical formulae for the mass 310 ($C_{22}H_{46}$, M 310.3599) is shown in Fig. 2.143. To differentiate between the individual signals a precision in the mass determination of 2 ppm would be necessary.

It is characteristic for a mass spectrum produced by a magnetic sector instrument that the resolution A is constant over the whole mass range. According to the formula for the resolution, the distance between the mass signals, Δm, for water (m = 17/18) is much greater than for signals in the upper mass range. The maximum resolution that can be achieved characterises the slit system and the quality of the ion optics of the magnetic sector instrument.

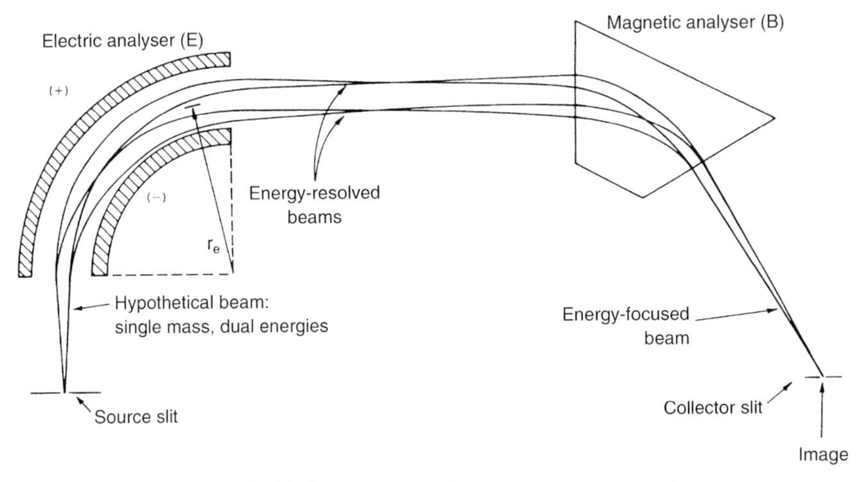

Fig. 2.139 Principle of the double focusing magnetic sector mass spectrometer (Nier Johnson geometry: EB).

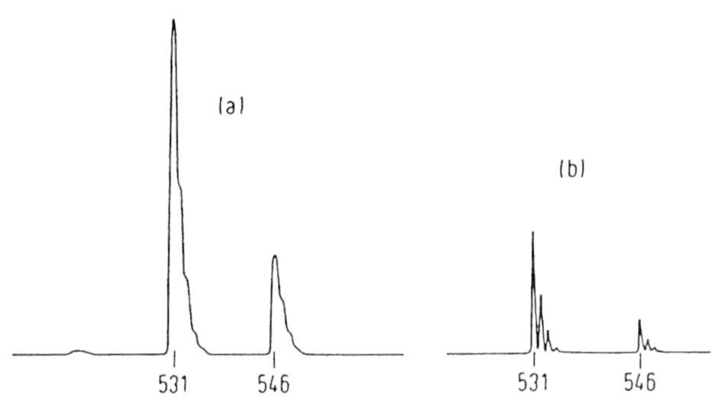

Fig. 2.140 Section of a poorly resolved spectrum (a) and of a better, more highly resolved one (b) which, however, results in a lowering of intensity of the signals (after Budzikiewicz 1980).

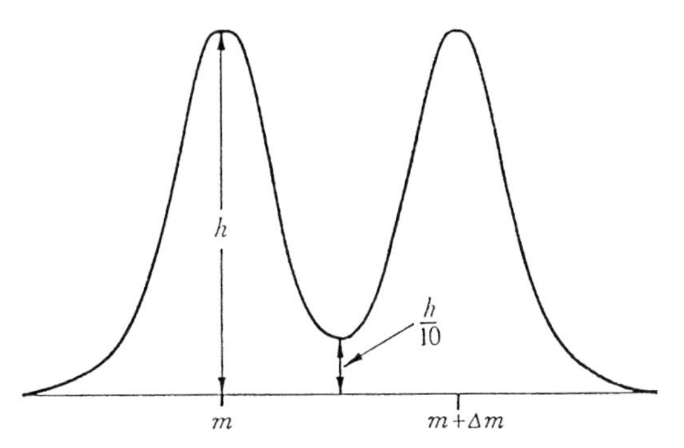

Fig. 2.141 General resolution conditions for two mass signals – 10% trough definition (after Budzikiewicz 1980).

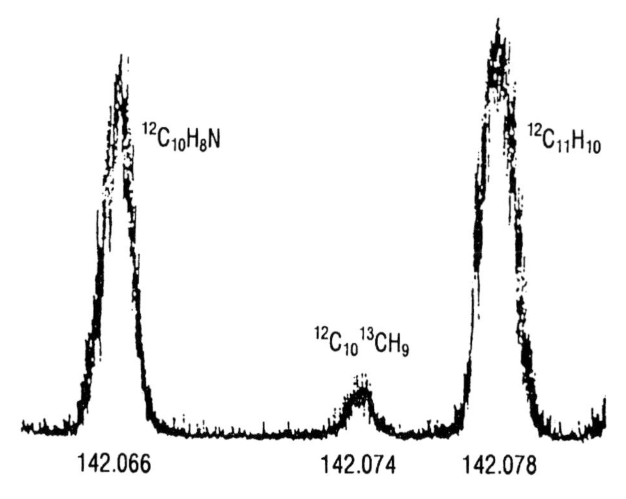

Fig. 2.142 Diagram showing mass signals in a high resolution spectrum at $A = 61\,000$ (Finnigan MAT 90).

Example of the Calculation of the Necessary Minimum Resolution

What mass spectroscopic resolution is required to obtain the signals of carbon monoxide (CO), nitrogen (N_2) and ethylene (C_2H_4) which are passed through suitable tubing into the ion source of a mass spectrometer?

Substance	Nominal mass	Exact mass
CO	m/z 28	m/z 27.994910
C_2H_4	m/z 28	m/z 28.006148
N_2	m/z 28	m/z 28.031296

For MS separation of CO and N_2, which appear at the same time in accordance with the formula given above, a resolving power of at least 2500 is necessary (see also Fig. 2.134).
All MS systems with low resolution need preliminary GC separation of the components (CO, C_2H_4 and N_2 would then arrive one after the other in the ion source of an MS). This is the case for all ion trap and quadrupole instruments.

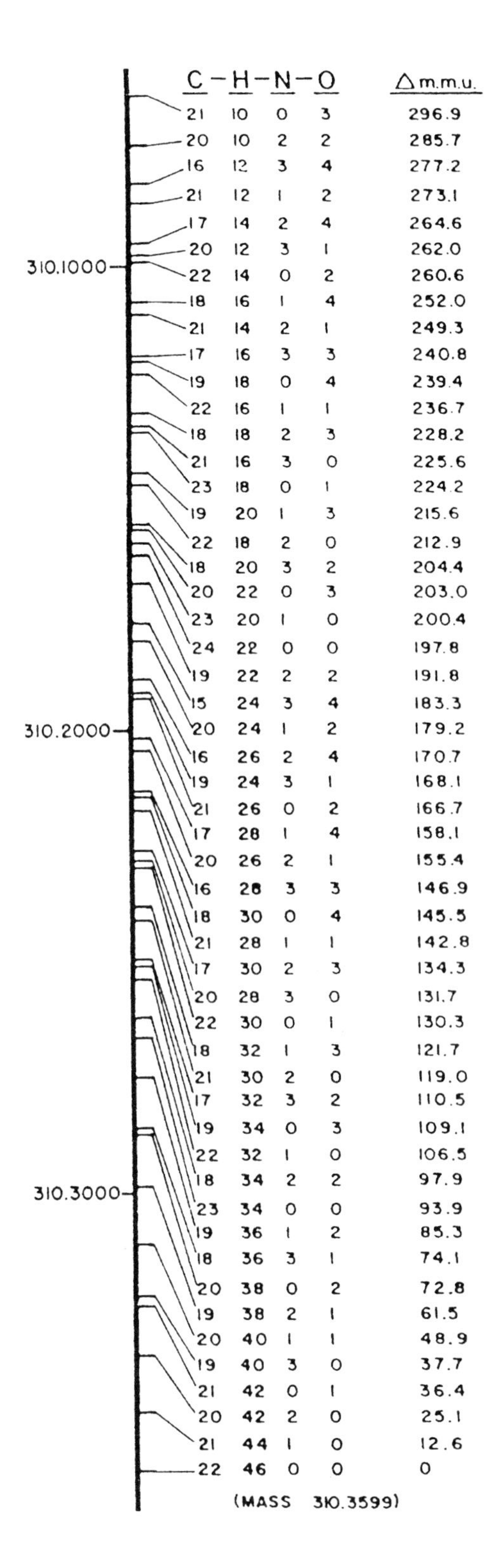

Fig. 2.143 Exact masses of chemically realistic empirical formulae consisting of C, H, N (≤ 3), O (≤ 5) given in deviations (Δ mu) from the molecular mass of $C_{22}H_{46}$ m/z 310 (after McLafferty 1993).

2.3.1.2 **Unit Mass Resolution**

A mass spectrum obtained with an ion trap or quadrupole mass spectrometer (Figs. 2.144 and 2.147) shows another characteristic: the distance between two mass signals and their signal widths are constant over the whole mass range! How far the instrument can scan to higher masses is therefore unimportant. The quadrupole/ion trap peaks of water (m/z = 17/18) are the same width and have the same separation as the masses in the upper mass range (Fig. 2.145).

Since the distance between two signals, Δm, is constant over the whole mass range for these instruments, the formula $A = m/\Delta m$ would have the result that the resolution A would

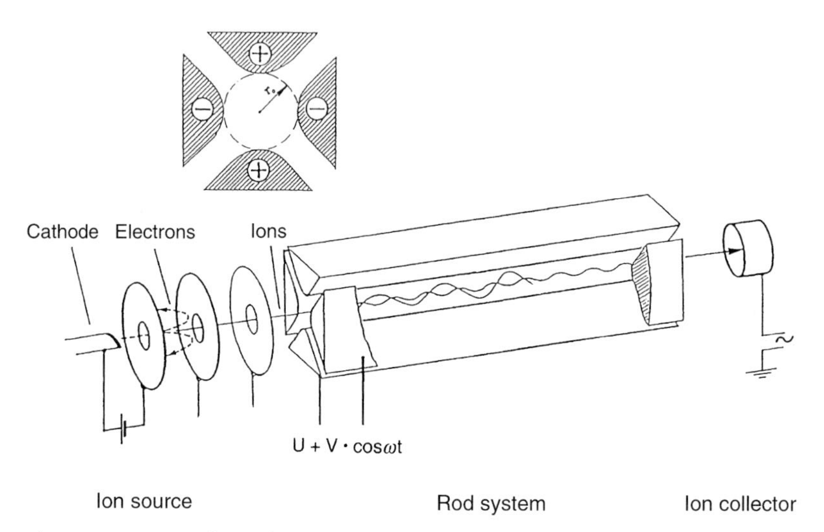

Fig. 2.144 Diagram of a quadrupole mass spectrometer.

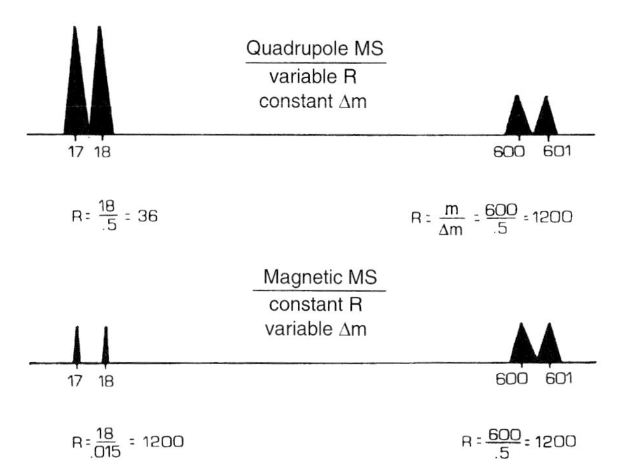

Fig. 2.145 Comparison of the principles of the mass spectra obtained from quadrupole/ion trap and magnetic sector analysers.

be directly proportional to the highest possible mass m (Fig. 2.146). At a peak separation of one mass unit the resolving power in the lower mass range would be small using the formula for the resolution (e.g. for water $A \approx 18$) and in the upper mass range higher (e.g. $A \approx 614$ for FC43).

The following conclusions may be drawn from these facts concerning the assessment of quadrupole and ion trap instruments:

1. The formula $A = m/\Delta m$ does not give any meaningful figures for quadrupole and ion trap instruments and therefore cannot be used in every case.
2. The visible optical resolution (on the screen) is the same in the upper and lower mass ranges. It can easily be seen that the signals corresponding to whole numbers are well separated. This corresponds in practice to the maximum possible resolution for quadrupole and ion trap instruments.
3. This resolving power, which is constant over the whole mass range, is set up by the manufacturer in the electronics of the instrument and is the same for all types and manufacturers. The peak width is chosen in such a way that the distance between two neighbouring nominal mass signals corresponds to one mass unit (1 u = 1000 mu). High resolution, as in the magnetic sector instrument, is not possible for quadrupole and ion trap instruments within the framework of the scan technique used. Because all quadrupole and ion trap instruments provide the same quality of nominal mass peak resolution they are said to work at unit mass resolution (see Fig. 2.146).

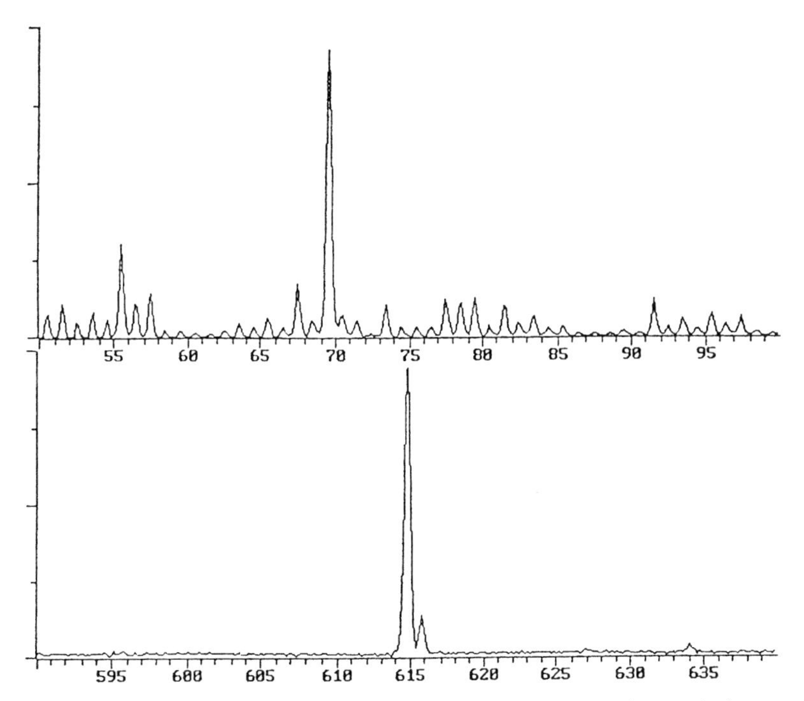

Fig. 2.146 Diagram of the mass signals obtained using an ion trap analyser in the lower and upper mass ranges (calibration standard FC43).

4. The mass range of quadrupole and ion trap instruments varies but still has no effect on the resolving power.

Frequently the terms mass range and unit mass resolution are mixed up when giving a quality criterion for a mass range above 1000 u for a quadrupole instrument (it is not obvious that a mass range up to 4000 u, for example, is always accompanied by unit mass resolution). The effective attainable resolution for a real measurable signal of a reference compound is accurate and meaningful.

The different types of analyser for quadrupole rod and quadrupole ion trap instruments function on the same mathematical basis (Paul/Steinwedel 1953 and patent specification 1956) and therefore show the same resolution properties (Fig. 2.147).

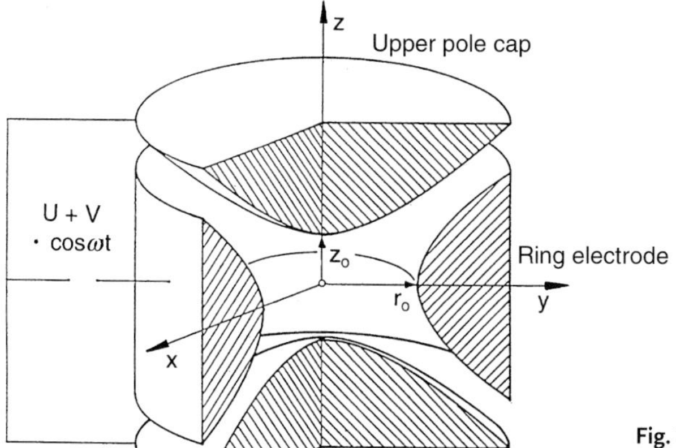

Fig. 2.147 Diagram of an ion trap analyser (Finnigan).

The display of line spectra on the screen must be considered completely separately from the resolution of the analyser. By definition a mass peak with unit mass resolution has a base width of one mass unit or 1000 mu. On the other hand the position of the top of the mass peak (centroid) can be calculated exactly. Data sometimes given with one or several digits of a mass unit gives the false impression of a resolution higher than unit mass resolution. Components appearing at the same time at the ion source with signals of the same nominal mass, which can naturally occur in GC/MS (as a result of co-eluates, the matrix, column bleed etc) cannot be separated at unit mass resolving power of the quadrupole and ion trap analysers (see Section 2.3.1.2). The position of the centroid can therefore not be used for any sensible evaluation. *In no case is this the basis for the calculation of a possible empirical formula*! Depending on the manufacturer, the labelling of the spectra can be found with pure nominal masses to several decimal places and can usually be altered by the user.

It has been observed with many benchtop instruments that the unavoidable contamination of ion volumes and lenses, the changes in the carrier gas flow as a result of temperature programs, and temperature drift of the ion source during the operation cause the position of the peak maximum to shift by several tenths of a mass during a short time. Adequate mass stability is therefore only given for the requirement of unit mass resolution. Each MS data system uses whole number mass spectra (nominal masses) for internal processing by background subtraction and library searching. The spectra of all MS libraries also only contain whole mass numbers!

2.3.1.3 High and Low Resolution in the Case of Dioxin Analysis

Can dioxin analysis be carried out with quadrupole and ion trap instruments?

Yes! But with what sort of quality? *The main problem is the resolving power.* The sensitivity of quadrupole and ion trap instruments is adequate for many applications. However, the selectivity is not.

It is possible to make mistakes on using selected ion monitoring techniques (SIM). These types of mistakes are smaller for full scan data collection with ion traps and on using MS/MS techniques, as usually a complete spectrum and a product ion spectrum are available for further confirmation. Errors caused by false positive peak detection and poor detection limits as a result of matrix overlap, which cannot be totally excluded, generally limit the use of low resolution mass spectrometers. For screening tests positive results should be followed by high resolution GC/MS for confirmation.

The high resolution and mass precision of the magnetic sector instrument allows the accurate mass to be recorded (e. g. 2,3,7,8-TCDD at m/z 321.8937 instead of a nominal mass of 322) and thus masks the known interference effects (Table 2.37). As a result very high selectivity with very low detection limits of < 10 fg are achieved, which gives the necessary assurance for making decisions with serious implications (Fig. 2.148).

Table 2.37 Possible interference with the masses m/z 319.8965 and 321.8936 relevant for 2,3,7,8-TCDD and the minimum analyser resolution required for separation.

Compound	Formula	m/z[a]	Resolution needed for separation
Tetrachlorobenzyltoluene	$C_{12}H_8Cl_4$	319.9508	5900
Nonachlorobiphenyl	$C_{12}HCl_9$	321.8491	7300
Pentachlorobiphenylene	$C_{12}H_3Cl_5$	321.8677	12500
Heptachlorobiphenyl	$C_{12}H_3Cl_7$	321.8678	13000
Hydroxytetrachlorodibenzofuran	$C_{12}H_4O_2Cl_4$	321.8936	cannot be resolved
DDE	$C_{14}H_9Cl_5$	321.9292	9100
DDT	$C_{14}H_9Cl_5$	321.9292	9100
Tetrachloromethoxybiphenyl	$C_{13}H_8OCl_4$	321.9299	8900

a) of the interfering ion

High resolution is therefore also required to verify positive screening results in dioxin analysis (see also Section 4.25). A comparison of the two spectrometric methods of the lower and higher resolution SIM techniques is shown in Fig. 2.149 for the mass traces in the detection of TCDF traces.

The resolution comparison with respect to different analyser technologies used in dioxin analysis is shown in Figs. 2.150 to 2.152 with simulated ideal mass spectra of the TCDD iso-

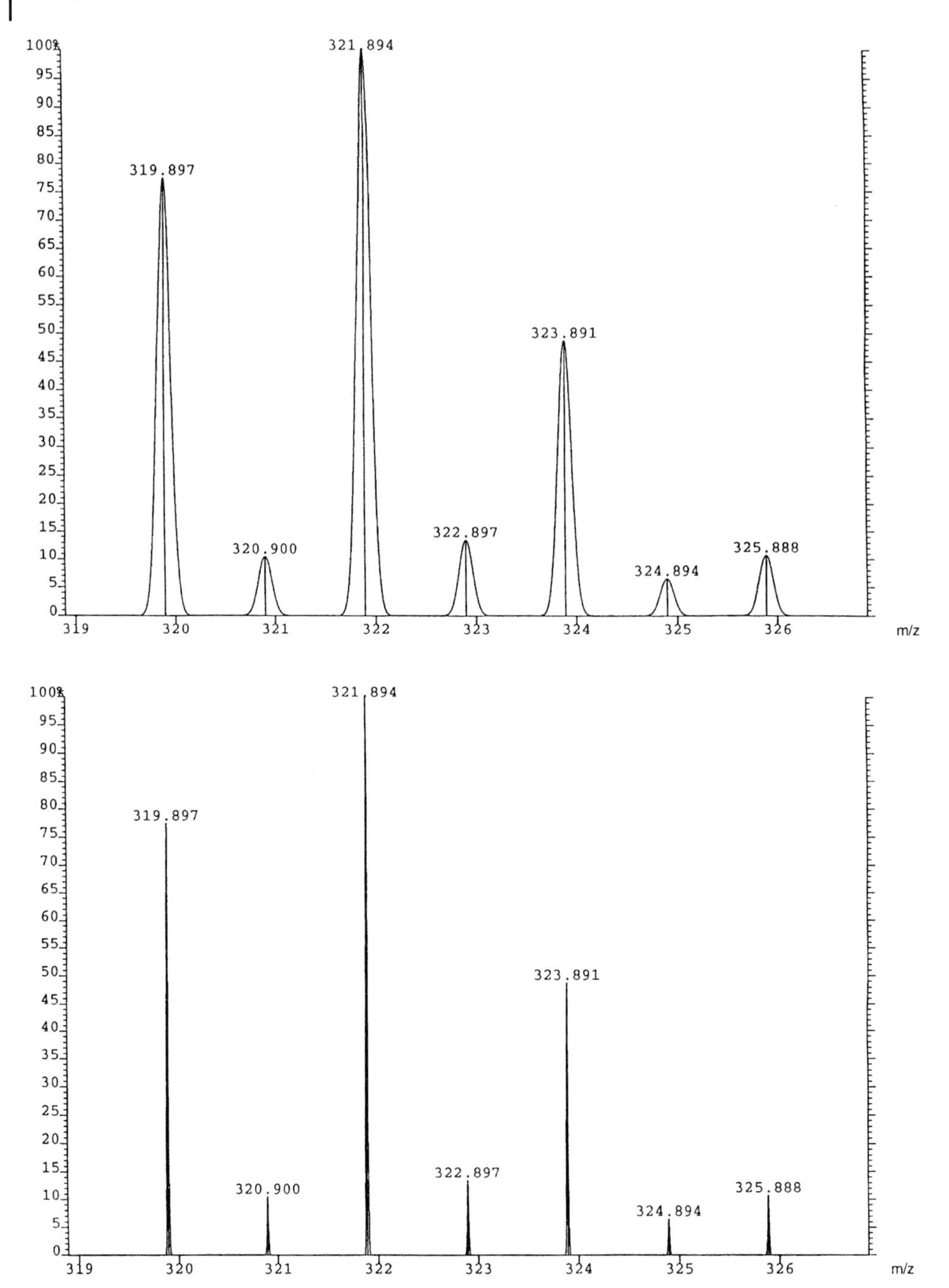

Fig. 2.148 The isotope pattern for 2,3,7,8-TCDD using a double focusing magnetic sector instrument with peak widths for resolutions of 1000 (above) and 10000 (below) (after Fürst).

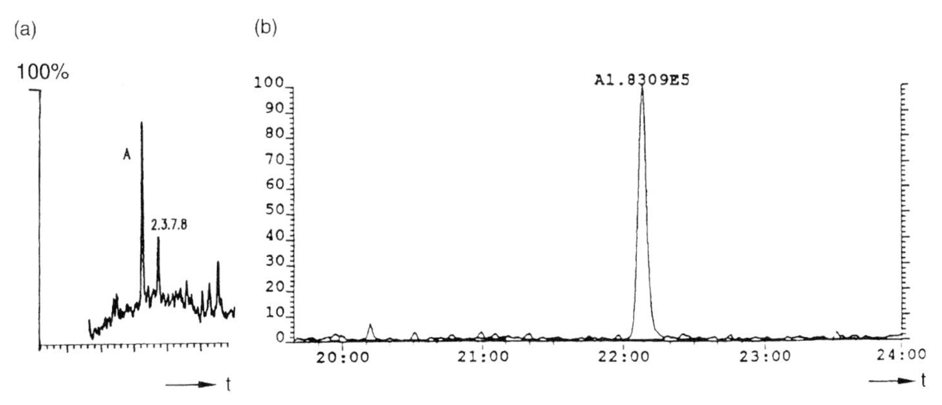

Fig. 2.149 Comparison of 2,3,7,8-TCDF traces in analyses using a low resolution quadrupole GC/MS system (a = m/z 319.9) and a high resolution GC/MS system (b = m/z 319.8965). Both chromatograms were run on an identical human milk sample. The component marked with A in the quadrupole chromatogram is an interfering component (after Fürst).

tope pattern in the molecular peak region. In practice, for real sample measurements a significant background contribution has to be kept in mind. The characteristic of the quadrupole analyser for unit mass resolution (Fig. 2.150) shows the mass peaks of the isotope pattern with the base peak width of 1 Da. The highly resolving quadrupole technology is able to reduce the peak width down to 0.4 Da with a remarkable good mass separation (Fig. 2.150). Time-of-flight analysers show very tall peaks but typically with a broad peak base. For that reason the TOF resolution power is measured at the half peak height. At the very good TOF resolution of 7,000 FPHW (frequently named hrTOF) the peak base still is wide over almost one full mass window (Fig. 2.151). Only the high resolution magnetic sector instruments provide with distinct difference the required resolution power of 10,000 for ultimate selectivity (Fig. 2.152).

2.3.2 Time-of-Flight Analyser

The concept of time-of-flight (TOF) mass spectrometry was proposed already in 1946 by William E. Stephens of the University of Pennsylvania (Borman 1998). In a TOF analyser, ions are separated by differences in their velocities as they fly down a field free drift region towards a collector in the order of their increasing mass-to-charge ratio (see Fig. 2.153). With that principle, TOF MS is probably the simplest method of mass measurement. TOF MS is fast, offers a high duty cycle with the parallel detection of ions and is theoretically unlimited in mass range. Due to the speed of detection and its inherent sensitivity by the parallel detection of a complete ion packet, it is specially suited for a full scan chromatographic detection, and, it is capable of running fast GC because of its high sampling rates of up to more than hundred spectra per second. TOF MS is also widely used for the determination of large biomolecules (electrospray ionisation, ESI; matrix assisted laser desorption, MALDI), among many other low and high molecular weight applications. TOF and Quadrupole-TOF analysers are inherently low resolution instruments. With current technologies the resolution power achieved is up to about 15,000 FWHM (Webb 2004).

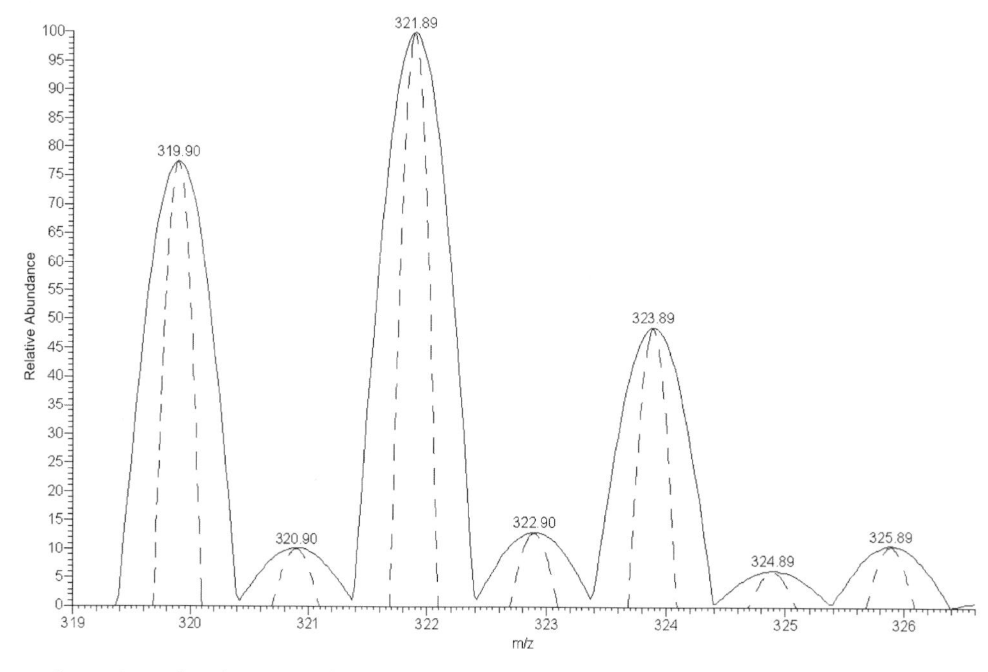

Fig. 2.150 Quadrupole mass resolution at 1 Da and 0.4 Da in H-SRM mode (TCDD isotope pattern).

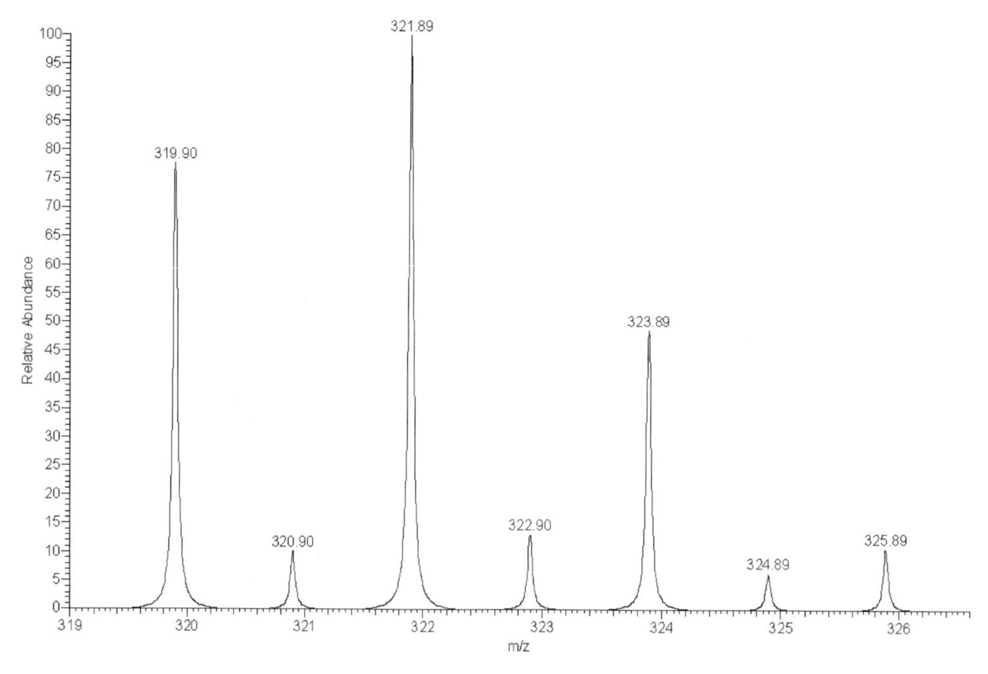

Fig. 2.151 TOF mass resolution at 7000 FPHW (TCDD isotope pattern)

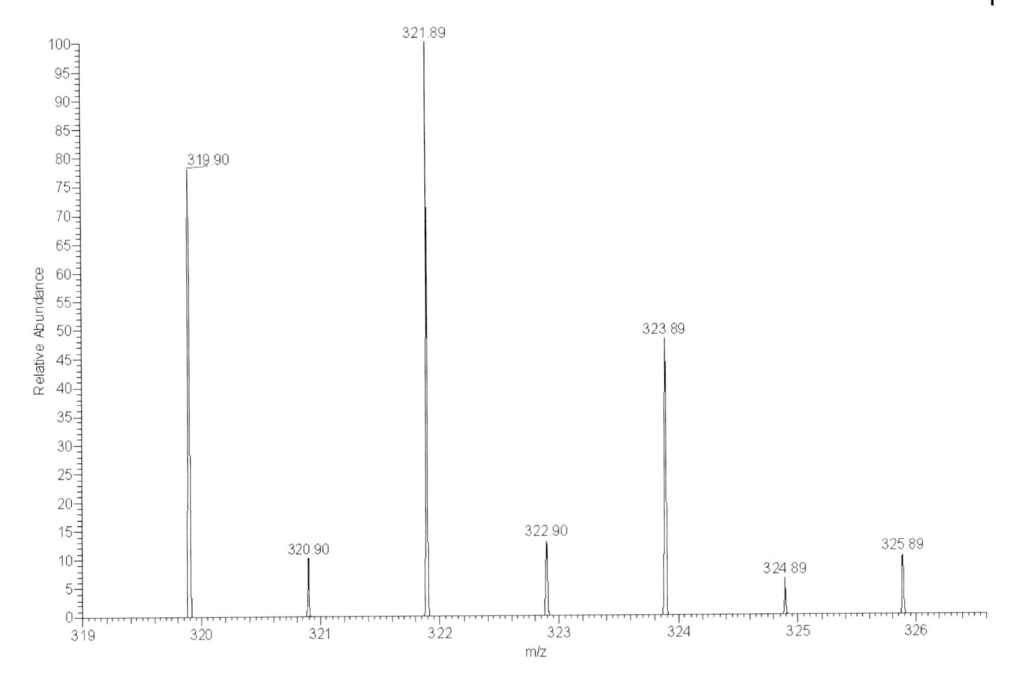

Fig. 2.152 High resolution magnetic sector analyser, mass resolution at 10,000 at 10% valley (TCDD isotope pattern).

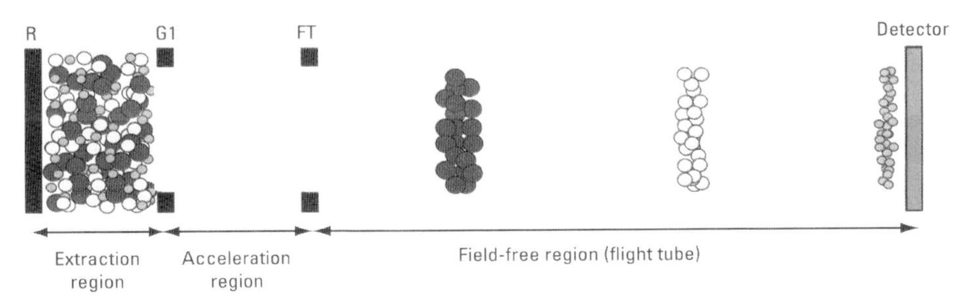

Fig. 2.153 Time-of-flight operating principle (McClenathan 2004) (reprinted with permission from the American Chemical Society)

TOF instruments were first designed and constructed starting in the late 1940s. Key advances were made by William C. Wiley and I.H. McLaren of the Bendix Corp., Detroit, USA, the first company to commercialize TOF mass spectrometers. According to pharmacology professor Robert J. Cotter of the Johns Hopkins University School of Medicine, Wiley and McLaren "devised a time-lag focusing scheme that improved mass resolution by simultaneously correcting for the initial spatial and kinetic energy distributions of the ions. Mass resolution was also greatly improved by the 1974 invention by Boris A. Mamyrin of the Physical-Technical Institute, Leningrad, Soviet Union of the reflectron, which corrects for the effects of the kinetic energy distribution of the ions." (Birmingham 2005).

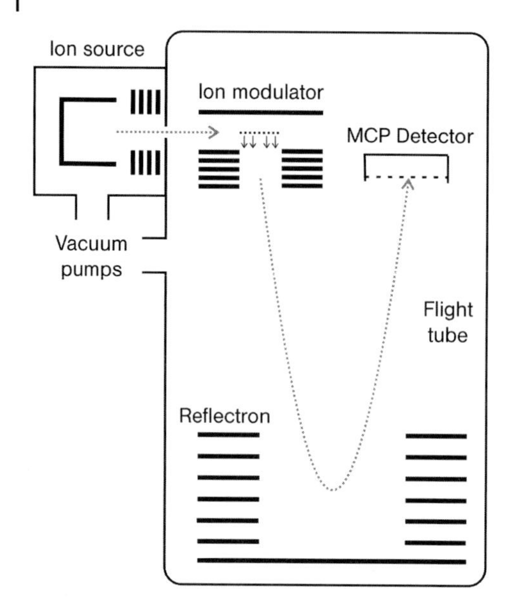

Fig. 2.154 TOF mass spectrometer with GC ion source, orthogonal ion modulation into the TOF analyser and MCP detection, dotted lines indicate the ion flight path.

The schematics of a TOF analyser is shown in Fig. 2.154. Ions are introduced as pulsed packets either directly from the GC or LC ion source of the instrument, or with hyphenated instruments, from a previous analyser (e.g. quadrupole-time-of-flight hybrid, Q-TOF). Ions from the ion source are accelerated and focused into a parallel beam that continuously enters the ion modulator region of the TOF analyser. Initially the modulator region is fieldfree and ions continue to move in their original direction. A pulsed electric field is applied at a frequency of several kilohertz (kHz) across a modulator gap of several mm width, pushing ions in an orthogonal direction (to their initial movement) into an accelerating section of several keV. The modulator pulses serve as trigger for recording spectra at the detector.

All ions in the pulsed ion packet receive the same initial kinetic energy, $E_{kin} = \frac{1}{2} mv^2$. Lighter ions travel faster and reach the detector earlier. As the ions enter and move down the fieldfree drift zone, they are separated by their mass to charge ratio in time. Commercial GC and LC/MS TOF instruments are usually equipped with a reflectron. This "ion mirror" is focusing the kinetic energy distribution, originating from the small initial velocity differences due to different spatial start positions of the ions when getting pulsed from the modulator into the drift tube (The mass resolution in general in TOF-MS is limited by the initial spatial and velocity spreads). The ratio of velocities components in the two orthogonal directions of movement from source and modulator is selected such that ions are directed to the centre of the ion mirror and get focused on a horizontal multichannel plate detector plane (MCP). The reflectron typically consists of a series of lenses with increasing potential pushing the ions back in a slight angle into the direction of the detector. Ions of higher velocity (energy) penetrate deeper into the mirror. Hence, the ion packet is getting focused in space and in time for increased mass resolution. All ions reaching the detector are recorded explaining the inherent high sensitivity in full scan mode of TOF analysers. Duty cycles reached with orthogonal ion acceleration instruments vary in a range of 5% to 30% depending on the mass range and limited by the slowest (heaviest) ion moving across the TOF mass analyser, covering the range between quadrupole and ion trap analysers.

The TOF mass separation is characterized by the following basic equation:

$$m/z = 2e \cdot E \cdot s\,(t/d)^2$$

with

m/z	mass-to-charge ratio
e	elementary charge
E	extraction pulse potential
s	length of ion acceleration, over which E is effective
t	measured flight time of an ion, triggered by pulse E
d	length of field free drift zone

The fast data acquisition rates make the TOF analyser the ideal mass detector for fast GC and comprehensive GC×GC. Despite of the described inherent sensitivity of TOF MS in full scan analysis some fundamental trade-offs in terms of response and spectral quality have to be considered when setting up TOF experiments. In contrast to the expectation that higher acquisition rates strongly support the deconvolution of coeluting compounds, the average ion abundance is dropping with increased scan rates and limiting the useful dynamic range. With the increase of the scan rate to 50 scans/s the ion abundance significantly drops to about 10% of a 5 scans/s intensity (see Fig. 2.153). Minor components may be left unrecognised. Hence, typical acquisition rates with current GC/TOFMS instrumentation are 20–50 Hz, with up to 100 Hz in fast GC or GC×GC applications.

Also the spectral quality has to be taken in account when setting up GC/TOF-MS methods for deconvolution experiments. Spectral skewing, as known from slow scanning quadrupole or sector mass spectrometers due to the increasing or decreasing substance intensity in the transient GC peak, does not occur in the ion package detection of TOF instruments. But, TOF spectral quality is limited by high data acquisition rates. There is still the general rule valid that increased measurement times support the spectrum dynamic range and hence its

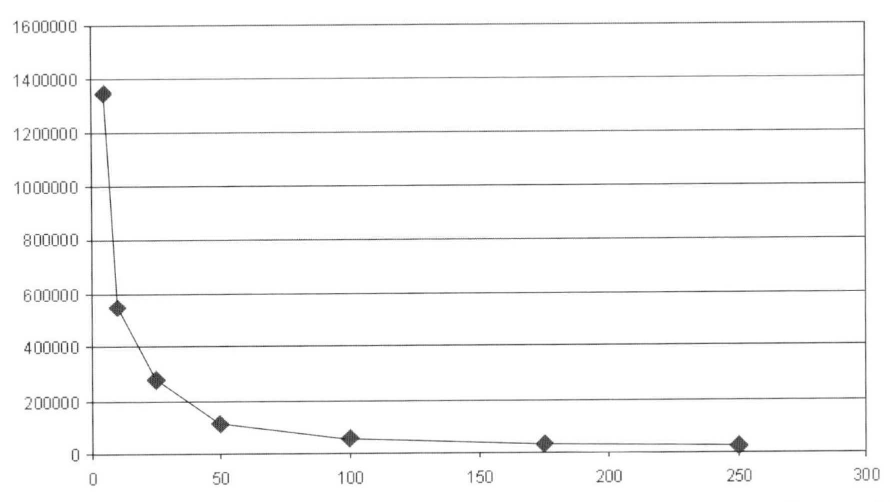

Fig. 2.155 Ion abundance dependence from the scan rate in TOF MS (counts vs. scan rate in spectra/s) (Meruva 2000).

quality. The effect can be demonstrated when comparing the acquired spectra at increasing scan rate against the reference spectra of the NIST library. The fit values of similarity match drop significantly with scan rates above 25 resp. 50 scan/s, see Table 2.38 for an experiment on two common pesticides.

Table 2.38 Average similarity match values depending on the data acquisition rate of TOF mass spectra vs. NIST library for disulfoton and diazinon. The average values of five replicate spectra from independent chromatographic runs are listed, except where low match quality did not provide identification.

Acquisition rate (spectra/s)	Disulfoton (fit value)	Diazinon (fit value)
5	884	852
10	876	851
25	837	827 [a]
50	820	623 [a]
100	735	583
175	697 [b]	626 [b]
250	660 [b]	586 [b]

a) Based on four replicate spectra.
b) Based on three replicate spectra.

2.3.3 Isotope Ratio Monitoring GC/MS

The measurement of isotope ratios was the first application of mass spectrometry. In 1907 Thomson for the first time showed the parabola mass spectrum of a Neon sample with his newly developed mass spectrometer. Later, in 1919, Aston concluded that the observed lines reveal the isotopes 20 and 22 of Neon, he later also discovered the isotope 21 of Ne with only 0.3 at% abundance. The term "isotope" was coined independently by Frederick Soddy. He observed substances with identical chemical behaviour but different atomic mass in the decay of natural radioactive elements and received 1921 the Nobel Prize in Chemistry for his investigations into the origin and nature of isotopes. The term "Isotope" is derived from the Greek "isos" for equal and "topos" for the place in the table of elements. Consequently the development in mass spectrometry in the first half of last century was dominated by elemental analysis with the determination of the elemental isotope ratios facilitated by further improved mass spectrometer systems with higher resolving power as introduced by Alfred Nier. He carried out the first measurements on $^{13}C/^{12}C$ abundance ratios in natural samples. Already in 1938 Nier studied bacterial metabolism by using ^{13}C as the tracer.

The following investigations by Alfred Nier formed the foundation for today's high precision isotope ratio mass spectrometry (IRMS). In contrast to mass spectrometric organic structure elucidation and target compound quantitation, IRMS is providing a different analytical dimension of precision data. Isotope ratio MS provides information on the physiochemical history, the origin or authenticity of a sample and is essential today in multiple areas of research and control such as food, life sciences, forensics, material quality control, geology or climate research, just to name the most important applications today.

Isotope ratio monitoring techniques using continuous flow sample introduction via gas chromatography had been introduced in the 1970s and developed into a mature analytical technology very quickly. Already in 1976, the first approach hyphenating a capillary GC with a magnetic sector MS for the systematic measurement of isotope ratios was published (Sano 1976). The continuous flow determination of individual compound $^{13}C/^{12}C$ ratios was introduced by Matthews and Hayes in 1978 and the term "isotope ratio monitoring GC/MS" was coined, today commonly abbreviated "irm-GC/MS" (Matthews 1978). The full range of $^{15}N/^{14}N$, $^{18}O/^{16}O$ and most importantly H/D determinations lasted until its publication in 1998 (Brand 1994, Heuer 1998, Hilkert 1998). Today irm-GC/MS is the established analytical methodology for delivering the precise ratios of the stable isotopes of the elements H, N, C, and O being the major constituents of organic matter. Sulfur is due to its only low abundance in organic molecules currently not amenable to irm-GC/MS analysis. A compelling example was presented by Ehleringer on the geographical origin of cocaine from South America applied to determine the distribution of illicit drugs (Ehleringer 2000, Bradley 2002). Organic compounds containing these elements can be quantitatively converted into simple gases for mass spectrometric analysis e. g. H_2, N_2, CO, CO_2, O_2, SO_2. For that conversion to simple gases, integrating and providing the full isotope information of a substance (in contrast to a fragmented organic mass spectrum) also the term "gas isotope ratio mass spectrometry" GIRMS is used frequently also including irm-GC/MS applications.

The Principles of Isotope Ratio Monitoring

Isotope ratios, although tabulated with average values for all elements, are not constant. All phase transition processes, transport mechanisms and enzymatic or chemical reactions are dependent on the physical properties of the reaction partners, i. e. most importantly the mass of the molecule involved in the process or chemical reaction, for their kinetic properties.

The high precision quantitative data on isotope ratios obtained are not absolute quantitation data. Natural isotope ratios exhibit only small but meaningful variations that are measured with highest precision relative to a known standard. In most of the applications in stable isotope analysis the differences in isotopic ratios between samples is of much more interest and significance than the absolute amount in a given sample. The isotope ratio is independent of the amount of material measured (and at the low end only determined by the available ion statistics). The relative ratio measurement can be accomplished in the required and even higher precision, which is at least one order of magnitude higher than the determination of absolute values.

Notations in irm-GC/MS

The small abundance differences of stable isotopes are best represented by the delta notation (4) in which the stable isotope abundance is expressed relative to an isotope standard, measured in the same run for reference:

$$\delta = (R_{sample}/R_{std} - 1) \times 1000\ [‰] \tag{4}$$

with R the molar ratio of the heavy to the common (light) isotope of an element e. g. R = $^{13}C/^{12}C$ or D/H or $^{18}O/^{16}O$. A graphical representation of meaning the δ-value is given in

Table 2.39 Natural abundances of light stable isotopes relevant to stable isotope ratio mass spectrometry.

Element	Isotope	Atomic weight	Relative abundance %	Elemental relative mass difference	Molecular relative mass difference %	Terrestrial range ‰	Terrestrial range ppm	Technical precision ‰	Technical precision ppm
Hydrogen	1H	1,0078	99,9840	D/H	1HD/1H1H				
Deuterium	2H (D)	2,0141	0,0156	(2/1)	(3/2)	700	109	1	0,16
				100%	**50%**				
Boron	10B	10,0129	19,7	11B/10B		60			
	11B	11,0093	8,3	**10%**					
Carbon	12C	12,0000	98,892	13C/12C	13C16O16O/12C16O16O				
	13C	13,0034	1,108	(13/12)	(45/44)	100	1123	0,05	0,56
				8,3%	**2,3%**				
Nitrogen	14N	14,0031	99.635	15N/14N	15N14N/14N14N				
	15N	15,0001	0,365	(15/14)	(29/28)	50	181	0,10	0,72
				7.1%	**3,6%**				
Oxygen	16O	15,9949	100	18O/16O	12C16O18O/12C16O16O				
	17O	16,9991	0,037	(18/16)	(46/44)				
	18O	17,9992	0,204	**12.5%**	**4.6%**	100	200	0,10	0,20
Silicon	28Si	28	92,21	29Si/28Si	29Si19F3/28Si19F3	6			
	29Si	29	4,7	(29 / 28)	(86 / 85)				
	30Si	30	3,09	**3,6%**	**1,2%**				
Sulphur	32S	32	95,02	34S/32S	34S16O16O/32S16O16O				
	33S	33	0,76						
	34S	34	4,22	(34/32)	(66/64)	100	4580	0,20	9,16
	35S	36	0,014	**6,3%**	**3,1%**				
Chlorine	35Cl	35	75,77	37Cl/35Cl	12CH337Cl/12CH335Cl	10		0,10	
	37Cl	37	24,23	(37/35)	(52/50)				
				5,7%	**4.0%**				

From Gilles St.Jean, Basic Principles in Stable Geochemistry, IRMS Short Course, 9th Canadian CF-IRMS Workshop 2002.

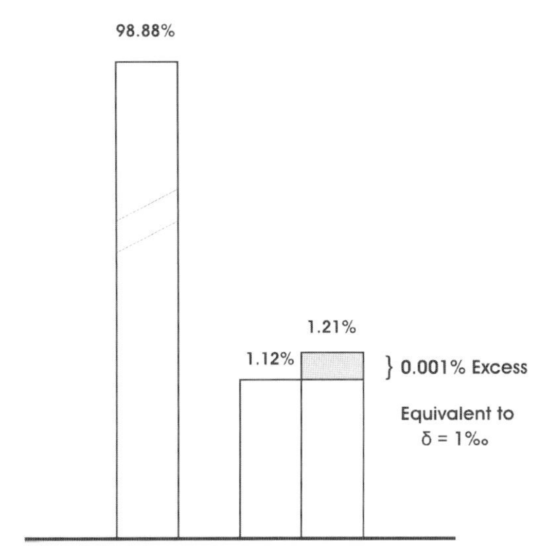

Fig. 2.156 Graphical representation of the δ-notation for a ^{13}C variation of 1‰.

Fig. 2.156 for the variation of the ^{13}C isotope proportion of carbon. δ-Values can be positive or negative indicating a higher or lower abundance of the major isotope in the sample than in the reference.

Other systems in use for expressing isotope ratios are ppm (part per million, relative value), at% (atom percent, absolute value), and APE (atom percent excess, relative value) as the most common terminology in biomedical tracer studies. The choice is dependent primarily on the specific field of application e.g. with high degrees of enrichments being very much different from natural abundances. Also typical in this field are historical traditions in the use of different notation systems. Values of at% and δ-notation can be converted as follows:

$$\text{at\%} = \frac{R_{st} \cdot (\delta/1000 + 1)}{1 + R_{st} \cdot (\delta/1000 + 1)} \cdot 100$$

with R_{st} absolute standard ratio

Isotopic Fractionation

Isotope effects, as they can be observed in phase transition or dissociation reactions are usually the result of incomplete processes in diffusive or equilibrium fractionation. This effect is caused by different translational velocities of the lighter and heavier molecule through a medium or across a phase boundary. As the kinetic energy at a given temperature is the same for all gas molecules, the kinetic energy equation

$$E_{kin} = {}^1\!/_2\, \mathrm{m} \cdot v^2$$

applies for both molecules $^{12}CO_2$ and $^{13}CO_2$ with their respective masses 44 and 45 u:

$$^1\!/_2\,(44 \cdot v_a^2) = {}^1\!/_2\,(45 \cdot v_b^2)$$

resulting in

$$\nu_a/\nu_b = \sqrt{(45/44)} = 1.0113$$

The velocity ratio of both CO_2 species explains that the average velocity of $^{12}CO_2$ is 1.13% higher than that of the heavier molecule.

Example: Boiling water will lose primarily the light $^1H_2{}^{16}O$ molecules as can be seen in Fig. 2.157. Heavier water molecules will consequently be concentrated in the liquid; the vapour will become depleted. The reversed processes are observed during condensation, e.g., during the formation of raindrops from humid air. As the extent of this process is temperature dependent isotopic "thermometers" are formed and ultimately isotopic „signatures“ of materials and processes are created.

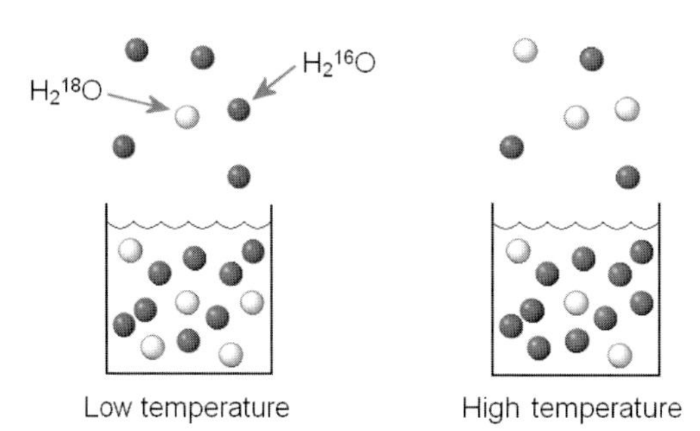

Fig. 2.157 Alteration of the isotope ratio during the evaporation of water resulting in a different $^{16}O/^{18}O$ ratio at low and high temperatures in vapour as well as in the liquid.

Equilibrium isotope effects usually are associated with phase transition processes like evaporation, diffusion or dissociation reactions. When occurring during sample preparation, incomplete phase transition processes lead to severe alteration of the initial isotope ratio. In fact, and that is of highest importance for all sample preparation steps in IRMS, only complete conversion reactions are acceptable to maintain the integrity of the original isotope ratio of the sample.

Isotope effects are also observed in chemical and biochemical reactions (kinetic fractionation). Of particular significance is the isotope effect occurring during enzymatic reactions with a general depletion in the heavier isotope being key to uncover metabolic processes. Chemical bonds to the heavier isotope are stronger, more stable and need higher dissociation energies in chemical reactions due to the different vibrational energy levels involved. Hence, the rate of enzymatic reaction is faster with the light isotope leading to differences in the abundance between substrate and product, unless the substrate is fully consumed, which is not the case in cellular steady state equilibria. Kinetic fractionation effects also have to be considered when employing compound derivatisation steps (e.g. silylation, methylation) during sample preparation for GC application (Meyer-Augenstein 1997).

The natural variations in isotopic abundances can be large, depending on the relative elemental mass differences: hydrogen (100%) > oxygen (12.5%) > carbon (8.3%) > nitrogen (7.1%), see also Table 2.39. An overview of the isotopic variations found in natural compounds is given in Figs. 2.158 to 2.161.

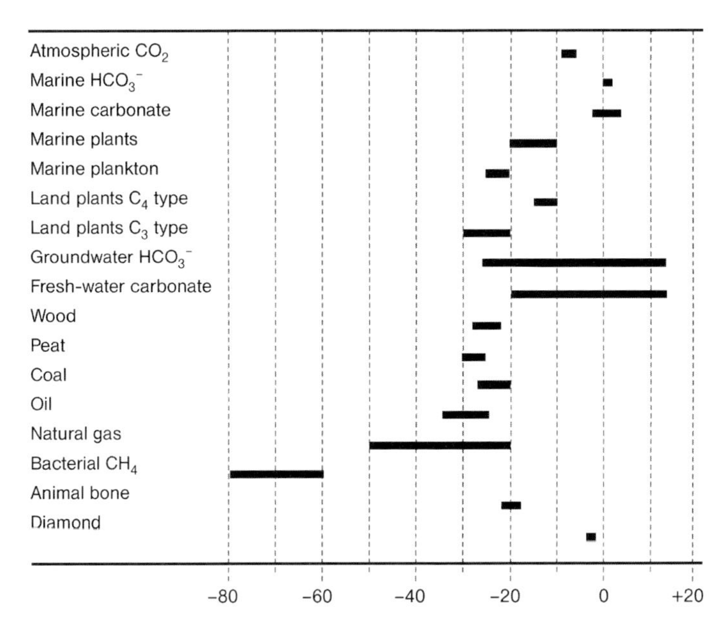

Fig. 2.158 $\delta^{13}C$ variations in natural compounds ($\delta^{13}C$ VPDB ‰ scale) (de Vries 2000, courtesy IAEA).

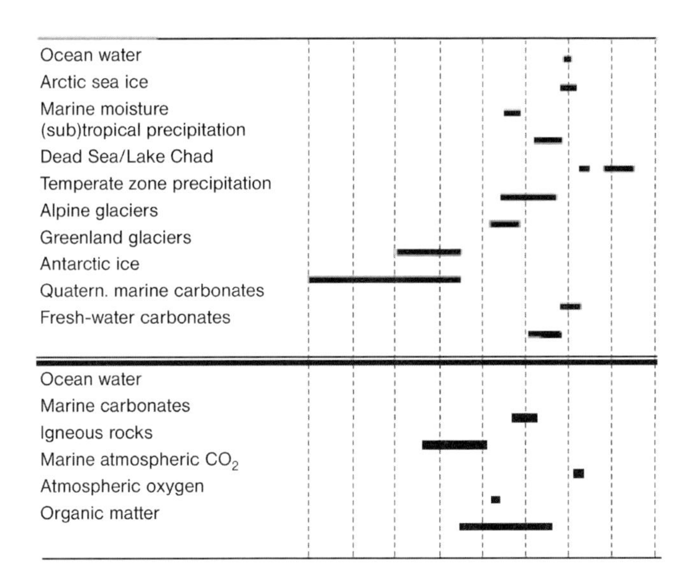

Fig. 2.159 $\delta^{18}O$ variations in natural compounds (top: $\delta^{18}O$ VSMOW ‰ scale for waters, $\delta^{18}O$ VPDB ‰ scale for carbonates, bottom: $\delta^{18}O$ VSMOW ‰ scale) (de Vries 2000, courtesy IAEA).

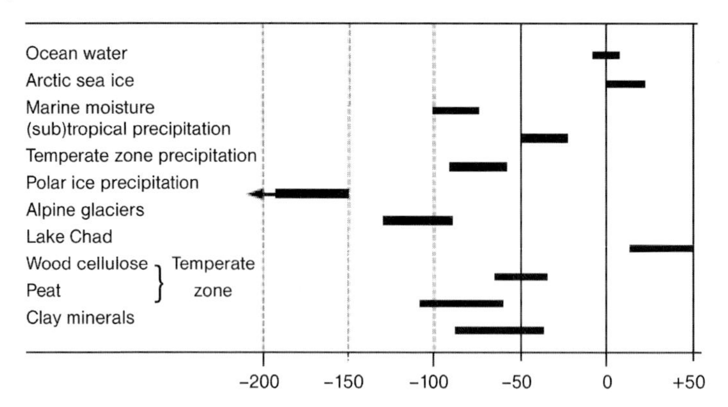

Fig. 2.160 δ^2H variations in natural compounds (δ^2H VSMOW ‰ scale) (de Vries 2000, courtesy IAEA).

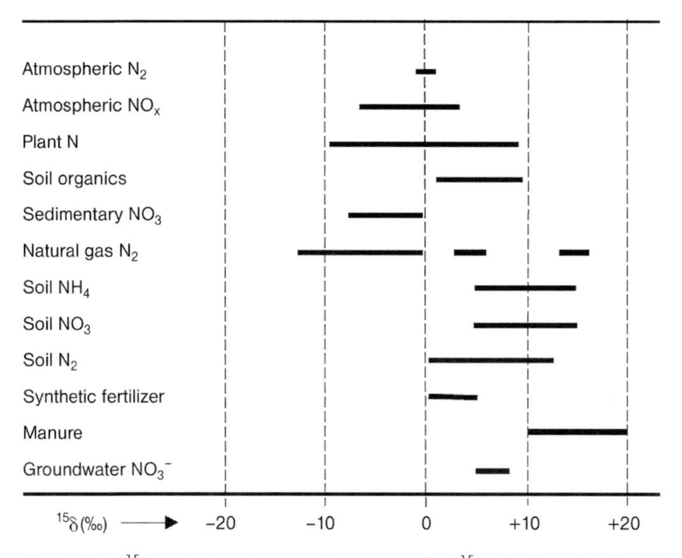

Fig. 2.161 $\delta^{15}N$ variations in natural compounds ($\delta^{15}N$ Air ‰ scale) (de Vries 2000, courtesy IAEA).

irm-GC/MS Technology

irm-GC/MS is applied to obtain compound specific data after a GC separation of mixtures in contrast to bulk analytical data from an elemental analyser system. The bulk analysis delivers the average isotope ratio within a certain volume of sample material. The entire sample is converted into simple gases using conventional elemental analysers (EA) or high temperature conversion elemental analysers (TC/EA). In contrast to a bulk analysis, irm-GC/MS delivers specific isotope ratio data in the low picomolar range of individual compounds after a conventional capillary GC separation (see Fig. 2.162).

The most important step in irm-GC/MS is the conversion of the eluting compounds into simple measurement gases like CO_2, N_2, CO, and H_2. In irm-GC/MS as a continuous flow application, this conversion is achieved on-line within the helium carrier gas stream while

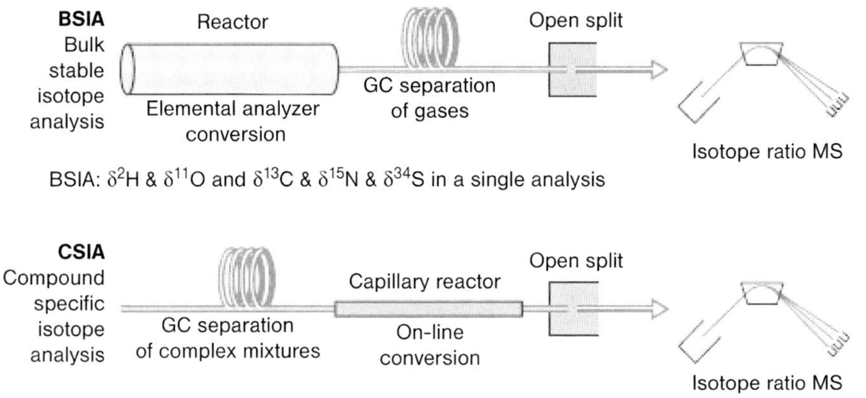

Fig. 2.162 Basic scheme of bulk sample (BSIA) vs. compound specific (CSIA) isotope analysis (courtesy Thermo Fisher Scientific).
BSIA: IRMS is coupled with an EA: The complete sample is first converted to simple gases followed by their chromatographic separation.
CSIA: IRMS is coupled with a GC: The sample components are first separated by capillary chromatography and then individually converted to the measurement gases.

preserving the chromatographic pattern of the sample. The products of the conversion are then fed by the carrier gas stream into the isotope ratio MS.

It is important to note that during chromatographic separation on regular fused silica capillary columns, a separation of the different isotopically substituted species of a compound already takes place. Due to the lower molar volume of the heavier components and the resulting differences in mobile/stationary phase interactions, the heavier components elute slightly earlier (Matucha 1991). This separation effect can be observed also in "organic" GC/MS analyses when having a closer look at the peak top retention times of the ^{13}C carbon isotope mass trace. irm-GC/MS makes special use of the chromatographic isotope separation effect by displaying the typical S-shaped ratio traces during analysis indicating the substances of interest (see Figs. 2.163 and 2.164).

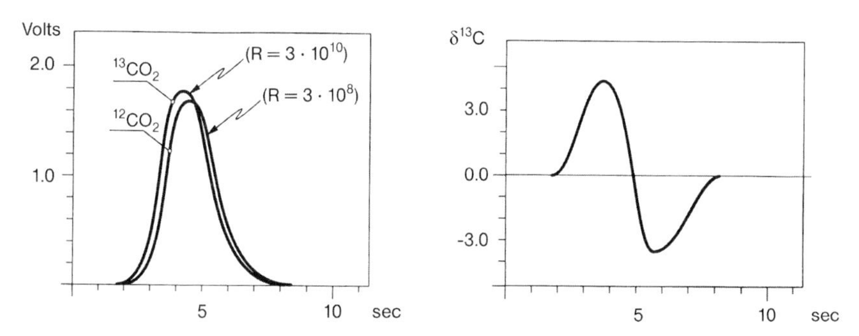

Fig. 2.163 Chromatographic elution profiles of CO_2 with natural ^{13}C abundance resulting in the typical S-shaped curvature when displaying the calculated δ-values (left: note the intensity difference of the isotopes with the choice of amplification resistors R in a difference of two orders of magnitude).

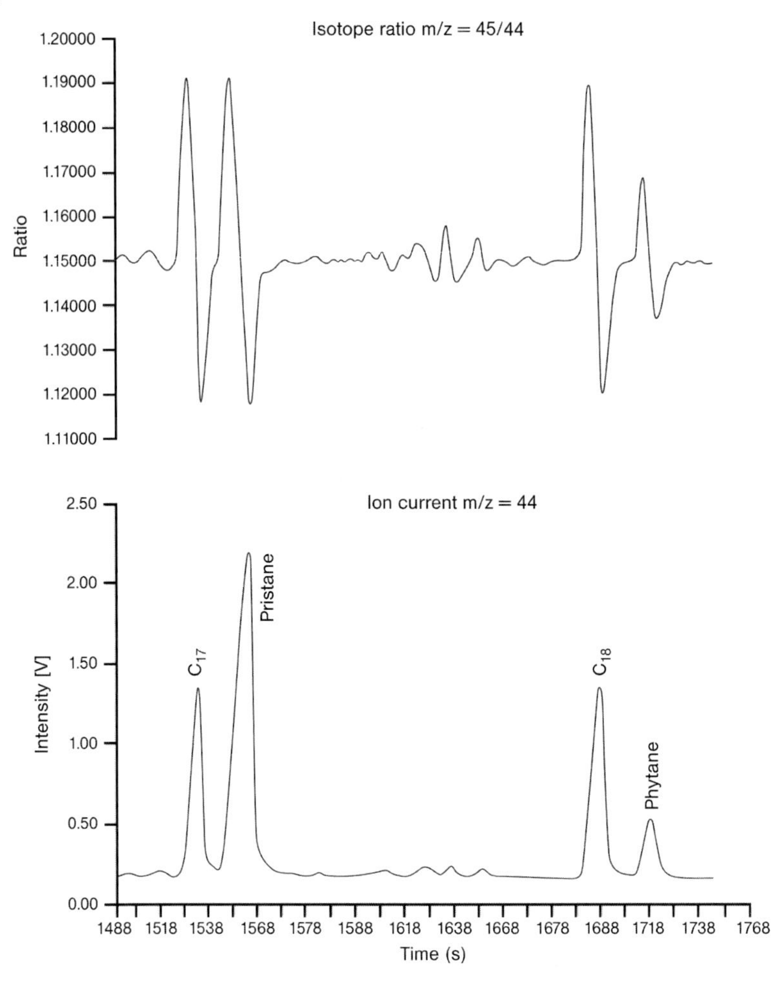

Fig. 2.164 irm-GC/MS analysis of the crude oil steroid biomarkers pristane and phytane with the $\delta^{13}C$ elution profile on top.

The Open Split Interface

For high precision isotope ratio determination, the IRMS ion source pressure must be kept absolutely constant. For this reason, it is mandatory in all continuous flow applications coupled to an isotope ratio MS, to keep the open split interface at atmospheric pressure. The open split coupling eliminates the isotopic fractionation due to an extended impact of the high vacuum of the mass spectrometer into the chromatography and conversion reactors. The basic principle of an open split is to pick up a part of the helium/sample stream by a transfer capillary from a pressureless environment, i. e. an open tube or a wide capillary, and transfer it into the IRMS (see Fig. 2.165). The He stream added by a separate capillary ensures protection from ambient air. Retracting the transfer capillary into a zone of pure helium allows cutting off parts of the chromatogram. In all modes, a constant flow of helium into the isotope ratio MS, and consequently constant ion source conditions are maintained.

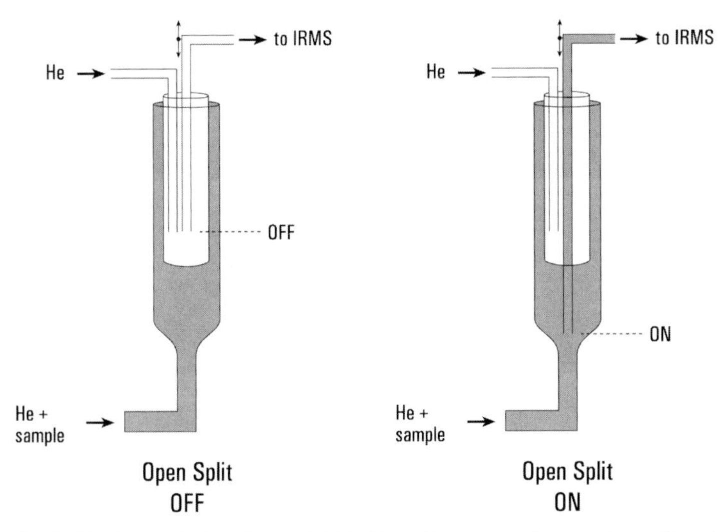

Fig. 2.165 Open split interface to IRMS, effected by moving the transfer capillary (to IRMS) from the column inlet to the He sample flow region (courtesy Thermo Fisher Scientific).

The valve-free open split is absolutely inert and does not create any pressure waves during switching.

In addition, the open split interface also offers an automatic peak dilution capability. Due to the transfer of the analyte gases in a helium stream more than one interface can be coupled in parallel for alternate use to an isotope ratio MS via separate needle valves.

Compound Specific Isotope Analysis by On-line Sample Combustion and Conversion

irm-GC/MS is amenable to all GC-volatile organic compounds down to the low picomole range. A capillary design of oxidation and high temperature conversion reactors is required to guarantee the integrity of GC resolution. Table 2.40 gives an overview of the techniques available today with the achievable precision values by state of the art IRMS technology. Compound specific isotope analysis (CSIA) is one of the newest (but today a mature and steadily expanding) fields of isotope analysis. The first GC combustion system for $\delta^{13}C$ determination was commercially introduced in 1988 by Finnigan MAT Bremen, Germany. This

Table 2.40 Analytical methods used in compound specific isotope analysis.

Element ratio	Measured species	Method	Temperature	Analytical precision [a)]
$\delta^{13}C$	CO_2	Combustion	up to 1000 °C	^{13}C 0.06 ‰, 0.02 ‰/nA
$\delta^{15}N$	N_2	Combustion	up to 1000 °C	^{15}N 0.06 ‰, 0.02 ‰/nA
$\delta^{18}O$	CO	Pyrolysis	≥ 1450 °C	^{18}O 0.15 ‰, 0.04 ‰/nA
$\delta^{2}H$	H_2	Pyrolysis	≥ 1280 °C	^{2}H 0.50 ‰, 0.20 ‰/nA

a) 10 pulses of reference gas (amplitude 3V, for H_2 5V) δ notation.

technique combines the resolution of capillary GC with the high precision of isotope ratio mass spectrometry (IRMS). In 1992, the capabilities for analysis of $\delta^{15}N$ and in 1996 for $\delta^{18}O$ were added. Quantitative pyrolysis by high temperature conversion (GC-TC IRMS) for δD analyses was introduced in 1998 introducing an energy discrimination filter in front of the HD collector at m/z 3 for the suppression of $^4He^+$ ions from the carrier gas interfering with the HD^+ signal (Hilkert 1999).

On-line Combustion for $\delta^{13}C$ and $\delta^{15}N$ Determination

All compounds eluting from a GC column are oxidized in a capillary reactor to form CO_2, N_2, and H_2O as a by-product at 940 to 1000°C. NO_x produced in the oxidation reactor is reduced to N_2 in a subsequent capillary reduction reactor. The H_2O formed in the oxidation process is removed by an on-line Nafion dryer, a maintenance-free water removal system. For the analysis of $\delta^{15}N$, all CO_2 is retained in a liquid nitrogen trap in order to avoid interferences from CO on the identical masses before transfer into the IRMS through the movable capillary open split. CO is generated in small amounts as a side reaction during the ionisation process from CO_2 in the ion source of the MS. A detailed schematic of the on-line combustion setup is given in Fig. 2.166.

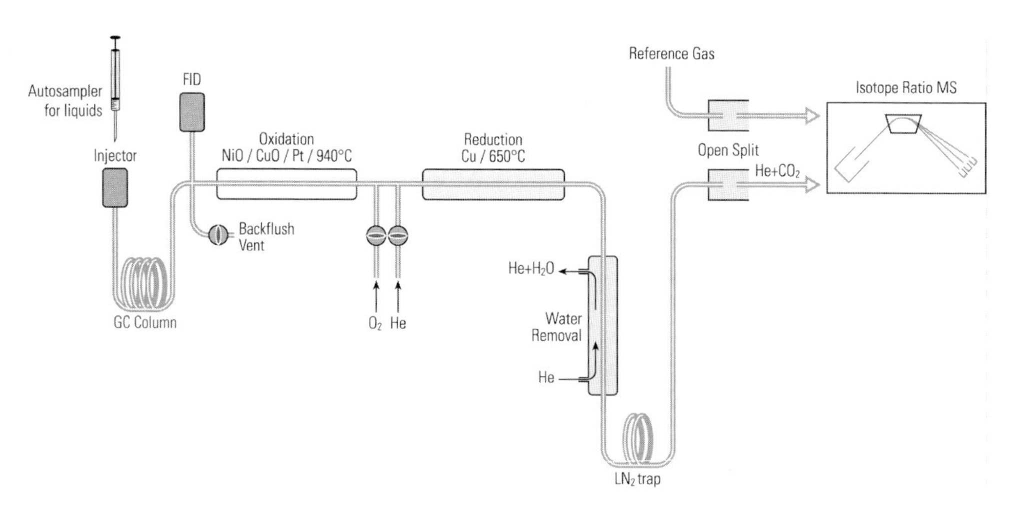

Fig. 2.166 On-line oxidation of compounds eluting from the GC for the production of CO_2 and N_2 for IRMS measurement (courtesy Thermo Fisher Scientific).

The Oxidation Reactor

The quantitative oxidation of all organic compounds eluting from the GC column, including the refractory methane, is performed at temperatures up to 1000 °C. The reactor consists of a capillary ceramic tube loaded with twisted Ni, Cu, and Pt wires. The resulting internal volume compares to a capillary column and secures the integrity of the chromatographic separation. The reactor can be charged and recharged automatically with O_2 added to a backflush flow every 2–3 days, depending on the operating conditions. The built-in backflush system reverses the flow through the oxidation reactor towards an exit directly after the GC column. The backflush is activated during the analysis to cut off an eluting solvent peak in front of the oxidation furnace by flow switching. All valves have to kept outside of the analytical flow path to maintain optimum GC performance.

The Reduction Reactor

The reduction reactor is typically comprising of copper material and operated at 650 °C to remove any O_2 bleed from the oxidation reactor and to convert any produced NO_x into N_2. It is of the same capillary design as the oxidation reactor.

Water Removal

Water produced during the oxidation reaction is removed through a 300 μm inner diameter Nafion capillary, which is dried by a counter current of He on the outside (Fig. 2.167). The water removal adds no dead volume and is maintenance-free.

On-line Water Removal From a He Stream Using Nafion

Water is removed from a He sample stream by a gas tight but hygroscopic Nafion tubing. The sample flow containing He, CO_2 and H_2O passes through the Nafion tubing which is mounted co-axially inside a glass tube. This glass tube, and therefore the outer surface of the Nafion tube, is constantly kept dry by a He flow of approximately 8 to 10 mL/min. Due to the water gradient through the Nafion membrane wall any water in the sample flow will move through the membrane. A dry gas comprising of only He and CO_2 results which is fed to the ion source of the mass spectrometer.

Structure of Nafion:

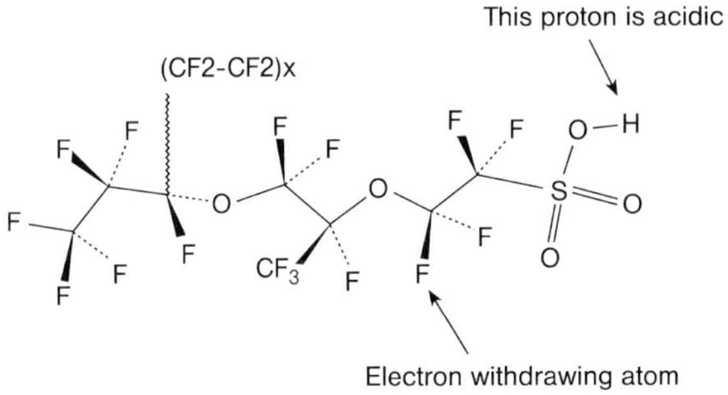

This proton is acidic

(CF2-CF2)x

Electron withdrawing atom

Properties:

Nafion is the combination of a stable Teflon backbone with acidic sulfonic groups. It is highly conductive to cations, making it ideal for many membrane applications. The Teflon backbone interlaced with ionic sulfonate groups gives Nafion a high operating temperature, e.g. up to 190 °C. It is selectively and highly permeable to water. The degree of hydration of the Nafion membrane directly affects its ion conductivity and overall morphology. See also: http://de.wikipedia.org/wiki/Nafion

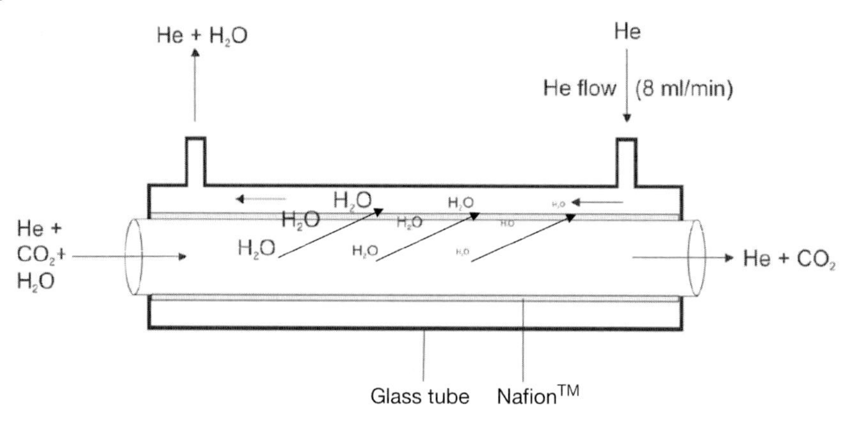

Fig. 2.167 Principle of on-line removal of water from a He stream using a Nafion membrane (courtesy Thermo Fisher Scientific).

The Liquid Nitrogen Trap

For the analysis of $\delta^{15}N$, all the CO_2 must be removed quantitatively to avoid an interference of CO^+ produced in from CO_2 in the ion source with the N^{2+} analyte. This is achieved by immersing the deactivated fused silica capillary between the water removal and the open split into a liquid nitrogen bath. The trapped CO_2 is easily released after the measurement series with no risk of CO_2 contamination of the ion source by using the movable open split.

On-line High Temperature Conversion for δ^2H and $\delta^{18}O$ Determination

A quantitative pyrolysis by high temperature conversion of organic matter is applied for the conversion of organic oxygen and hydrogen to form the measurement gases CO and H_2 for the determination of $\delta^{18}O$ or δD (see Fig. 2.169). This process requires an inert and reductive environment to prevent any O or H containing material from reacting or exchanging with the analyte.

For the determination of $\delta^{18}O$, the analyte must not contact the ceramic tube that is used to protect against air. For the conversion to CO the pyrolysis takes place in an inert platinum inlay of the reactor. Due to the catalytic properties of the platinum, the reaction can be performed at 1280 °C. For the determination of δD from organic compounds, the reaction is performed in an empty ceramic tube at 1450 °C. Such high temperatures are required to ensure a quantitative conversion.

Typically an on-line high temperature reactor is mounted in parallel to a combustion reactor at the GC oven. The complete setup for on-line high temperature conversion is given in Fig. 2.170. A water removal step has no effect here on the dry analyte gas.

Mass Spectrometer for Isotope Ratio Analysis

Mass spectrometers employed for isotope ratio measurements are dedicated non-scanning, static magnetic sector mass spectrometer systems. The ion source is particularly optimized by a "closed source" design for the ionisation of gaseous compounds at very high ion production efficiency of 500–1000 molecules/ion at a high response linearity (Brand 2004). After extraction from the source region, the ions are typically accelerated by 2.5 to 10 kV to form an ion beam which enters the magnetic sector analyser through the entrance slit (see Fig. 2.172).

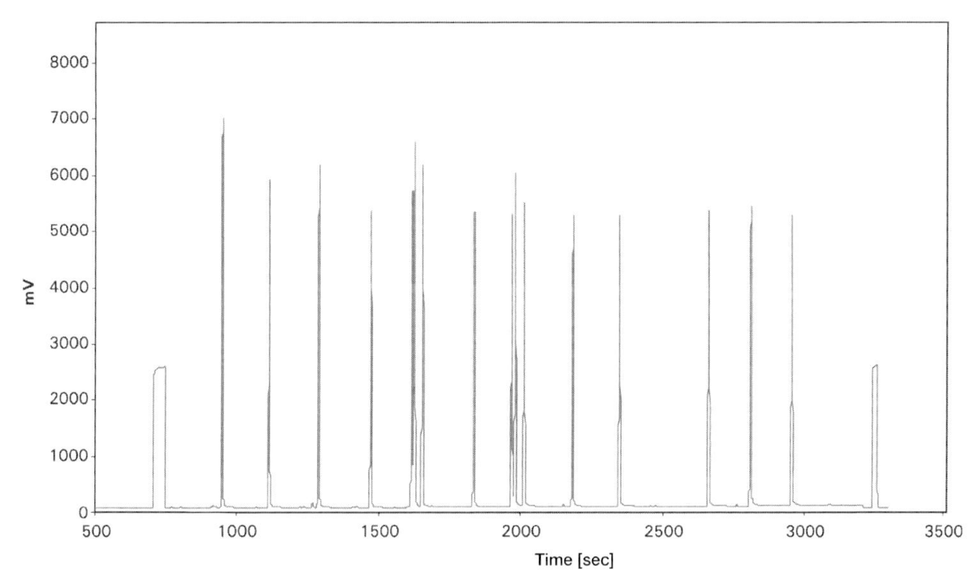

Fig. 2.168 irm-GC-C/MS chromatogram of a fatty acid methyl ester (FAME) sample after on-line combustion at 940 °C. The rectangular peaks in the beginning and at the end of the chromatogram are the CO_2 reference gas injection peaks (courtesy Thermo Fisher Scientific).

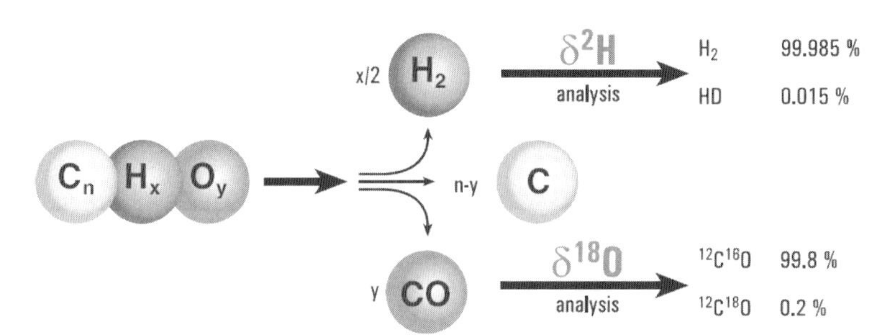

Fig. 2.169 Principle of high temperature conversion (courtesy Thermo Fisher Scientific).

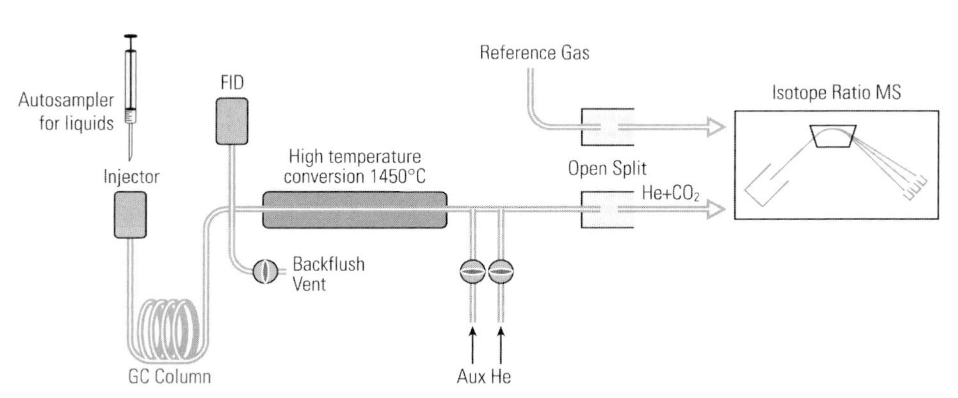

Fig. 2.170 On-line high temperature conversion of compounds eluting from the GC for the production of CO and H_2 for IRMS measurement (courtesy Thermo Fisher Scientific).

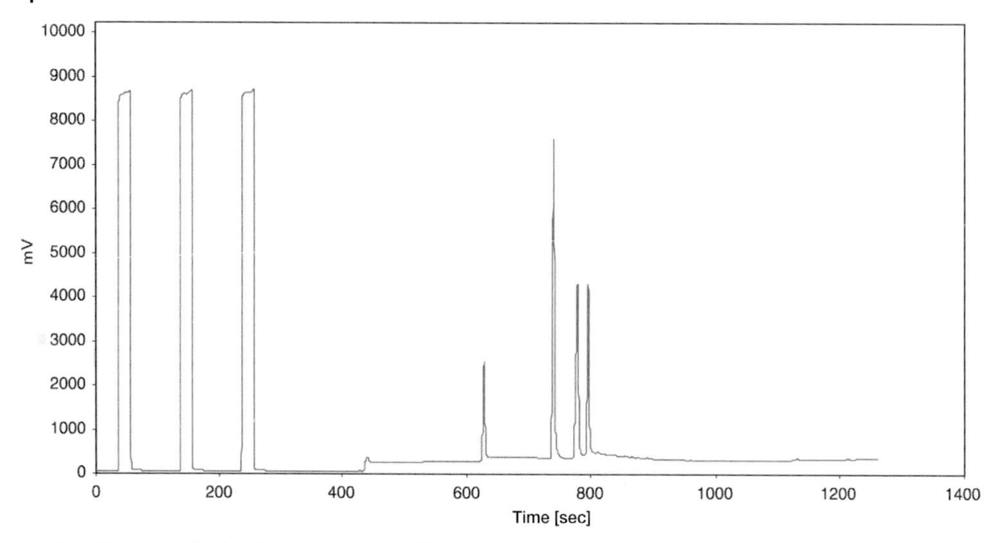

Fig. 2.171 irm-GC-TC/MS chromatogram of flavour components using high temperature conversion with three CO reference gas pulses at start (courtesy Thermo Fisher Scientific).

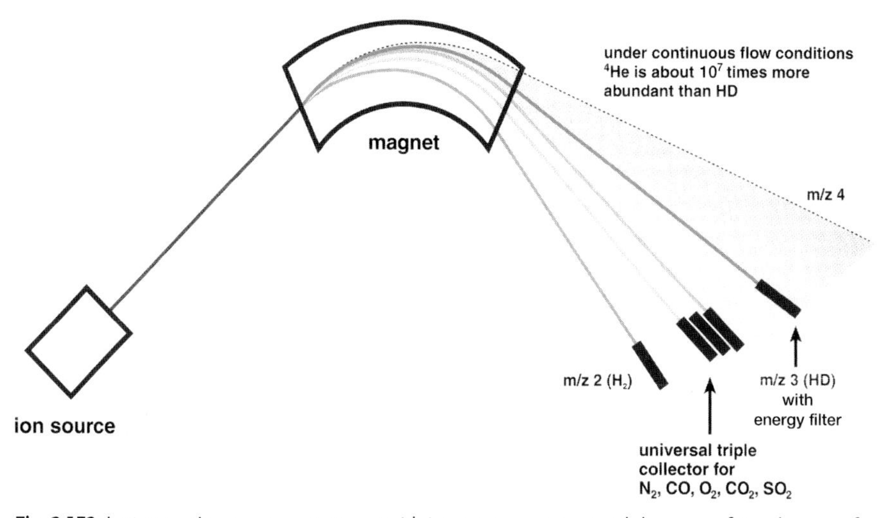

Fig. 2.172 Isotope ratio mass spectrometer with ion source, magnet and the array of Faraday cups for simultaneous isotope detection (courtesy Thermo Fisher Scientific).

The ion currents of the isotopes are measured simultaneously at the individual m/z values by discrete Faraday cups mounted behind a grounded slit (see Fig. 2.173). An array of specially designed deep Faraday cups for quantitative measurement is precisely positioned along the optical focus plane representing the typical isotope mass cluster to be determined.

Only the simultaneous measurement of the isotope ion currents with dedicated Faraday cups for each isotope using individual amplifier electronics, cancels out ion beam fluctuations due to temperature drifts or electron beam variations and provides the required preci-

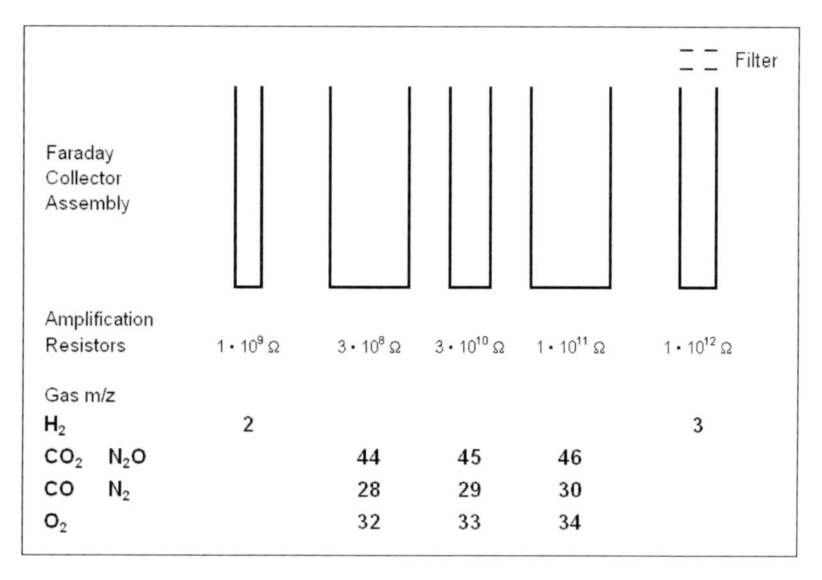

Fig. 2.173 Typical Faraday cup arrangement for measurement of the isotope ratios of the most common gas species (below of the cup symbols the values of the amplification resistors are given; the HD cup is shown to be equipped with a kinetic energy filter to prevent from excess $^4He^+$).

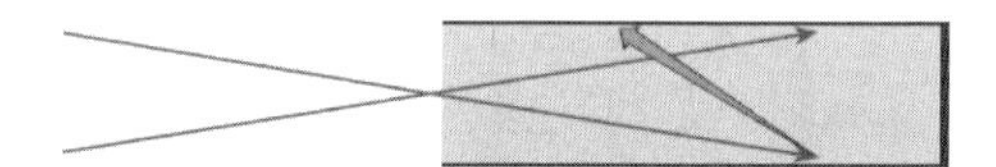

Fig. 2.174 Cross section of a Faraday cup for isotope ratio measurement, the large depth of the graphite cup prevents from losses caused by scattering. Arrows represent the ion beam with a focus point at the exit slit in front of the Faraday cup (courtesy Thermo Fisher Scientific).

sion (see Fig. 2.174, also refer to Table 2.39). Current instrumentation allows for irm-GC/MS analyses of organic compounds down to the low picomole range.

Isobaric interferences from other isotope species at the target masses require the measurement of more masses than just the targeted isotope ratio for a necessary correction. For the measurement of $\delta^{13}C$, three collectors at m/z 44, 45, and 46 are necessary (see Table 2.40). Algorithms for a fully automated correction of isobaric ion contributions are implemented in modern isotope ratio MS data systems e.g. for the isobaric interferences of ^{17}O and ^{13}C on m/z 46 of CO_2, see Table 2.41 (Craig 1957). Other possible interferences like CO on N_2, N_2O on CO_2 can be taken care of by the interface technology.

Table 2.41 Isobaric interferences when measuring $\delta^{13}C$ from CO_2.

m/z	Ion composition
44	$^{12}C^{16}O^{16}O$
45	$^{13}C^{16}O^{16}O$, $^{12}C^{16}O^{17}O$
46	$^{12}C^{16}O^{18}O$, $^{12}C^{17}O^{17}O$, $^{13}C^{16}O^{17}O$

The IT Principle

Standardisation in isotope ratio monitoring measurements should be done exclusively using the principle of "Identical Treatment of reference and sample material", the "IT Principle". Mostly, isotopic referencing is made with a co-injected peak of standard gas.

Injection of Reference Gases

The measurement of isotope ratios requires that sample gases be measured relative to a reference gas of a known isotope ratio. For the purpose of sample-standard referencing in irm-GC/MS, cylinders of calibrated reference gases, the laboratory standard of H_2, CO_2, N_2, or CO are used for an extended period of time. This referencing procedure turned out to be most economic and precise compared to the addition of an internal standard to the sample, also providing necessary quality assurance purposes. An inert, fused silica capillary supplies

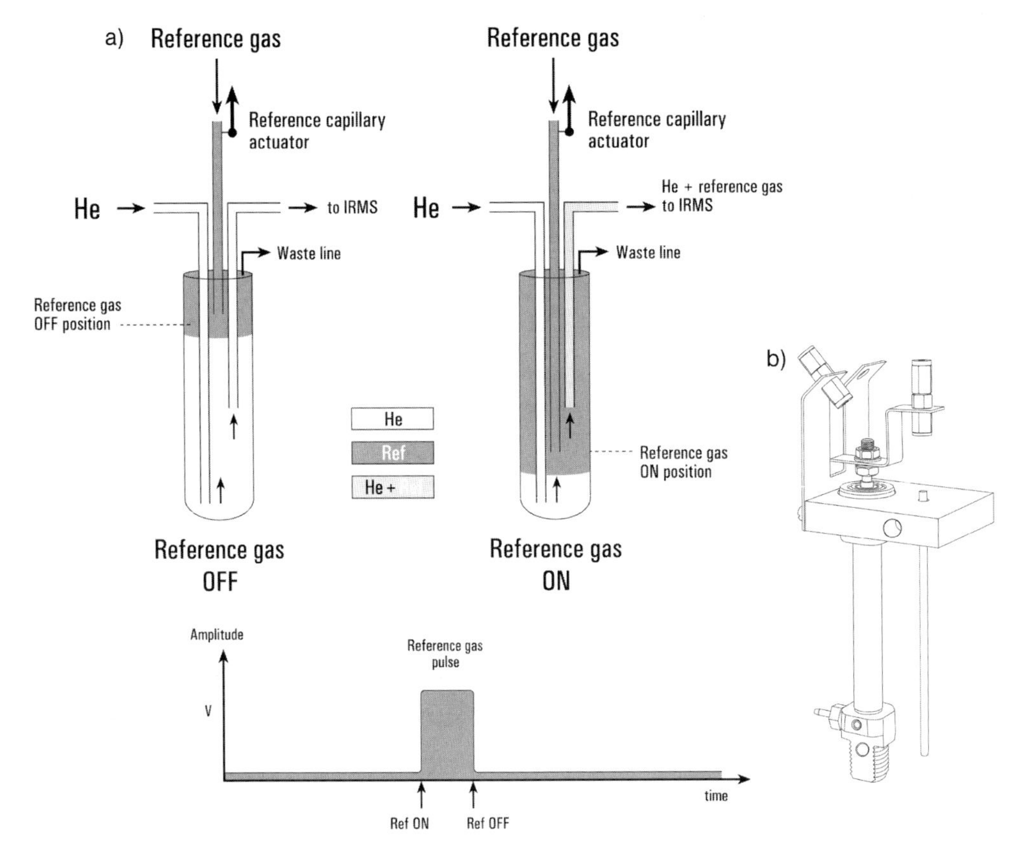

Fig. 2.175 (a) Reference gas injection port in irm-GC/MS implemented as an open coupling between GC column and IRMS ion source. (b) Reference gas injection port design: left side pneumatic drive, right side valve tube for fused silica columns, top column connectors for fixed (left) and movable (right) columns (courtesy Thermo Fisher Scientific).

the reference gas in the μL/min range into a miniaturized mixing chamber, the reference gas injection port (see Fig. 2.175 a and b). This capillary is lowered under software control into the mixing chamber for e. g. 20 s, creating a He reference gas mixture which is fed into the IRMS source via a second independent gas line. This generates a rectangular, flat top gas peak without changing any pressures or gas flows in the ion source. Reference gases used are pure nitrogen (N_2) carbon dioxide (CO_2), hydrogen (H_2), and carbon monoxide (CO).

Isotope Reference Materials

In IRMS the measurement of isotope ratios requires the samples to be measured relative to a reference of a known isotope ratio. This is the only means to achieve the required precision level of <1.5 ppm for e. g. the $^{13}C/^{12}C$ isotope ratio, which is 0.15 ‰ in the commonly used δ-notation. The employed isotopic reference scales (see Figs. 2.158 to 2.161) are arbitrarily defined by the community relative to the isotope ratio of a selected primary reference material for a given element.

Reference materials are available through the International Atomic Energy Agency (IAEA) in Vienna, Austria. The most commonly used and recognized international reference materials for stable isotope ratio analysis of natural abundances are given in Table 2.42. The abbreviations used with a preceding „V“ as in „“VSMOW“ refer to the reference materials prepared by the IAEA in Vienna compliant with the regular consultants meetings (for a detailed discussion see Groening 2004).

Table 2.42 IRMS primary reference materials for irm-GC/MS. Absolute isotope ratios R with 1σ standard uncertainties.

H	R = 0.00015576 ±0.000005	VSMOW	Standard Mean Ocean Water
C	R = 0.011224 ±0.000028	VPDB	Pee Dee Belemnita
N	R = 0.0036765	Air	Atmospheric Air
	R = 0.003663 ±0.000005	NSVEC	N standard by Jung and Svec
O	R = 0.0020052 ±0.000045	VSMOW	Standard Mean Ocean Water [a]

a) As a general rule $\delta^{18}O$ data of carbonates and CO_2 gas are reported against VPDB whereas $\delta^{18}O$ data of all other materials should be reported vs. VSMOW.

VSMOW

Oceans contain almost 97% of the water on Earth and have a uniform isotope distribution. Oceans are the major sink in the hydrological cycle. The SMOW standard was initially proposed by Craig 1961 as a concept for the origin of the scale and was calculated from an average of samples taken from different oceans. The reference water VSMOW was prepared by the IAEA in 1968 with the same $\delta^{18}O$ and a –0.2‰ lighter δD as defined by SMOW.

VPDB

Pee Dee Belemnita is $CaCO_3$ of the rostrum of a bellemnite (*Belemnita americana*) from the Pee Dee Formation of South Carolina, USA. The PDB reference material has been exhausted for a long time. The new VPDB reference material was anchored by the IAEA with a fixed δ value so that it corresponds nominally to the previous PDB scale.

Atmospheric Air
Nitrogen in atmospheric air has a very homogeneous isotope composition. The atmosphere is the main nitrogen sink and the largest terrestrial nitrogen reservoir. NSVEC is the reference material initially prepared by Jung and Svec, Iowa State University, USA.

The available IAEA reference materials are intended to calibrate local laboratory standards and not for continuous quality control purposes (Groening 2004). Every laboratory should prepare for routine work long lasting laboratory specific standards with similar characteristics to the used references, the working standards, which are calibrated against the primary reference materials.

For comparison to results of other laboratories, the raw data of the mass spectrometer are converted into VSMOW or VPDB (Boato 1960). Due to the relative measurement to a standard ratio, the formula contains a product term besides the sum of the δ-values:

$\delta_1 = \delta$ (sample/reference)	measured in the lab
$\delta_2 = \delta$ (reference/standard)	known lab calibration
$\delta_3 = \delta$ (sample/standard)	unknown

$$\delta_3 = \delta_1 + \delta_2 + 10^{-3} \cdot \delta_1 \cdot \delta_2$$

2.3.4 Ionisation Procedures

2.3.4.1 Electron Impact Ionisation

Electron impact ionisation (EI) is the standard process in all GC/MS instruments. An ionisation energy of 70 eV is currently used in all commercial instruments. Only a few benchtop instruments still allow the user to adjust the ionisation energy for specific purposes. In particular, for magnetic sector instruments the ionisation energy can be lowered to ca. 15 eV. This allows EI spectra with a high proportion of molecular information to be obtained (*low voltage ionisation*) (Fig. 2.176). The technique has decreased in importance due to its inherently low sensitivity since the introduction of chemical ionisation.

The process of electron impact ionisation can be explained by a wave or a particle model. The current theory is based on the interaction between the energy-rich electron beam with the outer electrons of a molecule. Energy absorption initially leads to the formation of a molecular ion M^+ by loss of an electron. The excess energy causes excitation in the rotational and vibrational energy levels of this radical cation. The subsequent processes of fragmentation depend on the amount of excess energy and the capacity of the molecule for internal stabilisation. The concept of localised charge according to Budzikiewicz empirically describes the fragmentations. The concept was developed from the observation that bonds near heteroatoms (N, O, S; Fig. 2.177) or π electron systems are cleaved preferentially in molecular ions. This is attributed to the fact that a positive or negative charge is stabilised by an electronegative structure element in the molecule or one favouring mesomerism. Bond breaking can be predicted by subsequent electron migrations or rearrangement. These types of process include α-cleavage, allyl cleavage, benzyl cleavage and the McLafferty rearrangement (see also Section 3.2.5).

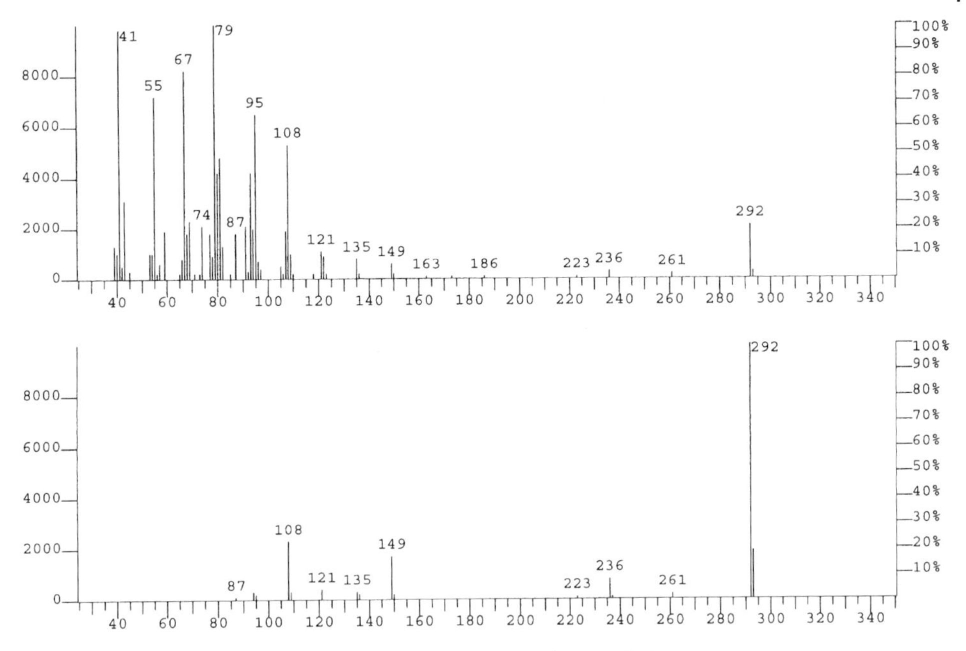

Fig. 2.176 EI spectra of methyl linolenate, $C_{17}H_{29}COOCH_3$ (after Spiteller). Recording conditions: direct inlet, above 70 eV, below 17 eV.

The energy necessary for ionisation of organic molecules is lower than the effective applied energy of 70 eV and is usually less than 15 eV (Table 2.43). The EI operation of all MS instruments at the high ionisation of 70 eV was established with regard to sensitivity (ion yield) and the comparability of the mass spectra obtained.

Table 2.43 First ionisation potentials [eV] of selected substances.

Helium	24.6	Pentane	10.34
Nitrogen	15.3	Nitrobenzene	10.18
Carbon dioxide	13.8	Benzene	9.56
Oxygen	12.5	Toluene	9.18
Propane	11.07	Chlorobenzene	9.07
1-Chloropropane	10.82	Propylamine	8.78
Butane	10.63	Aniline	8.32

Assuming a constant substance stream into an ion source, Fig. 2.178 shows the change in intensity of the signal with increasing ionisation energy. The steep rise of the signal intensity only begins when the ionisation potential (IP) is reached. Low measurable intensities just below it are produced as a result of the inhomogeneous composition of the electron beam. Generally the increase in signal intensity continues with increasing ionisation energy until a plateau is reached. A further increase in the ionisation energy is now indicated by a

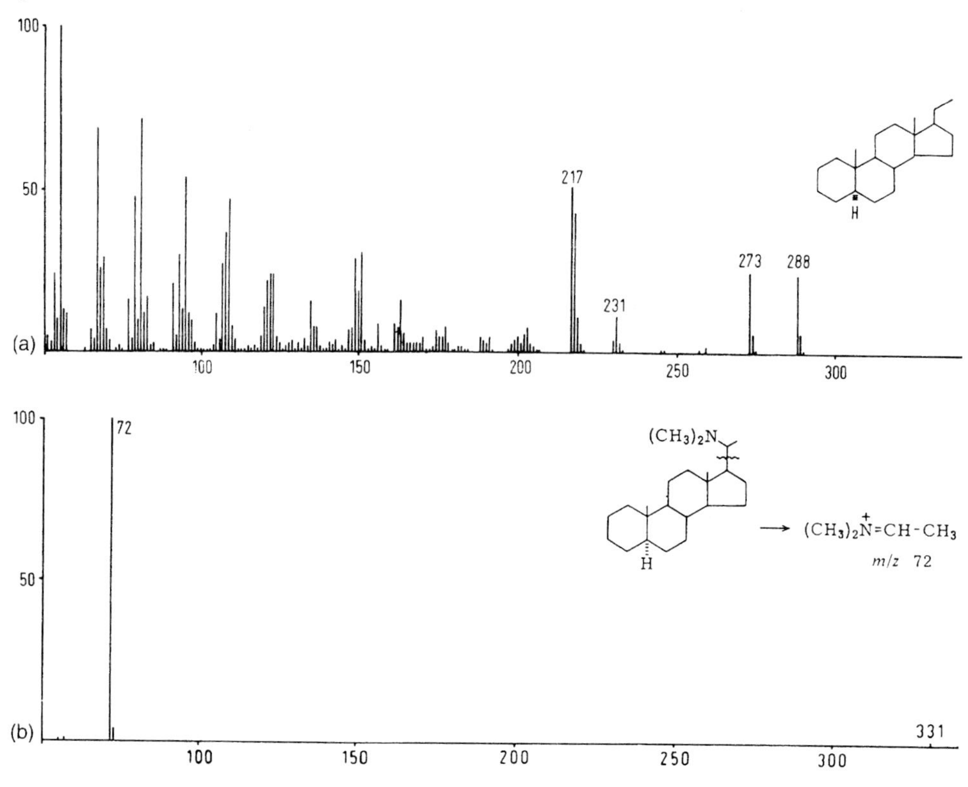

Fig. 2.177 Effect of structural features on the appearance of mass spectra – concept of localised charge (after Budzikiewicz).
(a) Mass spectrum of 5α-pregnane.
(b) Mass spectrum of 20-dimethylamino-5α-pregnane. The α-cleavage of the amino group dominates in the spectrum; information on the structure of the sterane unit is completely absent!

slight decrease in the signal intensity. An electron with an energy of 50 eV has a velocity of 4.2×10^6 m/s and crosses a molecular diameter of a few Ångstroms in ca. 10^{-16} s!

Further increases in the signal intensity are therefore not achieved via the ionisation energy with beam instruments, but by using measures to increase the density and the dispersion of the electron beam. The application of pairs of magnets to the ion source can be used, for example.

The standard ionisation energy of 70 eV which has been established for many years is also aimed at making mass spectra comparable. At an ionisation energy of 70 eV energy, which is in excess of that required for ionisation to M^+, remains as excess energy in the molecule, assuming maximum energy transfer. As a result fragmentation reactions occur and lead to an immediate decrease in the concentration of M^+ ions in the ion source. At the same time stable fragment ions are increasingly formed.

The fragmentation and rearrangement processes are now extensively known. They serve as fragmentation rules for the manual interpretation of mass spectra and thus for the identification of unknown substances. Each mass spectrum is the quantitative analysis by the analyser system of the processes occurring during ionisation. It is recorded as a line diagram.

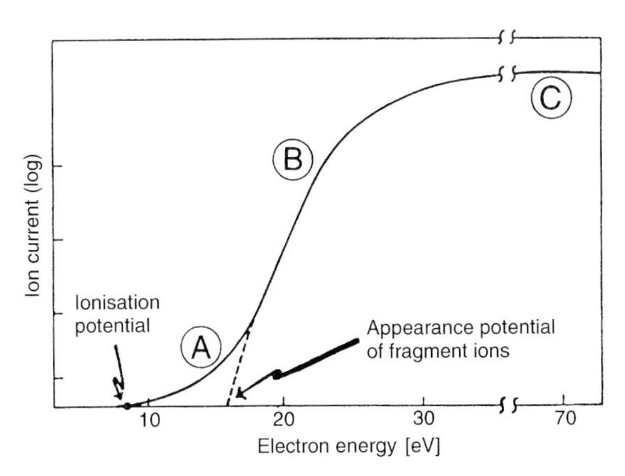

Fig. 2.178 Increase in the ion current with increasing electron energy (after Frigerio).
(A) Threshold region after reaching the appearance potential; molecular ions are mainly produced here.
(B) Build-up region with increasing production of fragment ions.
(C) Routine operation, stable formation of fragment ions.

What does a mass spectrum mean? In the graphical representation the mass to charge ratio m/z is plotted along the horizontal line. As ions with unit charge are generally involved in GC/MS with a few exceptions (e.g. polyaromatic hydrocarbons), this axis is generally taken as the mass scale and gives the molar mass of an ion (see Fig. 2.179). The intensity scale shows the frequency of occurrence of an ion under the chosen ionisation conditions. The scale is usually given both in percentages relative to the base peak (100% intensity) and in measured intensity values (counts). As neutral particles are lost in the fragmentation or rearrangement of a molecular ion M^+ and cannot be detected by the analyser, the mass of these neutral particles is deduced from the difference between the fragment ions and the molecular ion (or precursor fragments) (Fig. 2.180).

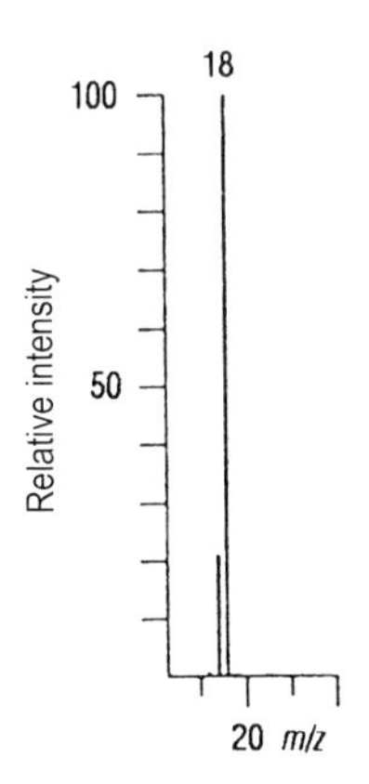

Introduction

Learning how to identify a simple molecule from its electron-ionization (EI) mass spectrum is much easier than from other types of spectra. The mass spectrum shows the mass of the molecule and the masses of pieces from it. Thus the chemist does not have to learn anything new – the approach is similar to an arithmetic brain-teaser. Try one and see.

In the bar-graph form of a spectrum the abscissa indicates the mass (actually m/z, the ratio of mass to the number of charges on the ions employed) and the ordinate indicates the relative intensity. If you need a hint, remember the atomic weigths of hydrogen and oxygen are 1 and 16 respectively."

Prof. McLafferty, Interpretation of Mass Spectra (1993)

Fig. 2.179 McLafferty's unknown spectrum.

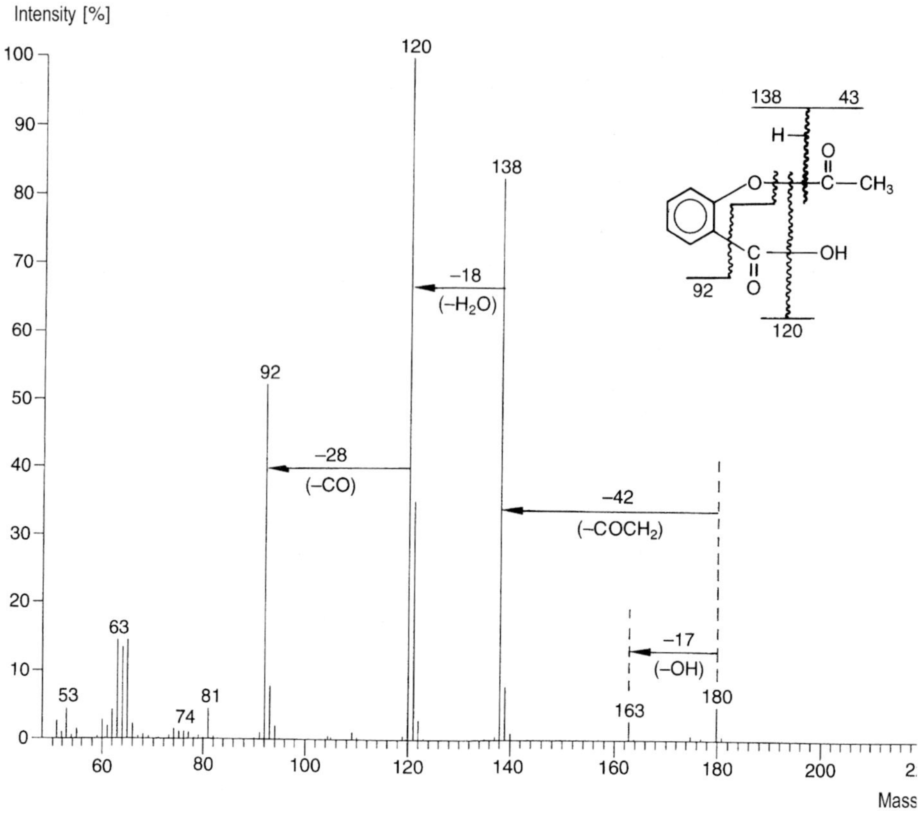

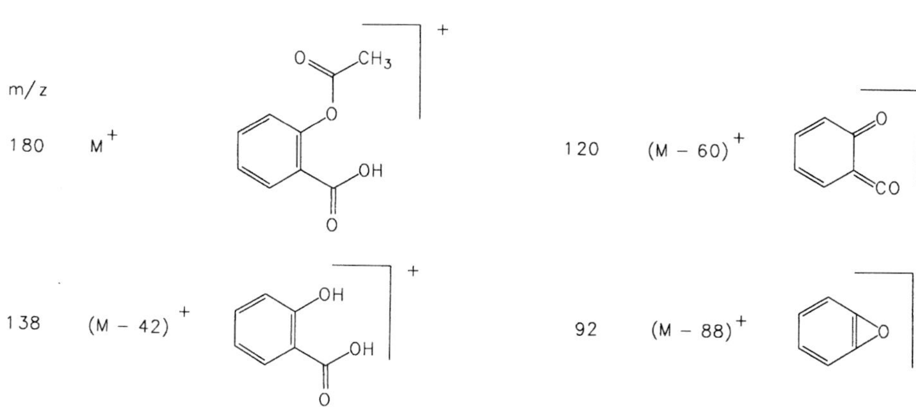

Fig. 2.180 A typical line spectrum in organic mass spectrometry with a molecular ion, fragment ions and the loss of neutral particles (acetylsalicylic acid, $C_9H_8O_4$, M 180).

In EI at 70 eV the extent of the fragmentation reactions observed for most organic compounds is independent of the construction of the ion source. For building up libraries of spectra the comparability of the mass spectra produced is thus ensured. All commercially available libraries of mass spectra are run under these standard conditions and allow the fragmentation pattern of an unknown substance to be compared with the spectra available in the library (see Section 3.2.4).

The Time Aspect in the Formation and Determination of Ions in Mass Spectrometry

- Flight time of an electron through an organic molecule (70 eV) — 10^{-16} s
- Formation of the molecular ion M^+ (EI) — 10^{-12} s
- Fragmentation reactions finished — 10^{-9} s
- Rearrangement reactions finished — 10^{-6}–10^{-7} s
- Lifetime of metastable ions — 10^{-3}–10^{-6} s

Flight times of ions:

- magnetic sector analyser — 10^{-5} s
- quadrupole analyser — 10^{-4} s
- ion trap analyser (storage times) — 10^{-2}–10^{-6} s

Types of Ions in Mass Spectrometry

- Molecular ion:
  The unfragmented positive or negatively charged ion with a mass equal to the molecular mass and of a radical nature because of the unpaired electron.
- Quasimolecular ion:
  Ions associated with the molecular mass which are formed through chemical ionisation e.g. as $(M + H)^+$, $(M - H)^+$ or $(M - H)^-$ and are not radicals.
- Adduct ions:
  Ions which are formed through addition of charged species e.g. $(M + NH_4)^+$ through chemical ionisation with ammonia as the CI gas.
- Fragment ions:
  Ions formed by cleavage of one or more bonds.
- Rearrangement products:
  Ions which are formed following bond cleavage and migration of an atom (see McLafferty rearrangement).
- Metastable ions:
  Ions (m_1) which lose neutral species (m_2) during the time of flight through the magnetic sector analyser and are detected with mass $m^* = (m_2)^2/m_1$.
- Base ion:
  This ion gives the highest signal (100%) (base peak) in a mass spectrum.

2.3.4.2 Chemical Ionisation

In electron impact ionisation (EI) molecular ions M^+ are first produced through bombardment of the molecule M by high energy electrons (70 eV). The high excess energy in M^+ (the ionisation potential of organic molecules is below 15 eV) leads to unimolecular fragmentation into fragment ions ($F1^+$, $F2^+$, ...) and uncharged species.

The EI mass spectrum shows the fragmentation pattern. The nature (m/z value) and frequency (intensity %) of the fragments can be read directly from the line spectrum. The loss of neutral particles is shown by the difference between the molecular ion and the fragments formed from it.

Which line in the EI spectrum is the molecular ion? Only a few molecules give dominant M^+ ions, e.g. aromatics and their derivatives, such as PCBs and dioxins. The molecular ion is frequently only present with a low intensity and with the small quantities of sample applied, as is the case with GC/MS, can only be identified with difficulty among the noise (matrix), or it fragments completely and cannot be seen in the spectrum.

Figure 2.181, which shows the EI/CI spectra of the phosphoric acid ester Tolclofos-methyl, is an example of this. The base peak in the EI spectrum shows a Cl atom. Loss of a methyl group $(M - 15)^+$ gives m/z 250. Is m/z 265 the nominal molecular mass? The CI spectrum shows m/z 301 for a protonated ion so the nominal molecular mass could be 300 u. The isotope pattern of two Cl atoms is also visible.

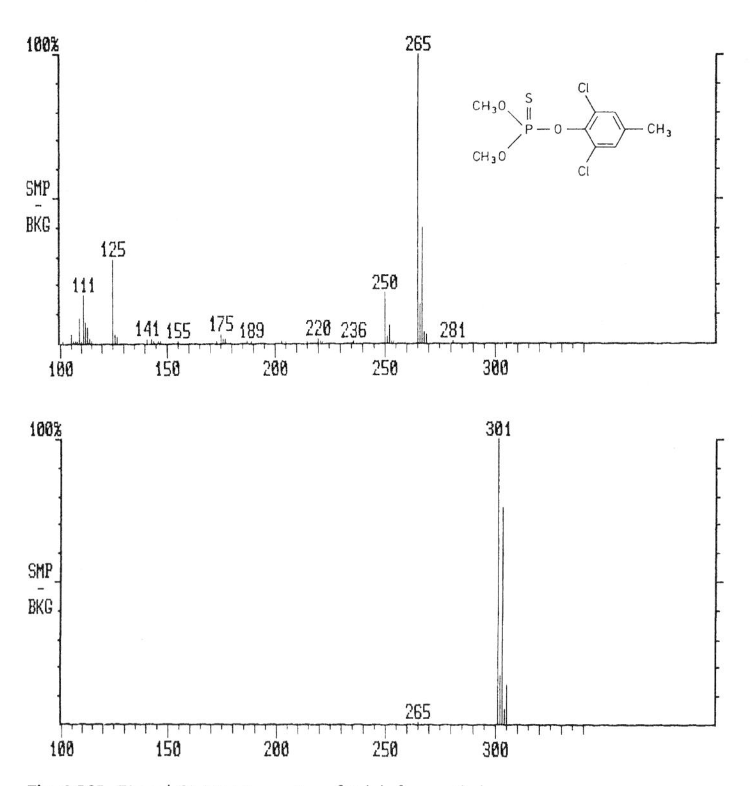

Fig. 2.181 EI and CI (NH_3) spectra of Tolclofos-methyl.

How can both EI and CI spectra be completed? Obviously Tolclofos fragments completely in EI by loss of a Cl atom to m/z 265 as $(M-35)^+$. With CI this fragmentation does not occur. The attachment of a proton retains the complete molecule with formation of the quasimolecular ion $(M+H)^+$.

The importance of EI spectra for identification and structure confirmation is due to the fragmentation pattern. All searches through libraries of spectra are based on EI spectra. With the introduction of the chemical ionisation capabilities for internal ionisation ion trap systems, a commercial CI library of spectra with more than 300 pesticides was first introduced by Finnigan.

The term chemical ionisation, unlike EI, covers all soft ionisation techniques which involve an exothermic chemical reaction in the gas phase mediated by a reagent gas and its reagent ions. Stable positive or negative ions are formed as products. Unlike the molecular ions of EI ionisation, the quasimolecular ions of CI are not radicals.

The principle of chemical ionisation was first described by Munson and Field in 1966. CI has now developed into a widely used technique for structural determination and quantitation in GC/MS. Instead of an open, easily evacuated ion source, a closed ion volume is necessary for carrying out chemical ionisation. In the high vacuum environment of the ion source, a reagent gas pressure of ca. 1 Torr must be maintained to achieve the desired CI reactions. Depending on the construction of the instrument, either changing the ion volume, an expensive change of the whole ion source or only a software switch is necessary (Fig. 2.182). Through the straightforward technical realisation in the case of combination ion sources and through the broadening of the use of ion trap mass spectrometers, CI has now become established in residue and environmental analysis, even for routine methods.

Chemical ionisation uses considerably less energy for ionising the molecule M. CI spectra therefore have fewer or no fragments and thus generally give important information on the molecule itself.

The use of chemical ionisation is helpful in structure determination, confirmation or determination of molecular weights, and also in the determination of significant substructures. Additional selectivity can be introduced into mass spectrometric detection by using the CI reaction of certain reagent gases, e.g. the detection of active substances with a transparent hydrocarbon matrix. Analyses can be quantified selectively, with high sensitivity and unaffected by the low molecular weight matrix, by the choice of a quantitation mass in the upper molecular weight range. The spectrum of analytical possibilities with CI is not limited to the basic reactions described briefly here. Furthermore, it opens up the whole field of chemical reactions in the gas phase.

The Principle of Chemical Ionisation

In CI two reaction steps are always necessary. In the *primary reaction* a stable cluster of reagent ions is produced from the reagent gas through electron bombardment. The composition of the reagent gas cluster is typical for the gas used. The cluster formed usually shows up on the screen for adjustment.

In the *secondary reaction* the molecule M in the GC eluate reacts with the ions in the reagent gas cluster. The ionic reaction products are detected and displayed as the CI spectrum. It is the secondary reaction which determines the appearance of the spectrum. Only exothermic reactions give CI spectra. In the case of protonation this means that the proton affinity PA of M must be higher than that of the reagent gas PA(R) (Fig. 2.183 and Table 2.44).

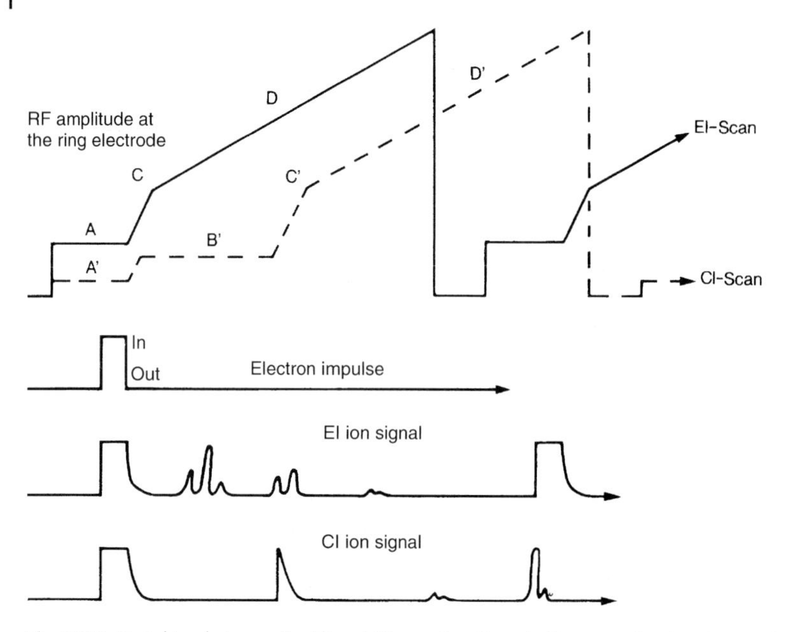

Fig. 2.182 Switching between the EI and CI scan functions in the case of an ion trap analyser with internal ionisation (Finnigan).
EI scan:
A Ionisation and storage of ions
C Starting mass
D Recording of an EI mass spectrum
CI scan:
A′ Ionisation and storage of reagent gas ions
B′ Reaction of reagent gas ions with neutral substance molecules
C′ Starting mass
D′ Recording of a CI mass spectrum

Through the choice of the reagent gas R, the quantity of energy transferred to the molecule M and thus the degree of possible fragmentation and the question of selectivity can be controlled. If PA(R) is higher than PA(M), no protonation occurs. When a nonspecific hydrocarbon matrix is present, this leads to transparency of the background, while active substances, such as plant protection agents, appear with high signal/noise ratios.

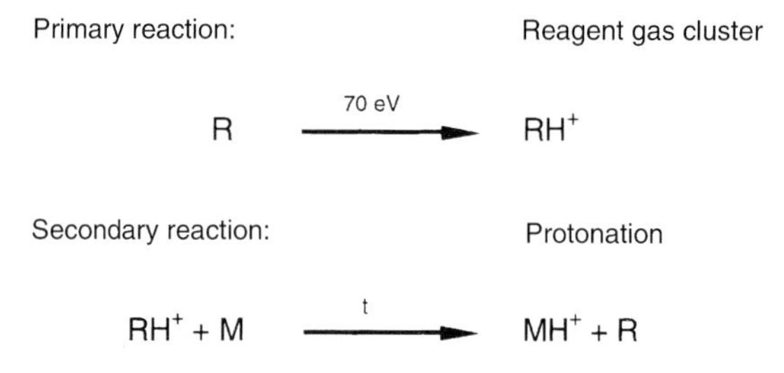

Fig. 2.183 Primary and secondary reactions in protonation.

Table 2.44 Proton affinities of some simple compounds.

Aliphatic amines			
NH_3	857	n-Pr_2NH	951
$MeNH_2$	895	i-Pr_2NH	957
$EtNH_2$	907	n-Bu_2NH	955
n-$PrNH_2$	913	i-Bu_2NH	956
i-$PrNH_2$	917	s-Bu_2NH	965
n-$BuNH_2$	915	Me_3N	938
i-$BuNH_2$	918	Et_3N	966
s-$BuNH_2$	922	n-Pr_3N	976
t-$BuNH_2$	925	n-Bu_3N	981
n-Amyl-NH_2	918	Me_2EtN	947
Neopentyl-NH_2	920	$MeEt_2N$	957
t-Amyl-NH_2	929	Et_2-n-PrN	970
n-Hexyl-NH_2	920	Pyrrolidine	938
Cyclohexyl-NH_2	925	Piperidine	942
Me_2NH	922	N-methylpyrrolidine	952
MeEtNH	930	N-methylpiperidine	956
Et_2NH	941	$Me_3Si(CH_2)_3NMe_2$	966
Oxides and sulfides			
H_2O	723	n-Bu_2O	852
MeOH	773	i-PrO-t-Bu	873
EtOH	795	n-$Pentyl_2O$	858
n-PrOH	800	Tetrahydrofuran	834
t-BuOH	815	Tetrahydropyran	839
Me_2O	807	H_2S	738
MeOEt	844	MeSH	788
Et_2O	838	Me_2S	839
i-PrOEt	850	MeSEt	851
n-Pr_2O	848	Et_2S	859
i-Pr_2O	861	i-Pr_2S	875
t-BuOMe	852	H_2Se	742
Disubstituted alkanes			
$NH_2CH_2CH_2NH_2$	947	$Me_2N(CH_2)_6NMe_2$	1041
$NH_2CH_2CH_2CH_2NH_2$	973	$NH_2CH_2CH_2OMe$	933
$NH_2CH_2CH_2CH_2CH_2NH_2$	995	$NH_2CH_2CH_2CH_2OH$	952
$NH_2(CH_2)_5NH_2$	986	$NH_2(CH_2)_6OH$	966
$NH_2(CH_2)_6NH_2$	989	$NH_2CH_2CH_2CH_2F$	928
$Me_2NCH_2CH_2NMe_2$	996	$NH_2CH_2CH_2CH_2Cl$	928
$Me_2NCH_2CH_2CH_2NH_2$	1002	$OHCH_2CH_2CH_2CH_2OH$	886
$Me_2NCH_2CH_2CH_2NMe_2$	1012	2,4-Pentanedione	886
$Me_2N(CH_2)_4NMe_2$	1028		

Table 2.44 (continued)

Substituted alkylamines and alcohols			
$CH_2FCH_2NH_2$	890	CF_3NMe_2	815
$CHF_2CH_2NH_2$	871	CHF_2CH_2OH	755
$CF_3CH_2NH_2$	850	CF_3CH_2OH	731
$(CF_3)_3CNH_2$	800	CCl_3CH_2OH	760
$CF_3CH_2NHCH_3$	880	Piperazine	936
$CF_3CH_2CH_2NH_2$	885	1,4-Dioxan	811
$CF_3CH_2CH_2CH_2NH_2$	900	Morpholine	915
$CH_2FCH_2CH_2NH_2$	914		
Unsaturated amines and anilines			
$CH_2{=}CHCH_2NH_2$	905	m-$CH_3OC_6H_4NH_2$	906
Cyclo-$C_3H_5NH_2$	899	p-$CH_3OC_6H_4NH_2$	899
$CH_2{=}C(CH_3)CH_2NH_2$	912	p-$ClC_6H_4NH_2$	876
$HC{\equiv}CCH_2NH_2$	884	m-$ClC_6H_4NH_2$	872
$(CH_2{=}CCH_3CH_2)_3N$	964	m-$FC_6H_4NH_2$	870
$(CH_2{=}CHCH_2)_3N$	958	C_6H_5NHMe	912
$HC{\equiv}CCH_2)_3N$	916	$C_6H_5NMe_2$	935
$C_6H_5NH_2$	884	$C_6H_5CH_2NH_2$	918
m-$CH_3C_6H_4NH_2$	896	$C_6H_5CH_2NMe_2$	953
p-$CH_3C_6H_4NH_2$	896		
Other N, O, P, S compounds			
Aziridine	902	$CH_2{=}CHCN$	802
Me_2NH	922	ClCN	759
N-Methylaziridine	926	BrCN	770
Me_3N	938	CCl_3CN	760
MeNHEt	930	CH_2ClCN	773
2-Methylaziridine	916	CH_2ClCH_2CN	795
Pyridine	921	$NCCH_2CN$	757
Piperidine	942	i-PrCN	819
MeCH=NEt	931	n-BuCN	818
n-PrCH=NEt	942	cyclo-C_3H_6CN	824
$Me_2C{=}NEt$	959	Ethylene oxide	793
HCN	748	Oxetane	823
CH_3NH_2	895	CH_2O	741
MeCN	798	MeCHO	790
EtCN	806	$Me_2C{=}O$	824
n-PrCN	810	Thiirane	818

Table 2.44 (continued)

Carbonyl compounds, iminoethers and hydrazines			
EtCHO	800	CF_3CO_2-n-Bu	782
n-PrCHO	809	$HCO_2CH_2CF_3$	767
n-BuCHO	808	$NCCO_2Et$	767
i-PrCHO	808	$CF_3CO_2CH_2CH_2F$	764
Cyclopentanone	835	$(MeO)_2C{=}O$	837
HCO_2Me	796	HCO_2H	764
HCO_2Et	812	$MeCO_2H$	797
HCO_2-n-Pr	816	$EtCO_2H$	808
HCO_2-n-Bu	818	FCH_2CO_2H	781
$MeCO_2Me$	828	$ClCH_2CO_2H$	779
$MeCO_2Et$	841	CF_3CO_2H	736
$MeCO_2$-n-Pr	844	n-PrNHCHO	878
CF_3CO_2Me	765	Me_2NCHO	888
CF_3CO_2Et	777	$MeNHNH_2$	895
CF_3C0_2-n-Pr	781		
Substituted pyridines			
Pyridine	921	4-CF_3-pyridine	890
4-Me-pyridine	935	4-CN-pyridine	880
4-Et-pyridine	939	4-CHO-pyridine	900
4-t-Bu-pyridine	945	4-$COCH_3$-pyridine	909
2,4-diMe-pyridine	948	4-Cl-pyridine	910
2,4-di-t-Bu-pyridine	967	4-MeO-pyridine	947
4-Vinylpyridine	933	4-NH_2 pyridine	961
Bases weaker than water			
HF	468	CO_2	530
H_2	422	CH_4	536
O_2	423	N_2O	567
Kr	424	CO	581
N_2	475	C_2H_6	551
Xe	477		

All data in [kJ/mol] calculated for proton affinities PA(M) at 25 °C corresponding to the reaction $M + H^+ \rightarrow MH^+$ (after Aue and Bowers 1979)

For chemical ionisation many types of reaction can be used analytically. In gas phase reactions, not only positive, but also negative ions can be formed. GC/MS systems currently commercially available usually only detect positive ions (positive chemical ionisation, PCI). To detect negative ions (negative chemical ionisation, NCI), special equipment to reverse the polarity of the analyser potential and a multiplier with a conversion dynode are required. Specially equipped instruments also allow the simultaneous detection of positive and negative ions produced by CI. The alternating reversal of polarity during scanning (pulsed positive ion negative ion chemical ionisation, PPINICI) produces two complementary data files from one analysis.

1. Positive Chemical Ionisation

Essentially four types of reaction contribute to the formation of positive ions. As in all CI reactions, reaction partners meet in the gas phase and form a transfer complex $M \cdot R^+$. In the following types of reaction the transfer complex is either retained or reacts further.

Protonation

Protonation is the most frequently used reaction in positive chemical ionisation. Protonation leads to the formation of the quasimolecular ion $(M + H)^+$, which can then undergo fragmentation:

$$M + RH^+ \rightarrow MH^+ + R$$

Normally methane, water, methanol, isobutane or ammonia are used as protonating reagent gases (Table 2.45). Methanol occupies a middle position with regard to fragmentation and selectivity. Methane is less selective and is designated a hard CI gas. Isobutane and ammonia are typical soft CI gases.

Table 2.45 Reagent gases for proton transfer.

Gas	Reagent ion	PA [kJ/mol]
H_2	H_3^+	422
CH_4	CH_5^+	527
H_2O	H_3O^+	706
CH_3OH	$CH_3OH_2^+$	761
i-C_4H_{10}	t-$C_4H_9^+$	807
NH_3	NH_4^+	840

The CI spectra formed through protonation show the quasimolecular ions $(M + H)^+$. Fragmentations start with this ion. For example, loss of water shows up as M-17 in the spectrum, formed through $(M + H)^+ - H_2O$!

The existence of the quasimolecular ion is often indicated by low signals from addition products of the reagent gas. In the case of methane, besides $(M + H)^+$, $(M + 29)^+$ and $(M + 41)^+$ appear (see methane), and for ammonia, besides $(M + H)^+$, $(M + 18)^+$ with varying intensity (see ammonia).

Hydride Abstraction

In this reaction a hydride ion (H^-) is transferred from the substance molecule to the reagent ion:

$$M + R^+ \rightarrow RH + (M-H)^+$$

This process is observed, for example, in the use of methane when the C_2H_5 ion (m/z 29) contained in the methane cluster abstracts hydride ions from alkyl chains.

With methane as the reagent gas, both protonation and hydride abstraction can occur, depending on the reaction partner M. The quasimolecular ion obtained is either $(M-H)^+$ or $(M + H)^+$. In charge exchange reactions M^+ is also formed.

Charge Exchange

The charge exchange reaction gives a radical molecular ion with an odd number of electrons as in electron impact ionisation. Accordingly, the quality of the fragmentation is comparable to that of an EI spectrum. The extent of fragmentation is determined by the ionisation potential IP of the reagent gas.

$$M + R^+ \rightarrow R + M^+$$

The ionisation potentials of most organic compounds are below 15 eV. Through the choice of reagent gas it can be controlled whether only the molecular ion appears in the spectrum or whether, and how extensively fragmentation occurs. In the extreme case spectra similar to those with EI are obtained. Common reagent gases for ionisation by charge exchange are benzene, nitrogen, carbon monoxide, nitric oxide or argon (Table 2.46).

Table 2.46 Reagent gases for charge exchange reactions.

Gas	Reagent ion	IP [eV]
C_6H_6	$C_6H_6^+$	9.3
Xe	Xe^+	12.1
CO_2	CO_2^+	13.8
CO	CO^+	14.0
N_2	N_2^+	15.3
Ar	Ar^+	15.8
He	He^+	24.6

In the use of methane, charge exchange reactions as well as protonations can be observed, in particular for molecules with low proton affinities.

Adduct Formation

If the transition complex described above does not dissociate, the adduct is visible in the spectrum:

$$M + R^+ \rightarrow (M + R)^+$$

This effect is seldom made use of in GC/MS analysis in contrast to being the prevailing reaction in ESI-LC/MS, but must be taken into account on evaluating CI spectra. The enhanced formation of adducts is always observed with intentional protonation reactions where differences in the proton affinity of the participating species are small. High reagent gas pressure in the ion source favours the effect by stabilising collisions.

Frequently an $(M + R)^+$ ion is not immediately recognised, but can give information which is as valuable as that from the quasimolecular ion formed by protonation. Cluster ions of this type nevertheless sometimes make interpretation of spectra more difficult, particularly when the transition complex does not lose immediately recognisable neutral species.

2. Negative Chemical Ionisation

Negative ions are also formed during ionisation in mass spectrometry even under EI conditions, but their yield is so extremely low that it is of no use analytically. The intentional production of negative ions can take place by addition of thermal electrons (analogous to an ECD), by charge exchange or by extraction of acidic hydrogen atoms.

Charge Transfer

The ionisation of the sample molecule is achieved by the transfer of an electron between the reagent ion and the molecule M

$$M + R^- \rightarrow M^- + R$$

The reaction can only take place if the electron affinity EA of the analyte M is greater than that of the electron donor R.

$$EA\ (M) > EA\ (R)$$

In practical analysis charge transfer to form negative ions is less important (unlike the formation of positive ions by charge exchange, see above).

Proton Abstraction

Proton transfer in negative chemical ionisation can be understood as proton abstraction from the sample molecule. In this way all substances with acidic hydrogens, e.g. alcohols, phenols or ketones, can undergo soft ionisation.

$$M + R^- \rightarrow (M-H)^- + RH$$

Proton abstraction only occurs if the proton affinity of the reagent gas ion is higher than that of the conjugate base of the analyte molecule. A strong base, e.g. OH^-, is used as the reagent for ionisation. Substances which are more basic than the reagent are not ionised.

Reagent gases and organic compounds were arranged in order of gas phase acidity by Bartmess and McIver in 1979. The order corresponds to the reaction enthalpy of the dissociation of their functional groups into a proton and the corresponding base. Table 2.47 can be used for controlling the selectivity via proton abstraction by choosing suitable reagent gases.

Extensive fragmentation reactions can be excluded by proton abstraction. The energy released in the exothermic reaction is essentially localised in the new compound RH. The anion formed does not contain any excess energy for extensive fragmentation.

Table 2.47 Scale of gas phase acidities (after Bartmess and McIver 1979).

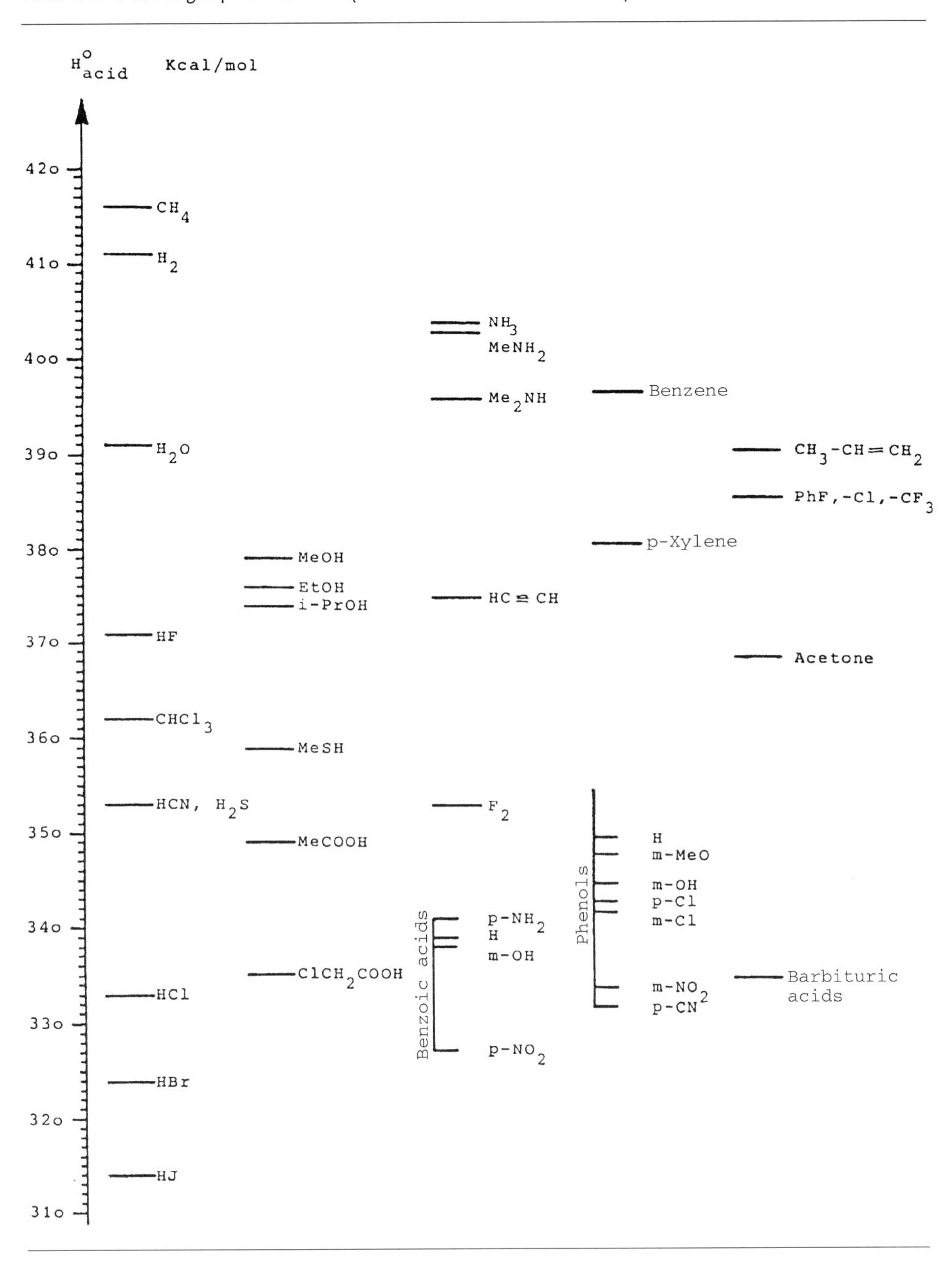

Reagent Ion Capture

The capture of negative reagent gas ions was described in the early 1970s by Manfred von Ardenne and coworkers for the analysis of long-chain aliphatic hydrocarbons with hydroxyl ions.

$$M + R^- \rightarrow M \cdot R^-$$

Besides associative addition, with weak bases adduct formation can lead to a new covalent bond. The ions formed are more stable than the comparable association products.

Substitution reactions, analogous to an S_N2 substitution in solution, occur more frequently in the gas phase because of poor solvation and low activation energy. Many aromatics give a peak at $(M + 15)^-$, which can be attributed to substitution of H by an O^- radical. Substitution reactions are also known for fluorides and chlorides. Fluoride is the stronger nucleophile and displaces chloride from alkyl halides.

Electron Capture

Electron capture with formation of negative ions is the NCI ionisation process most frequently used in GC/MS analysis. There is a direct analogy with the behaviour of substances in ECD and the areas of use may also be compared. The commonly used term ECD-MS indicates the parallel mechanisms and applications. With negative chemical ionisation the lowest detection limits in organic mass spectrometry have been reached (Fig. 2.184). The detection of 100 ag octafluoronaphthalene corresponds to the detection of ca. 200 000 molecules!

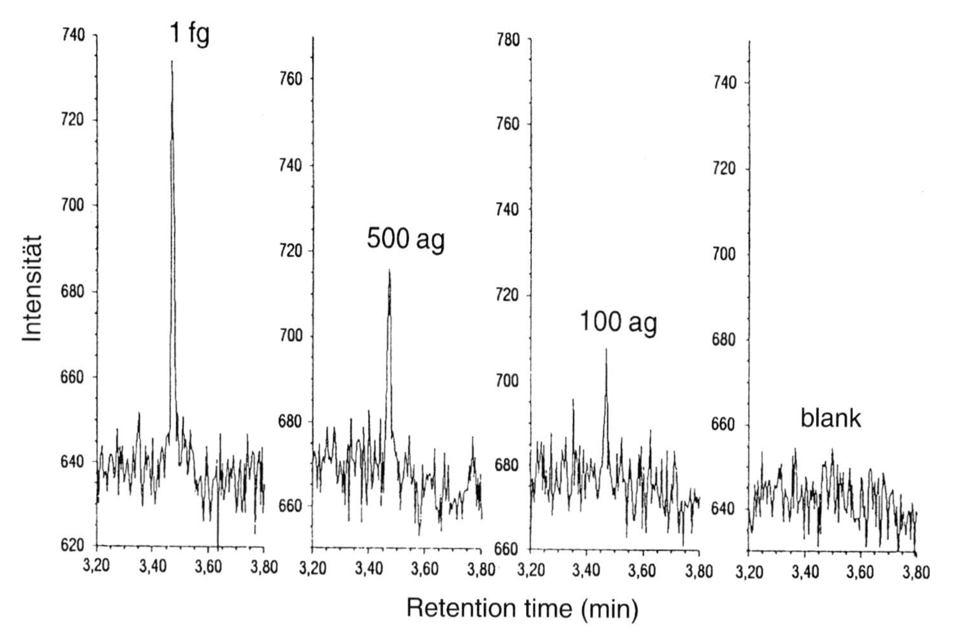

Fig. 2.184 GC/MS detection of traces of 10^{-15} to 10^{-16} g octafluoronaphthalene by NCI detection of the molecular ion m/z 272 (after McLafferty and Michnowicz 1992).

At the same energy electrons in the CI plasma have a much higher velocity (mobility) than those of the heavier positive reagent ions.

$$E = m/2 \cdot v^2 \qquad m(e^-) = 9.12 \cdot 10^{-28}\ g$$
$$m(CH_5^+) = 2.83 \cdot 10^{-22}\ g$$

Electron capture as an ionisation method is 100–1000 times more sensitive than ion/molecule reactions limited by diffusion. For substances with high electron affinities, higher sensitivities can be achieved than with positive chemical ionisation. NCI permits the detection of trace components in complex biological matrices (Fig. 2.185). Substances which have a high NCI response typically have a high proportion of halogen or nitro groups. In practice it has been shown that from ca. 5–6 halogen atoms in the molecule detection with NCI gives a higher specific response than that using EI (Fig. 2.186). For this reason, in analysis of polychlorinated dioxins and furans, although chlorinated, EI is the predominant method, in particular in the detection of 2,3,7,8-TCDD in trace analysis.

Another feature of NCI measurements is the fact that, like ECD, the response depends not only on the number of halogen atoms, but also on their position in the molecule. Precise quantitative determinations are therefore only possible with defined reference systems via the determination of specific response factors.

The key to sensitive detection of negative ions lies in the production of a sufficiently high population of thermal electrons. The extent of formation of M^- at sufficient electron density depends on the electron affinity of the sample molecule, the energy spectrum of the electron

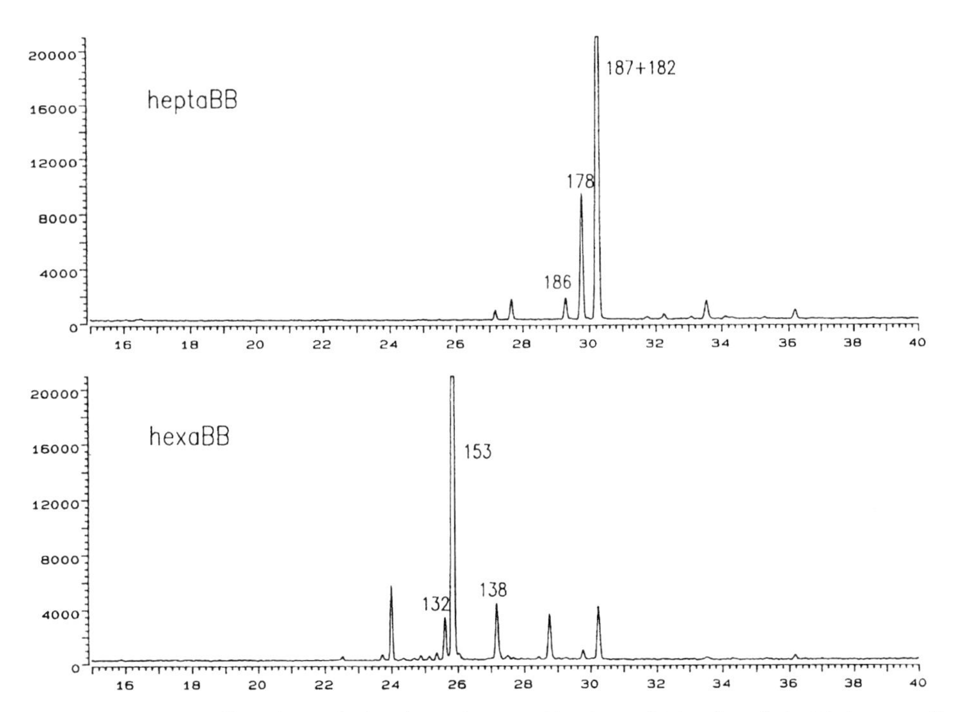

Fig. 2.185 Detection of heptabromobiphenylene (above) and hexabromobiphenylene (below) in human milk by NCI (after Fürst 1994).

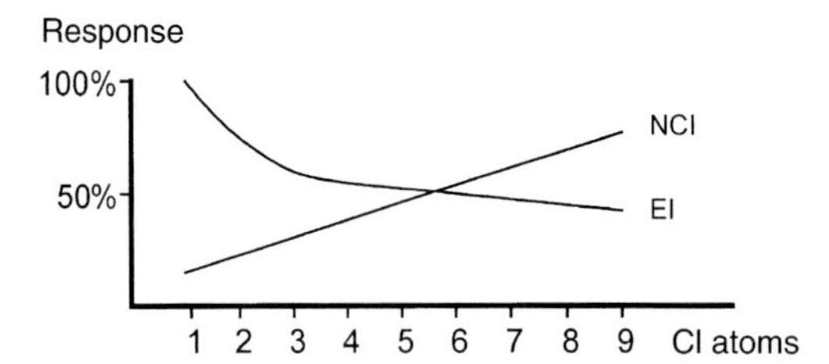

Fig. 2.186 Response dependence in the EI and NCI modes of the increasing number of Cl atoms in the molecule. The decrease in the response in the EI mode with increasing Cl content is caused by splitting of the overall signal by individual isotopic masses. In the NCI mode the response increases with increasing Cl content through the increase in the electronegativity for electrons capture (analogous to an ECD).

population and the frequency with which molecular anions collide with neutral particles and become stabilised (collision stabilisation). Even with an ion trap analyser, using external ionisation additions to give M^- can be utilised analytically. The storage of electrons in the ion trap itself is not possible because of their low mass (internal ionisation).

For residue analysis, derivatisation with perfluorinated reagents (e.g. heptafluorobutyric anhydride, perfluorobenzoyl chloride) in association with NCI is becoming important. Besides being easier to chromatograph, the compounds concerned have higher electron affinities allowing sensitive detection.

3. Reagent Gas Systems

Methane

Methane is one of the longest known and best studied reagent gases. As a hard reagent gas it has now been replaced by softer ones in many areas of analysis.

The reagent gas cluster of methane (Fig. 2.187) is formed by a multistep reaction, which gives two dominant reagent gas ions with m/z 17 and 29, and in lower intensity an ion with m/z 41.

CH_4 at 70 eV	→ CH_4^+, CH_3^+, CH_2, CH^+	and others		
CH_4^+ + CH_4	→ CH_5^+ + CH_3	m/z 17	50%	
CH_3^+ + CH_4	→ $C_2H_5^+$ + H_2	m/z 29	48%	
CH_2^+ + CH_4	→ $C_2H_3^+$ + H_2 + H			
$C_2H_3^+$ + CH_4	→ $C_3H_5^+$ + H_2	m/z 41	2%	

Good CI conditions are achieved if a ratio of m/z 17 to m/z 16 of 10:1 is set up. Experience shows that the correct methane pressure is that at which the ions m/z 17 and 29 dominate in the reagent gas cluster and have approximately the same height with good resolution. The ion m/z 41 should also be recognisable with lower intensity.

Methane is mainly used as the reagent gas in protonation reactions, charge exchange processes (PCI), and in pure form or as a mixture with N_2O in the formation of negative ions (NCI). In protonation methane is a hard reagent gas. For substances with lower proton affinity methane frequently provides the final possibility of obtaining CI spectra. The adduct ions $(M + C_2H_5)^+ = (M + 29)^+$ and $(M + C_3H_5)^+ = (M + 41)^+$ formed by the methane cluster help to confirm the molecular mass interpretation.

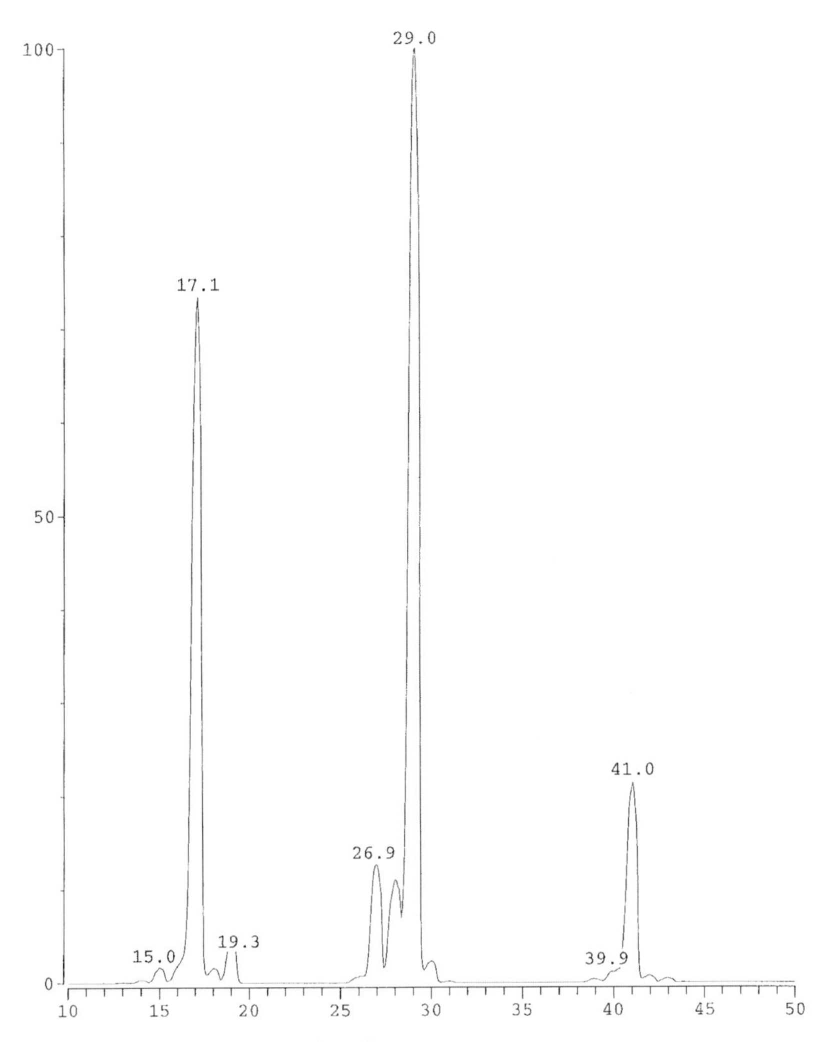

Fig. 2.187 Reagent gas cluster with methane.

Methanol

Because of its low vapour pressure, methanol is ideal for CI in ion trap instruments with internal ionisation. Neither pressure regulators nor cylinders or a long tubing system are required. The connection of a glass flask or a closed tube containing methanol directly on to the CI inlet is sufficient. In addition every laboratory has methanol available. Also for ion traps using external ionisation and quadrupole instruments liquid Cl devices are commercially available.

$(CH_3OH \cdot H)^+$ is formed as the reagent ion, which is adjusted to high intensity with good resolution (Fig. 2.188). The appearance of a peak at m/z 47 shows the dimer formed by loss of water (dimethyl ether), which is only produced at sufficiently high methanol concentrations. It does not function as a protonating reagent ion, but its appearance shows that the pressure adjustment is correct.

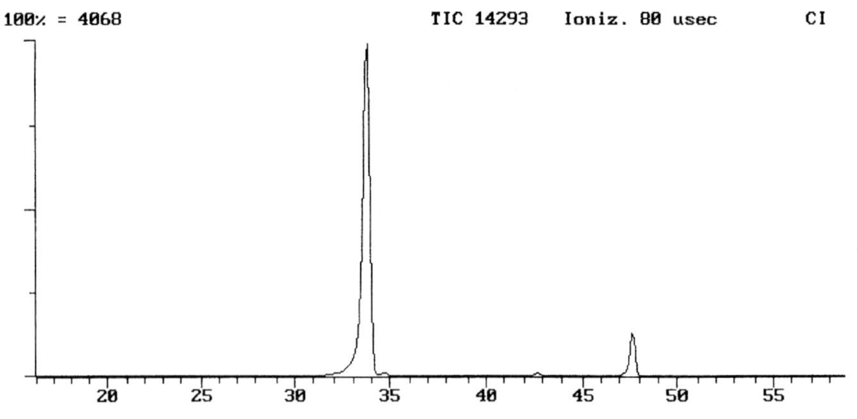

Fig. 2.188 Methanol as reagent gas (ion trap with internal ionisation).

Methanol is used exclusively for protonation. Because of its medium proton affinity, methanol allows a broad spectrum of classes of compounds to be determined. It is therefore suitable for a preliminary CI measurement of compounds not previously investigated. The medium proton affinity does not give any pronounced selectivity. However, substances with predominantly alkyl character remain transparent. Fragments have low intensities.

Water

For most mass spectroscopists water is a problematic substance. However, as a reagent gas, water has extraordinary properties. Because of the high conversion rate into H_3O^+ ions, and its low proton affinity, water achieves a high response for many compounds when used as the reagent gas. The spectra obtained usually have few fragments and concentrate the ion beam on a dominant ion.

When water is used as the reagent gas (Fig. 2.189), the intensity of the H_3O^+ ion should be as high as possible. With ion trap instruments with internal ionisation, no additional equipment is required. However a short tube length should be used for good adjustment.

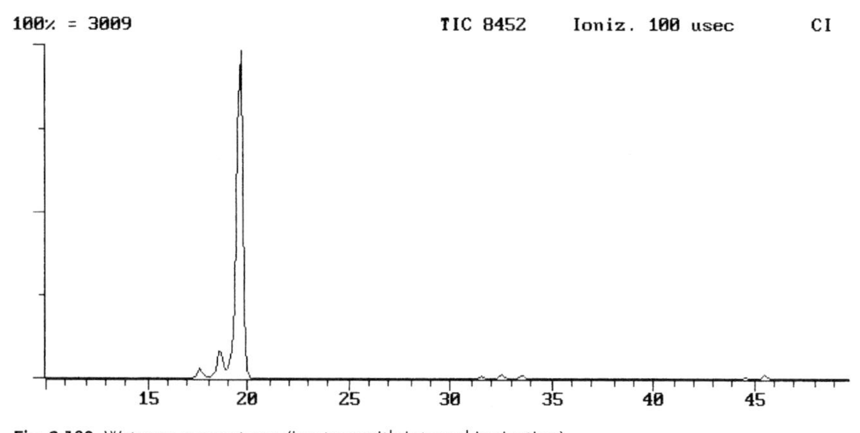

Fig. 2.189 Water as reagent gas (ion trap with internal ionisation).

For instruments with high pressure CI ion sources, the use of a heated reservoir and completely heated inlet tubing are imperative. Quadrupole instruments and ion traps using external ionisation require a special liquid CI device with heated CI gas reservoir and inlet lines.

The use of water is universal. In the determination of polyaromatic hydrocarbons considerable increases in response compared with EI detectors are found. Analytical procedures have even been published for nitroaromatics. Water can also be used successfully as a reagent gas for screening small molecules, e. g. volatile halogenated hydrocarbons (industrial solvents), as it does not interfere with the low scan range for these substances.

Isobutane

Like that of methane, chemical ionisation with isobutane has been known and well documented for years. The t-butyl cation (m/z 57) is formed in the reagent gas cluster and is responsible for the soft character of the reagent gas (Fig 2.190).

Isobutane is used for protonation reactions of multifunctional and polar compounds. Its selectivity is high and there is very little fragmentation. In practice, significant coating of the ion source through soot formation has been reported, which can even lead to dousing of the filament.

This effect depends on the adjustment and on the instrument. In such cases ammonia can be used instead.

Ammonia

To supply the CI system with ammonia a steel cylinder with a special reducing valve is necessary. Because of the aggressive properties of the gas, the entire tubing system must be made of stainless steel. In ion trap instruments only the ammonium ion NH_4^+ with mass m/z 18 is formed in the reagent gas cluster. At higher pressures adducts of the ammonium ion with ammonia $(NH_3)_n \cdot NH_4^+$ can be formed in the ion source.

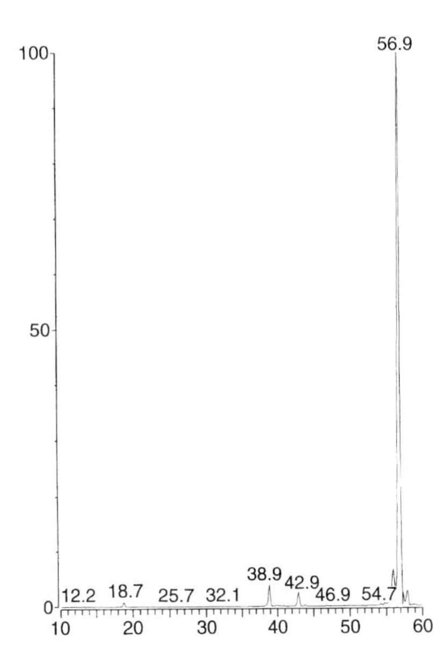

Fig. 2.190 iso-Butane as reagent gas.

Warning: Very often the ammonium ion is confused with water. Freshly installed reagent gas tubing generally has an intense water background with high intensities at m/z 18 and 19 as H_2O^+/H_3O^+. Clean tubing and correctly adjusted NH_3 CI gas shows no intensity at mass m/z 19 (Fig 2.191)!

Ammonia is a very soft reagent gas for protonation. The selectivity is correspondingly high, which is made use of in the residue analysis of many active substances. Fragmentation reactions only occur to a small extent with ammonia CI.

Adduct formation with NH_4^+ occurs with substances where the proton affinity differs little from that of NH_3 and can be used to confirm the molecular mass interpretation. In these cases the formation and addition of higher $(NH_3)_n \cdot NH_4^+$ clusters is observed with instruments with threshold pressures of ca. 1 Torr. Interpretation and quantitation can thus be impaired with such compounds.

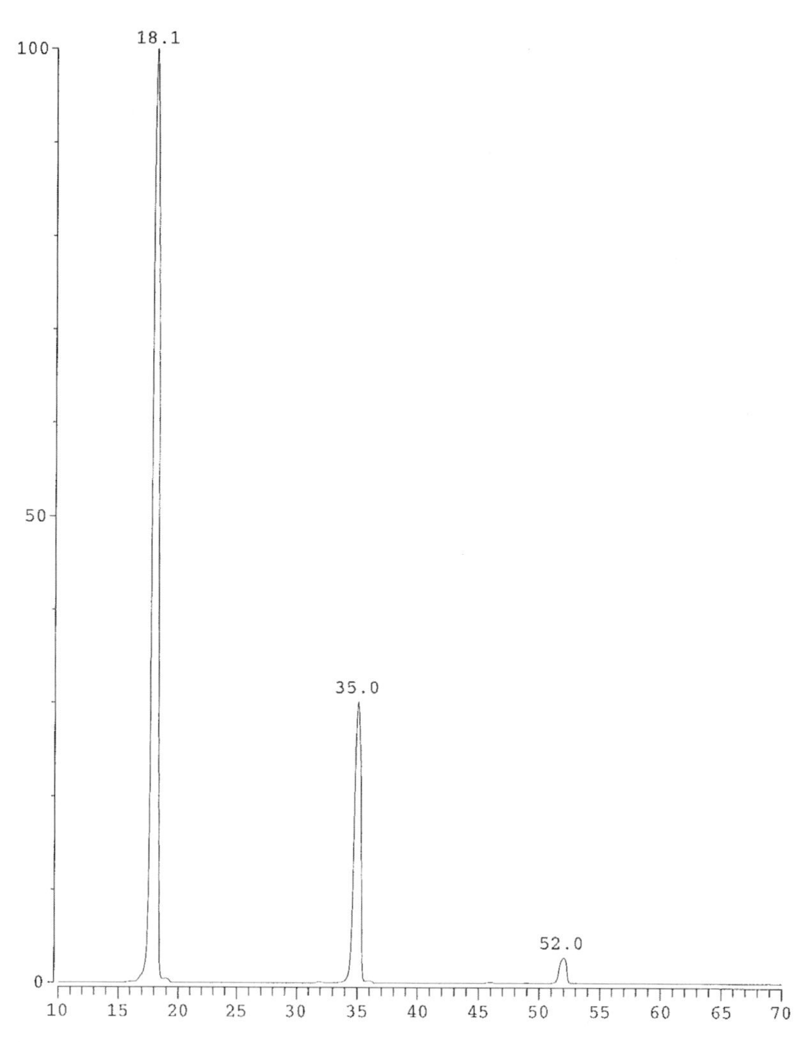

Fig. 2.191 Ammonia as reagent gas.

4. Aspects of Switching Between EI and CI

Quadrupole and Magnetic Sector Instruments

To initiate CI reactions and to guarantee a sufficient conversion rate, an ion source pressure of ca. 1 Torr in an analyser environment of 10^{-5}–10^{-7} Torr is necessary for beam instruments. For this, the EI ion source is replaced by a special CI source, which must have a gas-tight connection to the GC column, the electron beam and the ion exit in order to maintain the pressure in these areas.

Combination sources with mechanical devices for sealing the EI to the CI source have so far only proved successful with magnetic sector instruments. With the very small quadrupole sources, there is a significant danger of small leaks. As a consequence the response is below optimised sources and EI/CI mixed spectra may be produced.

Increased effort is required for conversion, pumping out and calibrating the CI source in beam instruments. Because of the high pressure, the reagent gas also leads to rapid contamination of the ion source and thus to additional cleaning measures in order to restore the original sensitivity of the EI system. Readily exchangeable ion volumes have been shown to be ideal for CI applications. This permits a high CI quality to be attained and, after a rapid exchange, unaffected EI conditions to be restored.

Ion Trap Instruments

Ion trap mass spectrometers with internal ionisation can be used immediately for CI without conversion. Because of their mode of operation as storage mass spectrometers, only an extremely low reagent gas pressure is necessary for instruments with internal ionisation. The pressure is adjusted by means of a special needle valve which is operated at low leakage rates and maintains a partial pressure of only about 10^{-5} Torr in the analyser. The overall pressure of the ion trap analyser of about 10^{-4}–10^{-3} Torr remains unaffected by it. CI conditions thus set up give rise to the term low pressure CI. Compared to the conventional ion source used in high pressure CI, in protonation reactions, for example, a clear dependence of the CI reaction on the proton affinities of the reaction partners is observed. Collision stabilisation of the products formed does not occur with low pressure CI. This explains why "high pressure" CI-typical adduct ions are not formed here, which would confirm the identification of the (quasi)molecular ion (e.g. with methane besides $(M + H)^+$, also M + 29 and M + 41 are expected). The determination of ECD-active substances by electron capture is not possible with low pressure CI.

Switching between EI and CI modes in an ion trap analyser with internal ionisation takes place with a keyboard command or through the scheduled data acquisition sequence in automatic operations. All mechanical devices necessary in beam instruments are dispensed with completely. The ion trap analyser is switched to a CI scan function internally without effecting mechanical changes to the analyser itself.

The CI reaction is initiated when the reagent ions are made ready by changing the operating parameters and a short reaction phase has taken place in the ion trap analyser. The scan function used in the CI mode with ion trap instruments (see Fig. 2.182) clearly shows two plateaus which directly correspond to the primary and secondary reactions. After the end of the secondary reaction the product ions, which have been produced and stored, are determined by the mass scan and the CI spectrum registered. In spite of the presence of the reagent gas, typical EI spectra can therefore be registered in the EI mode.

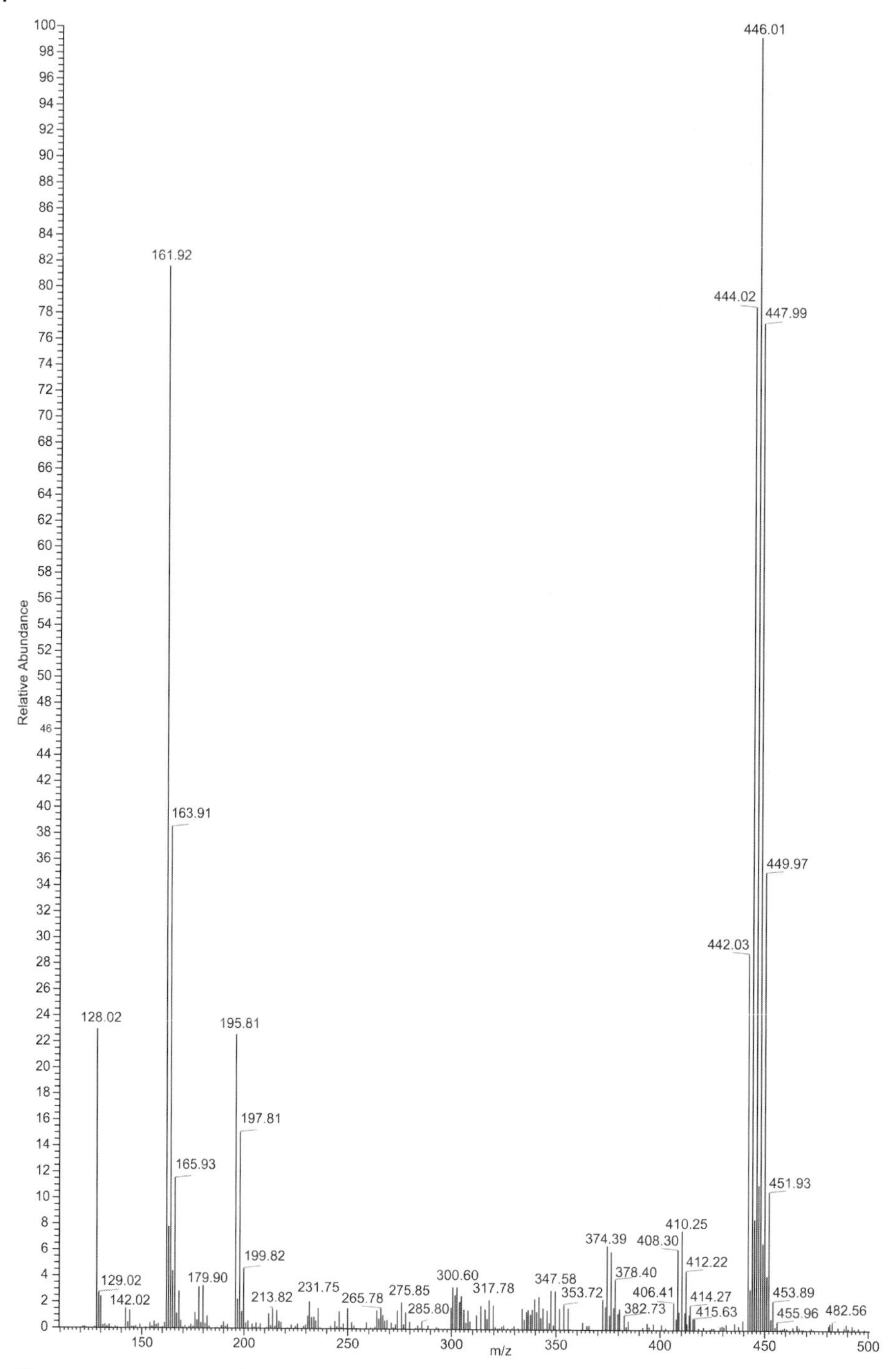

Fig. 2.192 NCI spectrum of a Toxaphen component (Parlar 69, polychlorinated bornane). The typical fragmentation known from EI ionisation is absent, the total ion current is concentrated on the molecular ion range (Theobald 2007).

The desired chemical ionisation is made possible by simply switching to the CI operating parameters.

On using autosamplers it is therefore possible to switch alternately between EI and CI data acquisition and thus use both ionisation processes routinely in automatic operations. The danger of additional contamination by CI gas does not occur with ion trap instruments because of the extremely small reagent gas input and allows this mode of operation to be run without impairing the quality.

Ion trap instruments with external ionisation have an ion source with a conventional construction. Changing between EI and CI ionisation takes place by changing the ion volume. Chemical ionisation in the classical manner of high pressure CI is thus carried out and thus the formation of negative ions by electron capture (NCI) in association with an ion trap analyser is made possible (Fig. 2.192).

2.3.5 Measuring Techniques in GC/MS

In data acquisition by the mass spectrometer there is a significant difference between detection of the complete spectrum (full scan) and the recording of individual masses (SIM, selected ion monitoring; MID multiple ion detection; SRM/MRM selected/multiple reaction monitoring). Particularly with continually operating spectrometers (ion beam instruments: magnetic sector MS, quadrupole MS) there are large differences between these two recording techniques with respect to selectivity, sensitivity and information content. For spectrometers with storage facilities (ion storage: ion trap MS, Orbitrap/ICR-MS) these differences are less strongly pronounced. Besides one-stage types of analyser (GC/MS), multistage mass spectrometers (GC/MS/MS) are playing an increasingly important role in residue target compound analysis and structure determination. With the MS/MS technique (multidimensional mass spectrometry), which is available in both beam instruments and ion storage mass spectrometers, much more analytical information and a high structure related selectivity for target compound quantitations can be obtained.

2.3.5.1 Detection of the Complete Spectrum (Full Scan)

The continuous recording of mass spectra (full scan) and the simultaneous determination of the retention time allow the identification of analytes by comparison with libraries of mass spectra. With beam instruments it should be noted that the sensitivity required for recording the spectrum depends on the efficiency of the ion source, the transmission through the analyser and, most particularly, on the dwell time of the ions. The dwell time per mass is given by the width of the mass scan (e.g. 50–550 u) and the scan rate of the chromatogram (e.g. 500 ms). From this a scan rate of 1000 u/s is calculated. Effective scan rates of modern quadrupole instruments exceed 11 000 u/s. Each mass from the selected mass range is measured only once during a scan over a short period (here 1 ms/u, Fig. 2.193). All other ions formed from the substance in parallel in the ion source are not detected during the mass scan (quadrupole as mass filter). Typical sensitivities for most compounds with benchtop quadrupole systems lie in the region of 1 ng of substance or less. Prolonging the scan time can increase the sensitivity of these systems for full scan operation (Fig. 2.193). However, there is an upper limit because of the rate of the chromatography. In practice for coupling

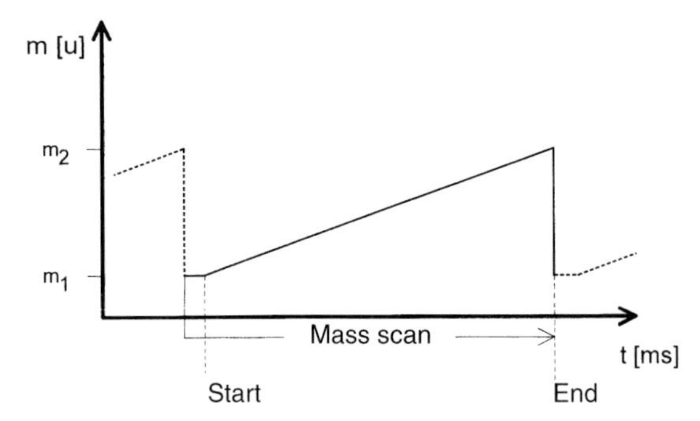

Fig. 2.193 Scan function of the quadrupole analyser: each mass between the start of the scan (m_1) and the end (m_2) is only registered once during the scan.

with capillary gas chromatography, scan rates of 0.5–1 s are used. For quantitative determinations it should be ensured that the scan rate of the chromatographic peak is adequate in order to determine the area and height correctly (see also Section 3.3). The SIM/MID mode is usually chosen to increase the sensitivity and scan rates of quadrupole systems for this reason (see also Section 2.3.5.2).

In ion storage mass spectrometers all the ions produced on ionisation of a substance are detected in parallel. The mode of function is opposite to the filter character of beam instruments and particularly strong when integrating weak ion beams (Table 2.48). All the ions formed are collected in a first step in the ion trap. At the end of the storage phase the ions, sorted according to mass during the scan, are directed to the multiplier. This process can take place very rapidly (Fig. 2.194). The scan rates of ion trap mass spectrometers are higher than 11 000 u/s. Typical sensitivities for full spectra in ion trap mass spectrometers are in the low pg range.

Table 2.48 Duty cycle for ion trap and beam instruments.

Scan range	Dwell time per mass		Sensitivity ratio
	Ion trap	Quadrupole	Ion trap/Quadrupole
[u]	[s/u]	[s/u]	
1	0.83	1	0.8
3	0.82	0.33	2.5
10	0.79	0.1	8
30	0.71	0.033	21
100	0.52	0.01	52
300	0.30	0.0033	91

The longer dwell time per mass leads to the highest sensitivity in the recording of complete mass spectra with ion trap instruments. Compared with beam instruments an increase in the duty cycle is achieved, depending on the mass range, of up to a factor of 100 and higher.

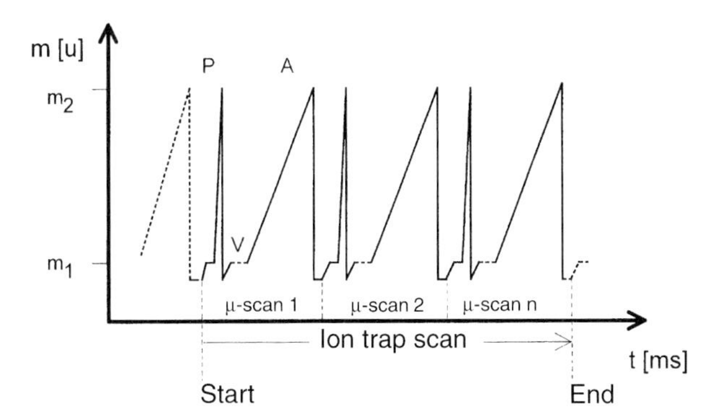

Fig. 2.194 Scan function of the ion trap analyser: within an ion trap scan, several μ-scans (three μ-scans shown here) are carried out and their spectra added before storage to disk.
P = pre-scan
A = analytical scan
V = variable ionisation time (AGC, automatic gain control)
(With ion trap instruments with external ionisation V stands for the length of the storage phase, ion injection time; for internal ionisation ion trap V stands for the ionisation time).

Dwell Times *t* Per Ion in Full Scan Acquisition With Ion Trap and Quadrupole MS

Mass range 50–550 u (500 masses wide), scan rate 500 ms

Ion trap MS	**Quadrupole MS**
Ion storage during ionisation. Ionisation time can vary, typically up to $t = 25$ ms at simultaneous storage of *all* ions formed.	$t = 500$ u/500 ms = 1 ms/u
Detection of *all* stored ions.	In a scan each type of ion is measured for only 1 ms; only a minute quantity of the ions formed are detected.

2.3.5.2 **Recording Individual Masses (SIM/MID)**

In the use of conventional mass spectrometers (beam instruments), the detection limit in the full scan mode is frequently insufficient for residue analysis because the analyser only has a very short dwell time per ion available during the scan. Additional sensitivity is achieved by dividing the same dwell time between a few selected ions by means of individual mass recording (SIM, MID) (Table 2.49 and Fig. 2.195).

At the same time a higher scan rate can be chosen so that chromatographic peaks can be plotted more precisely. The SIM technique is used exclusively for quantifying data on known target compounds, especially in trace analysis.

The mode of operation of a GC/MS system as a mass-selective detector requires the selection of certain ions (fragments, molecular ions), so that the desired analytes can be detected selectively. Other compounds contained in the sample besides those chosen for analysis re-

Table 2.49 Dwell times per ion and relative sensitivity in SIM analysis (for beam instruments) at constant scan rates.

Number of SIM ions	Total scan time[1)] [ms]	Total voltage setting time[2)] [ms]	Effective dwell time per ion[3)] [ms]	Relative sensitivity[4)] [%]
1	500	2	498	100
2	500	4	248	50
3	500	6	165	33
4	500	8	123	25
10	500	20	48	10
20	500	40	23	5
30	500	60	15	3
40	500	80	11	3
50	500	100	8	2
For comparison full scan:				
500	500	2	1	1

1) The total scan time is determined by the necessary scan rate of the chromatogram and is held constant.
2) Total voltage setting times are necessary in order to adjust the mass filter for the subsequent SIM masses. The actual times necessary can vary slightly depending on the type of instrument.
3) Duty cycle/ion.
4) The relative sensitivity is directly proportional to the effective dwell time per ion.

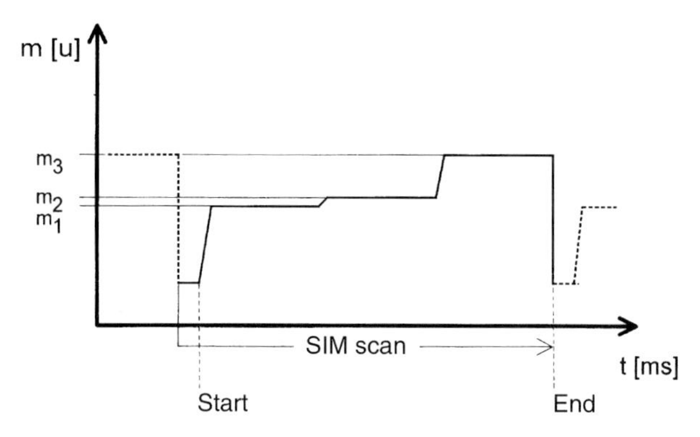

Fig. 2.195 SIM scan with a quadrupole analyser: the total scan time is divided here into the three individual masses m_1, m_2, and m_3 with correspondingly long dwell times.

main undetected. Thus the matrix present in large quantities in trace analysis is masked out, as are analytes whose appearance is not expected or planned. In the choice of masses required for detection it is assumed that for three selective signals in the fragmentation pattern per substance a secure basis for a yes/no decision can be found in spite of variations in the retention times (SIM, selected ion monitoring, MID, multiple ion detection). Identification of substances by comparison with spectral libraries is no longer possible. The relative intensities of the selected ions serve as quality criteria (qualifiers) (1 ion – no criterion, 2 ions – 1 criterion, 3 ions – 3 criteria!). This process for detecting compounds can be affected by errors through shifts in retention times caused by the matrix. In residue analysis it is known that with SIM analysis false positive findings occur in ca. 10% of the samples. Recently positive SIM data have been confirmed in the same way as positive results from classical GC detectors by running a complete mass spectrum of the analytes suspected. Confirmation of positive results, and statistically of negative results as well, is required by international directives either by full scan, ion ratios or HRMS.

Gain in Sensitivity Using SIM/MID

A typical SIM data acquisition of 5 selected masses at a scan rate of 0.5 s is given as a typical example:

	Ion trap MS	**Quadrupole MS**
Dwell time per ion:	Identical ionisation procedure to that with full scan, however selective and parallel or sequential storage of the selected SIM ions.	At a scan time of 500 ms the effective dwell time per SIM mass is divided up as $t = 500$ ms/5 masses = 100 ms/mass.
Function:	The ion trap is filled exclusively with the ions with the selected masses. If the capacity of the ion trap is not used up completely, the storage phase ends after a given time (ms).	To measure an SIM mass the quadrupole spends 100 times longer on one mass compared with full scan, and thus permits a dwell time which is 100 times longer for the selected ions to be achieved.
Sensitivity:	The gain in sensitivity is most marked with matrix-containing samples, as the length of the storage phase still mainly depends on the appearance of the selected SIM ions in the sample and is not shortened by a high concentration of matrix ions.	Theoretically the sensitivity increases by a factor of 100. In practice for real samples a factor of 30–50 compared with full scan is achieved.
Consequences for trace analysis:	Ion trap systems already give very high sensitivity in the full scan mode. Samples with high concentrations of matrix and detection limits below the pg level require the SIM technique (MS/MS is recommended).	Quadrupole systems require the SIM mode to achieve adequate sensitivity.

Confirmation:	For 3 SIM masses by 3 intensity criteria (qualifiers), with MS/MS by means of the product ion spectra.	For 3 SIM masses by 3 intensity criteria (qualifiers), check of positive results after further concentration by the full scan technique or external confirmation with an ion trap instrument (see Table 2.49).

SIM Set-up

1. Choice of column and program optimisation for optimal GC separation, paying particular attention to analytes with similar fragmentation patterns.
2. Full scan analysis of an average substance concentration to determine the selective ions (SIM masses, 2–3 ions/component); special matrix conditions are to be taken into account.
3. Determination of the retention times of the individual components.
4. Establishment of the data acquisition interval (time window) for the individual SIM descriptors.
5. Test analysis of a low standard (or better, a matrix spike) and possible optimisation (SIM masses, separation conditions).

Planning an analysis in the SIM/MID mode first requires a standard run in the full scan mode to determine both the retention times and the mass signals necessary for the SIM selection (Tables 2.50 and 2.51). As the gain in sensitivity achieved in individual mass recording with beam instruments is only possible on detection of a few ions, for the analysis of several compounds the group of masses detected must be adjusted. The more components there are to be detected, the more frequently and precisely must the descriptors be adjusted. Multicomponent analyses, such as the MAGIC-60 analysis with purge and trap (volatile halogenated hydrocarbons see Section 4.5), cannot be dealt easily with in the SIM mode.

The use of SIM analysis with ion trap mass spectrometers has also been developed. Through special control of the analyser (waveform ion isolation) during the ionisation phase only the preselected ions of analytical interest are stored (SIS, selective ion storage). This technique allows the detection of selected ions in ion storage mass spectrometers in spite of the presence of complex matrices or the co-elution of another component in high concentration. As the storage capacity of the ion trap analyser is only used for a few ions instead of for a full spectrum, extremely low detection limits are possible (< 1 pg/component) and the usable dynamic range of the analyser is extended considerably. Unlike conventional SIM operations with beam instruments, the detection sensitivity only alters slightly with the number of selected ions using the ion trap SIM technique (Fig. 2.196). For the SIM technique the sensitivity depends almost exclusively on the ionisation time. The SIM technique with ion trap instruments is regarded as a necessary prerequisite for carrying out MS/MS detection.

Table 2.50 Characteristic ions (m/z values) for selected polycondensed aromatics and their alkyl derivatives (in the elution sequence for methylsilicone phases).

Benzene-d_6	92, 94	Pyrene	202
Benzene	77, 78	Methylfluoranthene	215, 216
Toluene-d_8	98, 100	Benzofluorene	215, 216
Toluene	91, 92	Phenylanthracene	252, 253, 254
Ethylbenzene	91, 106	Benzanthracene	228
Dimethylbenzene	91, 106	Chrysene-d_{12}	240
Methylethylbenzene	105, 120	Chrysene	228
Trimethylbenzene	105, 120	Methylchrysene	242
Diethylbenzene	105, 119, 134	Dimethylbenz[a]anthracene	239, 241, 256
Naphthalene-d_8	136	Benzo[b]fluoranthene	252
Naphthalene	128	Benzo[j]fluoranthene	252
Methylnaphthalene	141, 142	Benzo[k]fluoranthene	252
Azulene	128	Benzo[e]pyrene	252
Acenaphthene	154	Benzo[a]pyrene	252
Biphenyl	154	Perylene-d_{12}	264
Dimethylnaphthalene	141, 155, 156	Perylene	252
Acenaphthene-d_{10}	162, 164	Methylcholanthrene	268
Acenaphthene	152	Diphenylanthracene	330
Dibenzofuran	139, 168	Indeno[1,2,3-cd]pyrene	276
Dibenzodioxin	184	Dibenzanthracene	278
Fluorene	165, 166	Benzo[b]chrysene	278
Dihydroanthracene	178, 179, 180	Benzo[g,h,i]perylene	276
Phenanthrene-d_{10}	188	Anthanthrene	276
Phenanthrene	178	Dibenzo[a,l]pyrene	302
Anthracene	178	Coronene	300
Methylphenanthrene	191, 192	Dibenzo[a,i]pyrene	302
Methylanthracene	191, 192	Dibenzo[a,h]pyrene	302
Phenylnaphthalene	204	Rubicene	326
Dimethylphenanthrene	191, 206	Hexaphene	328
Fluoranthene	202	Benzo[a]coronene	350

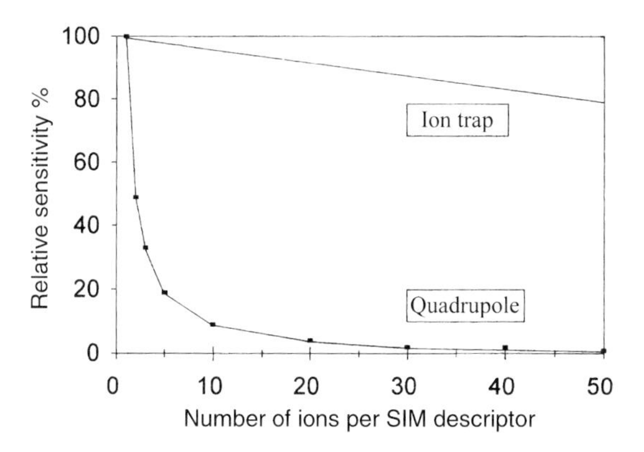

Fig. 2.196 Comparison of the SIS (ion trap analyser) and the SIM (quadrupole analyser) techniques based on the effective dwell time per ion (relative sensitivity).

Table 2.51 Main fragments and relative intensities for pesticides and some of their derivatives (DFG 1992).

Compound	Molar mass	1		2		3		4		5		6	
Acephate	183	43	(100)	44	(88)	136	(80)	94	(58)	47	(56)	95	(32)
Alaclor	269	45	(100)	188	(23)	160	(18)	77	(7)	146	(6)	224	(6)
Aldicarb	190	41	(100)	86	(89)	58	(85)	85	(61)	87	(50)	44	(50)
Aldrin	362	66	(100)	91	(50)	79	(47)	263	(42)	65	(35)	101	(34)
Allethrin	302	123	(100)	79	(40)	43	(32)	81	(31)	91	(29)	136	(27)
Atrazine	215	43	(100)	58	(84)	44	(75)	200	(69)	68	(43)	215	(40)
Azinphos-methyl	317	77	(100)	160	(77)	132	(67)	44	(30)	105	(29)	104	(27)
Barban	257	51	(100)	153	(76)	87	(66)	222	(44)	52	(43)	63	(43)
Benzazolin methyl ester	257	170	(100)	134	(75)	198	(74)	257	(73)	172	(40)	200	(31)
Bendiocarb	223	151	(100)	126	(58)	166	(48)	51	(19)	58	(18)	43	(17)
Bromacil	260	205	(100)	207	(75)	42	(25)	70	(16)	206	(16)	162	(12)
Bromacil N-methyl derivative	274	219	(100)	221	(68)	41	(45)	188	(41)	190	(40)	56	(37)
Bromophos	364	331	(100)	125	(91)	329	(80)	79	(57)	109	(53)	93	(45)
Bromophos-ethyl	392	97	(100)	65	(35)	303	(32)	125	(28)	359	(27)	109	(27)
Bromoxynil methyl ether	289	289	(100)	88	(77)	276	(67)	289	(55)	293	(53)	248	(50)
Captafol	347	79	(100)	80	(42)	77	(28)	78	(19)	151	(17)	51	(13)
Captan	299	79	(100)	80	(61)	77	(56)	44	(44)	78	(37)	149	(34)
Carbaryl	201	144	(100)	115	(82)	116	(48)	57	(31)	58	(20)	63	(20)
Carbendazim	191	159	(100)	191	(57)	103	(38)	104	(37)	52	(32)	51	(29)
Carbetamid	236	119	(100)	72	(54)	91	(44)	45	(38)	64	(37)	74	(29)
Carbofuran	221	164	(100)	149	(70)	41	(27)	58	(25)	131	(25)	122	(25)
Chlorbromuron	292	61	(100)	46	(24)	62	(11)	63	(10)	60	(9)	124	(8)
Chlorbufam	223	53	(100)	127	(20)	51	(13)	164	(13)	223	(13)	70	(10)
cis-Chlordane	406	373	(100)	375	(84)	377	(46)	371	(39)	44	(36)	109	(36)
trans-Chlordane	406	373	(100)	375	(93)	377	(53)	371	(47)	272	(36)	237	(30)
Chlorfenprop-methyl	232	125	(100)	165	(64)	75	(46)	196	(43)	51	(43)	101	(37)
Chlorfenvinphos	358	81	(100)	267	(73)	109	(55)	269	(47)	323	(26)	91	(23)
Chloridazon	221	77	(100)	221	(60)	88	(37)	220	(35)	51	(26)	105	(24)
Chloroneb	206	191	(100)	193	(61)	206	(60)	53	(57)	208	(39)	141	(35)
Chlorotoluron	212	72	(100)	44	(29)	167	(28)	132	(25)	45	(20)	77	(11)
3-Chloro-4-methylaniline (GC degradation product of Chlorotoluron)	141	141	(100)	140	(37)	106	(68)	142	(36)	143	(28)	77	(25)
Chloroxuron	290	72	(100)	245	(37)	44	(31)	75	(21)	45	(19)	63	(16)
Chloropropham	213	43	(100)	127	(49)	41	(35)	45	(20)	44	(18)	129	(16)
Chlorpyrifos	349	97	(100)	195	(59)	199	(53)	65	(27)	47	(23)	314	(21)
Chlorthal-dimethyl	330	301	(100)	299	(81)	303	(47)	332	(29)	142	(26)	221	(24)
Chlorthiamid	205	170	(100)	60	(61)	171	(50)	172	(49)	205	(35)	173	(29)
Cinerin I	316	123	(100)	43	(35)	93	(33)	121	(27)	81	(27)	150	(27)
Cinerin II	360	107	(100)	93	(57)	121	(53)	91	(50)	149	(35)	105	(33)
Cyanazine	240	44	(100)	43	(60)	68	(60)	212	(48)	41	(47)	42	(34)
Cypermethrin	415	163	(100)	181	(79)	165	(68)	91	(41)	77	(33)	51	(29)
2,4-DB methyl ester	262	101	(100)	59	(95)	41	(39)	162	(36)	69	(28)	63	(25)

Table 2.51 (continued)

Compound	Molar mass	Main fragment m/z (intensities) 1	2	3	4	5	6
Dalapon	142	43 (100)	61 (81)	62 (67)	97 (59)	45 (59)	44 (47)
Dazomet	162	162 (100)	42 (87)	89 (79)	44 (73)	76 (59)	43 (53)
Demetron-S-methyl	230	88 (100)	60 (50)	109 (24)	142 (17)	79 (14)	47 (11)
Desmetryn	213	213 (100)	57 (67)	58 (66)	198 (58)	82 (44)	171 (39)
Dialifos	393	208 (100)	210 (31)	76 (20)	173 (17)	209 (12)	357 (10)
Di-allate	269	43 (100)	86 (62)	41 (38)	44 (25)	42 (24)	70 (19)
Diazinon	304	137 (100)	179 (74)	152 (65)	93 (47)	153 (42)	199 (39)
Dicamba methyl ester	234	203 (100)	205 (60)	234 (27)	188 (26)	97 (21)	201 (20)
Dichlobenil	171	171 (100)	173 (62)	100 (31)	136 (24)	75 (24)	50 (19)
Dichlofenthion	314	97 (100)	279 (92)	223 (90)	109 (67)	162 (53)	251 (46)
Dichlofluanid	332	123 (100)	92 (33)	224 (29)	167 (27)	63 (23)	77 (22)
2,4-D isooctyl ester	332	43 (100)	57 (98)	41 (76)	55 (54)	71 (41)	69 (27)
2,4-D methyl ester	234	199 (100)	45 (97)	175 (94)	145 (70)	111 (69)	109 (68)
Dichlorprop isooctyl ester	346	43 (100)	57 (83)	41 (61)	71 (48)	55 (47)	162 (41)
Dichlorprop methyl ester	248	162 (100)	164 (80)	59 (62)	189 (56)	63 (39)	191 (35)
Dichlorvos	220	109 (100)	185 (18)	79 (17)	187 (6)	145 (6)	47 (5)
Dicofol	368	139 (100)	111 (39)	141 (33)	75 (18)	83 (17)	251 (16)
o,p′-DDT	352	235 (100)	237 (59)	165 (33)	236 (16)	199 (12)	75 (12)
p,p′-DDT	352	235 (100)	237 (58)	165 (37)	236 (16)	75 (12)	239 (11)
Dieldrin	378	79 (100)	82 (32)	81 (30)	263 (17)	77 (17)	108 (14)
Dimethirimol methyl ether	223	180 (100)	223 (23)	181 (10)	224 (3)	42 (2)	109 (2)
Dimethoate	229	87 (100)	93 (76)	125 (56)	58 (40)	47 (39)	63 (33)
DNOC methyl ether	212	182 (100)	165 (74)	89 (69)	90 (57)	212 (48)	51 (47)
Dinoterb methyl ether	254	239 (100)	209 (41)	43 (36)	91 (35)	77 (33)	254 (33)
Dioxacarb	223	121 (100)	122 (62)	166 (46)	165 (42)	73 (35)	45 (31)
Diphenamid	239	72 (100)	167 (86)	165 (42)	239 (21)	152 (17)	168 (14)
Disulfoton	274	88 (100)	89 (43)	61 (40)	60 (39)	97 (36)	65 (23)
Diuron	232	72 (100)	44 (34)	73 (25)	42 (20)	232 (19)	187 (13)
Dodine	227	43 (100)	73 (80)	59 (52)	55 (47)	72 (46)	100 (46)
Endosulfan	404	195 (100)	36 (95)	237 (91)	41 (89)	24 (79)	75 (78)
Endrin	378	67 (100)	81 (67)	263 (59)	36 (58)	79 (47)	82 (41)
Ethiofencarb	225	107 (100)	69 (48)	77 (29)	41 (26)	81 (21)	45 (17)
Ethirimol	209	166 (100)	209 (17)	167 (14)	96 (12)	194 (4)	55 (2)
Ethirimol methyl ether	223	180 (100)	223 (23)	85 (14)	181 (12)	55 (10)	96 (9)
Etrimfos	292	125 (100)	292 (91)	181 (90)	47 (84)	153 (84)	56 (73)
Fenarimol	330	139 (100)	107 (95)	111 (40)	219 (39)	141 (33)	251 (31)
Fenitrothion	277	125 (100)	109 (92)	79 (62)	47 (57)	63 (44)	93 (40)
Fenoprop isooctyl ester	380	57 (100)	43 (94)	41 (85)	196 (63)	71 (60)	198 (59)
Fenoprop methyl ester	282	196 (100)	198 (89)	59 (82)	55 (36)	87 (34)	223 (31)
Fenuron	164	72 (100)	164 (27)	119 (24)	91 (22)	42 (14)	44 (11)
Flamprop-isopropyl	363	105 (100)	77 (44)	276 (21)	106 (18)	278 (7)	51 (5)
Flamprop-methyl	335	105 (100)	77 (46)	276 (20)	106 (14)	230 (12)	44 (11)
Formothion	257	93 (100)	125 (89)	126 (68)	42 (49)	47 (48)	87 (40)

Table 2.51 (continued)

Compound	Molar mass	Main fragment m/z (intensities)											
		1		2		3		4		5		6	
Heptachlor	370	100	(100)	272	(81)	274	(42)	237	(33)	102	(33)		
Iodofenphos	412	125	(100)	377	(78)	47	(64)	79	(59)	93	(54)	109	(49)
loxynil isooctyl ether	483	127	(100)	57	(96)	41	(34)	43	(33)	55	(26)	37	(16)
loxynil methyl ether	385	385	(100)	243	(56)	370	(41)	127	(13)	386	(10)	88	(9)
Isoproturon	206	146	(100)	72	(54)	44	(35)	128	(29)	45	(28)	161	(25)
Jasmolin I	330	123	(100)	43	(52)	55	(34)	93	(25)	91	(24)	81	(23)
Jasmolin II	374	107	(100)	91	(69)	135	(69)	93	(67)	55	(66)	121	(58)
Lenacil	234	153	(100)	154	(20)	110	(15)	109	(15)	152	(13)	136	(10)
Lenacil N-methyl derivative	248	167	(100)	166	(45)	168	(12)	165	(12)	124	(9)	123	(6)
Lindane	288	181	(100)	183	(97)	109	(89)	219	(86)	111	(75)	217	(68)
Linuron	248	61	(100)	187	(43)	189	(29)	124	(28)	46	(28)	44	(23)
MCPB isooctyl ester	340	87	(100)	57	(81)	43	(62)	71	(45)	41	(42)	69	(29)
MCPB methyl ester	242	101	(100)	59	(70)	77	(40)	107	(25)	41	(22)	142	(20)
Malathion	330	125	(100)	93	(96)	127	(75)	173	(55)	158	(37)	99	(35)
Mecoprop isooctyl ester	326	43	(100)	57	(94)	169	(77)	41	(70)	142	(69)	55	(52)
Mecoprop methyl ester	228	169	(100)	143	(79)	59	(58)	141	(57)	228	(54)	107	(50)
Metamitron	202	104	(100)	202	(66)	42	(42)	174	(35)	77	(24)	103	(19)
Methabenzthiazuron	221	164	(100)	136	(73)	135	(69)	163	(42)	69	(30)	58	(25)
Methazole	260	44	(100)	161	(44)	124	(36)	187	(31)	159	(24)	163	(23)
Methidathion	302	85	(100)	145	(90)	93	(32)	125	(22)	47	(21)	58	(20)
Methiocarb	225	168	(100)	153	(84)	45	(40)	109	(37)	91	(31)	58	(21)
Methomyl	162	44	(100)	58	(81)	105	(69)	45	(59)	42	(55)	47	(52)
Metobromuron	258	61	(100)	46	(43)	60	(15)	91	(13)	258	(13)	170	(12)
Metoxuron	228	72	(100)	44	(27)	183	(23)	228	(22)	45	(21)	73	(15)
Metribuzin	214	198	(100)	41	(78)	57	(54)	43	(39)	47	(38)	74	(36)
Mevinphos	224	127	(100)	192	(30)	109	(27)	67	(20)	43	(8)	193	(7)
Monocrotophos	223	127	(100)	67	(25)	97	(23)	109	(14)	58	(14)	192	(13)
Monolinuron	214	61	(1003	126	(63)	153	(42)	214	(34)	46	(29)	125	(25)
Napropamide	271	72	(100)	100	(81)	128	(62)	44	(55)	115	(41)	127	(36)
Nicotine	162	84	(100)	133	(21)	42	(18)	162	(17)	161	(15)	105	(9)
Nitrofen	283	283	(100)	285	(67)	202	(55)	50	(55)	139	(37)	63	(37)
Nuarimol	314	107	(100)	235	(91)	203	(85)	139	(60)	123	(46)	95	(35)
Omethoat	213	110	(100)	156	(83)	79	(39)	109	(32)	58	(30)	47	(21)
Oxadiazon	344	43	(100)	175	(92)	57	(84)	177	(60)	42	(35)	258	(22)
Parathion	291	97	(100)	109	(90)	291	(57)	139	(47)	125	(41)	137	(39)
Parathion-methyl	263	109	(100)	125	(80)	263	(56)	79	(26)	63	(18)	93	(18)
Pendimethalin	281	252	(100)	43	(53)	57	(43)	41	(41)	281	(37)	253	(34)
Permethrin	390	183	(100)	163	(100)	165	(25)	44	(15)	184	(15)	91	(13)
Phenmedipham	300	133	(100)	104	(52)	132	(34)	91	(34)	165	(31)	44	(27)
Phosalone	367	182	(100)	121	(48)	97	(36)	184	(32)	154	(24)	111	(24)
Pirimicarb	238	72	(100)	166	(85)	42	(63)	44	(44)	43	(24)	238	(23)
Pirimiphos-ethyl	333	168	(100)	318	(94)	152	(88)	304	(79)	180	(73)	42	(71)
Pirimiphos-methyl	305	290	(100)	276	(93)	125	(69)	305	(53)	233	(44)	42	(41)
Propachlor	211	120	(100)	77	(66)	93	(36)	43	(35)	51	(30)	41	(27)
Propanil	217	161	(100)	163	(70)	57	(64)	217	(16)	165	(11)	219	(9)

Table 2.51 (continued)

Compound	Molar mass	Main fragment m/z (intensities) 1		2		3		4		5		6	
Propham	179	43	(100)	93	(88)	41	(42)	120	(24)	65	(24)	137	(23)
Propoxur	209	110	(100)	152	(47)	43	(28)	58	(27)	41	(21)	111	(20)
Pyrethrin I	328	123	(100)	43	(62)	91	(58)	81	(47)	105	(45)	55	(43)
Pyrethrin II	372	91	(100)	133	(70)	161	(55)	117	(48)	107	(47)	160	(43)
Quintozene	293	142	(100)	237	(96)	44	(75)	214	(67)	107	(62)	212	(61)
Resmethrin	338	123	(100)	171	(67)	128	(52)	143	(49)	81	(38)	91	(28)
Simazine	201	201	(100)	44	(96)	186	(72)	68	(63)	173	(57)	96	(40)
Tecnazene	259	203	(100)	201	(69)	108	(69)	215	(60)	44	(57)	213	(51)
Terbacil	216	160	(100)	161	(99)	117	(69)	42	(45)	41	(41)	162	(37)
Terbacil N-methyl derivative	230	56	(100)	174	(79)	175	(31)	57	(24)	176	(23)	41	(20)
Tetrachlorvinphos	364	109	(100)	329	(48)	331	(42)	79	(20)	333	(14)	93	(9)
Tetrasul	322	252	(100)	254	(67)	324	(51)	108	(49)	75	(40)	322	(40)
Thiabendazole	201	201	(100)	174	(72)	63	(12)	202	(11)	64	(11)	65	(9)
Thiofanox	218	57	(100)	42	(75)	68	(39)	61	(38)	55	(34)	47	(33)
Thiometon	246	88	(100)	60	(63)	125	(56)	61	(52)	47	(49)	93	(47)
Thiophanat-methyl	342	44	(100)	73	(97)	159	(89)	191	(80)	86	(72)	150	(71)
Thiram	240	88	(100)	42	(25)	44	(20)	208	(18)	73	(15)	45	(10)
Tri-allate	303	43	(100)	86	(73)	41	(43)	42	(31)	70	(23)	44	(21)
Trichlorfon	256	109	(100)	79	(34)	47	(26)	44	(20)	185	(17)	80	(8)
Tridemorph	297	128	(100)	43	(26)	42	(18)	44	(13)	129	(11)	55	(5)
Trietazine	229	200	(100)	43	(81)	186	(52)	229	(52)	214	(50)	42	(48)
Trifluralin	335	43	(100)	264	(33)	306	(32)	57	(7)	42	(6)	290	(5)
Vamidothion	287	87	(100)	58	(47)	44	(40)	61	(29)	59	(26)	60	(25)
Vinclozolin	285	54	(100)	53	(93)	43	(82)	124	(65)	212	(63)	187	(61)

The data refer to EI ionisation at 70 eV. The relative intensities can depend in individual cases an the type of mass spectrometer or mass-selective detector used. For confirmation mass spectra should be consulted which were run under identical instrumental conditions.

Example of Selected Ion Monitoring
for PCB analysis taking into account the PCB replacement product Ugilec T

The analysis strategy shown here has the aim of determining a PCB pattern as completely as possible in the relevant degrees of chlorination and to test in parallel for the possible presence of Ugilec T (tetrachlorobenzyltoluenes/trichlorobenzene). Three individual masses per time window for every two degrees of chlorination are planned for the selected SIM descriptors (scan width ±0.25 u based on the centroid determined in the full scan mode) to determine the overlapping retention ranges of the individual degrees of chlorination (Fig. 2.197). A staggered mass determination is thus obtained (Fig. 2.198).

No.	Substance	SIM masses [m/z]	Staggered time window [min:s]
1	Trichlorobenzenes	180/182/184	Start – 8:44
2	Cl_3-PCBs	256/258/260	8:44–14:30
3	Cl_4-PCBs	290/292/294	8:44–15:52
4	Cl_5-PCBs	324/326/328	14:30–18:34
5	Ugilec	318/320/322	15:52–20:15
6	Cl_6-PCBs	358/360/362	15:52–21:36
7	Cl_7-PCBs	392/394/396	18:34–23:38
8	Cl_8-PCBs	428/430/432	20:15–30:00
9	Cl_{10}-PCB	496/498/500	23:58–30:00

(The times refer to a capillary column J&W DB5 30 m × 0.25 mm × 0.25 μm.
Program: 60 °C – 2 min, 20 °/min to 180 °C, 5 °/min to 290 °C – 5 min. Splitless injection at 280 °C, column pressure 1 bar He. GC/MS system Finnigan INCOS 50. Sample: Aroclor 1240/1260 + Ugilec T, total concentration of the mixture ca. 10 ng/μl.)

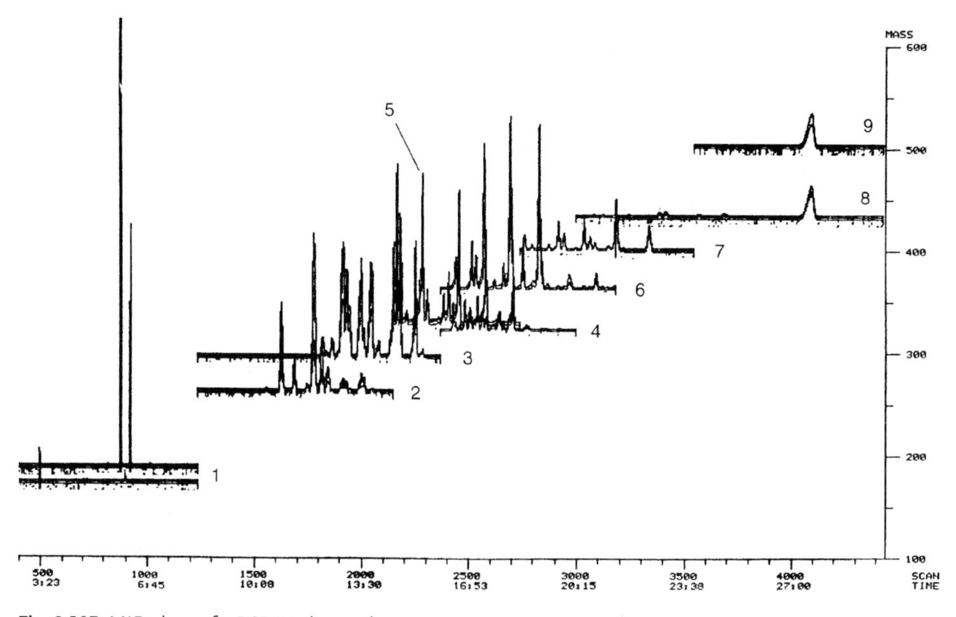

Fig. 2.197 MID chart of a PCB/Ugilec analysis.
In each case two degrees of chlorination of the PCBs and Ugilec T were detected in parallel, each with three masses. The overlapping MID descriptors were switched in such a way that each degree of chlorination was detected in two consecutive time windows (see text).

As an SIM/MID analysis is limited to the detection of certain ions in small mass ranges and the comparability of a complete spectrum for a library search is not required, on tuning the analyser a mass-based optimisation to a transmission which is as high as possible is un-

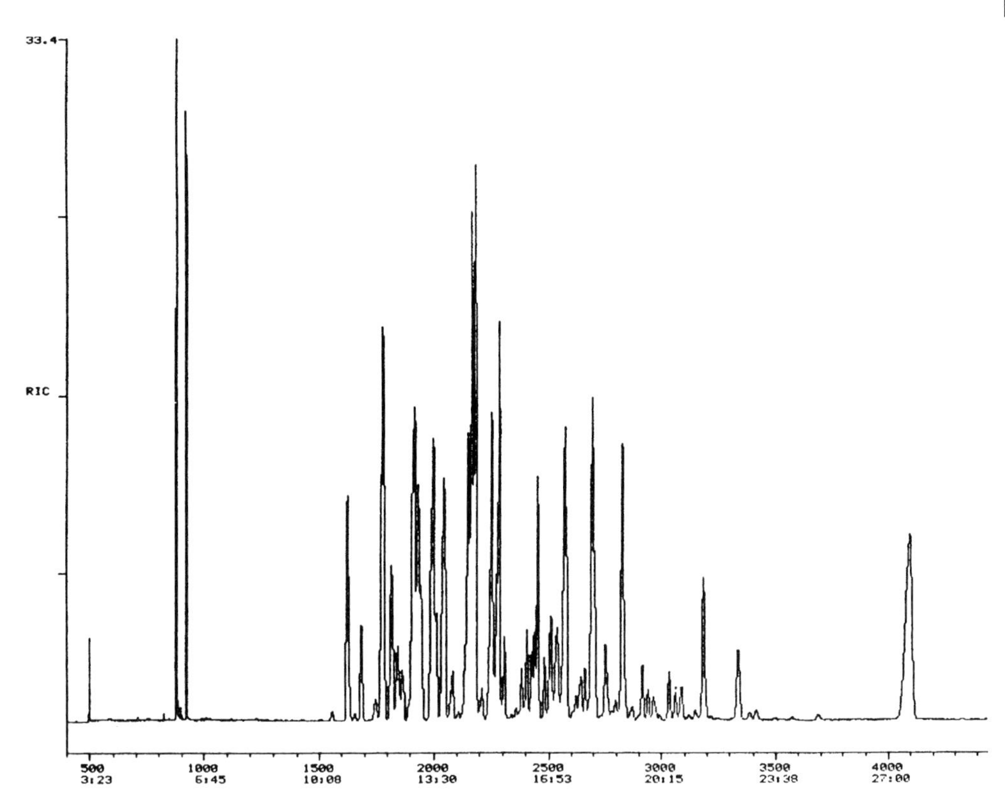

Fig. 2.198 Chromatogram (RIC) compiled after termination of the PCB/Ugilec T analysis in the MID mode from Fig. 2.197.

dertaken. Standard tuning aims to produce a balanced spectrum which corresponds to the data in a reference list for the reference substance FC43 (perfluorotributylamine), in order to guarantee good comparability of the spectra run with those in the library. Certain EPA methods require the source tuning using BFB (4-bromofluorobenzene) or DFTPP (decafluorotriphenylphosphine) for compliance with predefined mass intensities as set in the operating procedure. The position of a mass spectrum standard does not need to be adhered to in SIM analysis as no comparisons are made between spectra, but the relative intensities, e.g. of isotope patterns, must be evaluated. In the optimisation of the ion source special attention should therefore be paid to the masses (or the mass range) involved in SIM data acquisition. The source and lens potentials should then be selected manually so that a nearby fragment of the reference compound or an ion produced by column bleed (GC temperature ca. 200 °C) can be detected with the highest intensity but good resolution. In this way a significant additional increase in sensitivity can be achieved with quadrupole analysers for the SIM mode.

Figure 2.199 shows the chromatogram of a PCB standard as the result of a typical SIM routine analysis. In this case two masses are chosen as SIM masses for each PCB isomer. The switching points of the individual descriptors are recognised as steps in the base line. The different base line heights arise as a result of the different contributions of the chemical

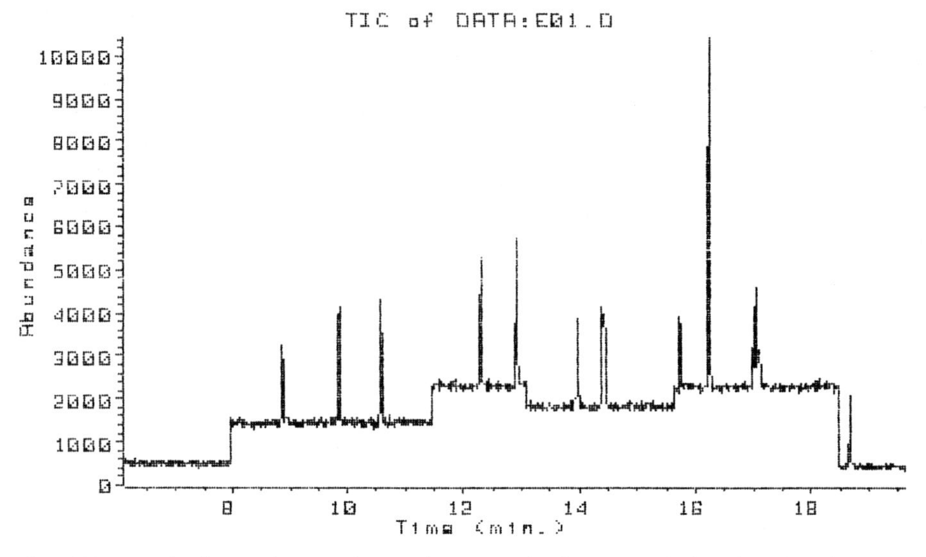

Fig. 2.199 Example of a typical PCB analysis in the SIM mode. The steps in the base line show the switching of the SIM descriptors.

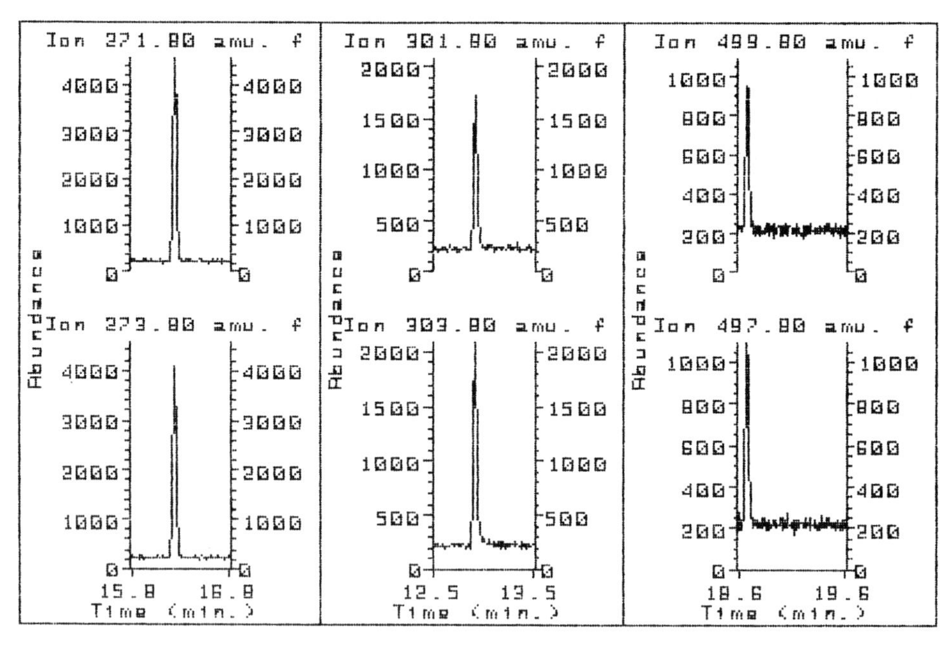

Fig. 2.200 Evaluation of the PCB analysis from Fig. 2-138 by showing the peaks at specific mass traces (see text).

noise to these signals. To control the evaluation, the substance signals can be represented as peaks in the expected retention time windows (Fig. 2.200). A deviation from the calibrated retention time (Fig. 2.200, right segment with the masses m/z 499.8 and 497.8) leads to a shift of the peak from the middle to the edge of the window and should be a reason for further checking. If qualifiers are present (e.g. isotope patterns), these should be checked using the relative intensities as a line spectrum, if possible (Fig. 2.201), or as superimposed mass traces (Fig. 2.202). In the region of the detection limit the noise width should be taken into account in the test of agreement.

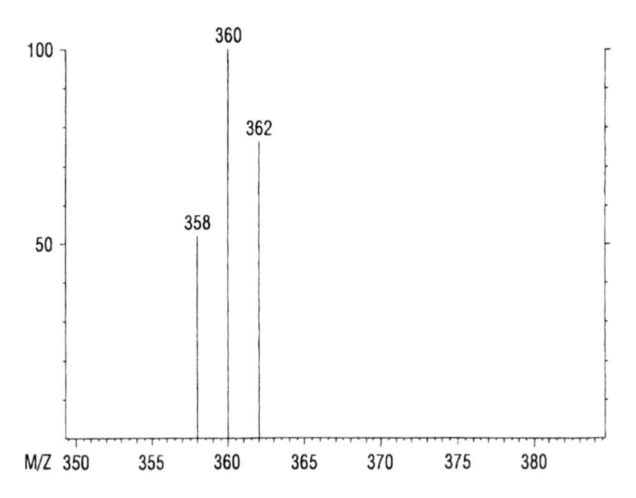

Fig. 2.201 Evaluation of isotope patterns from an MID analysis of hexachlorobiphenyl by comparison of relative intensities (shown as a bar graph spectrum).

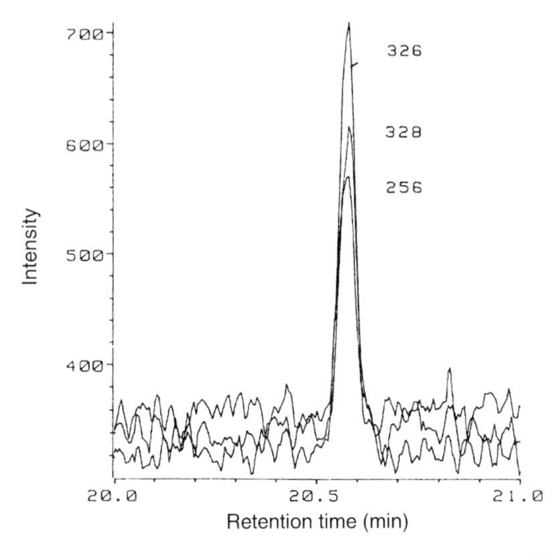

Fig. 2.202 Test of PCB isotope patterns (PCB 101, pentachlorobiphenyl) in the range of the detection limit (10 pg, S/N ca. 4:1) after SIM analysis (shown as mass traces).

2.3.5.3 High Resolution Accurate Mass MID Data Acquisition

Target compound analysis using high mass resolution e. g. for polychlorinated dioxins, furans (PCDD/F), pesticides, persistent organic pollutants (POP) or pharmaceutical residues are typically performed by monitoring the compound specific accurate mass ions at the expected retention time for each analyte. High Resolution GC/MS target compound applications benefit from a unique technical feature referred to as the lock-mass technique for performing multiple ion detection (MID) analyses. The lock mass technique provides ease of use, combined with a maximum quantitative precision and certainty in analyte confirmation.

The basic equation for sector mass spectrometers

$$m/z = c \cdot B^2/V$$

with
c = the instrument constant
B = magnetic field strength
V = acceleration voltage

shows that mass calibrations are feasible either at constant acceleration voltage by calibrating the magnetic field or vice versa.

For MID data acquisition, the fixed magnet setting with a variable acceleration voltage is typically used. The mass calibration for MID data acquisition follows a special procedure during the data acquisition in the form of a scan inherent mass calibration. This mass calibration is performed during MID analysis in every scan before acquisition of the target compound intensities. Most recent developments use two reference masses below and above the target masses referenced as the "lock-plus-cali mass technique" or in short "lock mode". It provides optimum mass accuracy for peak detection even in difficult chromatographic situations. The scan-to-scan mass calibration provides the best confidence for the acquired analytical data and basically is the accepted feature for the requirement of HRGC/HRMS as a confirmation method (EPA Method 1613 b).

The scan inherent mass calibration process is performed by the instrument control in the background without being noticed by the operator. In particular, it provides superior stability especially for high sample throughput with extended runtimes. A reference compound is leaked continuously from the reference inlet system during the GC run into the ion source. Typically perfluoro-tributylamine (FC43) is used as reference compound in HRGC/HRMS for dioxin analysis. Other reference compounds may be used to suit individual experimental conditions.

The exact ion masses of the reference compound are used in the MID acquisition windows for internal calibration. For the lock-and-cali-mass technique, two ions of the reference substance are individually selected for each MID window; one mass which is close, but below the analyte target mass, and the second, which is slightly above the analyte target masses. Although both reference masses are used for the internal calibration, it became common practice to name the lower reference mass the "lock mass" and the upper reference mass the "calibration mass".

During the MID scan the mass spectrometer is parking (locking) the magnetic field strength at the start of each MID window and then performing the mass calibration using the lock and calibration masses followed by the acquisition of the target and internal standard mass intensities.

Lock-Mass Technique

At the start of each MID retention time window, the magnet is automatically set to one mass (Da) below the lowest mass found in the MID descriptor. The magnet is parked, or "locked" and remains with this setting throughout the entire MID window. All analyzer jumps to the calibration and target compound masses are done by fast electrical jumps of the acceleration voltage.

Scan Inherent Calibration

The lock mass (L) is scanned in a small mass window starting below the mass peak by slowly decreasing the ion source acceleration voltage (see Fig. 2.203, ①). The mass resolution of the lock mass peak is calculated and written to the data file. Using the lock mass setting, a second reference mass is used for building the MID mass calibration. The calibration mass is checked by an electrical jump. A fine adjustment of the electrical calibration is made based on this measurement (see Fig. 2.203, ②). The electrical "jump" (see Fig. 2.203, ③) is very fast and takes only a few milliseconds. The dwell times for the sufficiently intense reference ions are very short.

The resulting electrical calibration is used for subsequent MID data acquisition.

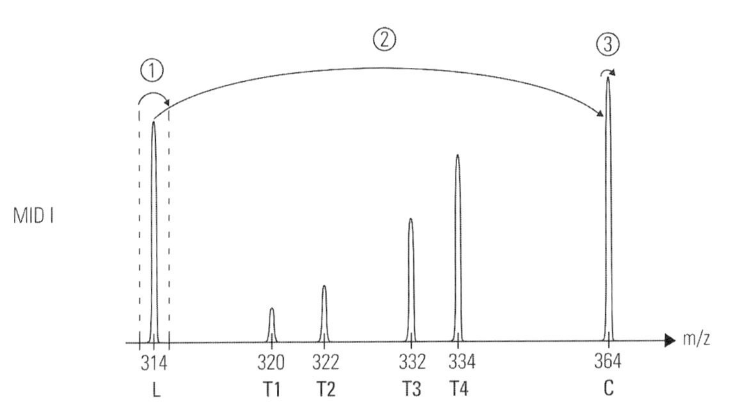

Fig. 2.203 Mass detection scheme in HRMS MID calibration. The arrows show the sequence of measurement in the mass calibration steps (the magnet is "locked" in this example at m/z 313).

①	Magnet locking and "lock mass" sweep, mass calibration and resolution determination	
②	Electrical jump to calibration mass	
③	Calibration mass sweep and mass calibration	
L	from FC43	lock mass (L), m/z 313.983364
C	from FC43	calibration mass (C), m/z 363.980170
T1, T2	from native TCDD	analyte target masses m/z 319.895992, 321.893042
T3, T4	from ^{13}C-TCDD	internal standard masses m/z 331.936250, 333.933300

Data Acquisition

With the updated and exact mass calibration settings, the analyzer now sets the acceleration voltage to the masses of the target ions. The intensity of each ion is measured based on a preset dwell time (see Fig. 2.204, ④). The dwell times to measure the analyte target ion intensities are significantly longer than the lock or calibration mass ions dwell times. This is done to achieve the optimum detection sensitivity for each analyte ion. The exact positioning on the top of the target ion mass peak allows for higher dwell times, significantly increased

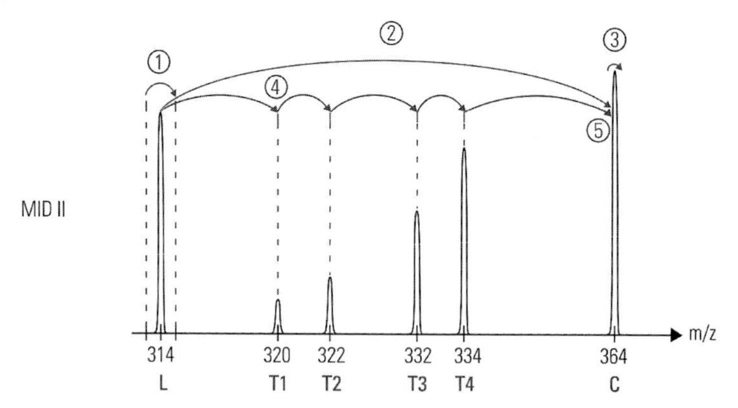

Fig. 2.204 Mass detection scheme in HRMS MID data acquisition. The arrows show the sequence of measurement in the target compound and internal standard data acquisition (legend for the mass scale see Fig. 2.203).
④ Consecutive electrical jumps to target and internal standard masses
⑤ Electrical jump to calibration mass, mass calibration

sensitivity, and higher S/N values compared to sweep scan techniques still used in older technology HRMS systems. It is important to note that the lock-plus-cali mass technique extends the dynamic range significantly into the lower concentration range.

Advantages of the Lock-plus-Cali Mass Technique

The lock-plus-cali mass calibration technique provides extremely stable conditions for data acquisitions of long sequences even over days, e.g., over the weekend and usually includes the performance documentation for quality control documented in the data file.

The electrical jumps of the acceleration voltage are very fast, and provide an excellent instrument duty cycle. Any outside influences from incidental background ions, long term drift or minute electronics fluctuations are taken care of and do not influence the result.

Both, the lock and cali masses are monitored in parallel during the run providing an excellent confirmation of system stability for data certainty. Together with the constant resolution monitoring, this technique provides the required traceability in MID data analysis.

Other acquisition techniques have been formerly used employing just one lock mass position. This technique requires a separate pre-run electrical mass calibration and does not allow a scan inherent correction of the mass position which may arise due to long term drifts of the analyzer during data acquisition.

As a consequence, with the one mass lock techniques, the mass jumps are less precise with increasing run times. Deviations from the peak top position when acquiring data at the peak slope result in less sensitivity, less reproducibility and poor isotope ratio confirmation.

The lock-plus-cali mass technique has proven to be superior in achieving lower LOQs and higher S/N values in high resolution MID. Figure 2.205 shows the typical chromatogram display of dioxin analysis with the TCDD target masses as well as the ^{13}C internal standard masses. In addition, the continuously monitored FC43 lock and cali masses are displayed as constant mass traces. Both traces are of valuable diagnostic use and confirm the correct measurement of the target compounds.

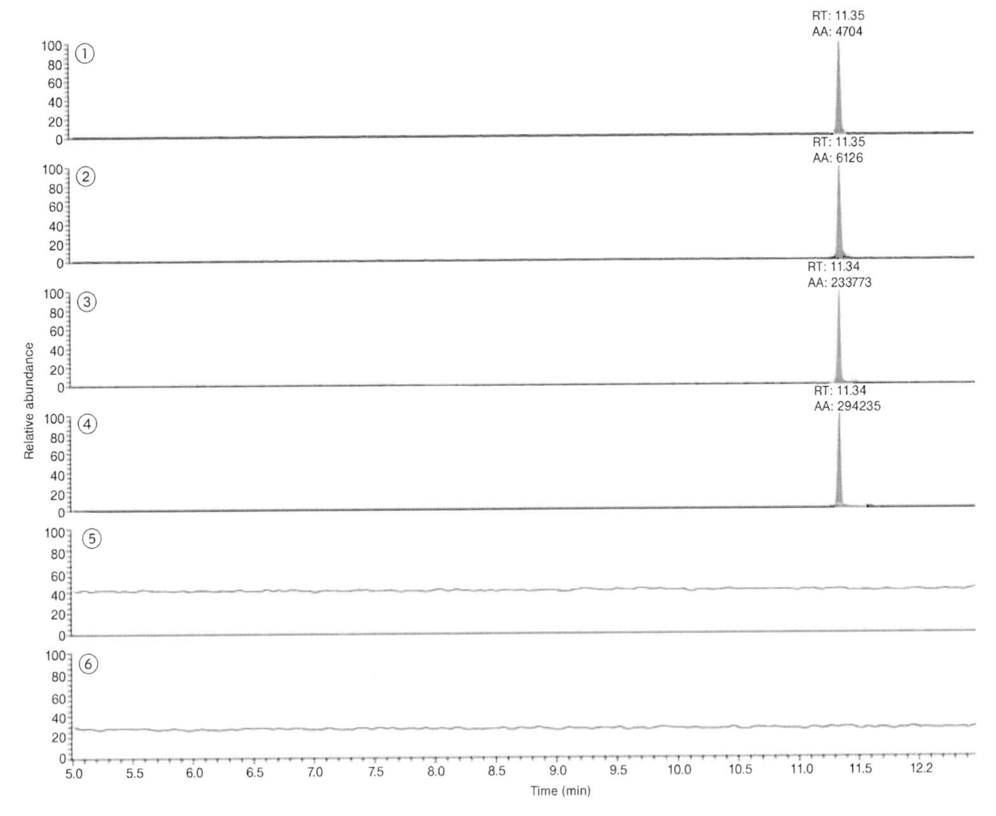

Fig. 2.205 Resulting mass chromatograms of a TCDD standard solution at 100 fg/µL (DB-5MS 60 m x 0.25 µm x 0.1 µm); ① Ratio mass of 2,3,7,8-tcdd (native) m/z 319.8960; ② Quan mass of 2,3,7,8-tcdd (native) m/z 321.8930; ③ Ratio mass of 2,3,7,8–13C12-TCDD (ISTD) m/z 331.9362; ④ Quan mass of 2,3,7,8–13C12-TCDD (ISTD) m/z 333.9333; ⑤ Lock mass of FC43 m/z 313.983364; ⑥ Cali mass of FC43 m/z 363.980170.

Setup of the MID Descriptor

The MID descriptor for the data acquisition contains all the information required by the HRMS mass analyzer. Included in each descriptor is the retention time information for switching between different target ions, the exact mass calibration, the target masses to be acquired, and the corresponding dwell times.

A sample chromatogram usually facilitates the setting of the retention time windows, as shown Fig. 2.206. The sample chromatogram is used to optimize the GC component separation and set the MID windows before data analysis, and usually consists of a higher concentration standard mix.

MID Cycle Time

In order to provide a representative and reproducible GC peak integration, the total MID cycle time on the chromatographic time scale should allow for the acquisition of 8 to 10 data points over a chromatographic peak. The cycle time has a direct influence on the available measurement time for each ion (dwell time). If the MID cycle time is too short,

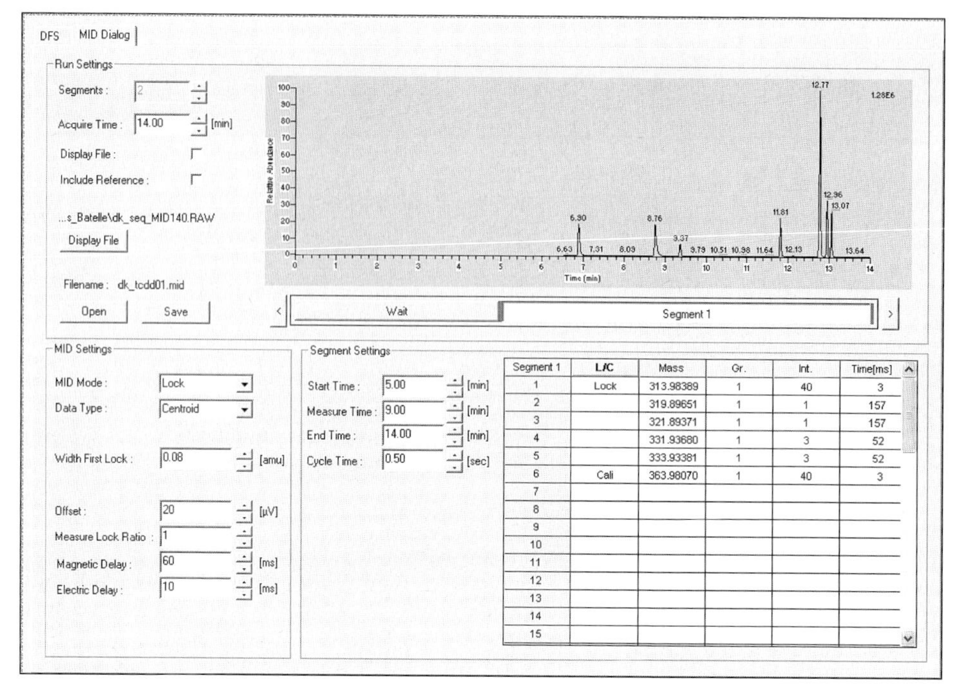

Fig. 2.206 MID editor with sample chromatogram (top) and target mass list and duty cycles for the highlighted retention time window (bottom right).

the sensitivity of the instrument is compromised, too high values lead to a poor GC peak definition.

2.3.6
MS/MS – Tandem Mass Spectrometry

"Can atomic particles be stored in a cage without material walls?

This question is already quite old. The physicist Lichtenberg from Göttingen wrote in his notebook at the end of the 18th century: "I think it is a sad situation that in the whole area of chemistry we cannot freely suspend the individual components of matter." This situation lasted until 1953. At that time we succeeded, in Bonn, in freely suspending electrically charged atoms, i.e. ions, and electrons using high frequency electric fields, so-called multipole fields. We called such an arrangement an ion cage."

Prof. Wolfgang Paul
in a lecture at the Cologne Lindenthal Institute in 1991

(from: Wolfgang Paul, A Cage for Atomic Particles – a basis for precision measurements in navigation, geophysics and chemistry, Frankfurter Allgemeine Zeitung, Wednesday 15th December 1993 (291) N4)

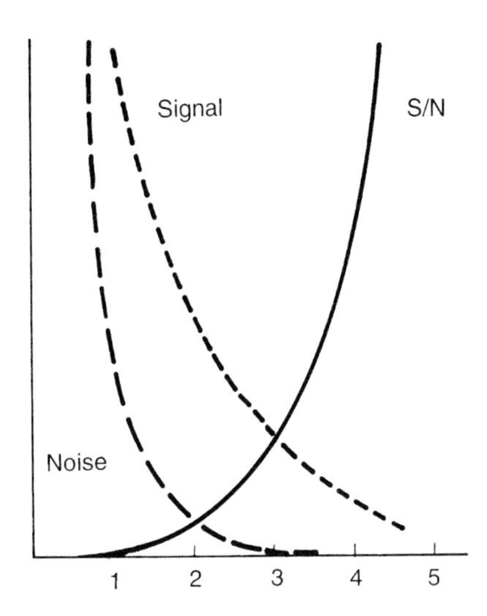

Fig. 2.207 Relationship between signal, noise and the number of analytical steps.

As part of the further development of instrumental techniques in mass spectrometry, MS/MS analysis has become the method of choice for trace analysis in complex matrices. Most of the current applications involve the determination of substances in the ppb and ppt ranges in samples of urine, blood and animal or plant tissues and in many environmental analyses. In addition the determination of molecular structures is an important area of application for multidimensional mass spectrometry.

As an analytical background to the use of the GC/MS/MS technique in residue analysis it should be noted that the signal/noise ratio increases with the number of analytical steps (Fig. 2.207). Clean-up steps lower the potential signal intensity. The sequence of wet chemical or instrumental sample preparation steps can easily lead to the situation whereby, as a consequence of processing losses, a substance can no longer be detected. From this consideration, the first separation step (MS^1) in a GC/MS/MS systen can be regarded as a mass-specific clean-up in the analysis of extracts with large quantities of matrix. After the subsequent induced fragmentation of selected ions, an analyte is identified using the characteristic mass spectrum of the product ions or it is quantified using the structure selective fragment ions for target compound quantitation in difficult matrices.

When is MS/MS Used?

- The sample matrix contributes significantly to the chemical background noise in SIM.
- Co-elution with isobaric impurities occurs.
- The structures of the compounds are unknown.
- Quantitation with the highest possible sensitivity in difficult matrices is necessary.
- The SIM analysis requires additional confirmation.

MS/MS Scan Techniques

Scan mode MS_1	MS_2	Result	Application
Single ion	Scan	Product ion spectrum (MS/MS spectrum)	Identification and confirmation of compounds, structure determination
Single ion	Single ion	Individual intensities of product ions (selected reaction monitoring, SRM)	Highly selective and highly sensitive target compound quantitation with complex matrices, e.g. pesticides
Scan	Single ion	Precursor masses of certain fragments (precursor ion scan)	Specific analysis of compounds (classes of substance) with common structural features, screening, e.g. crude oil biomarker
Scan	Scan-NL	Precursor ions, which undergo loss of neutral particles with NL u (neutral loss scan)	Specific analysis of compounds (classes of substance) with common functional groups/structural features, e.g. loss of COCl from PCDD/PCDF

Soft ionisation techniques instead of electron impact ionisation are the preferred ionisation processes for MS/MS analysis. Although fragmentation in the EI mode of GC/MS is desirable for substance identification, frequently only low selectivity and sensitivity are achieved in complex matrices. Soft ionisation techniques, such as chemical ionisation, concentrate the ion flow to a few intense ions which can form a good starting point for MS/MS analysis. For this reason the HPLC coupling techniques atmospheric pressure CI (APCI) and electrospray (ESI) are also in the forefront of the development and extension of MS/MS analysis. The use of the GC/MS/MS technique, which can be carried out with positive and negative chemical ionisation (PCI and NCI), will in the future dominate the multicomponent target compound methods for trace analyses.

The information content of the GC/MS/MS technique was already evaluated in 1983 by Richard A. Yost (University of Florida, codeveloper of the MS/MS technique). Based on the theoretical task of detecting one of the five million substances catalogued at that time by the Chemical Abstracts Service, a minimum information content of 23 bits ($\log_2(5 \cdot 10^6)$) was required for the result of the chosen analysis procedure and the MS procedures available were evaluated accordingly. The calculation showed that capillary GC/MS/MS can give 1000 times more information than the traditional GC/MS procedure (Table 2.52)!

For the instrumental technique required for tandem mass spectrometry, as in the consideration of the resolving power, the main differences lie between the performances of the magnetic sector and quadrupole ion trap instruments (see Section 2.3.1). For coupling with GC and HPLC, triple-quadrupole instruments have been used since the 1980s. Ion trap MS/MS instruments have been used in research since the middle of the 1980s and are now being used in routine residue analysis. Tandem magnetic sector instruments are mainly used in research and development for mass spectrometry. Much higher energies (keV range) can be used to induce fragmentation. For the selection of precursor ions and/or the detection of the product ion spectrum high resolution capabilities can be used.

Table 2.52 Information content of mass spectroscopic techniques (after Yost 1983, Kaiser 1978).

Technique	P	Factor
MS [a)]	$1.2 \cdot 10^4$	0.002
Packed GC/MS [b)]	$7.8 \cdot 10^5$	0.12
Capillary GC/MS [c)]	$\mathbf{6.6 \cdot 10^6}$	**1**
MS/MS	$1.2 \cdot 10^7$	2
Packed GC/MS/MS	$7.8 \cdot 10^8$	118
Capillary GC/MS/MS	$6.6 \cdot 10^9$	1000

a) MS: 1000 u, unit mass resolution, maximum intensity 2^{12}
b) Packed GC: $2 \cdot 10^3$ theoretical plates, 30 min separating time
c) Capillary GC: $1 \cdot 10^5$ theoretical plates, 60 min separating time

MS/MS Tandem Mass Spectrometry

- Ionisation of the sample (EI, CI and other methods).
- Selection of a precursor ion.
- Collision-induced dissociation (CID) to product ions.
- Mass analysis of product ions (product ion scan).
- Detection as a complete product ion spectrum or preselected individual masses (selected reaction monitoring, SRM).

Ion trap mass spectrometry offers a new extension to the instrumentation used in tandem mass spectrometry. The methods for carrying out MS/MS analyses differ significantly from those involving triple-quadrupole instruments and reflect the mode of operation as storage mass spectrometers rather than beam instruments.

Tandem mass spectrometry consists of several consecutive processes. The GC peak with all the components (analytes, co-eluents, matrix, column bleed etc) reaching the ion source is first ionised. From the resulting mixture of ions the precursor ion with a particular m/z value is selected in the first step (MS^1). This ion can in principle be formed from different molecules (structures) and, even with different empirical formulae, they can be of the same nominal m/z value. The MS^1 step is identical to SIM analysis in traditional GC/MS, which, for the reasons mentioned above, cannot rule out false positive signals. Fragmentation of the precursor ions to product ions occurs in a collisional behind MS^1 through collisions of the selected ions with neutral gas molecules (CID, collision induced dissociation, or CAD, collision activated decomposition). In this collision process the kinetic energy of the precursor ions (quadrupole: ca. 10–100 eV, magnetic sector: >1 keV) is converted into internal energy (in an ion trap typically <6 eV), which leads to a substance-specific fragmentation by cleavage or rearrangement of bonds and the loss of neutral particles. A mixture of product ions with lower m/z values is formed. Following this fragmentation step, a second mass spectrometric separation (MS^2) is necessary for the mass analysis of the product ions. The spectrum of the product ions is finally detected as the product ion or MS/MS spectrum of this compound.

With regard to existing analysis procedures, for target compound quantitation high speed and flexibility in the choice of precursor ions is necessary. For the analysis with internal standards (e.g. deuterated standards) and multi-component methods, the fast switch between multiple precursor ions is necessary to achieve a sufficiently high data rate for a reliable peak definition of all coeluting compounds, in particular for SRM quantitations.

Besides recording a spectrum for the product ions (product ion scan), or of an individual mass (SRM), two additional MS/MS scan techniques give valuable analysis data. By linking the scans in MS_2 and MS_1, very specific and targeted analysis routes are possible. In the precursor ion scan, the first mass analyser (MS_1) is scanned over a preselected mass range. All the ions in this mass range reach the collision chamber and form product ions (CID). The second mass analyser (MS_2) is held constant for a specific fragment. Only emerging ions of MS^1 which form the selected fragment are recorded. This recording technique makes the identification of substances of related structure, which lead to common fragments in the mass spectrometer, easier (e.g. biomarkers in crude oil characterisation, drug metabolites).

In neutral loss scan all precursor ions, which lose a particular neutral particle (that otherwise cannot be detected in MS), are detected. Both mass analysers scan, but with a constant preadjusted mass difference, which corresponds to the mass of the neutral particle lost. This analysis technique is particularly meaningful if molecules contain the same functional groups (e.g. metabolites as acids, glucuronides or sulfates). In this way it is possible to identify the starting ions which are characterised by the loss of a common structural element. Both MS/MS scan techniques can be used for substance-class-specific detection in triple-quadrupole systems. Ion trap systems allow the mapping of these processes by linking the scans between separate stages of MS in time.

Mode of Operation of Tandem Mass Spectrometers

In tandem mass spectrometry using quadrupole or magnetic sector instruments, the various steps take place in different locations in the beam path of the instrument. The term "tandem-in-space" (R. Yost) has been coined to show how they differ from ion storage mass spectrometers (Fig. 2.208). In the ion trap analyser, a typical storage mass spectrometer, these

Tandem-in-space

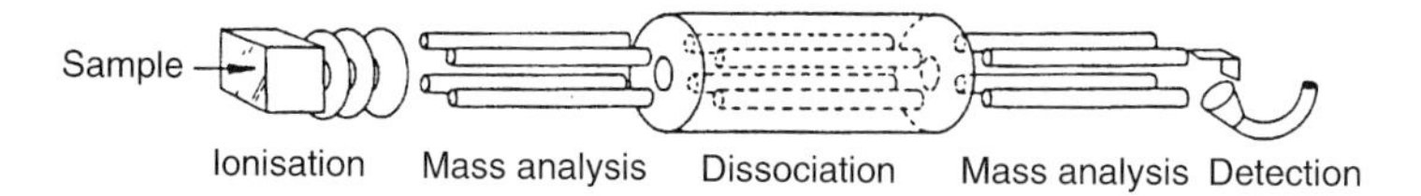

Tandem-in-time

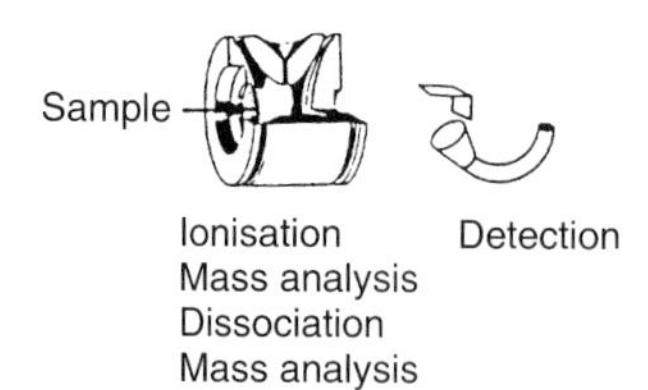

Fig. 2.208 GC/MS/MS techniques.
Tandem-in-space: triple-quadrupole technique
Tandem-in-time: ion trap technique

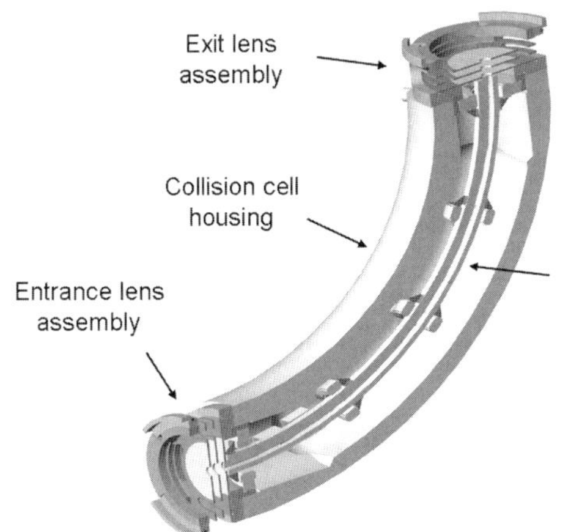

Fig. 2.209 Curved collision cell of a triple quadrupole system (90°) for reduction of non-specific noise by eliminating neutral particles and photons from the ion flight path (TSQ Quantum; courtesy Thermo Fisher Scientific).

processes take place in the same location, but consecutively. Richard Yost has described these as "tandem-in-time".

In a triple-quadrupole mass spectrometer (or other beam instruments, Fig. 2.209), an ion beam is passed continuously through the analyser from the ion source. The selection of precursor ions and collision-induced dissociation take place in dedicated devices of the analyser (Q_1, Q_2) independent of time. The only time-dependent process is the mass scan in the second mass analyser (Q_3, MS_2) for the recording of the product ion spectrum (Fig 2.210). An enclosed quadrupole device (Q_2) typically consisting of square rods is used. Formerly also hexapole or octapole rod systems have been employed. The collision cell is operated in an RF only made without mass seperating but ion focusing capabilities. (The mass filters MS_1 (Q_1) and MS_2 (Q_3) are operated at a constant RF/DC ratio!). Helium, nitrogen, argon or xenon at pressures of ca. 10^{-3} to 10^{-4} Torr are used as the collision gas. Heavier collision gases increase the yield of product ions. The collision energies are of the order of up to 100 eV and are typically controlled by a preceding lens stack, see Fig. 2.209.

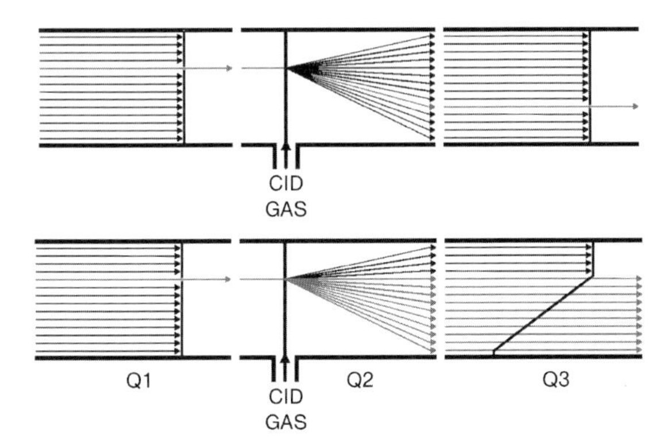

Fig. 2.210 Modes of operation of a triple quadrupole mass spectrometer.
Q_1 = first mass separating quad, mass selection of the precursor ion
Q_2 = collision cell, CID process
Q_3 = second mass separating quad, SRM detection (top) or product ion scan (bottom)

In instruments with ion trap analysers, the selection of the precursor ions, the collision-induced dissociation and the analysis of the product ions occur in the same place, but with time control via a sequence of frequency and voltage values at the end caps and ring electrode of the analyser (scan function see Fig. 2.211). The systems with internal or external ionisation differ in complexity and ease of calibration. In the case of internal ionisation the sample spectrum is first produced in the ion trap analyser and stored. The precursor ion m_p is then selected by ejecting ions above and below m_p from the trap by applying a multifrequency signal at the end caps (waveform). Ion trap systems with external ionisation can employ waveforms during injection of the ions into the analyser from the external source for isolation of the precursor ion in MS/MS mode or for isolation of the desired mass scan range in other scan modes. In this way a longer storage phase and a more rapid scan rate can be used for GC/MS in the MS/MS and SIM modes.

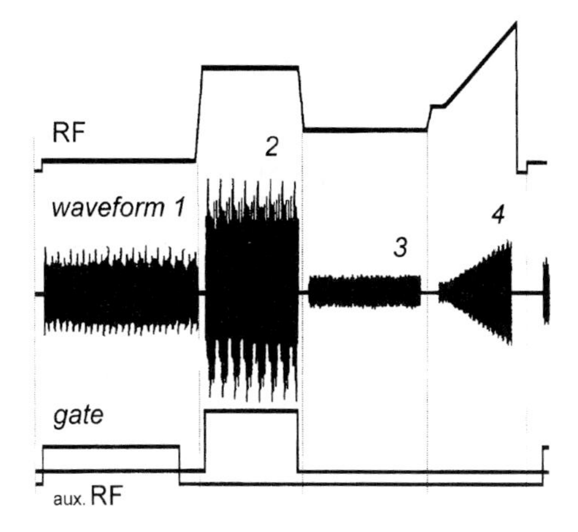

Fig. 2.211 MS/MS scan function of the ion trap analyser with external ionisation.

1 A gate switches the ion beam for transfer to the analyser, selection of the precursor ion m_p by a special frequency spectrum (ionisation waveform), variable ion injection time up to maximum use of the storage capacity.
2 One or more m/z values are then isolated using a synthesised frequency spectrum (isolation waveform). This phase corresponds to the SIM mode.
3 Collision induced dissociation (CID) by selective excitation to the secular frequency (activation waveform) of the precursor ion, storage of the product ions formed at a low RF value without exciting them further.
4 Product ion scan at a scan rate of ca. 5500 u/s (resonance ejection waveform), detection of the ions by the multiplier.

Collision-induced dissociation is initiated by an additional AC voltage at the end cap electrodes of the ion trap analyser. If the frequency of the AC voltage corresponds to the secular frequency of the selected ions, there is an uptake of kinetic energy by resonance. The collisions of the precursor ions m_p excited in this way with the helium molecules present are sufficiently energy-rich to effect fragmentation. The collision energy is determined by the level of the applied AC voltage. In the collisions the kinetic energy is converted into internal energy and used up in bond cleavage. The product ions formed are stored in this phase at low RF values of the ring electrode and are detected at the end of the CID phase by a normal mass scan. The time required for these processes is in the lower ms range and, at a high scan rate for the GC, allows the separate monitoring of the deuterated internal standard at the same time. With the realisation of tandem-in-time MS/MS an aspect of particular importance for the efficiency of the process lies in the reliable choice of the frequencies for precursor ion selection and excitation. Instruments with external ion sources permit a calibration of the frequency scale which is generally carried out during the automatic tuning. In this way the MS/MS operation is analogous to that of the SIM mode in that only the masses of the desired precursor ions m_p need to be known. In the case of instruments with internal ionisation, where the GC carrier can cause pressure fluctuations within the analyser, the empirical determination of excitation frequencies and the broad band excitation of a mass window is necessary.

The efficiency of the CID process is of critical importance for the use of GC/MS/MS in residue analysis. In beam instruments optimisation of the collision energy (via the potential of Q_2) and the collision gas pressure is necessary. The optimisation is limited by scattering effects at high chamber pressures and subsequent fragmentation of product ions. Modern collision cells are manufactured from square quadrupole rods which provide superior ion transmission characteristics even at extended length compared to former quadrupole or hexapole cells eliminating the formerly observed loss of ions due to the collision processes (see Fig. 2.209). The now possible extended length of collision cells provides an effective fragmentation process for the generation of the product ion spectrum. Typical collision energies for SRM quantitations are in the range of 10 to 25 eV. Cross-talk from different precursors to the same product ion is effectively eliminated by an active cleaning step between the scans (adds to the interscan time). Encapsulated collision cells using square quadrupoles are typically operated at up to 4 mbar Ar or N_2 pressure for highly sensitive target compound residue analysis.

With the ion trap analyser the energy absorbed by the ions depends on the duration of the resonance conditions for the absorption of kinetic energy and on the voltage level. The pressure of the helium buffer gas can play a role in the collision energy, but is not typically adjusted. Typical values for the induction phase are 3–15 ms and 500–1000 mV. With the ion trap technique there is higher efficiency in the fragmentation and transmission to product ions (Johnson and Yost 1990). In addition the ion trap technique has clear advantages because of the sensitivity resulting from the storage technique, which allows recording of complete product ion spectra even below the pg range.

Modern developments in ion trap technology include the automated determination of collision energies (ACE). This technique makes it easier to set up MS/MS methods by not requiring manual optimization of the collision energy for every compound of a large set of targets. The instrument uses an empirical calibration scheme to calculate an optimum collision energy comprising all of the known instrument parameters as precursor mass and damping gas flow. Three collision energy values can be provided to check results. The middle one typically is the optimum calculated from the instrument parameters. The found levels cover the range of optimum collision energies for compounds of varying ion chemistry. In practice a very good correlation with manual optimization e.g. for pesticides is achieved.

Low mass MS/MS fragments, that are cut from the spectrum by the former regular ion trap operation are stored and scanned for MS/MS spectra using the pulsed q-value dissociation technique (PQD). Normal CID in ion traps is done at a particular excitation q value defining the effective collision energy. Higher q values provide higher collision energy values by getting the precursor ions moving faster but cut the low mass end from the spectrum. PQD works by first exciting the precursor ion at high q value for a short period of time

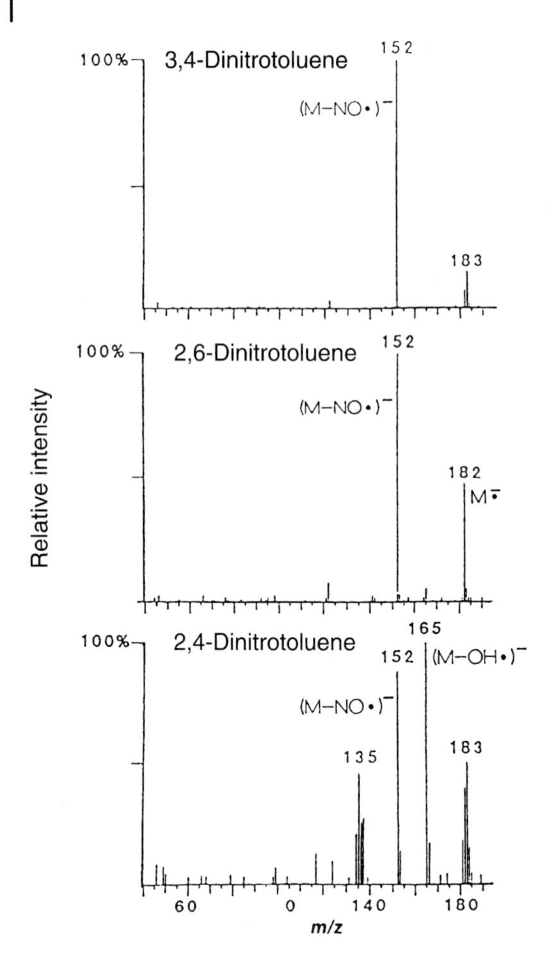

Fig. 2.212 NCI-MS/MS: Differentiation between dinitrotoluene isomers by comparison of the product ion spectra (ion trap analyser, reagent gas water, detection of negative ions) (after Brodbelt and Cooks 1988).

(50–100 μs) using a high collision energy as determined by ACE. During the induced fragmentation the q value is lowered significantly to store the full range of product ions. The result is that also low m/z fragments ions are being trapped. Therefore PQD is primarily useful as a qualitative tool for structure elucidation, but also finds utility in quantitative applications where mainly and intense low m/z fragment ions are observed as known for many N containing compounds e.g. amines. PQD is a patented technique (US 6,949,743 & 7,102,129) that was first used on the ion traps from Thermo Fisher Scientific, San Jose, CA, USA.

Today comparable results are available for both the triple-quadrupole and ion trap analysers. It is typical for the product ion spectra obtained with ion trap mass spectrometers that the intensities of the precursor ions are significantly reduced due to efficient CID. The product ion spectra show only a few well defined but intense product ions derived directly from the precursor. Further fragmentation of product ions usually does not occur as the m/z of the precursor ion is excited exclusively. Triple quadrupole instruments show less efficiency in the CID process resulting in a higher precursor ion signal. Primary product ions fragment consecutively in the collision cell and produce additional signals in the product ion spectrum. The efficiency of ion trap instruments is also reported to be higher than that of magnetic sector instruments. The mass range and the scan techniques of precursor scan and neutral loss scan make triple-quadrupole instruments suitable for applications outside the range of pure GC/MS use.

When evaluating MS/MS spectra it should be noted that no isotope intensities appear in the product ion spectrum (independent of the type of analyser used). During selection of the precursor ion for the CID process, naturally occurring isotope mixtures are separated and isolated. The formation of product ions is usually achieved by the loss of common neutral species. The interpretation of these spectra is generally straightforward and less complex compared with EI spectra. When comparing product ion spectra of different instruments the acquisition parameter used must be taken into account. In particular with beam instruments the recording parameters can be reflected in the relative intensities of the spectra. On the other hand, rearrangements and isomerisations are possible, which can lead to the same product ion spectra.

Structure Selective Detection Using MS/MS Transitions

Triple quadrupole as well as ion trap mass spectrometers besides structure elucidation find their major application in quantitation of target compounds in difficult matrices. The increased selectivity when observing specific transitions from a precursor ion to a structural related product ion provides highly confident analyses with excellent LOQs even in matrix samples. In the selected reaction monitoring mode (SRM) an intense precursor ion from the spectrum of the target compound is selected in the first quadrupole Q1, fragmented in the collision cell, and monitored on a selective product ion for quantitation, see the principle given graphically in Fig. 2.210. The high selectivity of the SRM method is controlled by the mass resolution of the first quadrupole Q1. While round rod quadrupoles are working at unit mass resolution, hyperbolic quadrupoles typically are set to 0.7 Da peak width at FPHW as standard and are operated in the highly resolved H-SRM mode for maximum selectivity even at narrow peak widths of 0.4 Da FPHW (patented by Thermo Fisher), also refer to Section 2.3.1.

Analogue to the SIM analysis mode used with single quadrupole instrumentation the SRM mode omits full spectral information for substance confirmation. A structure selective characteristic of the assay is given by the mass difference of the transition monitored. Typically, two independent transitions together with the chromatographic retention time provide the positive confirmation of the occurrence of a particular compound (Council Directive 96/23/EC). State-of-the-art triple quadrupole instrumentation provide transition times as low as 1–2 ms offering the potential for screening of a large number of compounds in a given chromatographic window e. g. for multicomponent pesticide analysis (see Section 4.14).

Data Dependant Data Acquisition

The advanced electronic capabilities of modern triple quadrupole instruments offer additional features for delivering compound specific information even during SRM quantitation analyses. Depending on the quality of a scan currently being acquired, the mode of the data acquisition for a following scan (data point) can be switched to acquire a full product ion spectrum. The level of product ion intensity that triggers the switch of acquisition mode is set method specific. As result one data point in the substance peak is used to generate and acquire the compound product ion spectrum (see Fig. 2.113). This spectrum provides the full structural information about the detected substance and is available for library search

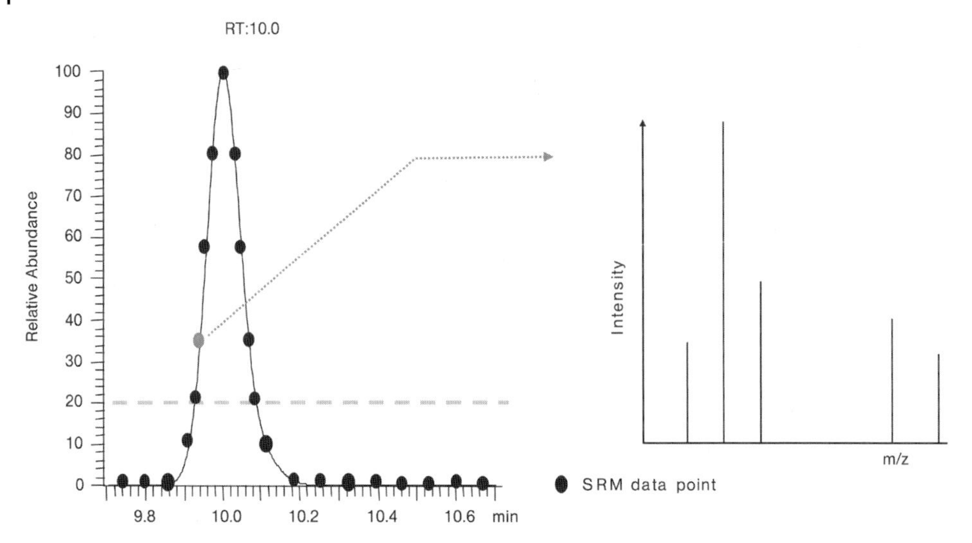

Fig. 2.213 Data dependent data acquisition scheme. The first SRM scan intensity above a user defined threshold (left, dotted line) is acquired as MS/MS product ion spectrum (right).

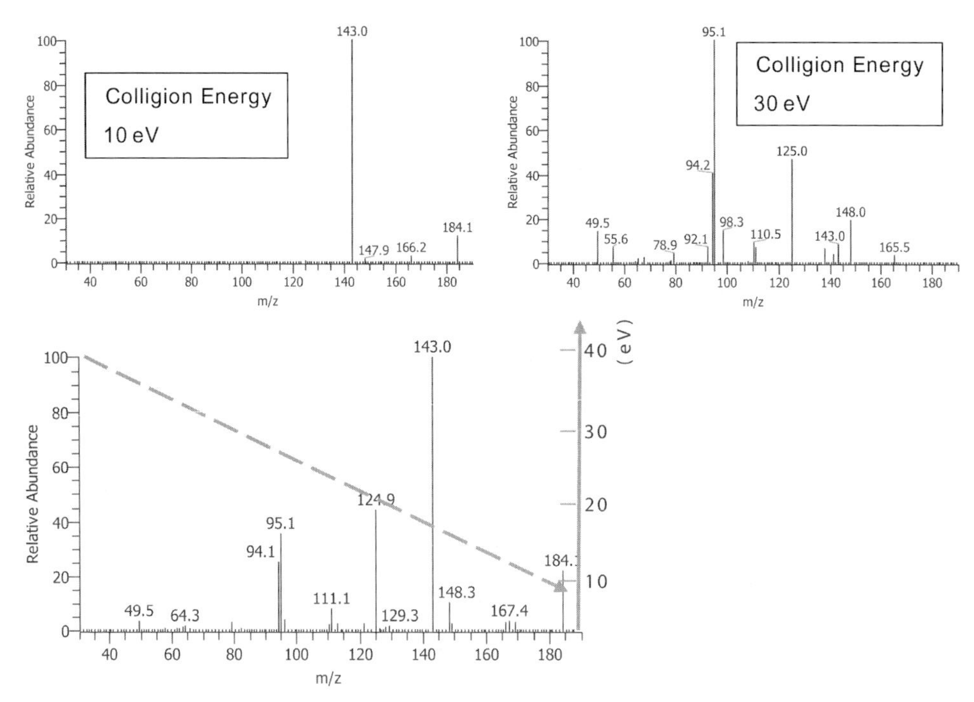

Fig. 2.214 Decreasing collision energy ramp in MS/MS mode provides information rich product ion spectrum.

and compound confirmation. Using the data dependent acquisition mode the final data file contains both quantitative as well as qualitative information.

For the generation of the MS/MS product ion spectrum the applied collision energy has substantial impact on the information content of the spectrum, represented by the occurrence and intensities of fragment and precursor ions. The maximum information on the compound structure is achieved by variation of the collision energy during the product ion scan. High collision energies lead to the generation of low mass fragments while lower collision energies provide the favoured main fragments and still some visible intensity of the precursor ion. The instrumental approach to this solution is the variation of the collision energy during the scan from a high energy to a low energy level by ramping down the ion acceleration voltage using the lenses located at the entrance of the ion beam into the collision cell (see Fig. 2.209). The resulting product ion spectrum consequently delivers the full available information about the structural characteristics of an unknown or the compound to be confirmed (see Fig. 2.114).

2.3.7 Mass Calibration

To operate a GC/MS system calibration of the mass scale is necessary. The calibration converts the voltage or time values into m/z values by controlling the analyser during a mass scan. For the calibration of the mass scale a mass spectrum of a known chemical compound is where both the fragments (m/z values) and their intensities are known and stored in the data system in the form of a reference table.

With modern GC/MS systems performing an up-to-date mass calibration is generally the final process in a tuning or autotuning program. This is preceded by a series of necessary adjustments and optimisations of the ion source, focusing, resolution and detection, which affect the position of the mass calibration. Tuning the lens potential particularly affects the transmission in individual mass areas. In particular, with beam instruments focusing must be adapted to the intensities of the reference substances in order to obtain the intensity pattern of the reference spectrum (see Section 2.3.5.2). The m/z values contained in a stored reference table are localised by the calibration program in the spectrum of the reference compound measured. The relevant centroid of the reference peak is calculated and correlated with the operation of the analyser. Using the stored reference table a precise calibration function for the whole mass range of the instrument can be calculated. The actual state of the mass spectrometer at the end of the tuning procedure is thus taken into account. The data are plotted graphically and are available to the user for assessment and documentation. For quadrupole and ion trap instruments the calibration graph is linear (Fig. 2.215), whereas with magnetic sector instruments the graph is exponential (Fig. 2.216).

Perfluorinated compounds are usually used as calibration standards (Table 2.53). Because of their high molecular weights the volatility of these compounds is sufficient to allow controllable leakage into the ion source. In addition, fluorine has a negative mass defect (^{19}F = 18.9984022), so that the fragments of these standards are below of the corresponding nominal mass and can easily be separated from a possible background of hydrocarbons with positive mass defects. The requirements of the reference substance are determined by the type of analyser. Quadrupole and ion trap instruments are calibrated with FC43 (PFTBA, perfluorotributylamine; Table 2.55 and Fig. 2.217) independent of their available mass

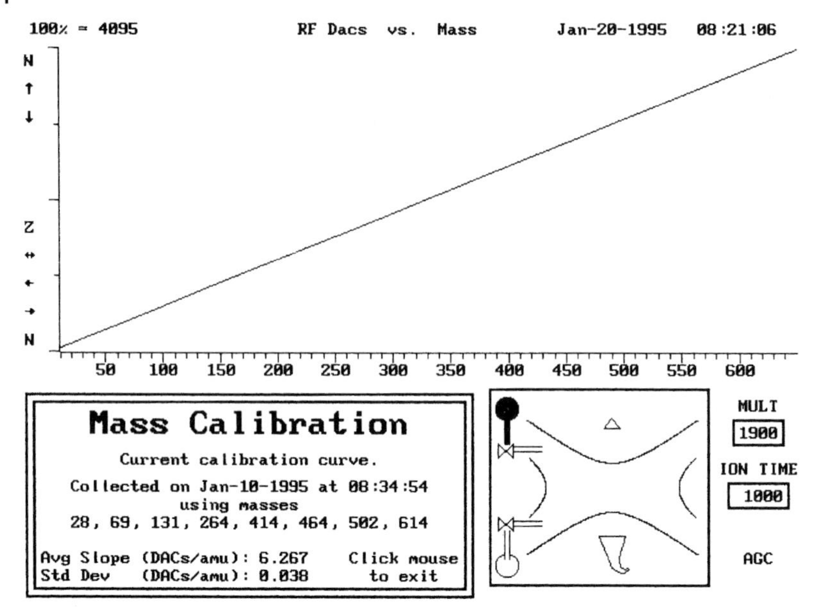

Fig. 2.215 Linear mass calibration of the quadrupol ion trap analyser (voltage of ring electrode against m/z value).

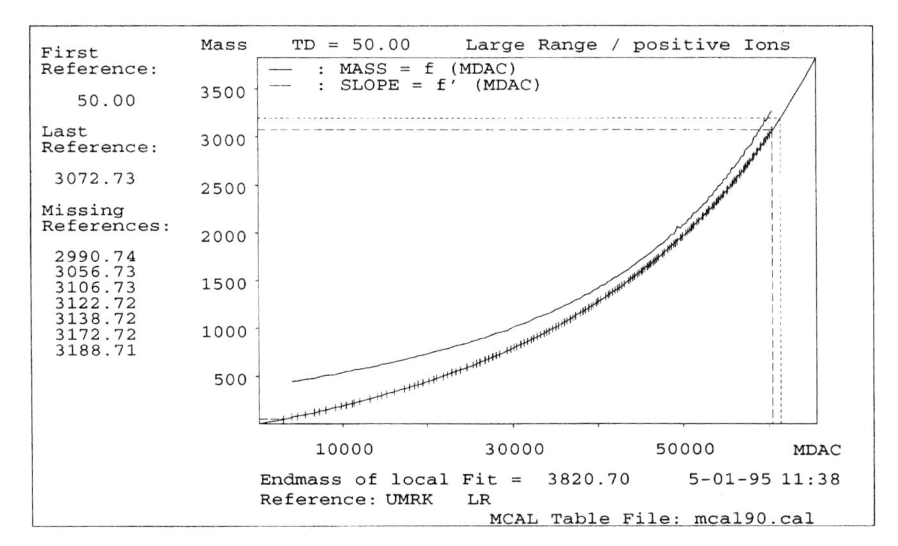

Fig. 2.216 Exponential mass calibration of the magnetic sector analyser.

range. For the compliance to a series of EPA methods and other regulations the targeted source tuning according to the specific manufacturers guidelines is required, see Table 2.54 (Eichelberger 1975). This for instance refers to EPA methods 501.3, 524.2, 8260B, CLP-SOW for the determination of volatiles using 4-bromofluorbenzene (BFB) and the EPA methods 625, 1625, 8250, 8270 on base/neutrals/acids or semivolatiles referring to difluorotriphenyl-

Table 2.53 Calibration substances and their areas of use.

Name	Formula	M	m/z max.	Instrument used	
				Magnetic sector	Quadrupole/ion trap
FC43 [1)]	$C_{12}F_{27}N$	671	614	Up to 620	Up to more than 1000
FC5311	$C_{14}F_{24}$	624	624	Up to 620	Up to more than 1000
PFK [2)]	–	–	1017	Up to 1000	Up to more than 1000
Perfluorinated triazines					
(C_7) [3)]	$C_{24}F_{45}N_3$	1185	1185	800–1200	–
(C_9) [4)]	$C_{30}F_{57}N_3$	1485	1485	800–1500	–
Fomblin [5)]	$(OCF(CF_3)CF_2)_x–(OCF_2)_y$			Up to 2500	–
CsI [6)]	–	–	15 981	High mass range	–

1) Perfluorotributylamine, 2) Perfluorokerosene, 3) Perfluorotriheptylazine, 4) Perfluorotrinonyltriazine, 5) Poly(perfluoropropylene oxide) (also used as diffusion pump oil), 6) Caesium iodide

Table 2.54 Ion abundance criteria for BFB and DFTPP target tuning.

Compound	m/z	Ion abundance criteria (relative abundance)	
BFB	50	15 to 40%	of mass 95
	75	30 to 60% (1)	of mass 95
	95	100%	base peak
	96	5 to 9%	of mass 95
	173	<2%	of mass 174
	174	>50	of mass 95
	175	5 to 9%	of mass 174
	176	>95% but <101%	of mass 174
	177	5 to 9%	of mass 176

(1) 30–80% of mass 95 for EPA 524.2

Compound	m/z	Ion abundance criteria (relative abundance)	
DFTPP	51	30–60%	of mass 198
	68	<2%	of mass 69
	70	<2%	of mass 69
	127	40–60%	of mass 198
	197	<1%	of mass 198
	198	100%	base peak
	199	5–9%	of mass 198
	275	10–30%	of mass 198
	365	> 1%	of mass 198
	441	present	of mass 443
	442	> 40%	of mass 198
	443	17–23%	of mass 442

Table 2.55 (a) Reference table FC43/PFTBA (EI, intensities >1%, quadrupole instrument).

Exact mass [u]	Intensity [%]	Formula	Exact mass [u]	Intensity [%]	Formula
68.9947	100.0	CF_3^+	225.9898	2.0	$C_5NF_8^+$
92.9947	2.0	$C_3F_3^+$	263.9866	27.0	$C_5NF_{10}^+$
99.9931	19.0	$C_2F_4^+$	313.9834	3.0	$C_6NF_{12}^+$
113.9961	11.0	$C_2NF_4^+$	351.9802	4.0	$C_6NF_{14}^+$
118.9915	16.0	$C_2F_5^+$	363.9802	1.0	$C_7NF_{14}^+$
130.9915	72.0	$C_3F_5^+$	375.9802	1.0	$C_8NF_{14}^+$
149.9899	4.0	$C_3F_6^+$	401.9770	2.0	$C_7NF_{16}^+$
168.9883	7.0	$C_3F_7^+$	413.9770	9.0	$C_8NF_{16}^+$
175.9929	3.0	$C_4NF_6^+$	425.9770	1.0	$C_9NF_{16}^+$
180.9883	3.0	$C_4F_7^+$	463.9738	3.0	$C_9NF_{18}^+$
213.9898	2.0	$C_4NF_8^+$	501.9706	8.0	$C_9NF_{20}^+$
218.9851	78.0	$C_4F_9^+$	613.9642	2.0	$C_{12}NF_{24}^+$

Table 2.55 (b) Reference table FC43/PFTBA (EI, intensities >0.1%, high resolution magnetic sector instrument).

Exact mass [u]	Intensity [%]	Formula	Exact mass [u]	Intensity [%]	Formula
4.00206		He^+	230.98508	1.1	$C_5F_9^+$
14.01510		CH_2^+	242.98508	0.0	$C_6F_9^+$
18.01002		H_2O^+	263.98656	37.6	$C_5NF_{10}^+$
28.00560		N_2^+	275.98656	0.2	$C_6NF_{10}^+$
30.99786		CF^+	280.98189	0.1	$C_6F_{11}^+$
31.98928		O_2^+	294.98496	0.2	$C_6NF_{11}^+$
39.96184		Ar^+	313.98336	2.7	$C_6NF_{12}^+$
43.98928		CO_2^+	325.98336	0.3	$C_7NF_{12}^+$
49.99626	3.0	CF_2^+	344.98177	0.0	$C_7NF_{13}^+$
68.99466	50.7	CF_3^+	351.98017	1.7	$C_6NF_{14}^+$
75.99933	1.0	$C_2NF_2^+$	363.98017	1.0	$C_7NF_{14}^+$
80.99466	1.2	$C_2F_3^+$	375.98017	1.2	$C_8NF_{14}^+$
92.99466	1.0	$C_3F_3^+$	401.97698	2.0	$C_7NF_{16}^+$
99.99306	3.9	$C_2F_4^+$	413.97698	10.8	$C_8NF_{16}^+$
113.99614	3.5	$C_2NF_4^+$	425.97698	4.0	$C_9NF_{16}^+$
118.99147	5.9	$C_2F_5^+$	451.97378	0.6	$C_8NF_{18}^+$
130.99147	49.8	$C_3F_5^+$	463.97378	7.3	$C_9NF_{18}^+$
149.98987	2.1	$C_3F_6^+$	475.97378	0.0	$C_{10}NF_{18}^+$
168.98827	5.1	$C_3F_7^+$	501.97059	22.7	$C_9NF_{20}^+$
175.99295	1.3	$C_4NF_6^+$	513.97059	0.1	$C_{10}NF_{20}^+$
180.98827	1.3	$C_4F_7^+$	525.97059	0.1	$C_{11}NF_{20}^+$
199.98668	0.2	$C_4F_8^+$	551.96740	0.0	$C_{10}NF_{22}^+$
213.98975	2.5	$C_4NF_8^+$	563.96740	0.1	$C_{11}NF_{22}^+$
218.98508	100.0	$C_4F_9^+$	575.96740	1.3	$C_{12}NF_{22}^+$
225.98975	1.4	$C_5NF_8^+$	613.96420	3.6	$C_{12}NF_{24}^+$

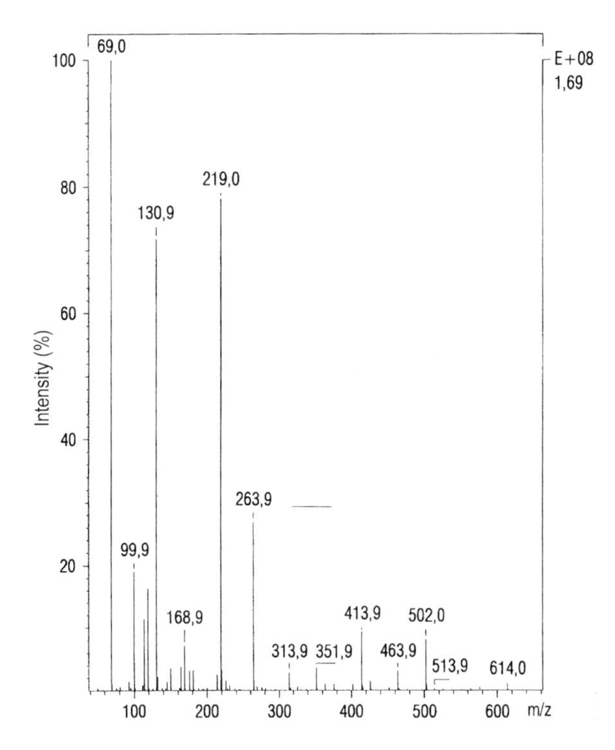

Fig. 2.217 FC43/PFTBA spectrum (Finnigan TSQ 700, EI ionisation, Q3).

phosphine (DFTPP) as tuning compounds providing consistent and instrument independent ion ratio profiles for quantitation. The precision of the mass scan and the linearity of the calibration allow the line to be extrapolated beyond the highest fragment which can be determined, which is m/z 614 for FC43. For magnetic sector instruments the use of PFK (perfluorokerosene; Table 2.56, Fig. 2.218) has proved successful besides FC43. It is particularly suitable for magnetic field calibration as it gives signals at regular intervals up to over m/z 1000. Also perfluorophenanthrene (FC5311) is frequently used as an alternative calibration compound to FC-43 which covers about the same mass range (see Table 2.57). Higher mass ranges can be calibrated using perfluorinated alkyltriazines or caesium iodide. Reference tables frequently take also account of the masses from the lower mass range, such as He, N_2, O_2, Ar or CO_2, which always form part of the background spectrum.

The calculated reference masses in the given tables are based on the following values for isotopic masses: ^{1}H 1.0078250321 u, ^{4}He 4.0026032497 u, ^{12}C 12.0000000000 u, ^{14}N 14.0030740052 u, ^{16}O 15.9949146221 u, ^{19}F 18.9984032000 u, and ^{40}Ar 39.9623831230 u. The mass of the electron 0.00054857991 u was taken into account for the calculation of the ionic masses (Audi 1995, Mohr 1999).

To record high resolution data for manual work peak matching is employed. At a given magnetic field strength one or two reference peaks and an ion of the substance being analysed are alternately shown on a screen. By changing the acceleration voltage (and the electric field coupled to it) the peaks are superimposed. From the known mass and voltage difference the exact mass of the substance peak is determined. This process is an area of the solid probe technique, as here the substance signal can be held constant over a long period.

Table 2.56 Reference table for PFK (perfluorokerosene), EI ionisation high resolution magnetic sector instrument.

Exact mass [u]	Intensity [%]	Formula	Exact mass [u]	Intensity [%]	Formula
4.002055		He^+	492.969112	1.10	$C_{11}F_{19}^+$
14.015101		CH_2^+	504.969112	0.65	$C_{12}F_{19}^+$
18.010016		H_2O^+	516.969112	0.50	$C_{13}F_{19}^+$
28.005599		N_2^+	530.965919	0.70	$C_{11}F_{21}^+$
30.007855	3.80	CF^+	542.965919	0.60	$C_{12}F_{21}^+$
31.989281		O_2^+	554.965919	0.50	$C_{13}F_{21}^+$
39.961835		Ar^+	566.965919	0.60	$C_{14}F_{21}^+$
51.004083	6.70	CHF_2^+	580.962725	0.70	$C_{12}F_{23}^+$
68.994661	100.00	CF_3^+	592.962725	0.65	$C_{13}F_{23}^+$
80.994661	0.50	$C_2F_3^+$	604.962725	0.60	$C_{14}F_{23}^+$
92.994661	3.30	$C_3F_3^+$	616.962725	0.50	$C_{15}F_{23}^+$
99.993064	5.60	$C_2F_4^+$	630.959531	0.50	$C_{13}F_{25}^+$
113.000889	0.02	$C_3HF_4^+$	642.959531	0.50	$C_{14}F_{25}^+$
118.991467	26.40	$C_2F_5^+$	654.959531	0.55	$C_{15}F_{25}^+$
130.991467	24.00	$C_3F_5^+$	666.959531	0.50	$C_{16}F_{25}^+$
142.991467	1.90	$C_4F_5^+$	680.956338	0.20	$C_{14}F_{27}^+$
154.991467	1.40	$C_5F_5^+$	692.956338	0.25	$C_{15}F_{27}^+$
168.988274	17.00	$C_3F_7^+$	704.956338	0.40	$C_{16}F_{27}^+$
180.988274	8.75	$C_4F_7^+$	716.956338	0.25	$C_{17}F_{27}^+$
192.988274	8.30	$C_5F_7^+$	730.953144	0.20	$C_{15}F_{29}^+$
204.988274	1.50	$C_6F_7^+$	742.953144	0.25	$C_{16}F_{29}^+$
218.985080	8.60	$C_4F_9^+$	754.953144	0.50	$C_{17}F_{29}^+$
230.985080	8.80	$C_5F_9^+$	766.953144	0.20	$C_{18}F_{29}^+$
242.985080	3.80	$C_6F_9^+$	780.949951	0.25	$C_{16}F_{31}^+$
254.985080	1.20	$C_7F_9^+$	792.949951	0.30	$C_{17}F_{31}^+$
268.981887	4.00	$C_5F_{11}^+$	804.949951	0.15	$C_{18}F_{31}^+$
280.981887	6.00	$C_6F_{11}^+$	816.949951	0.05	$C_{19}F_{31}^+$
292.981887	2.70	$C_7F_{11}^+$	830.946757	0.10	$C_{17}F_{33}^+$
304.981887	1.00	$C_8F_{11}^+$	842.946757	0.10	$C_{18}F_{33}^+$
318.978693	2.00	$C_6F_{13}^+$	854.946757	0.10	$C_{19}F_{33}^+$
330.978693	3.70	$C_7F_{13}^+$	866.946757	0.05	$C_{20}F_{33}^+$
342.978693	1.80	$C_8F_{13}^+$	880.943563	0.10	$C_{18}F_{35}^+$
354.978693	0.90	$C_9F_{13}^+$	892.943563	0.10	$C_{19}F_{35}^+$
368.975499	0.80	$C_7F_{15}^+$	904.943563	0.05	$C_{20}F_{35}^+$
380.975499	2.30	$C_8F_{15}^+$	916.943563	0.05	$C_{21}F_{35}^+$
392.975499	1.10	$C_9F_{15}^+$	930.940370	0.05	$C_{19}F_{37}^+$
404.975499	1.00	$C_{10}F_{15}^+$	942.940370	0.05	$C_{20}F_{37}^+$
416.975499	0.55	$C_{11}F_{15}^+$	954.940370	0.05	$C_{21}F_{37}^+$
430.972306	1.85	$C_9F_{17}^+$	966.940370	0.05	$C_{22}F_{37}^+$
442.972306	1.20	$C_{10}F_{17}^+$	980.937176	0.05	$C_{20}F_{39}^+$
454.972306	0.80	$C_{11}F_{17}^+$	992.937176	0.05	$C_{21}F_{39}^+$
466.972306	0.50	$C_{12}F_{17}^+$	1004.937176	0.05	$C_{22}F_{39}^+$
480.969112	1.40	$C_{10}F_{19}^+$	1016.937176	0.05	$C_{23}F_{39}^+$

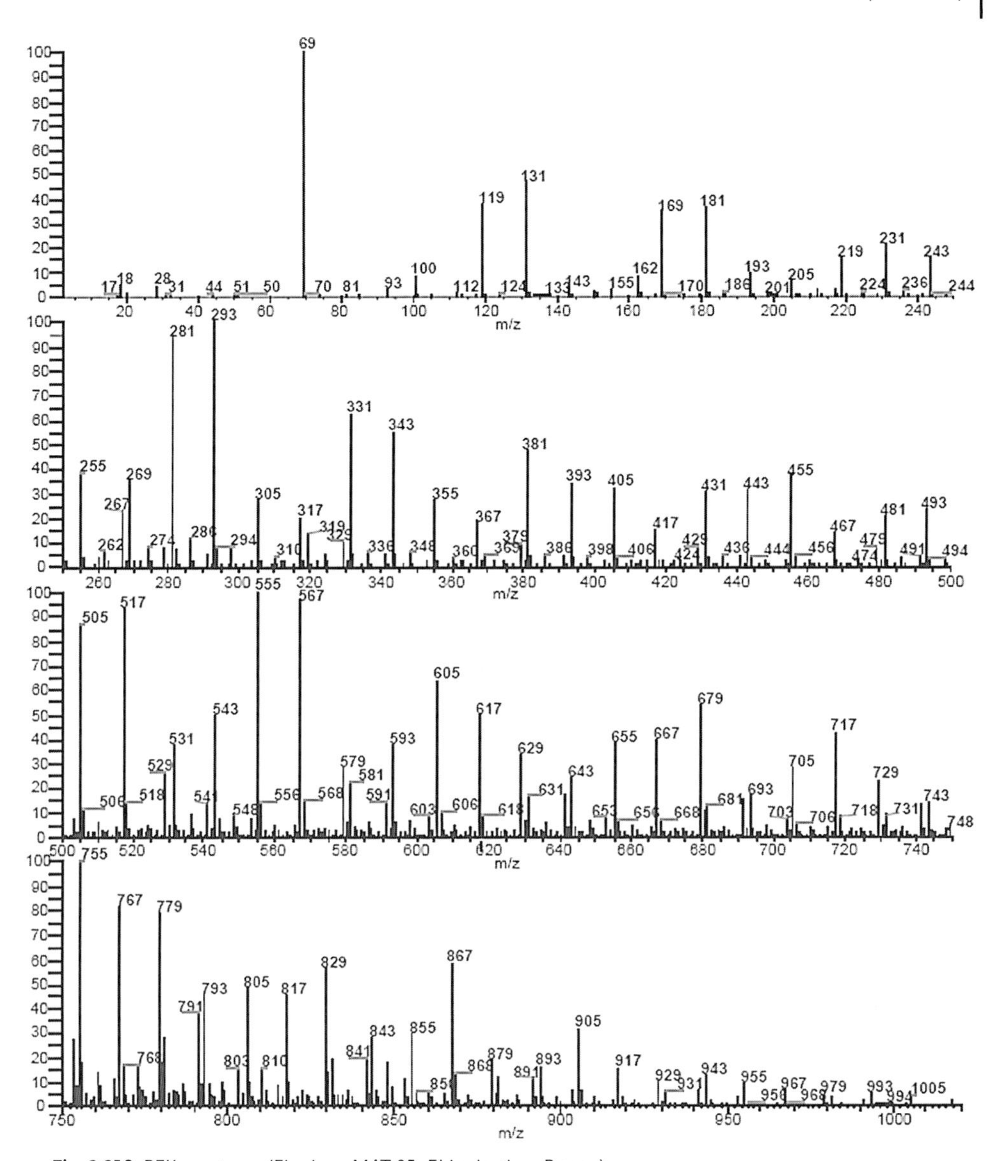

Fig. 2.218 PFK spectrum (Finnigan MAT 95, EI ionisation, B scan).

For GC/MS systems an internal scanwise calibration (scan-to-scan) by control of the data system is employed (internal mass calibration). At a given resolution (e.g. 10 000), a known reference is used which has been fed in at the same time. The analyser is positioned on the exact mass of the substance ion to be analysed relative to the measured centroid of the known reference. At the beginning of the next scan the exact position of the centroid of the reference mass is determined again and is used as a new basis for the next scan.

The usability of the calibration depends on the type of instrument and can last for a period of a few hours to several days or weeks. All tuning parameters, in particular the adjustment of the ion source, affect the calibration as described above. Special attention should be paid to the temperature of the source. Ion sources which do not contain any internal heating are heated by the cathode in the course of the operation. This results in a significant drift in the calibration over one day. Regular mass calibration is therefore recommended also to comply with the lab internal QA/QC procedures. Heated ion sources exhibit high stability which lasts for weeks and months.

The carrier gas flow of the GC has a pronounced effect on the position of the calibration. Ion sources with small volumes and also ion trap instruments with internal ionisation show a significant drift of several tenths of a mass unit if the carrier gas flow rate is significantly changed by a temperature program. Calibration at an average elution temperature, the use of an open split, or equipping the gas chromatograph with electronic pressure programming (EPC, electronic pressure control) for analysis at constant flow are imperative in this case. Severe contamination of the ion source or the ion optics also affects the calibration. However, reduced transmission of such an instrument should force cleaning to be carried out in good time.

For analyses with differing scan rates using magnetic instruments calibrations are carried out at the different rates which are required for the subsequent measurements. For scan rates which differ significantly, a mass drift can otherwise occur between calibration and measurement. Calibrations of quadrupole and ion trap instruments are practically independent of the scan rate.

Depending on the type of instrument, a new mass calibration is required if the ionisation process is changed. While for ion trap instruments switching from EI to CI ionisation is pos-

Table 2.57 EI positive ion spectra for perfluorophenanthrene (FC5311).

m/z	Rel. abundance %	m/z	Rel. abundance %
55	1.8	219	2.7
56	1.1	231	4.1
57	3.0	243	7.7
69	100.0	255	2.1
70	1.2	267	2.6
93	6.0	286	2.1
94	1.3	293	7.7
100	11.1	305	1.1
112	1.5	317	2.3
119	16.8	331	1.1
124	1.5	343	1.2
131	47.1	367	1.9
143	5.3	405	4.2
155	3.1	455	18.3
162	5.6	505	4.4
169	7.7	517	1.2
181	11.2	555	4.6
193	6.7	605	1.6
205	2.8	624	1.0
217	2.1	–	–

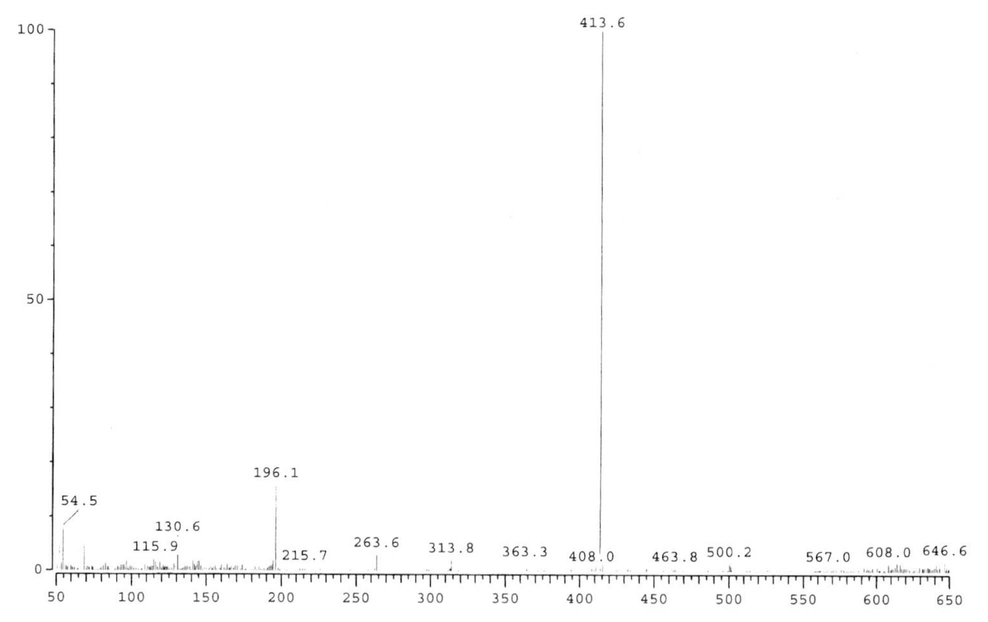

Fig. 2.219 FC43/PFTBA spectrum in PCI mode (CI gas methane, Finnigan GCQ).

sible without alterations to the analyser, with other types switching of the ion source or changing the ion volume are required. For an optimised CI reaction a lower source temperature is frequently used compared with the EI mode. After these changes have been made, a new mass calibration is necessary. This involves running a CI spectrum of the reference substance and consulting the CI reference table (Table 2.58). The perfluorinated reference substances can be used for both positive and negative chemical ionisation (PCI and NCI) (see Figs. 2.219 and 2.220). The intensities given in the CI reference tables can also be used here to optimise the adjustment of the reagent gas.

For reasons of quality control mass calibration should be carried out regularly and should be documented with a print-out (see Figs. 2.215 and 2.216).

2.4 Special Aspects of GC/MS Coupling

2.4.1 Vacuum Systems

Many benchtop quadrupole GC/MS systems are designed for flow rates of about 1 mL/min and are therefore suited for use with normal bore capillary columns (internal diameter 0.25 mm). Larger column diameters, however, can usually only be used with limitations. Even higher performance vacuum systems cannot improve on this value, as the design of the ion sources may be optimised for a particular carrier gas flow. Ion trap GC/MS systems with internal ionisation can tolerate a carrier gas flow of up to 3 mL/min. Normal and wide bore capillaries can be used with these instruments (see also Section 2.2.4).

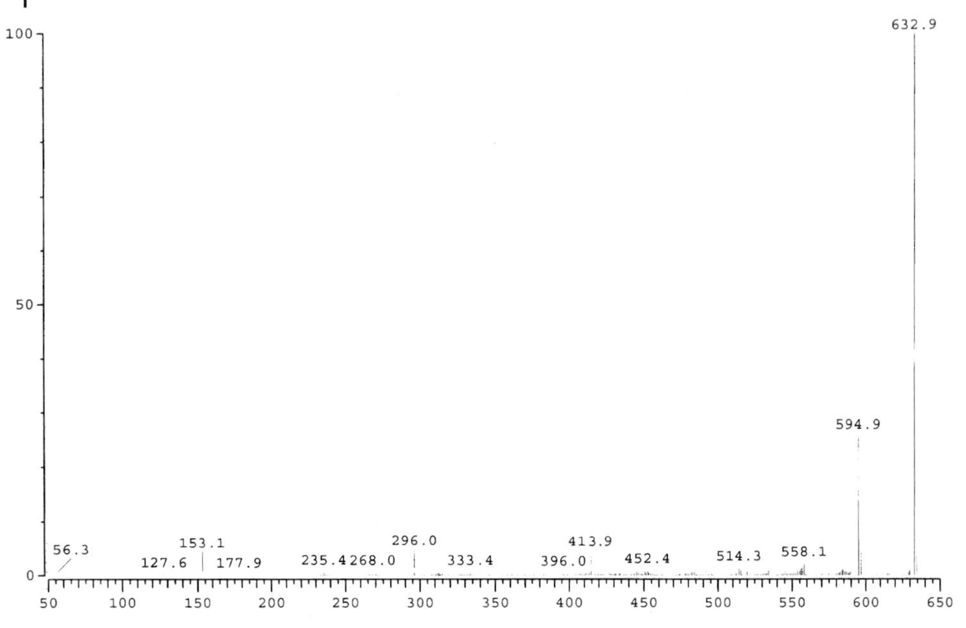

Fig. 2.220 FC43/PFTBA spectrum in NCI mode (CI gas methane, Finnigan GCQ).

Table 2.58 Reference table for PFK (perfluorokerosene) in NCI mode (CI gas ammonia)

Exact mass	Intensity [%]	Formula	Exact mass	Intensity [%]	Formula
168.9888	2	C_3F_7	530.9664	11	$C_{11}F_{21}$
211.9872	5	C_5F_8	535.9680	8	$C_{13}F_{20}$
218.9856	8	C_4F_9	542.9664	13	$C_{12}F_{21}$
230.9856	26	C_5F_9	554.9664	21	$C_{13}F_{21}$
249.9840	3	C_5F_{10}	573.9648	17	$C_{13}F_{22}$
261.9840	40	C_6F_{10}	585.9648	20	$C_{14}F_{22}$
280.9824	81	C_6F_{11}	604.9633	31	C_4F_{23}
292.9824	12	C_7F_{11}	611.9617	13	$C_{13}F_{24}$
311.9808	51	C_7F_{12}	623.9617	42	$C_{14}F_{24}$
330.9792	100	CF_{13}	635.9617	32	$C_{15}F_{24}$
342.9792	32	C_8F_{13}	654.9601	21	$C_{15}F_{25}$
361.9776	35	C_8F_{14}	661.9585	13	$C_{14}F_{26}$
380.9760	63	C_8F_{15}	673.9585	37	$C_{15}F_{26}$
392.9760	48	C_9F_{15}	685.9585	21	$C_{16}F_{26}$
411.9744	16	C_9F_{16}	699.9553	7	$C_{14}F_{28}$
423.9744	15	$C_{10}F_{16}$	704.9569	6	$C_{16}F_{27}$
430.9728	39	C_9F_{17}	711.9553	8	$C_{15}F_{28}$
442.9728	36	$C_{10}F_{17}$	723.9553	12	$C_{16}F_{28}$
454.9728	16	$C_{11}F_{17}$	735.9553	6	$C_{17}F_{28}$
473.9712	11	$C_{11}F_{18}$	750.9601	4	$C_{23}F_{25}$
480.9696	19	$C_{10}F_{19}$	761.9521	2	$C_{16}F_{30}$
492.9696	23	$C_{11}F_{19}$	773.9521	3	$C_{17}F_{30}$
504.9696	18	$C_{12}F_{19}$	787.9489	3	$C_{15}F_{32}$
516.9696	7	$C_{13}F_{19}$	799.9489	1	C_6F_{32}
523.9680	7	$C_{12}F_{20}$			

The carrier gas for GC/MS is either helium or hydrogen. It is well known that the use of hydrogen significantly improves the performance of the GC, lowers the elution temperatures of compounds, and permits shorter analysis times because of higher flow rates. The more favourable van Deemter curve for hydrogen (see Fig. 2.88) accounts for these improvements in analytical performance. As far as mass spectrometry is concerned, when hydrogen is used the mass spectrometer requires a higher vacuum capacity and thus a more powerful pumping system. The type of analyser and the pumping system determine the advantages and disadvantages.

For the turbo molecular pumps mostly used at present the given rating is defined for pumping out a nitrogen atmosphere. The important performance data of turbo molecular pumps is the compression ratio. The compression ratio describes the ratio of the outlet pressure (forepump) of a particular gas to the inlet pressure (MS). The turbo molecular pump gives a completely background-free high vacuum and exhibits excellent start-up properties, which is important for benchtop instruments or those for mobile use. The use of helium lowers the performance of the pump. The compression ratio (e.g. for the Balzer TPH 062) decreases from 10^8 for nitrogen to $7 \cdot 10^3$ for helium, then to $6 \cdot 10^2$ for hydrogen. The reason for the much lower performance with hydrogen is its low molecular weight and high diffusion rate.

A turbo molecular pump essentially consists of the rotor and a stator (Fig. 2.221). Rotating and stationary discs are arranged alternately. All the discs have diagonal channels, whereby the channels on the rotor disc are arranged so that they mirror the positions of the channels of the stator discs. Each channel of the disc forms an elementary molecular pump. All the channels on the disc are arranged in parallel. A rotor disc together with a stator disc forms a single pump stage which produces a certain compression. The pumping process is such that a gas molecule which meets the rotor acquires a velocity component in the direction of rotation of the rotor in addition to its existing velocity. The final velocity and the direction in which the molecule continues to move are determined from the vector sum of the two velo-

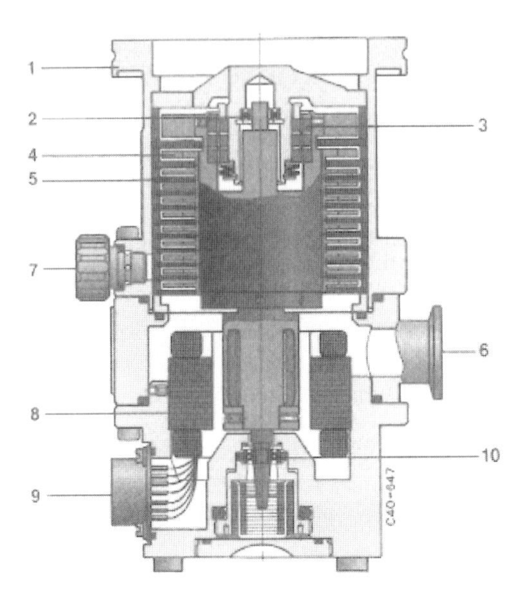

1 High vacuum connection (MS)
2 Emergency bearing
3 Permanent magnet bearing
4 Rotor
5 Stator
6 Forevacuum connection
7 Flooding connection
8 Motor
9 Electrical connection (control instrument)
10 High precision ball bearings with ceramic spheres

Fig. 2.221 Construction of a turbo molecular pump with ceramic ball bearings and permanent magnet bearings for use in mass spectrometry (after Balzers).

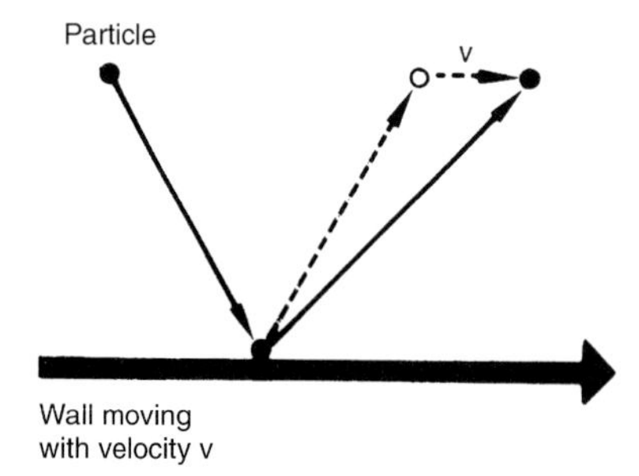

Fig. 2.222 Principle of the molecular pump (after Balzers).

cities (Fig. 2.222). The thermal motion of a molecule, which is initially undirected, is converted into directed motion when the molecule enters the pump. An individual pumping step only produces a compression of about 30. Several consecutive pumping steps, which reinforce each other's action, lead to very high compression rates.

The reduction in the performance of the pump on using hydrogen leads to a measurable increase in pressure in the analyser. Through collisions of substance ions with gas particles on their path through the analyser of the mass spectrometer, the transmission and thus the sensitivity of the instrument is reduced in the case of beam instruments (Fig. 2.223). The mean free path L of an ion is calculated according to:

$$L = p^{-1} \cdot 5 \cdot 10^{-3} \text{ [cm]} \qquad (23)$$

The effect can be compensated for by using higher performance pumps or additional pumps (differentially pumped systems for source and analyser). Ion trap instruments do not exhibit this behaviour because of their ion storage principle.

When hydrogen is used as the carrier gas in gas chromatography the use of oil diffusion pumps is an advantage. The pump capacity is largely independent of the molecular weight

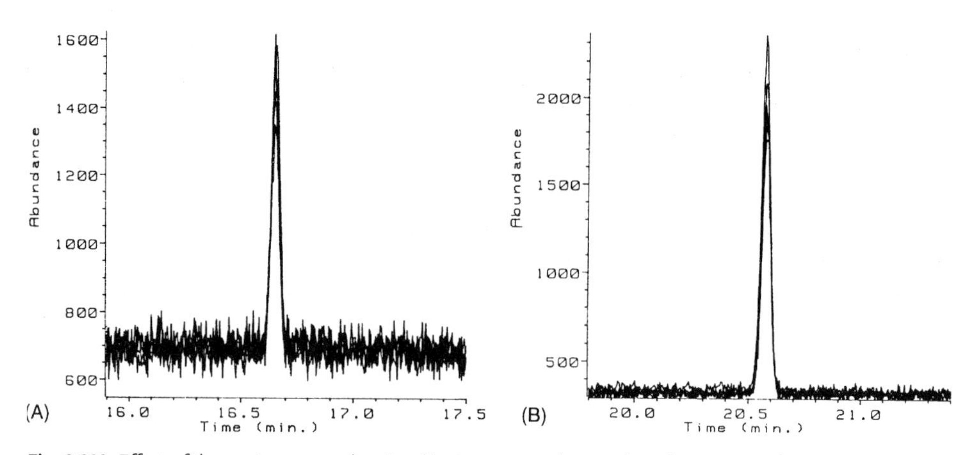

Fig. 2.223 Effect of the carrier gas on the signal/noise ratio in the quadrupole GC/MS (after Schulz). PCB 101, 50 pg, SIM m/z 256, 326, 328.
(A) Carrier gas hydrogen, S/N 4:1. (B) Carrier gas helium, S/N 19:1

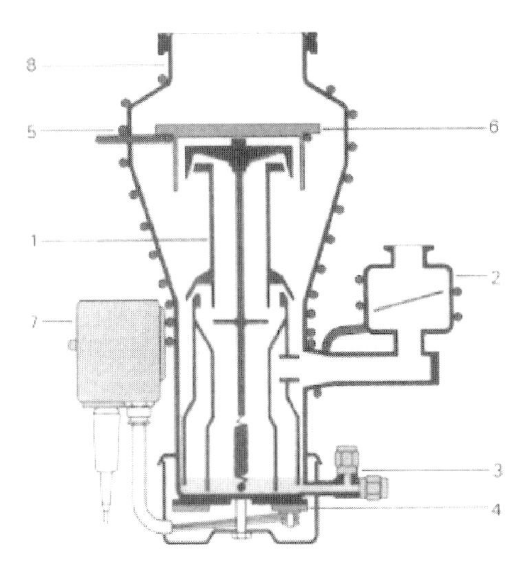

1 Jet system made of pressed aluminium components
2 Forevacuum baffle
3 Fuel top-up/measurement
4 Heating
5 Water cooling
6 Optically sealed baffle
7 Thermal protection switch/connection
8 Stainless steel pump casing

Fig. 2.224 Construction of an oil diffusion pump (after Balzers).

and is therefore very suitable for hydrogen and helium. Oil diffusion pumps also have long service lives and are economical as no movable parts are required. The pump operates using a propellant which is evaporated on a heating plate (Fig. 2.224). The propellant vapour is forced downwards via a baffle and back into a fluid reservoir. Gas molecules diffuse into the propellant stream and are conveyed deeper into the pump. They are finally evacuated by a forepump. Perfluoropolyethers (e.g. Fomblin) or polyphenyl ethers (e.g. Santovac S) are now used exclusively as diffusion pump fluids in mass spectrometry. However, the favourable operation of the pump results in disadvantages for the operation of the mass spectrometer. Because a heating plate is used, the diffusion pump starts sluggishly and can only be vented again after cooling. While older models require water cooling, modern diffusion pumps for GC/MS instruments are air-cooled by a ventilator so that heating and cool-down times of 30 min and longer are unavoidable. Propellant vapour can easily lead to a permanent background in the mass spectrometer, which makes the use of a cooled baffle necessary, depending on the construction. The detection of negative ions in particular can be affected by the use of fluorinated polymers.

Furthermore, it should be noted that, as a reactive gas, hydrogen can hydrogenate the substances being analysed. This effect has already been known from gas chromatography for many years. Reactions in hot injectors can lead to the appearance of unexpected by-products. In GC/MS ion sources reactions leading to hydrogenation products are also known. These cases are easy to identify by changing to helium as the carrier gas.

Both turbo molecular pumps and oil diffusion pumps require a mechanical forepump, as the compression is not sufficient to work against atmospheric pressure. Rotating vane pumps are generally used as forepumps (Fig. 2.225). Mineral oils are used as the operating fluid. Oil vapours from the rotating vane pump passing into the vacuum tubing to the turbo pump are visible in the mass spectrometer as a hydrocarbon background, in particular, on frequent venting of the system. Special devices for separation or removal are necessary. As an alternative for the production of a hydrocarbon-free forevacuum, spiro molecular pumps can be used. These pumps can be run dry without the use of oil. Their function involves centrifugal acceleration of the gas molecules through several pumping steps like the turbo mo-

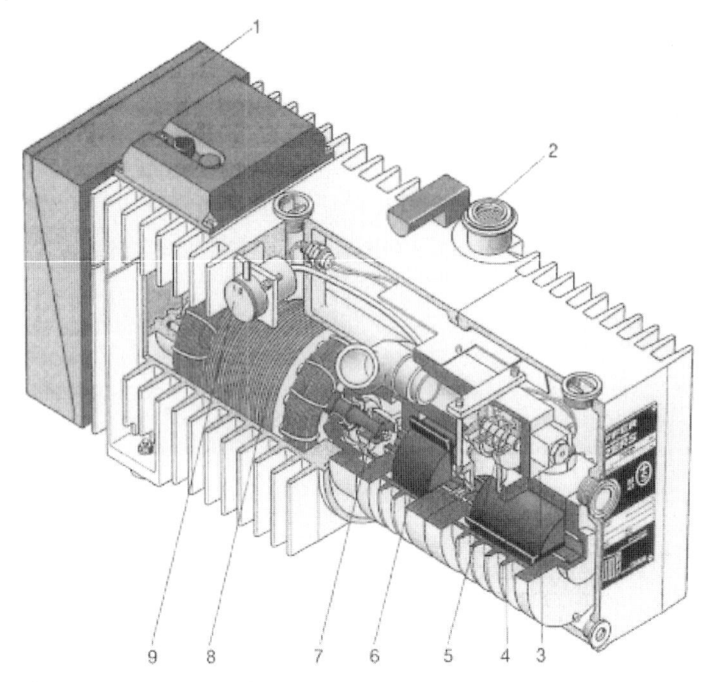

Fig. 2.225 Cross-section of a two-stage rotary vane vacuum pump (after Balzers).
1 Start-up control
2 Exhaust
3 Aeration valve
4 Pump stage 1
5 High vacuum safety valve
6 Pump stage 2
7 Motor coupling
8 Motor
9 Gas ballast valve

lecular pump. For the start-up an integrated membrane pump is used. In particular for mobile use of GC/MS systems and in other cases where systems are frequently disconnected from the mains electricity, spiro molecular pumps have proved useful.

2.4.2 GC/MS Interface Solutions

2.4.2.1 Open Split Coupling

Open split coupling has been known to many spectroscopists from the earliest days of the GC/MS technique. The challenge of open split coupling lay in the balancing of incompatible flow rates, particularly in the cutting out of the solvent peak or other unwanted main components (Fig. 2.226). The cathode was thus protected from damage and the ion source from contamination and the associated reduction in sensitivity thus prevented. These initial problems have now receded completely into the background thanks to fused silica capillaries and easily maintained ion source constructions.

Many advantages of open split coupling are nevertheless frequently exploited in various applications.

The rapid change of GC columns and the desire of the analyst to have a further detector besides the mass spectrometer after the split are now priorities. Because of the different pressure ratios a makeup gas now usually has to be added (Fig. 2.227). If the carrier gas

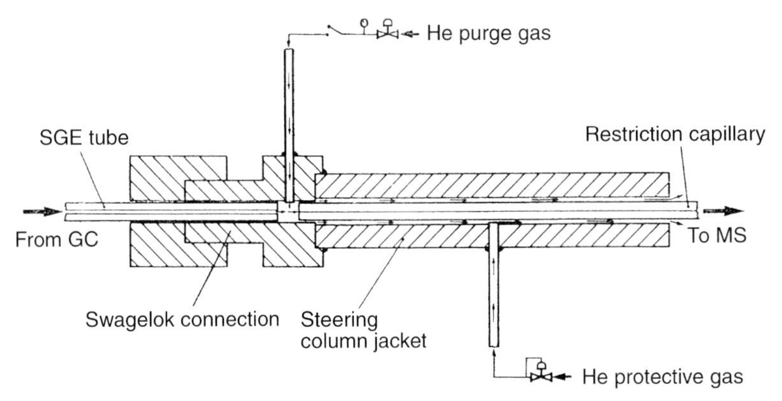

Fig. 2.226 Lengthways section of an open coupling for the expulsion of solvents (after Abraham).

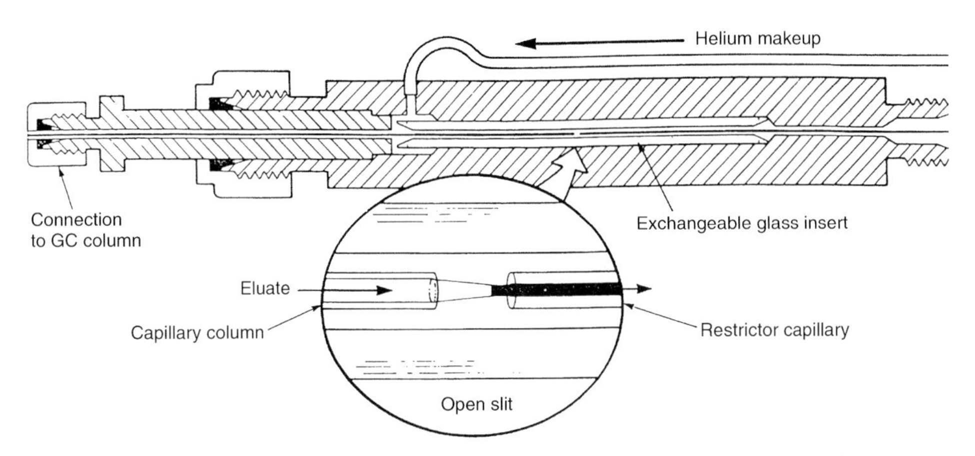

Fig. 2.227 Lengthways section of an open coupling for complete transfer of substance with controllable He makeup (ITD 800, Finnigan).

stream through the column is not sufficient, a considerable air leak can appear if the detector is only coupled by means of a T-piece. In this case the mass spectrometer sucks air through the detector into the ion source via the T-piece. By adding a makeup gas at the split point, these situations can be reliably prevented. The split situated after the column can thus be adapted rapidly to different column types and split ratios.

Advantages of Open Split Coupling

- The retention times of classical detectors, such as FID or ECD, are retained in the GC/MS and allow the direct comparison of chromatograms.
- At the end of the column a split can allow the use of an additional element-specific detector, such as ECD or NPD, and thus give information additional to that given by the mass spectrometer.
- The capillary column can be readily exchanged, usually without venting the mass spectrometer, allowing rapid resumption of work.

- The choice of GC conditions (column length, diameter, flow rate) can be optimised to give the best possible separation independent of the mass spectrometer.
- The use of wide bore, megabore and even packed columns (gas analysis) is possible. The excess eluate not sucked into the mass spectrometer is passed out via the split.
- A constant flow of material, which is independent of the oven temperature of the gas chromatograph, reaches the mass spectrometer. This allows precise optimisation of the ion source.

Disadvantages of Open Split Coupling

- The split point is at atmospheric pressure. To prevent the penetration of air the split point must be flushed with carrier gas. The additional screw joints can give rise to damaging leaks.
- If, because of the balance of column flow and suction capacity of the mass spectrometer, there is a positive split ratio, the sensitivity of the system decreases.
- On improper handling, e.g. on penetration of particles from the sealing ferrule, or a poor cut area at the end of the column, the quality of the GC is impaired (tailing).

2.4.2.2 Direct Coupling

If the end of the column is inserted right into the ion source of the mass spectrometer, the eluate reaches the mass spectrometer directly and unhindered. The entire eluate thus reaches the ion source undivided (Fig. 2.228). For this reason direct coupling is regarded as ideal for residue analysis. The pumping capacity of modern mass spectrometers is adjusted to be compatible with the usual flow rates of regular 0.25 mm and 0.32 mm ID capillary columns.

The end part of the capillary column which is inserted into the ion source requires particular attention to be paid to the construction of the GC/MS interface. The effect of the vacuum on the open end of the column causes molecular flow (as opposed to viscous flow in the column itself) with an increased number of wall collisions. According to the systematic investigations of Henneberg and Schomburg, there is possible adsorption on the column walls at the entry of the capillary into the ion source. It should therefore be ensured that there is uniform heating at this location in particular. Heating the interface too strongly can cause thermal decomposition caused by an activation of the column wall and a reduction in chromatographic performance.

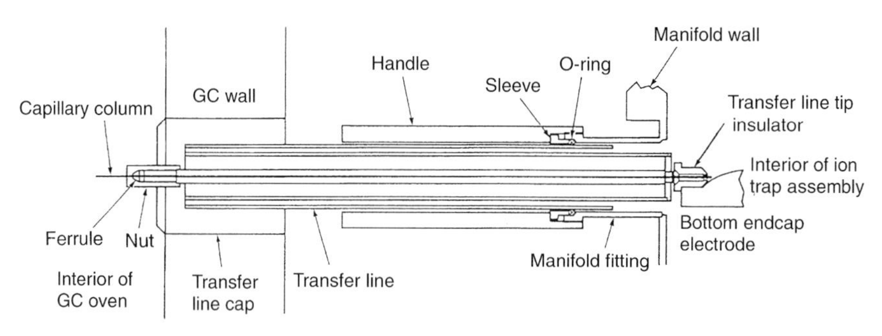

Fig. 2.228 Cross-section of a direct coupling (Finnigan).

Advantages of Direct Coupling

- Uncomplicated construction and handling.
- Single route for the substance from the GC injector to the ion source via a capillary column.
- Reliable vacuum seal at the point where the column is screwed in.

Disadvantages of Direct Coupling

- To change the column the mass spectrometer must be cooled and vented.
- The carrier gas flow into the ion source is not constant as it depends on the GC temperature program chosen and the column dimensions. Working under constant flow conditions (EPC) is recommended.
- The vacuum in the MS system affects the GC column and shortens the retention times in comparison with FID or ECD, for example. A direct comparison with these chromatograms is not straightforward.
- Constantly high interface temperatures limit the lifetime and inertness of sensitive types of column film, such as Carbowax 1701. It is essential to be aware of and adhere to the maximum continuous temperature of a particular GC column.
- The choice of column and adjustment of optimal flow rate is limited by the maximum carrier gas flow for the mass spectrometer.

2.4.2.3 Separator Techniques

For the coupling of capillary gas chromatography with mass spectrometry, special separators are no longer necessary. Only GC systems, which still involve packed columns with flow rates of ca. 10 mL/min or the use of halfmil capillary columns (internal diameter 0.53 mm), cannot be coupled to mass spectrometers without a separator. Separators are inserted between the column and the mass spectrometer to selectively lower the carrier gas load. This procedure, however, involves considerable loss of analyte.

The one-step jet separator (Biemann-Watson separator) frequently used with packed columns, operates according to the principle of diffusion of the smaller carrier gas molecules away from the transmission axis. These are evacuated by an additional rotating vane pump. A reduction in the carrier gas flow of up to ca. 1:100 can be used. Heavier molecules reach the ion source of the mass spectrometer via a transfer capillary. The use of separators involves the loss of ca. 40–80% of the sample material and since capillary column techniques have become widely used for GC/MS residue analysis, they should only be regarded as having historical interest.

References for Chapter 2

Section 2.1.1

Arthur, C. L., Potter, D. W., Buchholz, K. D., Motlagh, S., Pawliszyn, J., Solid-Phase Microextraction for the Direkt Analysis of Water: Theory and Practice, LCGC 10 (1992), 656–661.

Berlardi, R., Pawliszyn, J., The Application of Chemically Modified Fused Silica Fibers in the Extraction of Organics from Water Matrix Samples and Their Rapid Transfer to Capillary Columns, Water Pollution Research J. of Canada, 1989, 24, 179.

Boyd-Boland, A., Magdic, S., Pawleszyn, J., Simultaneous Determination of 60 Pecticides in Water Using Solid Phase Microextraction and Gas Chromatography-Mass Spectrometry, Analyst 121, 929–937 (1996).

Boyd-Boland, A., Pawleszyn, J., Solid Phase Microextraction of Nitrogen Containing Herbicides, J. Chromatogra., 1995, 704, 163–172.

Brenk, F. R., Bestimmung von polychlorierten Biphenylen in Ölproben, LaborPraxis 3 (1986), 222–224.

Brenk, F. R., GC-Auswertung und Quantifizierung PCB-haltiger Öle. LaborPraxis 4 (1986), 332–340.

Bundt, J., Herbel, W., Steinhart, H., Franke, S., Francke, W., Structure Type Separation of Diesel Fuels by Solid Phase Extraction and Identification of the Two- and Three-Ring Aromatics by Capillary GC-Mass Spectrometry, J. High Res. Chrom. 14 (1991), 91–98.

Ciupe, R., Spangenberg, J., Meyer, T., Wild, G., Festphasenextraktionstrennsystem zur gaschromatographischen Bestimmung von PAK's in Mineralöl, GIT Fachz. Lab. 8 (1994), 825–829.

Deutsche Gesellschaft für Mineralölwissenschaft und Kohlechemie e.V., DGMK-Projekt 387 – Bestimmung von PCB's in Ölproben, DGMK-Berichte, Hamburg 1985.

Deußing, G., Efficient Multi-residue Analysis of EDCs in Drinking Water, GIT Lab.J. 5, (2005) 17–19.

Dünges, W., Muno, H., Unckell, F., Rationelle prächromatographische Mikromethoden für die ppb- und ppt-Analytik, DVGW-Schriftenreihe Wasser Nr. 108, Eschborn 1990.

Eisert, R., Pawliszyn, J., New Trends in Solid Phase Microextraction, Critical Reviews in Analytical Chemistry 27, 103–135 (1997).

Farajzadeh, M.A., Hatami, M., A new selective SPME fiber for some n-alkanes and its use for headspace sampling of aqueous samples, J. Sep. Sci. 26, (2003) 802–808.

Festphasen-Mikroextraktion – Extraktion organischer Komponenten aus Wasser ohne Lösungsmittel, SUPELCO International 1992 (Firmenschrift), 3–5.

Gorecki, T., Martos, P., Pawliszyn, J., Strategies for the Analysis of Polar Solvents in Liquid Matrices, Anal. Chem. 70, 19–27 (1998).

Gorecki, T., Pawleszyn, J., Sample introduction approaches for solid phase microextraction/rapid GC, Anal. Chem., 1995, 34, 3265–3274.

Gorecki, T., Pawliszyn, J., The Effect of Sample Volume on Quantitative Analysis by SPME. Part I: Theoretical Considerations, Analyst 122, 1079–1086 (1997).

Hein, H., Kunze, W., Umweltanalytik mit Spektrometrie und Chromatographie, VCH: Weinheim 1994.

Kicinski, H. G., Adamek, S., Kettrup, A., Trace Enrichment and HPLC Analysis of Polycyclic Aromatic Hydrocarbons in Environmental Samples Using Solid Phase Extraction in Connection with UV/VIS Diode-Array and Fluorescence Detection. Chromatographia 28 (1989), 203–208.

Kicinski, H. G., Kettrup, A., Festphasenextraktion und HPLC-Bestimmung von polycyclischen Aromaten aus Trinkwasser, Vom Wasser 71 (1988), 245–254.

Ligor, M., Buszewski, B., The comparison of solid phase microextraction-GC and static headspace-GC for determination of solvent residues in vegetable oils, J. Sep. Sci. 31 (2008) 364–371.

Lord, H.L., Pawliszyn, J., Recent Advances in Solid Phase Microextraction and Membrane Extraction with a Sorbent Interface, see at http://www.science.uwaterloo.ca/chemistry/pawliszyn/Mesi/Recentadvances.htm.

Lord, H. L., Pawliszyn, J., Recent Advances in Solid Phase Microextraction, LCGC International 1998 (12) 776–785.

Luo, Y., Pan, L., Pawliszyn, J., Determination of Five Benzodiazepines in Aqueous Solution and Biological Fluids using SPME with Carbowax/DVB Fibre Coating, J. Microcolumn. Sep. 10, 193–201 (1998).

MacGillivray, B., Pawleszyn, J., Fowlei, P., Sagara, C., Headspace Solid-Phase Microextraction Versus Purge and Trap for the Determination of Substituted Benzene Compounds in Water, J. Chromatogr. Sci., 1994, 32, 317–322.

Magdic, S., Boyd-Boland, A., Jinno, K., Pawleszyn, J., Analysis of Organophosphorus Insecticides from Environmental Samples by Solid Phase Micro-extraction, J. Chromatogr. A 736, 219–228 (1996).

Martos, P., Saraullo, A., Pawliszyn, J., Estimation of Air/Coating Distribution Coefficients for Solid Phase Microextraction Using Retention Indexes from Linear Temperature-Programmed Capillary Gas Chromatography. Application to the Sampling and Analysis of Total Petroleum Hydrocarbons in Air, Anal. Chem. 69, 402–408 (1997).

Nielsson, T., Pelusio, F., et al., A Critical Examination of Solid Phase Micro-Extraction for Water Analysis, In: 16th Int. Symp. Cap. Chrom. Riva del Garda, P. Sandra (Ed.), Huethig 1994, 1148–1158.

Nolan, L., Ziegler, H., Shirey, R., Hastenteufel, S., Festphasenextraktion von Pestiziden aus Trinkwasserproben, LaborPraxis 11 (1991), 958–960.

Pan, L., Chong, J. M., Pawliszyn, J., Determination of Amines in Air and Water using Derivatization Combined with SPME, J. Chromatogr. 773, 249–260 (1997).

Pawlyszyn, J., Solid Phase Microextraction: Theory and Practice, Wiley-VCH: New York 1997.

Poerschmann, Z., Zhang, Z., Kopinke, F.-D., Pawliszyn, J., Solid Phase Microextraction for Determining the Distribution of Chemicals in Aqueous Matrices, Anal. Chem. 69, 597–600 (1997).

Potter, D., Pawliszyn, J., Detection of Substituted Benzenes in Water at the pg/ml Level Using Solid Phase Microextraction, J. Chromatogr., 1992, 625, 247–255.

Potter, D., Pawliszyn, J., J. Chromatogr. 625 (1992), 247–255.

Reupert, R., Brausen, G., Trennung von PAK's durch HPLC und Nachweis durch Fluoreszenzdetektion, GIT Fachz. Lab. 11 (1991), 1219–1221.

Reupert, R., Plöger, E., Bestimmung stickstoffhaltiger Pflanzenbehandlungsmittel in Trink-, Grund- und Oberflächenwasser: Analytik und Ergebnisse, Vom Wasser 72 (1989), 211–233.

Saraullo, A., Martos, P., Pawliszyn, J., Water Analysis by SPME Based on Physical Chemical Properties of the Coating, Anal. Chem. 69, 1992–1998 (1997).

SBSE background information, see www.gerstel.de/produkt_twister.htm.

Serodio, P., Nogueira, J.M.F., Multi-residue screening of Endocrine Disrupting Chemicals in water samples by stir bar sorptive extraction gas chromatography mass spectrometry detection, Anal. Chim. Acta 517, (2004) 21–32.

Shirey, R. E., Eine neue Polyacrylatfaser zur Festphasenmikroextraktion von polaren halbflüchtigen Verbindungen aus Wasser, SUPELCO Reporter 13 (1994), Firmenschrift, 8–9.

Shirey, R. E., Mani, Venkatachalem, New Carbon-coated Solid Phase Microextraction (SPME) Fibers for Improved Analyte Recovery, SUPELCO talk 497015, given at the 1997 Pittcon conference in Atlanta, Georgia, USA.

Shutao, W. , Wang, Y., Hong, Y., Jie, Y, Preparation of a Carbon-Coated SPME Fiber and Application to the Analysis of BTEX and Halocarbons in Water, Chromatographia 63, (2006) 365–371.

Steffen, A., Pawleszyn, J., The Analysis of Flavour Volatiles using Headspace Solid Phase Microextraction, J. Agric. Food Chem. 44, 2187-2193 (1996).

SUPELCO, Bulletin 928, SPME Troubleshooting Guide, Supelco T101928, see www.sigma-aldrich.com.

SUPELCO, Festphasen-Mikroextraktion organischer Komponenten aus wäßrigen Proben ohne Lösungsmittel, Die SUPELCO-Reporter 12 (1993), Firmenschrift, 3–5.

Van der Kooi, M. M. E., Noij, Th. H. M., Evaluation of Solid Phase Micro-Extraction for the Analysis of Various Priority Pollutants in Water, In: 16th Int. Symp. Cap. Chrom. Riva del Garda, P. Sandra (Ed.). Huethig 1994. 1087–1098.

Van Horne, K. C., Handbuch der Festphasenextraktion, ICT: Frankfurt, Basel, Wien 1993.

Yang, M., Orton, M., Pawliszyn, J., Quantitative Determination of Caffeine in Beverages Using a Combined SPME-GC/MS Method, J. Chem. Ed. 74, 1130–1132 (1997).

Zhang, Z., Pawliszyn, J., Headspace Solid Phase Microextraction, Anal. Chem. 65, 1843–1852, 1993.

Zhang, Z., Yang, M. J., Pawleszyn, J., Solid Phase Microextraction – A New Solvent-Free Alternative for Sample Preparation, Anal. Chem. 66, 844A–853A, 1994.

Zief, M., Kieser, R., Solid Phase Extraction for Sample Preparation, A technical guide to theory, method development and use, J.T. Baker Technical Library, see http://www.mallbaker.com/Literature/Documents/8008.pdf.

Section 2.1.2

Abraham, B., Rückstandsanalyse von anabolen Wirkstoffen in Fleisch mit Gaschromatographie-Massenspektrometrie, Dissertation, Technische Universität Berlin, 1980.

Bartle, K., Pullen, F., A happy coupling? – Aspects of the ongoing debate about the place of supercritical fluid chromatography coupled to MS, Analysis Europa 4 (1995) 44–45.

Bittdorf, H., Kernforschungszentrum Karlsruhe, Institut für Radiochemie Abt. Wassertechnologie, Überprüfung der Einsetzbarkeit der Supercritical Fluid Extraction (SFE) bei der Probenvorbereitung von Nitroaromaten, (pers. Mitteilung).

Blanch, G. P., Ibanez, E., Herraiz, M., Reglero, G., Use of a Programmed Temperature Vaporizer for Off-Line SFE/GC Analysis in Food Composition Studies, Anal. Chem. 66 (1994) 888–892.

Bowadt, S., Johansson, B., et al., Independent Comparison of Soxhlet and Supercritical Fluid Extraction for the Determination of PCBs in an Industrial Soil, Anal. Chem. 67 (1995) 2424–2430.

Burford, M. D., Hawthorne. S. B., Miller, D. J., Extraction Rates of Spiked versus Native PAH's from Heterogeneous Environmental Samples Using Supercritical Fluid Extraction and Sonication in Methylene Chloride, Anal. Chem. 65 (1993), 1497–1505.

Cammann, K., Kleiböhmer, W., Meyer, A., SFE of PAH's from Soil with a High Carbon Content and Analyte Collection via Combined Liquid/ Solid Trapping, In: P. Sandra (Ed.), 15th Int. Symp. Cap. Chrom., Riva del Garda May 1993, Huethig Verlag 1993, 105–110.

Croft, M. Y., Murby, E. J., Wells, R. J., Simultaneous Extraction and Methylation of Chlorophenoxyacetic Acids from Aqueous Solution Using Supercritical Carbon Dioxide as a Phase Transfer Solvent, Anal. Chem. 66 (1994), 4459–4465.

Hartonen, K., Riekkola, M.-L., Determination of Beta-Blockers from Urine by SFE and GC/MS, In: 16th Int. Symp. Cap. Chrom. Riva del Garda, P. Sandra (Ed.), Huethig 1994, 1729–1739.

Häufel, J., Creutznacher, H., Weisweiler, W., Analyse von PAK in Bodenproben – Extraktion mit organischen Lösungsmitteln gegen überkritische Fluide, GIT Fachz. Lab. 7 (1994). 764–768.

Hawthorne, S. B., Analytical-Scale Supercritical Fluid Extraction, Anal. Chem. 62 (1990), 633A.

Hawthorne, S. B., Methodology for Off-Line Supercritical Fluid Extraction, In: S. A. Westwood (Ed.), Supercritical Fluid Extraction and its Use in Chromatographic Sample Preparation, Blackie Academic & Professional: London 1993, 39–64.

Klein, E., Hahn, J., Gottesmann, P., Extraktion mit überkritischem CO_2 (SFE) – Applikationen aus der Lebensmittelüberwachung, Lebensmittelchemie 47 (1993), 84.

Langenfeld, J. J., Hawthorne, S. B., Miller, D. J., Pawliszyn, J., Role of Modifiers for Analytical-Scale Supercritical Fluid Extraction of Environmental Samples, Anal. Chem. 66 (1994), 909–916.

Lehotay, S. J., Eller, K. I., Development of a Method of Analysis for 46 Pesticides in Fruits and Vegetables by Supercritical Fluid Extraction and Gas Chromatography/Ion Trap Mass Spectrometry, J. AOAC Int. 78 (1995) 821–830.

Levy, J. M., Interfacing SFE On-Line to GC, SUPREX Technology Focus, SFE 91–6. Suprex Corporation: Pittsburg 1991.

Levy, J. M., Houck, R. K., Developments in Off-Line Collection for Supercritical Fluid Extraction, Am. Laboratory 4 (1993).

Lopez-Avila, V., Young, R., er al., Mini-round-robin study of a supercritical fluid extraction method for polynuclear aromatic hydrocarbons in soil with dichloromethane as a static modifier, J. Chromatogr. A 672 (1994) 167–175.

Luque de Castro, M. D., Valcarel, M., Tena, M. T., Analytical Supercritical Fluid Extraction, Springer: Berlin 1994.

Meyer, A., Kleiböhmer. W., Rapid Determination of PCP in Leather Using SFE with in Situ Derivatization, In: 16th Int. Symp. Cap. Chrom. Riva del Garda, P. Sandra (Ed.), Huethig 1994, 1752–1753.

Onuska, F. I., Terry, K. A., Supercritical Fluid Extraction of Polychlorinated Dibenzo-p-dioxins from Municipal Fly Ash, In: P. Sandra (Ed.). 13th Int. Symp. Cap. Chrom., Riva del Garda, May 1991, Huethig Verlag 1991.

Onuska, F. I., Terry, K. A., Supercritical Fluid Extraction of 2,3,7,8-Tetrachlorodibenzo-p-dioxin from Sediment Samples, J. High Res. Chrom. (1989), 537.

Oostdyk, T. S., Grob, R. L., Snyder, J. L., McNally, M. E., Study of Sonication and Supercritical Fluid Extraction of Primary Aromatic Amines, Anal. Chem. 65 (1993), 596–600.

Paschke, Th., SFE zur einfachen Fett- und Ölbestimmung, LaborPraxis 7 (1995) 24–26.

Paschke, Th., Untersuchungen zur Umweltrelevanz nitrierter polyzyklischer aromatischer Verbindungen in der Luft. Dissertation, Universität Gesamthochschule Siegen, Fachbereich 8, Siegen 1994.

Paschke, Th., Hawthorne, S. B., Miller, D. J., Supercritical Fluid Extraction of Nitrated Polycyclic Aromatic Hydrocarbons and Polycyclic Aromatic Hydrocarbons from Diesel Exhaust Particulate Matter. J. Chromatogr. 609 (1992), 333–340.

Pawliszyn, J., Kinetic Model of SFE Based on Packed Tube Extractor, Proc. 3rd Int. Symp. Supercritical Fluids, Strasbourg (France) 17.–19. Oct. 1994.
Raynie, D. E., Warning Concerning the Use of Nitrous Oxide in Supercritical Fluid Extractions, Anal. Chem. 65 (1993), 3127–3128.
Sachs, H., Uhl, M., Opiat-Nachweis in Haar-Extrakten mit Hilfe der GC/MS/MS und Supercritical Fluid Extraction (SFE), Toxichem + Krimichem 59 (1992), 114–120.
Smith, R. M., Nomenclature for Supercritical Fluid Chromatography and Extraction (IUPAC Recommendations 1993), Pure & Appl. Chem. 65 (1993), 2397–2403.
Snyder, J. L., Grob, R. L., McNally, M. E., Oostdyk, T. S., Comparison of Supercritical Extraction with Classical Sonication and Soxhlet Extractions for Selected Pesticides, Anal. Chem. 64 (1992), 1940–1946.
Taylor, L. T., Strategies for Analytical SFE, Anal. Chem. (1995) 364A-370A.
Towara, J., Extraktion von polychlorierten Dibenzo-p-Dioxinen und Dibenzofuranen (PCDD/F) aus Klärschlammproben mit Supercritical Fluid Extraction (SFE), Lehrstuhl für Ökologische Chemie und Geochemie der Universität Bayreuth, Studienarbeit 1993.
US EPA Methode Nr. 3560, Revision 1, December 1992.
Wenclawiak, B., (Ed.): Analysis with Supercritical Fluids: Extraction and Chromatography, Springer: Berlin 1992.

Section 2.1.3

Curren, M. S. S. , King, J.W., New sample preparation technique for the determination of avoparcin in pressurized hot water extracts from kidney samples, J. Chromatogr. A 954, (2002) 41–49.
Dionex ASE 300 Product Brochure, Dionex Corporation 2000.
Dionex product information, 2006, see www.dionex.com.
Hubert, Popp, P., Wenzel, K.-D., Engewald, W., Schüürmann, G., Accelerated Solvent Extraction – More efficient Extraktion of POPs and PAHs from real contaminated plant and soil samples, Rev. Anal. Chem. 20, (2001) 101–144.
Kaltenecker, M., Schwind, K.H., Hecht, H., Petz, M, Extraktion von Toxaphen aus tierischen Geweben mit Accelerated Solvent Extraction (ASE), GIT Labor Fachz. 7, (1999) 742–745.
Li, M.K., Landriault, M., Fingas, M., Llompart, M., Accelerated Solvent Extraction (ASE) of environmental organic compounds in soils using a modified supercritical fluid extractor, Journal of Hazardous Materials 102, (2003) 93–104.
US EPA, Method 3545A, January 1998, see www.epa.gov/sw-846/pdfs/3545a.pdf.
Wenzel, K.-D., A. Hubert, M. Manz, L. Weissflog, W. Engewald, G. Schüürmann. Accelerated solvent extraction of semivolatile organic compounds from biomonitoring samples of pine needles and mosses. Anal. Chem. 70, (1998) 4827–4835.

Section 2.1.4

Kerkdijk, H., Mol, H.G.J. , van der Nagel, B., Volume Overload Cleanup: An Approach for On-Line SPE-GC, GPC-GC, and GPC-SPE-GC, Anal. Chem. 79 (2007) 7975-7983.
Majors, R.E., Multidimensional HPLC, J. Chromatogr. Sci. 18 (1980) 571-577.
Munari, F., Grob, K., Coupling HPLC to GC: Why? How? With what instrumentation?, J. Chrom. Sci. 28 (1990) 61-66.
Jongenotter, G.A., Kerkhoff, M.A.T., Van der Knaap, H.C.M., Vandeginste, B.G.M., Automated On-Line GPC-GC-FID Involving Co-Solvent Trapping and the On-Column Interface for the Determination of Organophosphorus Pesticides in Olive Oils, J. High. Resol. Chromatogr. 22 (1999) 17-23.
De Paoli, M., et al., Determination of organophosphorus pesticides in fruit by on-lin automated LC-GC, J. Chromatogr. 626 (1992) 145-150.
Trestianu, S., Munari, F., Grob, K., Riva 85 - Riva 96, Ten years experience on creating instrumental solutions for large volume sample injections into capillary GC columns, CE Instruments publication, Rodano 1996.

Section 2.1.5

Ashley, D. L., Bonin, M. A., et al., Determining Volatile Organic Compounds in Human Blood from a Large Sample Population by Using Purge and Trap Gas Chromatography/Mass Spectrometry, Anal. Chem. 64 (1992), 1021–1029.
Belouschek, P., Brand, H., Lönz, P., Bestimmung von chlorierten Kohlenwasserstoffen mit kombinierter Headspace- und GC/MS-Technik, Vom Wasser 79 (1992), 1–8.

Deutsche Einheitsverfahren zur Wasser-, Abwasser- und Schlammuntersuchung, Gemeinsam erfaßbare Stoffgruppen (Gruppe F), DIN 38407 (Teil 5), Beuth Verlag: Berlin 1990.
Eichelberger, J. W., Bellar, T. A., Behymer, T. D., Budde, W. L., Analysis of Organic Wastes from Solvent Recycling Operations, Finnigan MAT Application Report No. 212.
Ettre, L. S., Kolb, B., Headspace Gas Chromatography: The Influence of Sample Volume on Analytical Results, Chromatographia 32(1/2), July 1991, 5–12.
Hachenberg, H., Die Headspace Gaschromatographie als Analysen- und Meßmethode, DANI-Analysentechnik: Mainz 1988.
Hachenberg, H., Schmidt, A. P., Gaschromatographic Headspace Analysis, J. Wiley and Sons, Chichester 1977.
Johnson, E., Madden, A., Efficient Water Removal for GC/MS Analysis of Volatile Organic Compounds with Tekmar's Moisture Control Module, Finnigan MAT Technical Report No. 616, 1990.
Kolb, B., Ettre, L.S., Static Headspace–Gas Chromatography: Theory and Practice, New York: Wiley, 1997.
Kolb, B., Angewandte Gaschromatographie, Bodenseewerk Perkin Elmer 1981 Heft 38.
Kolb, B., Applications of Headspace Gas Chromatography, Perkin Elmer GC Applications Laboratory 1991.
Kolb, B., Applied Headspace Gas Chromatography, J. Wiley and Sons, Chichester 1980.
Kolb, B., Die Bestimmung des Wassergehaltes in Lebensmitteln und Pharmaka mittels der gaschromatographischen Headspace-Technik, Lebensmittel- & Biotechnol. 1 (1993), 17–20.
Kolb, B., HSGC mit Kapillar-Trennsäulen, Labor-Praxis 1986.
Kolb, B., Bichler, Chr., Auer, M., Simultaneous Determination of Volatile Aromatic and Halogenated Hydrocarbons in Water and Soil Samples by Dual-Channel ECD/PID Equilibrium Headspace Analysis, In: P. Sandra: 15th Int. Symp. Cap. Chrom., Riva del Garda, May 24–27 1993, Huethig Verlag 1993, 358–364.
Kolb, B., Ettre, L.S., Static Headspace-Gas Chromatography: Theory and Practice, John Wiley & Sons Inc: Hoboken, New Jersey 2006.
Krebs, G., Schneider, E., Schumann, A., Head Space GC – Analytik flüchtiger aromatischer und halogenierter Kohlenwasserstoffe aus Bodenluft, GIT Fachz. Lab. 1 (1991) 19–22.
Lin, D. P., Falkenberg, C., Payne, D. A., et al., Kinetics of Purging for the Priority Volatile Organic Compounds in Water, Anal. Chem. 1993 (65), 999–1002.
Madden, A. T., Lehan, H. J., The Effects of Condensate Traps on Polar Compounds in Purge & Trap Analysis, Pittsburgh Conference 1991.
Maggio, A., Milana, M. R., et. al., Multiple Headspace Extraction Capillary Gas Chromatography (MHE-CGC) for the Quantitative Determination of Volatiles in Contaminated Soils, In: P. Sandra: 13th Int. Symp. Cap. Chrom., Riva del Garda, May 1991, Huethig Verlag 1991, 394–405.
Matz, G., Kesners, P., Spray and Trap Method for Water Analysis by Thermal Desorption Gas Chromatography/Mass Spectrometry in Field Applications, Anal. Chem, 65 (1993), 2366–2371.
Pigozzo, F., Munari, F., Trestianu ,S., Sample Transfer from Head Space into Capillary Columns, In: P. Sandra: 13th Int. Symp. Cap. Chrom., Riva del Garda, May 1991, Huethig Verlag 1991, 409–416.
TekData, Fundamentals of Purge and Trap, Tekmar Technical Documentation B 121988, ohne Jahresangabe.
Westendorf, R. G., Automatic Sampler Concepts for Purge and Trap Gas Chromatography, American Laboratory 2 (1989).
Westendorf, R. G., Design and Performance of a Microprozessor-based Purge and Trap Concentrator, American Laboratory 10 (1987).
Westendorf, R. G., Performance of a Third Generation Cryofocussing Trap for Purge and Trap Gas Chromatography, Pittsburgh Conference 1989.
Westendorf, R. G., A Quantitation Method for Dynamic Headspace Analysis Using Multiple Runs, J. Chrom.Science 11 (1985).
Willemsen, H. G., Gerke, Th., Krabbe, M. L., Die Analytik von LHKW und BTX im Rahmen eines DK-ARW(AWBR)-IKSR-Projektes, 17. Aachener Werkstattgespräch am 28. und 29. Sept. 1993, Zentrum für Aus- und Weiterbildung in der Wasser- und Abfallwirtschaft Nordrhein-Westfalen GmbH (ZAWA): Essen 1993.
Wylie, P. L., Comparing Headspace with Purge and Trap for Analysis of Volatile Priority Pollutants, Research and Technology 8 (1988), 65–72.
Wylie, P. L., Comparison of Headspace with Purge and Trap Techniques for the Analysis of Volatile Priority Pollutants, In: P. Sandra (Ed.), 8th Int. Symp. Cap. Chrom., Riva del Garda, May 19th–21st 1987, Huethig 1987, 482–499.

Zhu, J.Y., Chai, X.-S., Zhu, Some Recent Developments in Headspace Gas Chromatography, Curr. Anal. Chem. 1, (2005) 79–83.

Section 2.1.6

Betz, W. R., Hazard, S. A., Yearick, E. M., Characterization and Utilization of Carbon-Based Adsorbents for Adsorption and Thermal Desorption of Volatile, Semivolatile and Non-Volatile Organic Contaminants in Air, Water and Soil Sample Matrices, Int. Labmate XV (1989), 1.

Brown, R. H., Purnell, C. J., Collection and Analysis of Trace Vapour Pollutants in Ainbient Atmospheres, J. Chromatogr. 178 (1979), 79–90.

Carbotrap – an Excellent Adsorbent for Sampling Many Airbome Contaminants, GC Bulletin 846C, Supelco Firmenschrift 1986.

Efficently Monitor Toxic Compounds by Thermal Desorption, GC Bulletin 849C, Supelco Firmenschrift 1988.

Figg, K., Rubel, W., Wieck, A., Adsorptionsmittel zur Anreicherung von organischen Luftinhaltsstoffen, Fres. Z. Anal. Chem. 327 (1987), 261–278.

Föhl, A., Basnier, P., Untersuchung der Löschverfahren und Löschmittel zur Bekämpfung von Bränden gefährlicher Güter – GC/MS Rauchgasanalyse –, Forschungsbericht Nr. 81, Forschungsstelle für Brandschutztechnik an der Universität Karlsruhe (TH) 1992.

Knobloch, Th., Efer, J., Engewald, W., Adsorptive Anreicherung an Kohlenstoffadsorbentien und Thermodesorption in: SUPELCO Deutschland GmbH (Hrsg), Themen der Umweltanalytik, VCH: Weinheim 1993, 103–110.

Knobloch, Th., Engewald, W., Sampling and Gas Chromatographic Analysis of Volatile Organic Compounds (VOCs) in Hot and Extremely Humid Emissions, In: 16th Int. Symp. Cap. Chrom. Riva del Garda. P. Sandra (Ed.), Huethig 1994, 472–484.

Manura, J.J., Adsorbent Resins - Calculation and Use of Breakthrough Volume Data, Scientific Instrument Services, Ringoes NJ, USA, Application Note, see http://www.sisweb.com/index/referenc/resins.htm

Measurement of gases; calibration gas mixtures; preparation by continuous injection method, VDI Directive 3490 Blatt 8, Kommission Reinhaltung der Luft im VDI und DIN – Normenausschuss KRdL, Jan 2001, see also: www.vdi.de

MDHS 3 – Generation of Test Atmospheres of Organic Vapours by Syringe Injection Technique, September 1984 (HSE-Methode).

Mülle, A., Tschickard, M., Herstellung von Kalibriergasen im Emissions- und Immissionsbereich, Symposium Probenahme und Analytik flüchtiger organischer Gefahrstoffe, München/Neuberberg, 2./3. Dez.1993, Abstracts.

New Adsorbent Trap for Monitoring Volatile Organic Compounds in Wastewater, Environmental Notes, Supelco Firmenschrift: Bellafonte (USA) 1992.

Niebel, J., Analyse von Bodenluft durch Konzentration auf TENAX und thermischer Desorption mit dem mobilen GC/MS-System SpektraTrak 620, Applikationsschrift. Axel Semrau GmbH: Sprockhövel 1993.

Niebel, J., Analyse von Bodenproben auf BTX durch Thermodesorption mit dem mobilen GC/MSSystem SpektraTrak 620, Applikationsschrift, Axel Semrau GmbH: Sprockhövel 1993.

Niebel, J., Untersuchung von lösungsmittelbelasteten Bodenproben mit dem mobilen GC/MSD VIKING SpektraTrak 620. Applikationsschrift, Axel Semrau GmbH: Sprockhövel 1994.

Packing Traps for Tekmar Automatic Desorbers, Tekmar Firmenschrift, TekData B012589.

Perkin Elmer, Zusammenfassung des User Meetings Thermodesorption 18./19.9.91 in Mainz, Tagungsunterlagen.

Thermal Desorption, Technique and Applications, Perkin Elmer Applikation Report No. L-1360, 1990.

Tschickard, M., Analyse von polychlorierten Biphenylen in Innenraumluft mit Thermodesorption und Ion Trap Detektor, Perkin Elmer User Meeting, Landesamt für Umweltschutz und Gewerbeaufsicht, Mainz 1991, Tagungsunterlagen.

Tschickard, M., Analytik von leicht- und schwerflüchtigen Luftschadstoffen mit GC-Thermodesorption, Symposium Probenahme und Analytik flüchtiger organischer Gefahrstoffe, München/Neuherberg, 2./3. Dez. 1993. Abstracts.

Tschickard, M., Bericht über eine Prüfgasapparatur zur Herstellung von Kalibriergasen nach dem Verfahren der kontinuierlichen Injektion, Gesch.Zchn: 35–820 Tsch, Mainz: Landesamt für Umweltschutz und Gewerbeaufsicht 25.5.1993.

Tschickard, M., Routineeinsatz des Thermodesorbers ATD50 in der Gefahrstoffanalytik, Methodenbericht des Landesamtes für Umweltschutz und Gewerbeaufsicht Rheinland-Pfalz, Mainz 1991, In: Perkin Elmer, Zusammenfassung des

User Meetings Thermodesorption 18./19.9.91 in Mainz, Tagungsunterlagen.

Tschickard, M., Hübschmann, H.-J., Bestimmung von MAK-Werten und TRK-Werten mit dem Ion Trap-Detektor, Finnigan MAT Application Report No. 67, Bremen 1988.

Vorholz, P., Hübschmann, H.-J., Einsatz des mobilen VIKING SpektraTrak 600 GC/MS-Systems zur Vor-Ort-Analyse von Geruchsemissionen auf einer Kompostierungsanlage. Applikationsschrift, Axel Semrau GmbH: Sprockhövel 1994.

Section 2.1.7

Brodda, B.-G., Dix, S., Fachinger, J., Investigation of the Pyrolytic Degradation of Ion Exchange Resins by Means of Foil Pulse Pyrolysis Coupled with Gas Chromatography/Mass Spectrometry, Sep. Science Technol. 28 (1993), 653–673.

EPA Method 8275A, Semivolatile organic compounds (PAHs and PCBs) in soil/sludges and solid wastes using thermal extraxtion/gas chromatography mass spectrometry (TE/GC/MS), Rev 1, December 1996.

Ericsson, I., Determination of the Temperature-Time Profile of Filament Pyrolyzers, J. Anal. Appl. Pyrolysis 2 (1980), 187–194.

Ericsson, I., Influence of Pyrolysis Parameters on Results in Pyrolysis-Gas Chromatography, J. Anal. Appl. Pyrolysis 8 (1985), 73–86.

Ericsson, I., Sequential Pyrolysis Gas Chromatographic Study of the Decomposition Kinetics of cis-1,4-Polybutadiene. J. Chrom. Science 16 (1978), 340–344.

Ericsson, I., Trace Determination of High Molecular Weight Polyvinylpyrrolidone by Pyrolysis-Gas-Chromatography, J. Anal. Appl. Pyrolysis 17 (1990), 251–260.

Ericsson, I., Lattimer, R. P., Pyrolysis Nomenclature, J. Anal. Appl. Pyrolysis 14 (1989), 219–221.

Fischer, W. G., Kusch, P., An Automated Curie-Point Pyrolysis-High Resolution Gas Chromatography System, LCGC 6 (1993), 760–763.

Galletti, G. C., Reeves, J. B., Pyrolysis-Gas Chromatography/Mass Spectrometry of Lignocellulosis in Forages and By-Products, J. Anal. Appl. Pyrolysis 19 (1991), 203–212.

Hancox, R. N., Lamb, G. D., Lehrle, R. S., Sample Size Dependence in Pyrolysis: An Embarrassment or an Utility?, J. Anal. Appl. Pyrolysis 19 (1991), 333–347.

Hardell, H. L., Characterization of Spots and Specks in Paper Using PY-GC-MS Including SPM, Poster at the 10th Int. Conf. Fund. Aspects Proc. Appl. Pyrolysis, Hamburg 28.9.–2.10.1992.

Irwin, W. J., Analytical Pyrolysis, Marcel Dekker: New York 1982.

Klusmeier, W., Vögler, P., Ohrbach, K. H., Weber, H., Kettrup, A., Thermal Decomposition of Pentachlorobenzene, Hexachlorobenzene and Octachlorostyrene in Air, J. Anal. Appl. Pyrolysis 14 (1988), 25–36.

Matney, M. L., Limero, Th. F., James, J. T., Pyrolysis-Gas Chromatography/Mass Spectrometry Analyses of Biological Particulates Collected during Recent Space Shuttle Missions, Anal. Chem. 66 (1994), 2820–2828.

Richards, J. M., McClennen, W. H., Bunger, J. A., Meuzelaar, H. L. C., Pyrolysis Short-Column GC/MS Using the Ion Trap Detektor (ITD) and Ion Trap Mass Spectrometer (ITMS). Finnigan MAT Application Report No. 214, 1988.

Roussis, S.G., Fedora, J.W., Use of a Thermal Extraction Unit for Furnace-type Pyrolysis: Suitability for the Analysis of Polymers by Pyrolysis/GC/MS, Rapid Comm. Mass Spectrom. 10, (1998), 82–90.

Schulten, H.-R., Fischer, W., Wallstab, HRC & CC 10 (1987), 467.

Simon, W., Giacobbo, H., Chem. Ing. Techn. 37 (1965), 709.

Snelling, R. D., King, D. B., Worden, R., An Automated Pyrolysis System for the Analysis of Polymers. Poster, 10P, Proc. Pittsburgh Conference, Chicago 1994.

Uden, P. C., Nomenclature and Terminology for Analytical Pyrolysis (IUPAC Recommendations 1993), Pure & Appl. Chem. 65 (1983), 2405–2409.

Zaikin, V. G., Mardanov, R. G., et al., J. Anal. Appl. Pyrolysis 17 (1990), 291.

Section 2.2.1

Donato, P., Tranchida, P.Q., Dugo, P., Dugo, G., Mondello, L., Review: Rapid analysis of food products by means of high speed gas chromatography, J. Sep. Sci. 30 (2007), 508–526.

EPA Method 610, see www.epa.gov/epahome/index/.

Facchetti, R., Galli, S., Magni, P., Optimization of Analytical Conditions to Maximize Separation Power in Ultra-Fast GC, Proceedings of 25th International Symposium of Capillary Chromatography, KNL05, Riva del Garda, Italy, May 13–17, 2002, ed. P. Sandra.

Facchetti, R., Cadoppi, A., Ultra Fast Chromatography: A Viable Solution for the Separation of Essential Oil Samples, The Column, Nov 2005, 8–11 (www.thecolumn.eu.com).
Magni, P., Facchetti, R., Cavagnino, D., Trestianu, S., Proceedings of 25th International Symposium of Capillary Chromatography, KNL05, Riva del Garda, Italy, May 13–17, 2002, ed. P. Sandra.
Trapp, O., Maximale Information bei minimaler Analysenzeit, Nachr. Chem. 54, (2006) 1111–1114.
Trapp, O., Kimmel, J.R., et al., Angew. Chem. 43, (2004) 6541 – 6544. (for HT-TOF-MS).
Warden, J., Magni, P., Wells, A., Pereira, L., Increasing Throughput in GC Environmental Methods, Int. Chrom. Symp., Riva del Garda 2004.
Zimmermann, R., Welthagen, W., Multidimenionale Analytik mit GC und MS, Nachr. Chem. 54, (2006) 1115 – 1119.

Section 2.2.2

2DGC: Multidimensional Gas Chromatography on a single GC or dual GC with or without MS using a Moving Capillary Stream Switching system, see www.brechbuehler.ch/usa/mcss.htm.
Beens, H., Boelwns, R., Tijssen, R., Blomberg, Simple, non-moving modulation interface for comprehensive two-dimensional gas chromatography, J., J. High Res. Chromatogr. 21, (1998) 47.
Beens, J., Adahchour, M., Vreuls, R.J.J., van Altena, K., Brinkman, U.A.Th., Simple, non-moving modulation interface for comprehensive two-dimensional gas chromatography, J. Chromatogr. A, 919/1 (2001), 127–132.
Beens, J., Blomberg, J., Schoenmakers, P.J., Proper Tuning of Comprehensive Two-Dimensional Gas Chromatography (GCxGC) to Optimize the Separation of Complex Oil Fractions, J. High Resol. Chromatogr. 23(3) (2000), 182–188.
Bertsch, W.J., Two-Dimensional Gas Chromatography. Concepts, Instrumentation, and Applications – Part 1, J. High Res. Chromatogr., 22, (1999) 647–665.
Bertsch, W.J., Two-Dimensional Gas Chromatography. Concepts, Instrumentation, and Applications – Part 2: Comprehensive Two-Dimensional Gas Chromatography, J. High Res. Chromatogr., 23, (2000) 167–181.
Cavagnino, D., Bedini, F., Zilioli, G., Trestianu, S., Improving sensitivity and separation power by using LVSL-GCxGC-FID technique for pollutants detection at low ppb level, Poster at the Gulf Coast Conference, Comprehensive Two-Dimensional GC Symposium, 2003.
De Alencastro, L.F., Grandjean, D., Tarradellas, J., Application of Multidimensional (Heart-Cut) Gas Chromatography to the Analysis of Complex Mixtures of Organic Pollutants in Environmental Samples, Chimia 57, (2003) 499–504.
Deans, D.R., A new technique for heart cut-ting in gas chromatography, Chromatographia 1, (1968) 18–22.
Dimandja, J.M.D., GCxGC, Anal. Chem. 76, 9 (2004) 167A-174A.
Focant, J., Sjödin, A., Turner, W., Patterson, D., Measurement of Selected Polybrominated Diphenyl Ethers, Polybrominated and Polychlorinated Biphenyls, and Organochlorine Pesticides in Human Serum and Milk Using Comprehensive Two-Dimensional Gas Chromatography Isotope Dilution Time-of-Flight Mass Spectrometry, Anal. Chem. 76, 21 (2004) 6313–6320.
Guth, H., Use of the Moving Capillary Switching System (MCSS) in Combination with Stable Isotope Dilution Analysis (IDA) for the Quantification of a Trace Component in Wine, Poster at the 18th International Symposium on Capillary Chromatography, May 20–24, 1996.
Hamilton, J., Webb, P., Lewis, A., Hopkins, J., Smith, S., Davy, P., Partially oxidised organic components in urban aerosol using GCXGC-TOF/MS, Atmos. Chem. Phys. Discuss. 4, (2004) 1393–1423.
Hollingsworth, B.V., Reichenbach, S.E., Tao, Q., Visvanathan, A., Comparative visualization for comprehensive two-dimensional gas chromatography, J. Chromatography A, 1105, (2006) 51–58 (see also www.csl.unl.edu).
Horii, Y., Petrick, G., Katase, T., Gamo, T., Yamashita, N., Congener-specific carbon isotope analysis of technical PCN and PCB preparations using 2DGC IRMS, Organohalogen Comp. 66, (2004) 341–348.
Kellner, R., Mermet, J.-M., Otto, M., Valcarcel, M., Widmer, H.M., Analytical Chemistry, 2nd Ed., Wiley-VCH: Weinheim 2004.
Leco Corp. The Use of Resample, a New ChromaTOF® Feature, to Improve Data Processing for GCxGC-TOFMS, Separation Science Application Note 02/05, (2005).
Liu, Z., Phillips, J.B., J. Chrom. Sci., 29, (1991) 227.
Majors, R.E., Multidimensional and Comprehensive Liquid Chromatography, LCGC North America 10 (2005).
Marriott, P.J., Morrison, P.D., Shellie, R.A., Dunn, M.S., Sari, E., Ryan, D., Multidimensional and

Comprehensive Two-Dimensional Gas Chromatography, LCGC Europe 12 (2003) 2–10.
McNamara, K., Leardib, R., Hoffmann, A., Developments in 2D GC with Heartcutting, LCGC Europe 12 (2003) 14–22.
Mondello, L., Tranchida, P.Q., Dugo, P., Dugo, G., Comprehensive Two-Dimensional Gas Chromatography-Mass Spectrometry: A Review, Mass Spectrom. Rev. 27 (2008), 101–124.
Patterson, D.G., Welch, S.M., Focant J.F., Turner, W.E., The use of various gas chromatography and mass spectrometry techniques for human biomonitoring studies, Presented at the Dioxin 2006 Conference, FCC-2602–409677, Oslo 2006.
Reichenbach, S.E., Ni, M., Kottapalli, V., Visvanathan, A., Information technologies for comprehensive two-dimensional gas chromatography, Chemometrics Intel. Lab. Systems 71, (2004) 107– 120.
Sulzbach, H., Controlling chromatograph gas streams – by adjusting relative positioning off supply columns in connector leading to detector, Carlo Erba Strumentazione GmbH, Germany, patent DE 4017909, 1991.
Sulzbach, H., Multidimensionale Kapillarsäulen-Gaschromatographie. GIT 40, (1996) 131.
Welthagen, W., Schnelle-Kreis, J., Zimmerman, R., Search criteria and rules for comprehensive two-dimensional gas chromatography-time-of-flight mass spectrometry analysis of airborne particulates, J. Chromatogr. A., 1019, 233–249, 2003.

Section 2.2.3

Bergna, M., Banfi, S., Cobelli, L., The Use of a Temperature Vaporizer as Preconcentrator Device in the Introduction of Large Amount of Sample, In: 12th Int. Symp. Cap. Chrom. Riva del Garda, P. Sandra, G. Redant (Eds.), Huethig 1989. 300–309.
Blanch, G. P., Ibanez, E., Herraiz, M., Reglero, G., Use of a Programmed Temperature Vaporizer for Off-Line SFE/GC Analysis in Food Composition Studies. Anal. Chem. 66 (1994), 888–892.
Cavagna, B., Pelagatti, S., Cadoppi, A., Cavagnino, D., GC-ECD Determination of Decabromodiphenylether using Direct On-Column Injection into Capillary Columns, Poster Thermo Fisher Scientific Milan, PittCon 2008, New Orleans.
David, F., Sandra. P., Large Volume Sampling by PTV Injection. Application in Dioxine Analysis, In: 14th Int. Symp. Cal). Chrom. Riva del Garda, P. Sandra (Ed.), Huethig 1991, 380–382.
David, W., Gaschromatographie: Alte Ideen – erfolgreich neu kombiniert! LABO 7–8 (1994), 62–68.
Donike, M., Die temperaturprogrammierte Analyse von Fettsäuremethylsilylestern: Ein kritischer Qualitätstest für gas-chromatographische Trennsäulen, Chromatographia 6 (1973), 190.
Efer, J., Müller, S., Engewald, W., Möglichkeiten der PTV-Technik für die gaschromatographische Trinkwasseranalytik, GIT Fachz. Lab. 7 (1995), 639–646.
Färber, H., Peldszus, S., Schöler, H. F., Gaschromatographische Bestimmung von aziden Pestiziden in Wasser nach Methylierung mit Trimethylsulfoniumhydroxid, Vom Wasser 76 (1991), 13–20.
Färber, H., Schöler, H. F., Gaschromatographische Bestimmung von Harnstoffherbiziden in Wasser nach Methylierung mit Trimethylaniliniumhydroxid oder Trimethylsulfoniumhydroxid, Vom Wasser 77 (1991), 249–262.
Färber, H., Schöler, H. F., Gaschromatographische Bestimmung von OH- und NH-aziden Pestiziden nach Methylierung mit Trimethylsulfoniumhydroxid im „Programmed Temperature Vaporizer" (PTV), Lebensmittelchemie. 46 (1992), 93–100.
Grob, K., Classical Split and Splitless Injection in Capillary Gas Chromatography with some Remarks on PTV Injection, Huethig: Heidelberg 1986.
Grob, K., Einspritztechniken in der Kapillar-Gaschromatographie, Huethig: Heidelberg 1995.
Grob, K., Guidelines on How to Carry Out On-Column Injections, HRC & CC, 6 (1983), 581–582.
Grob, K., Injection Techniques in Capillary GC, Anal. Chem. 66 (1994), 1009A-1019A.
Grob, K., Grob, K. jun., HRC & CC, 1 (1978), 1.
Grob, K., Grob, K. jun., J. Chromatogr., 151 (1978), 31 1.
Grob, K., Split and Splitless Injection for Quantitative Gas Chromatography. Concepts, Processes, Practical Guidelines, Sources of Error, 4th Ed., Wiley-VCH: Weinheim 2001.
Hinshaw, J. W., Splitless Injection: Corrections and Further Information, LCGC Int., 5 (1992), 20–22.
Hoh, E., Mastovska, K., Large volume injection techniques in capillary gas chromatography, J. Chromatogr. A, 1186 (2008) 2-15
Karasek, F. W., Clement, R. E., Basic Gas Chromatography-Mass Spectrometry, Elsevier: Amsterdam 1988.
Klemp, M. A., Akard, M. L., Sacks, R. D., Cryofocussing Inlet with Reversed Flow Sample Col-

lection for Gas Chromatography, Anal. Chem. 65 (1993), 2516–2521.
Magni & Porzano, Journal of Separation Science 26 (2003), 1491–1498.
Matter, L., Lebensmittel- und Umweltanalytik mit der Kapillar-GC, VCH: Weinheim 1994.
Matter, L., Poeck, M., Kalte Injektionsmethoden, GIT Fachz.Labor, 1 1 (1987), 1031–1039.
Matter, L., Poeck, M., Probeaufgabetechniken in der Kapillar-GC, GIT Suppl. Chrom. 3 (1987), 81–86.
Mol, H. G. J., Janssen, H. G. M., Crainers, C. A., Use of a Temperature Programmed Injector with a Packed Liner for Direct Water Analysis and On-Line Reversed Phase LC-GC, In: 15th Int. Symp. Cap. Chrom. Riva del Garda, P. Sandra (Ed.), Huethig 1993, 798–807.
Mol, H. G. J., Janssen, H. G. M., Cramers, C. A., Brinkman, K. A. Th., Large Volume Sample Introduction Using Temperature Programmable Injectors-Implication of Line Diameter, In: 16th Int. Symp. Cap. Chrom. Riva del Garda, P. Sandra (Ed.), Huethig 1994, 1124–1136.
Müller, H.-M., Stan, H.-J., Pesticide Residue Analysis in Food with CGC. Study of Long-Time Stability by the Use of Different Injection Techniques, In: 12th Int. Symp. Cap. Chrom. Riva del Garda, P. Sandra, G. Redant (Eds.), Huethig 1989, 582–587.
Müller, H.-M., Stan, H.-J., Thermal Degradation Observed with Varying Injection Techniques: Quantitative Estimation by the Use of Thermolabile Carbamate Pesticides, In: 8th Int. Symp. Cap. Chrom. Riva del Garda, P. Sandra (Ed.), Huethig 1987, 588–596.
Müller, S., Efer, J., Wennrich. L., Engewald, W., Levsen, K., Gaschromatographische Spurenanalytik von Methamidophos und Buminafos im Trinkwasser – Einflußgrößen bei der PTV-Dosierung großer Probenvolumina, Vom Wasser 81 (1993), 135–150.
Pretorius, V., Bertsch, W., HRC & CC 6 (1983), 64.
Saravalle, C. A., Munari, F., Trestianu, S., Multipurpose Cold Injector for High Resolution Gas Chromatography. HRC & CC 10 (1987) 288–296.
Schomburg, G., In: Capillary Chromatography, 4th Int. Symp. Hindelang, R. Kaiser (Ed.). Inst. F. Chromatographie, 371 und A921 (1981).
Schomburg, G., Praktikumsversuch GDCh-Fortbildungskurs 305/89 Nr. 1, Erhöhung der Nachweisgrenze von Spurenkomponenten mit Hilfe des PTV-Injektors durch Anreicherung mittels Mehrfachaufgabe, Mühlheim 1989.
Schomburg, G., Gaschromatographie, VCH: Weinheim 1987.
Schomburg, G., Praktikumsversuch GDCh-Fortbildungskurs 305/89 Nr. 5, Gaschromatographie in Kapillarsäulen, Mühlheim 1989.
Schomburg, G., Probenaufgabe in der Kapillargaschromatographie, LaBo 7 (1983), 2–6.
Schomburg, G., Praktikumsversuch GDCh-Fortbildungskurs 305/89 Nr. 7, Schnelle automatische Split-Injektion mit der Spritze von Proben mit größerem Flüchtigkeitsbereich, Mühlheim 1989.
Schomburg, G., Behlau, H., Dielmann, R., Weeke, F., Husmann, H., J. Chromatogr. 142 (1977), 87.
Stan, H.-J., Müller, H.-M., Evaluation of Automated and Manual Hot-Splitless, Cold-Splitless (PTV) and On-Column Injektion Technique Using Capillar Gas Chromatography for the Analysis of Organophosphorus Pesticides, In: 8th Int. Symp. Cap. Chrom. Riva del Garda, P. Sandra (Ed.), Huethig 1987, 406–415.
Staniewski, J., Rijks, J. A., Potentials and Limitations of the Liner Design for Cold Temperature Programmed Large Volume Injection in Capillary GC and for LC-GC Interfacing, In: 14th Int. Symp. Cap. Chrom. Riva del Garda, P. Sandra (Ed.), Huethig 1991, 1334–1347.
Sulzbach, H., Magni, P., Quantitative Ultraspurengaschromatographie, Huethig: Heidelberg 1995.
Tipler, A., Johnson, G. L., Optimization of Conditions for High Temperature Capillary Gas Chromatography Using a Split-Mode Programmable Temperature Vaporizing Injektion System, In: 12th Int. Symp. Cap. Chrom. Riva del Garda, P. Sandra, G. Redant (Eds.), Huethig 1989, 986–1000.
Poy, F., Chromatographia, 16 (1982), 345.
Poy, F., 4th Int. Symp. Cap. Chrom., Hindelang 1981, Vorführung.
Poy, F., Cobelli, L., In: Sample Introduction in Capillar Gas Chromatography Vol 1, P. Sandra (Ed.), Huethig: Heidelberg 1985, 77–97.
Poy, F., Visani, S., Terrosi, F., HRC & CC 4 (1982), 355.
Poy, F., Visani, S., Terrosi, F., J. Chromatogr. 217 (1981), 81.
Vogt, W., Jacob, K., Obwexer, H. W., J. Chromatogr. 174 (1979), 437.
Vogt, W., Jacob, K., Ohnesorge, A. B., Obwexer, H. W., J. Chromatogr. 186 (1979), 197.

Section 2.2.4

Jennings, W., Gas Chromatography with Glass Capillary Columns, Academic Press: New York 1978.
Knapp, D. R., Handbook of Analytical Derivatization Reactions, John Wiley & Sons: New York 1979.
Matter, L., Anwendung der chromatographischen Regeln in der Kapillar-GC, In: Matter L., Lebensmittel- und Umweltanalytik mit der Kapillar-GC, VCH: Weinheim 1994, 1–27.
Pierce, A. E., Silylation of Organic Compounds, Pierce Chemical Company: Rockford, III, 1982.
Restek Corporation, A Capillary Chromatography Seminar, Restek: Bellafonte PA (USA), 1993 (Firmenschrift).
Schomburg, G., Gaschromatographie, VCH: Weinheim 1987.
van Ysacker, P. G., Janssen, H. G., Snijders, H. M. J., van Cruchten, H. J. M., Leclercq, P. A., Cramers, C. A., High-Speed-Narrow-Bore-Capillary Gas Chromatography with Ion-Trap Mass Spectrometric Detection. 15th Int. Symp. Cap. Chromatogr. Riva del Garda 1993, In: P. Sandra (Ed.), Huethig 1993.

Section 2.2.5

Bock, R., Methoden der analytischen Chemie 1, Verlag Chemie, Weinheim 1974.
Desty, D. H., LC-GC Intl. *4(5)*, 32 (1991).
Martin, A. J. P., Synge R. L. M., J. Biol. Chem. 35, 1358 (1941).
Schomburg, G., Gaschromatographie, Verlag Chemie, Weinheim 1987.

Section 2.2.6

Bretschneider, W., Werkhoff, P., Progress in All-Glass Stream Splitting Systems in Capillary Gas Chromatography Part I, HRC & CC 11 (1988), 543–546.
Bretschneider, W., Werkhoff, P., Progress in All-Glass Stream Splitting Systems in Capillary Gas Chromatography Part 11, HRC & CC 11 (1988), 589–592.
Ewender, J., Piringer O., Gaschroniatographische Analyse flüchtiger aliphatischer Amine unter Verwendung eines Amin-spezifischen Elektrolytleitfähigkeitsdetektors, Dt.Lebensm.Rundschau 87 (1991), 5–7.
Hill, H. H., McMinn, D. G., Detectors for Capillary Gas Chromatography, John Wiley & Sons Inc., New York 1992.
Kolb, B., Otte, E., Gaschromatographische Detektoren. Manuskript. Technische Schule Bodenseewerk Perkin Elmer GmbH.
Piringer, O., Wolff, E., New Electrolytic Conductivity Detector for Capillary Gas Chromatography – Analysis of Chlorinated Hydrocarbons. J Chromatogr. 284 (1984), 373–380.
Schneider, W., Frohne, J. Ch., Brudderreck, H., Selektive gaschromatographische Messung sauerstoffhaltiger Verbindungen mittels Flammenionisationsdetektor, J. Chromatogr. 245 (1982), 71.
Wentworth, W. E., Chen, E. C. M. in: Electron Capture Theory and Practice in Chromatography, Zlatkis, A., Poole, C. F. (Eds.). Elsevier, New York 1981, 27.

Section 2.3.1

Balogh, M.P., Debating Resolution and Mass Accuracy, LCGC Asia Pacific 7/4 Nov (2004), 16–20.
Balzers Fachbericht. Das Funktionsprinzip des Quadrupol-Massenspektrometers, Firmenschrift DN 9272.
Brubaker, W. M.,An improved quadrupole mass spectrometer analyser. J. Adv. Mass Spectr. 4, (1968), 293–299.
Brunnée, C., The Ideal Mass Analyzer: Fact or Fiction?, Int. J. Mass Spectrom. Ion Proc. 76 (1987), 121–237.
Brunnée, C., Voshage, H., Massenspektroskopie, Thiemig: München 1964.
Dawson, P. H., Whetten, N.R., Mass Spectrometry Using Radio-Frequency Quadrupole Fields, N. R., J. Vac. Sci. Technol. 5 (1968), 1.
Dawson, P. H., Whetten, N. R., Quadrupole mass spectrometry, Dynamic Mass Spectrometry, 2, (1971) 1–60.
Dawson, P.H., Whetten, N.R., The three dimensional quadrupole trap. Naturwissenschaften 56, (1969), 109–112.
Eljarrat, E., Barcelœ, D., Congener-specific determination of dioxins and related compounds by gas chromatography coupled to LRMS, HRMS, MS/MS and TOFMS, J. Mass Spectrom. 37(11) (2002), 1105–1117.
Feser, K., Kögler, W., The Quadrupole Mass Filter for GC/MS Applications, J. Chromatogr. Science 17 (1979), 57–63.

Grange, A.H., Determining Ion Compositions Using an Accurate Mass Triple Quadrupole Mass Spectrometer, 2005, http://www.epa.gov/nerlesd1/chemistry/ice/default.htm

Jemal, M., Quyang, Z., Rapid Commun. Mass Spectrom. 17 (2003), 24–38.

Liu,Y.-M., Akervik, K., Maljers, L., Optimized high resolution SRM quantitative analysis using a calibration correction method on a triple quadrupole system, ASMS 2006 Poster Presentation, TP08, #115.

Malavia, J., Santos, F.J., Galceran, M.T., Comparison of gas chromatography-ion-trap tandem mass spectrometry systems for the determination of polychlorinated dibenzo-p-dioxins, dibenzofurans and dioxin-like polychlorinated biphenyls, J. Chromatogr. A, 1186 (2008), 302–311.

Miller, P. E., Denton, M. B., The Quadrupol Mass Filter: Basic Operating Concepts, J.Chem.Education 63 (1986), 617–622.

Paul, W., Reinhard, H. P., von Zahn, U., Das elektrische Massenfilter als Massenspektrometer und Isotopentrenner. Z. Physik 152 (1958), 143–182.

Paul, W., Steinwedel, H., Apparat zur Trennung von geladenen Teilchen mit unterschiedlicher spezifischer Ladung, Deutsches Patent 944 900, 1956 (U. S. Patent 2 939 952 v. 7. 6. 1960).

Paul, W., Steinwedel, H., Ein neues Massenspektrometer ohne Magnetfeld. Z. Naturforschg. 8 a (1953), 448–450.

Sparkman, D., Mass Spectrometry Desk Reference, Global View Publishing, Pittsburgh, 2000.

Todd, J. F. J., Instrumentation in Mass Spectrometry, In: Advances in Mass Spectrometry, J. F. J. Todd (Ed.), Wiley: New York 1986, 35–70.

Van Ysacker, P. G., Janssen, H. M., Leclerq, P. A., Wollnik, H., Cramers, C. A., Comparison of Different Mass Spectrometers in Combination with High-Speed Narrow-Bore Capillary Gas Chromatography, In: 16th Int. Symp. Cap. Chrom. Riva del Garda, P. Sandra (Ed.), Huethig 1994, 785–796.

Wang, J., The Determination of Elemental Composition by Low Resolution MS/MS, Proceedings of the 43rd ASMS Conference. Atlanta 21–26 May 1995, 722.

Webb., K., Resolving Power and Resolution in Mass Spectrometry, Best Practice Guide, VIMMMS/2004/01, Teddington UK: LGC November 2004, see also www.lgc.co.uk.

Section 2.3.2

Borman, Stu, A Brief History in Mass Spectrometry, May 26, 1998, http://masspec.scripps.edu/information/history/index.html

Cotter, R.J., Time-of-Flight Mass Spectrometry, American Chemical Society: Washington DC, 1997.

Cotter, R.J., Analytical Chemistry 64 (1992), 1027A.

Dimandja, J.M., A new tool for the optimized analysis of complex volatile mixtures: Comprehensive two-dimensional gas chromatography/time-of-flight mass spectrometry, Am. Laboratory 2, (2003) 42–53.

GC Image, LLC, and the University of Nebraska, GC Image™ Users' Guide © 2001–2004 by, www.gcimage.com.

Guilhaus, M., Review of TOF-MS Journal of Mass Spectrometry, 30, (1995) 1519.

Mamyrin, B.A., et al., Soviet Physics – JETP, 37, (1973), 45.

Mamyrin, B.A., et al., Int. J. Mass Spectrom. Ion Proc., 131, (1994) 1.

McClenathan, D, Ray, S.J., Plasma Source TOFMS, Anal Chem 76, 9 (2004) 159A-166A.

Meruva, N.K., Sellers, K.W., Brewer, W.E., Goode, S.R., Morgan, S.L., Comparisons of chromatographic performance and data quality using fast gas chromatography, paper 1397, Pittcon 2000, New Orleans, 17 March 2000.

Webb, K., Methodology for Accurate Mass Measurement of Small Molecules, VIMMS/2004/01, Teddington, UK: LGC Ltd. November 2004

Wiley, W.C., MacLaren, I.H., The Review of Scientific Instruments, 26, (1955) 1150.

Section 2.3.3

Bradley, D., Tracking cocaine to its roots, Today's Chemist at Work 5, (2002) 15–16.

Brand, A., Mass Spectrometer Hardware for Analyzing Stable Isotope Ratios, in: de Groot, P.A. (Ed.), Handbook of Stable Isotope Analytical Techniques, Vol. 1, Elsevier: Amsterdam 2004, 835 – 856.

Brand, W.A., Tegtmeyer, A.R., Hilkert, A., Org. Geochem. 21, (1994), 585.

Coplan, T., Isotope Ratio Data, JUPAC.

Coplen, T.B., Explanatory Glossary of Terms Used in Expression of Relative Isotope Ratios and Gas Ratios, International Union of Pure and Applied Chemistry Inorganic Chemistry Division Com-

mission on Isotopic Abundances and Atomic Weights, Peer Review, January 16, 2008, see http://www.iupac.org.
Craig, H. Isotopic standards for carbon and oxygen and correction factors for mass-spectrometric analysis of carbon dioxide. Geochimica Cosmochimica Acta 12, (1957) 133–149.
de Vries, J.J., Natural abundance of the stable isotopes of C, O and H, in: Mook,W.G. (Ed.), Environmental Isotopes in the Hydrological Cycle, Vol. 1: Introduction – Theory, methods, review, see: www.iaea.org.
Ehleringer, J.R., Casale, J.F. ,Lott, M.J., Ford,V.L., Tracing the geographical origin of cocaine, Nature (2000) 408–311.
Ehleringer, J.R., Cerling, T.E., Stable Isotopes, in: Mooney, H.A., Canadell, J.G. (Eds.): Encyclopaedia of Global Environmental Change,Vol.2. The Earth System: Biological and ecological dimensions of global environmental change, Wiley: Chichester 2002, 544–550.
Fry, B., Stable Isotope Ecology, Springer: New York 2006.
Groening, M., International Stable Isotope Reference Materials, in: de Groot, P.A. (Ed.), Handbook of Stable Isotope Analytical Techniques Vol. 1, Elsevier: Amsterdam 2004, 875 – 906.
Heuer, K., Brand, W.A., Hilkert, A.W., Juchelka, D., Mosandl, A., Podebred, F. , Z. Lebensm. Unters. Forsch. 206, (1998), 230.
Hilkert, A., Douthitt, C.B., Schlüter, H.J., Brand, W.A., Isotope Ratio Monitoring Gas Chromatography/Mass Spectrometry of D/H by High Temperature Conversion Isotope Ratio Mass Spectrometry, Rapid Commun. Mass Spectrom. 13, (1999), 1226–1230.
Matthews D.E., J.M. Hayes, Anal. Chem. 50, (1978), 1465.
Matucha, M., J. Chromatogr. 588, (1991) 251–258.
Meyer-Augenstein, W., The chromatographic side of isotope ratio mass spectrometry: Pitfalls and answers, LCGC Int. Jan. 1997, 17 – 25.
Rosman, K.J.R., Taylor, P.D.P., Isotopic Compositions of the Elements 1997, Pure & Appl. Chem., 70, (1) (1998), 217–235.
Rossmann, A., Determination of stable isotope ratios in food analysis, Food Reviews Int. 17, (3) (2001), 347–381.
Sano, M., et al., A new technique for the detection of metabolites labelled by the isotope ^{13}C using mass fragmentography, Biomed. Mass Spectrom. 3, (1976), 1–3.
Sharp, Z., Principles of Stable Isotope Geochemistry, Pearson Prentice Hall: Upper Saddle River 2007.
St. Jean, G., Basic Principles of Stable Isotope Geochemistry, Short Course Manuscript of the 9th Canadian CF-IRMS Workshop, August 25, 2002.

Section 2.3.4

Ardenne, M., Steinfelder, K., Tümmler, R., Elektronenanlagerungsmassenspektrometrie organischer Substanzen, Springer Verlag: Berlin 1971.
Aue, D. H., Bowers, M. T., Stability of Positive Ions from Equilibrium Gas-Phase Basicity Measurements, in: M. T. Bowers (Ed.): Gas Phase Ion Chemistry, Vol. 2, Academic Press: New York 1979.
Bartmess, J., Mclver, R., The Gas Phase Acidity Scale, in: M. T. Bowers (Ed.): Gas Phase Ion Chemistry, Vol. 2, Academic Press: New York 1979.
Beck, H., Eckart, K., Mathar, W., Wittkowski, R., Bestimmung von polychlorierten Dibenzofuranen (PCDF) und Dibenzodioxinen (PCDD) in Lebensmitteln im ppq-Bereich, Lebensmittelchem. Gerichtl. Chem. 42 (1988), 101–105.
Bowadt, Frandsen, E., et al., Combined Positive and Negative Ion Chemical Ionisation for the Analysis of PCB'S, In: P.Sandra (Ed.), 15th Int. Symp. Cap. Chrom., Riva del Garda, May 1993, Huethig 1993.
Budzikiewicz, H., Massenspektrometrie, 3. erw. Aufl., VCH: Weinheim 1993.
Budzikiewicz, H., Massenspektrometrie negativer Ionen, Angew.Chem. 93 (1981), 635–649.
Buser, H.-R., Müller, M., Isomer- and Enantiomer-Selective Analyses of Toxaphene Components Using Chiral High-Resolution Gas Chromatography and Detection by Mass Spectrometry/Mass Spectrometry, Environ. Sci. Technol. 28 (1994), 119–128.
Class, T. J., Determination of Pyrethroids and their Degradation Products in Indoor Air and on Surfaces by HRGC-ECD and HRGC-MS (NCI), In: P. Sandra (Ed.), 12th Int. Symp. Cap. Chrom., Riva del Garda, May 1991, Huethig 1991.
Crow, F. W., Bjorseth, A., et al., Anal. Chem. 53 (1981), 619.
DePuy, C. H., Grabowski, J. J., Bierbaum, V. M., Chemical Reactions of Anions in the Gas Phase, Science 218 (1982), 955–960.
Dorey, R. C., Williams, K., Rhodes, C. L., Fossler, C. L., Heinze, T. M., Freeman, J. P., High Kinetic Energy Chemical Ionization in the Quadrupole Ion Trap: Methylamine CIMS of Amines, Presented at the 42nd ASMS Conference on Mass

Spectrometry and Allied Topics, Chicago. June 1–6, 1994.
Dougherty, R. C., Detection and Identification of Toxic Substances in the Environment, In: C. Merritt, C. McEwen (Eds.): Mass Spectrometry Part B, Marcel Dekker: New York 1980, 327.
Dougherty, R. C., Negative Chemical Ionization Mass Spectrometry, Anal.Chem. 53 (1981), 625A–636A.
Frigerio, A., Essential Aspects of Mass Spectrometry, Halsted Press: New York 1974.
Gross, J. H., Mass Spectrometry, Springer: Heidelberg 2004.
Hainzl, D., Burhenne, J., Parlar, H., Isolierung von Einzelsubstanzen für die Toxaphenanalytik, GIT Fachz. Lab. 4 (1994), 285–294.
Harrison, A. G., Chemical Ionization Mass Spectrometry, CRC Press: Boca Raton 1992.
Horning, E. C., Caroll, D. I., Dzidic, I., Stillwell, R. N., Negative Ion Atmospheric Pressure Ionization Mass Spectrometry and the Electron Capture Detector. In: A. Zlatkis, C. F. Poole (Eds.): Electron Capture, J. Chromatogr. Library, Vol. 20, Elsevier: Amsterdam 1981.
Howe, I., Williams, D. H., Bowen, R., Mass Spectrometry, 2nd Ed., McGraw Hill: New York 1981.
Hübschmann, H.-J., Einsatz der chemischen Ionisierung zur Analyse von Pflanzenschutzmitteln, Finnigan MAT Application Report No. 75, 1990.
Hunt, D. F., Stafford, G. C., Crow, F., Russel, J., Pulsed Positive Negative Ion Chemical Ionization Mass Spectrometry, Anal. Chem. 48 (1976), 2098–2105.
Keller, P. R., Harvey, G. J., Foltz, D. J., GC/MS Analysis of Fragrances Using Chemical Ionization on the Ion Trap Detector: An Easy-to-Use Method for Molecular Weight Information and Low Level Detection, Finnigan MAT Application Report No. 220, 1989.
McLafferty, F. W., Michnowicz, J. A., Chem. Tech. 22 (1992), 182.
McLafferty, F. W., Turecek, F., Interpretation of Mass Spectra. 4th Ed., University Science Books: Mill Valley 1993.
Munson, M. S. B., Field, F. H., J. Am. Chem. Soc. 88 (1966), 4337.
Schröder, E., Massenspektroskopie, Springer: Berlin 1991.
Smit, A. L. C., Field, F. H., Gaseous Anion Chemistry. Formation and Reaction of OH^-; Reactions of Anions with N_2O; OH^- Negative Chemical Ionization, J Am. Chem. Soc. 99 (1977), 6471–6483.
SPECTRA, Analytical Applications of Ion Trap Mass Spectrometry, Vol. 11 (2), 1988.
Spiteller, M., Spiteller, G., Massenspektrensammlung von Lösungsmitteln, Verunreinigungen. Säulenbelegmaterialien und einfachen aliphatischen Verbindungen, Springer: Wien 1973.
Stan, H.-J., Kellner, G., Analysis of Organophosphoric Pesticide Residues in Food by GC/MS Using Positive and Negative Chemical Ionisation, In: W. Baltes, P. B. Czedik-Eysenberg, W. Pfannhauser (Eds.), Recent Developments in Food Analysis, Verlag Chemie: Weinheim 1981.
Stout, S. J., Steller, W. A., Application of Gas Chromatography Negative Ion Chemical Ionization Mass Spectrometry in Confirmatory Procedures for Pesticide Residues, Biomed. Mass Spectrom. 11 (1984), 207–210.
Theobald, F., Huebschmann, H.J., High Sensitive MID Detection Method for Toxaphenes by High Resolution GC/MS, Application Note, Thermo Fisher Scientific, Bremen, Germany, AN30128, 2007.

Section 2.3.5

Brodbelt, J. S., Cooks, R. G., Ion Trap Tandem Mass Spectrometry, Spectra 11/2 (1988) 30–40.
Brunée, C., New Instrumentation in Mass Spectrometry, Spectra 9, 2/3 (1983) 10–36.
Brunée, C., The Ideal Mass Analyzer: Fact or Fiction?, Int. J. Mass Spec. Ion Proc. 76 (1987) 121–237.
Busch, K. L., Glish, G. L., McLuckey, S. A., Mass Spectrometry/Mass Spectrometry: Techniques and Applications of Tandem Mass Spectrometry, VCH Publishers: New York 1988.
Dawson, P. H., In: Dawson, P. H., Ed.: Quadrupole Mass Spectrometry and its Applications, Elsevier: Amsterdam 1976, 19–70.
DFG Deutsche Forschungsgemeinschaft, Manual of Pesticide Residue Analysis Vol. II, Thier, H.P., Kirchhoff, J., Eds.,VCH: Weinheim 1992, 25–28.
Johnson, J. V., Yost, R. A., Anal. Chem. 57 (1985) 758A.
Johnson, J. V., Yost, R. A., Kelley, P. E., Bradford, D. C., Tandem-in-Space and Tandem-in-Time Mass Spectrometry: Triple Quadrupoles and Quadrupole Ion Traps, Anal. Chem. 62 (1990) 2162–2172.
Julian, R. K., Nappi, M., Weil, C., Cooks, R. G., Multiparticle Simulation of Ion Motion in the Ion Trap Mass Spectrometer: Resonant and Direct Current Pulse Excitation, J. Am. Soc. Mass spectrom. 6 (1995) 57–70.

Kaiser, H., Foundations for the Critical Discussion of Analytical Methods, Spectrochim. Acta Part B, 33b (1978) 551.
March, R. E., Hughes, R. J., Quadrupole Storage Mass Spectrometry, John Wiley: New York 1989.
McLafferty, F. E., Ed.: Tandem Mass Spectrometry, John Wiley: New York 1983.
Noble, D., MS/MS, Flexes Ist Muscles, Anal. Chem. 67 (1995) 265A–269A.
Perfluorotributylamine (PFTBA, FC43) Reference Table, Thermo Fisher Scientific Data Sheet PS30040_E.
Plomley, J. B., Koester, C. J., March, R. E., Determination of N-Nitrosodimethylamine in Complex Environmental Matrices by Quadrupole Ion Storage Tandem Mass Spectrometry Enhanced by Unidirectional Ion Ejection, Anal. Chem. 66 (1994) 4437–4443.
Polychlorinated Dibenzodioxins and -furans, Thermo Fisher Scientific Data Sheet PS30042_E.
Slayback, J. R. B., Taylor, P. A., Analysis of 2.3.7,8-TCDD and 2,3,7,8-TCDF in Environmental Matrices Using GC/MS/MS Techniques. Spectra 9/4 (1983) 18–24.
Soni, M. H., Cooks, R. G., Selective Injection and Isolation of Ions in Quadrupole Ion Trap Mass Spectrometry Using Notched Waveforms Created Using the Inverse Fourier Transform. Anal. Chem. 66 (1994) 2488–2496.
Strife, R. J., Tandem Mass Spectrometry of Prostaglandins: A Comparison of an Ion Trap and a Reversed Geometry Sector Instrument, Rap. Comm. Mass Spec. 2 (1988) 105–109.
Wagner-Redeker, W., Schubert, R., Hübschmann, H.-J., Analytik von Pestiziden und polychlorierten Biphenylen mit dem Finnigan 5100 GC/MS-System, Finnigan MAT Application Report No. 58, 1985.
Yasek, E., Advances in Ion Trap Mass Spectrometry, Finnigan MAT Technical Report No. 614, Finnigan Corporation: San Jose USA 1989.
Yost, R. A., MS/MS: Tandem Mass Spectrometry, Spectra 9/4 (1983) 3–6.

Section 2.3.6

Council Directive 96/23/EC concerning the performance of analytical methods and the interpretation of results, Off. J. Europ. Comm. L221/8, 17, 8. 2002.

Section 2.3.7

Audi, G., Wapstra, A.H., The 1995 update to the atomic mass evaluation, Nuclear Phys. A 595 (1995), 409–480.
Eichelberger, J.W., Harris, L.E., Budde, W.L., DFTPP Tuning, Anal. Chem., 47 (1975) 995.
Mohr, P.J., Taylor, B.N., CODATA Recommended Values of the Fundamental Physical Constants: 1998, J. Phys. Chem. Ref. Data 28(6) (1999), 1713–1852.

Section 2.4.2

Abraham, B., Rückstandsanalyse von anabolen Wirkstoffen in Fleisch mit Gaschromatographie-Massenspektrometrie, Dissertation, Technische Universität Berlin, 1980.
Cramers, C. A., Scherpenzeel, G. J., Leclercq, P. A., J. Chromatogr. 203 (1981), 207.
Gohlke, R. S., Time-of-Flight Mass Spectrometry and Gas-Liquid Partition Chromatography, Anal. Chem. 31 (1959), 535–541.
Henneberg, D., Henrichs, U., Husmann, H., Schomburg, G., J. Chromatogr. 187 (1978), 139.
Schulz, J., Nachweis und Quantifizieren von PCB mit dem Massenselektiven Detektor, LaborPraxis 6 (1987), 648–667.
Tiebach, R., Blaas, W., Direct Coupling of a Gas Chromatograph to an Ion Trap Detectoir, J. Chromatogr. 454 (1988), 372–381.

3
Evaluation of GC/MS Analyses

3.1
Display of Chromatograms

Chromatograms obtained by GC/MS are plots of the signal intensity against the retention time, as with classical GC detectors. Nevertheless, there are considerable differences between the two types of chromatogram arising from the fact that data from GC/MS analyses are in three dimensions. Figure 3.1 shows a section of the chromatogram of the total ion current in the analysis of volatile halogenated hydrocarbons. The retention time axis also shows the number of continually registered mass spectra (scan no.). The mass axis is drawn above the time axis at an angle. The elution of each individual substance can be detected by evaluating the mass spectra using a 'maximising masses peak finder' program and can be shown by a marker. Each substance-specific ion shows a local maximum at these positions, which

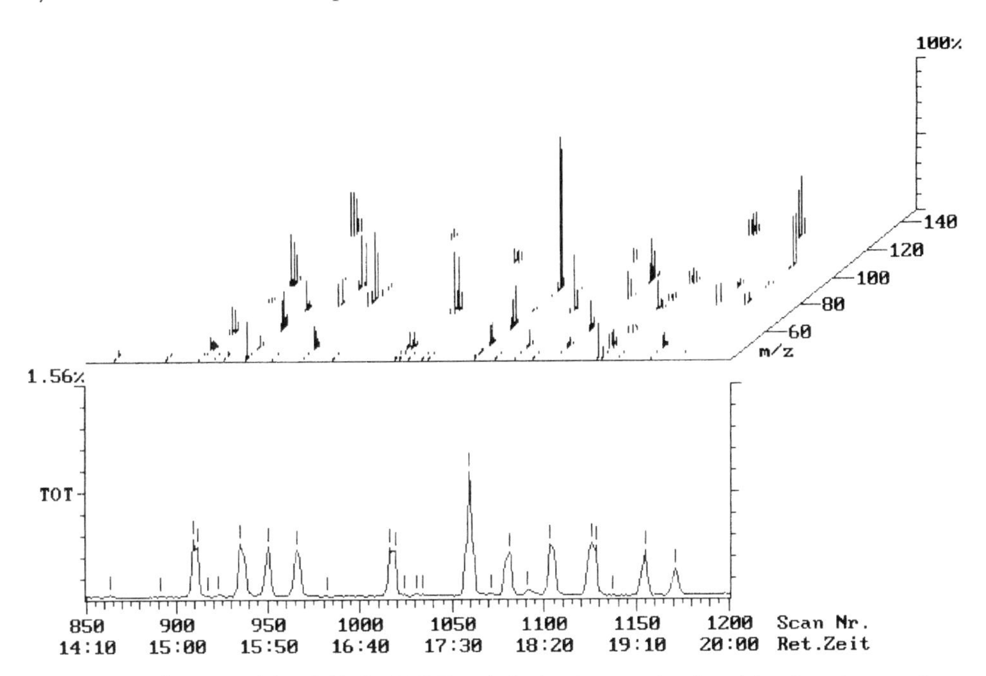

Fig. 3.1 Three-dimensional data field of a GC/MS analysis showing retention time, intensity and mass axis

Handbook of GC/MS: Fundamentals and Applications, Second Edition. Hans-Joachim Hübschmann

ISBN: 978-3-527-31427-0

can be determined by the peak finder. The mass spectra of all the analytes detected are shown in a three dimensional representation for the purposes of screening. For further evaluation the spectra can be examined individually.

3.1.1 Total Ion Current Chromatograms

The intensity axis in GC/MS analysis is shown as a total ion current (TIC) or as a calculated ion chromatogram (reconstructed ion chromatogram, RIC). The intensity scale may be given in absolute values, but a percentage scale is more frequently used. Both terms describe the mode of representation characteristic of the recording technique. At constant scan rates the mass spectrometer plots spectra over the pre-selected mass range and thus gives a three-dimensional data field arising from the retention time, mass scale and intensity. A signal parameter equivalent to FID detection is not directly available. (Magnetic sector mass spectrometers were equipped with a total ion current detector directly at the ion source until the end of the 1970s!). A total signal intensity comparable to the FID signal at a particular point in the scan can, however, be calculated from the sum of the intensities of all the ions at this point. All the ion intensities of a mass spectrum are added together by the data system and stored as a total intensity value (total ion current) together with the spectrum. The total ion current chromatogram thus constructed is therefore dependent on the scan range used for data acquisition. When making comparisons it is essential to take the data acquisition conditions into consideration.

SIM/MID analyses give a chromatogram in the same way but no mass spectrum is retrievable. The total ion current in this case is composed of the intensities of the selected ions. Analyses where switching from individual masses to fixed retention times is planned often show clear jumps in the base line (see Fig. 2.138).

The appearance of a GC/MS chromatogram (TIC/RIC) showing the peak intensities is therefore strongly dependent on the mass range shown. The repeated GC/MS analysis of one particular sample employing mass scans of different widths leads to peaks of different heights above the base line of the total ion current. The starting mass of the scan has a significant effect here. The result is a more or less strong recording of an unspecific background which manifests itself in a higher or lower base line of the TIC chromatogram. Peaks of the same concentration are therefore shown with different signal/noise ratios in the total ion current at different scan ranges. In spite of differing representation of the substance peaks, the detection limit of the GC/MS system naturally does not change. Particularly in trace analysis the concentration of the analytes is usually of the same order of magnitude or even below that of the chemical noise (matrix) in spite of good sample processing so that the total ion current cannot represent the elution of these analytes. Only the use of selective information from the mass chromatogram (see Section 3.1.2) brings the substance peak sought on to the screen for further evaluation.

In the case of data acquisition using selected individual masses (SIM/MID/SRM), only the changes in intensity of the masses selected before the analysis are shown. Already during data acquisition only those signals (ion intensities) are recorded from the total ion current which correspond to the prescriptions of the user. The greater part of the total ion current is therefore not detected using the SIM/MID/SRM technique (see Section 2.3.3). Only substances which give signals in the region of the selected masses as a result of fragment or molecular ions are shown as peaks. A mass spectrum for the purpose of checking identity is therefore not available. For confirmation this should be measured using an alternating full scan/SIM mode or separately in a subsequent analysis. The retention time and the relative intensities of two or three specific lines are used as qualifying features. In trace analysis unambiguous detection is never possible using this method. Positive results of an SIM/MID analysis basically require additional confirmation by a mass spectrum. SRM analyses offer the recording of a MS/MS spectrum or require the monitoring of multiple transitions as additional qualifiers.

3.1.2
Mass Chromatograms

A meaningful assessment of signal/noise ratios of certain substance peaks can only be carried out using mass chromatograms of substance-specific ions (fragment/molecular ions). The three-dimensional data field of GC/MS analyses in the full scan mode does not only allow the determination of the total ion intensity at a point in the scan. To show individual analytes selectively the intensities of selected ions (masses) from the total ion current are shown and plotted as an intensity/time trace (chromatogram).

The evaluation of these mass chromatograms allows the exact determination of the detection limit above the signal/noise ratio of the substance-specific ion produced by a compound. With the SIM/MID mode this ion would be detected exclusively, but a complete mass spectrum for confirmation would not be available. In the case of complex chromatograms of real samples mass chromatograms offer the key to the isolation of co-eluting components so that they can be integrated perfectly and quantified.

In the analysis of lemons for residues from plant protection agents a co-elution situation was discovered by data acquisition in the full scan mode of the ion trap detector and was evaluated using a mass chromatogram.

The routine testing with an ion trap GC/MS system gives a trace which differs from that using an element-specific NPD detector (Fig. 3.2). A large number of different peaks appear in the retention region which indicates the presence of Quinalphos as the active substance in the NPD evaluation (Fig. 3.3). The Quinalphos peak has a shoulder on the left side and is

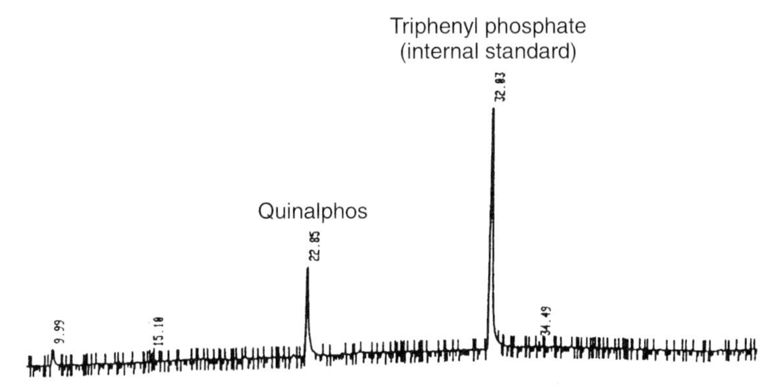

Fig. 3.2 Analysis of a lemon extract using the NPD as detector. The chromatogram shows the elution of a plant protection agent component as well as the internal standard

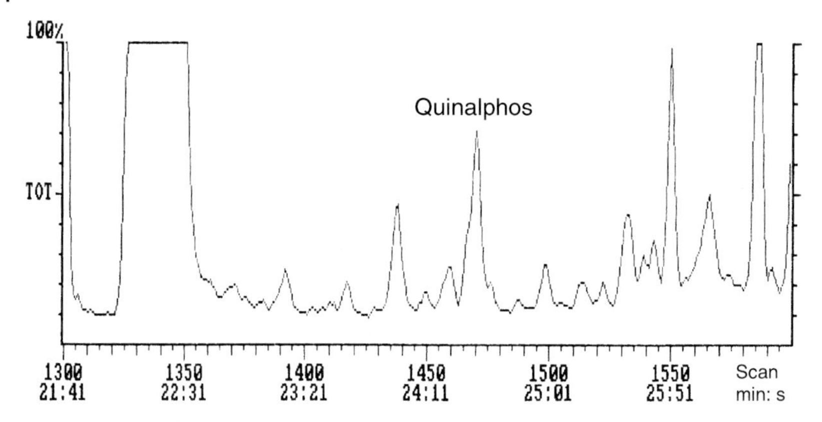

Fig. 3.3 Confirmation of the identity of a lemon extract by GC/MS. The total ion current clearly shows the questionable peak with shoulder mass spectra. A co-eluting second active substance, Chlorfenvinphos, gives rise to the shoulder.

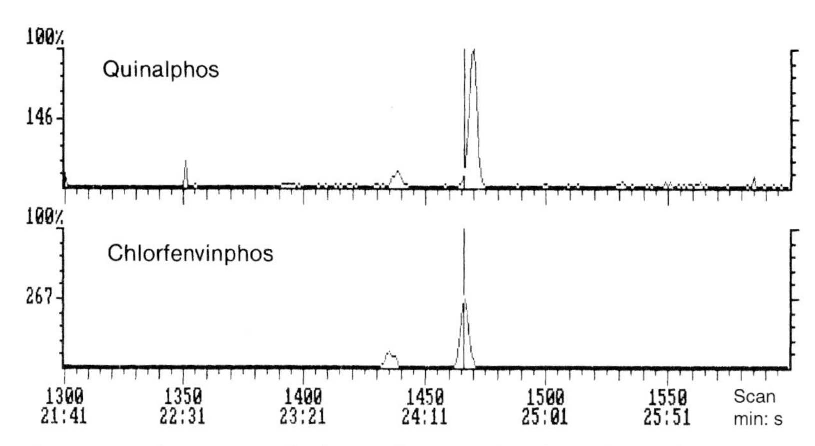

Fig. 3.4 Mass chromatograms for the specific masses show the co-elution of Quinalphos (m/z 146) and Chlorfenvinphos (m/z 267). The retention time range is identical with that in Fig. 3.3. The selective plot of the mass signals can be identified as part of the total ion current.

closely followed by another less intense component. In the mass chromatogram of the characteristic individual masses (fragment ions) it can be deduced from the total ion current that another eluting active substance is present (Fig. 3.4). Unlike NPD detection, with GC/MS analysis it becomes clear after evaluating the mass chromatogram and mass spectra that the co-eluting substance is Chlorfenvinphos.

In routine analysis this evaluation is carried out by the data system. If the information on the retention time of an analyte, the mass spectrum, the selective quantifying mass and a valid calibration are supplied, a chromatogram can be evaluated in a very short time for a large number of components (Fig. 3.5, see also Section 3.3).

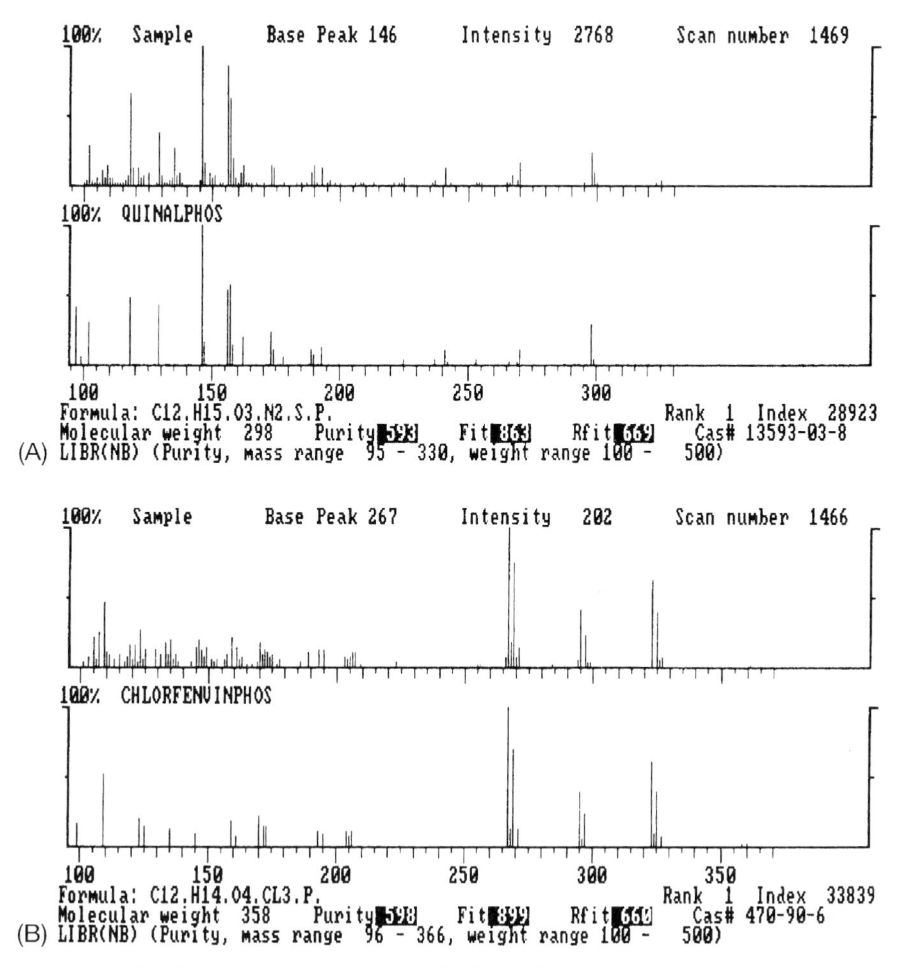

Fig. 3.5 The phosphoric acid esters are successfully identified by a library comparison after extraction of the spectra by a background subtraction.
(A) Quinalphos is confirmed by comparison with the NBS library (FIT value 863)
(B) Chlorfenvinphos is confirmed by a comparison with the NBS library (FIT value 899)

3.2 Substance Identification

3.2.1 Extraction of Mass Spectra

One of the great strengths of mass spectrometry is the immediate provision of direct information about an eluting component. The careful extraction of the substance-specific signals from the chromatogram is critical for reliable identity determination. For identification or confirmation of individual GC peaks recording mass spectra which are as complete as possible is an important basic prerequisite.

By plotting mass chromatograms co-elution situations can be discovered, as shown in Section 3.1.2. The mass chromatograms of selected ions give important information via their maximising behaviour. Only when maxima are shown at exactly the same time can it be assumed that the fragments observed originate from a single substance, i.e. from the same chemical structure. The only exception is the ideal simultaneous co-elution of compounds. If peak maxima with different retention times are shown by various ions, it must be assumed that there are co-eluting components (see Figs. 3.7 and 3.17).

Manual Spectrum Subtraction

By subtraction of the background or the co-elution spectra before or behind a questionable GC peak the mass spectrum of the substance sought is extracted from the chromatogram as free as possible from other signals. All substances co-eluting with an unknown substance including the matrix components and column bleed are described in this context as chemical background. The differentiation between the substance signals and the background and its elimination from the substance spectrum is of particular importance for successful spectroscopic comparison in a library search. In the example of the GC/MS analysis of lemons for plant protection agents described above this procedure is used to determine the identity of the active substances.

The possibilities for subtraction of mass spectra are shown in the following real example of the analysis of volatile halogenated hydrocarbons by purge and trap-GC/MS. Figure 3.6 shows part of a total ion current chromatogram. The peak marked with X shows a larger width than that of the neighbouring components. On closer inspection of the individual spectra in the peak it can be seen that in the rising slope of the peak ions with m/z 39, 75,

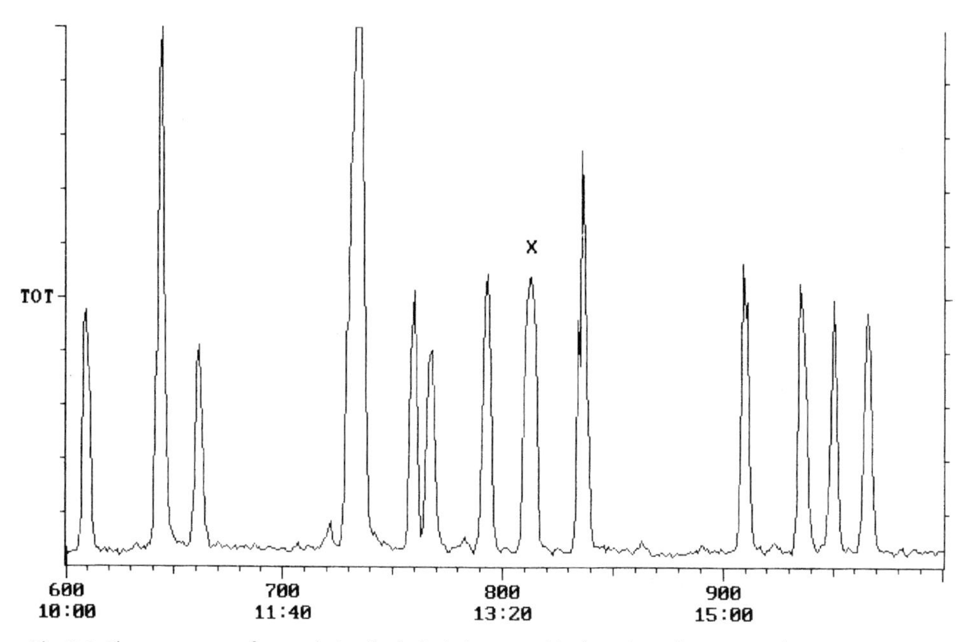

Fig. 3.6 Chromatogram of an analysis of volatile halogenated hydrocarbons by purge and trap GC/MS. The component marked with X has a larger half width than the neighbouring peaks

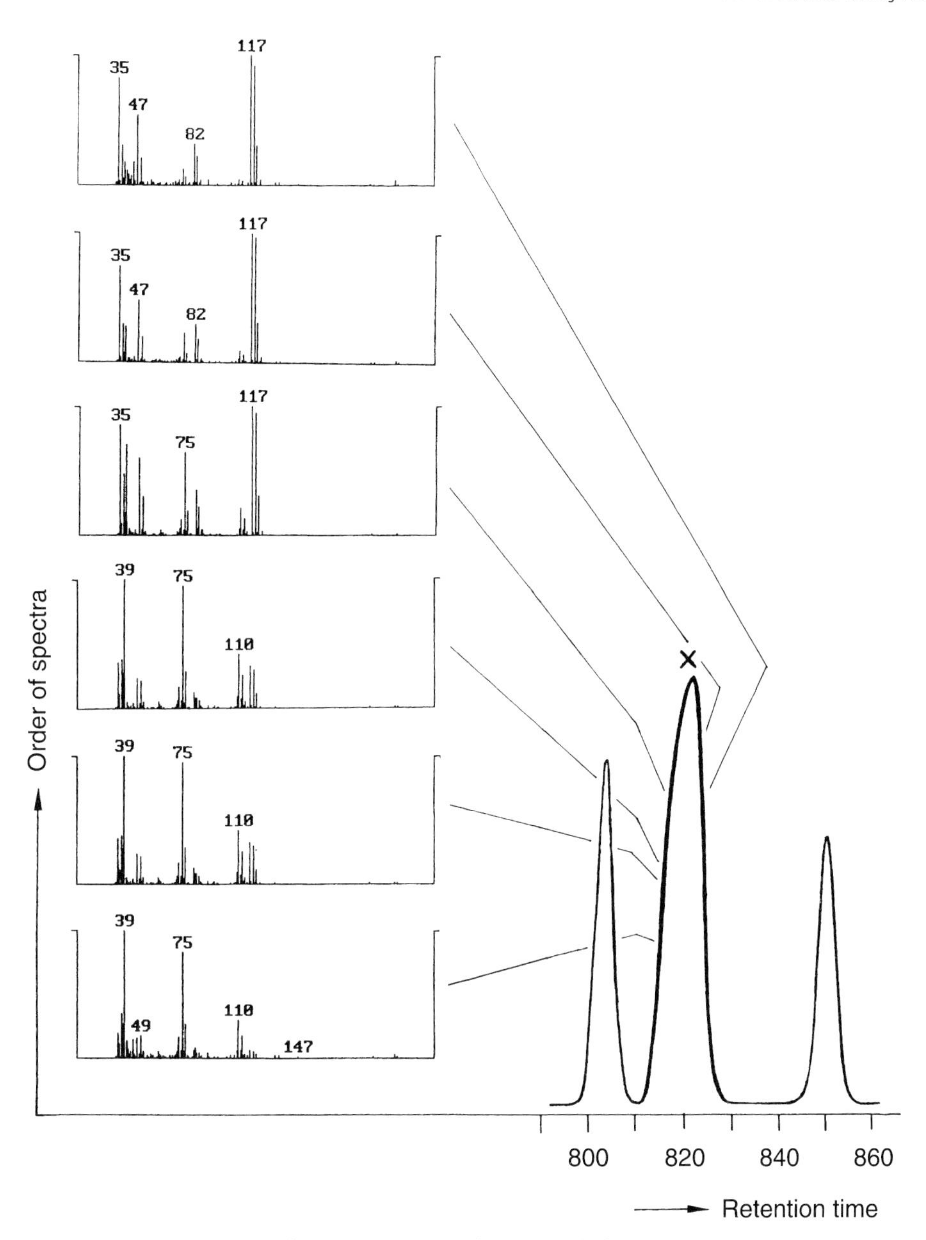

Fig. 3.7 Continuous plotting of the mass spectra in the peak marked X in Fig. 3-6

110 and 112 dominate. As the elution of the peak continues other ion signals appear. The ions with m/z 35, 37, 82, 84, 117, 119 and 121 appear in increased strength, while the previously dominant signals decrease. Figure 3.7 shows this situation using the continuing presentation of individual mass spectra in a characterised GC peak.

From the individual mass spectra (Fig. 3.7) it can be recognised that some signals obviously belong together. Figure 3.8 shows the mass chromatograms of the ions m/z 110/112

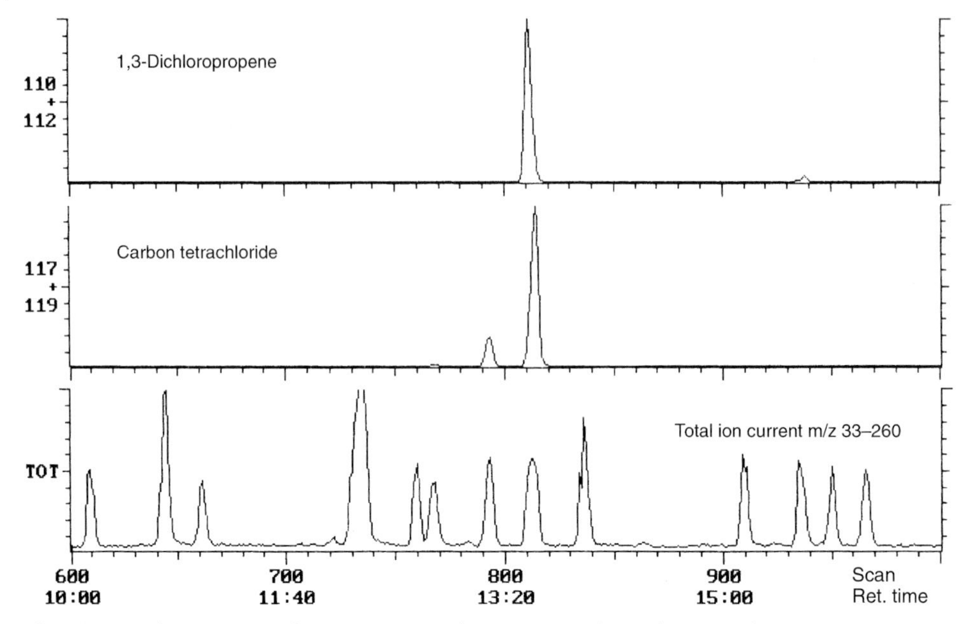

Fig. 3.8 Mass chromatograms for m/z 110/112 and m/z 117/119 shown above a total ion current chromatogram

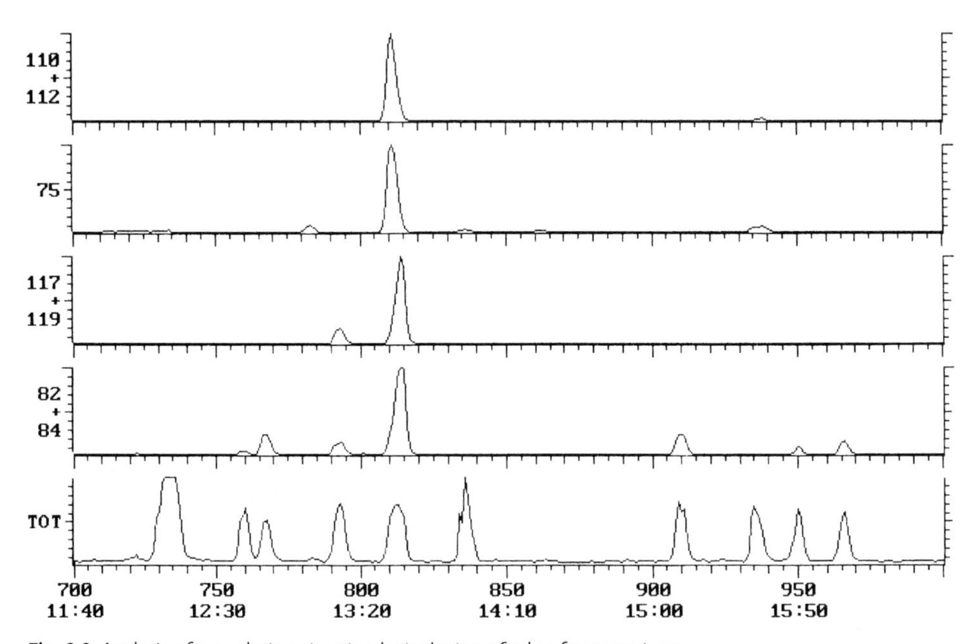

Fig. 3.9 Analysis of a co-elution situation by inclusion of other fragment ions

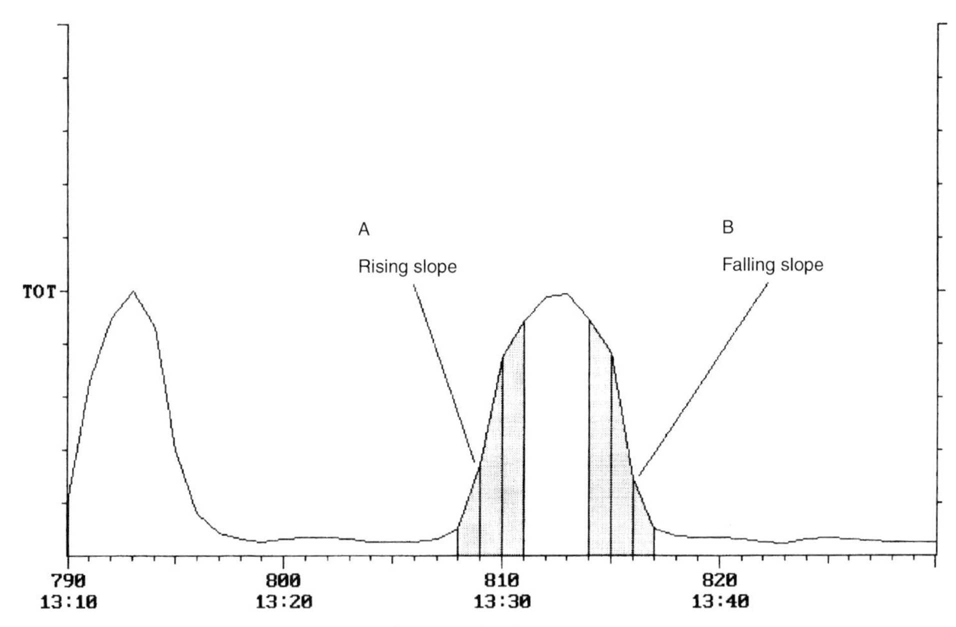

Fig. 3.10 Plot of a peak with selected areas for spectral subtraction

and m/z 117/119 (as a sum in each case) above the total ion current from the detected mass range of m/z 33-260. The mass chromatograms show an intense GC peak at the questionable retention time in each case. The peak maxima are not superimposed and are slightly shifted towards each other. This is an important indication of the co-elution of two components (Fig. 3.9).

If other ions are included in this first mass analysis, it can be concluded that the fragments belong together from their common maximising behaviour.

After the individual mass signals have been assigned to the two components, the extraction of the spectrum of each compound can be performed. Figure 3.10 shows the division of the peak into the front peak slope A and the back peak slope B. With the background subtraction function contained in all data systems the spectra in the areas A and B are added and subtracted from one another.

The subtraction of the areas A and B gives the clean spectra of the co-eluting analytes. In Fig. 3.11 the subtraction A – B shows the spectrum of the first component, which is shown to be 1,3-dichloropropene (Fig. 3.12) from a library comparison. The reverse procedure, i.e. the subtraction B – A, gives the identity of the second component (Figs. 3.13 and 3.14).

A further frequent use of spectrum subtraction allows the removal of background signals caused by the matrix or column bleed. Figure 3.15 shows the elution of a minor component from the analysis of volatile halogenated hydrocarbons, which elutes in the region where column bleed begins. For background subtraction the spectra in the peak and from the region of increasing column bleed are subtracted from one another.

The result of the background subtraction is shown in Fig. 3.16. While the spectrum clearly shows column bleed from the substance peak with m/z 73, 207 and a weak CO_2 background at m/z 44, the resulting substance spectrum is free from the signals of the interfering che-

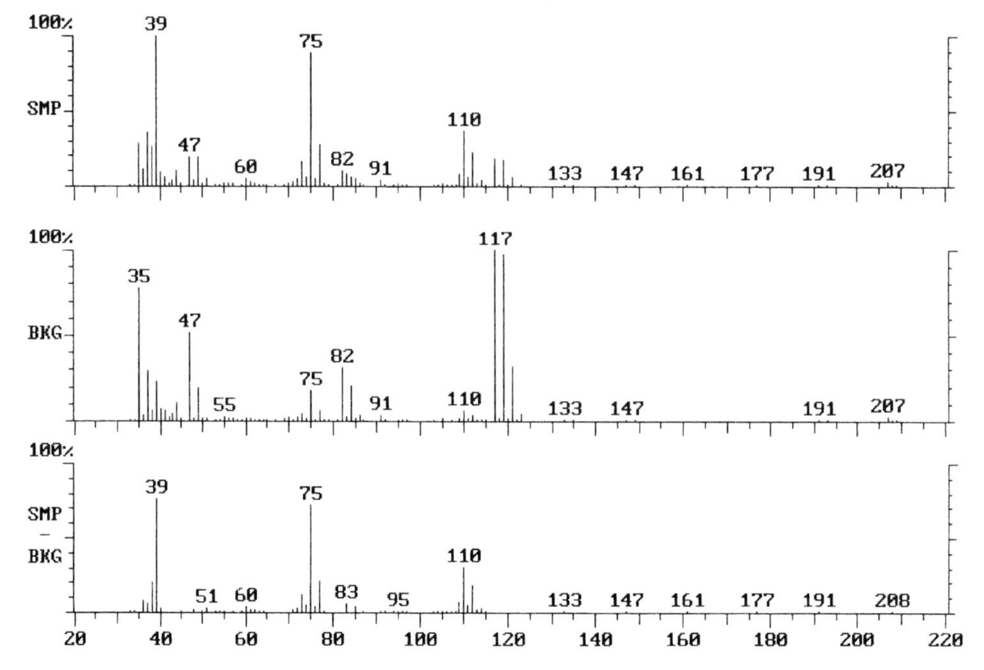

Fig. 3.11 Spectral subtraction of the areas A–B:
SMP = Spectra of the rising peak slope (A), sample
BKG = Spectra of the falling peak slope (B), background
SMP-BKG = Resulting spectrum of the component eluting first

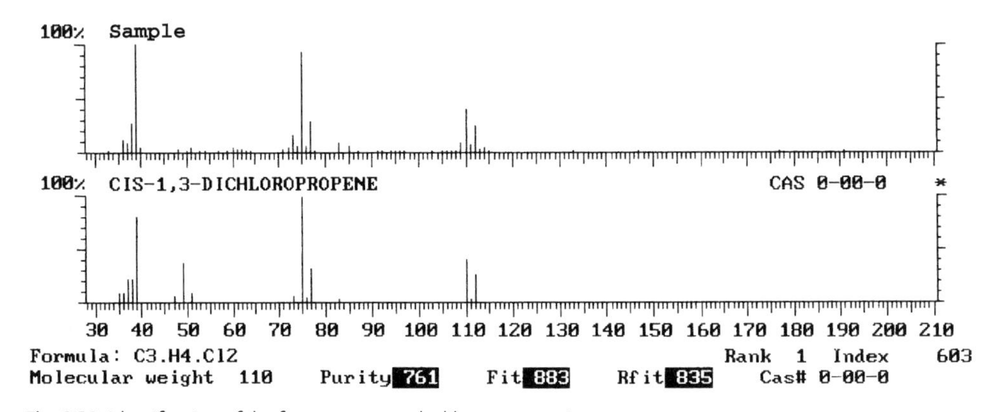

Fig. 3.12 Identification of the first component by library comparison

mical background after the subtraction. This clean spectrum can then be used for a library search in which it can be confirmed as 1,2-dibromo-3-chloropropane.

In the subtraction of mass spectra it should generally be noted that in certain cases substance signals can also be reduced. In these cases it is necessary to choose another background area. If changes in the substance spectrum cannot be prevented in this way, the

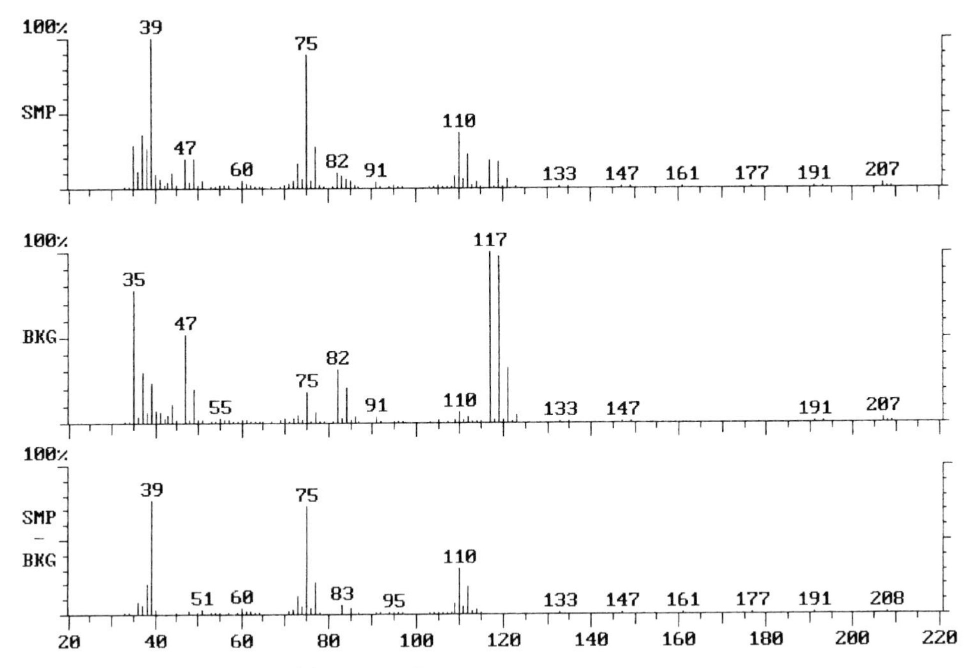

Fig. 3.13 Spectral subtraction of the areas B–A:
SMP = Spectra of the falling peak slope (B), sample
BKG = Spectra of the rising peak slope (A), background
SMP-BKG = Resulting spectrum of the component eluting second

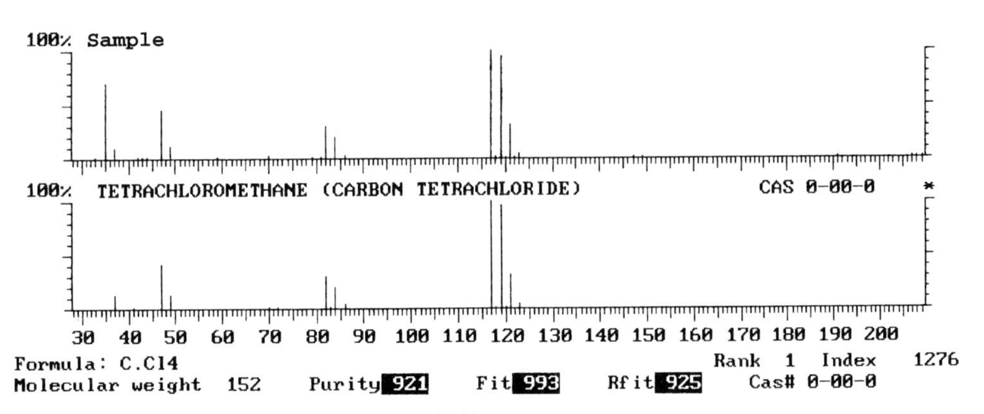

Fig. 3.14 Identification of the second component by library comparison

library search should be carried out with a small proportion of chemical noise. In the library search programs of individual manufacturers there is also the possibility of editing the spectrum before the start of the search. In critical cases this option should also be followed to remove known interfering signals resulting from the chemical noise from the substance spectrum.

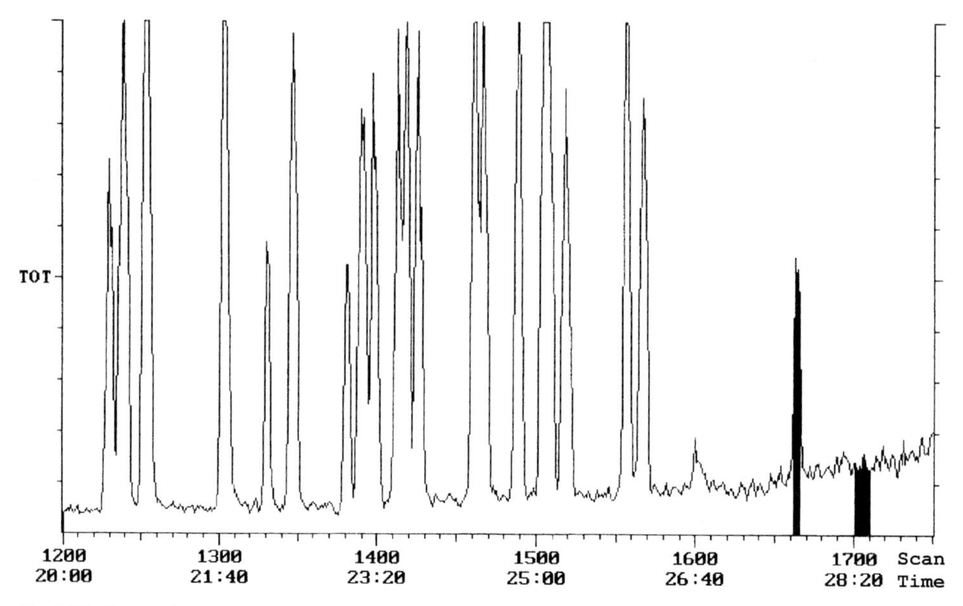

Fig. 3.15 The total ion current chromatogram of an analysis of volatile halogenated hydrocarbons shows the elution of a minor component at the beginning of column bleed (the areas of background subtraction are shown in black)

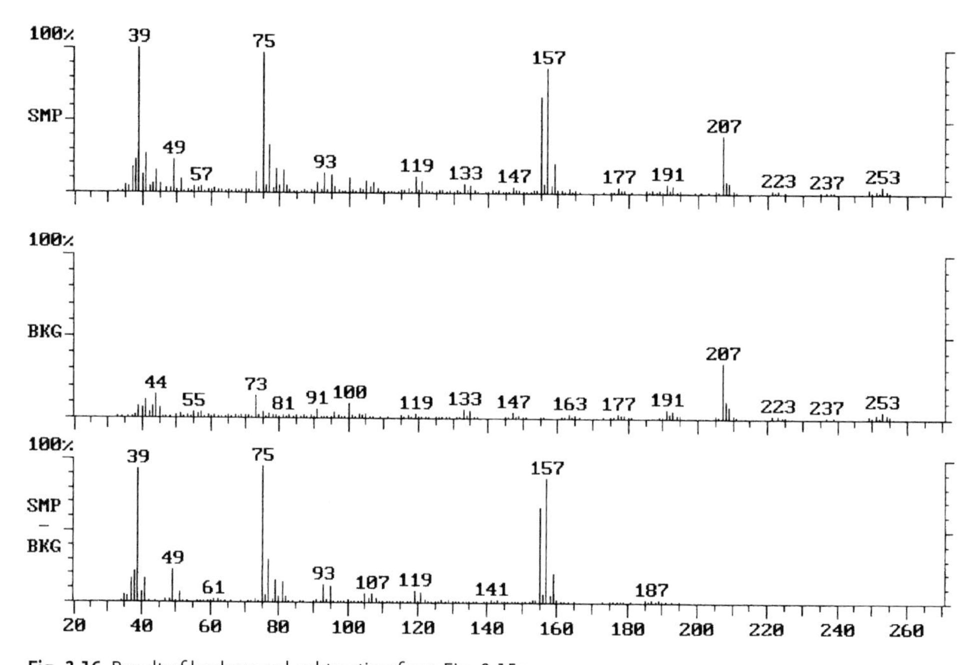

Fig. 3.16 Result of background subtraction from Fig. 3-15:
SMP = Spectra from the substance peak, sample
BKG = Spectra from the background (column bleed)
SMP-BKG = Resulting substance spectrum

Deconvolution of Mass Spectra

The advancements in full scan sensitivity especially in ion trap and time-of-flight MS instrumentation as well as the increased application of fast and two-dimensional GC methods is creating a strong demand for post-acquisition deconvolution methods. The extraction of pure spectra from compounds co-eluting with other analytes or interferences with conventional background subtraction methods of GC/MS data systems is of very limited use and cannot recognize the transient dependence of ion intensities of multiple compounds eluting close together.

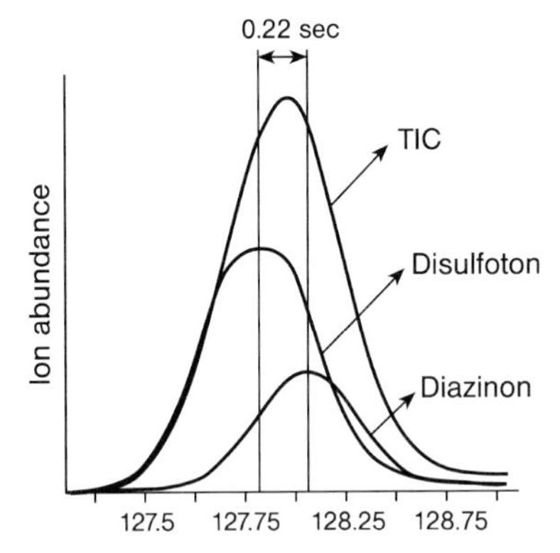

Fig. 3.17 Deconvolution example (Meruva 2006)

An automated mass spectra deconvolution and identification system (AMDIS) was developed at the National Institute of Standards and Technology (NIST) with the support of the Special Weapons Agency of the Department of Defense, for the critical task of verifying the Chemical Weapons Convention ratified by the United States Senate in 1997. In order to meet the rigorous requirements for this purpose, AMDIS was tested against more than 30,000 GC/MS data files accumulated by the EPA Contract Laboratory Program without a single false positive for the target set of known chemical warfare agents. While this level of reliability may not be required for all laboratories, this shows the degree to which the algorithms have been tested. After two years of development and extensive testing it has been made available to the general analytical chemistry community for download on the Internet.

The AMDIS program analyses the individual ion signals and extracts and identifies the spectrum of each component in a mixture analyzed by GC/MS. The software comprises by an integrated set of procedures for first extracting the pure component spectra from the chromatogram and then to identify the compound by a reference library.

The overall process involves four sequential steps in spectrum purification and identification:

1. Noise analysis by a complete analysis of noise signals with the use of this information for component perception. A correction for baseline drift is done for each component in case the chromatogram does not have to have a flat baseline.
2. Component perception identifies the location of each of the eluted components on the retention time scale by investigating the elution peak profile.

3. True spectral "deconvolution" of the data. Even if there is no available constant background for subtraction, AMDIS extracts clean spectra. The extraction of closely coeluting components is possible even for analytes that peak within a single scan of each other in a wide range of each component's concentration.
4. Library search for compound identification to match each deconvoluted spectrum to a reference library spectrum.

Unlike a traditional identification algorithm, AMDIS includes uncertainties in the deconvolution, purity, and retention times in the match factor. The final match factor is a measure of both the quality of the match and of the confidence in the identification.

AMDIS can operate as a "black box" chemical identifier, displaying all identifications that meet a user-selectable degree of confidence. Identification can be aided by internal standards and retention times. Also employed can be retention index windows when identifying target compounds and internal and external standards as maintained in separate libraries. AMDIS

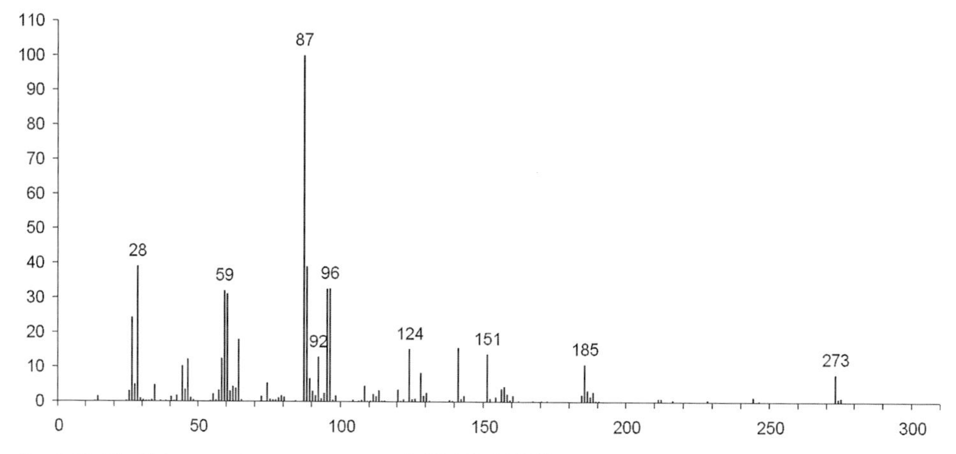

Fig. 3.18 Disulfoton, spectrum pur compound, NIST#: 118988

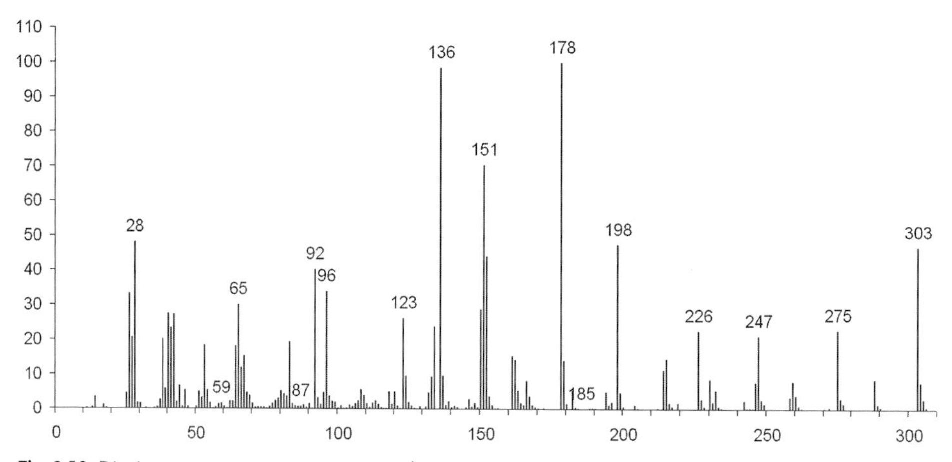

Fig. 3.19 Diazinone, spectrum pure compound, NIST#: 118996

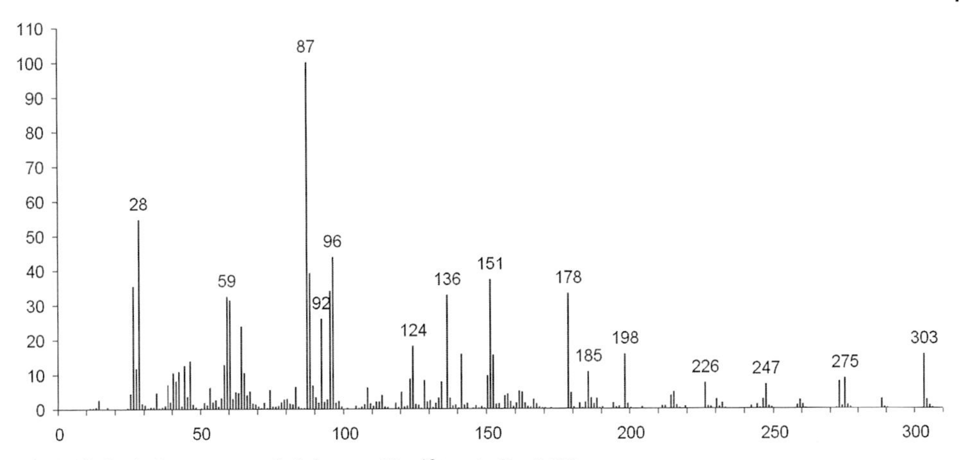

Fig. 3.20 Coelution spectrum 3 : 1 (at max Disulfoton in Fig. 3.17)

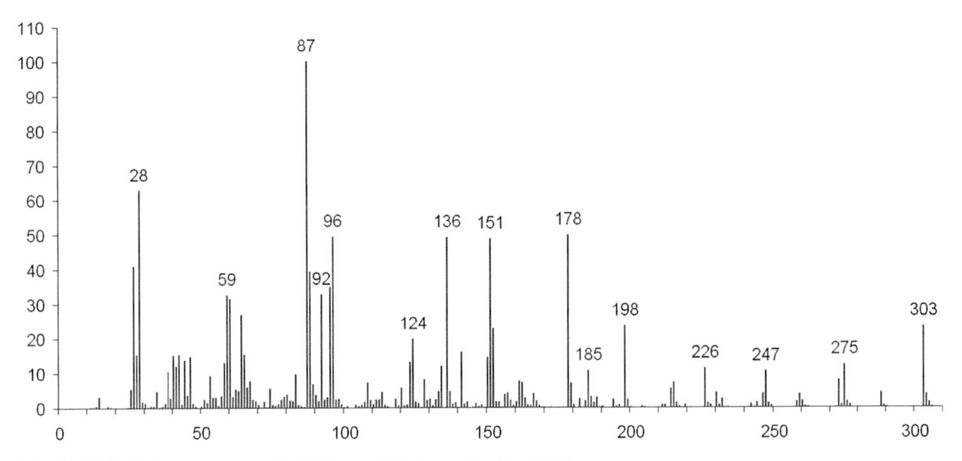

Fig. 3.21 Coelution spectrum 2 : 1 (at max Diazinone in Fig. 3.17)

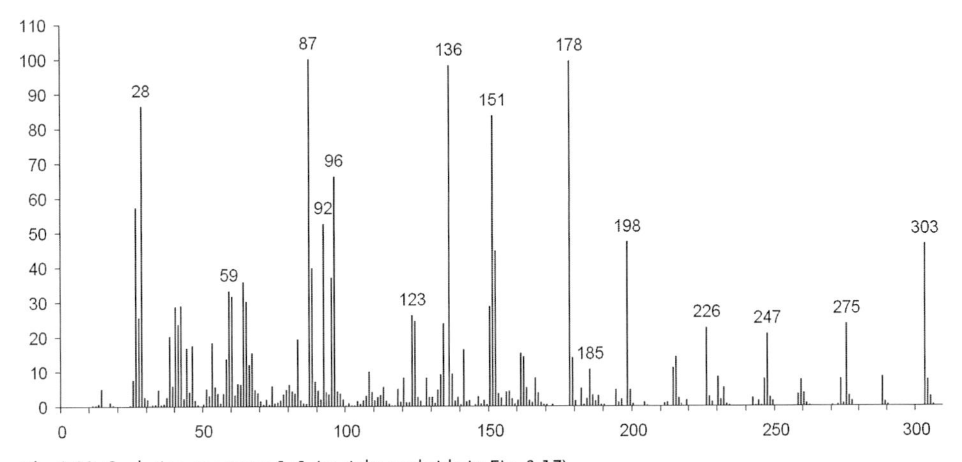

Fig. 3.22 Coelution spectrum 1 : 1 (at right peakside in Fig. 3.17)

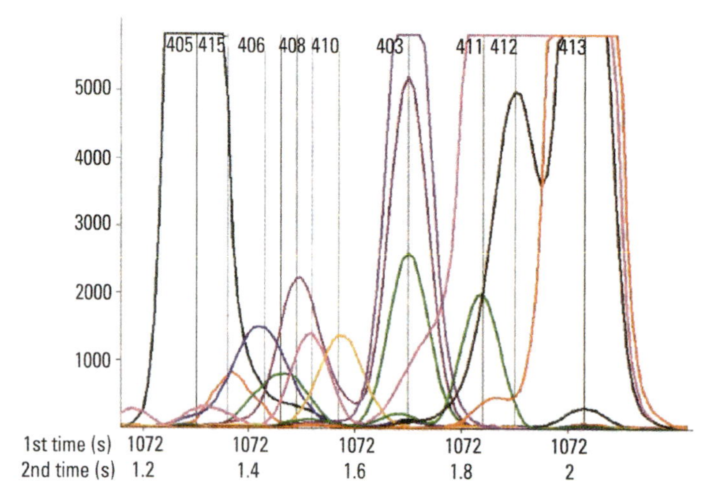

Fig. 3.23 Deconvolution of overlapping peaks in GCxGC/TOF-MS. Every vertical line indicates the peak of an identified component (Dimandja 2004, reprinted with permission from Analytical Chemistry, Copyright 2004 by American Chemical Society).

reads GC/MS raw data files in the formats of the leading GC/MS manufacturer or is already integrated in the instrument data systems.

With its unique deconvolution algorithms AMDIS has proven its capabilities for the efficient removal of overlapping interferences in many GC/MS applications. The deconvolution process turned out to be independent from the type of analyzer and scan rate used to resolve overlapping peaks for substance identification as well as multi component residue analysis (Dimandja 2004, Mallard 2005, Zhang 2006). Without time consuming manual data evaluation AMDIS provides sensitive compound information even with complex background present.

AMDIS has been designed to reconstruct "pure component" spectra from complex GC/MS chromatograms even when components are present at trace levels. For this purpose, observed chromatographic behavior is used along with a range of noise-reduction methods. AMDIS is distributed with specialized libraries (environmental, flavor and fragrance, and drugs and toxins) that were derived from the NIST Library. AMDIS has a range of other features, including the ability to search the entire NIST Library with any of the spectra extracted from the original data file. It can also employ retention index windows when identifying target compounds and can make use of internal and external standards maintained in separate libraries. A history list of selected performance standards is also maintained.

As of version 2.62, AMDIS reads data files in the following formats:

- Bruker (*.MSF)
- Finnigan GCQ (*.MS)
- Finnigan INCON (*.MI)
- Finnigan ITDS (*.DAT)
- HP ChemStation (*.D)
- HP MS Engine (*.MS)
- INFICON GCMS (*.acq)
- JEOL/Shrader (*.lrp)

- Kratos Mach3 (*.run)
- MassLynx NT (*.*)
- Micromass (*.)
- NetCDF (*.CDF)
- PE Turbo Mass (*.raw)
- Saturn SMS (*.sms)
- Shimadzu MS (*.R##)
- Shrader/GCMate (*.lrp)
- Thermo Xcalibur Raw (*.raw)
- Varian Saturn (*.MS)

3.2.2
The Retention Index

If the chromatographic conditions are kept constant, the retention times of the compounds remain the same. All identification concepts using classical detectors function on this basis. The retention times of compounds, however, can change through ageing of the column and more particularly through differing matrix effects.

The measurement of the retention times relative to a co-injected standard can help to overcome these difficulties. Fixed retention indices (*RI*) are assigned to these standards. An analyte is included in a retention index system with the *RI* values of the standards eluting before and after it. It is assumed small variances in the retention times affect both the analyte and the standards so that the *RI* values calculated remain constant.

The first retention index system to become widely used was developed by Kovats. In this system a series of n-alkanes is used as the standard. Each n-alkane is assigned the value of the number of carbon atoms multiplied by 100 as the retention index (pentane 500, hexane 600, heptane 700 etc.). For isothermal operations the *RI* values for other substances are calculated as follows:

$$\text{Kovats index} \quad RI = 100 \cdot c + 100 \, \frac{\log{(t'_R)_x} - \log{(t'_R)_c}}{\log{(t'_R)_{c+1}} - \log{(t'_R)_c}} \tag{24a}$$

The t'_R values give the retention times of the standards and the substance corrected for the dead time t_0 ($t'_R = t_R - t_0$). As the dead time is constant in the cases considered, uncorrected retention times are mostly used. The determination of the Kovats indices (Fig. 3.24) can be carried out very precisely and on comparison between various different laboratories is reproducible within ± 10 units. In libraries of mass spectra the retention indices are also given (see the terpene library by Adams, the pesticide library by Ockels, the toxicology library by Pfleger/Maurer/Weber).

On working with linear temperature programs a simplification is used which was introduced by Van den Dool and Kratz, whereby direct retention times are used instead of the logarithmic terms used by Kovats:

$$\text{Modified Kovats index} \quad RI = 100 \cdot c + 100 \, \frac{(t'_R)_x - (t'_R)_c}{(t'_R)_{c+1} - (t'_R)_c} \tag{24b}$$

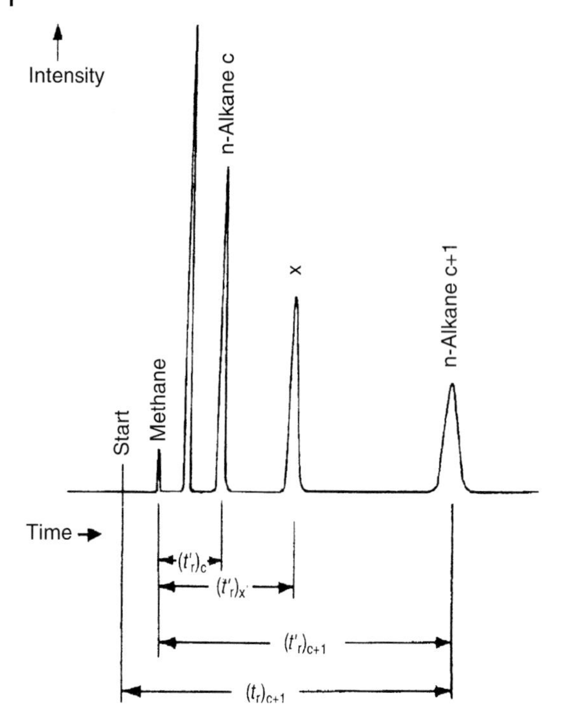

Fig. 3.24 Determination of the Kovats index for a substance X by interpolation between two n-alkanes (after Schomburg)

The weakness of retention index systems lies in the fact that not all analytes are affected by variances in the measuring system to the same extent. For these special purposes homologous series of substances which are as closely related as possible have been developed. For use in trace analysis in environmental chemistry and particularly for the analysis of plant protection agents and chemical weapons, the homologous M-series (Fig. 3.25) of n-alkylbis(trifluoromethyl)phosphine sulfides has been synthesised.

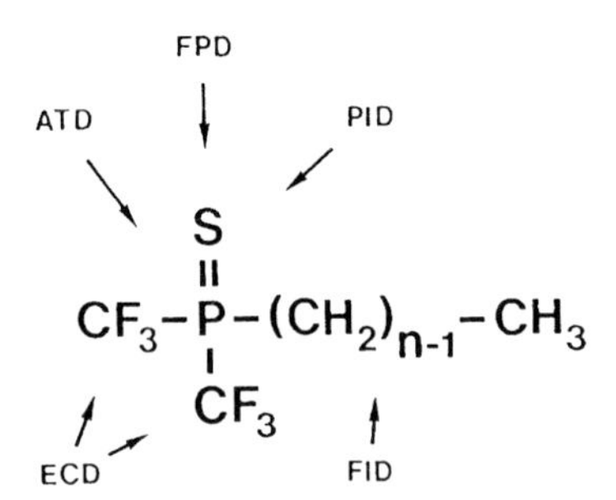

Fig. 3.25 n-Alkylbis(trifluoromethyl)phosphine sulfides (M series with n = 6, 8, 10, ≤20)

The molecule in the M-series contains active groups which also respond to the selective detectors ECD, NPD, FPD and PID and naturally also give good responses in FID and MS (detection limits: ECD ca. 1 pg, FID ca. 300 pg, Fig. 3.26). In the mass spectrometer all components of the M-series show intense characteristic ions at M-69 and M-101 and a typical fragment at m/z 147 (Fig. 3.27). The M-series can be used with positive and negative chemical ionisation.

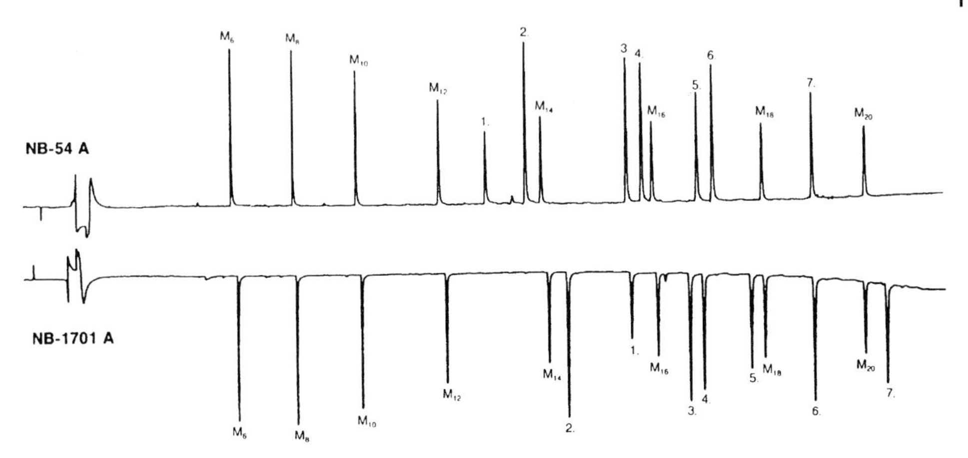

Fig. 3.26 Chromatograms of the M series and of pesticides (phosphoric acid esters) on columns of different polarities (after HNU/Nordion).
Carrier gas He, detector NPD, program: 50 °C (2 min), 150 °C (20°/min), 270 °C (6°/min).
Components: M series M_6, M_8, <: M_{20}, 1 Dimethoate, 2 Diazinon, 3 Fenthion, 4 Trichloronate, 5 Bromophos-ethyl, 6 Ditalimphos, 7 Carbophenothion

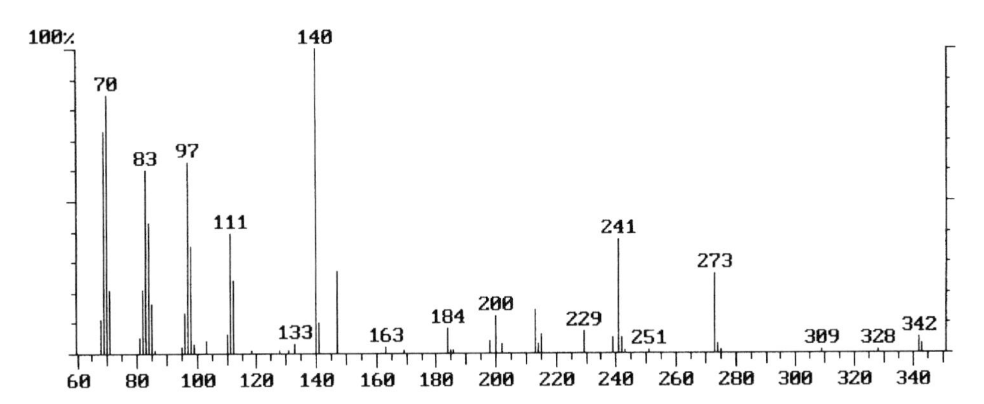

Fig. 3.27 M series: EI Mass Spectrum of the component M 10 (HNU)

The use of retention indices in spite of, or perhaps because of, the wide use of GC/MS systems is now becoming more important again as a result of the outstanding stability of fused silica capillaries and the good reproducibility of gas chromatographs now available. The broadening of chromatography data systems with optional evaluation routines is just beginning. These are especially dedicated to the processing of retention indices, e.g. for two-column systems.

If the retention index of a compound is not known, it can also be estimated from empirical considerations of the elements and partial structures present in the molecule (Tables 3.1 and 3.2). A first approximation can already be made using the empirical formula of an analyte. This is particularly valuable for assessing suggestions from the GC/MS library search because, besides good correspondence to a spectrum, plausibility with regard to the reten-

Comparison of calculated retention indices with empirically determined values

Substance	Calculated RI	Determined RI
Atrazine	1500	1716
Parathion-ethyl	2000	1970
Triadimefon	2000	1979

Examples of retention index calculation

Parathion-ethyl: $C_{10}H_{14}NO_5PS$

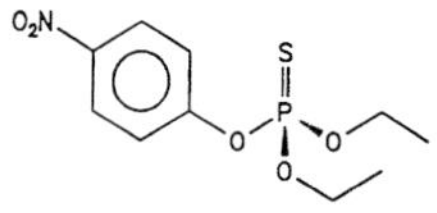

Number	Element	Contribution	Total
10	C	100	1000
14	H	–	–
1	N	100	100
5	O	100	500
1	P	200	200
1	S	200	200
		Sum	2000

Atrazine: $C_8H_{14}ClN_5$

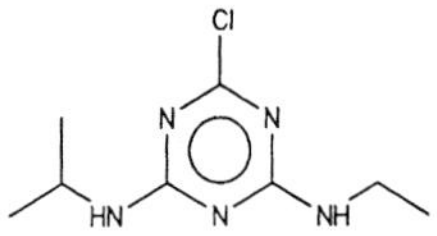

Number	Element	Contribution	Total
8	C	100	800
14	H	–	–
1	Cl	200	200
5	N	100	500
		Sum	1500

Triadimefon: $C_{14}H_{14}ClN_3O_2$

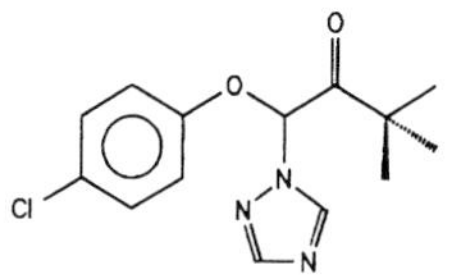

Number	Element	Contribution	Total
14	C	100	1400
14	H	–	–
1	Cl	200	200
3	N	100	300
2	O	100	200
1	tert-C	–100	–100
		Sum	2000

tion behaviour can be tested (see also Section 4.14). According to Weber (1992) the values determined give a correct estimation within 10%.

Polar groups with hydrogen bonding increase the boiling point of a compound and are thus responsible for stronger retention. For the second and every additional polar group the retention index increases by 150 units. Branches in the molecule increase the volatility. For each quaternary carbon atom present in a t-butyl group the retention index is reduced by 100 units. Values can be estimated with higher precision from retention indices for known structures by calculating structure elements according to Tables 3.1 and 3.2.

Table 3.1 Contributions for the determination of the retention index from the empirical formula (Weber 1992).

Element	Index contribution
H, F	0
C, N, O	100
Si in $Si(CH_3)_3$	0
P, S, Cl	200
Br	300
J	400

Table 3.2 Retention behaviour of structural isomers (Weber 1992).

Alkyl branches:	tertiary < secondary < n-alkyl
Disubstituted aromatics:	ortho < meta, para

3.2.3 Libraries of Mass Spectra

In electron impact ionisation (70 eV) a large number of fragmentation reactions take place with organic compounds. These are independent of the manufacturer's design of the ion source. The focusing of the ion source has a greater effect on the characteristics of a mass spectrum, which leads to a particularly wide adjustment range, especially in the case of quadrupole analysers. The relative intensities of the higher and lower mass ranges can easily be reversed. In the early days of use of quadrupole instruments this possibility was highly criticised by those using the established magnetic instruments. The problem is resolved in so far as both the manual and the automatic tuning of the instruments are aimed at giving the intensities of a reference compound (in contrast SIM tuning aims to give high sensitivity within a specific mass range, see Section 2.3.5.2). Perfluorotributylamine (FC43) is used as the reference substance in all GC/MS systems. Other influences on the mass spectrum in GC/MS systems are caused by the changing substance concentration during the mass scan (beam instruments). On running spectra over a large mass range (e.g. in the case of methyl stearate with a scan of 50 to 350 u in 1 s) sharp GC peaks lead to a mismatch of intensities (skewing) between the front and back slopes of the peak. The skew of intensities is thus the opposite of the true situation. This effect can only be counteracted by the use of fast scan rates, which, however, result in lower sensitivity. In practice standardised spectra must be used for these systems in order to calculate the compensation (for background subtraction see Section 3.2.1). Ion trap mass spectrometers do not show this reversal of intensities because there is parallel storage of all the ions formed. A mass spectrum should therefore not be regarded as naturally constant, but the result of an extremely complex process.

In practice the variations observed affect the relative intensities of particular groups of ions in the mass spectrum. The fragmentation processes itself are not affected (the same fragments are found with all GC/MS instruments) nor are the isotope ratios which result from natural distribution. Only adherence to a parameter window which is as narrow as possible (so-called standard conditions) during data acquisition creates the desired independence from the external influences described.

The comparability of the mass spectra produced is thus ensured for building up libraries of mass spectra. All commercially available libraries were run under the standard conditions mentioned and allow the comparison of the fragmentation pattern of an unknown substance with those available from the library. For the large universal libraries it should be assumed that most of the spectra were initially not run with GC/MS systems, and that still today many reference spectra are run using a solid sample inlet or similar inlet techniques. For example, the reference spectrum of Aroclor 1260 (a mixture of PCBs with a 60% degree of chlorination) can only be explained in this way. Information on the inlet system used is rarely found in library entries.

EI spectra are particularly informative because of their fragmentation patterns. All search processes through libraries of spectra are mainly based on EI spectra. With the introduction of the highly reproducible advanced chemical ionisation into ion trap mass spectrometers, the first commercial CI library with over 300 pesticides was produced (Finnigan, 1992). The introduction of substructure libraries (MS/MS product ion spectra) is currently ongoing (NIST). The commercially available libraries are divided into general extensive collections and special task-related collections with a narrow range of applications.

3.2.3.1 **Universal Mass Spectral Libraries**

NIST/EPA/NIH Mass Spectral Library

The NIST/EPA/NIH Mass Spectral Library is probably the most popular and most widely distributed library for GC/MS instruments. The 2005 edition has been largely expanded by the number of EI mass spectra with the addition of Kovats retention indices. MS/MS mass spectra are increasingly included. Extensive spectra evaluation and quality control has been involved in the new edition of the NIST database. Each spectrum was critically examined by experienced mass spectrometrists, and each chemical structure has been examined for correctness and consistency, using both human and computer methods. Spectra of stereoisomers have been intercompared, chemical names have been examined by experts and IUPAC names provided. CAS registry numbers have been verified.

The NIST library is available with the full-featured NIST MS Search Program, which also includes integrated tools for GC/MS spectral deconvolution (see Section 3.2.1,

AMDIS), mass spectral interpretation tools with thermodynamics-based interpretation of fragmentation and chemical substructure analysis. The binary format has not changed from the 2002 version, although several new files have been added that associate equivalent compounds and link individual compounds to the retention index library. Raw data files are provided in both an SDFile format (structure and data together) as well in earlier formats. The SDFile format holds the chemical structure as a MOLFile and the data in a simple ASCII format. The NIST MS Search Program is also part of many commercial instrumental GC/MS software suites.

The 2005 edition of the NIST database is characterized by (NIST 2005):

- 190,825 EI spectra of 163,198 unique compounds
- 111 average peaks/spectrum
- 98 median peaks/spectrum
- 163,198 compounds with EI spectra
- 18,592 compounds with replicate spectra
- 27,627 replicate spectra from high quality sources
- 163,195 chemical structures
- 11% increase in coverage to earlier version
- 5,191 MS/MS spectra of 1,943 unique ions
- 1,920 ions (1,628 cations and 292 anions)
- 25,728 compounds with 121,112 retention indices
- 120,786 experimental values with references
- Structure-based RI estimates

The increase in the number of spectra was accomplished primarily by the addition of complete, high quality spectra either measured specifically for the library or taken from major practical collections, including:

- Chemical Concepts – including Prof Henneberg's industrial chemicals collection (Max-Planck-Institute for Coal Research, Muehlheim, Germany), see details below
- Georgia and Virginia Crime Laboratories
- TNO Flavors and Fragrances
- AAFS Toxicology Section, Drug Library, see details below
- Association of Official Racing Chemists
- St. Louis University Urinary Acids
- VERIFIN & CBDCOM Chemical Weapons

The new addition of Kovats retention index values contains 121,112 Kovats retention index values for 25,893 compounds on non-polar columns, 12,452 of which are compounds represented in the electron ionization library. Full annotation is provided, including literature source and measurement conditions. These are provided in a format accessible by the NIST Search Program and separately as an ASCII SDFile.

The new addition of MS/MS Spectra provides 5,191 spectra of 1,943 different ions (1,671 positive and 341 negative ions). A range of instruments is represented, including ion trap and triple quadrupole mass spectrometers. Spectra have been provided by contributors, measured at NIST and extracted from the literature. It also documents spectrum variations between instrument classes at different conditions. It was found that at sufficiently high signal-to-noise measurement conditions, modern instruments are capable of providing very reproducible, library searchable spectra. While collision energy can be an important variable, spectra varies in an understandable way depending on compound and instrument class and conditions. This library is provided in formats equivalent to the electron ionization library but with new fields added to describe the instrument and analysis conditions. A small number of MS1 spectra are also included for reference purposes. These generally contain the ions used for MS/MS.

Wiley Registry of Mass Spectral Data

The Wiley Registry$^{(TM)}$ of Mass Spectral Data has recently been published in its 8th Edition (Wiley 2006). Is is one of the most comprehensive mass spectral libraries available in the file format for many mass spectral data systems for applications in forensics, environmental analysis, toxicology, and homeland security.

The 8th edition of the Wiley Registry contains nearly 400,000 mass spectra with over 183,000 searchable structures sourced from leading laboratories throughout the world. Most spectra are accompanied by the structure and trivial name, molecular formula, molecular weight, nominal mass and base peak. New in the edition are:

- Chemical warfare precursors
- Combinatorial library compounds
- High molecular diversity for fragmentation analysis
- New high resolution organics
- Structure and substructure searchable
- Spectra with retention indices

Also available is the combination of the large Wiley Registry with the current NIST database. The Wiley Registry 8th Edition/NIST 2005 (W8/N05) provides the most extensive mass spectral library with:

- 532,573 mass spectra
- 319,256 mass spectra with assigned, searchable structures
- More than 2 million names and synonyms
- High resolution spectra with most new spectra containing over 125 peaks per spectrum

The W8/N05 provides comprehensive coverage of small molecule organics, pharmaceutical drugs, illegal drugs, poisons, pesticides, steroids, natural products, organic compounds, and chemical warfare agents for different applications:

- Toxicology/Forensics/Public Health: The library contains a wide scope of spectra covering drugs, poisons, pesticides, and metabolites.
- Industrial R&D/Quality Assurance: The library contains a comprehensive collection of small organic compounds and their metabolites, including a combinatorial library appropriate for fragmentation analysis.
- Research/Teaching: The library contains data for fragmentation analysis as well as comprehensive coverage of most compounds measurable by GC/MS.
- Environmental: The library contains most known pesticides and includes precursors being used in the production of new pesticide classes.

- Available formats and compatibility: The library is available in two formats: Chemstation (Agilent) and the NIST MS Search (Bruker, JEOL, LECO, Perkin Elmer, Thermo, Varian, Waters,). Other formats are available on request.

The Palisade Complete Mass Spectra Library

The world's largest commercial mass spectral database is offered by Palisade with more than 600,000 mass spectra. All commercially available reference spectra – including those in the NIST and Wiley libraries – are contained in a single library. New spectra and structure updates will be distributed annually and new spectral collections are announced on the Internet for downloading by subscribers.

The Palisade Complete 600K includes all spectra of the NIST 2002 and Wiley Registry collections plus over 150,000 new spectra available only through Palisade:

- 606,000 total spectra
- Spectra type EI 70 eV
- 495,000 unique compounds
- 327,000 CAS assignments
- 437,000 compounds with structures
- 985,000 chemical names
- 350 Mb Disk storage space

3.2.3.2 Application Libraries of Mass Spectra

Mass Spectra of Geochemicals, Petrochemicals and Biomarkers

This database is focused on organic, geochemical, and petrochemical applications and comprises (De Leeuw 2004):

- 1,100 mass spectra of well-defined compounds.
- Information including mass spectra, chemical structure, chemical name, molecular formula, molecular weight (nominal mass), base peak, reference, and measurement condition.
- Chemical structures elucidated, if necessary, by a variety of techniques including NMR spectroscopy and single-crystal X-ray structure analysis (Wiley).

Chemical Concepts Library of Mass Spectra

The CC Mass Spectral Data collection (Chemical Concepts 2006) has been recently updated and consists of mass spectra of more than 40,000 compounds. It is included in the new release of the NIST 2005 library.

The main part of this mass spectra reference library comes from the Industrial Chemicals Collection of Prof. Henneberg, Max-Planck-Institut for Coal Research, Muelheim, Germany. Also universities and institutes such as ETH, Zürich, Switzerland and ISAS, Dortmund, Germany have contributed their research spectra to this collection. Prior to being included into the library the data pass consistency and quality checks performed at the Max-Planck-Institut.

Additional information included with the mass spectra are (Wiley):

- Chemical structure
- Chemical name
- Molecular formula
- Molecular weight (Nominal mass)
- Base peak
- Reference
- Measurement condition

Alexander Yarkov – Mass Spectra of Organic Compounds

The new specialized data collection contains 37,055 mass spectra of physiologically active organic compounds. The data resulted from quality control in combinatorial synthesis and cover a wide range of compound classes.

Additional information included with the mass spectra are (Wiley):

- Chemical structure
- Chemical name
- Molecular formula
- Molecular weight (nominal mass)
- Base peak
- Reference
- Measurement condition

Mass Spectra of Designer Drugs

This mass spectrum collection edited by Peter Rösner covers the entire range of designer drugs up to December 2006. It is the first database featuring systematic structures in depth. Carefully compiled by the mass spectral experts at the Regional Departments of Criminal Investigation in Kiel, Hamburg, and the Federal Criminal Laboratory in Wiesbaden, Germany, this database includes 67,321 mass spectra of 5,789 chemical compounds like designer drugs and medicinal drugs. Chemical warfare agents are added due to the recent interest in homeland security. All data has been taken from both legal and underground literature, providing the most comprehensive picture of these compounds available worldwide. Highly potential hallucinogens like the Bromo-DragonFLY are covered (Wiley).

Mass Spectra of Volatiles in Food

This mass spectral database is dedicated to the application areas of the food and flavour industries, and was selected and quality controlled by the mass spectral experts at the Central Institute of Nutrition and Food Research in the Netherlands. The database is now available in its 2nd edition (Wiley).

Mass Spectral and GC Data of Drugs, Poisons, Pesticides, Pollutants and Their Metabolites

This specialized collection is dedicated to environmental and forensic analysis, occupational toxicology and food analysis and contains data obtained from clinical samples over the course of more than 20 years. It encompasses 7,500 potentially harmful substances, from simple analgesics to designer drugs, and from pesticides and pollutants to chemical warfare agents, including metabolites to allow the identification of the mother substance. Karl Pfleger is the former, and Hans H. Maurer the current, head of the Clinical Toxicology Laboratory at the clinical campus of Saarland University in Homburg, Germany. Together with Armin Weber, they have developed this unique and most comprehensive toxicological database.

In 2000, the 2nd edition was expanded by 2,000 new mass spectra to more than 6,300. Covered also in the printed hard cover edition parts 1–4:

- Data of nearly all the new drugs relevant to clinical and forensic toxicology, doping control, food chemistry, etc.
- Nearly complete coverage of trimethylsilylated, perfluoroacylated, perfluoroalkylated and methylated compounds.
- Sections on sample preparation and GC-MS methods.

The new 3rd edition 2007 gives all spectra in order of their molecular mass, since this has become the prime benchmark criterion for the identification of unknown substances (Wiley).

Mass Spectra of Drugs, Pharmaceuticals and Metabolites

The collection edited by Rolf Kuehnle contains 2,200 mass spectra of Drugs, Pharmaceuticals and Metabolites. The inclusion of the silylated derivatives as used for GC/MS analyses is of special value in this collection. Additional information included are chemical structure, chemical name, molecular formula, molecular weight (nominal mass), base peak and the reference (Wiley).

Mass Spectra of Pharmaceuticals and Agrochemicals 2006

The collection of 4,563 unreduced spectra includes compounds that are subject to the drug trafficking laws as well as their precursors, by-products and metabolites. Other compound groups included are medical drugs, drugs with psychotropic effect, anabolics and pesticides. Chemical structures, synonym and systematic name, molecular weight, formula and experimental conditions complete the data record (Wiley).

Mass Spectra of Androgenes, Estrogens and other Steroids

The collection edited by Hugh L. J. Makin contains 2,979 EI mass spectra of androgens and estrogens and their trimethylsilyl-, O-methoxyoxime- and acetal derivatives. Each spectrum is accompanied by the structure and trivial name, molecular formula, molecular weight, nominal mass and base peak. All spectra of androgens and estrogens have been obtained on the same mass spectrometer under identical conditions (Wiley).

AAFS Drug Library

The American Academy of Forensic Sciences, Toxicology Section, committee was set up to coordinate the generation of reliable mass spectra of new drugs and metabolite standards, and to make these available to the profession on a timely basis. The mass spectral database as zip file and the list of entries is available for download from the Internet.

This library is a "subset" of one that has been compiled over a period of many years by Dr. Graham Jones and colleagues in Edmonton, Alberta, Canada. Pure drug spectra, plus GC breakdown products and pure metabolite standards have been edited into this compilation of over 2,300 mass spectra, including many replicate entries. All spectra were run on Agilent quadrupole GC/MS instruments tuned against PFTBA. The current version of the full spectra library was last updated March 2006 (AAFS).

The Lipid Library

The Archives of Mass Spectra by W.W. Christie comprise approx. 1670 mass spectra in total. They are made available on the web for study but without interpretation for the following compound groups:

- Methyl esters of fatty acids
- Picolinyl esters
- DMOX derivatives
- Pyrrolidine derivatives
- Miscellaneous fatty acid derivatives, lipids, artefacts, etc.

All the mass spectra illustrated in these pages were obtained by electron-impact ionization at an ionization potential of 70 eV on quadrupole mass spectrometers. The website also offers the Bibliography of Mass Spectra with lists of references mainly concerning the use of mass spectrometry for structural analysis of fatty acids mainly.

3.2.4
Library Search Procedures

In general it is expected that the identity of an unknown compound will be found in a library search procedure. However it is better to consider the results of a search procedure from the aspect of similarity between the reference and the unknown spectrum. Other information for confirmation of identity, such as retention time, processing procedure, and other spectroscopic data should always be consulted. A short review in the journal Analytical Chemistry (W. Warr, 1993) began with the sentence 'Library searching has limitations and can be dangerous in novice hands'. Examples of critical cases are different compounds which have the same spectra (isomers), the same compounds with different spectra (measuring conditions, reactivity, decomposition) or the fact that a substance being searched for is not in the library but similar spectra are suggested. In particular the limited scope of the libraries must be taken into account. According to a short press publication by the Chemical Abstract Service in 1994 the total of CAS registry numbers had passed the 12 million mark. Every year ca. 600 000 new compounds are added!

As a result of the different software equipment used in current benchtop GC/MS systems two search procedures have become widely established: INCOS and PBM. The SISCOM procedure (Search for Identical and Similar Compounds) developed by Henneberg/Weimann is also available on stand-alone work-stations. It stands out on account of its excellent performance for data-system-supported interpretation of mass spectra. The procedures for determination of similarity between spectra are based on very different considerations. The INCOS and PBM procedures aim to give suggestions of possible substances to explain an unknown spectrum. Both algorithms dominate in the qualitative evaluation using magnetic sector, quadrupole and ion trap GC/MS systems. Other search procedures, such as the Biemann search, have been replaced by newer developments and broadening of the algorithms by the manufacturers of spectrometers.

The newest development in the area of computer-supported library searches, the further development of the INCOS procedure, has been presented by Steven Stein (NIST) through targeted optimisation of the weighting and combination with probability values. An improvement in the hit rate was demonstrated in a comparison of test procedures with more than 12 000 spectra.

3.2.4.1 The INCOS/NIST Search Procedure

At the beginning of the 1970s the INCOS company (Integrated Control Systems) presented a search procedure which operated both on the principles of pattern recognition and with the components of classical interpretation techniques and which could reliably process data from different types of mass spectrometer. The early years of GC/MS were characterised by the rapid development of quadrupole instruments which were ideal for coupling with gas chromatographs, because of their scan rates, which were high compared with the magnetic sector instruments of that time. The spectral libraries then available had been drawn up from spectra run on magnetic sector instruments.

From the beginning the INCOS procedure was able to take into account the relatively low intensities of the higher masses in spectra run on quadrupole systems, besides the typical high mass intensity magnetic sector spectra. The INCOS search has remained virtually unchanged since the 1970s. The search is known for its high hit probability, even with mass spectra with a high proportion of matrix noise obtained in residue analysis, and its complete independence from the type of instrument.

After a significance weighting (square root of the product of the mass and the intensity) and data reduction by a noise filter and a redundancy filter, the extensive reference database is searched for suitable candidates for a pattern comparison in a rapid pre-search. The presence of up to eight of the most significant masses counts as an important starting criterion. The intensity ratios are not yet considered. It is required that only those reference spectra which contain at least eight of the most significant masses of the unknown spectrum are considered. Depending on the requirements of the user, spectra with less than eight matching masses are also further processed (Fig. 3.28 E: pre-search report, the number of candidates is marked with **). At the start of the search the parameter 'minimum number to search' must be adapted by the user. Reference spectra which only contain a small number or no matching masses or whose molecular weight does not match an optional suggestion, are excluded from the list of possible candidates and are not further processed.

The main search is the critical step in the INCOS algorithm, in which the candidates found in the pre-search are compared with the unknown spectrum and arranged in a prioritised list of suggestions. Of critical importance for the tolerance of the INCOS procedure for different types of mass spectrometer and marginal conditions of data acquisition (and thus for the high hit rate) is a process known as local normalisation.

Local normalisation introduces an important component into the search procedure which is comparable to the visual comparison of two patterns (Fig. 3.29). Individual clusters of ions and isotope patterns are compared with one another in a local mass window. The central mass of such a window from the reference spectrum is compared with the intensity of this mass in the unknown spectrum in order to assess the matching of the line pattern in windows a few masses to the left and right. In this way the nearby region of each mass signal is examined and, for example, the matching of isotope patterns (Cl, Br, S, Si for example) and cleavage reactions are assessed.

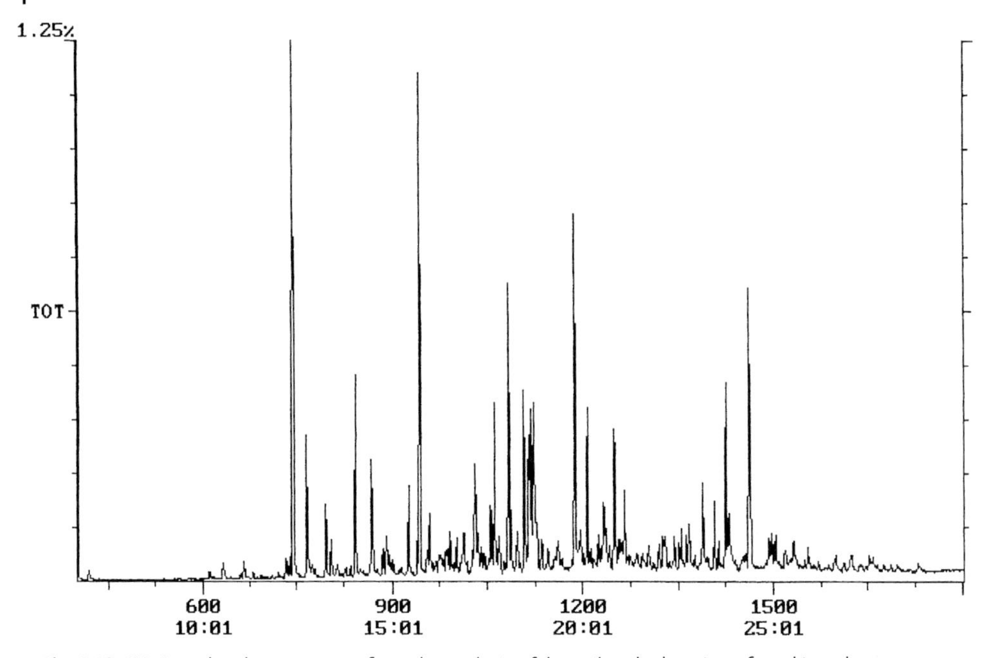

Fig. 3.28 (A) Complex chromatogram from the analysis of the soil at the location of a coking plant

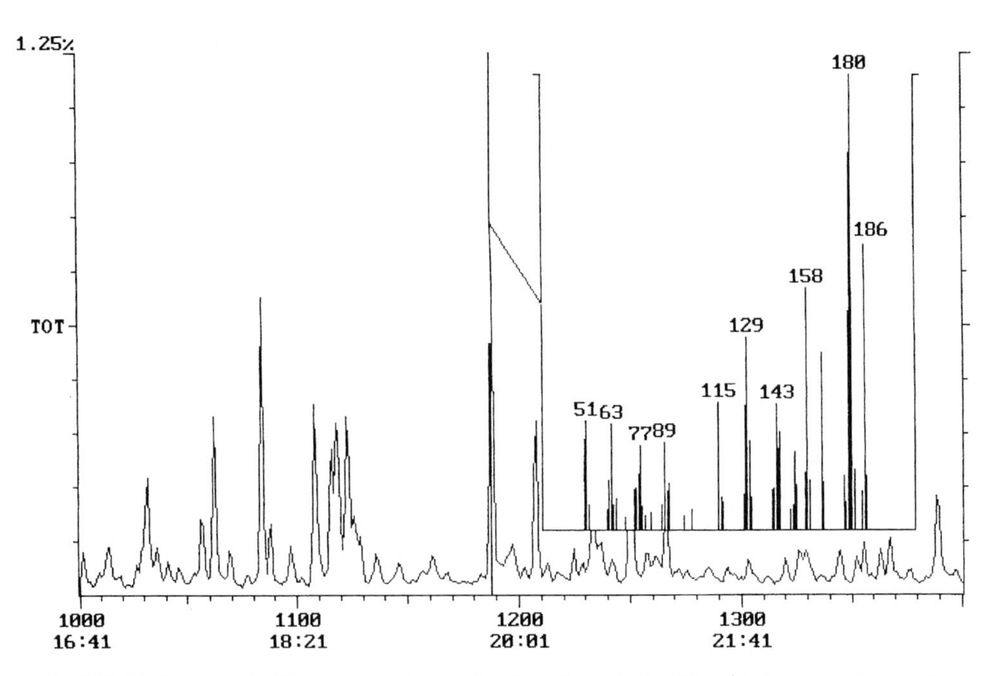

Fig. 3.28 (B) A large peak with a mass spectrum with intense ions at m/z 180 and 186 appears at a retention time of ca. 20 min

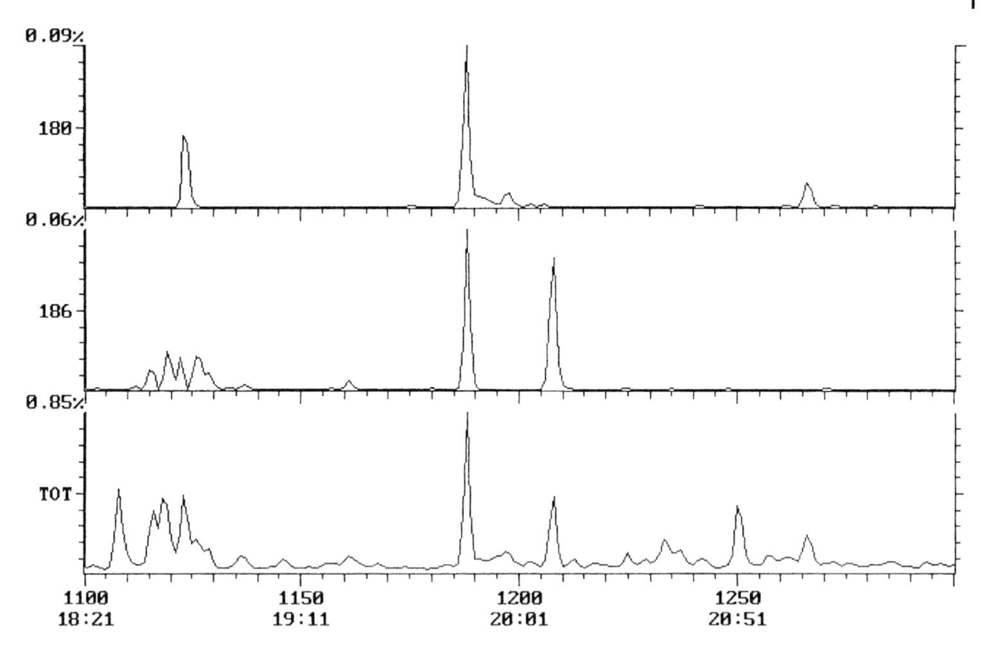

Fig. 3.28 (C) The mass chromatograms m/z 180 and 186 show peak maxima at the same retention time. Both peaks show the same intensity pattern

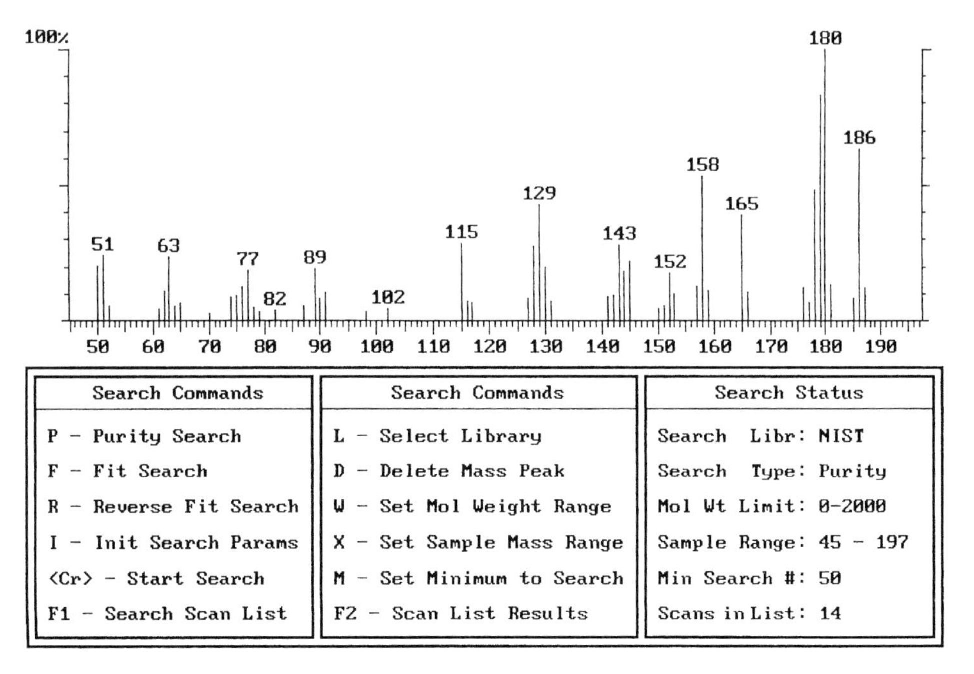

Fig. 3.28 (D) The spectrum of the peak is compared with the NIST library: INCOS sorting according to purity, all molecular weights permitted

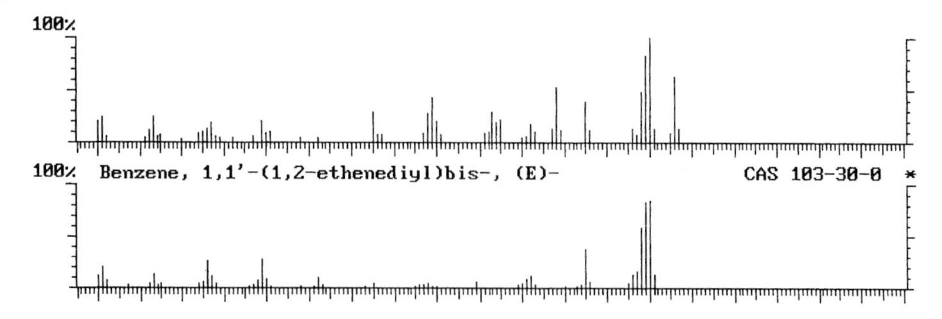

MATCHES	COUNT	TOTAL
8	0	0
7	8	8
6	37	45
5	162	207**
4	476	683
3	1446	2129
2	3874	6003
1	11352	17355
0	36638	53993

RANK	INDEX	PURITY	FIT	RFIT
1	15297	562	821	599
2	15292	558	900	586
3	15293	556	806	590
4	40669	545	875	575
5	15296	539	806	582
6	15300	535	865	555
7	15291	534	844	555
8	15295	532	855	578
9	15298	526	847	566
10	16674	524	790	552

Fig. 3.28 (E) The first suggestion of the INCOS search with tables showing the pre-search (left) and the spectrum comparison (main search, right). The pre-search table shows that 207 candidates were taken over into the main search. The main search table shows the order of rank of the first 10 hits sorted according to purity. All suggestions have high FIT values and low RFIT values

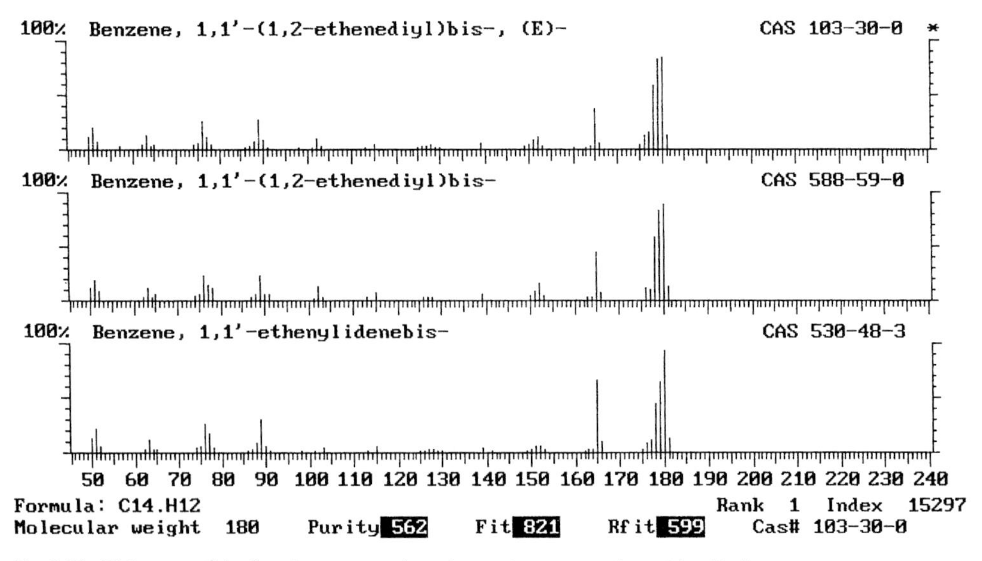

Fig. 3.28 (F) Spectra of the first three suggestions. Isomeric compounds are identified

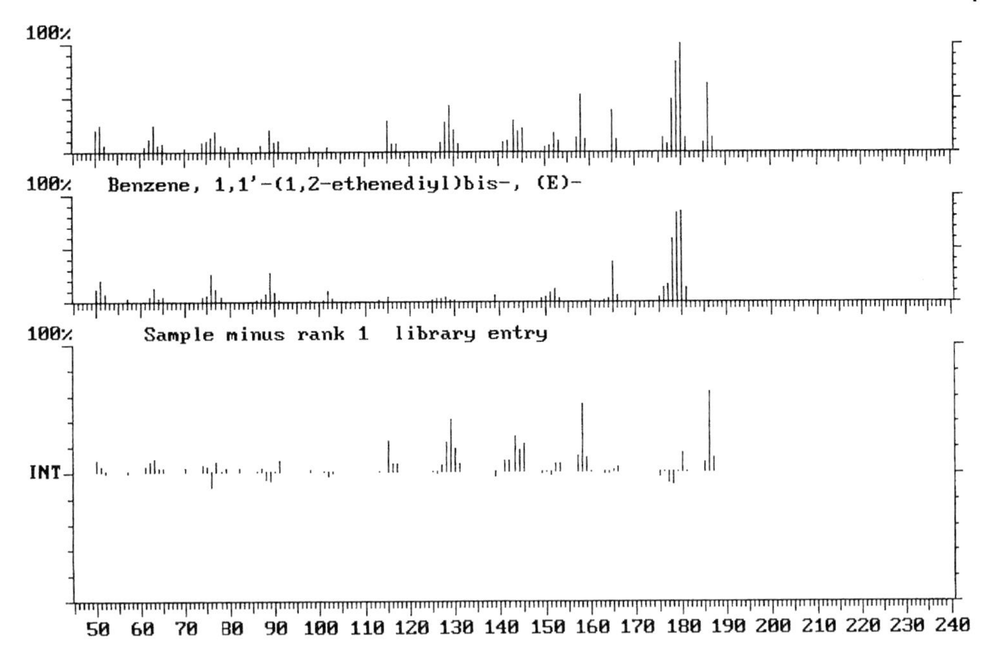

Fig. 3.28 (G) Difference spectrum of the first suggestion compared with the unknown spectrum. The positive part of the difference can be used for a new search

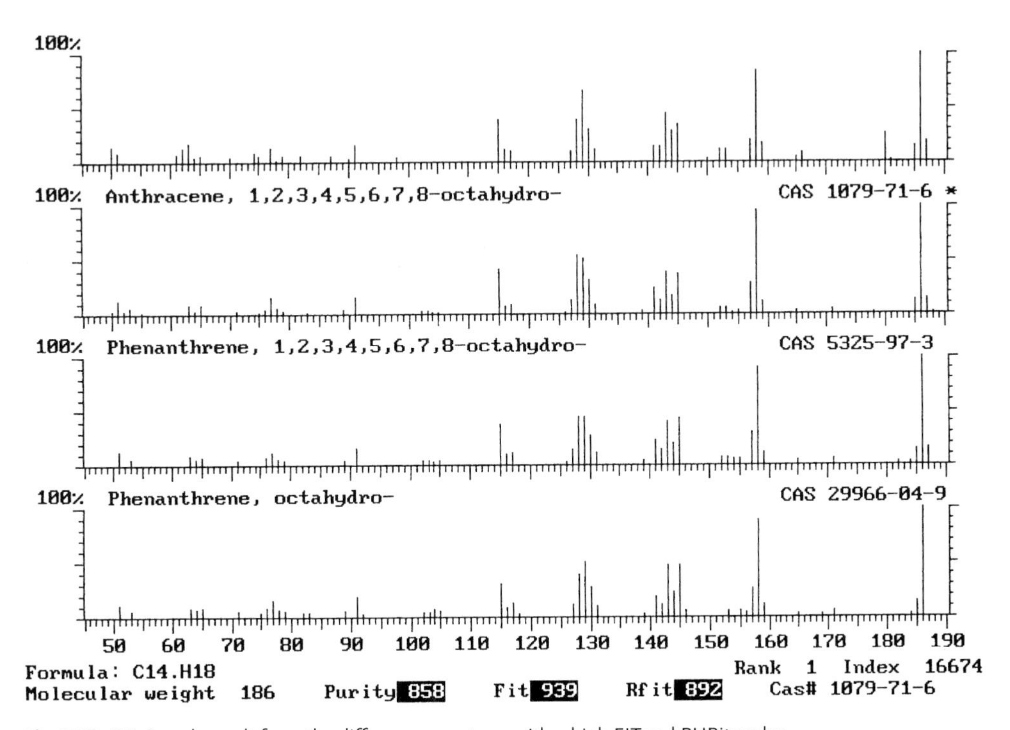

Fig. 3.28 (H) Search result from the difference spectrum with a high FITand PURity value

The advantage of this procedure lies in the fact that deviating relative intensities caused by a high proportion of chemical noise or the type of data acquisition do not have any effect on the result of the search. A variance in the relative signal intensities in a mass spectrum can be caused by varying the choice of spectra from the rising or falling slopes of the peak in the case of quadrupole and magnetic sector instruments and by changes in the tuning parameters of the ion source or its increasing contamination. Furthermore, local normalisation has a positive effect on spectra with a high proportion of noise (trace analysis, chemical background).

Local normalisation is the reason why spectral libraries which are searched using INCOS only require one mass spectrum per substance entry.

Two values are determined for spectral comparison as a result of local normalisation. The FIT value gives a measure of how well the reference spectrum is represented in terms of its masses in the unknown spectrum (reverse search procedure). The reversed mode of viewing, whereby the presence of the unknown spectrum in the reference spectrum is examined (forward search procedure), is expressed as the RFIT value (reversed fit). The combination of the two values gives information on the purity of the unknown spectrum (Fig. 3.29). If the FIT value is high and the RFIT value much lower, it can be assumed that the spectrum measured contains considerably more lines than the reference spectrum used for comparison. Using mass chromatograms or background subtraction it would be necessary to find out whether a co-eluate, chemical noise, the presence of a homologous substance, or another reason is responsible for the appearance of the additional lines.

All the candidates found in the pre-search are processed in the main search as described. As a result sorted lists according to PURity, FIT and RFIT are available (Table 3.3). The initial sorting according to purity is recommended because with this value the best estimation of the possible identity is achieved. Further sorting according to FIT values gives additional solutions which generally supplement the further steps towards identification, with valuable information on partial structures or identifying a particular class of compound.

For the subsequent manual processing the difference between a reference spectrum and the measured unknown spectrum can be established (see Fig. 3.28 G).

A new library search is possible with the remaining portion of the spectrum. In certain cases the co-elution of components at identical times can thus be established which, even

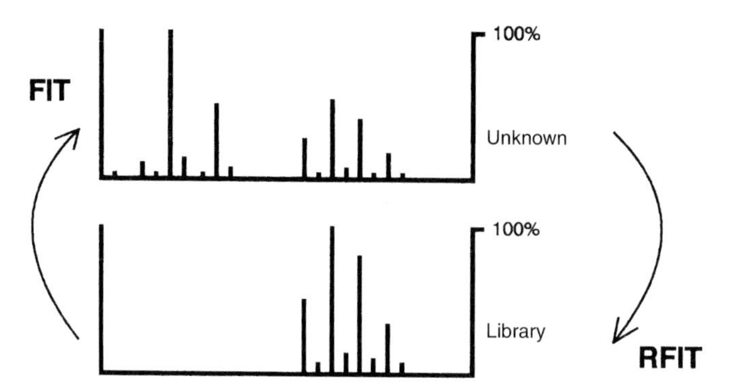

Fig. 3.29 Diagram showing local normalisation.
FIT value high: All masses in the library spectrum are present in the unknown spectrum and the isotope pattern also fits after 'local' normalisation of the intensities.
RFIT value low: Only a few masses from the unknown spectrum are present in the library spectrum.

Table 3.3 Results of the INCOS spectral comparison

FIT	RFIT	PUR	Assessment
high	high	high	Identification or that of an isomer very probable
high	low	low	Identification possible, but homologues, co-elution, noise present
low	high	low	Possibly an incomplete spectrum

Sorting of the suggestions should **first be carried out according to PUR** and then according to FIT

with careful capillary gas chromatographic procedures, is observed with complex samples, for example in environmental analysis. Using the example of a particular case (Figs. 3.28 A–H) the individual steps of the INCOS spectral comparison are shown.

In the 1990s Steven Stein from NIST took the INCOS approach and extended it to the most common situations when unknown compounds are not present in the library. Structurally similar compounds can appear in the NIST library search hit list. By using the results of the library search Stein added probabilities to the hit list that give information about common substructures which may be present or absent in the unknown compound. Based on this advanced performance, the NIST library search is recommended as the first step in the structural elucidation of compounds not found in reference libraries.

INCOS Library Search

Principle Pattern recognition (after Joel Karnovsky, INCOS)

Course

1. Significance weighting	$\sqrt{m \cdot I}$
2. Noise filter	Window ± 50 u, ≥ 40 masses
3. Redundancy filter	Window ± 7 u, ≥ 6 masses
4. Pre-search	8 masses + molecular weight
5. Main search	local normalisation FIT, RFIT and PUR calculation
6. Sorting and display	

Advantages

+ high hit probability
+ secure identification even with spectra with high noise levels
+ search is independent of scan times and type of instrument because of local normalisation
+ only 1 spectrum per substance necessary in the library
+ very fast
+ can be used for a variety of data systems in the form of NIST algorithm

Limitations

– manual difference calculation necessary for co-eluates
– with spectra with many equally distributed fragments

3.2.4.2 The PBM Search Procedure

A completely different strategy forms the basis of the PBM algorithm (probability based match). The statistical mathematical treatment by Prof. McLafferty allows predictions to be made on the probable identity of a substance suggestion. The search procedure was also developed in the 1970s at Cornell University as part of the Cornell algorithm (STIRS, the self-training interpretative and retrieval system was developed as an interpretative system). In the subsequent years parts of the PBM procedure were adapted for PCs, also under the name PBM. These only contained the less powerful mode, 'pure search'. At the beginning of the 1990s a PC version known as benchtop/PBM (Palisade) was released, which now also provides the mode 'mixture search' for the data systems of commercial GC/MS systems.

A significance weighting of the mass signals (here, a sum of the mass m and the intensity I) is also carried out first with the PBM search procedure. In addition the frequency of individual mass signals based on their appearance in the whole spectral library is also taken into account (the reference is the latest version of the Wiley library). The pre-search of the PBM procedure is also orientated towards the method of significance weighting. In the database used for PBM the reference spectra are sorted according to the values of maximum significance. For a given maximum significance of the weighted masses of an unknown spectrum a set of reference spectra can thus be selected. The choice of several sets of mass spectra is specified as the search depth, which, starting from $(m + I)_{max}$, can reach search depths of 3 to $(m + I)_{max} - 3$. The number of possible candidates for the main search is thus broadened.

The main search in the PBM procedure can be carried out in two ways. In the pure search mode only the fragments of the unknown spectrum are searched for in the reference spectra and compared (forward search). This procedure requires mass spectra which are free from overlap and matrix signals, which is the case with simple separations at a medium concentration range.

The mixture search mode first tests whether the mass signals of the reference spectrum are present in the unknown spectrum (reverse search). Local normalisation analogous to the INCOS procedure is also carried out from version 3.0 (1993) onwards. With every spectrum selected from the pre-search, a subtraction from the unknown spectrum is carried out in the course of the procedure. The result of the subtraction is, in turn, compared with the candidates remaining from the pre-search and matching criteria are met. In this way account is taken of the possibility that, even with high resolution gas chromatography, mixed spectra are detected because of the matrix or co-elution. A successful search requires that the second (or third) component of a mixture is contained in the pre-search result (hit list). Combinations of spectra with markedly different highest significances do not appear together in the pre-search hit list because the pre-search is limited by the depth of the search. Using this procedure only the probability values of the hit list are improved. This constitutes a major difference between this and the SISCOM search (see Section 3.2.4.3).

The sorting of results from the PBM search procedure takes place on the basis of probabilities which are determined in the course of spectrum comparison (Table 3.4). In the forefront is the aim of giving a statement about the identity of a suggestion. The assignment to class I gives information as to which degree a spectrum is the same or stereoisomeric with the suggestion. Here maximum values can reach 80%, but not higher. An extension of the definition to the ring positions of positional isomers, homologous compounds, the change in position of individual C-atoms or of double bonds gives a probability defined as class IV. A higher value implies that a compound with structural features is present which has little or no effect on the appearance of the mass spectrum. In such cases the mass spectroscopic

Table 3.4 Results of the benchtop/PBM spectral comparison

Class I	The probability that the suggestion is **identical** or a *stereoisomer*
Class IV	Extension of the probability from class I to compounds having *structural differences compared with the reference,* which only have a small effect on the mass spectrum (homologues, positional isomers).
% Contamination	Gives the proportion of ions which are *not present in the reference*

procedure shows itself to be insensitive to the differences in structural details between individual molecules. The user is thus given a value for assessment which indicates that the mass spectrum being considered matches the spectrum well on account of its few specific fragments, but probably cannot be identified conclusively. The literature should be consulted for details of the calculation of the probability values in PBM.

The clear grading of the probabilities is typical for the PBM procedure (Fig. 3.30). Sensible suggestions with high values of 70 to over 90% (class IV) are given. Then in the hit list there

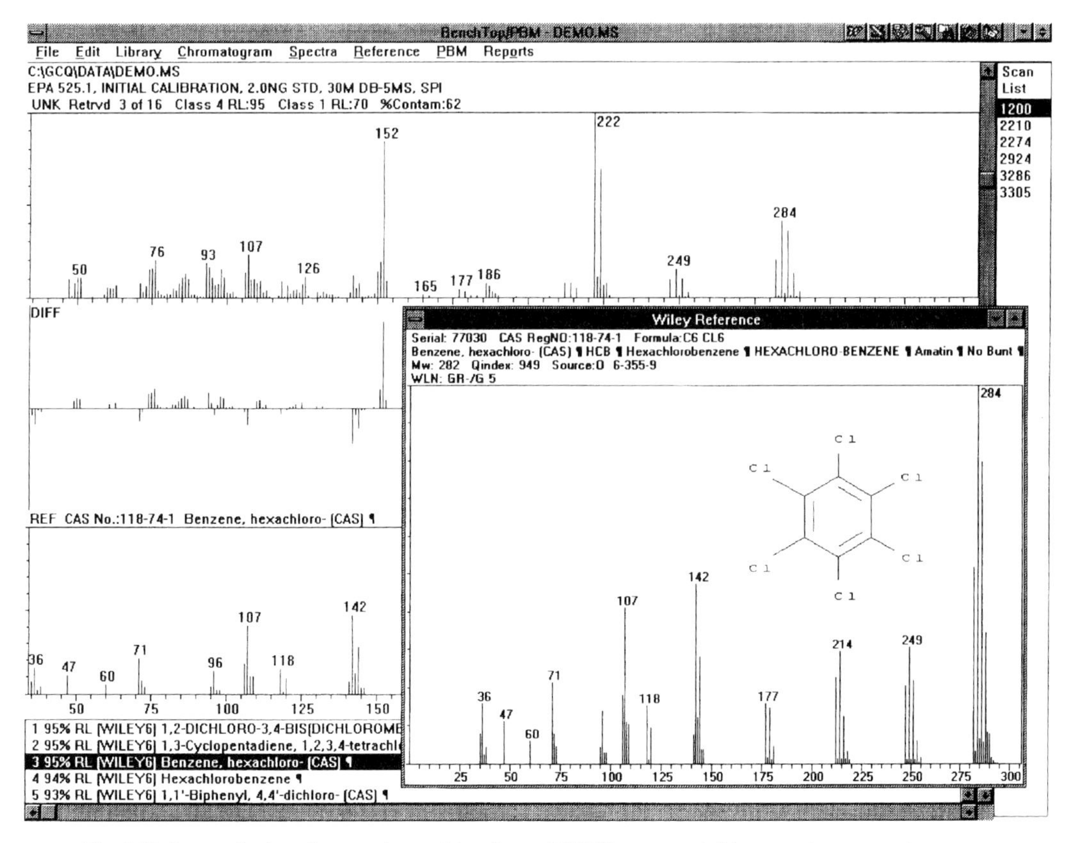

Fig. 3.30 Screen display of suggestions arising from a PBM library search (above: unknown spectrum; middle: difference spectrum; below and in the reference window: library suggestion; underneath: list of suggestions with probabilities)

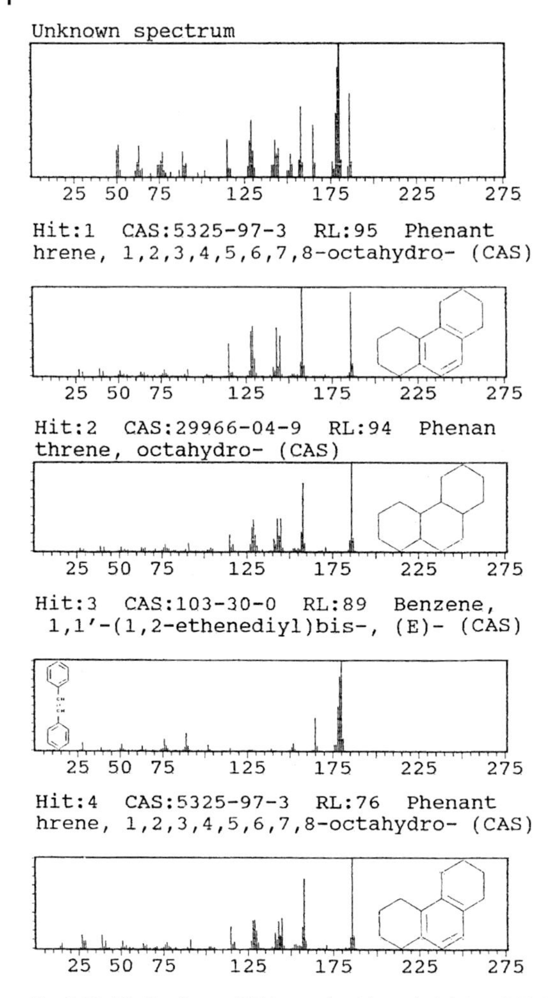

Fig. 3.31 Hit list from a PBM search with probabilities (RL). The unknown spectrum is the same as that shown in Fig. 3-25 B. The mixture search mode gives suggestions for both co-eluting compounds

is the rapid lowering of the values to below 20%, which avoids the occurrence of false positive suggestions. It is obvious that, as a result of the procedure in individual cases, recognisably correct suggestions are given with low probability values (false negative suggestions). Therefore poorly placed suggestions must also be included in the discussion of the search results (Fig. 3.31). In such cases the framework conditions, such as measuring range, quality of the library spectrum or possibly a large quantity of noise in the extracted spectrum should be investigated.

In practice the hit quota of the PBM search depends on the spectrum quality, which is also affected by the data acquisition parameters of the instrument, the choice of the spectrum in the peak (rising/falling slope, maximum), scan time and on the type of instrument. In this context it is useful that in PBM libraries several spectra per substance are available from different sources.

PBM Library Search

Principle Statistical mathematical tests after Prof. McLafferty

Course

Modi:	Pure search or mixture search
Both modi:	Significance weighting $(m + I)$
Both modi:	Pre-search by search depths 1 to 3 $(m + I)_{max}$ to $(m + I)_{max} - 3$

Pure Search Mode	**Mixture Search Mode**
Forward search	Reverse search
Fragments of the unknown spectrum must be present in the reference	Fragments of the reference spectrum must be present in the unknown spectrum
	Spectrum subtraction with each reference from the pre-search
	Comparison of results by forward search in the remaining references
Sorting	Sorting

Advantages

+ the mixture search is more meaningful
+ information on multiple occurrence of similar spectra
+ suggestions with regard to structural formulae and trivial names
+ available as complete additional software for many MS and GC/MS systems

Limitations

- pure search is less useful for complex GC/MS analyses
- very dependent on the quality of spectra and thus on recording parameters, choice of spectrum in the GC peak and on the type of instrument
- therefore more spectra per substance in PBM libraries

Note

There are different PBM programs among the data systems of MS producers, some are more powerful than others; the mixture search mode is often omitted for reasons of speed.

3.2.4.3 The SISCOM Procedure

The SISCOM procedure developed by Henneberg and Weimann (Max Planck Institute for Coal Research, Mülheim a. d. Ruhr, Germany) was targeted to structural determination in industrial MS laboratories. The procedure is commercially available in the form of MassLib in association with various libraries. Its primary goal is to give the most plausible suggestion for the structure of an unknown component and offers information to support the interpretation of the spectra from a large basis of knowledge, i.e. the collection of spectra which have already been explained: the Search for Identical and Similar Compounds.

One of the great strengths of the procedure is the deconvolution of the mass spectrum of an analyte from chromatographic data using an automatic background correction. Within this process the variation of mass intensities during a GC peak is used to determine a reliable substance spectrum (Fig. 3.32). In addition to the identity search, the search procedure offers a similarity search. Through this, account is taken of the fact that mass spectra can differ somewhat from a reference spectrum depending on the conditions of measurement, the compound can be contaminated by other compounds or by the matrix or the quality of the reference spectrum can be in doubt. A similarity search also allows the limitation caused by the limited scope of spectral libraries to be counteracted. All the above-named search systems are limited by this handicap. If the aim of the search is only identification, the results of the similarity search can be rearranged in the direction of identity using a special algorithm. Complete spectra are used here and weightings with regard to mass and intensity are undertaken. A mixture correction procedure is also used to determine the nature of co-eluates. The retention index can also be included in the evaluation of GC/MS analyses.

An essential part of the recognition of similarities between an unknown spectrum and a reference spectrum is the simultaneous use of different methods of comparison, which eval-

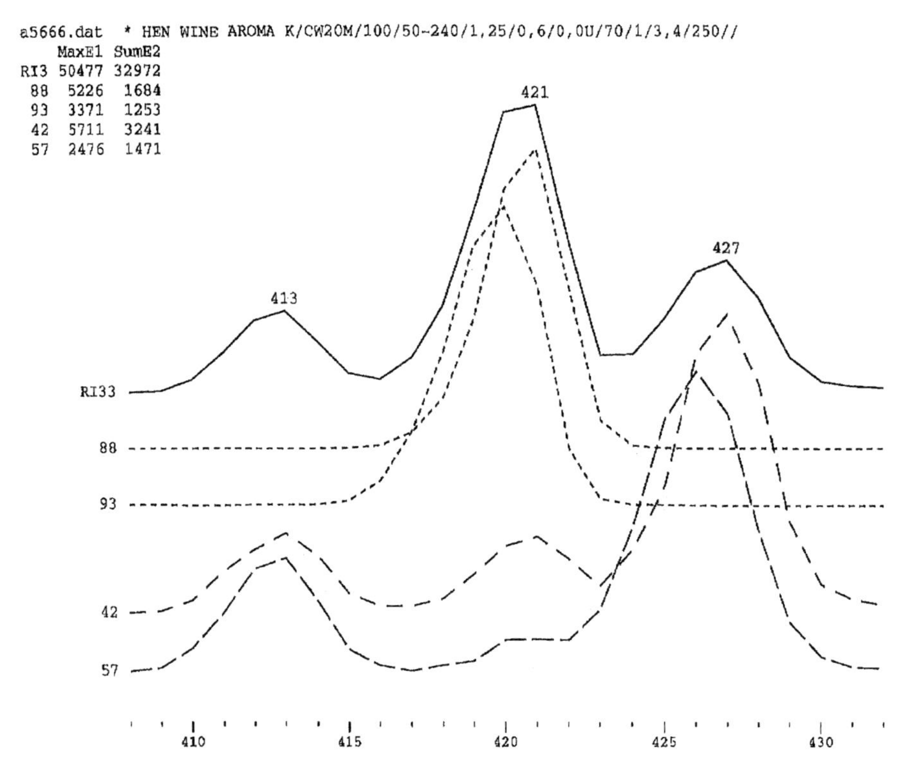

Fig. 3.32 Automatic peak purity control showing the relevant mass chromatograms on co-elution using the SISCOM procedure (MassLib, Chemical Concepts)

uate different aspects of the similarities between two spectra. In the SISCOM procedure the results of these methods are combined finally in such a way that the contribution of the part of the comparison which does not agree can be compensated for by the part which does agree. The differences between an unknown spectrum and the reference spectrum can thus be tolerated and do not hinder the selection of a relevant reference. The results of the different methods of comparison are represented by factors. The order of the hit list is based on a weighted function of these comparison factors. The calculation of a comparison value as a combination of factors with different relative importance is a multidimensional consideration. The ordering of references in a one-dimensional list of suggestions can therefore only be a compromise. The recognition of relevant reference spectra, which are arranged as subsequent hits in the similarity evaluation, is possible for an experienced user when employing the whole range of comparison factors.

The libraries used for the SISCOM search are divided into 14 sections. The assignment of a reference spectrum to one of these sections is based on the strongest peak in the ion series spectrum. Ion series spectra consist of 14 peaks. Each of these peaks (or ion series) represents the sum of all the intensities of the original spectrum and with a mass of m/z modulo 14 = n. Thus, for example, the ion series 1 corresponds to the sum of the intensities of the masses 15, 29, 43, ... The strongest ion series in the spectrum of the unknown substance decides which section of the library is to be used for further searching. As the ion series spectra of related compounds (similar structures) are almost identical (even when the intensity ratios in the spectra differ), the ion series spectra are particularly suitable as a filter for a similarity search.

To compare the unknown spectrum (U) with the reference (R) from the library, SISCOM uses coded spectra, which are calculated from the original spectra by reduction to characteristics. To be determined as characteristic in the SISCOM sense, a mass peak must be larger than the arithmetic mean of its neighbouring homologues (in ± 14 u intervals). The characteristics used in a spectroscopic comparison are divided into three groups. One group contains the characteristics present in U and R; and the other two groups contain characteristics present only in U and only in R respectively. Three comparison factors arise from these three groups:

NC: The number of common characteristics contained in both the unknown spectrum and the reference spectrum. High NC values indicate similar structures.

NR: The number of characteristics from the reference spectrum which are not present in the unknown spectrum. NR is a measure in relation to NC of the extent to which the reference spectrum is represented in the unknown spectrum or is part of it.

NU: The number of characteristics in the unknown spectrum which are not present in the reference spectrum. NU gives the part of the unknown spectrum which is not explained by the reference spectrum. Contamination or a mixture may be present.

These three factors NC, NR, and NU do not contain any intensity parameters. 150 candidates result from them (pre-search).

Three other factors are derived from these first three which use intensities. All six factors are used together to order the 150 candidates in the pre-search for the hit list.

IR: The relative intensity in % of the characteristics of NR. IR completes or differentiates the importance of NR. A few characteristics not contained can have high peak intensities or many characteristics which are absent only consist of peaks of low intensity.

IU: The relative intensity as a % of the characteristics of NC. IU is equivalent to IR and is a measure of the purity of the spectrum, in the case where the suggestion is contained as part of a mixture.

PC: The measure of the correlation of the peak patterns of the relative intensities in the characteristics of NC. As PC is essentially calculated from the occurrence of intense peaks, high values are of only limited importance in spectra with few dominant masses. The PC only makes a limited contribution to the similarity index SI, i. e. deviations, for example as a result of measuring conditions or different types of instrument, are tolerated better.

The evaluation of the hit list requires experience and sound knowledge of mass spectrometry. In the case of identical reference spectra NR and IR are small but PC is high. For a pure spectrum IU is small. If a mixture is present IU is large. Components of a mixture can also appear in the suggestion list using SISCOM, provided that NR and IR are very small and PC correspondingly high. Substances with similar structures are assigned comparable factors, as are components of mixtures. Particularly high NC values indicate high similarity even when PC is negligibly small. Isomeric compounds are an example of this. In the evaluation it should also be noted that a search result depends on the start of the mass range measured in the unknown mass spectrum and on the reference spectrum as well as on the presence of reduced reference spectra. For optimal search results the unknown and the reference spectra should both contain the lower mass region from m/z 25 and be present in nonreduced form. Many important characteristics from this region (e. g. m/z 27, 30, 31, 35 ...) can contribute to similarity searches or identification of spectra.

SISCOM provides a powerful automatic mixture correction. The subtraction procedure is based on the assumption that specific ions exist for a substance which is removed. These ions disappear completely if the exact percentage proportion is removed. If larger proportions are removed negative peaks are formed. The mixture correction uses an iterative procedure based on these which is orientated towards the sum of the negative intensities.

The SISCOM search with mixture correction subtracts the spectrum of the best hits from the unknown spectrum. According to experience this is almost always one of the components present (if a corresponding spectrum is present in the library) or an isomer with almost the same spectrum. With the spectrum thus corrected a completely new search is carried out! In this way a second component is then found if it was not present in the first hit list. The best hit in this second search is also subtracted from the original unknown spectrum and a third complete search is carried out so that in the case of a mixture a purified spectrum is now also present for the first component, which leads to higher identity values. This threefold search gives high identity values from the purified spectra. For identity values of less than 80% identification is very improbable. This makes it easier for only the relevant results to be produced in automatic evaluation processes. A further advantage of this method is that the result of the correction procedure is independent of whether the unknown spectrum is pure or the spectrum of a mixture. In the case of mixtures the MassLib program package (Chemical Concepts, Weinheim, Germany) has a module available for determining the ratio of components in a GC peak using regression analysis (Beynon 1982).

3.2.5
Interpretation of Mass Spectra

There are no hard and fast rules about fully interpreting a mass spectrum. Unlike spectra obtained using other spectroscopic procedures, such as UV, IR, NMR, or fluorescence, mass spectra do not show uptake or emission of energy by the compound (i. e. the intact molecule), but reflect the qualitative and quantitative analysis of the processes accompanying ionisation (fragment formation, rearrangements, chemical reaction) (see Section 2.3.2). The time factor and the energy required for ionisation (electron beam, temperature, pressure) also play a role. With all other spectroscopic procedures the features of certain functional groups or other structural elements always appear in the same way. In mass spectrometry, however, the appearance of certain details of a structure always depends on the total structure of the compound and may not occur at all or only under certain specific conditions. The failure of expected signals to appear generally does not prove in mass spectrometry that certain structural elements are not present; only positive signals count. It is also true that a mass spectrum cannot be associated with a particular chemical structure without additional information.

Nevertheless procedures are recommended for deciphering information hidden in a mass spectrum. (Certain users accuse experienced mass spectrometrists of having a criminological feel for the subject – and they are justified!). In this spectroscopic discipline the experience of a frequently investigated class of substances is rapidly built up. New groups of substances usually require new methods of resolution. It is therefore of extremely great importance that other parameters relating to the substance besides mass spectrometry (spectrum and high resolution data), such as UV, IR, NMR spectra, solubility, elution temperature, acidic or basic clean-up, synthesis reaction equations or those of conversion processes, should be incorporated into the interpretation of the spectrum.

The procedure shown in the following scheme has proved to be effective:

1. Spectrum display
   Does the mass spectrum originate from a single substance or are there signals which do not appear to belong to it? Can the subtraction of the background give a clearer representation or is the spectrum falsified by the subtraction? What information do the mass chromatograms give about the most significant ions?

2. Library search
   As all GC/MS systems are connected to very powerful computer systems, each interpretation process for EI spectra should begin with a search through available spectral libraries (see Section 3.2.4). Spectral libraries are an inestimable source of knowledge, which can give information as to whether the substance belongs to a particular class or on the appearance of clear structural features, even when identification seems improbable. Careful use of the database spares time and gives important suggestions. The different search procedures can all help, although to differing degrees.

3. Molecular ion
   Which signal could be that of the molecular ion? Is an $M^{+/-}$, $(M+H)^+$ or an $(M-H)^{+/-}$ present? Which signals are considered to be noise or chemical background? Mass chromatograms can also help here. If the molecular weight is known, e.g. from CI results, the library search should be carried out again limited to the molecular mass.

4. Isotope pattern
   Is there an obvious isotope pattern, e.g. for chlorine, bromine, silicon or sulfur? The molecular ion shows all elements with stable isotopes in the compound. Is it possible to find out the maximum number of carbon atoms? This is always a problem with residue analysis as usually it is not possible to detect ^{13}C signal intensities with certainty. Also a noticeable absence of isotope signals, particularly with individual fragments, can be important for identifying the presence of phosphorus, fluorine, iodine, arsenic and other monoisotopic elements. Only the molecular or quasimolecular ions give complete information on isotopes.

5. Nitrogen rule
   Is nitrogen present? An uneven molecular mass indicates an uneven number of nitrogen atoms in the empirical formula.

6. Fragmentation pattern
   What information does the fragmentation pattern give? Are there pairs of fragments, the sum of which give the molecular weight? Which fragments could be formed from an α-cleavage? Here the use of a table giving details of fragmentation of molecular ions $(M-X)^+$ is advisable (see Table 3.12).

7. Key fragments
   What can be said about characteristic fragments in the lower mass region? Is there information on aromatic building blocks or ions formed through rearrangements (McLafferty, retro-Diels-Alder)? Here also tables with appropriate explanations are helpful (see Table 3.12).

8. Structure postulate
   Bringing together information rapidly leads to a rough interpretation, which initially gives a partial structure and finally it postulates the molecular structure. Two possibilities test this postulated structure: which fragmentation pattern would the proposed structure give? Is the proposed substance available as a reference and does this correspond to the unknown spectrum?

Within this interpretation scheme spectroscopic comparison through library searching is definitely placed at the beginning of the interpretation. Confirmation of a suggestion from a library search is achieved through an assessment using the above scheme. Interpretation has never finished simply with the print-out of the list of suggestions!

Steps to the Interpretation of Mass Spectra

1. Library search!
2. Only one substance?
3. Molecular ion?
4. Isotope pattern?
5. Nitrogen?
6. Fragmentation pattern?
7. Fragments?
8. Reference spectrum?

Why is ^{12}C the Official Reference Mass for Atomic Mass Units?

Prior to the 1970s, two conventions were used for determining relative atomic masses. Physicists related their mass spectrometric determinations to the mass of ^{16}O (i.e., ^{16}O has a mass of exactly 16 on the amu scale), the most abundant isotope of oxygen, and chemists used the weighted mass of all three isotopes of oxygen: ^{16}O, ^{17}O, and ^{18}O. At an international congress devoted to the standardization of scientific weights and measures, the redoubtable A.O. Nier proposed a solution to these disparate conventions whose negative consequences were becoming serious. He suggested that the carbon-12 isotope (^{12}C) be the reference for the atomic mass unit (amu). By definition, its mass would be exactly 12 amu, a convention that would be acceptable to the physicists. In accordance with this convention, the average mass for oxygen (the weighted sum of the three naturally occurring isotopes) becomes 15.9994 amu, a number close enough to 16 to satisfy the chemists (Sharp 2007).

3.2.5.1 Isotope Patterns

For organic mass spectrometry only a few elements with noticeable isotope patterns are important, while in inorganic mass spectrometry there are many isotope patterns of metals, some of them very complex. From Table 3.5 it can be seen that the elements carbon, sulfur, chlorine, bromine and silicon consist of naturally occurring non-radioactive isotopes. The elements fluorine, phosphorus and iodine are among the few monoisotopic elements in the periodic table.

If isotope signals appear in mass spectra there is the possibility that these elements can be recognised by their typical pattern and also the number of them in molecular and fragment ions can be determined. With carbon there are limitations to this procedure in trace analysis, because usually the quantity of substance available is too small for a sufficiently stable analysable signal to be obtained. In order to allow conclusions to be drawn on the maximum number of carbon atoms, a larger quantity of substance is necessary. For this the technique of individual mass registration (SIM, MID) with longer dwell times is particularly suitable for giving good ion statistics. In the evaluation in the case of carbon only the maximum number of carbon atoms (isotope intensity/1.1%) can be calculated, as contributions from other elements must be taken into account.

Table 3.5 Exact masses and natural isotope frequencies

Element	Isotope	Nominal mass[1] (g/mol)	Exact mass (g/mol)	Abundance[2] (%)	Factors for calculating the isotope intensity[3] M+1	M+2
Hydrogen	^{1}H	1	1.007825	99.99		
	D, ^{2}H	2	2.014102	0.01		
Carbon	^{12}C	12	12.000000[5]	98.9		
	^{13}C	13	13.003354	1.1	1.1	0.006
Nitrogen	^{14}N	14	14.003074	99.6	0.4	
	^{15}N	15	15.000108	0.4	0.4	
Oxygen	^{16}O	16	15.994915	99.76		
	^{17}O	17	16.999133	0.04	0.04	
	^{18}O	18	17.999160	0.20		0.20
Fluorine[4]	F	19	18.998405	100		
Silicon	^{28}Si	28	27.976927	92.2		
	^{29}Si	29	28.976491	4.7	5.1	
	^{30}Si	30	29.973761	3.1		3.4
Phosphorus	P	31	30.973763	100		
Sulfur	^{32}S	32	31.972074	95.02		
	^{33}S	33	32.971461	0.76	0.8	
	^{34}S	34	33.976865	4.22		4.4
Chlorine	^{35}Cl	35	34.968855	75.77		
	^{37}Cl	37	36.965896	24.23		32.5
Bromine	^{79}Br	79	78.918348	50.5		
	^{81}Br	81	80.916344	49.5		98.0
Iodine	I	127	126.904352	100		

1) The calculation of the nominal mass of an empirical formula is carried out using the mass numbers of the most frequently occurring isotope e. g. Lindane $^{12}C_6{}^{1}H_6{}^{35}Cl_6$ M 288.
2) The isotope abundance is a relative parameter. The abundances of an element add up to 100%.
3) In the isotope pattern of the ion the intensity of the first mass peak (nominal mass) is assumed to be 100%. The intensities of the isotope peaks (satellites) M+1 and M2 are given by multiplying the factors with the number of atoms of an element in the ion.
Example: $C_{10}H_{22}$ M^+ 142 intensity m/z 143: 10 · 1.1 = 11%
C_6Cl_6 M^+ 282 intensity m/z 284: 6 · 32.5 = 195%
S_6 M^+ 192 intensity m/z 194: 6 · 4.4 = 26.4%
4) Fluorine, sodium, aluminium, phosphorus, manganese, arsenic and iodine, for example, appear as monoisotopic elements in mass spectrometry.
5) See box on p. 331 on ^{12}C as the reference mass.

Some elements, in particular the halogens chlorine and bromine, which are contained in many active substances, plastics, and other technical products, can be recognised by the typical isotope patterns. These easily recognised patterns are shown in Figs. 3.33 to 3.39. The intensities shown are scaled down to a unit ion stream of the isotope pattern. The lowering of the specific response of the compound as a function of, for example, the degree of chlorination, is shown. The simple occurrence of the elements shown as a series is used as a reference in each case. The relative intensities within an isotope pattern are given as a percentage in the

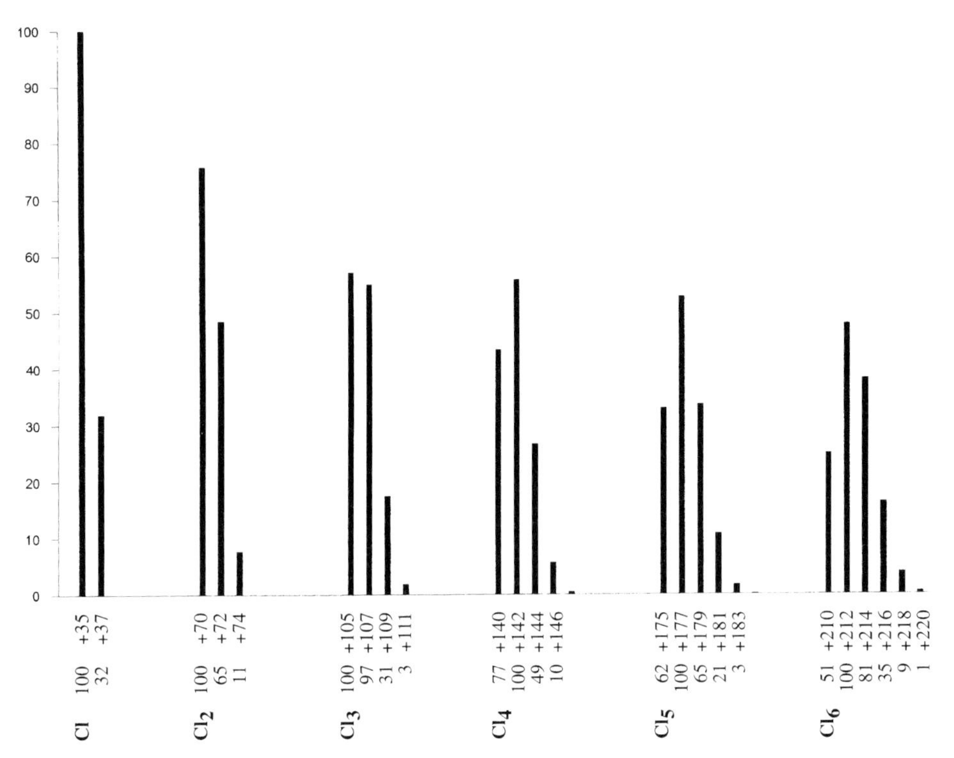

Fig. 3.33 Isotope pattern of chlorine (Cl to Cl_6)

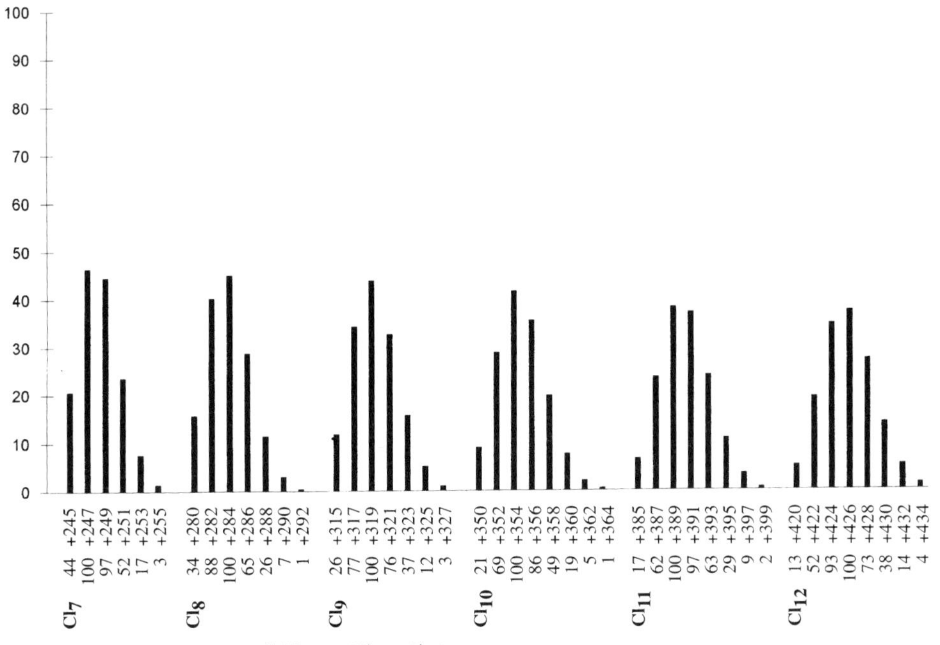

Fig. 3.34 Isotope pattern of chlorine (Cl_7 to Cl_{12})

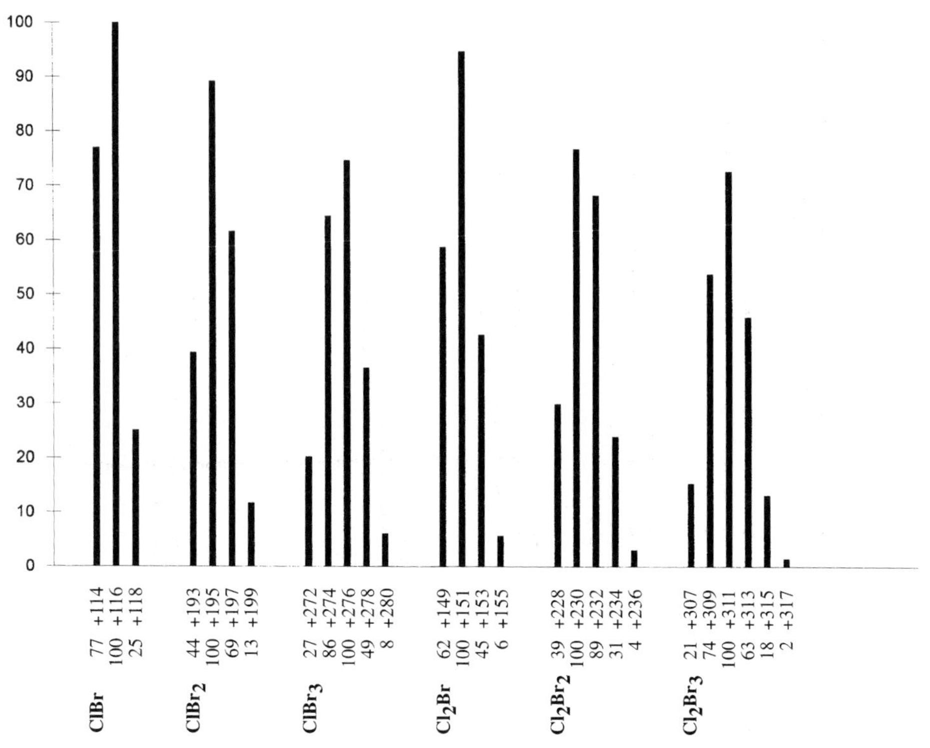

Fig. 3.35 Isotope pattern of chlorine/bromine ($ClBr$ to Cl_2Br_3)

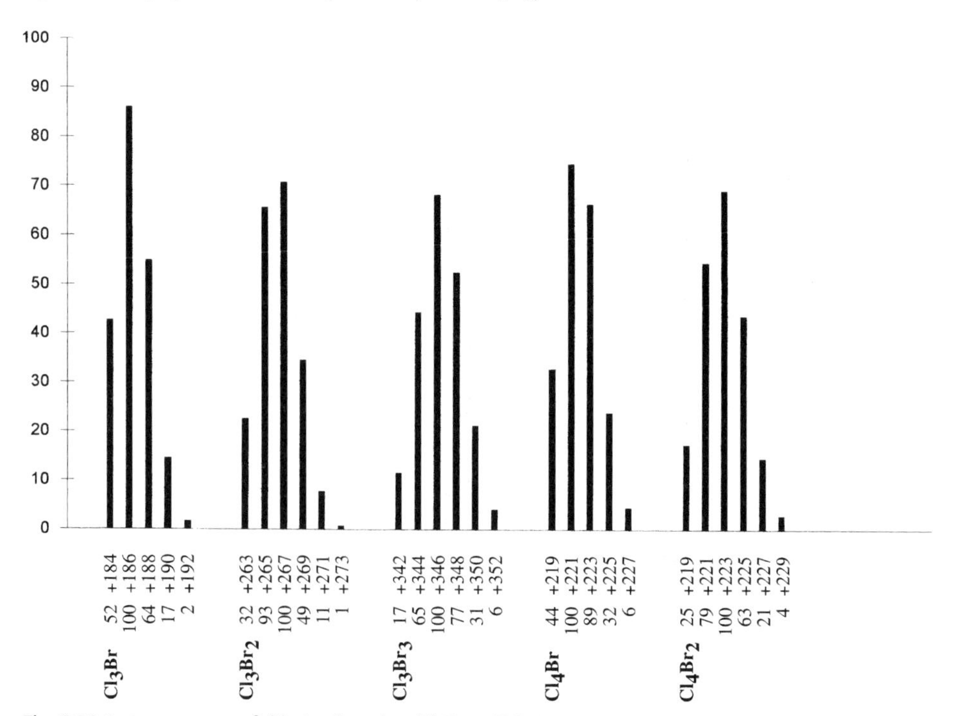

Fig. 3.36 Isotope pattern of chlorine/bromine (Cl_3Br to Cl_4Br_2)

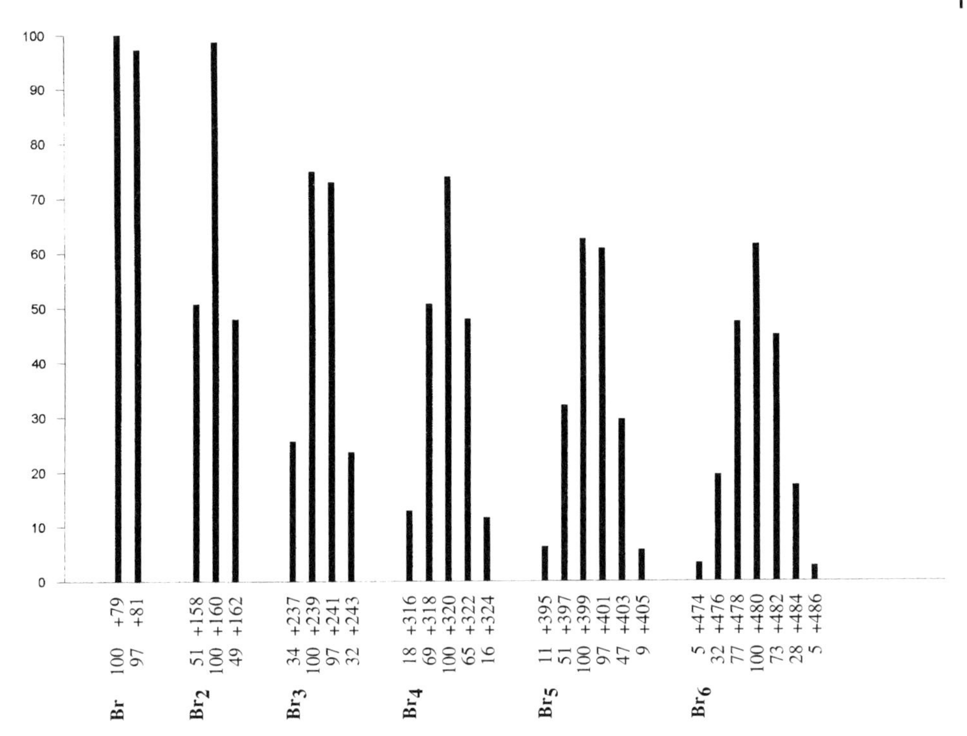

Fig. 3.37 Isotope pattern of bromine (Br to Br_6)

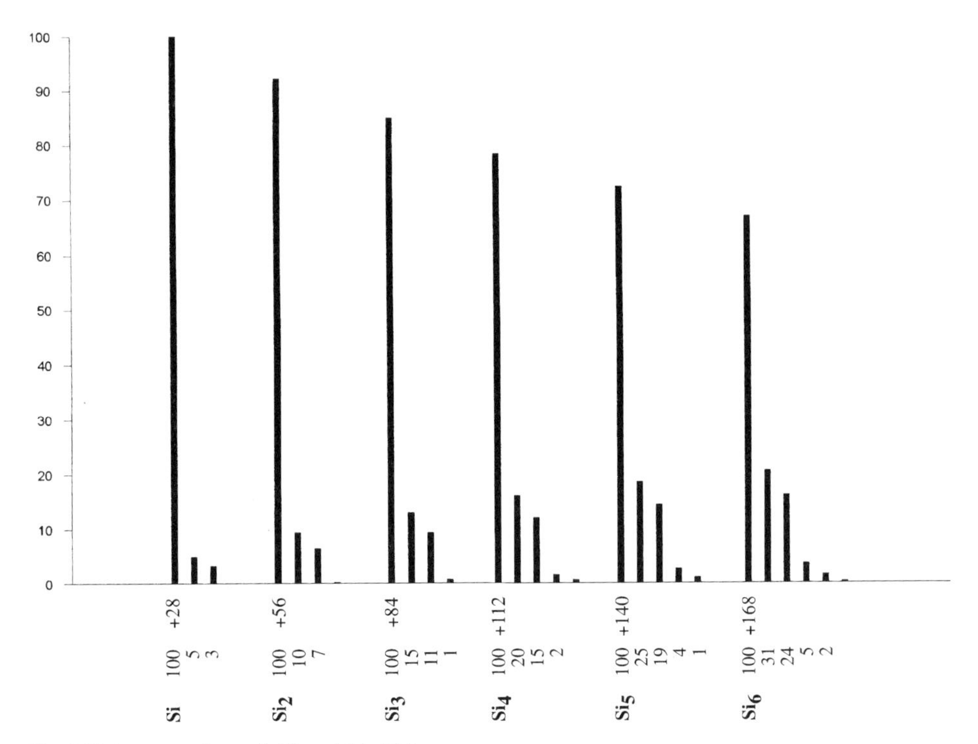

Fig. 3.38 Isotope pattern of silicon (Si to Si_6)

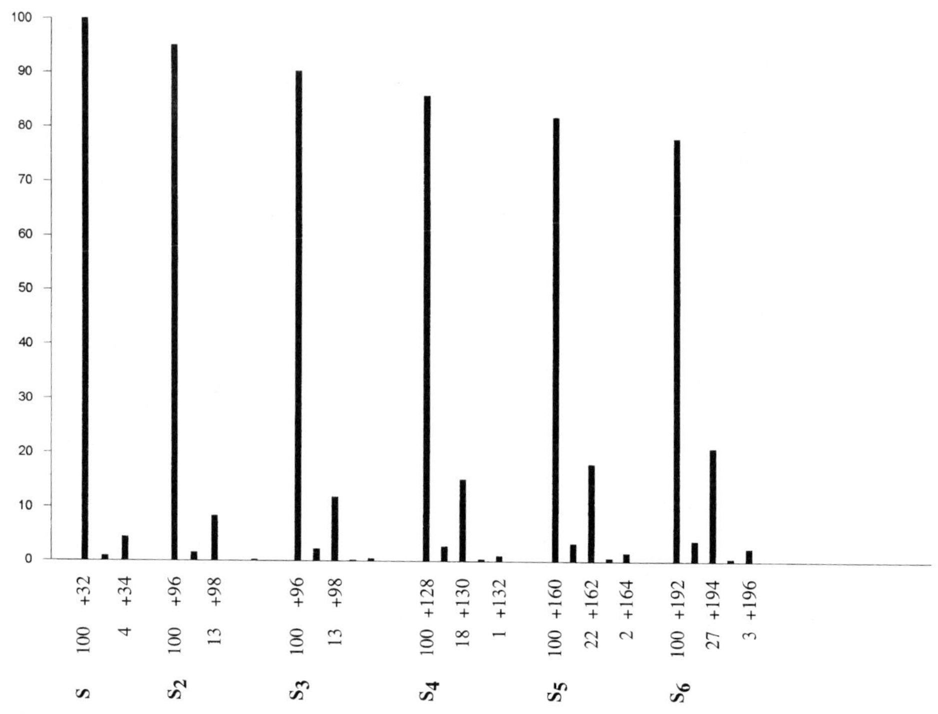

Fig. 3.39 Isotope pattern of sulfur (S to S_6)

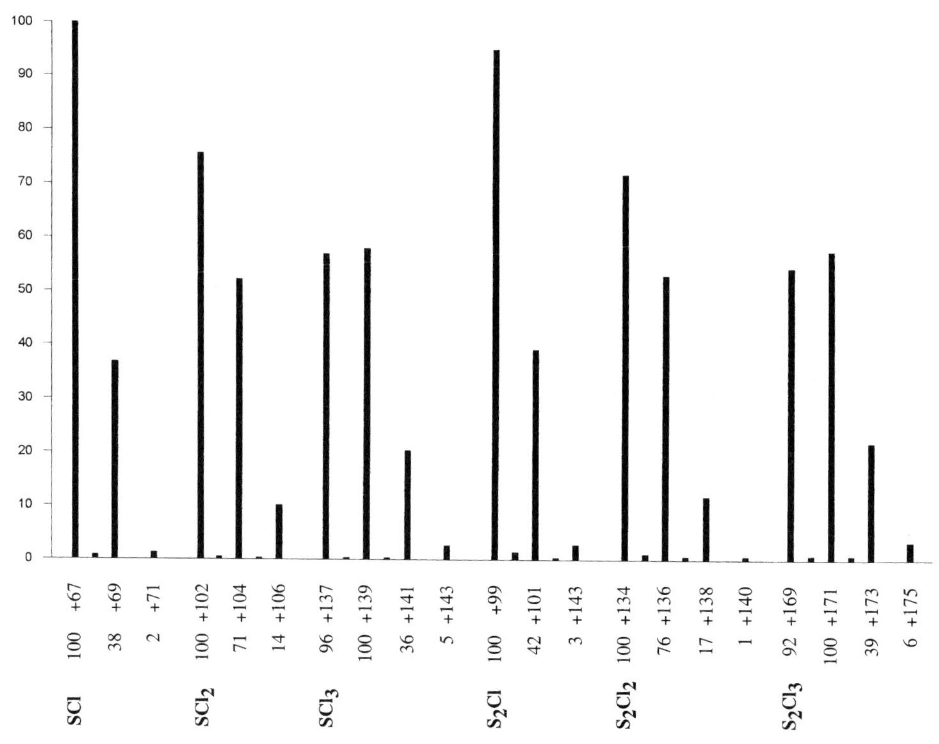

Fig. 3.40 Isotope pattern of sulfur/chlorine (SCl to S_2Cl_3)

caption underneath the isotope lines with the mass contribution to the molecular weight based on the most frequent occurrence in each case. (Source: isoMass – PC program for calculation and analysis of isotope patterns by Ockels, SpectralService). What is noticeable for chlorine and bromine is the distance between the isotopes of two mass units. Provided that chlorine and bromine occur separately, the degree of chlorination or bromination can be easily determined by comparison of the variation in the intensities. For compounds which contain both chlorine and bromine the degree of substitution cannot be determined by comparison of the patterns alone. In these cases the high atomic weight of bromine is a help (for calculation of retention indices from the empirical formula see Section 3.2.2). In GC/MS coupling relatively high molecular weights (fragment ions) are detected through the presence of bromine in a molecule even at short retention times. In library searches mixed isotope patterns of chlorine and bromine are reliably recognised.

With sulfur the distance between the isotope peaks is also two mass units. However, when the proportion of sulfur in the molecule is low (e.g. in the case of phosphoric acid esters) it is difficult to be sure of the presence of sulfur (Fig. 3.39). Here a detailed investigation of the fragmentation is necessary. Higher sulfur contents (see also Fig. 3.164) give clear information. The combinations of sulfur and chlorine in which chlorine is clearly dominant are also shown (Fig. 3.40). Differences are extremely difficult to see with the naked eye and with computers only when measurements are obtainable with good ion statistics.

Silicon occurs very frequently in trace analysis. Silicones get into the analysis through derivatisation (silylation), partly through clean-up (joint grease), but more frequently through bleeding from the septum or the column (septa of autosampler vials, silicone phases). The typical isotope pattern of all silicone masses can be recognised rapidly and excluded from further evaluation measures (Fig. 3.38).

3.2.5.2 Fragmentation and Rearrangement Reactions

The starting point for a fragmentation is the molecular ion (EI) or the quasimolecular ion (CI). A large number of reactions which follow the primary ionisation need to be described here. All the reactions follow the thermodynamic aim of achieving the most favourable energy balance possible. The basic mechanisms which are involved in the production of spectra of organic compounds will be discussed briefly here (Fig. 3.41). For more depth the references cited should be consulted (Budzikiewicz, Pretsch et al., Howe et al., McLafferty et al.).

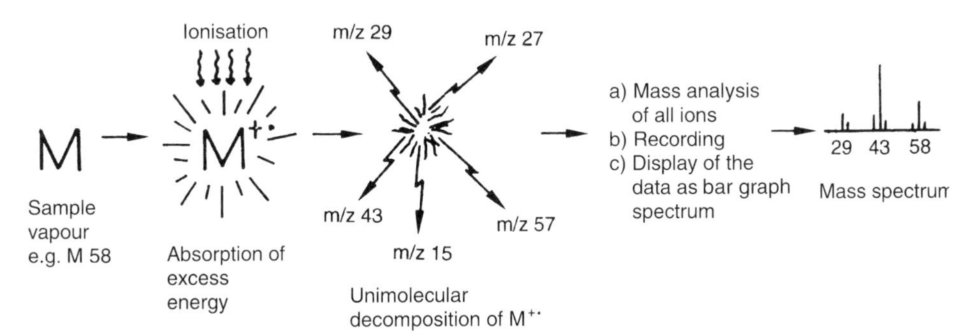

Fig. 3.41 Principle of mass spectrometry: formation of molecular ions as the starting point of a fragmentation (after Frigerio)

There are two possible mechanisms for the cleavage of carbon chains following ionisation, known as α-cleavage and formation of carbenium ions. The starting point in each case is the localisation of positive charge on electron-rich structures in the molecule.

α-Cleavage

α-cleavage takes place after ionisation by loss of one nonbonding electron from a heteroatom (e.g. in amines, ethers, ketones, see Tables 3.6 to 3.8) or on formation of an allylic or benzylic carbenium ion (from alkenes, alkylaromatics):

Amines

$$R_1\text{-}\overset{+\bullet}{N}(R_2)\text{-}CH_2\text{-}R \longrightarrow R_1\text{-}\overset{+}{N}(R_2){=}CH_2 + R^\bullet$$

Table 3.6 Characteristic ions from the α-cleavage of amines

m/z	R_1	R_2
30	H	H
44	CH_3	H
58	CH_3	CH_3

Ethers

$$R_1\text{-}\overset{+\bullet}{\underline{O}}\text{-}CH_2\text{-}R \longrightarrow R_1\text{-}\overset{+}{\underline{O}}{=}CH_2 + R^\bullet$$

Table 3.7 Characteristic ions from the α-cleavage of ethers

m/z	R_1
31	H
45	CH_3
59	C_2H_5
73	C_3H_7
87	C_4H_9
etc.	

Ketones

$$R_1\text{-}C(=\overset{+\bullet}{O})\text{-}R \longrightarrow R_1\text{-}C{\equiv}\overset{+}{O}| + R^\bullet$$

Table 3.8 Characteristic ions from the α-cleavage of ketones

m/z	R_1
29	H
43	CH_3
57	C_2H_5
71	C_3H_7
85	C_4H_9

Formation of Carbenium Ions

Fragmentation involving formation of carbenium ions takes place at a double bond in the case of aliphatic carbon chains (allylic carbenium ion) or at a branch. With alkylaromatics side chains are cleaved giving a benzylic carbenium ion (benzyl cleavage), which dominates as the tropylium ion m/z 91 in many spectra of aromatics.

Alkenes

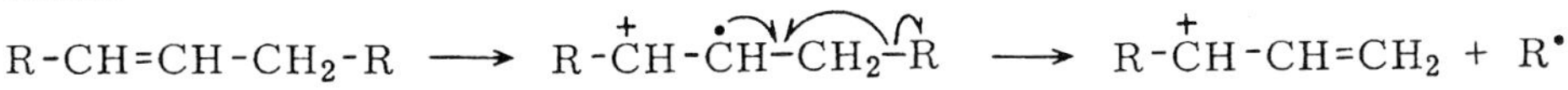

$$R\text{-}CH{=}CH\text{-}CH_2\text{-}R \longrightarrow R\text{-}\overset{+}{C}H\text{-}\overset{\bullet}{C}H\text{-}CH_2\text{-}R \longrightarrow R\text{-}\overset{+}{C}H\text{-}CH{=}CH_2 + R^{\bullet}$$

Branched carbon chains

$$R_3C \overset{+}{\bullet} R \longrightarrow R_3C^{+} + R^{\bullet}$$

The formation of carbenium ions occurs preferentially at tertiary branches rather than secondary ones.

Alkylaromatics (benzyl cleavage)

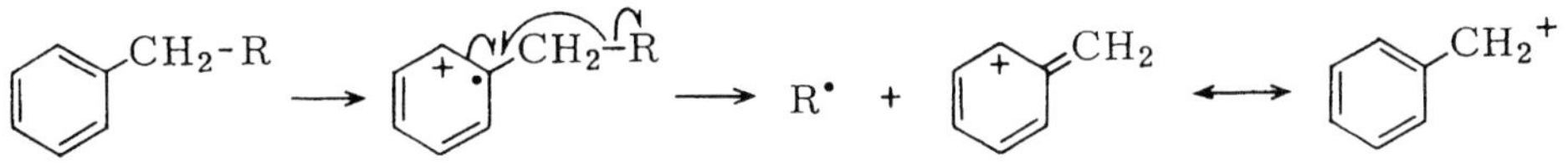

$$C_6H_5\text{-}CH_2\text{-}R \longrightarrow [C_6H_5\text{-}CH_2\text{-}R]^{+\bullet} \longrightarrow R^{\bullet} + C_6H_5{=}CH_2{}^{+} \longleftrightarrow C_6H_5\text{-}CH_2{}^{+}$$

Loss of Neutral Particles

The elimination of stable neutral particles is a common fragmentation reaction. These include H_2O, CO, CO_2, NO, HCN, HCl, RCOOH, and alkenes (see also McLafferty rearrangement). These reactions can usually be recognised from the corresponding ions and mass differences in the spectra. Eliminations are particularly likely to occur when α-cleavage is impossible. MS/MS mass spectrometer providing the neutral loss scan mode allow a substance class specific detection based on the elimination of neutrals.

Alcohols (loss of water)

$$[C_nH_{2n+1}OH]^{+\bullet} \longrightarrow [C_nH_{2n}]^{+\bullet} + H_2O$$

Carbonyl Functions (loss of CO)

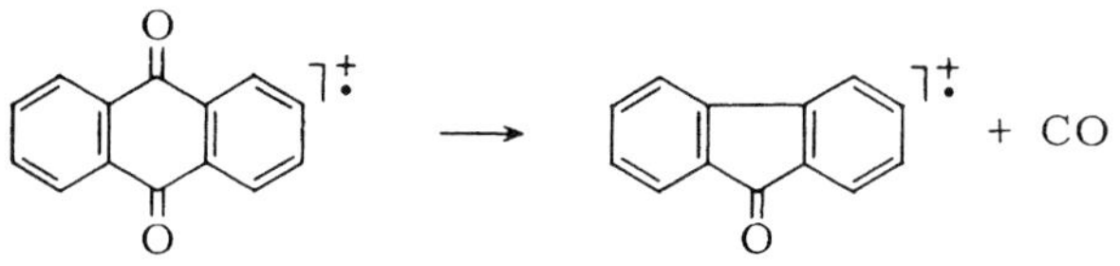

$$[\text{anthraquinone}]^{+\bullet} \longrightarrow [\text{fluorenone}]^{+\bullet} + CO$$

Heterocycles (loss of HCN)

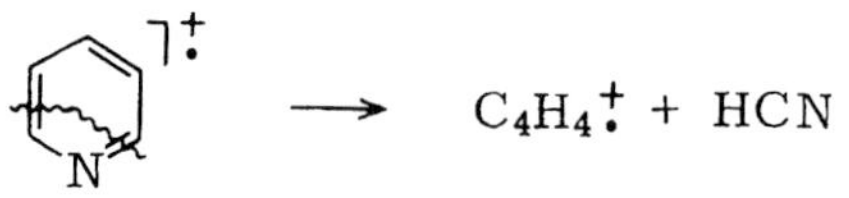

$$\longrightarrow \quad C_4H_4^{+\bullet} + HCN$$

Retro-Diels-Alder (loss of alkenes)

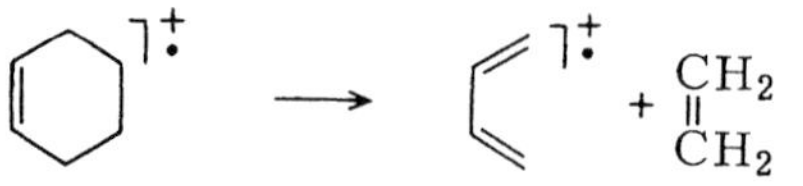

$$\longrightarrow \quad + \; CH_2{=}CH_2$$

The McLafferty Rearrangement

The McLafferty rearrangement involves the migration of an H-atom in a six-membered ring transition state. The following conditions must be fulfilled for the rearrangement to take place:

- The double bond C=X is C=C, C=O or C=N.
- There is a chain of three σ-bonds ending in a double bond.
- There is an H-atom in the γ-position relative to the double bond which can be abstracted by the element X of the double bond.

According to convention the McLafferty rearrangement is classified as the loss of neutral particles (alkene elimination with a positive charge remaining on the fragment formed from the double bond, see Table 3.9).

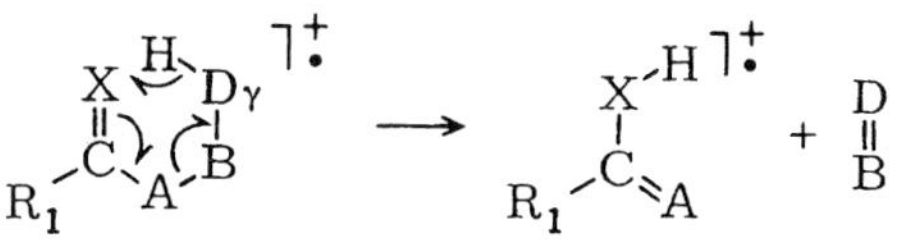

Table 3.9 Characteristic ions formed in the McLafferty rearrangement

m/z	R_1	found
44	H	in aldehydes
60	OH	in organic acids
74	$O{-}CH_3$	in methyl esters

Table 3-10 Mass correlations to explain cleavage reactions $(M - X)^+$ and key fragments X^+

m/z	Fragment X^+	$M^+ - X$	Explanations
12	C		
13	CH		
14	CH_2, N, N_2^{++}		
15	CH_3	$M^+ - 15$	Nonspecific, CH_3 at high intensity
16	O, NH_2, O_2^{++}, CH_4	$M^+ - 16$	Rarely CH_4, (but frequently $R^+ - CH_4$ in alkyl fragments), O from N-oxides and nitro compounds, NH_2 from anilines
17	OH, NH_3	$M^+ - 17$	Nonspecific O-indication, NH_3 from primary amines
18	H_2O, NH_4	$M^+ - 18$	Nonspecific O-indication, strong for many alcohols, some acids, ethers and lactones
19	H_3O, F	$M^+ - 19$	F-indication
20	HF, Ar^{++}, CH_2CN^{++}	$M^+ - 20$	F-indication
21	$C_2H_2O^{++}$ (rarely)		
22	CO_2^{++}		
23	Na (rarely)		
24	C_2		
25	C_2H	$M^+ - 25$	Rarely with a terminal $C{\equiv}CH$ group
26	C_2H_2, CN	$M^+ - 26$	From purely aromatic compounds, rarely from cyanides
27	C_2H_3, HCN	$M^+ - 27$	CN from cyanides, C_2H_3 from terminal vinyl groups and some ethyl esters
28	C_2H_4, N_2, CO	$M^+ - 28$	CO from aromatically bonded O, ethylene through RDA from cyclohexenes, by H-migration from alkyl groups, nonspecific from alicyclic compounds
29	C_2H_5, CHO	$M^+ - 29$	Aromatically bonded O, nonspecific with hydrocarbons
30	C_2H_6, H_2NCH_2, NO, CH_2O, BF (N-fragment)	$M^+ - 30$	CH_2O from cyclic ethers and aromatic methyl ethers, NO from nitro compounds and nitro esters
31	CH_3O, CH_2OH, CH_3NH_2, CF, (O-fragment)	$M^+ - 31$	Methyl esters, methyl ethers, alcohols
32	O_2, CH_3OH, S	$M^+ - 32$	Methyl esters, some sulfides and methyl ethers
33	CH_3OH_2, SH, CH_2F	$M^+ - 33$	SH nonspecific S-indication, $M^+ - 18–15$ nonspecific O-indication, strong with alcohols
34	SH_2, (S-fragment)	$M^+ - 34$	Nonspecific S-indication, strong with thiols
35	^{35}Cl, SH_3	$M^+ - 35$	Chlorides, nitrophenyl-compounds $(M^+ - 17 - 18)$
36	HCl, C_3	$M^+ - 36$	Chlorides

Table 3-10 (continued)

m/z	Fragment X^+	$M^+ - X$	Explanations
37	^{37}Cl, C_3H		
38	$H^{37}Cl$, C_3H_2		
39	C_3H_3	$M^+ - 39$	Weak with aromatic hydrocarbons
40	Ar, C_3H_4	$M^+ - 40$	Rarely with CH_2CN
41	C_3H_5, CH_3CN	$M^+ - 41$	C_3H_5 from alicyclic compounds, CH_3CN from aromatic N-methyl and o-C-methyl heterocycles
42	$CH_2{=}C{=}O$, C_3H_6, C_2H_4N	$M^+ - 42$	Nonspecific with aliphatic and alicyclic systems, strong through RDA from cyclohexenes, by rearrangement from α-, β-cyclohexenones, enol and enamine acetates: McLafferty product
43	CH_3CO, C_3H_7, C_2H_4N, CONH	$M^+ - 43$	Acetyl, propyl, aromatic methyl ethers ($M^+ - 15{-}28$), nonspecific with aliphatic and alicyclic systems
44	CO_2, CH_3NHCH_2, CH_2CHOH, (N-fragment)	$M^+ - 44$	CO_2 from acids, esters, butane from aliphatic hydrocarbons
45	C_2H_5O, HCS, (Sulfides)	$M^+ - 45$	Ethyl esters, ethyl ethers, lactones, acids, CO_2H from some esters; CH_3NHCH_3 from dimethylamines
46	C_2H_5OH, NO_2	$M^+ - 46$	Ethyl esters, ethyl ethers, rarely acids, nitro compounds, n-alkanols ($M^+ - 18{-}28$)
47	CH_3S, $C^{35}Cl$, $C_2H_5OH_2$, $CH(OH)_2$, (S-fragment)		
48	CH_3SH, $CH^{35}Cl$		
49	$C^{37}Cl$, $CH_2{}^{35}Cl$		
50	C_4H_2, $CH_3{}^{35}Cl$		
51	C_4H_3, (Aromatic fragment)		
52	C_4H_4, $CH_3{}^{37}Cl$, (Aromatic fragment)		
53	C_4H_5		
54	//‾\\, C_2H_4CN	$M^+ - 54$	Cyclohexene (RDA)
55	C_4H_7, C_2H_3CO	$M^+ - 55$	C_4H_7 from alicyclic systems and butyl esters
56	C_4H_8, C_2H_4CO	$M^+ - 56$	Nonspecific with alkanes and alicyclic systems
57	C_4H_9, C_2H_5CO, C_3H_2F	$M^+ - 57$	Nonspecific with alkanes and alicyclic systems
58	CH_3COHCH_2, C_2H_5-$CHNH_2$, $C_2H_6NCH_2$	$M^+ - 58$	C_3H_6O from α-methylaldehydes and acetonides
59	C_2H_6COH, $C_2H_5OCH_2$, CO_2CH_3, CH_3CONH_2	$M^+ - 59$	Methyl esters
60	$CH_2CO_2H_2$, CH_2ONO	$M^+ - 60$	O-acetates (M^+ – AcOH), methyl esters ($M^+ - CH_3OH{-}CO$)

Table 3-10 (continued)

m/z	Fragment X^+	$M^+ - X$	Explanations
61	$CH_3CO_2H_2$, C_2H_4SH		
62	$HOCH_2CH_2OH$	$M^+ - 62$	Ethylene ketals
63	C_5H_3		
64	SO_2	$M^+ - 64$	SO_2 cleavage from sulfonic acids
65	C_5H_5		
67			
68	C_4H_4O, C_3H_6CN		
69	C_5H_9, C_3H_5CO, CF_3, C_3HO_2 (1,3-Dioxyaromatics)		
70	C_5H_{10}		
71	C_5H_{11}, C_4H_7CO		
72	$C_4H_{10}N$, $C_3N_7NHCH_2$, $C_2H_5COHCH_2$		
73	$CO_2C_2H_5$, $C_3H_7OCH_2$, $CH_2CO_2CH_3$, C_4H_8OH, (O-fragments)		
74	$CH_2{=}COHOCH_3$, $CH_3CH{=}COHOH$		
75	$C_2H_5CO_2H_2$, $C_2H_5SCH_2$, $CH_3OCHOCH_3$, (Dimethyl acetates)		
76	C_6H_4		
77	C_6H_5		
78	C_6H_6		
79	C_6H_7, ^{79}Br		
80	C_6H_8, $H^{79}Br$, CH_3S_2H		
81	C_6H_9, ^{81}Br		
82	C_6H_{10}, $H^{81}Br$		
83	C_6H_{11}, C_4H_7CO		
84			
85	C_6H_{13}, C_4H_9CO		
86	$C_3H_7COH{=}CH_2$		
87	$CO_2C_3H_7$, $CH_2CO_2C_2H_5$, $CH_2CH_2CO_2CH_3$, (O-fragments)		
88	$CH_2{=}COHOC_2H_5$, $CH_3CH{=}COHOCH_3$		
91	n-Alkyl chlorides		
92			
93	$CH_2{}^{79}Br$		

Table 3-10 (continued)

m/z	Fragment X^+	m/z	Fragment X^+
94	$CH_3{}^{79}Br$, 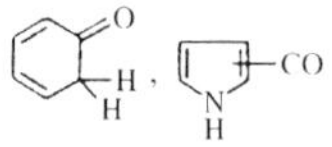	120	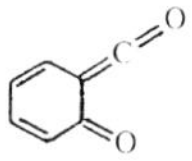
95	$CH_2{}^{81}Br$, 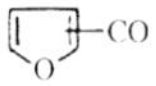	121	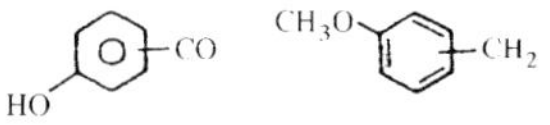
96	$C_5H_{10}CN$, $CH_3{}^{81}Br$	127	J
97	C_7H_{13}, 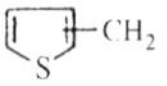	128	HJ, 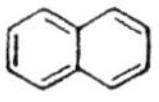
98	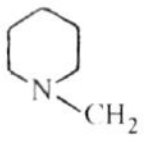	130	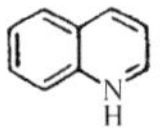
99	C_7H_{15}, 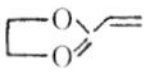 (Ethylene ketals)	131	C_3F_5
104	$C_2H_5CHONO_2$, 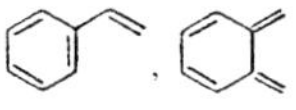	135	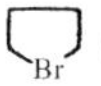 (n-Alkyl bromides)
105	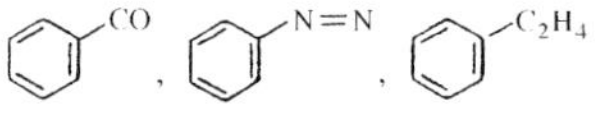	141	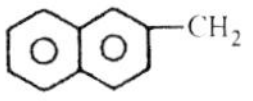
111	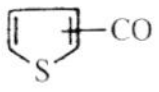	142	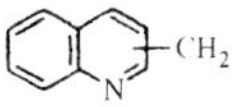
115	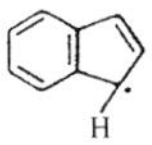	149	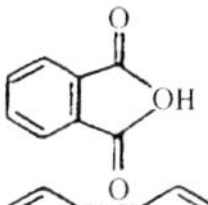
119	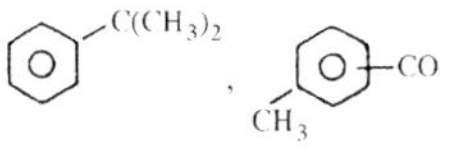	152	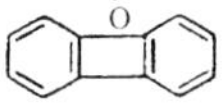

3.2.5.3 DMOX Derivatives for Location of Double Bond Positions

The location of double bonds in polyunsaturated fatty acids by GC/MS involves many different methodologies either by suitable derivatisation or chemical ionisation techniques. Derivatisation reactions include the reaction with 4,4-dimethyloxazoline (DMOX) (Dobson 2002, Fay 1991), dimethyl disulfide (DMDS) (Moss 1991), as well as methoxy and methoxybromo derivatives (Shantha 1984) besides other known derivatives. Chemical ionisation (CI) allows specific reactions with C,C-double and triple bonds with suitable reagent gases which in many cases permit a location of the sites of unsaturation in organic molecules (Budzikiewicz 1985).

Methoxybromo derivatives of unsaturated fatty acids including conjugated acids yield simple mass spectra to locate the position of double bonds in these acids. Unlike other methods using methoxy derivatives, the methoxybromo derivatives yield fewer ions, the diagnostic peaks forming the most intense ions of the spectra with the characteristic appearance of fragments corresponding to $[CH_3(CH_2)_nCH(OMe)CH(Br)CH_22H]^+$.

The preparation of 4,4-dimethyloxazoline (DMOX) derivatives is most widely applied for routine identification of fatty acids in unknown samples. A relatively mild reaction minimizes possible isomerisation reactions. The DMOX derivatives are comparable to FAMES in volatility and hence in chromatographic resolution. When ionised under regular EI conditions, radical induced cleavage processes give rise to mass spectra that are easy to interpret

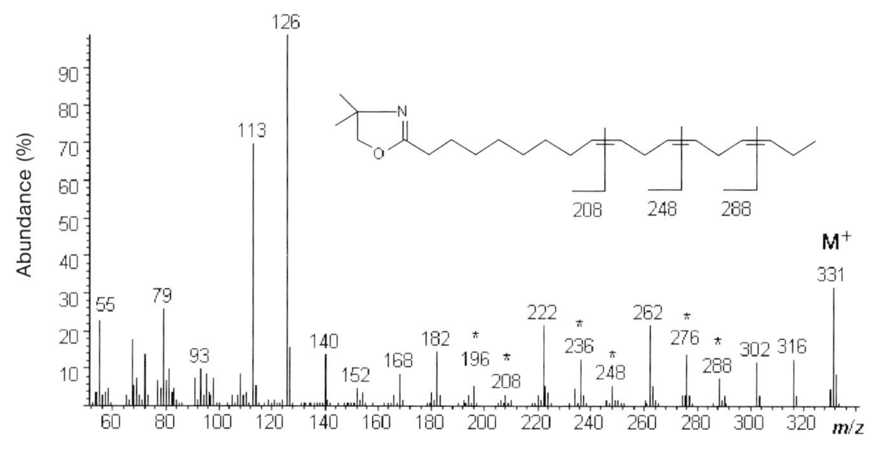

Fig. 3.42 Spectrum of the DMOX derivative of 9,12,15-Octadecatrieenoate (C18:3, n-3). Masses indicated by * are the diagnostic ions with 12 u mass distance for locating double bonds (Christie – The Lipid Library).

in terms of locating double bond positions in the hydrocarbon chain. The total number of carbons and the degree of unsaturation can be taken from the molecular ion information. Typical mass distances of 12 u unveil double bonds corresponding to fragments containing n and n-1 carbons (see the mass peaks marked by * with the pairs m/z 196/208, 236/248, and 276/288 in Fig. 3.42). Monoenes with double bonds between C7 and C15 follow these rules and exhibit intense allylic ions. Especially with an increasing degree of unsaturation and for conjugated double bonds, DMOX spectra are more informative compared to other derivatives. With double bonds closer to the carboxyl end, the spectra show at C4, C5 and C6 characteristic odd-numbered ions at m/z 139, 153 and 167 respectively. At C3 the base peak m/z 152 and at C2 m/z 110 are prominent ions.

3.2.6
Mass Spectroscopic Features of Selected Substance Classes

3.2.6.1 Volatile Halogenated Hydrocarbons

This group of compounds does not belong to a single class of compounds (Figs. 3.43 to 3.52). In a single analysis more than 60 aliphatic and aromatic compounds (Magic 60) can be determined by headspace GC/MS (static/purge and trap). A common feature is the appearance of chlorine and bromine isotope patterns in the mass spectra. With aliphatic compounds molecular ions do not always appear. With increasing molecular size the M^+ intensities decrease. Usually the loss of Cl (and also Br, F) as a radical from the molecular ion occurs. Fluorine can be recognised as HF from the difference of 20 u or as the CF fragment m/z 31, and bromine from signals with significantly higher masses but relatively short retention times in the GC. For detection it is a good recommendation to include the masses m/z 35/37 in the scan to guarantee ease of identification during spectroscopic comparison.

Aromatic halogenated hydrocarbons generally show an intense molecular ion. There is successive radical cleavage of chlorine. In the lower mass range the characteristic aromatic fragments appear with lower intensity.

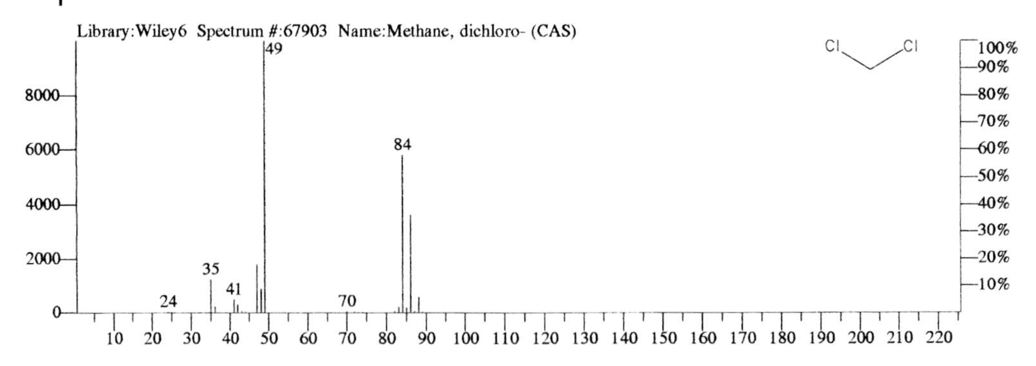

Fig. 3.43 Dichloromethane (R30) CH_2Cl_2, M: 84, CAS Reg. No.: 75-09-02

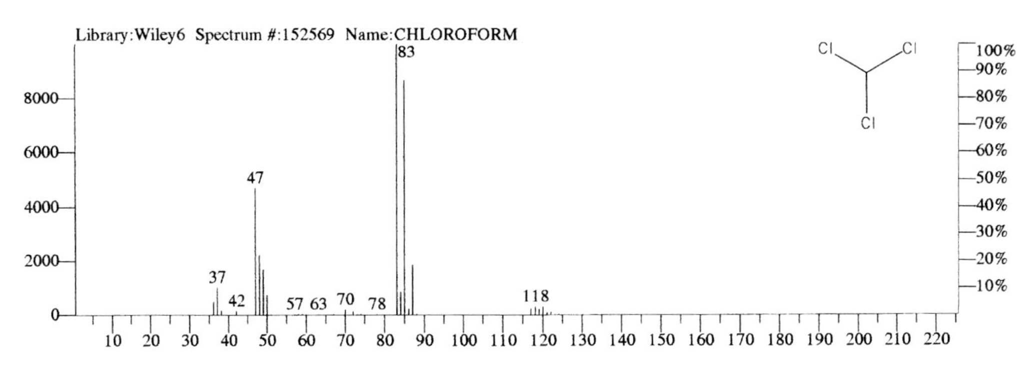

Fig. 3.44 Chloroform $CHCl_3$, M: 118, CAS Reg. No.: 67-66-3

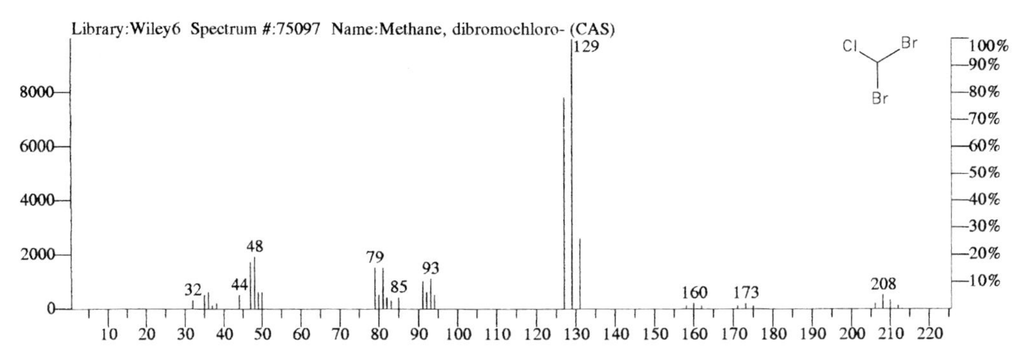

Fig. 3.45 Dibromochloromethane $CHBr_2Cl$, M: 206, CAS Reg. No.: 124-48-1

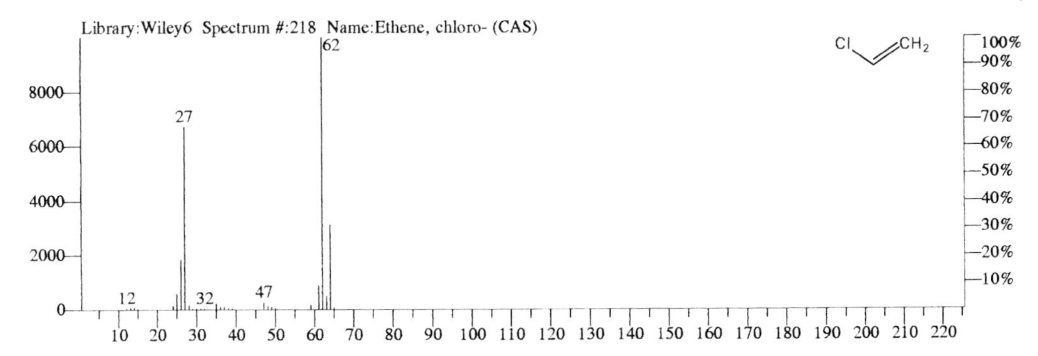

Fig. 3.46 Vinyl chloride C_2H_3Cl, M: 62, CAS Reg. No.: 75-01-4

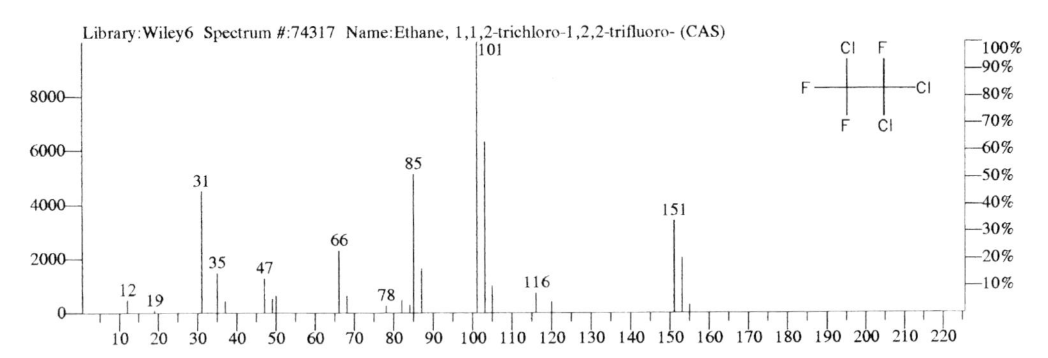

Fig. 3.47 1,1,2-Trifluoro-1,2,2-trichloroethane (R113) $C_2Cl_3F_3$, M: 186, CAS Reg. No.: 76-13-1

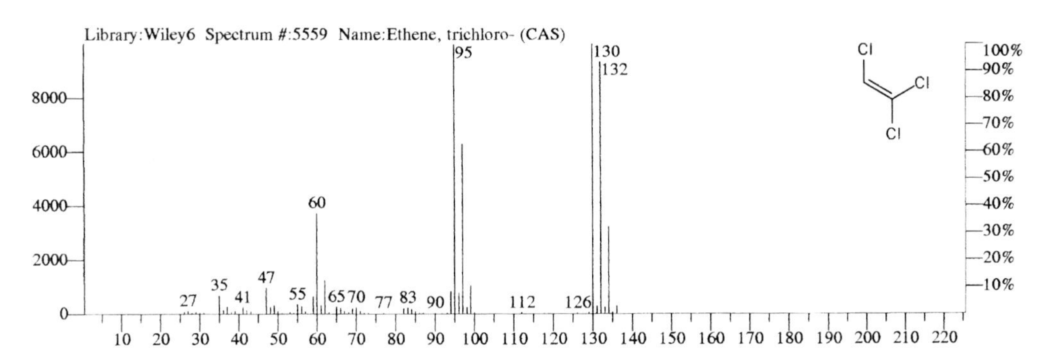

Fig. 3.48 Trichloroethylene C_2HCl_3, M: 130, CAS Reg. No.: 79-01-6

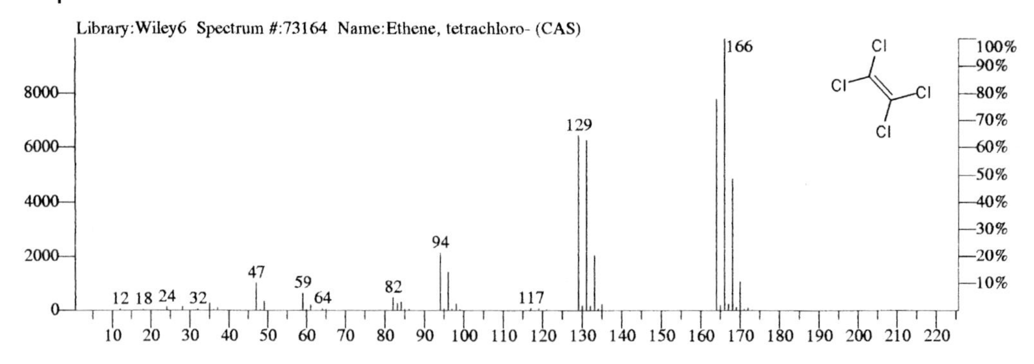

Fig. 3.49 Tetrachloroethylene (Per) C_2Cl_4, M: 164, CAS Reg. No.: 127-18-4

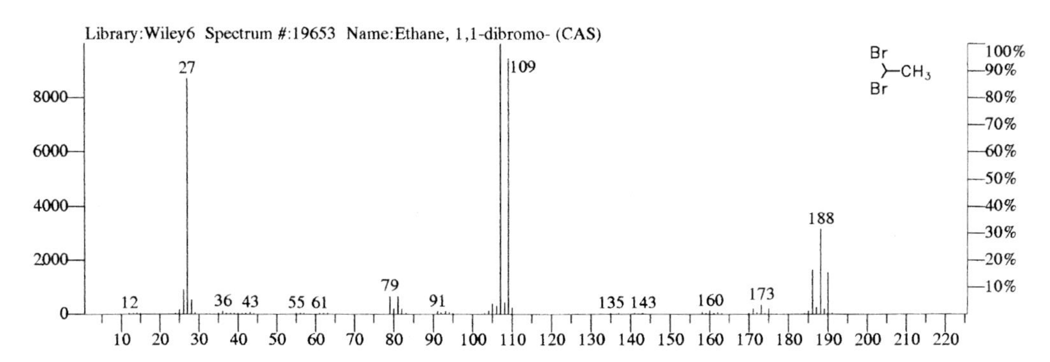

Fig. 3.50 1,1-Dibromoethane $C_2H_4Br_2$, M: 186, CAS Reg. No.: 557-91-5

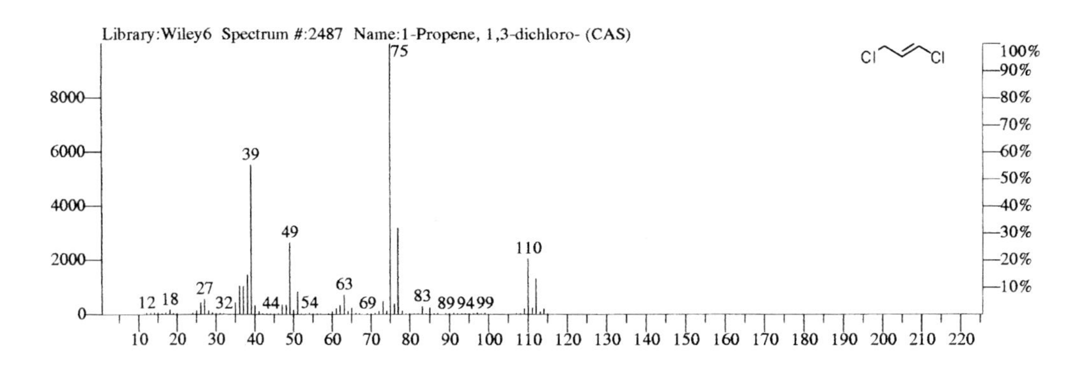

Fig. 3.51 Dichloropropene $C_3H_4Cl_2$, M: 110, CAS Reg. No.: 542-75-6

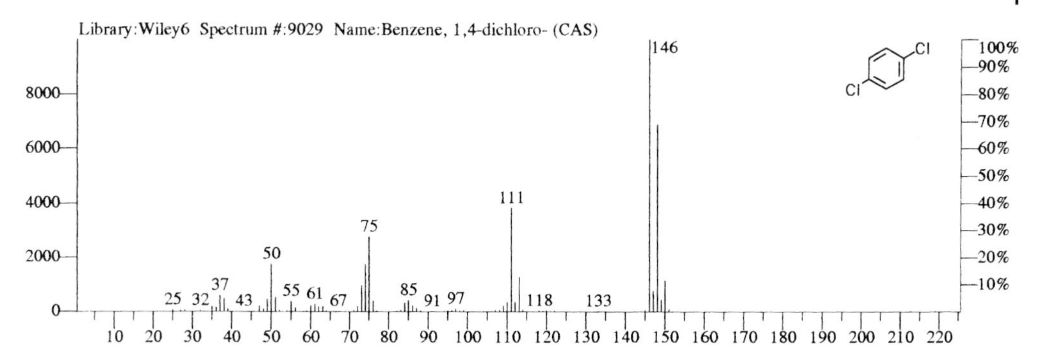

Fig. 3.52 p-Dichlorobenzene $C_6H_4Cl_2$, M: 146, CAS Reg. No.: 106-46-7

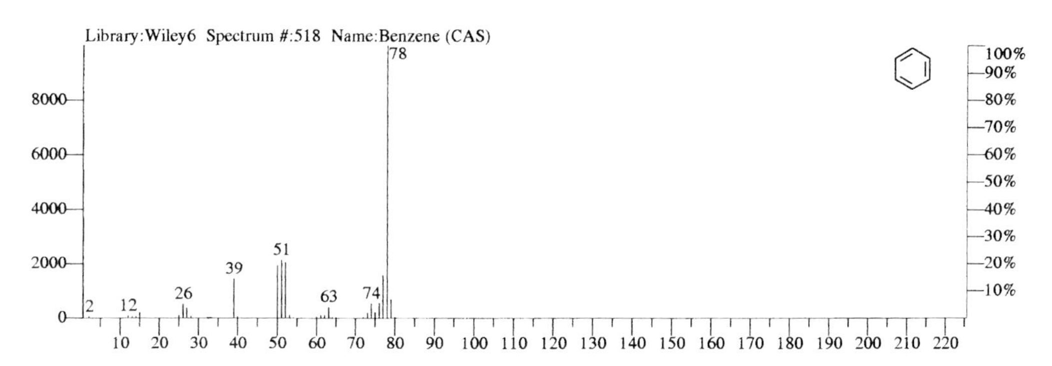

Fig. 3.53 Benzene C_6H_6, M: 78, CAS Reg. No.: 71-43-2

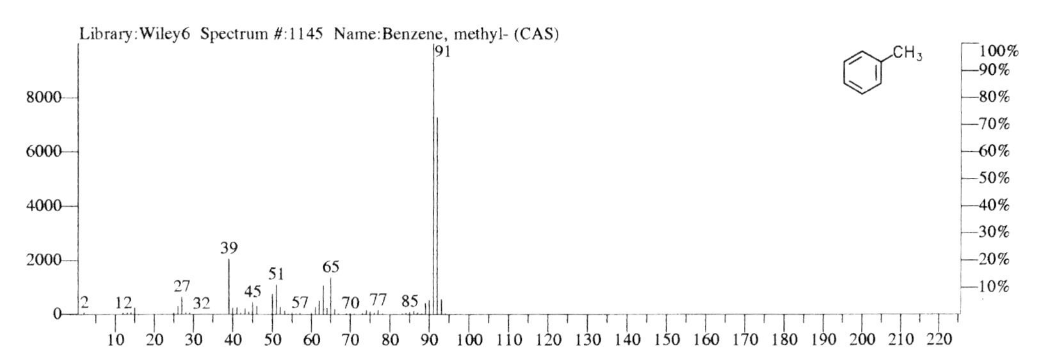

Fig. 3.54 Toluene C_7H_8, M: 92, CAS Reg. No.: 108-88-3

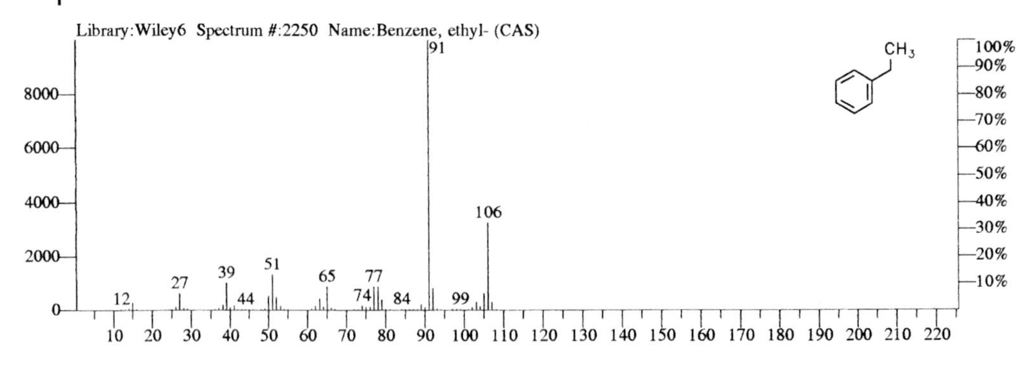

Fig. 3.55 Ethylbenzene C_8H_{10}, M: 106, CAS Reg. No.: 100-41-4

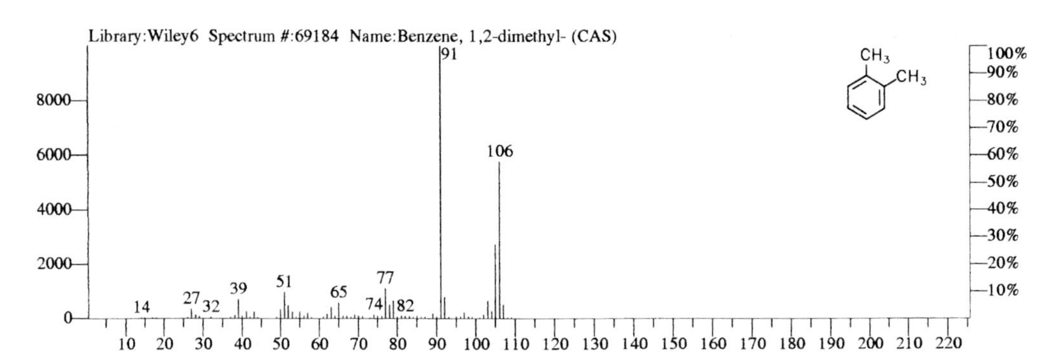

Fig. 3.56 o-Xylene C_8H_{10}, M: 106, CAS Reg. No.: 95-47-6

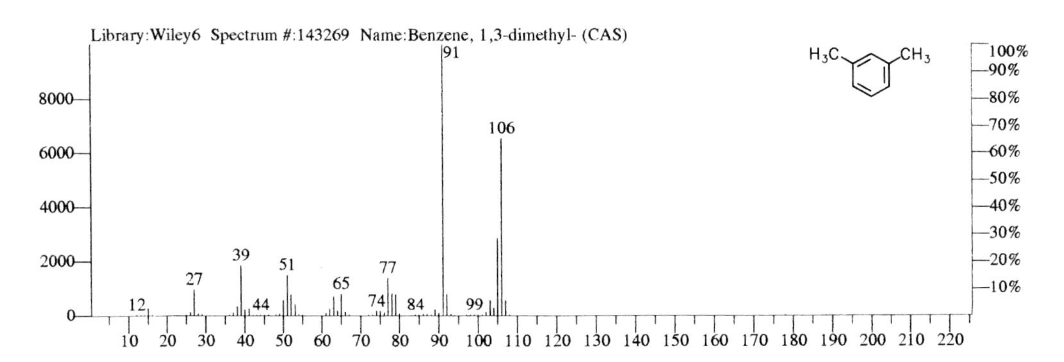

Fig. 3.57 m-Xylene C_8H_{10}, M: 106, CAS Reg. No.: 108-38-3

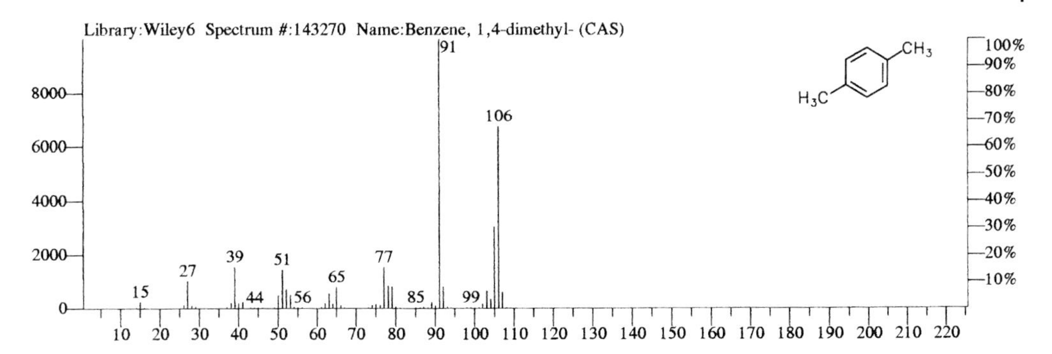

Fig. 3.58 p-Xylene C_8H_{10}, M: 106, CAS Reg. No.: 106-42-3

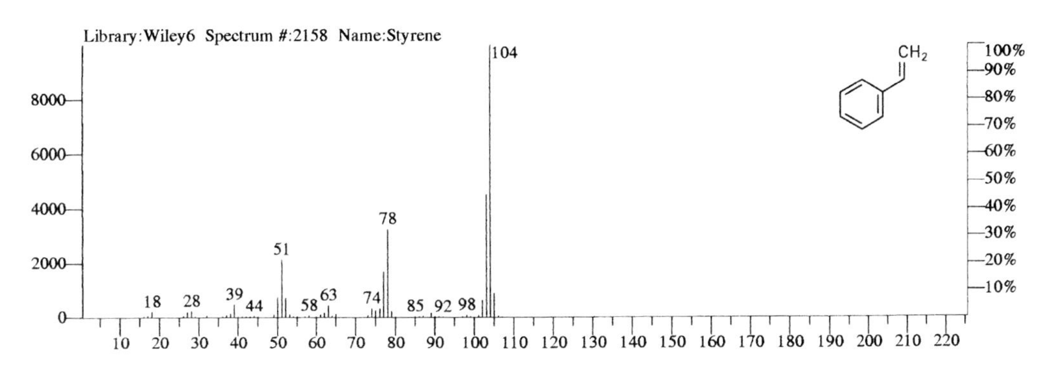

Fig. 3.59 Styrene C_8H_8, M: 104, CAS Reg. No.: 100-42-5

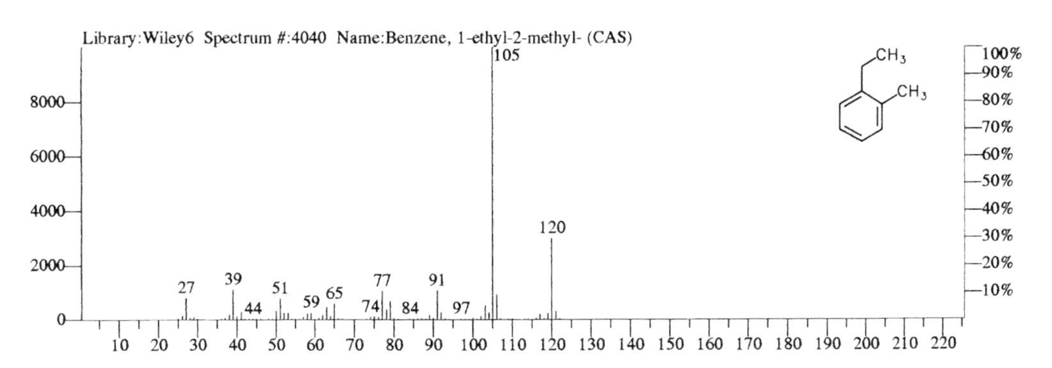

Fig. 3.60 1-Ethyl-2-methylbenzene C_9H_{12}, M: 120, CAS Reg. No.: 611-14-3

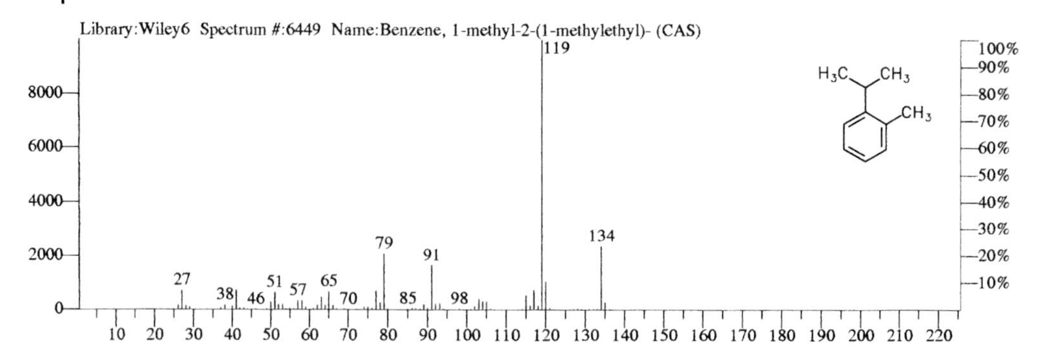

Fig. 3.61 1-Methyl-2-isopropylbenzene $C_{10}H_{14}$, M: 134, CAS Reg. No.: 1074-17-5

3.2.6.2 Benzene/Toluene/Ethylbenzene/Xylenes (BTEX, Alkylaromatics)

Alkylaromatics form very stable molecular ions which can be detected with very high sensitivity (Figs. 3.53 to 3.61). The tropylium ion occurs at m/z 91 as the base peak, which is, for example, responsible for the uneven base peak in the toluene spectrum (M 92). The fragmentation of the aromatic skeleton leads to typical series of ions with m/z 38–40, 50–52, 63-67, 77–79 ('aromatic rubble'). Ethylbenzene and the xylenes cannot be differentiated from their spectra because they are isomers. In these cases the retention times of the components are more meaningful.

With alkyl side chains the problem of isomerism must be taken into account. For dimethylnaphthalene, for example, there are 10 isomers! Alkylaromatics fragment through benzyl cleavage. From a propyl side chain onwards benzyl cleavage can take place with H-transfer to the aromatic ring (to $C_8H_{10}^+$, m/z 106) or without transfer (to $C_8H_9^+$, m/z 105) depending on the steric or electronic conditions.

3.2.6.3 Polyaromatic Hydrocarbons (PAH)

Polyaromatic hydrocarbons (PAH) form very stable molecular ions (Figs. 3.62 to 3.70). They can be recognised easily from a 'half mass' signal caused by doubly charged molecular ions as m/2z = 1/2 m/z. (Be aware of heterocycles with odd numbers of nitrogen, the doubly charged ion appears on two masses in CRMS, see Fig. 3.68). Masses in the range m/z 100 to m/z 320 should be scanned for the analysis. In this range all polycondensed aromatics from naphthalene to coronene (including all 16 EPA components) can be determined and a possible matrix background of hydrocarbons with aliphatic character can be almost completely excluded from detection.

3.2.6.4 Phenols

In mass spectrometry phenolic substances are determined by their aromatic character (Figs. 3.71 to 3.76). Depending on the side chains, intense molecular ions and less intense fragments appear. In GC/MS phenols are usually chromatographed as their methyl esters or acetates. Phenols especially halogenated phenols are acidic. In trace analysis chlorinated and brominated phenols are the most important and can be recognised by their clear isotope patterns. The loss of CO (M-28) gives a less intense signal but is a clear indication of the presence of phenols. Halogenated phenols clearly show the loss of HCl (M-36) and HBr (M-80) in their spectra. With phenols isomers are also best recognised from their retention times rather than their mass spectra.

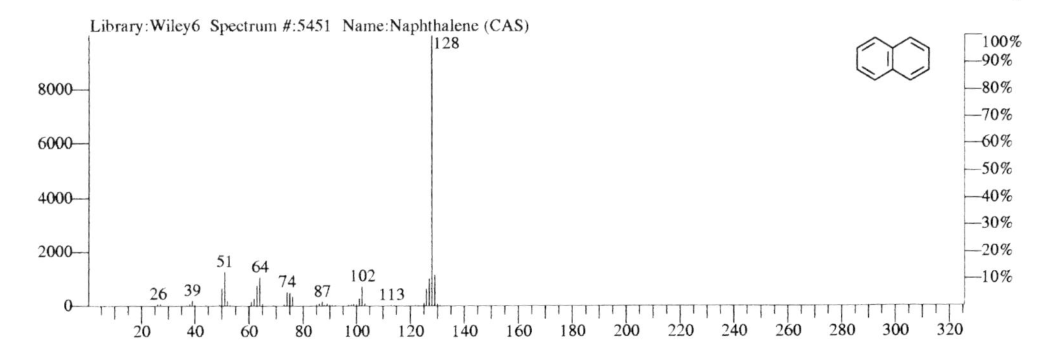

Fig. 3.62 Naphthalene $C_{10}H_8$, M: 128, CAS Reg. No.: 91-20-3

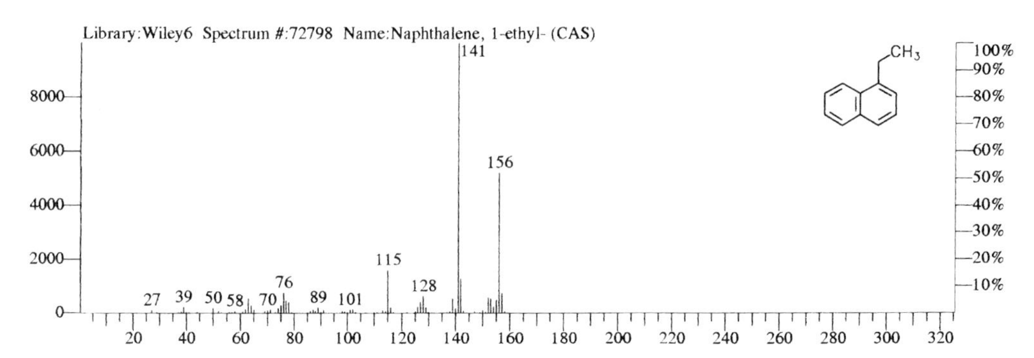

Fig. 3.63 1-Ethylnaphthalene $C_{12}H_{12}$, M: 156, CAS Reg. No.: 1127-76-0

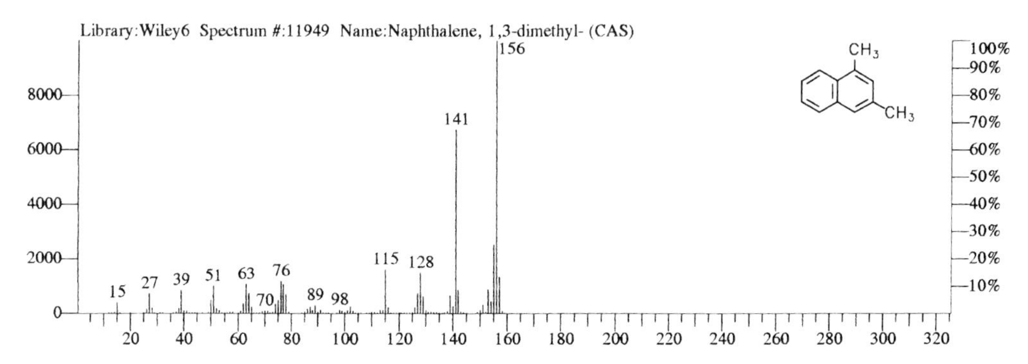

Fig. 3.64 1,3-Dimethylnaphthalene $C_{12}H_{12}$, M: 156, CAS Reg. No.: 575-41-7

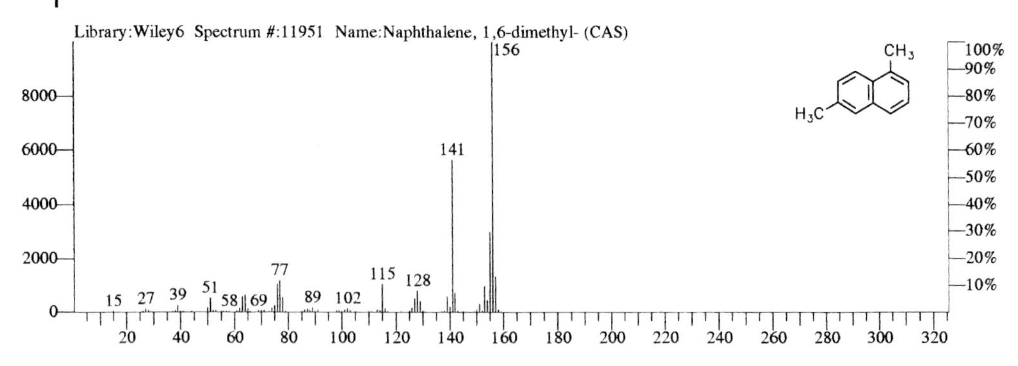

Fig. 3.65 1,6-Dimethylnaphthalene $C_{12}H_{12}$, M: 156, CAS Reg. No.: 575-43-9

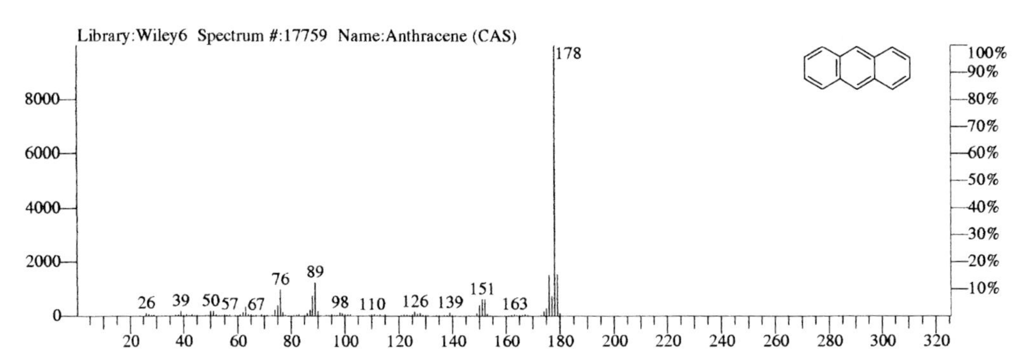

Fig. 3.66 Anthracene $C_{14}H_{10}$, M: 178, CAS Reg. No.: 120-12-7

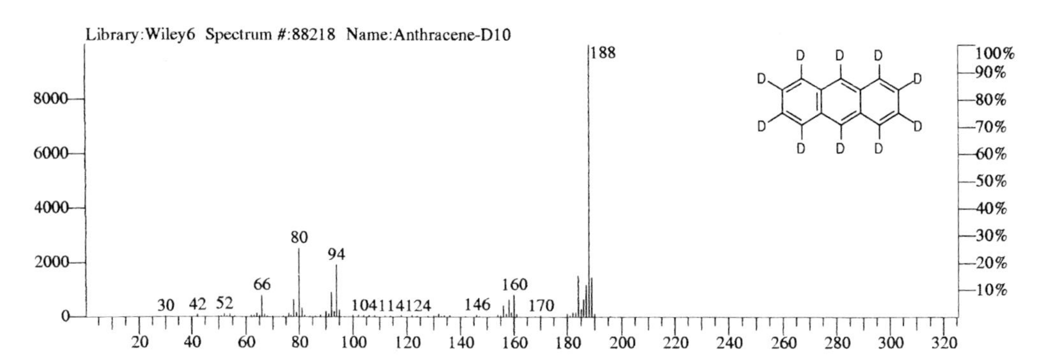

Fig. 3.67 Anthracene-d_{10} $C_{14}D_{10}$, M: 188, CAS Reg. No.: 1719-06-8

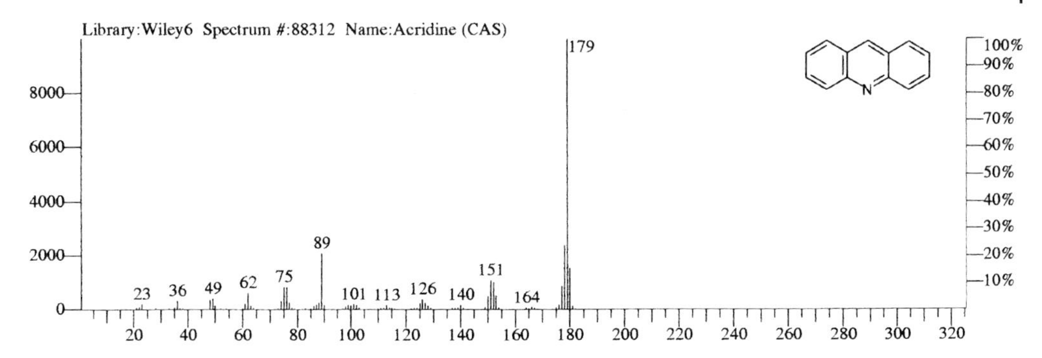

Fig. 3.68 Acridine $C_{13}H_9N$, M: 179, CAS Reg. No.: 260-94-6

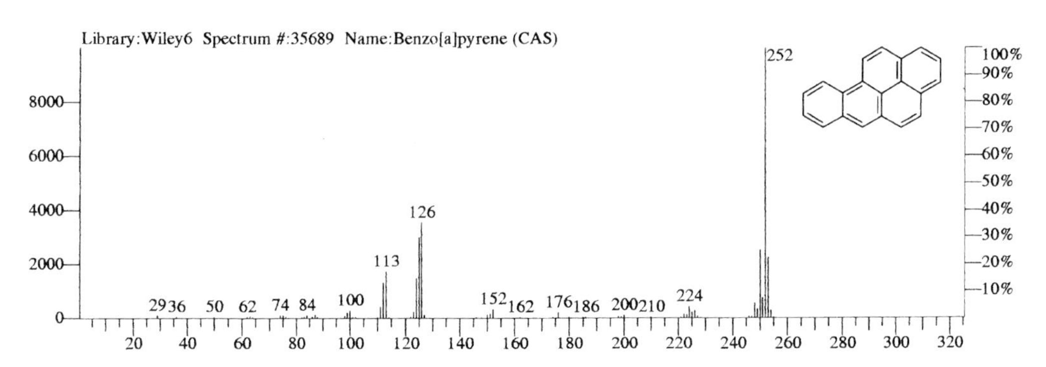

Fig. 3.69 Benzo[a]pyrene $C_{20}H_{12}$, M: 252, CAS Reg. No.: 50-32-8

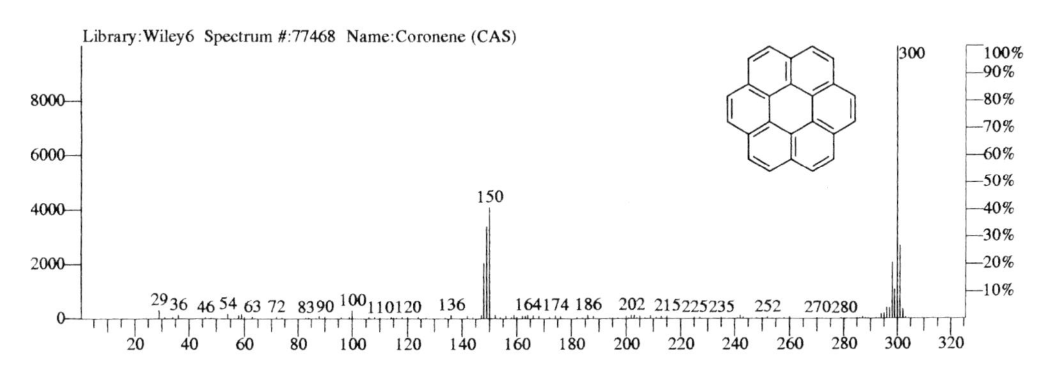

Fig. 3.70 Coronene $C_{24}H_{12}$, M: 300, CAS Reg. No.: 191-07-1

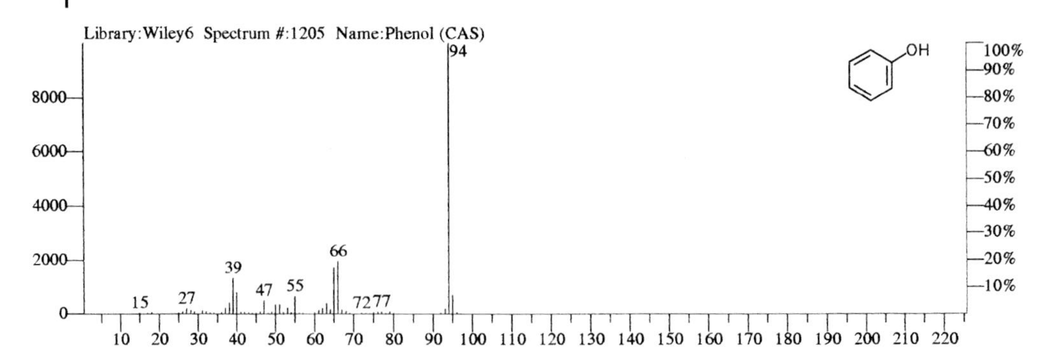

Fig. 3.71 Phenol C_6H_6O, M: 94, CAS Reg. No.: 108-95-2

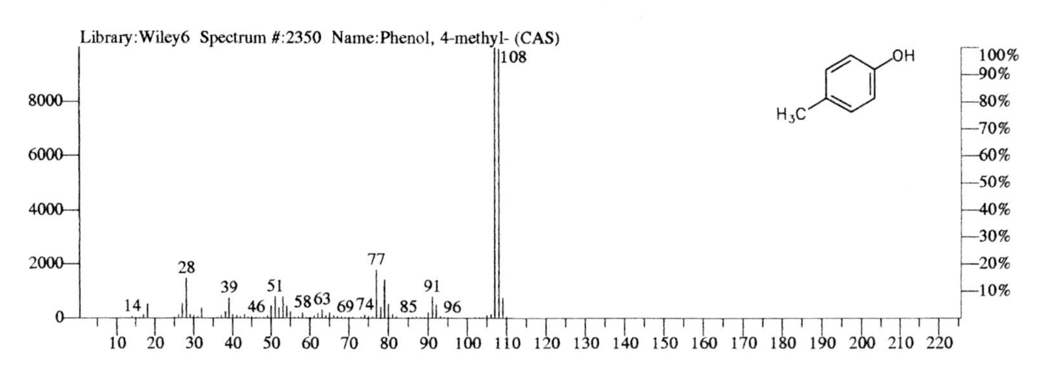

Fig. 3.72 p-Cresol C_7H_8O, M: 108, CAS Reg. No.: 106-44-5

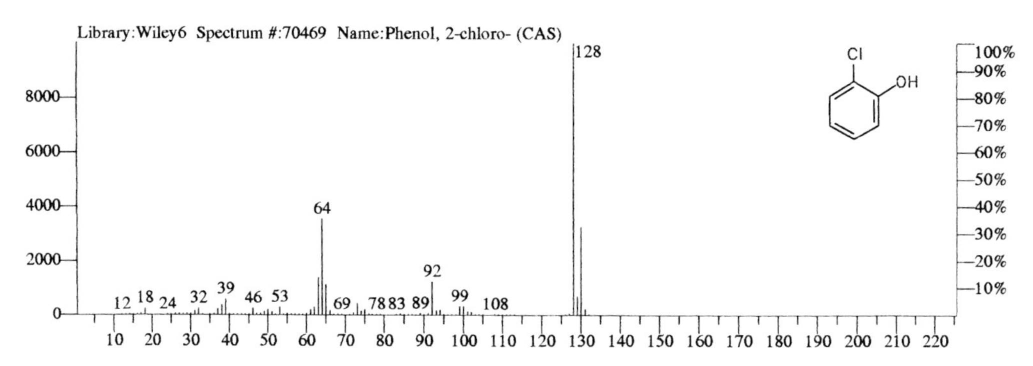

Fig. 3.73 o-Chlorophenol C_6H_5ClO, M: 128, CAS Reg. No.: 95-57-8

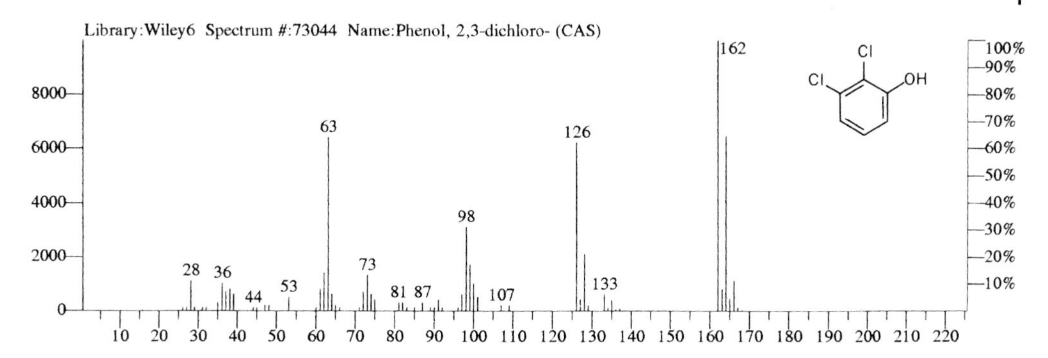

Fig. 3.74 2,3-Dichlorophenol $C_6H_4Cl_2O$, M: 162, CAS Reg. No.: 576-24-9

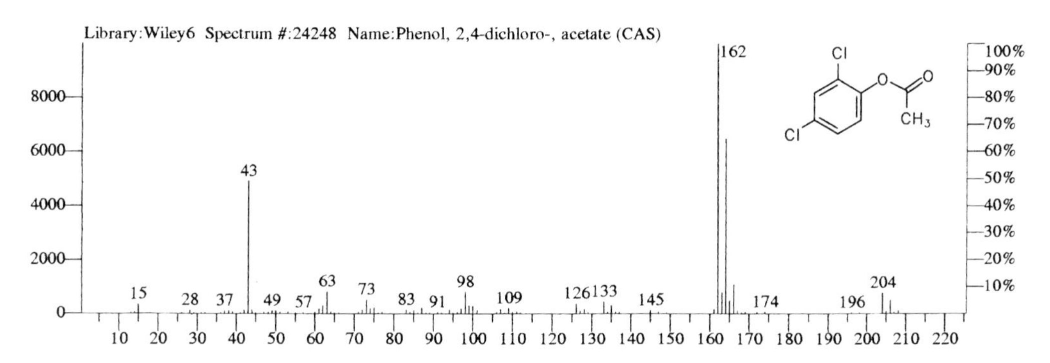

Fig. 3.75 2,4-Dichlorophenyl acetate $C_8H_6Cl_2O_2$, M: 204, CAS Reg. No.: 6341-97-5

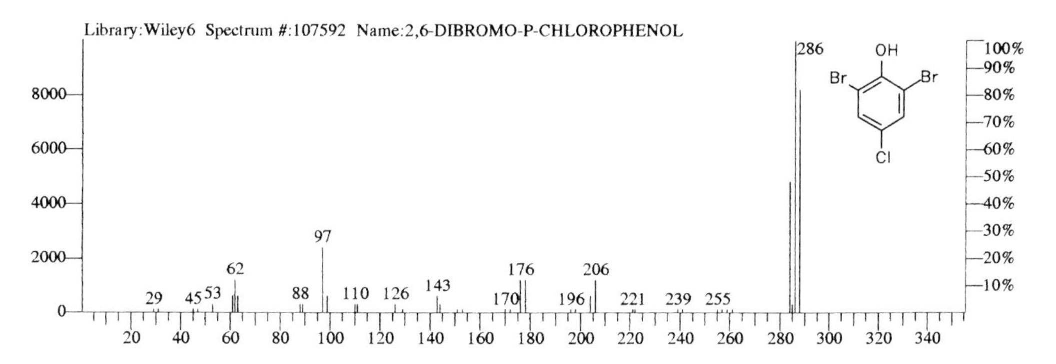

Fig. 3.76 2,6-Dibromo-4-chlorophenol $C_6H_3Br_2ClO$, M: 284

3.2.6.5 **Pesticides**

The character of the active substance forms the basis of the classification of these compounds (Figs. 3.77 to 3.108). In a collection of pesticide spectra there is therefore a wide variety of compound classes, which are covered to some extent by the other substance classes described here. Even when considering what appears to be a single group, such as phosphoric acid esters, it is virtually impossible to make any generalisations on fragmentation. Usually only stable compounds with aromatic character form intense molecular ions. In other cases molecular ions are of lower intensity and in trace analyses cannot be isolated from the matrix. For many plant protection agents (phosphoric acid esters, triazines, phenylureas etc), the use of chemical ionisation is advantageous for confirming identities or allowing selective detection.

Chlorinated Hydrocarbons

In this group of organochlorine pesticides there is a large number of different types of compound (Figs. 3.77 to 3.86). In the example of Lindane and HCB the difference between compounds with saturated and aromatic character can clearly be seen (molecular ion, fragmentation).

The high proportion of chlorine in these analytes leads to intense and characteristic isotope patterns. With the nonaromatic compounds (polycyclic polychlorinated alkanes made by Diels-Alder reactions, e.g. Dieldrin, Aldrin) spectra with large numbers of lines are formed through extensive fragmentation of the molecule. These compounds can be readily analysed using negative chemical ionisation, whereby the fragmentation is prevented.

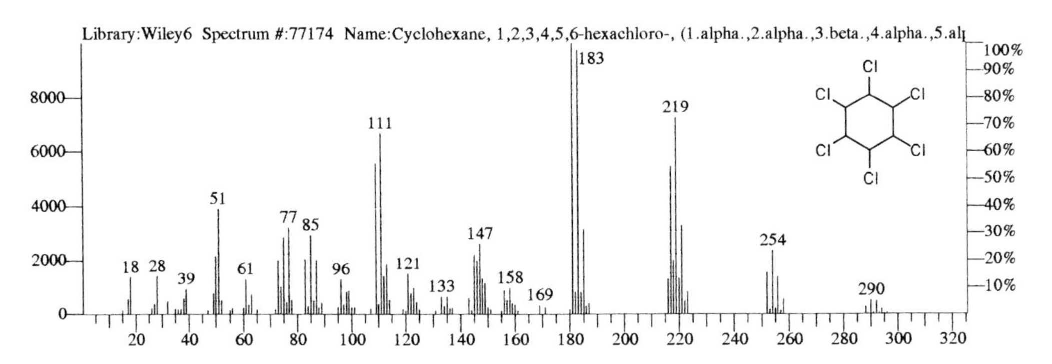

Fig. 3.77 Lindane $C_6H_6Cl_6$, M: 288, CAS Reg. No.: 58-89-9

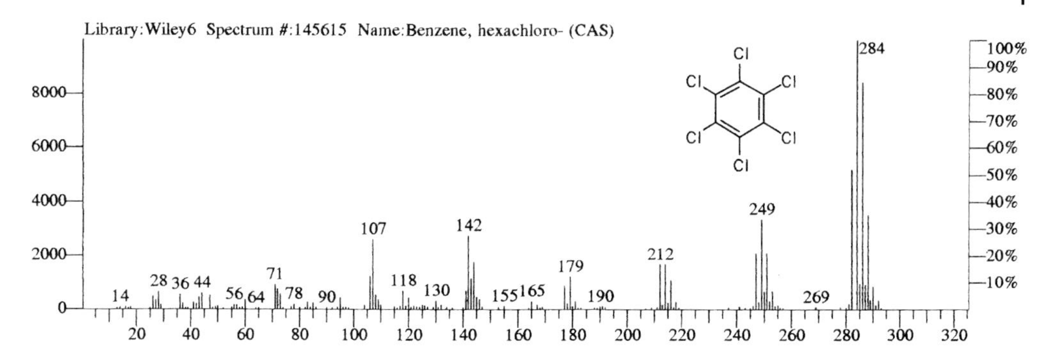

Fig. 3.78 Hexachlorobenzene (HCB) C_6Cl_6, M: 282, CAS Reg. No.: 118-74-1

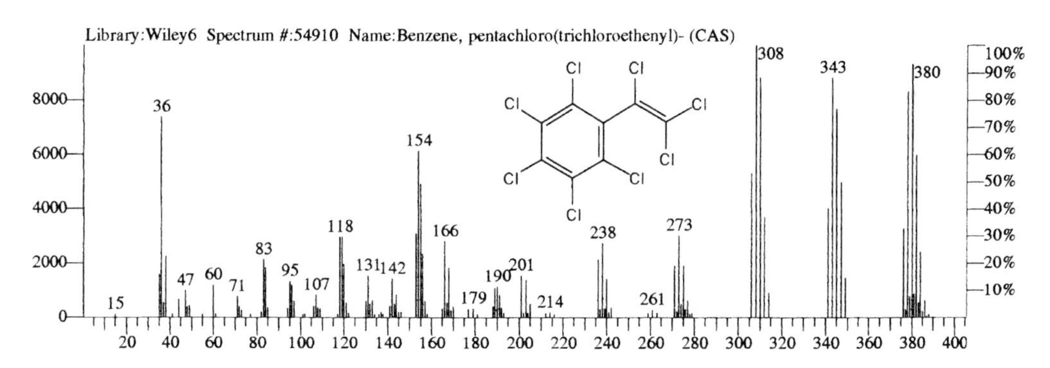

Fig. 3.79 Octachlorostyrene C_8Cl_8, M: 376, CAS Reg. No.: 29082-74-4

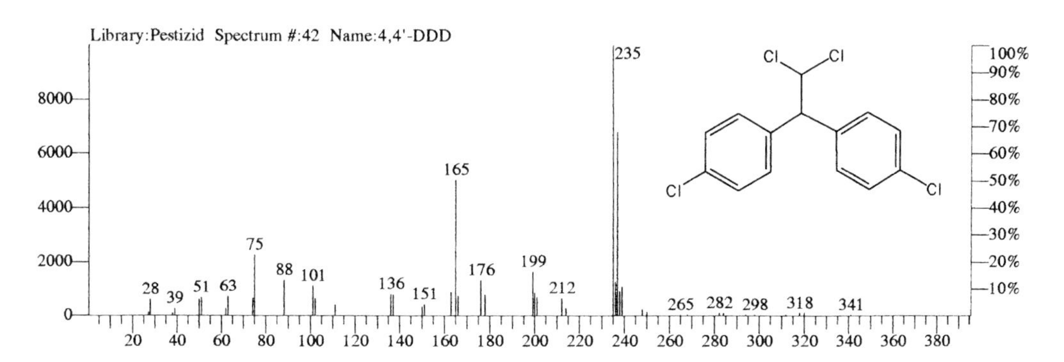

Fig. 3.80 4,4′-DDD $C_{14}H_{10}Cl_4$, M: 318, CAS Reg. No.: 72-54-8

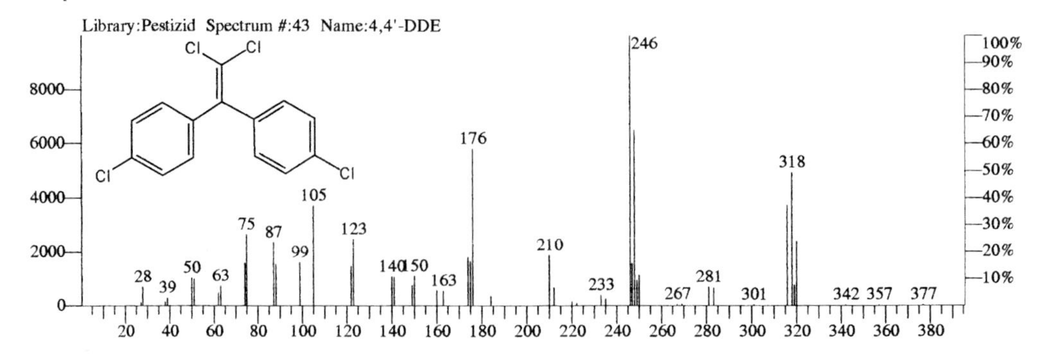

Fig. 3.81 4,4′-DDE $C_{14}H_8Cl_4$, M: 316, CAS Reg. No.: 72-55-9

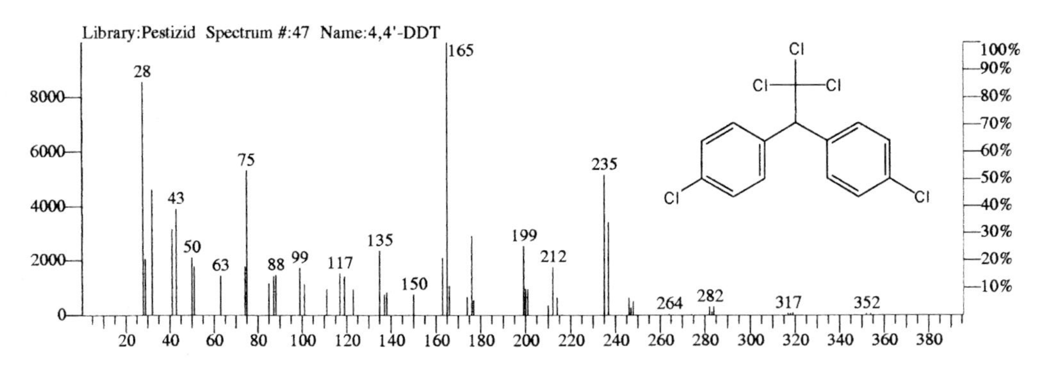

Fig. 3.82 4,4′-DDT $C_{14}H_9Cl_5$, M: 352, CAS Reg. No.: 50-29-3

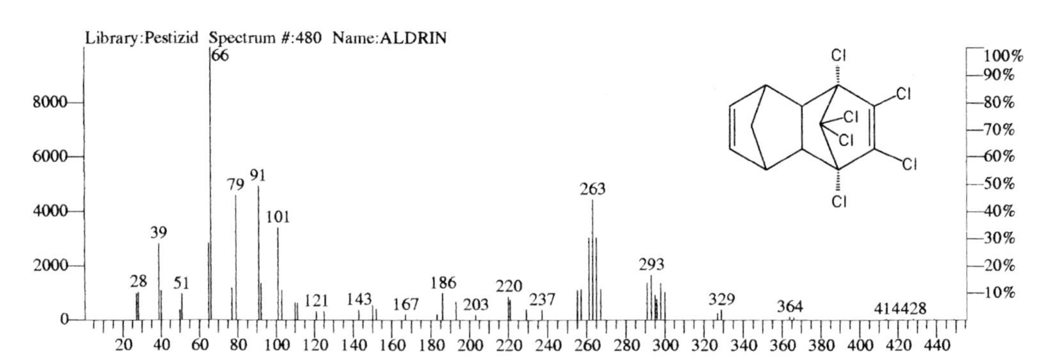

Fig. 3.83 Aldrin $C_{12}H_8Cl_6$, M: 362, CAS Reg. No.: 309-00-2

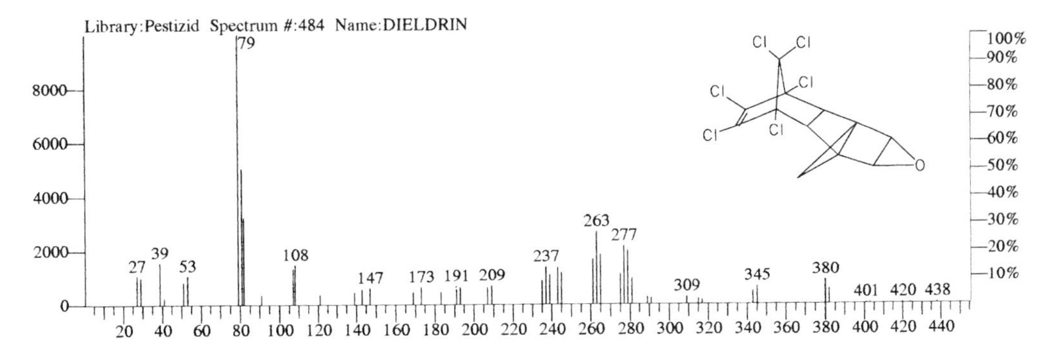

Fig. 3.84 Dieldrin $C_{12}H_8Cl_6O$, M: 378, CAS Reg. No.: 60-57-1

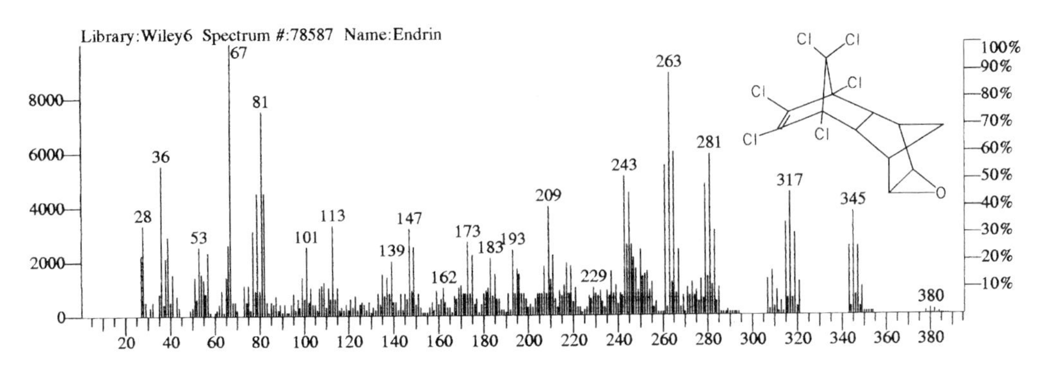

Fig. 3.85 Endrin $C_{12}H_8Cl_6O$, M: 378, CAS Reg. No.: 72-20-8

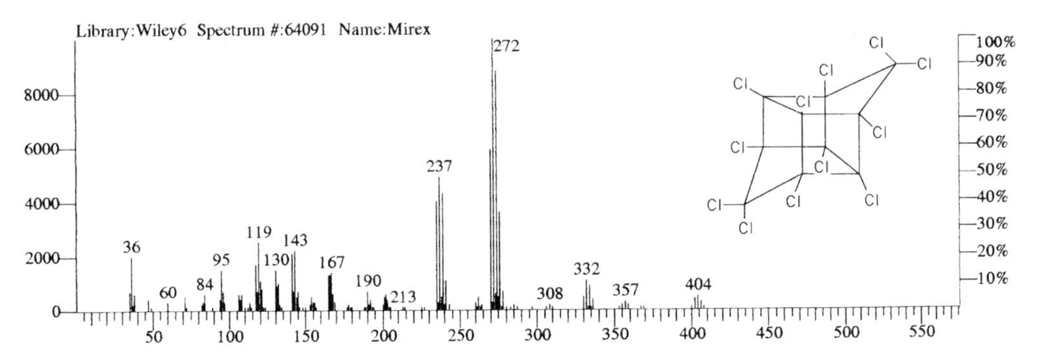

Fig. 3.86 Mirex $C_{10}Cl_{12}$, M: 540, CAS Reg. No.: 2385-85-5

Triazines

Triazine herbicides are substitution products of 1,3,5-triazines and thus belong to a single series of substances (Figs. 3.87 to 3.89). Hexazinon is also determined together with the triazine group in analysis (Fig. 3.90). Without exception the EI spectra of triazines show a large number of fragment ions and usually also contain the molecular ion with varying intensity. The high degree of fragmentation is responsible for the low specific response of triazines in trace analysis. All triazine analyses can be confirmed and quantified readily by positive chemical ionisation (e.g. with NH_3 as the reagent gas).

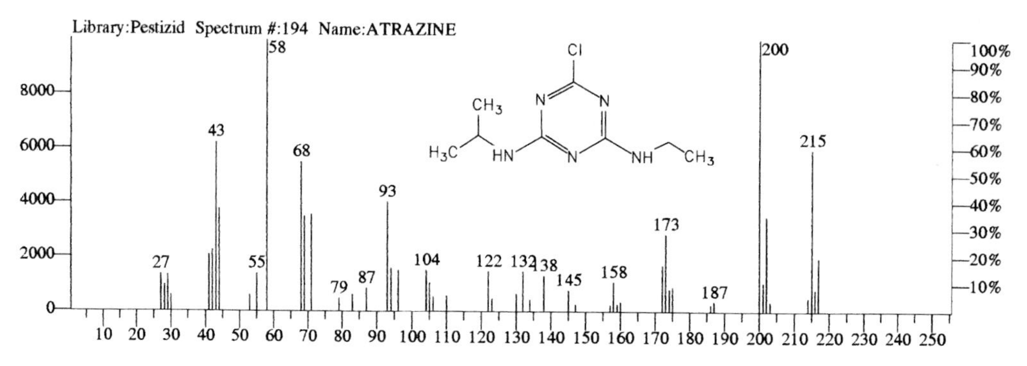

Fig. 3.87 Atrazine $C_8H_{14}ClN_5$, M: 215, CAS Reg. No.: 1912-24-9

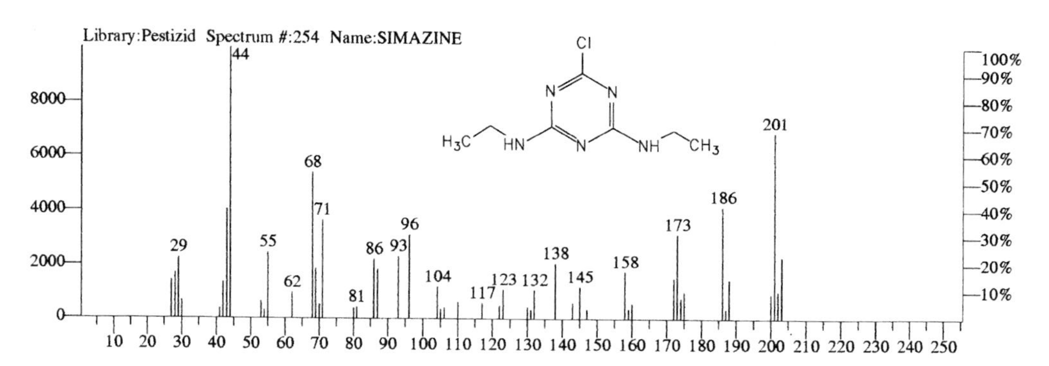

Fig. 3.88 Simazine $C_7H_{12}ClN_5$, M: 210, CAS Reg. No.: 122-34-9

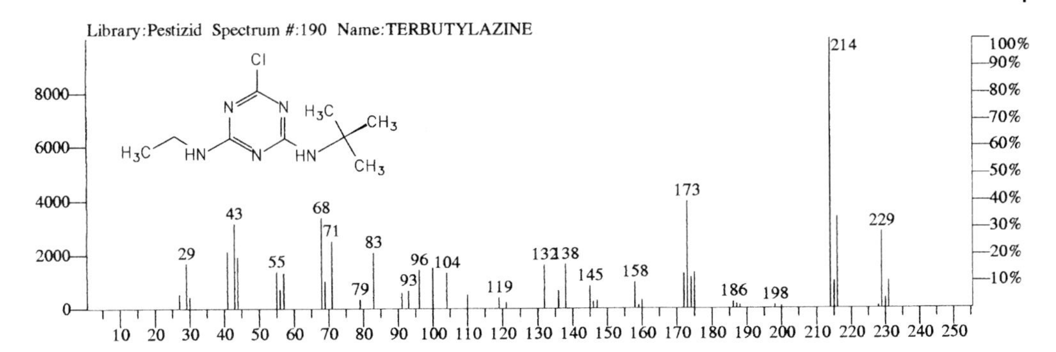

Fig. 3.89 Terbutylazine $C_9H_{16}ClN_5$, M: 229, CAS Reg. No.: 5915-41-3

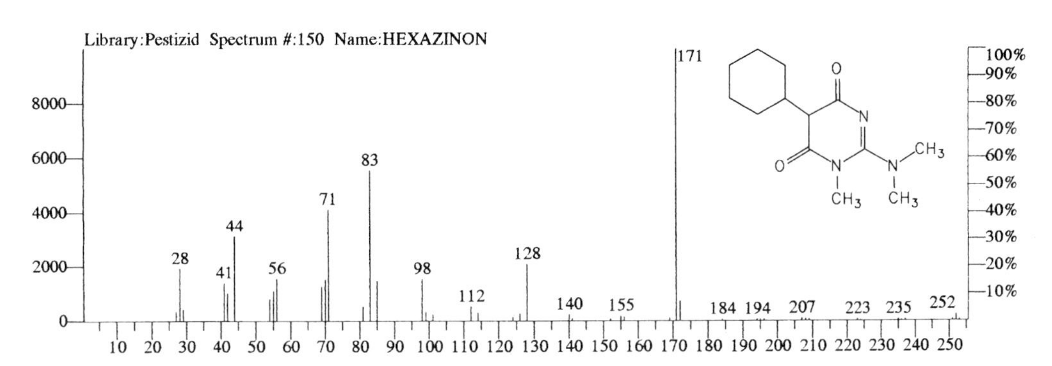

Fig. 3.90 Hexazinon $C_{12}H_{20}N_4O_2$, M: 252, CAS Reg. No.: 51235-04-2

Carbamates

The highly polar carbamate pesticides cannot always be analysed by GC/MS (Figs. 3.91 to 3.93). The low thermal stability leads to decomposition even in the injector. Substances with definite aromatic character, however, form stable intense molecular ions.

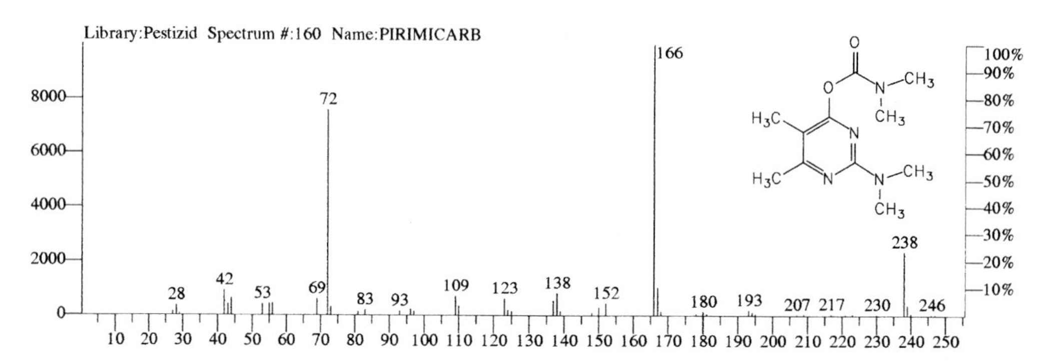

Fig. 3.91 Pirimicarb $C_{11}H_{18}N_4O_2$, M: 238, CAS Reg. No.: 23103-98-2

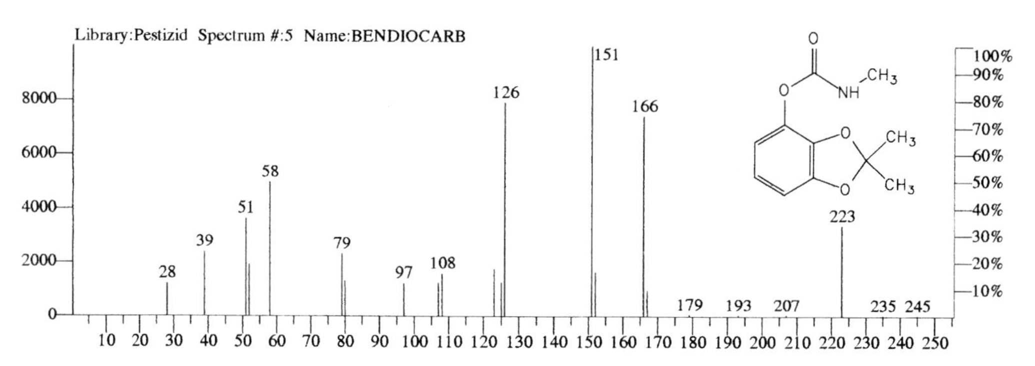

Fig. 3.92 Bendiocarb $C_{11}H_{13}NO_4$, M: 223, CAS Reg. No.: 22781-23-3

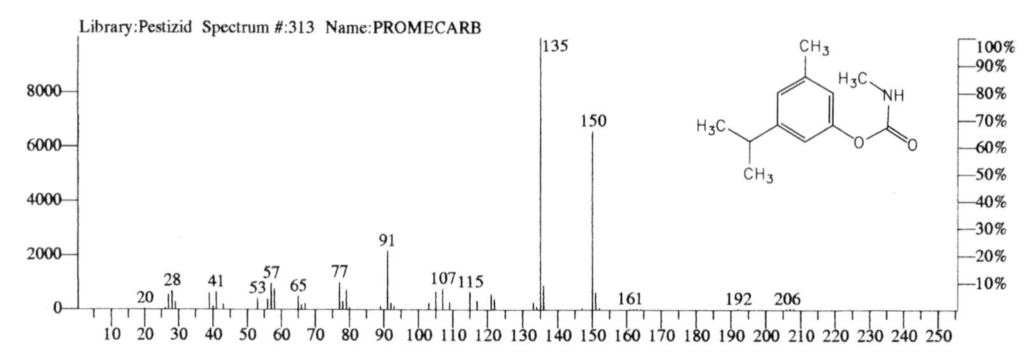

Fig. 3.93 Promecarb $C_{12}H_{17}NO_2$, M: 207, CAS Reg. No.: 2631-37-0

Phosphoric Acid Esters

This large group of pesticides does not exhibit uniform behaviour in mass spectrometry (Figs. 3.94 to 3.99). In trace analysis the detection of molecular ions is usually difficult, except in the case of aromatic compounds (e. g. Parathion). The high degree of fragmentation frequently extends into the area of matrix noise. Because of this it is more difficult to detect individual compounds, as in full scan analysis low starting masses must be used. Positive chemical ionisation is suitable for phosphoric acid esters because generally a strong CI reaction can be expected as a result of the large number of functional groups. The presence of the halogens Cl or Br can only be determined with certainty from the (quasi)molecular ions.

Phosphoric acid esters are also used as highly toxic chemical warfare agents (Tabun, Sarin and Soman, see Section 3.2.6.10).

The occurrence of fragments belonging to a particular group in the spectra of phosphoric acid esters has been intensively investigated (Table 3.11). Phosphoric acid esters are subdivided in the usual way as follows:

I	Dithiophosphoric acid esters	$(RO)_2$-P(S)-S-Z
II	Thionophosphoric acid esters	$(RO)_2$-P(S)-O-Z
III	Thiophosphoric acid esters	$(RO)_2$-P(O)-S-Z
IV	Phosphoric acid esters	$(RO)_2$-P(O)-O-Z

Table 3.11 Fragments typical of various groups from phosphoric acid ester (PAE) pesticides (after Stan)

Group	R	m/z 93	m/z 97	m/z 109	m/z 121	m/z 125
Dithio-PAE	Ia CH_3	+	–	–	–	+
	Ib C_2H_5	+	+	–	+	+
Thiono-PAE	IIa CH_3	+	–	+	–	+
	IIb C_2H_5	+	+	+	+	+
Thiol-PAE	IIIa CH_3	(+)	–	+	–	+
	IIIb C_2H_5	–	+	+	+	–
PAE	IVa CH_3	(+)	–	+	–	–
	IVb C_2H_5	+/–	–	+	–	–

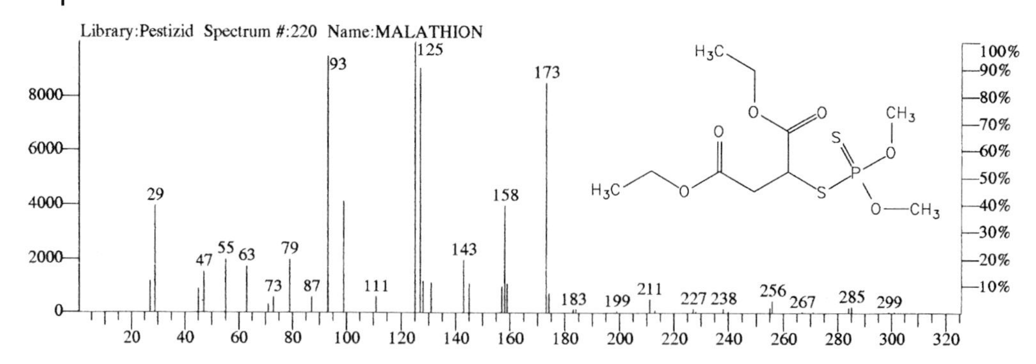

Fig. 3.94 Malathion $C_{10}H_{19}O_6PS_2$, M: 330, CAS Reg. No.: 121-75-5

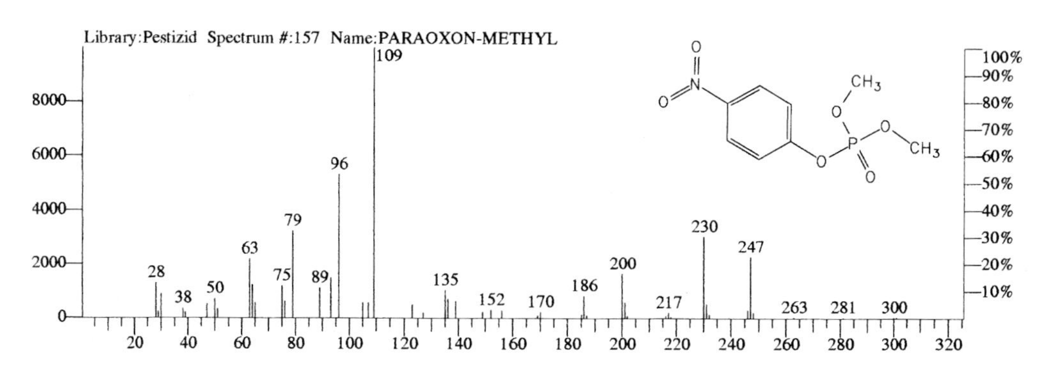

Fig. 3.95 Paraoxon-methyl $C_8H_{10}NO_6P$, M: 247, CAS Reg. No.: 950-35-6

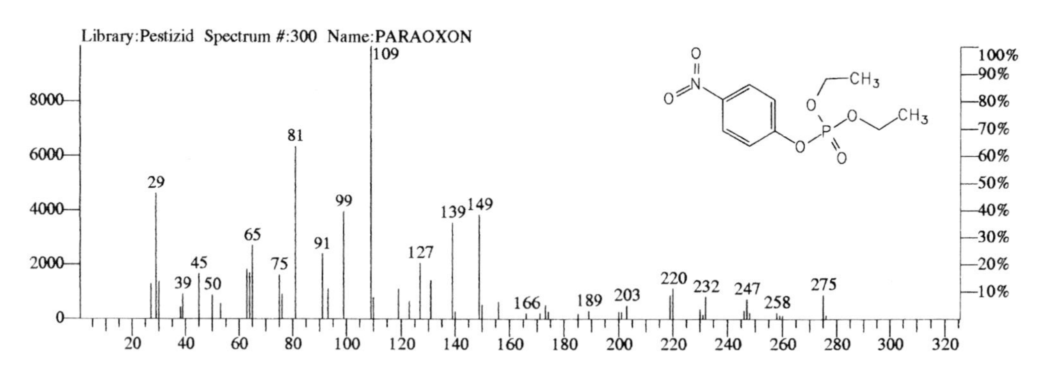

Fig. 3.96 Paraoxon(-ethyl) $C_{10}H_{14}NO_6P$, M: 275, CAS Reg. No.: 311-45-5

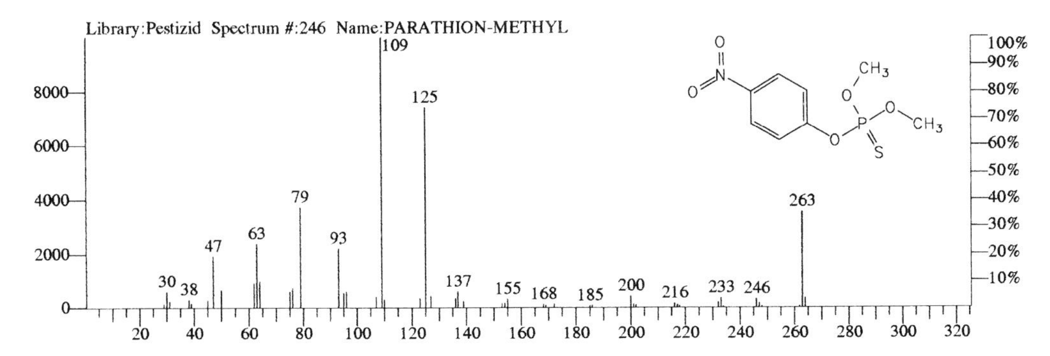

Fig. 3.97 Parathion-methyl $C_8H_{10}NO_5PS$, M: 263, CAS Reg. No.: 298-00-0

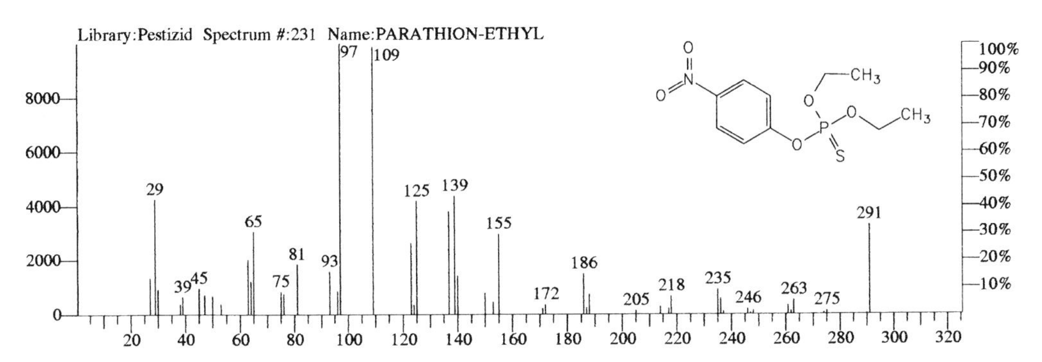

Fig. 3.98 Parathion-ethyl $C_{10}H_{14}NO_5PS$, M: 291, CAS Reg. No.: 56-38-2

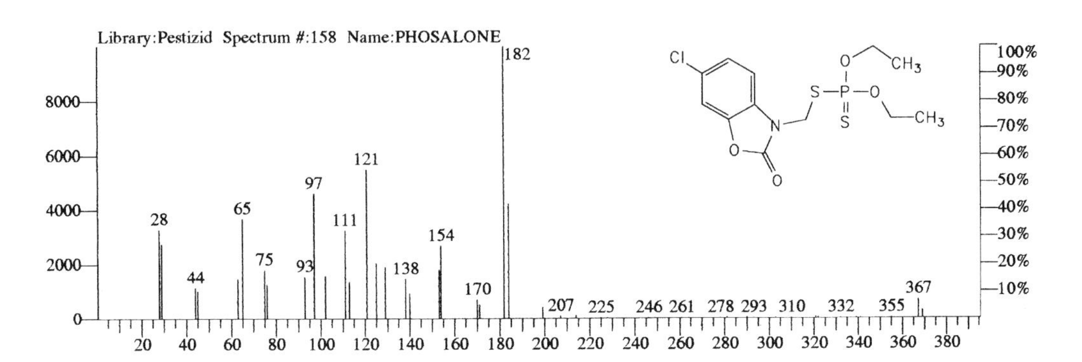

Fig. 3.99 Phosalone $C_{12}H_{15}ClNO_4PS_2$, M: 367, CAS Reg. No.: 2310-17-0

Phenylureas

GC/MS can only be used for the determination of phenylureas after derivatisation of the active substances. HPLC or HPLC/MS are currently the most suitable analytical methods because of the thermal lability and polarity of these compounds. A spectrum with few lines is frequently obtained in GC/MS analyses, which is dominated by the dimethylisocyanate ion m/z 72 which is specific to the group. The molecular ion region is of higher specificity but usually lower intensity (Figs. 3.100 to 3.105).

The phenylureas are rendered more suitable for GC by methylation of the azide hydrogen (e.g. with trimethylsulfonium hydroxide (TMSH) in the PTV injector after Färber). The mass spectra of the methyl derivatives correspond to those of the parent substances, except that the molecular ions are 14 masses higher with the same fragmentation pattern (Figs. 3.101, 3.103 and 3.105).

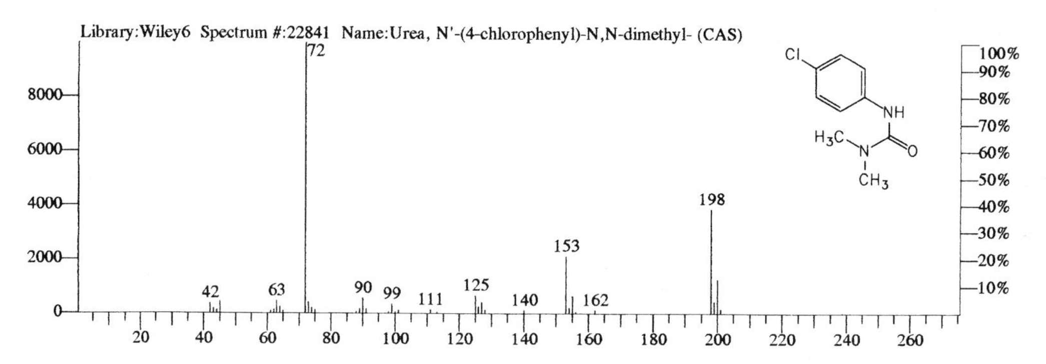

Fig. 3.100 Monuron $C_9H_{11}ClN_2O$, M: 198, CAS Reg. No.: 150-68-5

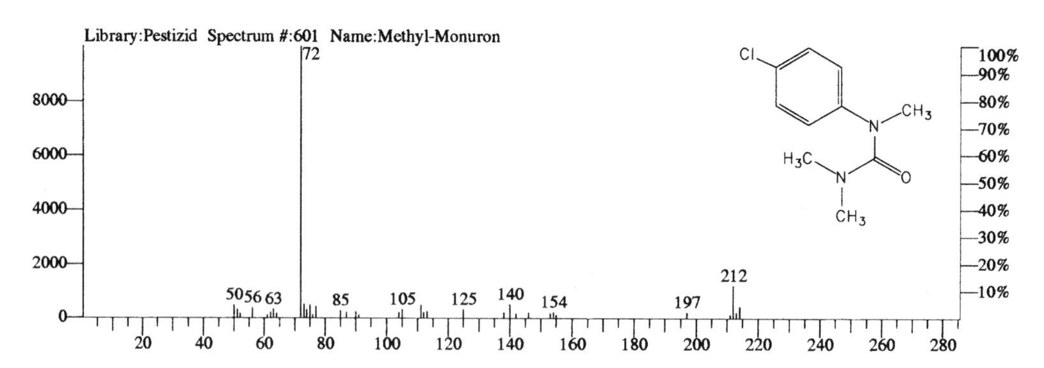

Fig. 3.101 Methyl-Monuron $C_{10}H_{13}ClN_2O$, M: 212

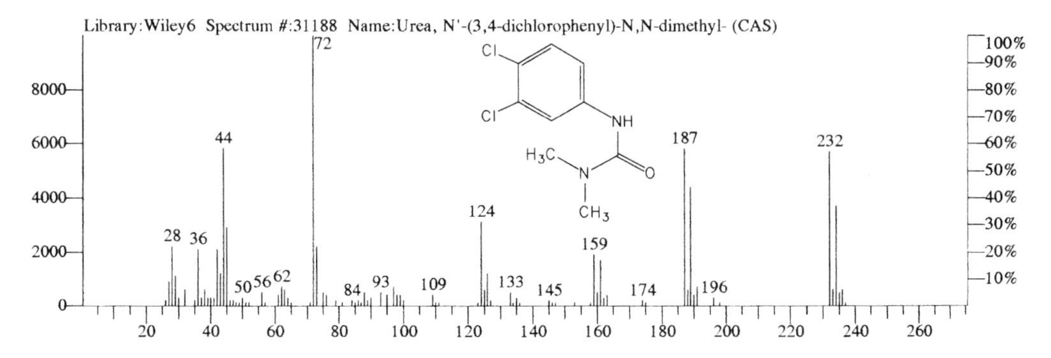

Fig. 3.102 Diuron $C_9H_{10}Cl_2N_2O$, M: 232, CAS Reg. No.330-54-1

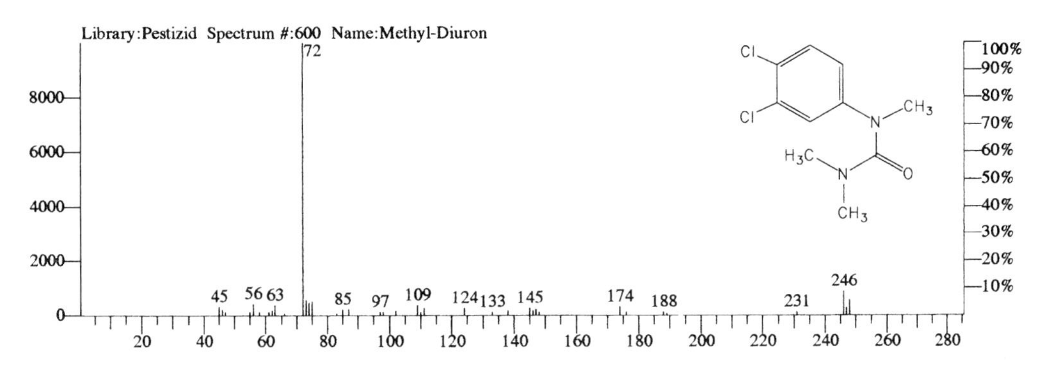

Fig. 3.103 Methyl-Diuron $C_{10}H_{12}Cl_2N_2O$, M: 246

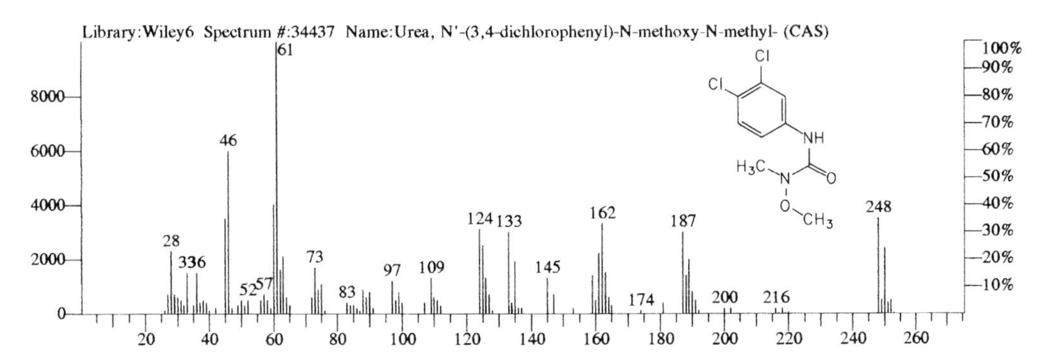

Fig. 3.104 Linuron $C_9H_{10}Cl_2N_2O_2$, M: 248, CAS Reg. No.: 330-55-2

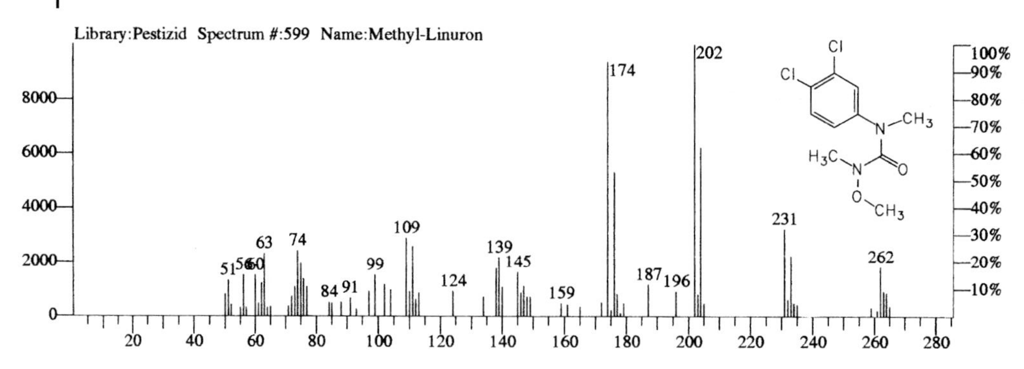

Fig. 3.105 Methyl-Linuron $C_{10}H_{12}Cl_2N_2O_2$, M: 262

Phenoxyalkylcarboxylic Acids

The free acids cannot be used with GC/MS in the case of trace analysis. They are determined using the methyl ester (Figs. 3.106 to 3.108). If the aromatic character predominates, intense molecular ions occur in the upper mass range. Increasing the length of the side chains significantly reduces the intensity of the molecular ion and leads to signals in the lower mass range. It should be noted that the presence of Cl or Br can only be determined with certainty from the (quasi)molecular ion. With EI the molecular ion fragments losing a Cl radical. Because of this the isotope signals of the fragments cannot be evaluated conclusively (see MCPB methyl ester). A final confirmation can be achieved through chemical ionisation.

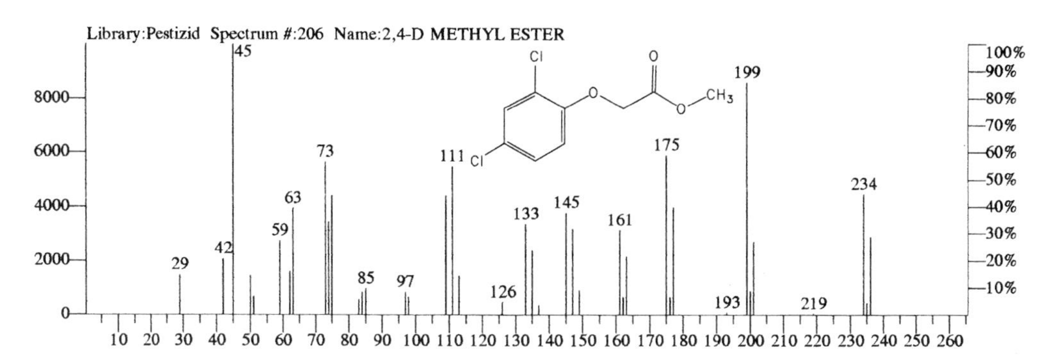

Fig. 3.106 2,4-D Methylester $C_9H_8Cl_2O_3$, M: 234, CAS Reg. No.: 1928-38-7

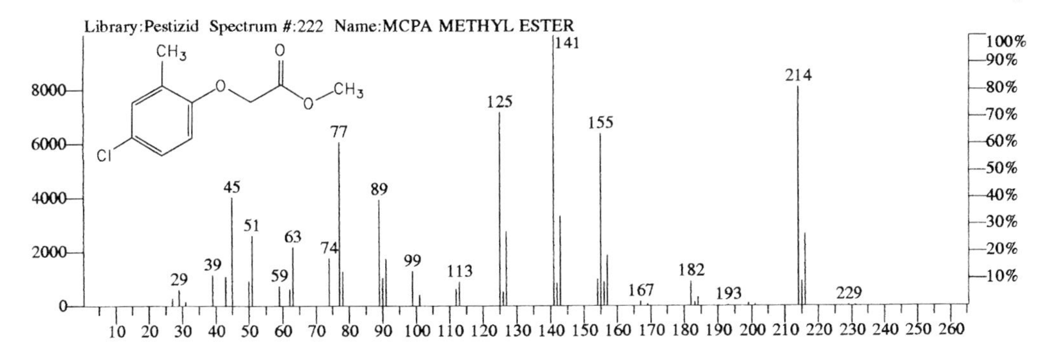

Fig. 3.107 MCPA Methylester $C_{10}H_{11}ClO_3$, M: 214, CAS Reg. No.: 2436-73-9

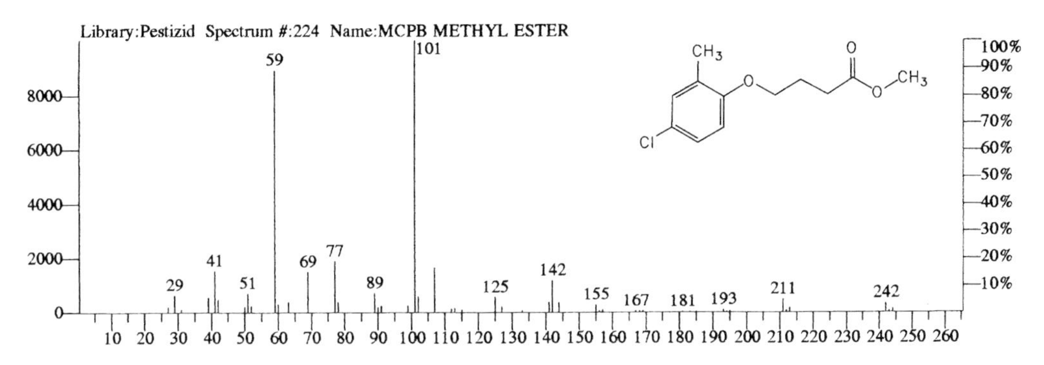

Fig. 3.108 MCPB Methylester $C_{12}H_{15}ClO_3$, M: 242, CAS Reg. No.: 57153-18-1

3.2.6.6 Polychlorinated Biphenyls (PCBs)

The spectra of polychlorinated biphenyls have similar features independent of the degree of chlorination (Figs. 3.109 to 3.118). As they are aromatic, their molecular ions are strongly pronounced. The degree of chlorination can clearly be determined from the isotope pattern. Fragmentation involves successive loss of Cl radicals and, in the lower mass range, degradation of the basic skeleton. For data acquisition the mass range above m/z 100 or 150 is required, so that detection of PCBs is usually possible above an accompanying matrix background. Individual isomers with a particular degree of chlorination have almost identical mass spectra. They can be differentiated on the basis of their retention times and therefore good gas chromatographic separation is a prerequisite for the determination of PCBs. The isomers 31 and 28 (nomenclature after Ballschmitter and Zell) are used as resolution criteria (Table 3.12). The spectra of compounds with different degrees of chlorination are shown in the following figures.

Table 3.12 PCB congeners for quantitation (after Ballschmitter)

PCB No.	Structure	
28	2,4,4′	Cl_3–PCB
52	2,2′,5,5′	Cl_4–PCB
101	2,2′,4,5,5′	Cl_5–PCB
118	2,3′,4,4′,5	Cl_5–PCB
138	2,2′,3,4,4′,5	Cl_6–PCB
153	2,2′,4,4′,5,5′	Cl_6–PCB
180	2,2′,3,4,4′,5,5′	Cl_7–PCB
209	2,2′,3,3′,4,4′,5,5′,6,6′	Cl_{10}–PCB [a]

a) Used as internal standard

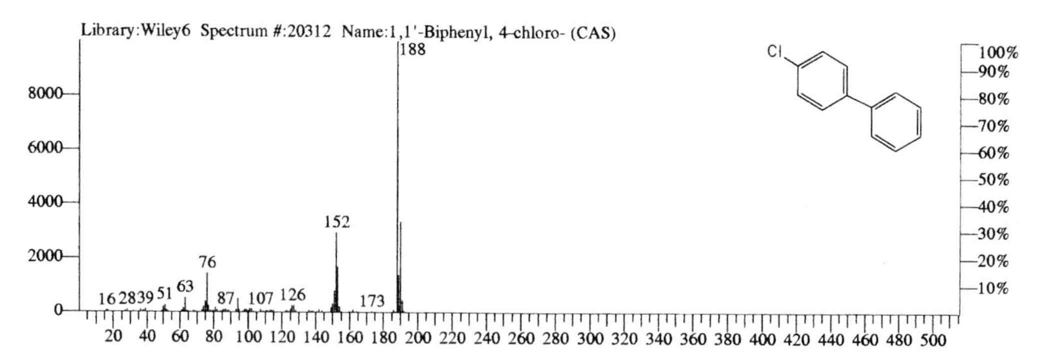

Fig. 3.109 Monochlorobiphenyl $C_{12}H_9Cl$, M: 188

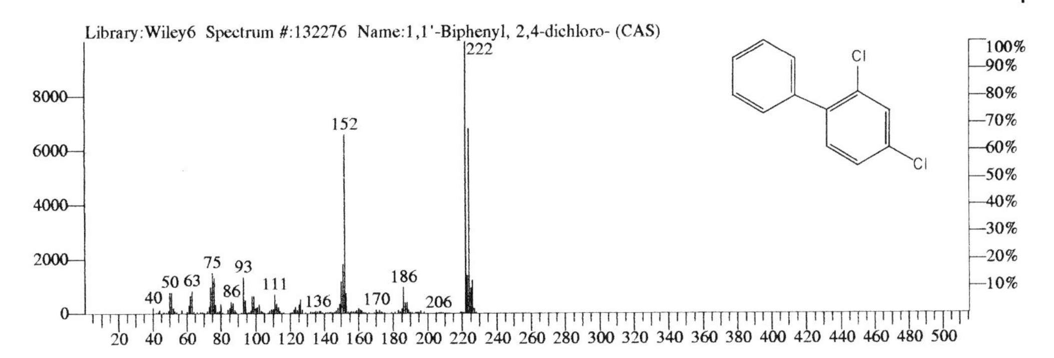

Fig. 3.110 Dichlorobiphenyl $C_{12}H_8Cl_2$, M: 222

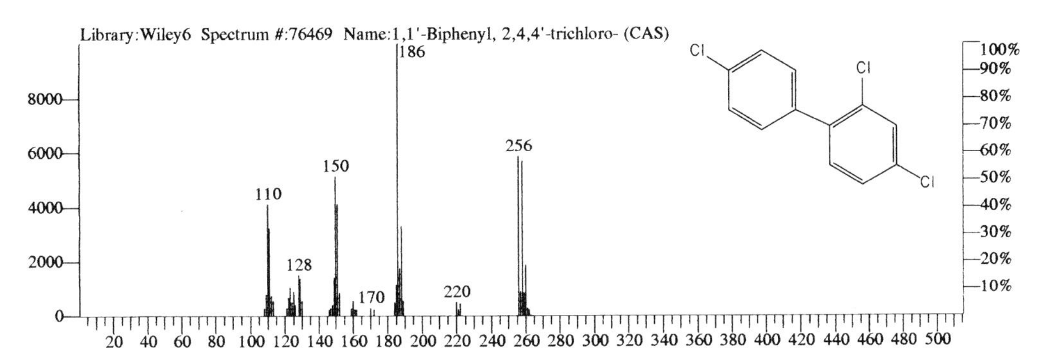

Fig. 3.111 Trichlorobiphenyl (e.g., PCB 28, 31) $C_{12}H_7Cl_3$, M: 256

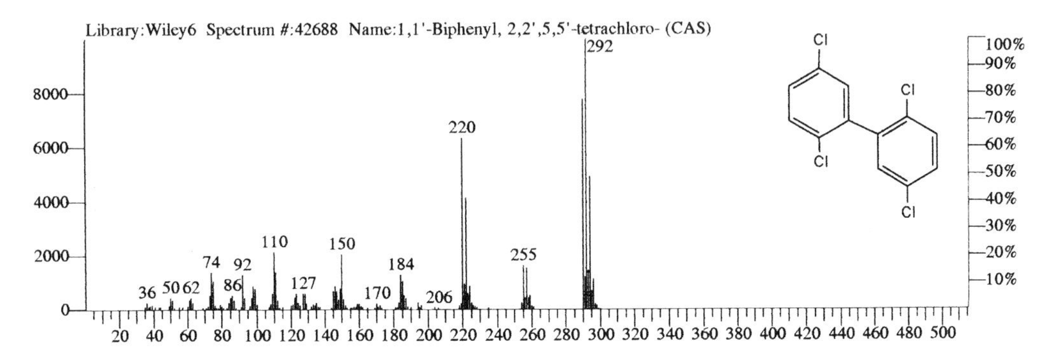

Fig. 3.112 Tetrachlorobiphenyl (e.g., PCB 52) $C_{12}H_6Cl_4$, M: 290

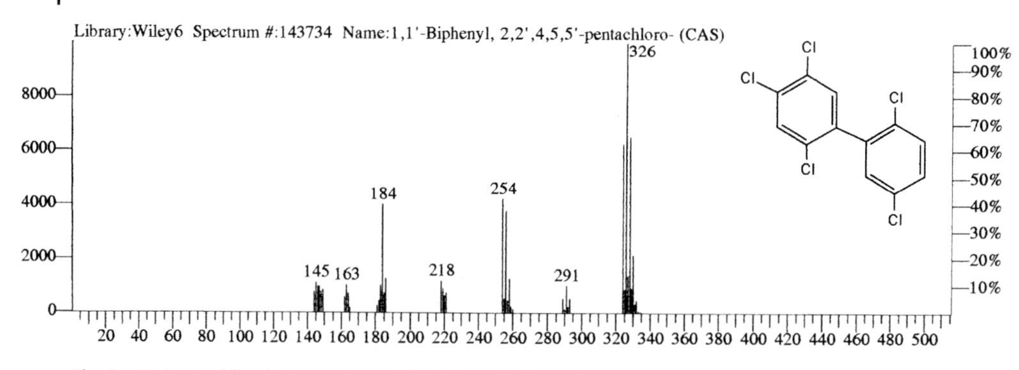

Fig. 3.113 Pentachlorobiphenyl (e.g., PCB 101, 118) $C_{12}H_5Cl_5$, M: 324

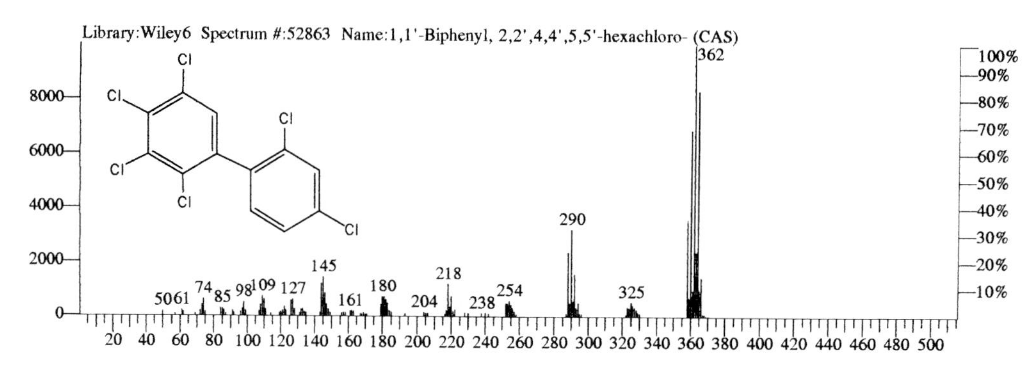

Fig. 3.114 Hexachlorobiphenyl (e.g., PCB 138, 153) $C_{12}H_4Cl_6$, M: 358

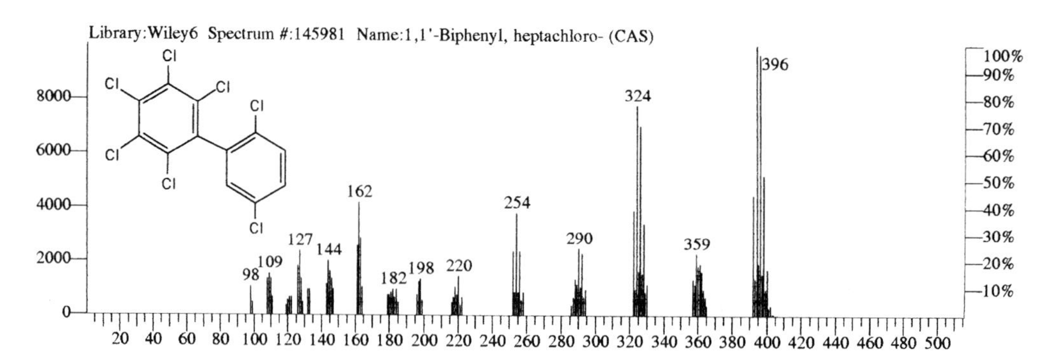

Fig. 3.115 Heptachlorobiphenyl (e.g., PCB 180) $C_{12}H_3Cl_7$, M: 392

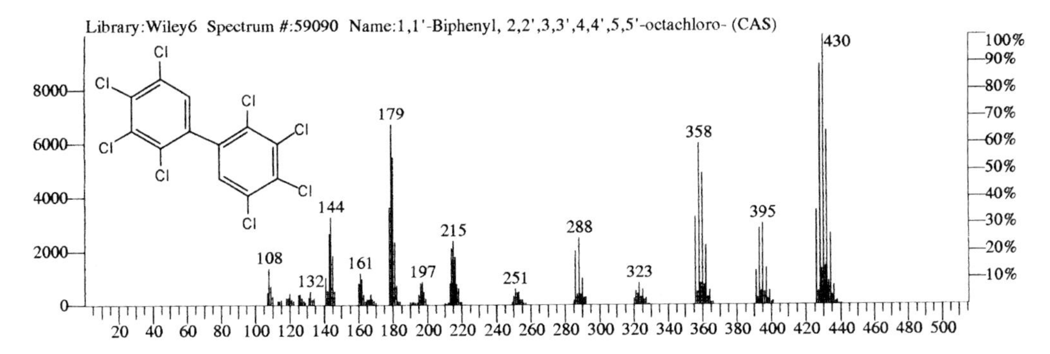

Fig. 3.116 Octachlorobiphenyl $C_{12}H_2Cl_8$, M: 426

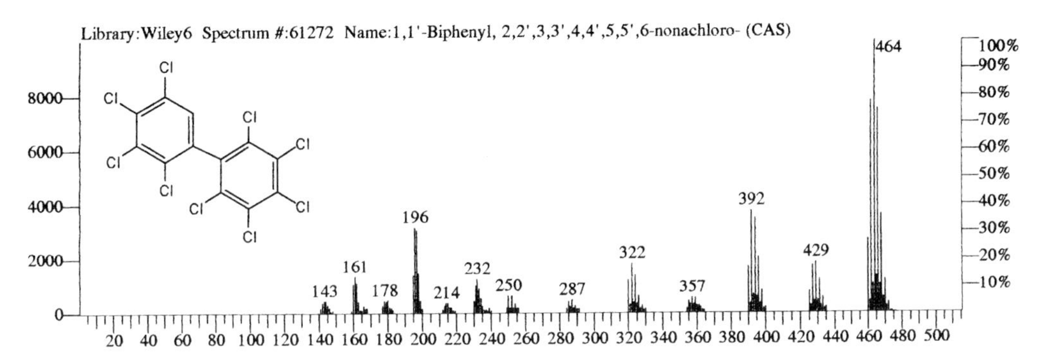

Fig. 3.117 Nonachlorobiphenyl $C_{12}HCl_9$, M: 460

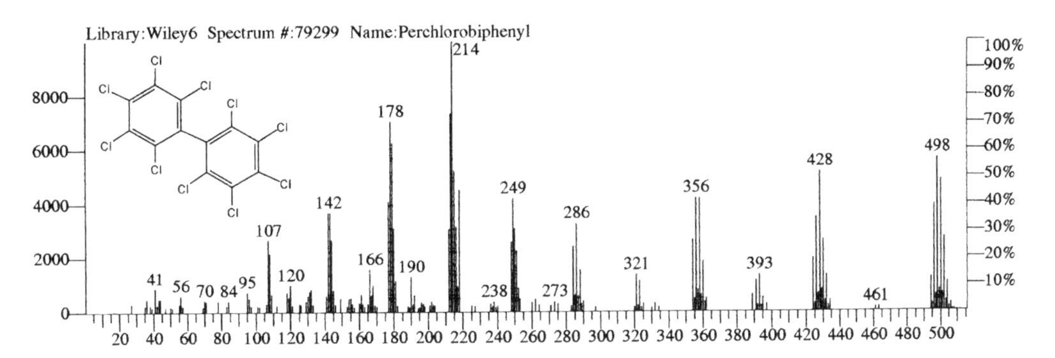

Fig. 3.118 Decachlorobiphenyl (PCB 209) $C_{12}Cl_0$, M: 494

3.2.6.7 **Polychlorinated Dioxins/Furans (PCDDs/PCDFs)**

The persistence of this class of substance in the environment parallels their mass spectroscopic stability. As they are aromatic, dioxins and furans give good molecular ion intensities with pronounced isotope patterns and only low fragmentation (Figs. 3.119 and 3.120). Internal standards with 6-fold and 12-fold isotopic labelling (^{13}C) are used for quantitation. High levels of labelling are necessary in order to obtain mass signals for the standard above the native Cl isotope pattern.

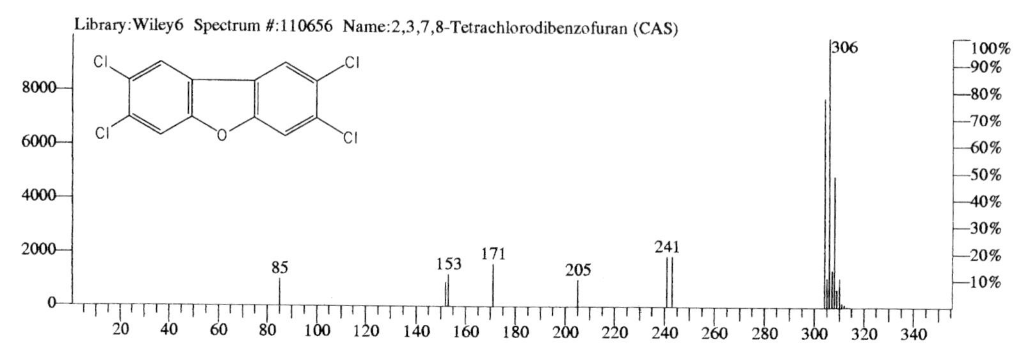

Fig. 3.119 2,3,7,8-Tetrachlorodibenzofuran (2,3,7,8-TCDF) $C_{12}H_4Cl_4O$, M: 304, CAS Reg. No.: 51207-31-9

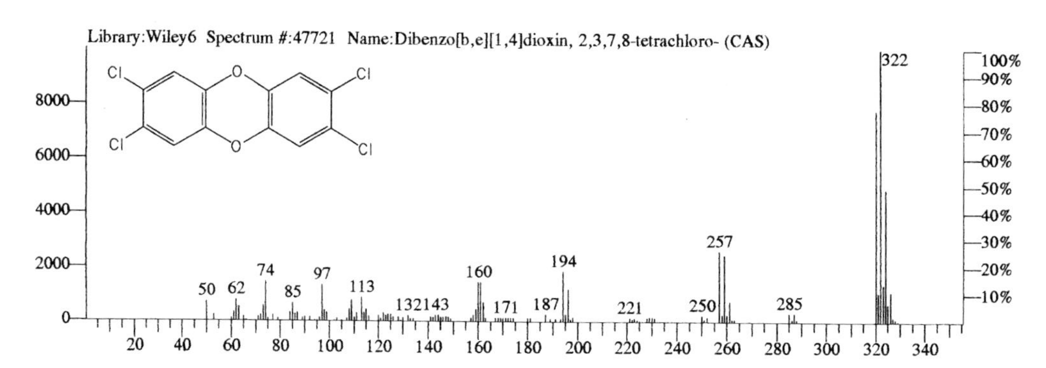

Fig. 3.120 2,3,7,8-Tetrachlorodibenzodioxin (2,3,7,8-TCDD) $C_{12}H_4Cl_4O_2$, M: 320, CAS Reg. No.: 1746-01-6

3.2.6.8 Drugs

Various different classes of active substance are assigned to the group of drugs. The amphetamines are of particular interest with regard to mass spectrometry because their EI spectra are dominated by α-cleavage (Figs. 3.121 to 3.123). In this case the mass scan must be started at a correspondingly low mass to determine the ions m/z 44 and 58. The drugs of the morphine group (morphine, heroin, codeine and cocaine) give stable molecular ions and can be recognised with certainty on the basis of their fragmentation pattern (Figs. 3.124 to 3.127). The selective detection and confirmation of the identity of drugs is also possible in complex matrices using chemical ionisation with ammonia or isobutane as the CI gas.

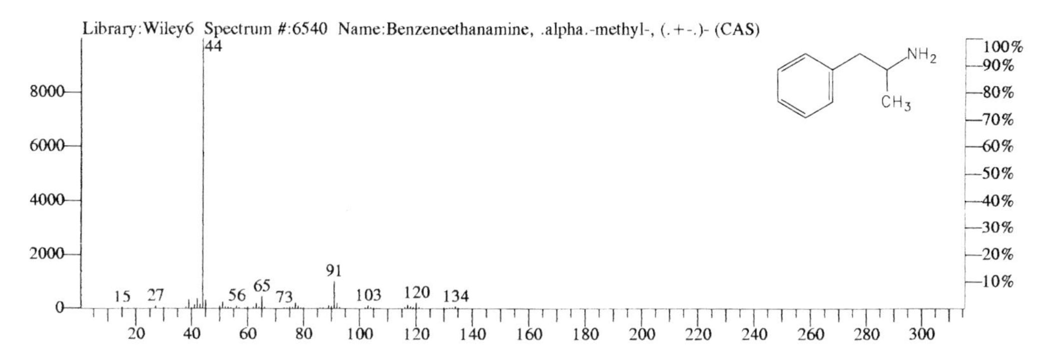

Fig. 3.121 Amphetamine $C_9H_{13}N$, M: 135, CAS Reg. No.: 300-62-9

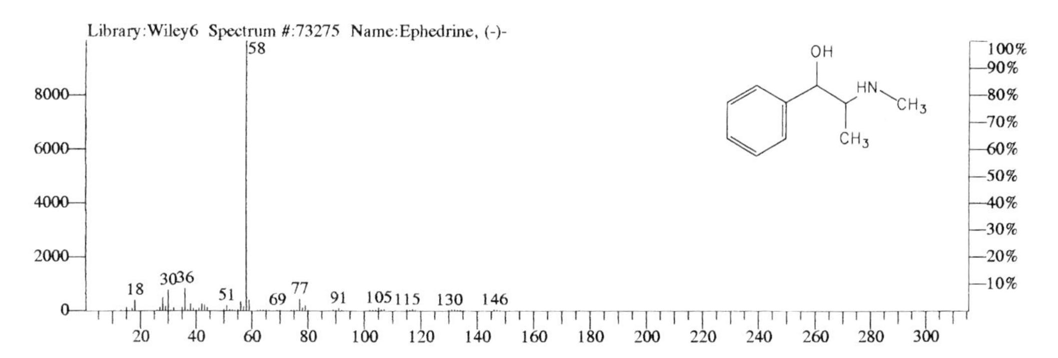

Fig. 3.122 Ephedrine $C_{10}H_{15}NO$, M: 165, CAS Reg. No.: 299-42-3

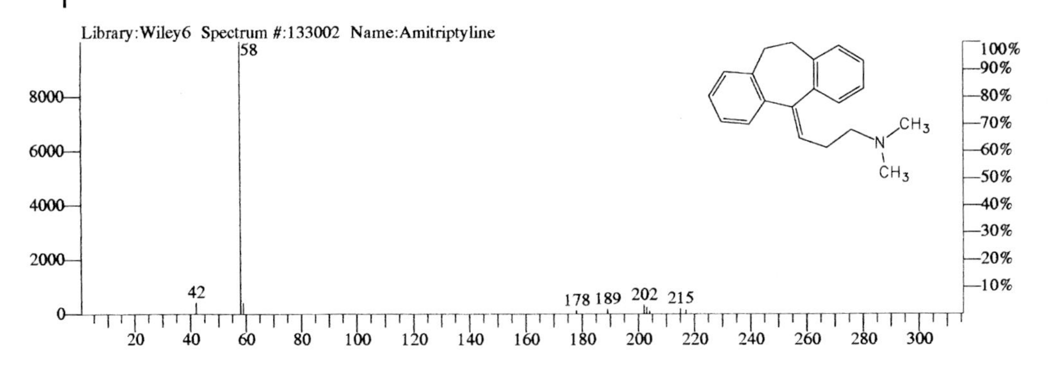

Fig. 3.123 Amitriptyline $C_{20}H_{23}N$, M: 277, CAS Reg. No.: 50-48-6

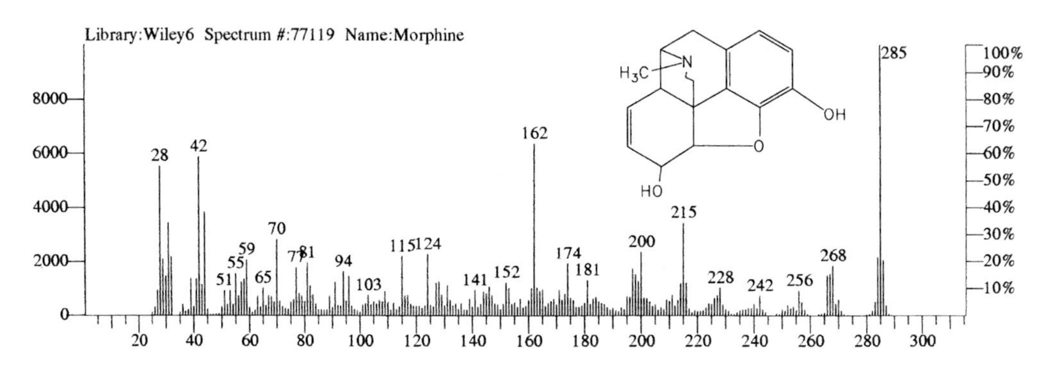

Fig. 3.124 Morphine $C_{17}H_{19}NO_3$, M: 285, CAS Reg. No.: 57-27-2

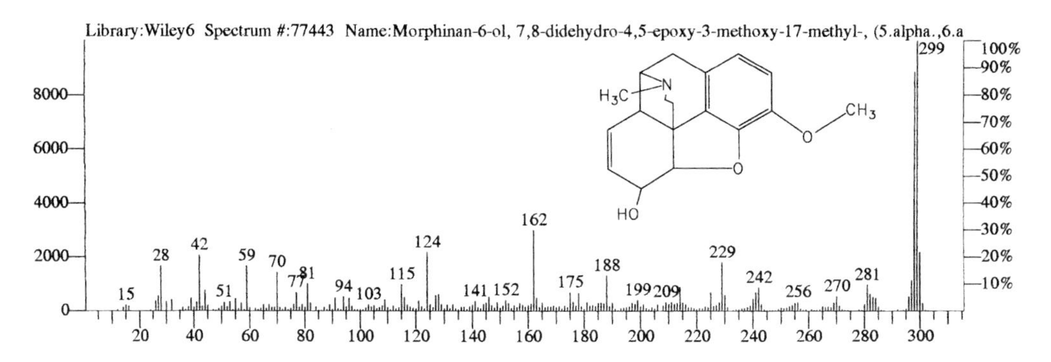

Fig. 3.125 Codeine $C_{18}H_{21}NO_3$, M: 299, CAS Reg. No.: 76-57-3

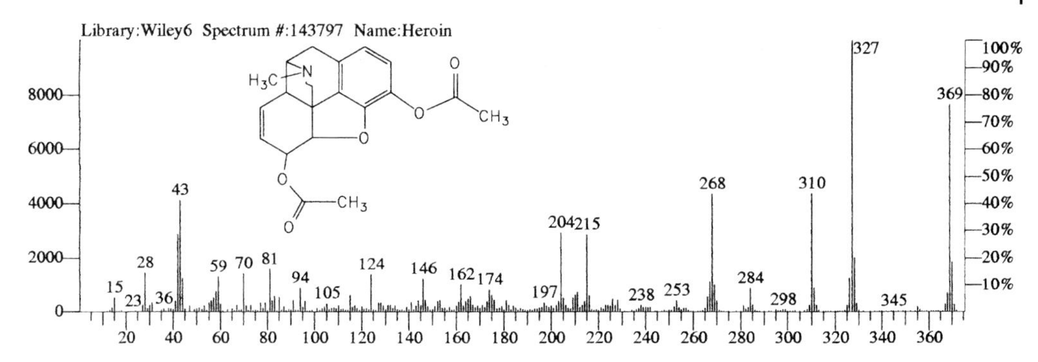

Fig. 3.126 Heroin $C_{21}H_{23}NO_5$, M: 369, CAS Reg. No.: 561-27-3

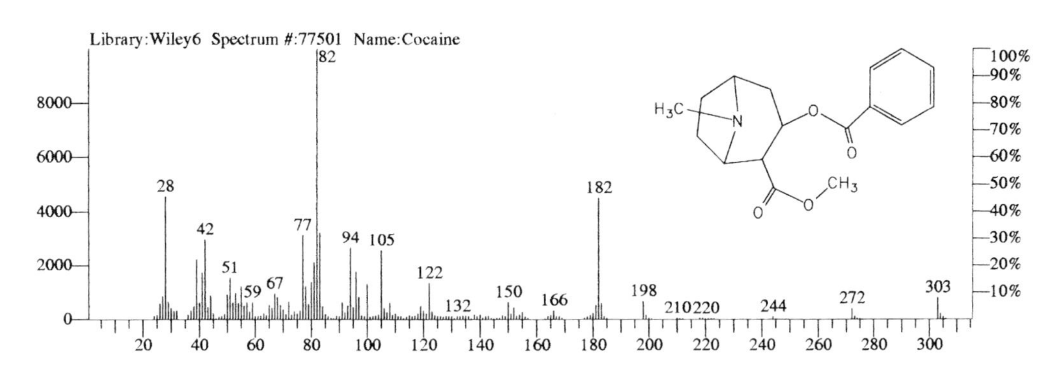

Fig. 3.127 Cocaine $C_{17}H_{21}NO_4$, M: 303, CAS Reg. No.: 50-36-2

3.2.6.9 Explosives

All explosives show a high proportion of oxygen in the form of nitro groups. GC/MS analysis of the nonaromatic compounds (Hexogen, Octogen, Nitropenta etc.) is problematic because of decomposition, even in the injector. However, aromatic nitro compounds and their metabolites (aromatic amines) can be detected with extremely high sensitivity because of their stability (Figs. 3.128 to 3.143). With EI molecular ions usually appear at low intensity because nitro compounds eliminate NO (M-30). o-Nitrotoluenes are an exception because a stable ion is formed after loss of an OH radical (M-17) as a result of the proximity of the two groups (ortho effect) (Figs. 3.131 and 3.132). If the scan is run including the mass m/z 30, the mass chromatogram can show the general elution of the nitro compounds. In this area chemical ionisation is useful for confirming and quantifying results. In particular water has proved to be a useful CI gas in ion trap systems for residue analysis of old munitions.

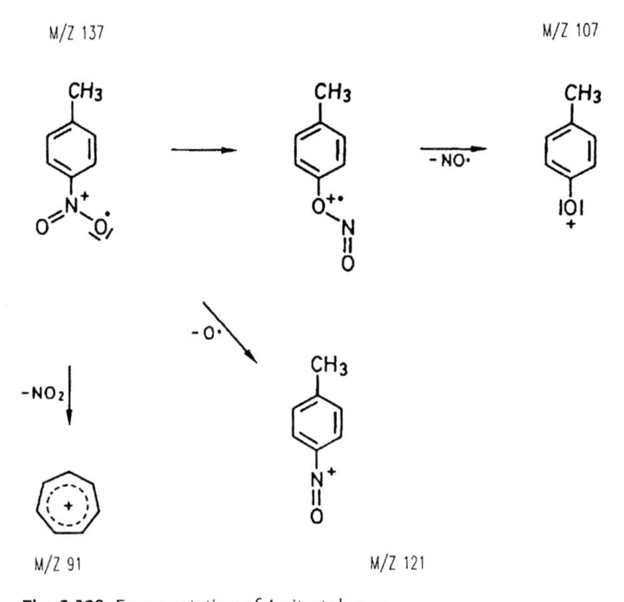

Fig. 3.128 Fragmentation of 4-nitrotoluene

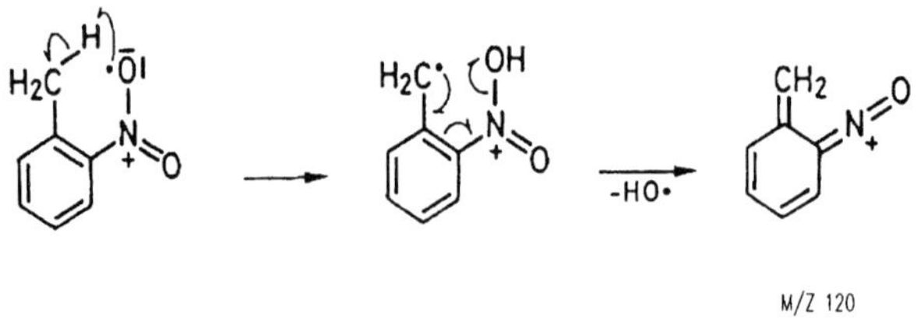

Fig. 3.129 Fragmentation of 2-nitrotoluene (ortho effect)

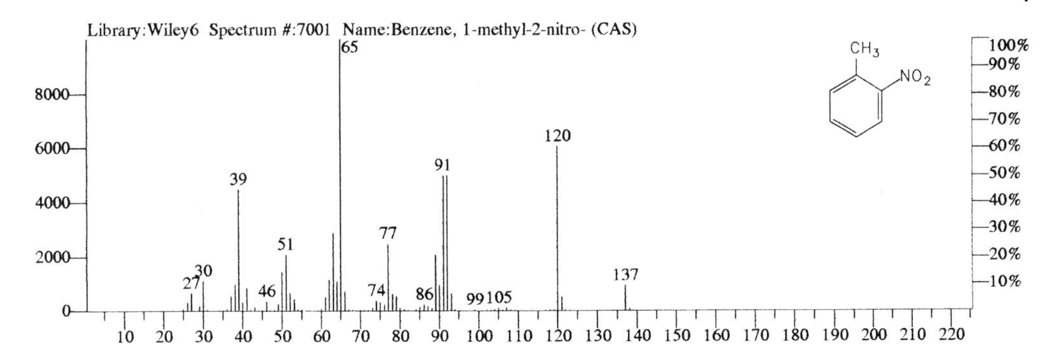

Fig. 3.130 2-Nitrotoluene $C_7H_7NO_2$, M: 137, CAS Reg. No.: 88-72-2

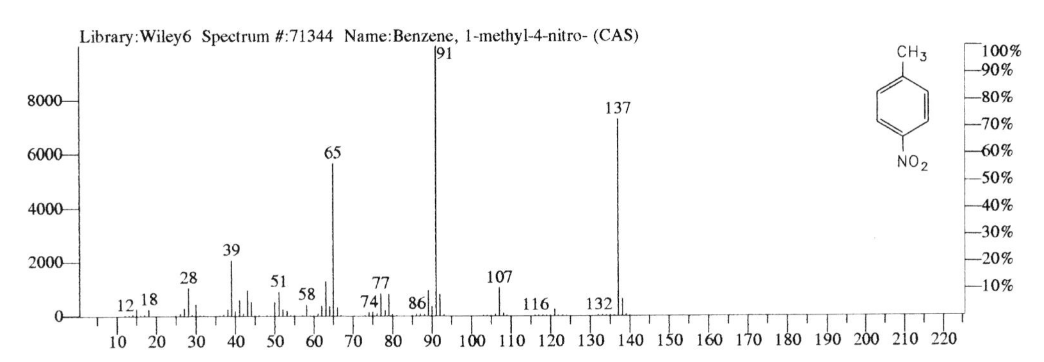

Fig. 3.131 4-Nitrotoluene $C_7H_7NO_2$, M: 137, CAS Reg. No.: 99-99-0

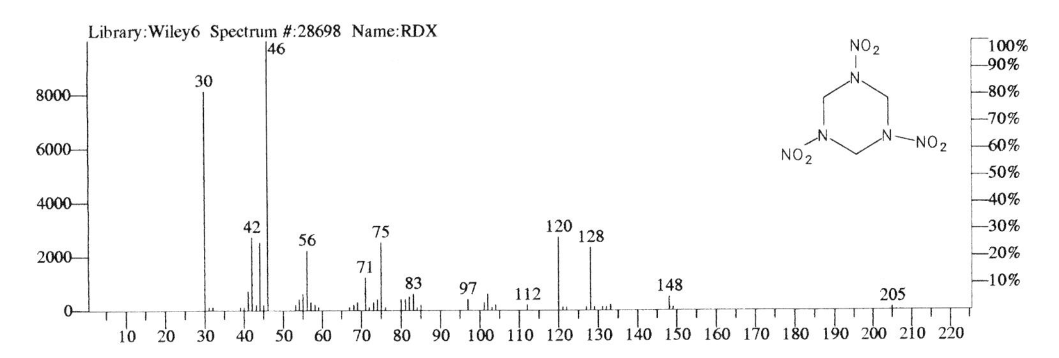

Fig. 3.132 Hexogen (RDX) $C_3H_6N_6O_6$, M: 222, CAS Reg. No.: 121-82-4

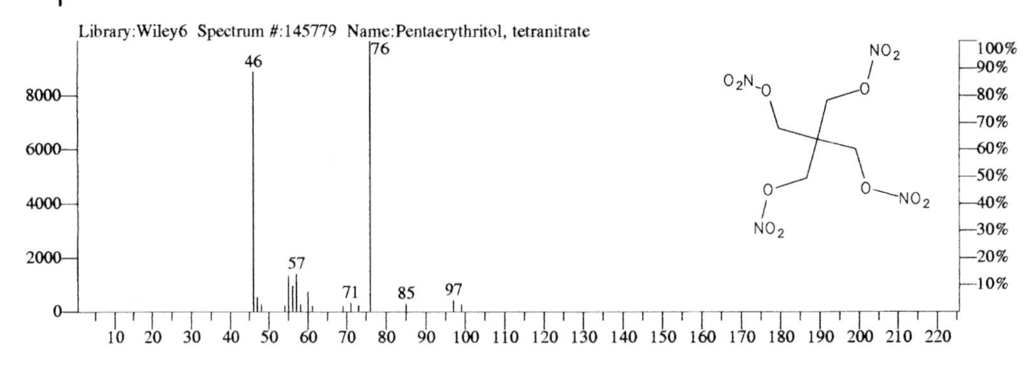

Fig. 3.133 Nitropenta (PETN) $C_5H_8N_4O_{12}$, M: 316, CAS Reg. No.: 78-11-5

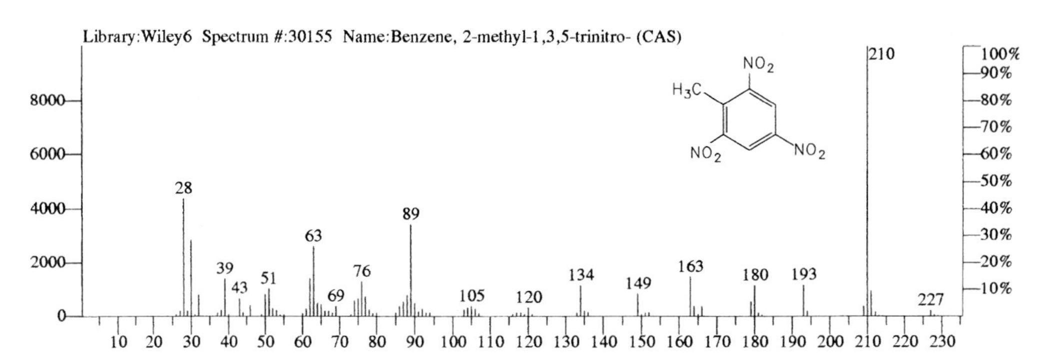

Fig. 3.134 Trinitrotoluene (TNT) $C_7H_5N_3O_6$, M: 227, CAS Reg. No.: 118-96-7

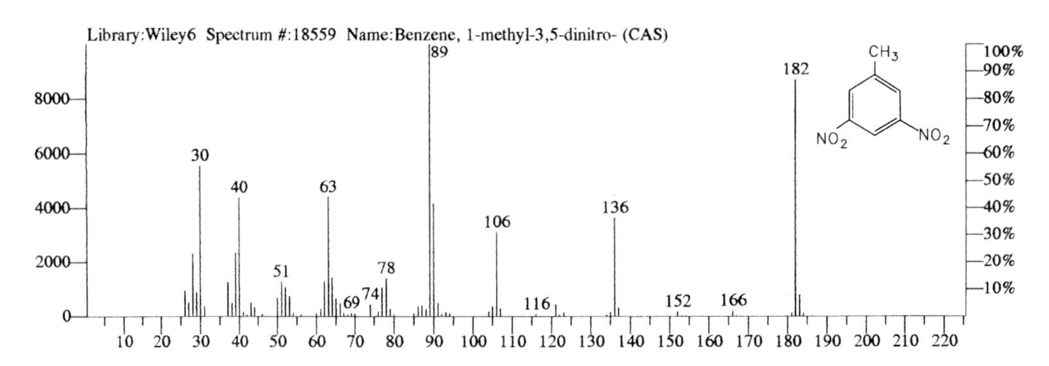

Fig. 3.135 3,5-Dinitrotoluene (3,5-DNT) $C_7H_6N_2O_4$, M: 182, CAS Reg. No.: 618-85-9

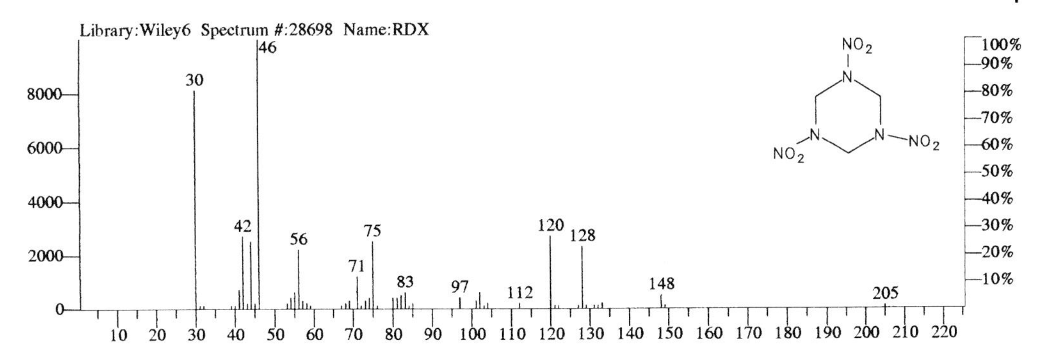

Fig. 3.136 2,6-Dinitrotoluene (2,6-DNT) $C_7H_6N_2O_4$, M: 182, CAS Reg. No.: 606-20-2

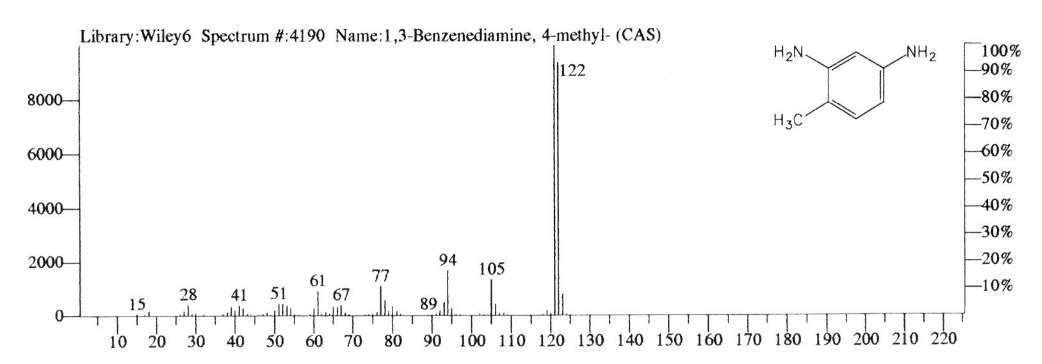

Fig. 3.137 2,4-Diaminotoluene (2,4-DAT) $C_7H_{10}N_2$, M: 122, CAS Reg. No.: 95-80-7
(Note: 2,4-DAT and 2,6-DAT form a critical pair during GC separation)

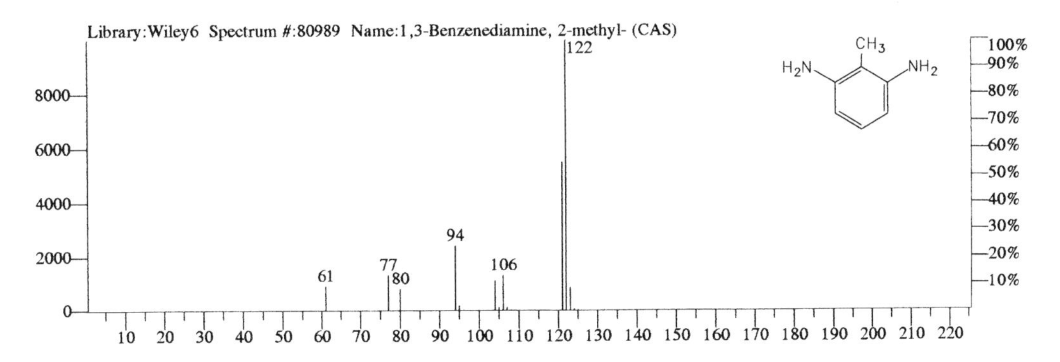

Fig. 3.138 2,6-Diaminotoluene (2,6-DAT) $C_7H_{10}N_2$, M: 122, CAS Reg. No.: 823-40-5

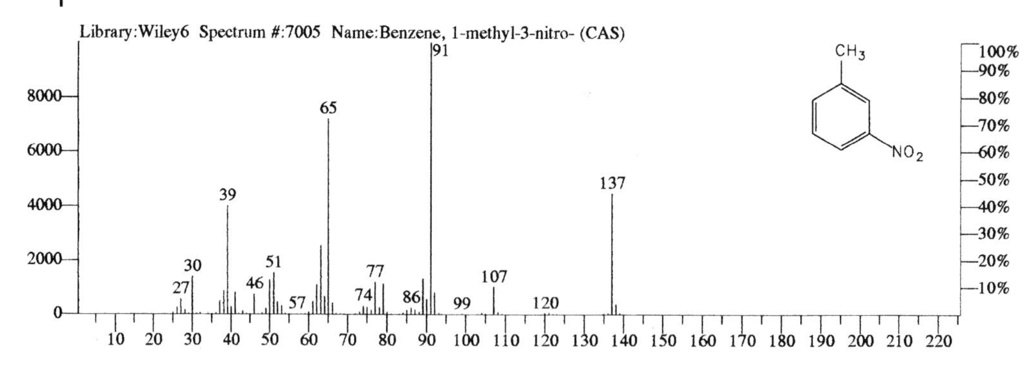

Fig. 3.139 (Mono-)3-nitrotoluene (3-MNT) $C_7H_7NO_2$, M: 137, CAS Reg. No.: 99-08-1

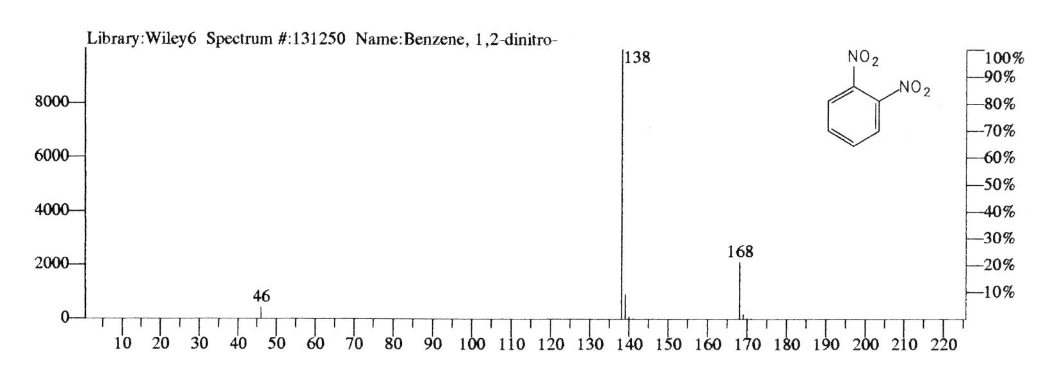

Fig. 3.140 1,2-Dinitrobenzene (1,2-DNB) $C_6H_4N_2O_4$, M: 168, CAS Reg. No.: 528-29-0

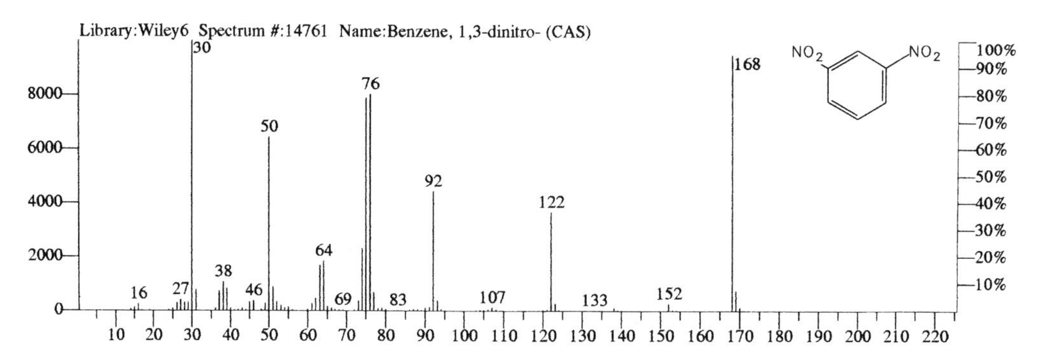

Fig. 3.141 1,3-Dinitrobenzene (1,3-DNB) $C_6H_4N_2O_4$, M: 168, CAS Reg. No.: 99-65-0

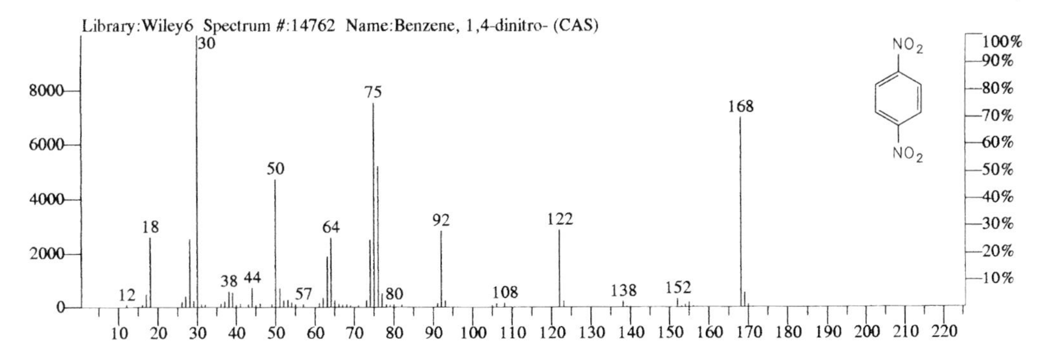

Fig. 3.142 1,4-Dinitrobenzene (1,4-DNB) $C_6H_4N_2O_4$, M: 168. CAS Reg. No.: 100-25-4

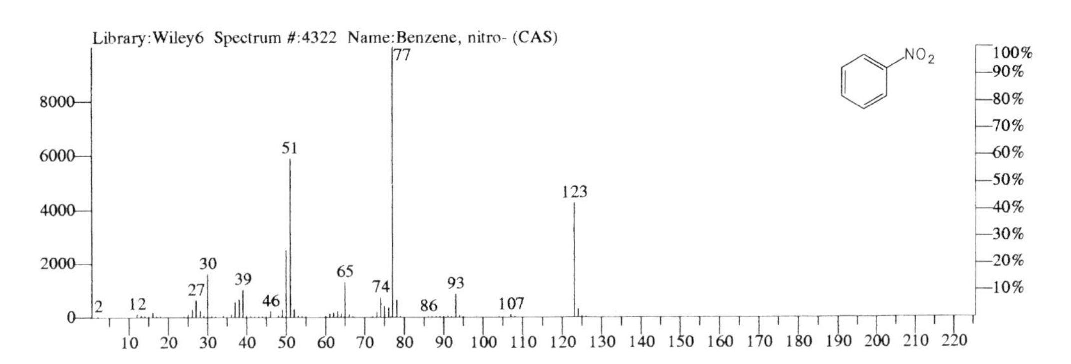

Fig. 3.143 (Mono-)nitrobenzene (MNB) $C_6H_5NO_2$, M: 123, CAS Reg. No.: 98-95-3

3.2.6.10 **Chemical Warfare Agents**

The identification of chemical warfare agents is important for testing disarmament measures and for checking disused military sites. Here also there is no single chemical class of substances. Volatile phosphoric acid esters and organoarsenic compounds belong to this group (Figs. 3.144 to 3.149). Chemical ionisation is the method of choice for identification and confirmation of identity.

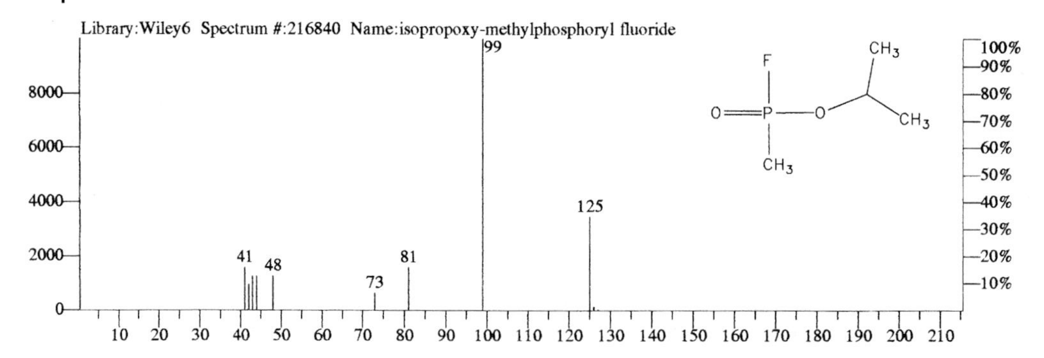

Fig. 3.144 Sarin $C_4H_{10}FO_2P$, M: 140, CAS Reg. No.: 107-44-8

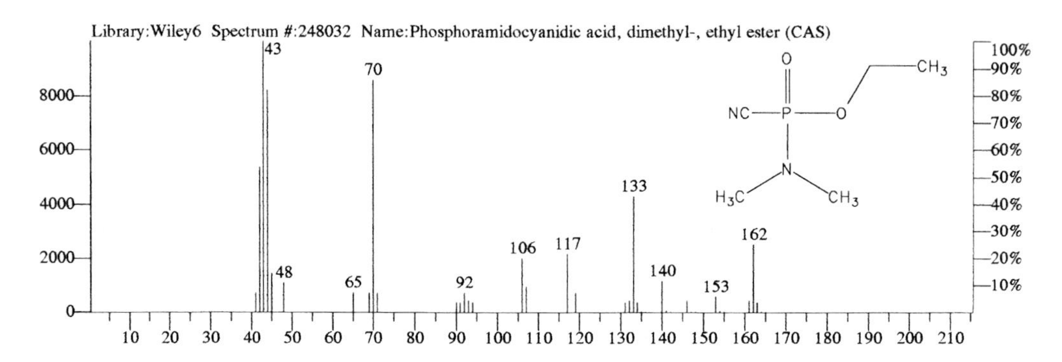

Fig. 3.145 Tabun $C_5H_{11}N_2O_2P$, M: 162, CAS Reg. No.: 77-81-6

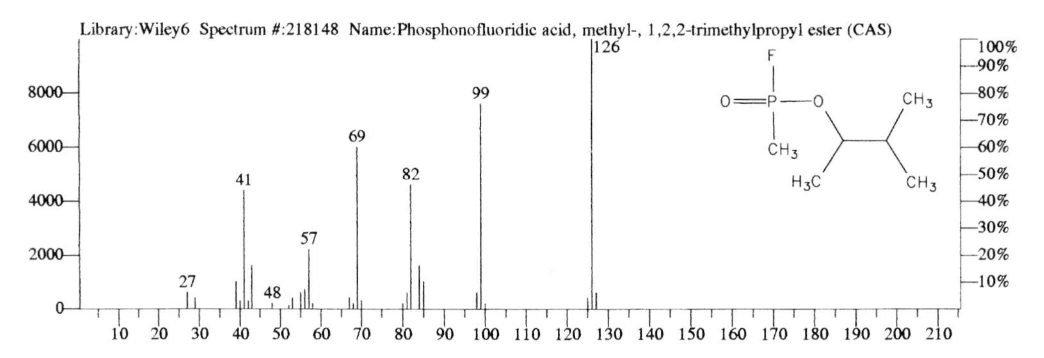

Fig. 3.146 Soman $C_7H_{16}FO_2P$, M: 182, CAS Reg. No.: 96-64-0

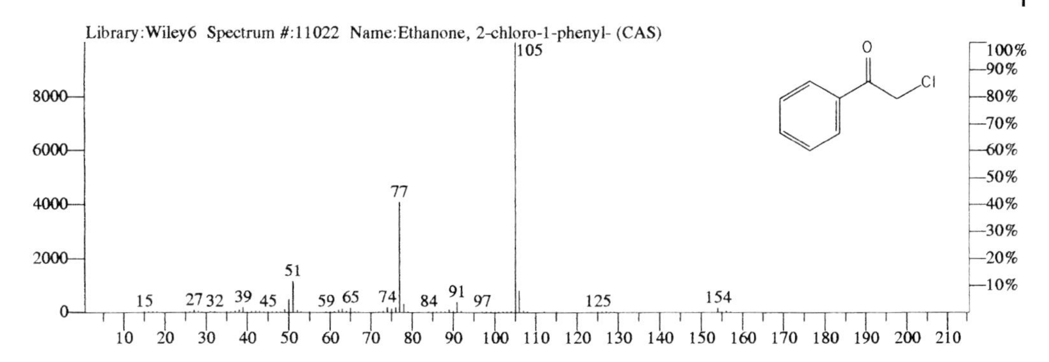

Fig. 3.147 Chloroacetophenone (CN) C_8H_7ClO, M: 154, CAS Reg. No.: 532-27-4

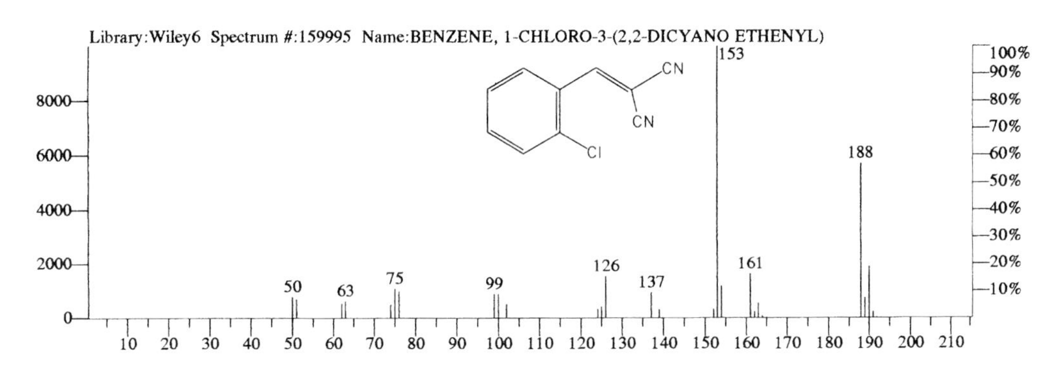

Fig. 3.148 o-Chlorobenzylidenemalnonitrile (CS) $C_{10}H_5ClN_2$, M: 188, CAS Reg. No.: 2698-41-1

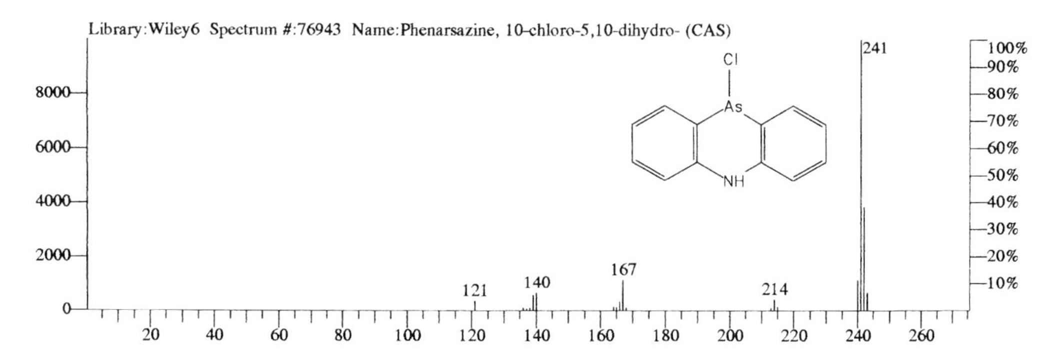

Fig. 3.149 Adamsite (DM) $C_{12}H_9AsClN$, M: 277, CAS Reg. No.: 578-94-9

3.2.6.11 Brominated Flame Retardants (BFR)

Multiply brominated aromatic compounds, which can only be chromatographed in a few cases because of their high molecular weights, are used as flameproofing agents (Figs. 3.150 and 3.151). The molecular weights of the polybrominated biphenyls (PBB) and polybrominated diphenylether (PBDE), which are mostly used, have molecular weights of up to 1000 (decabromobiphenyl M 950). Brominated flame retardants have recently become of interest to analysts because burning materials containing PBB and PBDE can lead to formation of polybrominated dibenzodioxins and -furans. Decabromodiphenylether is suspected to be accumulated in the human body (PBDE 209), leading to the requirement of trace analysis. The spectra of PBB and PBDE are characterised by the symmetrical isotope pattern of the bromine and the high stability of the aromatic molecular ion. Flameproofing agents based on brominated alkyl phosphates have a greater tendency to fragment.

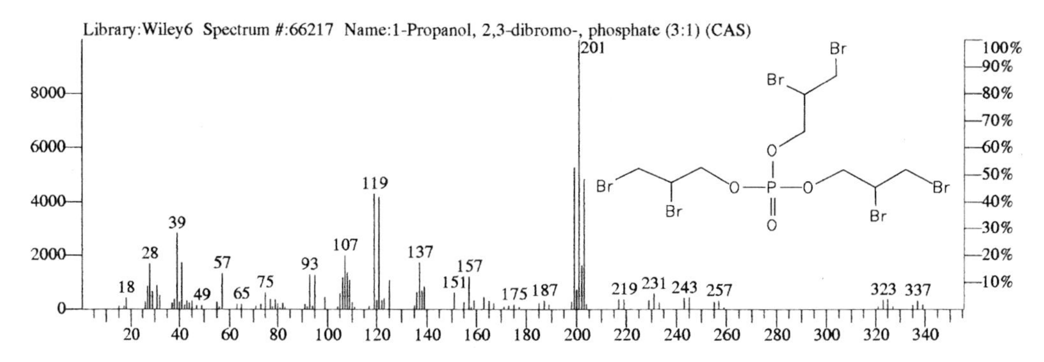

Fig. 3.150 Bromkal P67–6HP (Tris) $C_9H_{15}Br_6O_4P$, M: 692, CAS Reg. No.: 126-72-7

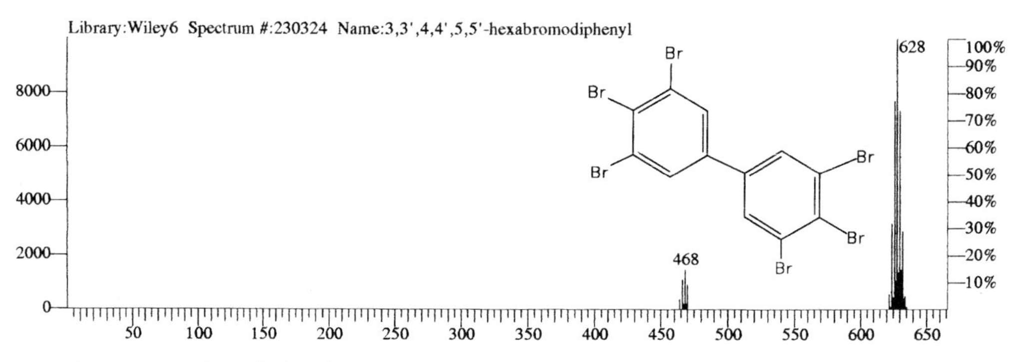

Fig. 3.151 Hexabromobiphenyl (HBB) $C_{12}H_4Br_6$, M: 622

3.3 Quantitation

It is important to point out that accuracy assessment is a continuous process, which should be implemented in the routine work as a part of the QA/GC set-up of the laboratory.

Richard Boqué
Universitat Rovira i Virgili, Tarragona, Spain (Boqué 2008)

Besides the identification of components of a mixture, the use of GC/MS systems to determine the concentration of target compounds is a most important application in governmental, industrial as well as private control labs. The need to determine quantitatively an increasing number of components in complicated matrices in ever smaller concentrations makes the use of GC/MS systems in routine analysis appropriate for economic reasons. Gone are the days when only positive results from classical GC were confirmed by GC/MS. In many areas of trace analysis the development of routine multicomponent methods has only become possible through the selectivity of detection and the specificity of identification of GC/MS and GC/MS/MS systems. The development of GC/MS data systems has therefore been successful in recent years, particularly in areas where the use of the mass spectrometric substance information coincides with the integration of chromatographic peaks and thus increases the certainty of quantitation. Compared with chromatography data systems for stand-alone GC and HPLC, there are therefore differences and additional possibilities which arise from the use of the mass spectrometer as the detector.

Scan Rate of the Chromatogram

Mass spectrometers do not continuously record the substance stream arriving in the detector (as for example with FID, ECD etc.). The chromatogram is comprised of a series of measurement points which are represented by mass spectra. The scan rate chosen by the user establishes the time interval between the data points. The maximum possible scan rate depends on the scan speed of the spectrometer. It is determined by the width of the mass range to be acquired and the necessity of achieving an analytical detection capacity of the instrument which is as high as possible. For routine measurements scan rates of 0.5 to 1.0 s/scan are usually chosen. Compared with a sharp concentration change in the slope of a GC peak, these scan rates are only slow.

In the integration of GC/MS chromatograms, the choice of scan rate is particularly important for the distribution of the peak area values. With the scan rate of 1s/scan, which is fre-

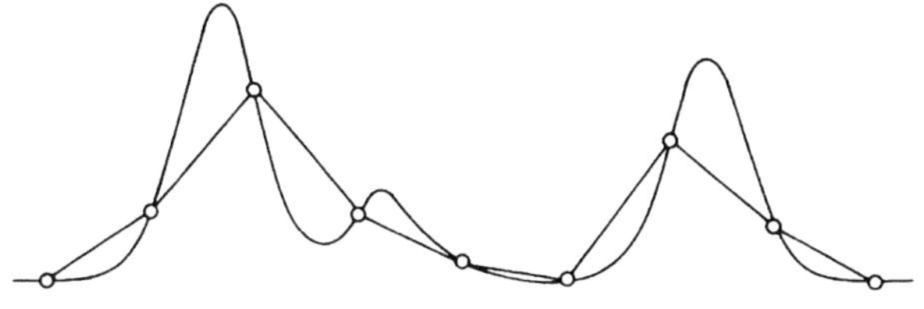

Fig. 3.152 Comparison of an actual chromatogram with one reconstructed from data points at a low scan rate (after Chang)

quently used, the peak area of rapidly eluting components cannot be determined reliably as too few data points for the correct plotting of the GC peak are recorded (Fig. 3.152). The incorporation of the top of the peak in the calculation of height and area cannot be carried out correctly in these cases. Under certain circumstances small peaks next to sharp peaks can be lost in the plot. In the case of a distribution of area values in quantitative determinations, special attention should first be paid to the recording parameters before a possible cause is looked for in the detector itself.

To determine the optimal scan rate, the base width of the components eluting early should first be investigated. In practice it has been shown that the peak area of a symmetrical peak can be described well by ca. 10 measuring points. Therefore, the base width (elution time) of the most rapidly eluting component should be a measure of the scan rate in quantitative GC/MS analysis. From the statistical view on determining a reliable peak area a number of 6 to 8 data points has proven to be a good practical compromise. There is no problem with base widths in the region of ca. 10 s, but volatile compounds elute with peak widths of 5 s or even lower in good chromatography. Here the limits of scanning instruments become clear, as faster scan rates (<0.5 s/scan) lead to a decrease in the dwell time per ion and thus to a significant loss in sensitivity. Individual mass recording (SIM/MID/SRM) which allows high sensitivity at high scan rates, or parallel detection of ions by ion trap or time-of-flight mass spectrometers are possible solutions.

3.3.1 Decision Limit

The question of when a substance can be said to be detected cannot be answered differently for quantitative GC/MS compared with all other chromatographic systems. The answer lies in the determination of the signal/noise ratio.

In the basic adjustment of mass spectrometers, unlike classical GC detectors, the zero point is adjusted correctly (electrometer zero) to ensure the exact plot of isotope patterns. For this the adjustment is chosen in such a way that minimum noise of the electronics is determined which can then be removed during data acquisition by software filters. Besides electrical noise, there is also chemical noise (matrix, column bleed, leaks etc.), particularly in trace analysis.

The decision as to whether a substance has been detected or not is usually assessed in the signal domain. Here it is established that the decision limit is such that the smallest detectable signal from the substance can be clearly differentiated from a blank value (critical value).

In the measurement of a substance-free sample (blank sample) an average signal is obtained which corresponds to a so-called blank value. Multiple measurements confirm this blank value statistically and give its standard deviation. With GC/MS this blank value determination is carried out in practice in the immediate vicinity to the substance peak, whereby the noise widths before and after the peak are taken into consideration. The average noise and the signal intensity can also be determined manually from the print-out, also suitable S/N algorithms are available in data systems. A substance can be said to be detected if the substance signal exceeds a certain multiple of the noise width (standard deviation). This value is chosen arbitrarily as 2, 3 or higher (depending on the laboratory SOP or on the application) and is used as the deciding criterion in most routine evaluations of GC/MS data systems (Fig. 3.153).

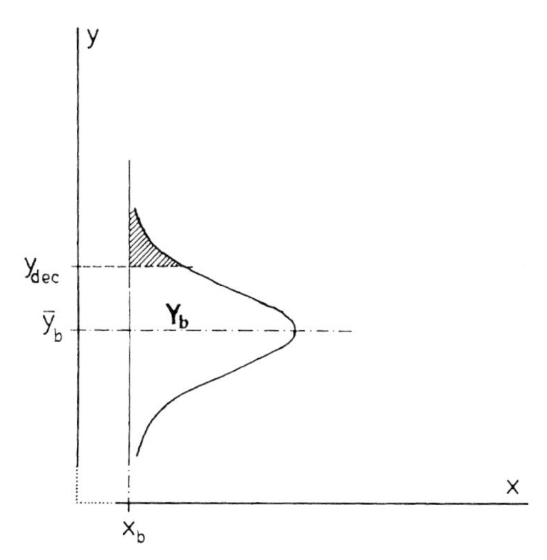

Fig. 3.153 Statistical interpretation of the decision limit.
y_b = random sample of blank values, y_{dec} = decision limit.
In a statistically defined system the decision limit, i. e. the smallest detectable signal which can be differentiated from a blank value, can only be obtained from multiple measurements of the blank value and the analysis at a given error probability (after Ebel)

3.3.2
Limit of Detection

How does the decision limit relate to the limit of detection (LOD)? Usually both terms are used synonymously, which is totally incorrect! The LOD of a method is never given in counts or parts of a scale but is given a dimension, such as pg/µl, ng/l, or ppb etc., in any case in the substance domain! The transfer from the signal domain of the decision limit into the substance domain of the LOD is effected using a valid calibration function (Fig. 3.154)! Nevertheless, the LOD can only be regarded as a qualitative parameter as here the uncertainty in the calibration function is not taken into account.

3.3.3
Limit of Quantitation

The limit of quantitation (LOQ) of a method is also given as a quantity of substance or concentration in the substance domain. This limit incorporates the calibration and thus also the uncertainty (error consideration) of the measurements. Unlike the LOD it is guaranteed statistically and gives the lower limiting concentration which can be unambiguously determined quantitatively. It can differ significantly from the blank value.

As a component can only be determined after it has been detected, the LOQ cannot be lower than the LOD. As there is a relative uncertainty in the result of ca. 100 % at a concentration of a substance corresponding to the LOD in an analysis sample, the LOQ must be correspondingly higher than the LOD, depending on the requirement.

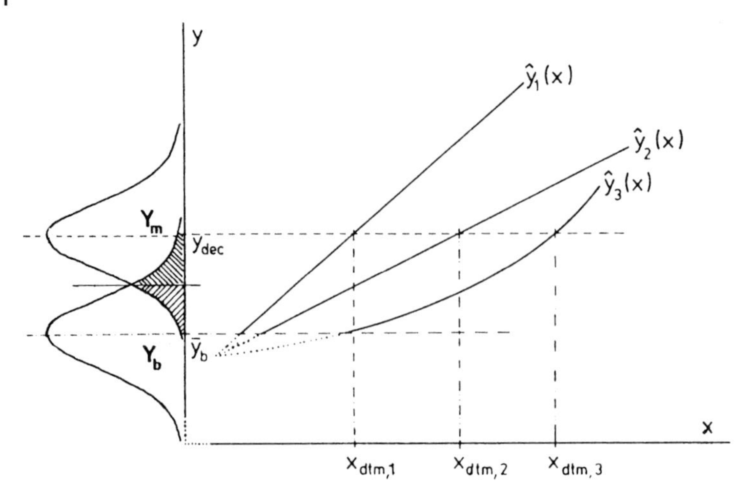

Fig. 3.154 Definition of the limit of detection (LOD) from the decision limit.
The limit of detection (LOD) is defined as that quantity of substance, concentration or content X_{dtm} which is given using the calibration function from the smallest detectable signal y_{dec} (decision limit). Calibration functions with different sensitivities (slopes) lead to different limits of detection at the same signal height (after Ebel)

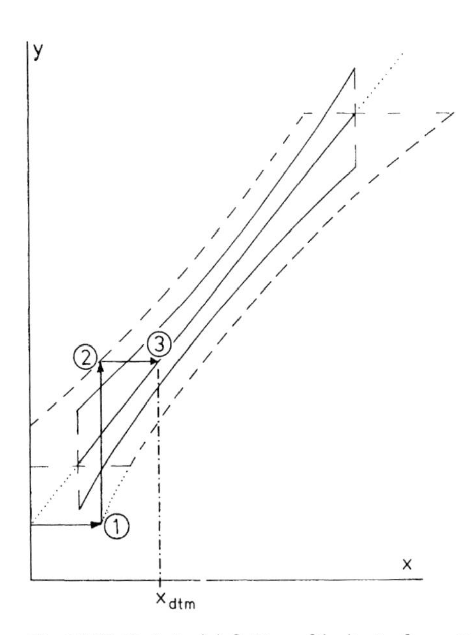

Fig. 3.155 Statistical definition of the limit of quantitation.
1 Measured value at quantity of substance = 0 (signal domain).
2 Signal size which can be differentiated significantly taking the standard deviation into account.
3 Limit of quantitation x_{dtm} from the calibration function.
The limit of quantitation is the lower limiting concentration x_{dtm} at a fixed statistical error probability which can definitely be quantitatively determined and be differentiated significantly from zero concentration (after Ebel)

3.3.4 Sensitivity

The term sensitivity, which is frequently used to describe the quality of a residue analysis, or the current state of a measuring instrument, is often used incorrectly as a synonym for the lowest possible LOD or LOQ. A sensitive analysis procedure, however, exhibits a large change in signal with a small change in substance concentration. The sensitivity of a procedure thus describes the increase a of a linear calibration function (see Section 3.3.5). At the same confidence interval of the measured points of a calibration function (Fig. 3.156), sensitive analysis procedures give a narrower confidence interval than less sensitive ones! The LOD is independent of the sensitivity of an analytical method.

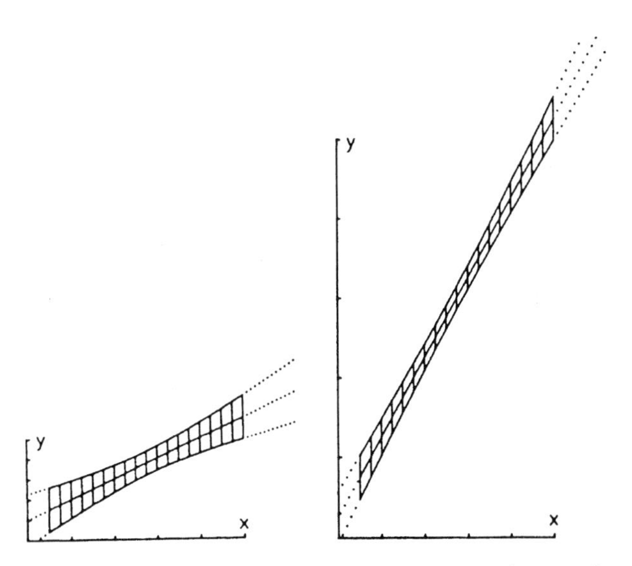

Fig. 3.156 Confidence intervals for calibration curves at the same distribution of the measured values but with different slopes (after Ebel)

3.3.5 The Calibration Function

For an analysis procedure that has been worked out, the calibration function is constructed from the measurement of known concentrations as, because of the dependence of the height of the signal on the operating parameters and the current state of the mass spectrometer, fixed response factors can only be used to a limited degree in GC/MS. The calibration function generally describes the dependence of the signal on the substance concentration. In the case of a linear dependence, the regression calculation gives a straight line for the calibration function, the equation for which contains the blank value a_0 and the sensitivity a:

$$f(x) = a_0 + a \cdot x \tag{25}$$

The calibration function is defined exclusively within the working range given by the experimental calibration. GC/MS systems achieve very low LODs so that, at a correspondingly dense collection of calibration points near the blank value, a nonlinear area is described. This area can be caused by unavoidable active sites (residual activities) in the system and 'swallows up' a small but constant quantity of substance. Such a calibration function tends to approach the x axis before reaching the origin.

In the upper concentration range the signal hardly increases at all with increasing substance concentration because of increasing saturation of the detector. The calibration function stops increasing and tends to be asymptotic (Fig. 3.157).

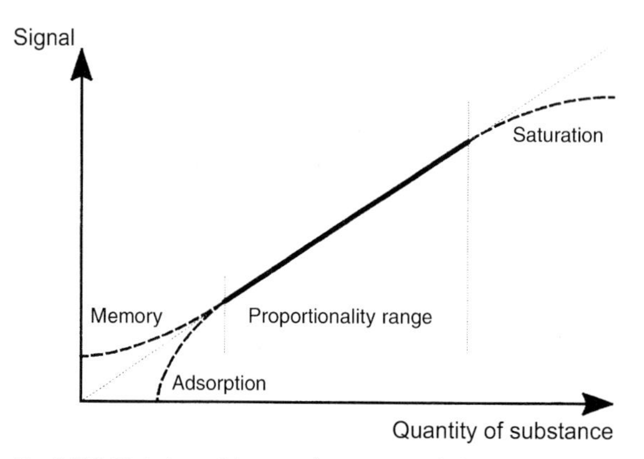

Fig. 3.157 Variation of the signal intensity with the quantity of substance

The best fit of the calibration curve to the measured points is determined by a regression calculation. The regression coefficient gives information on the quality of the curve fit. Linear regressions are not always suitable for the best fit in GC/MS analysis; quadratic regressions frequently give better results. Particularly in the area of trace analysis the type of fit within the calibration can change through the nonlinear effects described above. In such cases it is helpful to limit the regression calculation to that of a sample concentration lying close to the calibration level (local linearisation, next three to four data points). If individual data systems do not allow this, a point to point calibration can be carried out instead.

Important aspects regarding the optimisation of a calibration can be derived from the above:

- One-point calibrations have no statistical confirmation and can therefore only be used as an orientation.
- Multipoint calibrations must cover the expected concentration range. For regulated methods a factor of 10 below the regulated concentration (maximum residue level, MRC) is required.
- Extrapolation beyond the experimentally measured points is not allowed. (Calibration functions do not have to pass through the origin).
- Multiple measurements at an individual calibration level define the confidence interval which can be achieved.
- At the same number of calibration points those near the LOQ give an improvement in the fit (unlike the equidistant position of the calibration level).

3.3.6
Quantitation and Standardisation

In order to determine the substance concentration in an unknown analysis sample, the peak areas of the sample are calculated using the calibration function and the results are given in terms of quantity or concentration. Many data systems also take into account the sample amount as weighed out and dilution or concentration steps in order to be able to give the concentration in the original sample.

For GC/MS the methods of external or internal standard calibration or standard addition are used as standardisation procedures.

3.3.6.1 External Standardization

External standardisation corresponds to the classical calibration procedure (Fig. 3.158). The substance to be determined is used to make a standard solution with a known concentration. Measurements are made on standard solutions of different concentrations (calibration steps, calibration levels). For calibration the peak areas determined are plotted against the concentrations of the different calibration levels.

With external standardisation, standard deviations of 5–10% are obtained in GC/MS, which occur even using highly reproducible injection techniques involving an autosampler. The causes of this distribution are injection errors and small changes in the mass spectrometer. As absolute values from different analysis runs are used for external standardisation, the calibration function and the sample measurement show all the effects which can cause a change in response between the mass spectrometric measurements. Various factors can contribute to this, e.g. slightly different ionisation efficiency through geometric changes in the filament, slightly varying transmission due to contamination of the lens systems with the matrix, or a contribu-

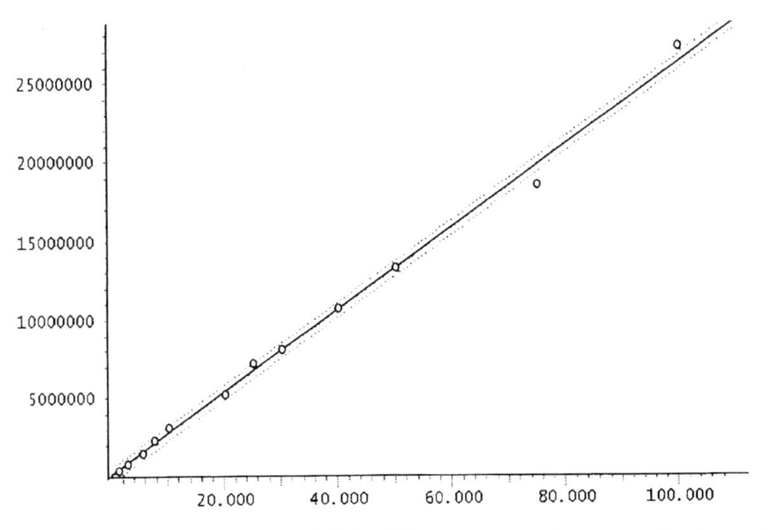

Fig. 3.158 External calibration for PCB 28 (GCQ data system, Finnigan, on-line thermodesorption GC/MS coupling, after Tschickard, Environmental Protection Office, Mainz)

tion from the multiplier on signal production. These factors can be compensated for by internal standardisation.

3.3.6.2 Internal Standardisation

The principle of the internal standard is based on the calculation of relative values which are determined within the same analysis. One or more additional substances are introduced as a fixed reference parameter, the concentration of which is kept constant in the standard solutions and is always added to the analysis sample at the same concentration. For the calculation the peak area values (or peak heights) of the substance being analysed relative to the peak area (or height) of the internal standard are used. In this way all volume errors and variations in the function of the instrument are compensated for and quantitative determinations of the highest precision are achieved. Standard deviations of less than 5% can be achieved with internal standardisation.

The time at which the internal standard is added during the analysis depends on the analysis requirement. For example, the internal standard can be added to the sample at an early stage (surrogate standard) to simplify the clean-up. The addition of different standards at different individual stages of the clean-up allows the efficiency of individual clean-up steps to be monitored. Addition to the extract directly before the measurement can serve to confirm which instrument specification is required and to monitor the minimum signal/noise ratio necessary.

The choice of the internal standard is particularly important also in GC/MS. Basically the internal standard should behave as far as possible in the same way as the substance being analysed. Unlike classical GC detectors, the GC/MS procedure offers the unique possibility of using isotopically labelled, but nonradioactive analogues of the substances being analysed. Deuterated standards are frequently used for this purpose as they fulfil the require-

Requirements of the Internal Standard in GC/MS Analysis

- The internal standard chosen must be stable to clean-up and analysis and as inactive as possible.
- As far as possible the properties of the standard should be comparable to those of the analyte with regard to sample preparation and analysis; therefore isotopically labelled standards should ideally be used.
- The standard itself must not be present in the original sample.
- The retention behaviour in the GC should be adjusted so that elution occurs in the same section of the program (isothermal or heating ramp).
- The use of several standards allows them to be used as retention time standards.
- The retention behaviour of the internal standard should be adjusted to ensure that overlap with matrix peaks or other components to be determined is avoided and faultless integration is possible.
- The fragmentation behaviour (mass spectrometric response) should be comparable.
- The choice of the quantifying mass of the standard should exclude interference by the matrix or other components.

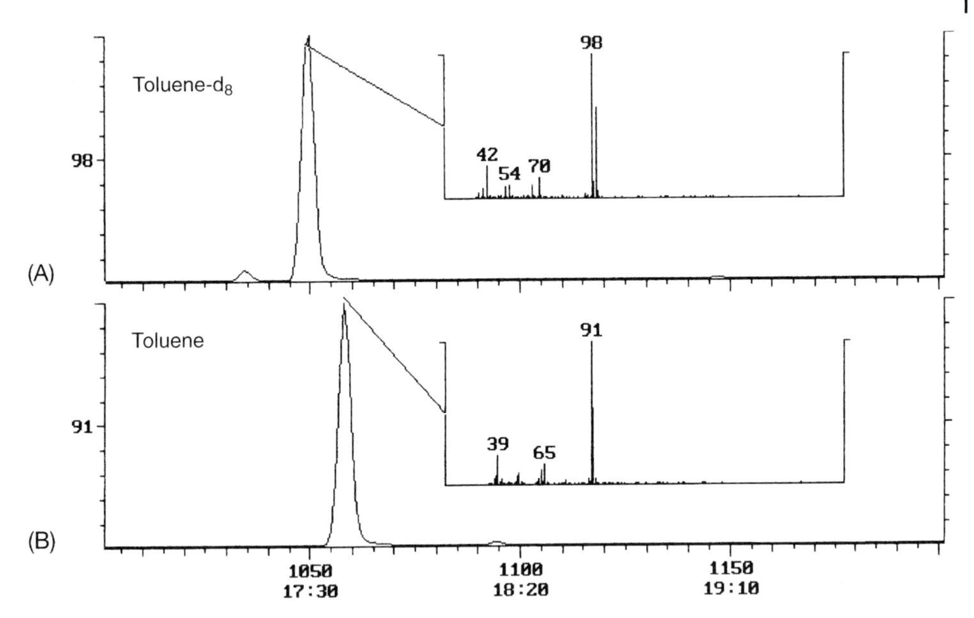

Fig. 3.159 Elution of toluene-d_8 as the internal standard using purge and trap GC/MS (mass chromatograms). (A) Toluene-d_8, m/z 98, (B) Toluene m/z 91

ment of comparable behaviour during clean-up and analysis to the greatest degree. The extent of deuteration should always be sufficiently high for interference with the natural isotope intensities of the unlabelled substance to be excluded. The internal deuterated standard thus chosen can thus be detected selectively through its own mass trace and integrated (Fig. 3.159), see also section "Isotope Dilution".

In the preparation of standard solutions for calibration with internal standards, their concentration must be kept constant in all the calibration levels. Pipetting in separately the same volumes of the internal standard followed by different volumes of a mixed standard stock solution and then making up with solvent to a fixed volume in the sample vial has proved to be a good method (Fig. 3.160). The addition of the internal standard can be performed in routine analysis by program control of the liquid autosampler. Keeping the mixed standard and the internal standard storage vessels separate simplifies many manual steps.

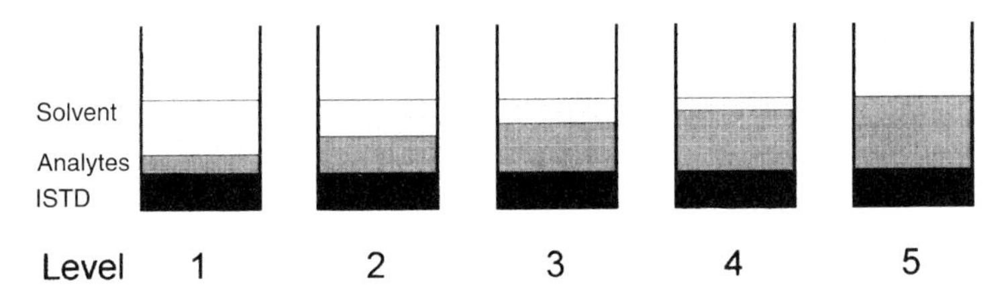

Fig. 3.160 Preparation of solutions for calibration with internal standards (c (ISTD) = constant, c (analytes) = variable, total volume = constant, fill with solvent, calculate concentrations on the basis of total volumes)

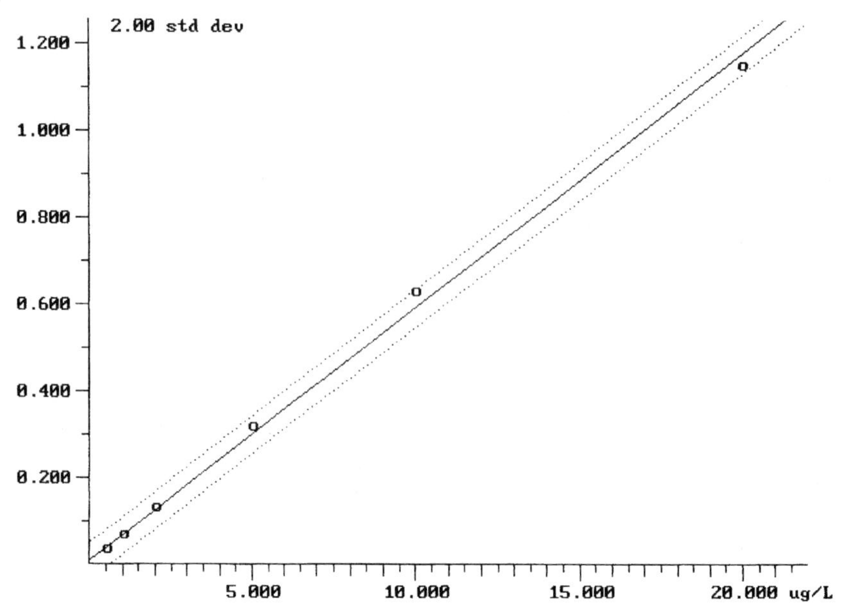

Fig. 3.161 Calibration function using the internal standard procedure.
Plot: area of the sample/area of the standard against concentration in the sample

To calculate the results of analyses using internal standards, after integration of the standard chromatograms the areas of the analysis substance relative to the area of the internal standard are plotted against the concentration in the sample (Fig. 3.161). The parameter determined relative to the internal standard is thus independent of deviations in the injection volume and possible variations in the performance of the detector, as all these influences affect the analyte and the internal standard to the same extent. To calculate the analysis results the ratio of the peak area for the analyte to that for the internal standard is determined and the concentration calculated using the calibration function (Table 3.13).

Isotope Dilution

The "isotope dilution" quantitation method is the most sophisticated and provides the highest accuracy of the internal standard quantification methodology. As internal standard compounds the most similar analogues of the native analytes labelled with stable isotopes are administered. Widely used are ^{13}C-labelled or deuterated compounds, e.g., $^{13}C_{12}$-TCDD or d_{12}-Benzo[a]pyrene. The chemical and physical behavior of the native and labeled compound is almost identical as ideally required by the concept of internal standardization.

The term "isotope dilution" was borrowed from isotope ratio mass spectrometry for quantitative organic analysis. The basic definition asserts that it is not necessary to carry out a standard curve by different standard dilutions. Instead a known quantity of a rare isotope is added as a spike to each sample. The measurement of the isotope ratio of the resulting mix compared to the known isotope ratio of the spike and natural abundances allows the calculation of the concentration of the unknown in the individual sample. In organic quantitation methods typically the average relative response of standards in different concentrations (native ag. labeled ISTD) is used as the basis for quantitative calibration.

Table 3.13 List of volatile halogenated hydrocarbons found in a drinking water analysis using the ISTD procedure (I: internal standard, S: surrogate standard, A: analyte)

No.	Substance	Type	Scan #	Retention time	Me	Calculated conc.	Unit
1	Fluorobenzene IS1	I	680	17:00	BB	5.000	µg/L
2	1,2-Dichlorobenzene IS2	S	1050	26:15	BB	5.001	µg/L
3	Bromofluorobenzene SS1	S	943	23:34	BB	5.000	µg/L
9	Trichlorofluoromethane	A	424	10:36	BB	0.045	µg/L
11	Dichloromethane	A	522	13:03	BB	0.518	µg/L
17	Chloroform	A	630	15:45	BB	8.714	µg/L
18	1,1,1-Trichloroethane	A	641	16:01	BB	0.071	µg/L
21	Benzene	A	664	16:36	BB	0.056	µg/L
22	Carbon tetrachloride	A	651	16:17	BB	0.023	µg/L
24	Trichloroethylene	A	702	17:33	BV	0.060	µg/L
26	Dichlorobromomethane	A	732	18:18	BB	7.415	µg/L
29	Toluene	A	779	19:28	BB	0.167	µg/L
32	Dibromochloromethane	A	829	20:44	BB	4.860	µg/L
34	Tetrachloroethylene	A	813	20:19	BV	0.058	µg/L
35	Chlorobenzene	A	868	21:42	BB	0.032	µg/L
37	Ethylbenzene	A	874	21:51	BV	0.037	µg/L
38	meta, para-Xylene	A	881	22:02	VB	0.054	µg/L
39	Bromoform	A	923	23:05	BB	0.635	µg/L
45	Bromobenzene	A	954	23:51	BB	0.034	µg/L
51	1,2,4-Trimethylbenzene	A	998	24:57	BB	0.084	µg/L
55	Isopropyltoluene	A	1019	25:28	BB	0.057	µg/L
59	Naphthalene	A	1182	29:33	BB	0.155	µg/L
61	Hexachlorobutadiene	A	1176	29:24	BV	0.126	µg/L

A typical example for extended use of the isotope dilution method is the quantification of polychlorinated dioxins. Six-fold and twelve-fold labeled dioxins and furans in the required chlorination degrees are used as recovery and surrogate standards during sample preparation and GC/MS analysis, as described in the well known and widely applied EPA 1613 method. The nearly identical compound characteristics during clean-up, the chromatographic process and detection, guarantees most reliable quantitation results based on the sample individual recovery values in one analysis run (see details in the application Section 4.27 on dioxin analysis).

In the clean-up of biological material the carrier effect can be exploited through the addition of an internal standard. The standard added at comparatively high concentrations can cover up active sites in the matrix and thus improve the extraction of the substance being analysed. The result is a significantly improved extraction recovery from active matrices for the native compound for a reliable representation of the given content in the sample. In chromatography deuterated standards have a slightly shorter retention time than the native analytes. This retention time difference, though small, is always visible in the mass chromatogram, which is used for selective integration of the individual components.

3.3.6.3 The Standard Addition Procedure

Matrix effects frequently lead to varying extraction yields. The headspace and purge and trap techniques in particular are affected by this type of problem. If measures for standardising the matrix are unsuccessful, the standard addition procedure can be used, as in, for example, atomic absorption (AAS) for the same reason. Here the calibration is analogous to the external standardisation described above and involves addition of known quantities of the analyte to be determined. The calibration samples, however, are prepared with constant quantities of the analyte by addition of corresponding volumes of the standard solution. One sample is left as it is, i. e. no standard is added.

The analysis results are calculated by plotting the peak areas against the quantity added. The calibration function cuts the y axis at a height corresponding to the concentration of analyte in the sample, expressed as a peak area. The concentration in the sample can be read off by extrapolation of the calibration function to the point where it cuts the x axis (Fig. 3.162).

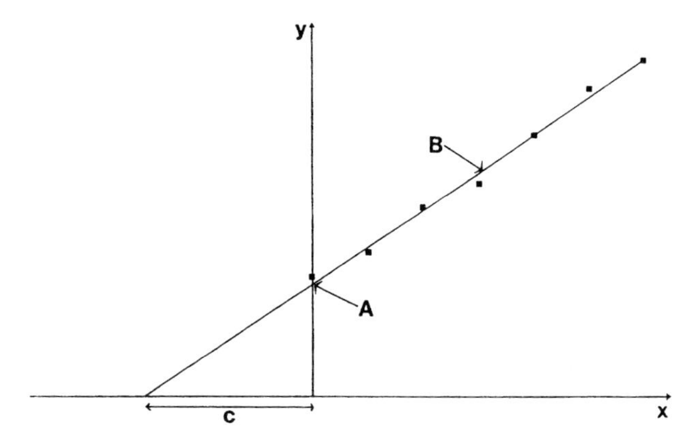

Fig. 3.162 Calibration using standard addition. x axis: quantity of substance, y axis: signal intensity.
(A) Intercept with the y axis: signal of the unaltered sample, quantity added = 0
(B) At this point the signal intensity has been doubled. If there is a direct dependence between concentration and signal, a quantity has been given at this point which corresponds to the content in the sample
(C) Content in the sample

In spite of the good results obtained with this procedure, in practice there is one major disadvantage. A separate calibration must be carried out for every sample. On the other hand a calibration for a series of samples can be used with external or internal standards. A further criticism of the addition procedure arises from the statistical aspect. The linear extrapolation of the calibration function is only carried out assuming its validity for this area also. A possible nonlinear deviation in this lower concentration range would give rise to considerable errors. To allow a better fit of the calibration function, a larger number of additions should be analysed. Nevertheless, the reservation still exists that the calibration curve must be extrapolated beyond the valid area for the values to be obtained.

The Accuracy of Analytical Data

In spite of precise measuring procedures and careful evaluation of the data, the accuracy of analytical data is highly dependent on the sampling procedure, transportation to the laboratory, choice of a representative laboratory sample, and the sample preparation procedure (Fig. 3-163)! Even powerful instrumental analysis cannot correct errors that have already occurred.

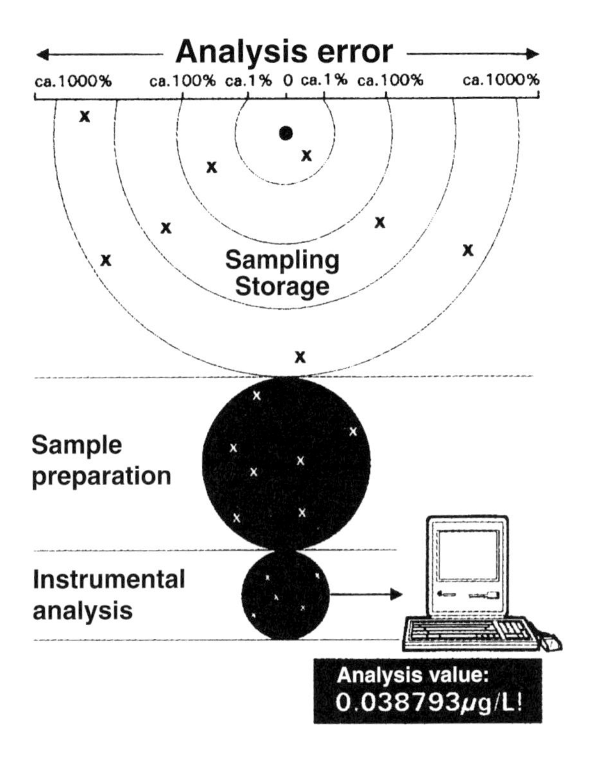

Fig. 3.163 Sources of error in analysis (after Hein/Kunze)

3.4 Frequently Occurring Impurities

With increasing concentration during sample preparation and increasing sensitivity of GC/MS systems, the question of what are the necessary laboratory conditions for trace analysis is becoming more important. Interfering signals from impurities are increasingly appearing (Table 3.14). These arise from 'cleaned' glass apparatus (e.g. rotary evaporators, pipettes), through contact with polymers e.g. cartridges from solid-phase extraction, septa of autosampler, solvent and sample bottles), from the solvents themselves (e.g. stabilisers) or even from the laboratory surroundings (e.g. dust, solvent vapours etc). There are also sources of interfering signals in the GC/MS system itself. These are often in the surroundings of the GC injector. They range from septum bleed or decomposing septa to impurities which have be-

Table 3.14 Mass signals of frequently occurring contaminants

m/z values	Possible cause
149, 167, 279	Phthalate plasticisers, various derivatives
129, 185, 259, 329	Tri-n-butyl acetyl citrate plasticiser
99, 155, 211, 266	Tributyl phosphate plasticiser
91, 165, 198, 261, 368	Tricresyl phosphate plasticiser
108, 183, 262	Triphenylphosphine (synthesis by-product)
51, 77, 183, 201, 277	Triphenylphosphine oxide (synthesis by-product)
41, 55, 69, 83. ... 43, 57, 71, 85, ... 99, 113, 127, 141, ... 285, 299, 313, 327, ... 339, 353, 367, 381, ... 407, 421, 435, 449, ... 409, 423, 437, 451, ...	Hydrocarbon background (forepump oil, greasy fingers, tap grease etc)
64, 96, 128, 160, 192, 224, 256	Sulfur (as S_8)
205, 220	Antioxidant 2,6-di-t-butyl-4-methylphenol (BHT, Ionol) and isomers (technical mixture)
115, 141, 168, 260, 354, 446	Poly(phenyl ether) (diffusion pump oil)
262, 296, 298	Chlorophenyl ether (impurity in diffusion pump oil)
43, 59, 73, 87, 89, 101, 103, 117, 133	Poly(ethylene glycol) (all Carbowax phases)
73, 147, 207, 221, 281, 355, 429	Silicone rubber (all silicone phases)
133, 207, 281, 355, 429	Silicone grease
233, 235	Rhenium oxide ReO_3^- (from the cathode in NCl)
217, 219	Rhenium oxide ReO_2^-
250, 252	Rhenium oxide $HReO_4^-$

come deposited in the split vent and get into the measuring system on subsequent injections. The capillary column used is well known for sometimes high background noise caused by column bleed (silicone phases). As these typical mass signals constantly occur in trace analyses, the structures of the most frequently occurring ions are listed (Fig. 3.164). Sources of contamination in the mass spectrometer are also known. Among these are background signals arising from pump oil (fore pump, diffusion pump) and degassing products from sealing materials (e.g. Teflon) and ceramic parts after cleaning.

For trace analysis the carrier gas used (helium, hydrogen) should be of the highest possible purity from the beginning of installation of the instrument. Contamination from the gas supply tubes (e.g. cleaned up for ECD operations!) leads to ongoing interference and can only be removed at great expense. With central gas supply plants in particular gas purification (irreversible binding of organic contaminants to getter materials) directly at the entry to the GC/MS system is recommended for trace analysis. Leaks in the vacuum system, in the GC system and in the carrier gas supply always lead to secondary effects and should therefore be carefully eliminated. Under no circumstances should plastic tubing be used for the carrier gas supply to GC/MS systems (Table 3.15).

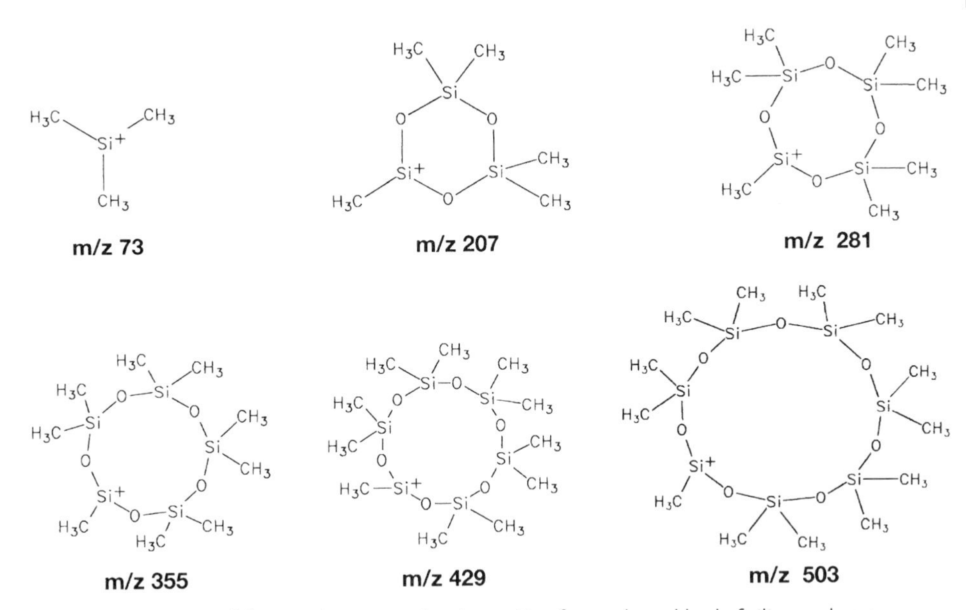

Fig. 3.164 Structures of the most important signals resulting from column bleed of silicone phases (after Spiteller)

Table 3.15 Diffusion of oxygen through various line materials (air products)

Line material	Contamination by O_2 [ppm]
Copper	0
Stainless steel	0
Kel-F	0.6
Neoprene	6.9
Polyethylene (PE)	11
Teflon (PTFE)	13
Rubber	40

Note: measured in argon 6.0 at a line diameter of 6 mm and 1 m length with a flow rate of 5 l/h.

Butyl Rubber Septa

In GC/NCI-MS an intense contamination can be observed exhibiting m/z 166 with a significant isotope cluster. This compound was found to migrate from butyl rubber septa of autosampler vials into the sample and can be explained as artifacts from vulcanization facilitators (Kapp 2006).

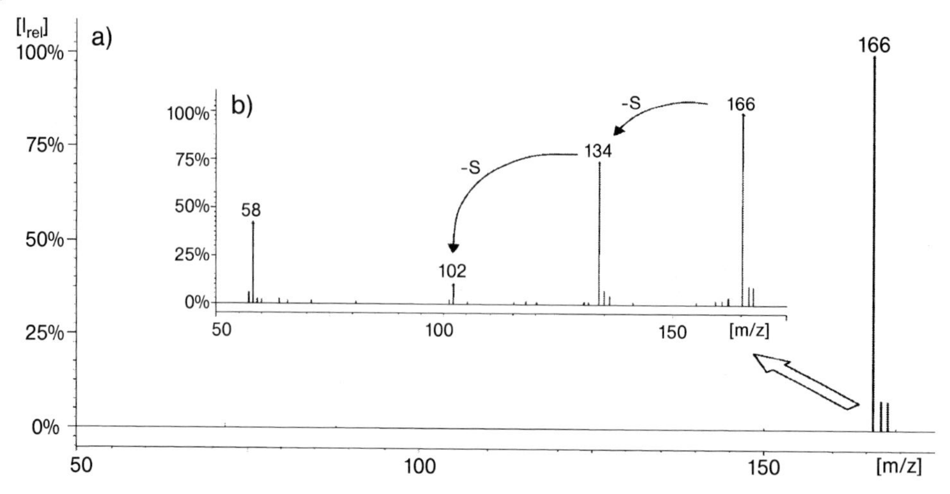

Fig. 3.165 (a) GC-NCI-MS spectrum of a contaminant from buthyl rubber septa.
(b) Confirmation of two S atoms by GC-MS/MS

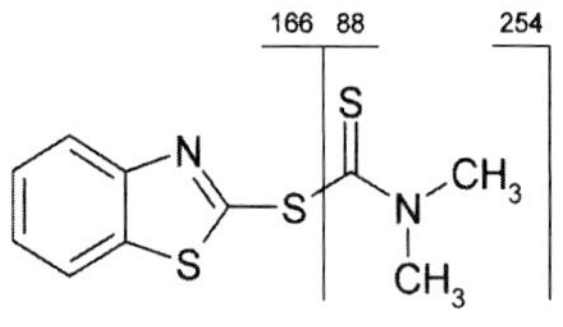

Fig. 3.166 Structure of the artefact 2-benzothiazolyl-N,N-dimethyldithiocarbamate

Some typical interfering components frequently occurring in residue analysis are listed together with their spectra (Figs. 3.167 to 3.179). They are listed in order of the mass of the base peak.

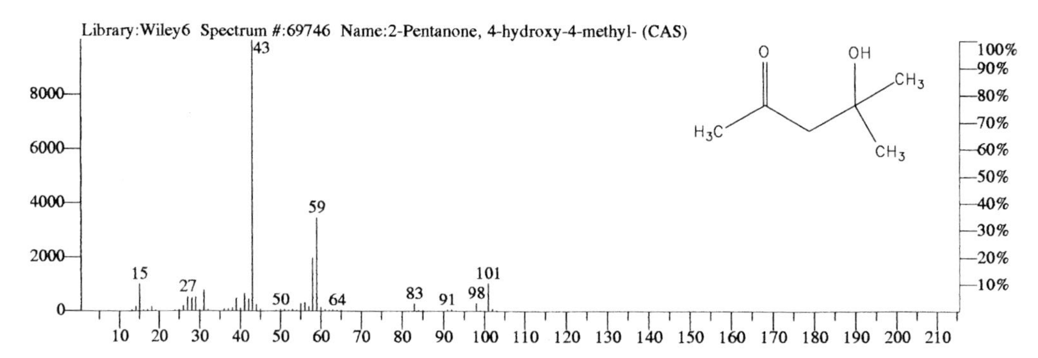

Fig. 3.167 Diacetone alcohol $C_6H_{12}O_2$, M: 116, CAS: 123-42-2
Occurrence: acetone dimer, forms from the solvent under basic conditions

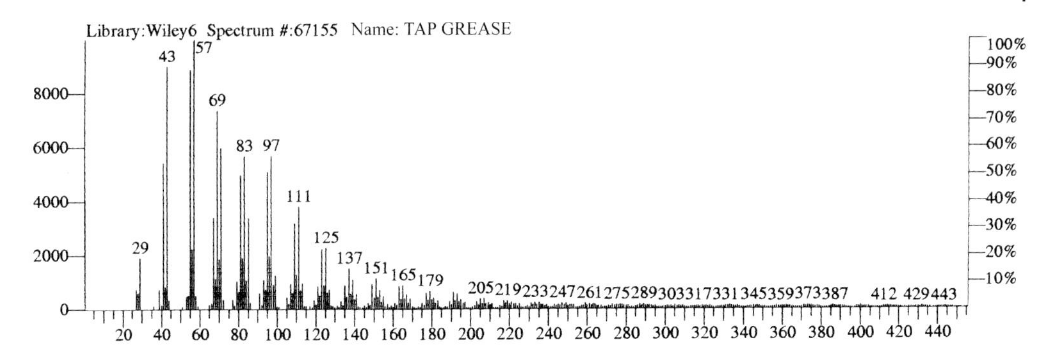

Fig. 3.168 Hydrocarbon background
Occurrence: greasy fingers as a result of maintenance work on the analyser, the ion source or after changing the column, background of forepump oil (rotatory slide valve pumps), joint grease

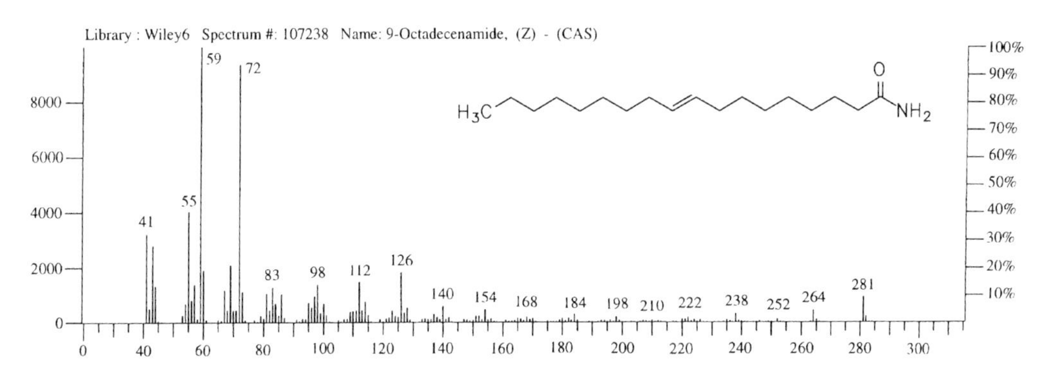

Fig. 3.169 Oleic acid amide
Occurrence: lubricant for plastic sheeting, erucamide occurs just as frequently

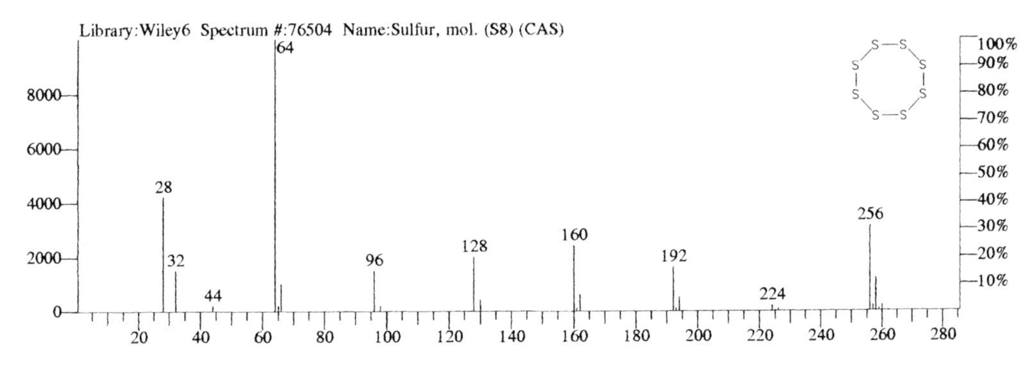

Fig. 3.170 Molecular sulfur S_8, M: 256, CAS: 10544-50-0
Occurrence: from soil samples (microbiological decomposition), impurities from rubber objects, and by-product in the synthesis of thio-compounds

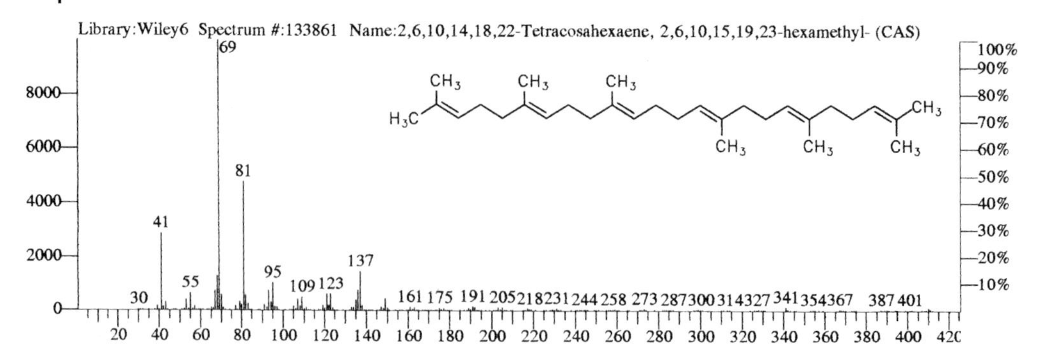

Fig. 3.171 Squalene $C_{30}H_{50}$, M: 410, CAS: 7683-64-9
Occurrence: stationary phase in gas chromatography

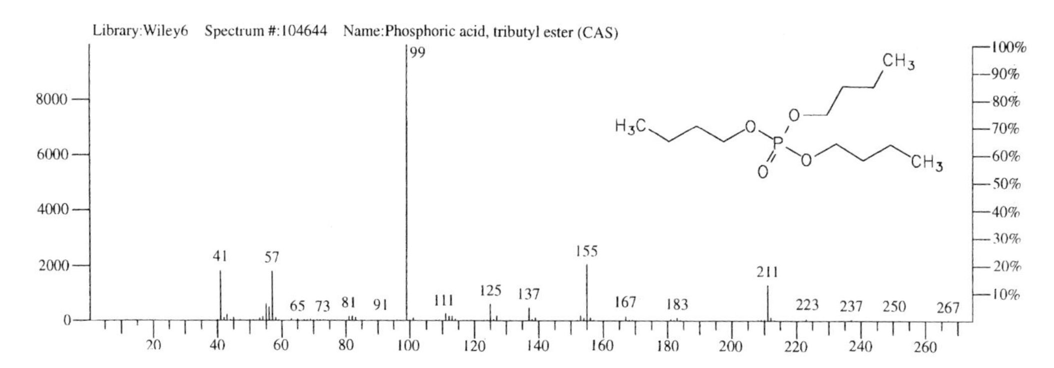

Fig. 3.172 Tributyl phosphate $C_{12}H_{27}O_4P$, M: 266, CAS: 126-73-8
Occurrence: widely used plasticiser

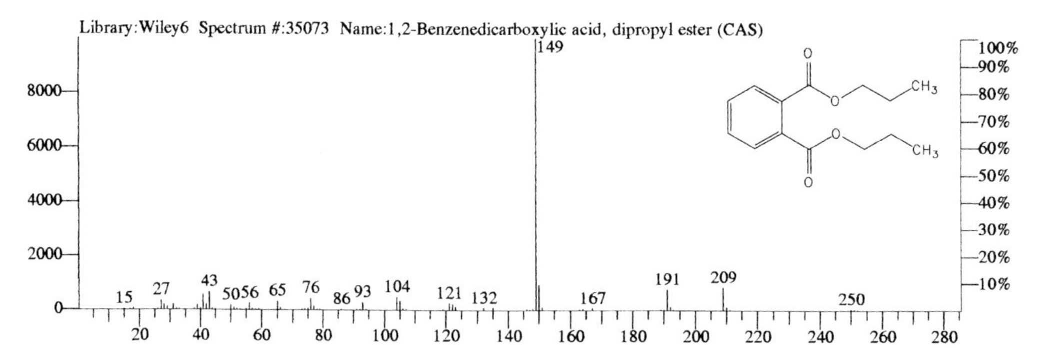

Fig. 3.173 Propyl phthalate $C_{14}H_{18}O_4$, M: 250, CAS: 131-16-8
Occurrence: plasticiser, widely used in many plastics (also SPE cartridges, bottle tops etc), typical mass fragment of all dialkyl phthalates m/z 149

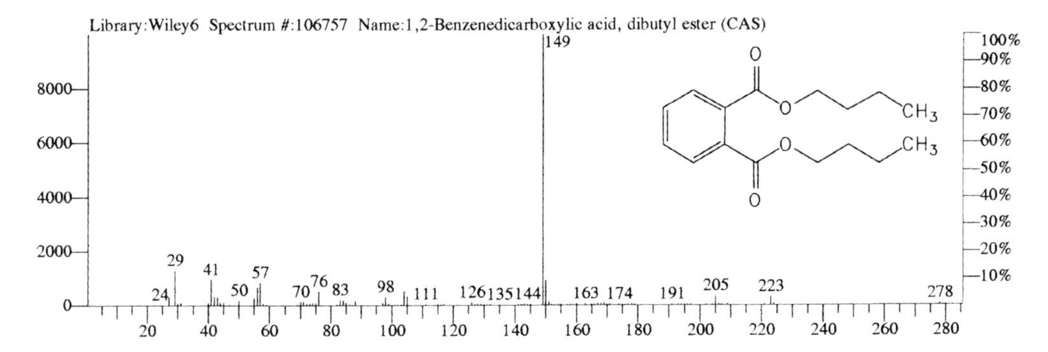

Fig. 3.174 Dibutyl phthalate $C_{16}H_{22}O_4$, M: 278, CAS: 84-74-2

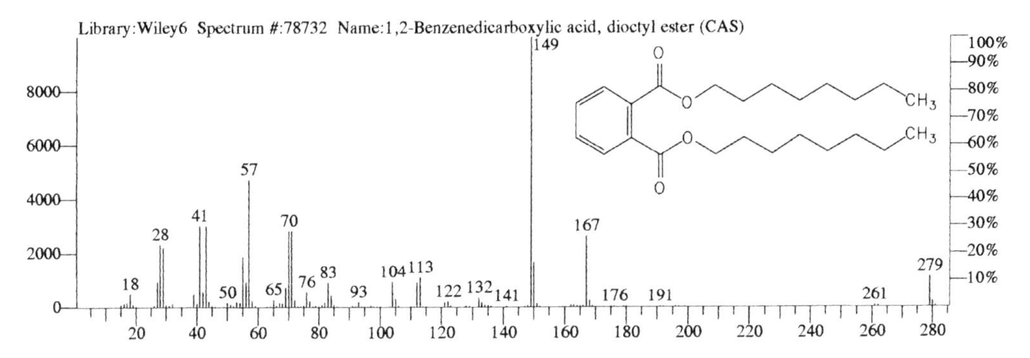

Fig. 3.175 Dioctyl phthalate $C_{24}H_{38}O_4$, M: 390, CAS: 117-84-0

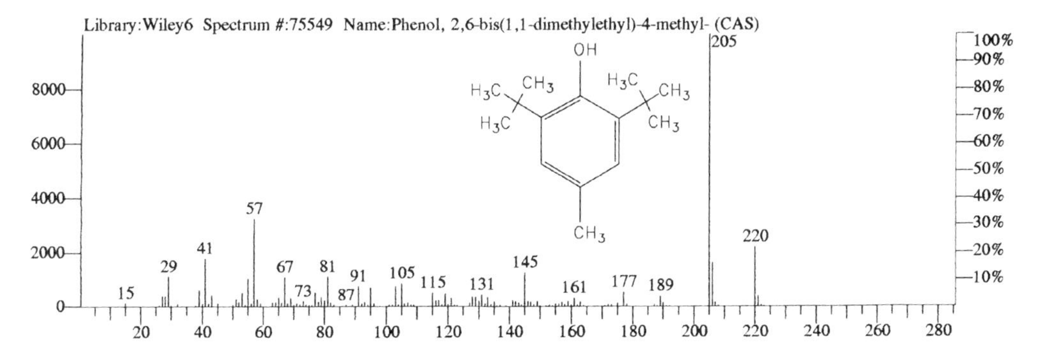

Fig. 3.176 Ionol (BHT) $C_{15}H_{24}O$, M: 220, CAS: 128-37-0
Occurrence: antioxidant in plastics, stabiliser (radical scavenger) for ethers, THF, dioxan, technical mixture of isomers

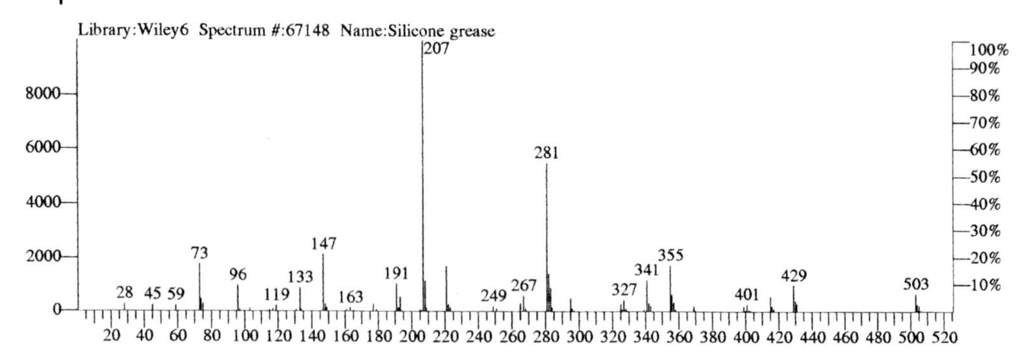

Fig. 3.177 Silicones, silicone grease
Occurrence: typical column bleed from many silicone phases, from septum caps of sample bottles, injectors etc (see also Fig. 3.160)

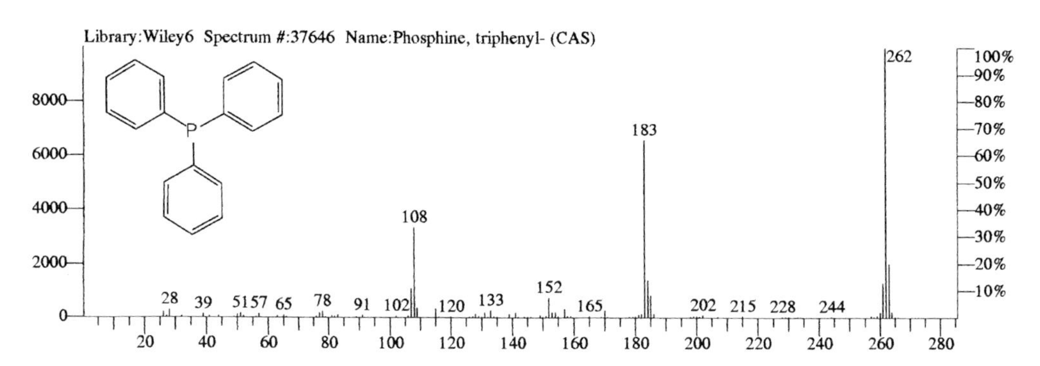

Fig. 3.178 Triphenylphosphine $C_{18}H_{15}P$, M: 262, CAS: 603-35-0
Occurrence: catalyst for syntheses

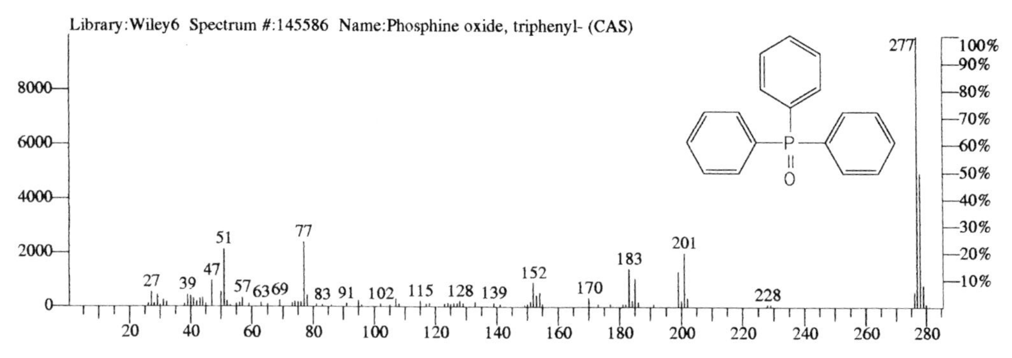

Fig. 3.179 Triphenylphosphine oxide $C_{18}H_{15}OP$, M: 278, CAS: 791-28-6
Occurrence: forms from the catalyst in syntheses (e.g. the Wittig reaction)

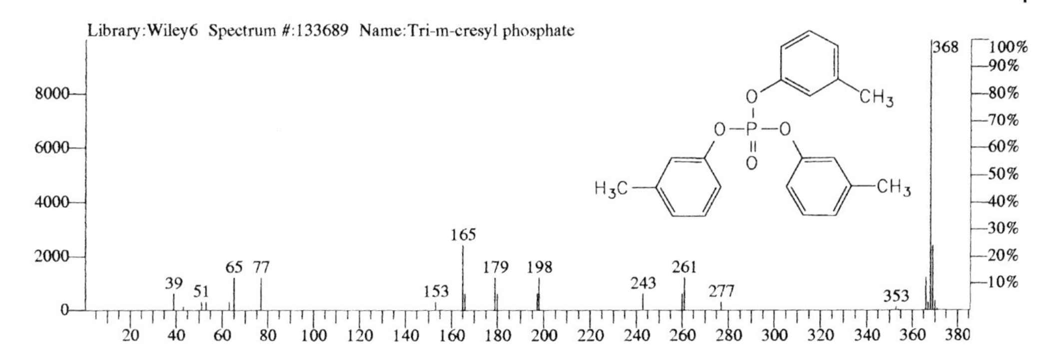

Fig. 3.180 Tri-m-cresyl phosphate $C_{21}H_{21}O_4P$, M: 368, CAS: 563-04-2
Occurrence: plasticiser (e.g. in PVC, nitrocellulose etc)

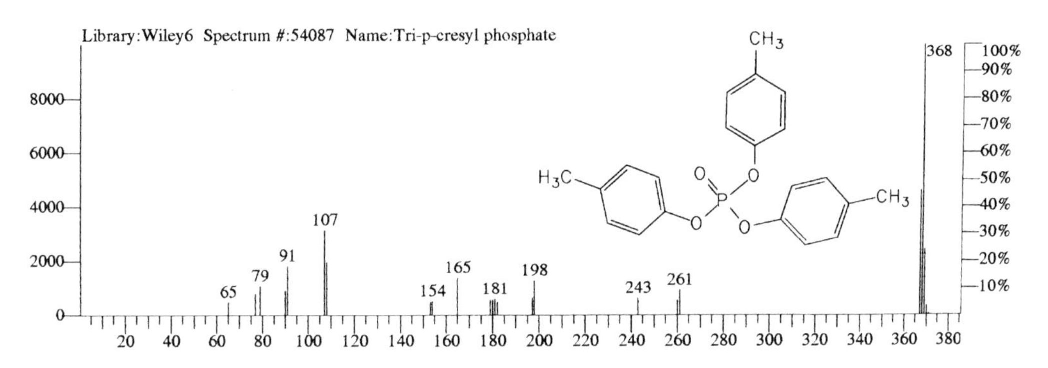

Fig. 3.181 Tri-p-cresyl phosphate $C_{21}H_{21}O_4P$, M: 368, CAS: 78-32-0
Occurrence: plasticiser (e.g. in PVC, nitrocellulose etc)

References for Chapter 3

Section 3.1.1

Barwick, V., et al., Best practice guide for generating mass spectra, LGC: Teddington UK, December 2006, see also www.lgc.co.uk.

Bemgard, A., Colmsjö, A., Wrangskog, K., Prediction of Temperature-Programmed Retention Indexes for Polynuclear Aromatic Hydrocarbons in Gas Chromatography, Anal. Chem. 66 (1994), 4288–4294.

Ciccioli, P., Brancaleoni, E., Cecinato, A., Frattoni, M., A Method for the Selective Identification of Volatile Organic Compounds (VOC) in Air by HRGC-MS, In: 15th Int. Symp. Cap. Chromatogr., Riva del Garda, May 1993, P. Sandra (Ed.), Heidelberg: Huethig 1993, 1029–1042.

Deutsche Forschungsgemeinschaft, Gaschromatographische Retentionsindizes toxikologisch relevanter Verbindungen auf SE-30 oder OV-1, Mitteilung 1 der Kommission für Klinisch-toxikologische Analytik, Verlag Chemie: Weinheim 1982.

Hall, G. L., Whitehead, W. E., Mourer, C. R., Shibamoto, T., A New Gas Chromatographic Retention Index for Pesticides and Related Compounds, HRC & CC 9 (1986), 266–271.

Katritzky, A. L., Ignatchenko, E. S., Barcock, R. A., Lobanov, V. S., Karelson, M., Prediction of Gas Chromatographic Retention Times and Response Factors Using a General Quantitative StructureProperty Relationship Treatment, Anal. Chem. 66 (1994), 1799–1807.

Kostiainen, R., Nokelainen, S., Use of M-Series Retention Index Standards in the Identification of Trichothecenes by Electron Impact Mass Spectrometry, J. Chromatogr. 513 (1990), 31–37.

Lipinski, J., Stan, H.-J., Compilation of Retention Data for 270 Pesticides on Three Different Capillary Columns, In: 10th Int.Symp.Cap.Chromatogr., Riva del Garda, May 1989, P. Sandra (Ed.), Heidelberg: Huethig 1989, 597–611.

Manninen, A., et al., Gas Chromatographic Properties of the M-Series of Universal Retention Index Standards and their Application to Pesticide Analysis, J.Chromatogr. 394 (1987), 465–471.

Schomburg, G., Gaschromatographie, 2. Aufl., Weinheim: VCH 1987, 54–62.

Weber, E., Weber, R., Buch der Umweltanalytik, Band 4, Methodik und Applikationen in der Kapillargaschromatographie, GIT Verlag: Darmstadt 1992, 64–65.

Zenkevich, I. G., The Exhaustive Database for Gaschromatographic Retention Indizes of Low-Boiling Halogenhydrocarbons at PLOT Alumina Columns, In: 15th Int.Symp.Cap.Chromatogr., Riva del Garda, May 1993, P. Sandra (Ed.), Heidelberg: Huethig 1993, 181–186.

Section 3.2.1

Ausloos, P., Clifton, C.L., Lias, S.G., Mikaya, A.I., Stein, S.E.,Tchekhovskoi, D.V., Sparkman, O.D., Zaikin,V., Zhu, D., The Critical Evaluation of a Comprehensive Mass Spectral Library, J. Am. Soc. Mass Spectrom. 10, (1999) 287–299.

Dimandja, J.M.D., GCxGC, Anal. Chem. 76, 9 (2004) 167A-174A.

Halket, J.M., Przyborowska, A., Stein, S.E., Mallard, W.G., Down, S., Chalmers, R., Deconvolution Gas Chromatography/Mass Spectrometry of Urinary Organic Acids – Potential for Pattern Recognition and Automated Identification of Metabolic Disorders, Rapid Commun. Mass Spectrom. 13, (1999) 279–284.

Mallard, W.G., Reed, J., AMDIS – USER GUIDE, U.S. Department of Commerce, Technology Administration, National Institute of Standards and Technology (NIST), Standard Reference Data Program, Gaithersburg, MD 20899. This user guide is available in PDF format: www.chemdata.nist.gov/mass-spc/amdis/AMDIS.pdf.

Mallard, G., Stein, S., Toropov, O., AMDIS – Automatic Mass Spectral Deconvolution and Identification Software, Agilent Technologies Product Information Deconvolution Reporting Software 2005.

Meng, C.K., Szelewski, M., Find Hidden Target Compounds, Agilent Technologies Newsletter 2005.

Shao, X., Wang, G., Wang, S., Su, Q., Extraction of Mass Spectra and Chromatographic Profiles from Overlapping GC/MS Signals with Background, Anal. Chem. 76, (2004) 5143–5148.

Stein, S.E., Estimating Probabilities of Correct Identification from Results of Mass Spectral Library Searches, J. Am. Soc. Mass Spectrom. 5, (1994) 316–323.

Stein, S.E., Scott, D.R., Optimization and Testing of Mass Spectral Library Search Algorithms for Compound Identification, J. Am. Soc. Mass Spectrom. 5, (1994) 859–866.

Stein, S.E., Chemical Substructure Identification by Mass Spectral Library Searching, J. Am. Soc. Mass Spectrom. 6, (1995) 644–655.

Stein, S.E., An Integrated Method for Spectrum Extraction and Compound Identification from Gas Chromatography/Mass Spectrometry Data, J. Am. Soc. Mass Spectrom. 10, (1999) 770–781.

Zhang, W., Wu, P., Li, C., Study of automated mass spectral deconvolution and identification system (AMDIS) in pesticide residue analysis, Rapid Comm. Mass Spectrom. 20, (2006) 1563–1568.

General References

The complete manual for AMDIS is available as a PDF file (http://chemdata.nist.gov/mass-spc/amdis/AMDIS_PDF263.zip). It is also available as a Microsoft Word © document along with additional test data for AMDIS. The manual and sample data package are archived in a ZIP files, with additional instructions in a readme file. This data is supplied for use with the "Quick Start AMDIS" located in Appendix A of the AMDIS Help system.

Algorithms: Details of algorithms employed in AMDIS are described in the paper "An Integrated Method for Spectrum Extraction and Compound Identification from GC/MS Data" which appeared in the Journal of the American Society of Mass Spectrometry in Volume 10, 1999, pages 770–781. This paper is available in PDF format: http://chemdata.nist.gov/mass-spc/amdis/method.pdf

AMDIS for download see: http://chemdata.nist.gov/mass-spc/amdis/AMDIS32_FULLSETUP_264.zip

Mallard, W.G., Reed, J., AMDIS – USER GUIDE, U.S. Department of Commerce, Technology Administration, National Institute of Standards and Technology (NIST), Standard Reference Data Program, Gaithersburg, MD 20899 (www.chemdata.nist.gov/mass-spc/amdis/AMDIS.pdf)

Section 3.2.2

Bianchi, F., Careri, M., Mangia, A., Musci, M., Retention indices in the analysis of food volatiles in temperature-programmed gas chromatography: Database creation and evaluation, J. Separation Sci. 30, (2007) 563–572.

Weber, E., Weber, R., Buch der Umweltanalytik, Band 4, Methodik und Applikationen in der Kapillargaschromatographie, GIT Verlag: Darmstadt 1992.

Section 3.2.3

AAFS Drug Library, see: http://www.ualberta.ca/~gjones/mslib.html.

Adams, R.P., Identification of Essential Oils by Ion Trap Mass Spectrometry, Academic Press: San Diego 1989.

Ausloos, P., Clifton, C.L., et al., The Critical Evaluation of a Comprehensive Mass Spectral Library, J. Am. Soc. Mass Spectrom. 10, (1999) 287–299.

Budzikiewicz, H., Massenspektrometrie, 4. Auflage, Wiley-VCH: Weinheim 1998.

Cairns, T., Siegmund, E., Jacobsen, R., Mass Spectral Data Compilation of Pesticides and Industrial Chemicals, Dpt. Health and Human Services, Food and Drug Administration, Office of Regulatory Affairs, Los Angeles CA. 1985.

Central Institute of Nutrition and Food Research, Mass Spectra of Volatiles in Food (SpecData), 2nd Edition, December 2003, John Wiley & Sons, ISBN: 978–0-471–64825-3.

Chemical Concepts, Mass Spectra Chemical Concepts (SpecInfo), December 2006, John Wiley & Sons, ISBN: 978–0-471–66229-7.

Christie, W.W., Mass Spectrometry of Fatty Acid Derivatives, see http://www.lipidlibrary.co.uk/masspec.html.

De Leeuw, J.W., Mass Spectra of Geochemicals, Petrochemicals and Biomarkers (SpecData), August 2004, John Wiley & Sons, ISBN: 978-0-471-64798-0.

Heller, C., Mass Spectral and GC Data of Drugs, Poisons, Pesticides, Pollutants and Their Metabolites, Toxichem 60, 3 (1993).

Kuehnle, R., Mass Spectra of Drugs, Pharmaceuticals and Metabolites (SpecInfo), December 2006, John Wiley & Sons.

Makin, H.L.J., Mass Spectra of Androgenes, Estrogens and other Steroids 2005, October 2005, John Wiley & Sons, ISBN: 978-0-471-74893-9.

McLafferty, F. W., Stauffer, D. B., The Wiley/NBS Registry of Mass Spectral Data, Wiley & Sons: New York 1989.

National Institute of Science and Technology (NIST), NIST Standard Reference Database 1A, NIST/EPA/NIH Mass Spectral Library with Search Program, see: http://www.nist.gov/srd/nist1a.html.

Palisade Mass Spectrometry, The Palisade Complete Mass Spectral Library, see: http://www.palisade-ms.com.

Pfleger, K., Maurer, H. H., Weber, A., Mass Spectra and GC Data of Drugs, Poisons, Pesticides, Pollutants and Their Metabolites, Parts 1,2,3, 2. erw. Aufl., VCH: Weinheim 1992.

Pfleger, K., Maurer, H.H., Weber, A., Mass Spectral and GC Data of Drugs, Poisons, Pesticides, Pollutants and Their Metabolites, 3rd Edition, May 2007, John Wiley & Sons, ISBN: 978-3-527-31538-3.

Pfleger, K., Maurer, H.H., Weber, A., Mass Spectral and GC Data of Drugs, Poisons, Pesticides, Pollutants and Their Metabolites, Mass Spectral and GC Data of Drugs, Poisons, Pesticides, Pollutants and Their Metabolites: Parts I-IV, 2nd, Revised and Enlarged Edition, October 2000, John Wiley & Sons, ISBN: 978-3-527-29793-1.

Rösner, P., Mass Spectra of Designer Drugs, August 2005, John Wiley & Sons, ISBN: 978-0-471-74385-9.

Stein, S.E., Estimating Probabilities of Correct Identification From Results of Mass Spectral Library Searches, J. Am. Soc. Mass Spectrom. 5, (1994) 316–323.

Stein, S.E., An Integrated Method for Spectrum Extraction and Compound Identification from Gas Chromatography/Mass Spectrometry Data, J. Am. Soc. Mass Spectrom. 10, (1999) 770–781.

Stein, S.E., Scott, D.R., Optimization and Testing of Mass Spectral Library Search Algorithms for Compound Identification, JASMS in press.

U.S. Department of Commerce, NIST/EPA/NIH Mass Spectral Library 2005, September 2005, John Wiley & Sons, ISBN: 978-0-471-75594-4.

Wiley Registry of Mass Spectral Data, 8th Edition, July 2006, John Wiley & Sons, ISBN: 978-0-470-04785-9.

Wiley Registry 8th Edition/NIST 2005 Mass Spectral Library, September 2006, ISBN: 978-0-470-04786-6.

Yarkov, A., Mass Spectra of Organic Compounds (SpecInfo), July 2006, John Wiley & Sons, ISBN: 978-0-471-66231-0.

Section 3.2.4

Atwater, B. L., Stauffer, D. B., McLafferty, F. W., Peterson, D. W., Reliability Ranking and Scaling Improvements to the Probability Based Matching System for Unknown Mass Spectra, Anal. Chem. 57 (1985), 899–903.

Beynon, J. H., Brenton. A. G., Introduction to Mass Spectrometry. University Wales Press: Swansea 1982.

CLB-Redaktion, Die zwölfmillionste chemische Verbindung beschrieben, CLB 45 (1994), 215.

Davies, A. N., Mass Spectrometric Data Systems, Spectroscopy Europe 5 (1993), 34–38.

Fachinformationszentrum Chemie GmbH. MS Online Programm Mass-Lib-Retrieval Programm for Use with Mass Spectra Libraries, User Manual V 3.0, Fachinformationszentrum Chemie: Berlin 1988.

Henneberg, D., Weimann, B, Search for Identical and Similar Compounds in Mass Spectral Data Bases, Spectra 1984, 11–14.

Hübschmann, H. J., MS-Bibliothekssuche mittels INCOS und PBM, LABO Analytica 4 (1992), 102–118.

McLafferty, F. W., Turecek F., Interpretation of Mass Spectra, 4th. Ed., University Science Books: Mill Valley, CA 1993.

Neudert, B., Bremser, W., Wagner, H., Multidimensional Computer Evaluation, Org. Mass Spectrom. 22 (1987) 321–329.

Palisade Corporation, *Benchtop/PBM Users Guide*, 1994.

Sokolow, S., Karnovsky, J., Gustafson, P., The Finnigan Library Search Program, Finnigan MAT Applikation No. 2, 1978.

Stauffer, D. B., McLafferty, F. W., Ellis, R. D., Peterson, D. W., Adding Forward Searching Capabilities to a Reverse Search Algorithm for Unknown Mass Spectra, Anal. Chem. 57 (1985), 771–773.

Stauffer, D. B., McLafferty, F. W., Ellis. R. D., Peterson, D. W., Probability-Based-Matching Algorithm with Forward Searching Capabilities for Matching Unknown Mass Spectra of Mixtures, Anal. Chem. 57 (1985), 1056–1060.

Stein, S. E., Estimating Probabilities of Correct Identification From Results of Mass Spectral Library Search, J. Am. Soc. Mass Spectrom. 5 (1994) 316–323.

Stein, S. E., Scott, D. R., Optimization and Testing of Mass Spectral Library Search Algorithms for Compound Identification, J. Am. Soc. Mass Spectrom. 5 (1994), 859–866.

Warr, W. A., Computer-Assisted Structure Elucidation-Part 1: Library Search and Spectral Data Collections, Anal. Chem. 65 (1993) 1045A–1050A.

Warr, W. A., Computer-Assisted Structure Elucidation – Part 2: Indirect Database Approaches and Established Systems, Anal. Chem. 65 (1993) 1087A–1095A.

Yang, L., Martin, M., Glazner, M., et al., Comparison of Three Benchtop Mass Spectrometers: Initial Development Phase of a Volatile Organic Analyzer for Space Station Freedom, presented at the ASMS, Tucson 1990.

Section 3.2.5

Budzikiewicz, H., Massenspektrometrie, 4. Auflage, Wiley-VCH: Weinheim 1998.

Budzikiewicz, H., Structure elucidation by ion-molecule reactions in the gas phase: The location of C, C-double and triple bonds, Fres. J. Anal. Chem., 321/2, (1985) 150–158.

Christie, W.W., Mass spectrometry of fatty acids – Part 1, first published in Lipid Technology, 8, (1996) 18–20; substantially re-written see: www.lipidlibrary.co.uk/topics/ms_fa_1/file.pdf

Christie, W.W., A practical guide to the analysis of conjugated linoleic acid, inform 12/2, (2001) 147–152.

Dobson, G., Christie, W.W., Spectroscopy and spectrometry of lipids – part 2, Mass spectrometry of fatty acid derivatives, Eur. J. Lipid Sci. Technol. 104, (2002) 36–43.

Fay, L., Richli, U., Location of double bonds in polyunsaturated fatty acids by gas chromatography-mass spectrometry after 4,4-dimethyloxazoline derivatization, J.Chromatogr. 541, (1991) 89–98.

IAEA, Environmental Isotopes in the Hydrological Cycle, Principles and Applications, Ed. Mook, W.G., Volume I: Introduction – Theory, Methods, Review, http://www.iaea.org/programmes/ripc/ih/volumes/ volumes.html.

Jhama, G.N., Attygalleb, A.B., Meinwalda, J., Location of double bonds in diene and triene acetates by partial reduction followed by methylthiolation, J. Chromatogr. A 1077/1, (2005) 57–67.

López, J.F., Grimalt, J.O., Phenyl- and cyclopentyl-imino derivatization for double bond location in unsaturated C (37)-C, (40) alkenones by GC-MS, J Am Soc Mass Spectrom. 15(8), (2004) 1161–1172.
McLafferty, F. W., Turecek, F., Interpretation von Massenspektren. Heidelberg: Springer 1995.
Moss, C.W., Lambert-Fair, M.A., Location of double bonds in monounsaturated fatty acids of Campylobacter cryaerophila with dimethyl disulfide derivatives and combined gas chromatography-mass spectrometry, J. Clin. Microbiol. 27(7), (1989) 1467–1470.
Ockels, W., Hübschmann, H. J., Interpretation von Massenspektren I/II, Manuskript zum Fortbildungskursus der Axel Semrau GmbH, Sprockhövel 1994.
Rosman, K.J.R., Taylor, P.D.P., Isotopic Compositions of the Elements 1997, Pure Appl. Chem. 70/1, (1998) 217–235.
Shantha, N. C., Kaimal, T. N. B., Mass spectrometric location of double bonds in unsaturated fatty acids including conjugated acids as their methoxybromo derivatives, Lipids 19/12, (1984) 971–974.
Sharp, Z., Principles of Stable Isotope Geochemistry, Pearson Prentice Hall Upper Saddle River 2007.

Section 3.2.6

Benz, W., Henneberg, D., Massenspektrometrie organischer Verbindungen, Akad. Verlagsgesellschaft: Frankfurt 1969.
Budzikiewicz, H., Massenspektrometrie, 4. Auflage, Wiley-VCH: Weinheim 1998.
Färber, H., Schöler, F., Gaschromatographische Bestimmung von Harnstoffherbiziden in Wasser nach Methylierung mit Trimethylanilinium-hydroxid oder Trimethylsulfoniumhydroxid, Vom Wasser 77 (1991,), 249–262.
Howe, I., et al., Mass Spectrometry, McGraw-Hill: New York 1981.
McLafferty, F. W., Mass Spectrometric Analysis – Molecular Rearrangements, Anal. Chem. 31 (1959), 82–87.
McLafferty, F. W., Stauffer, D. B., The Wiley/NBS Registry of Mass Spectral Data, 5th Ed., Wiley-Interscience: New York.
Ockels, W., Pestizid-Bibliothek, Axel Semrau GmbH: Sprockhövel 1993.
Ockels, W., Hübschmann, H. J., Interpretation von Massenspektren, Fortbildungskurs, Axel Semrau GmbH: Sprockhövel 1993.
Ockels, W., Hübschmann, H. J., Interpretation von Massenspektren II – Substanzidentifizierung in der Rückstandsanalytik, Fortbildungskurs, Axel Semrau GmbH: Sprockhövel 1994.
Pretsch, E., et al., Tabellen zur Strukturaufklärung organischer Verbindungen mit spektrometrischen Methoden, 3. Auflage, 1. korr. Nachdruck, Springer: Berlin 1990.
Schröder, E., Massenspektrometrie, Springer: Berlin 1991.
Seibl, J., Massenspektroskopie, Akad. Verlagsgesellschaft: Frankfurt 1970.
Stan, H.-J., Abraham, B., Jung, J., Kellert, M., Steinland, K., Nachweis von Organophosphor-insecticiden durch Gas-Chromatographie-Massenspektroskopie, Fres. Z. Anal. Chem. 287 (1977), 271–285.
Stan, H.-J., Kellner, G., Analysis of Organophosphoric Pesticide Residues in Food by GC/MS Using Positive and Negative Chemical Ionisation, In: Baltes, W., Czodik-Eysenberg, P. B., Pfannhauser, W. (Eds.), Recent Development in Food Analysis, Proceedings of the First European Conference on Food Chemistry (EURO FOOD CHEM I), Vienna 17–20 February 1981, Verlag Chemie: Weinheim 1981, 183–189.

Section 3.3

Boqué, R., Marato, A., Vander Heyden, Y., Assessment of Accuracy in Chromatographic Analysis, LCGC Europe May (2008) 264–267.
ISO 5725:1994. Accuracy (trueness and precision) of measurement methods and results (Parts 1–4), International Organization for Standardisation, Geneva 1994.

Section 3.3.6

Barker, J., Mass Spectrometry, 2nd Ed., Wiley: Chichester 1999.
Chang, C., Parallel Mass Spectrometry for High Performance GC and LC Detection, Int. Laboratory 5 (1985) 58–68.
Doerffel, K., Statistik in der analytischen Chemie, VCH: Weinheim 1984.
Ebel, S., Dorner, W., Jahrbuch Chemielabor 1987, VCH: Weinheim 1987.
Ebel, S., Kamm, U., Fres. Z. Anal. Chem. 316 (1983) 382–385.
Funk, W., Dammann, V., Donnevert, G., Qualitätssicherung in der Analytischen Chemie, VCH: Weinheim 1992.

Guichon, G., Guillemin, C. L., Quantitative Gas Chromatography, Elsevier: Amsterdam, Oxford, New York, Tokyo 1988.

Hein, H., Kunze, W., Umweltanalytik mit Spektrometrie und Chromatographie, VCH: Weinheim 1994.

Meyer, V. R., Richtigkeit bei der Peakflächenbestimmung, GIT Fachz. Lab. 1 (1994) 4–5.

Miller, J. N., The Method of Standard Additions, Spectroscopy Europe 1992, 4/6, 26–27.

Montag, A., Beitrag zur Ermittlung der Nachweis- und Bestimmungsgrenze analytischer Meßverfahren, Fres. Z.Anal.Chem. 1982 (312), 96–100.

Nachweis-, Erfassungs- und Bestimmungsgrenze, DIN 32645, Beuth-Verlag, Berlin.

Naes, T., Isakson, T., The Importance of Outlier Detection in Spectroscopy, Spectroscopy Europe 1992, 4/4, 32–33.

Vogelsang, J., Höchstmengenüberschreitung und Streuung der Analysenwerte bei der Rückstandsanalyse von Pestiziden in Trinkwasser, Lebensmittelchem. 1990 (44), 7.

Section 3.4

Kapp, T., Vetter, W., Migration von Additiven aus Buthylgummihaltigen Verschlüssen von Probegläschen, Lebensmittelchemie 60 (2006) 152.

Spiteller, M., Spiteller, G., Massenspektrensammlung von Lösemitteln, Verunreinigungen, Säulenbelegmaterialien und einfachen aliphatischen Verbindungen, Wien: Springer 1973.

4 Applications

The applications given here have been chosen in order to describe typical areas of use of GC/MS, such as air, water, soil, foodstuffs, the environment, waste materials, drugs and pharmaceutical products. Special emphasis has been placed on current and reproducible examples which give an idea of what is going on in routine laboratories. The selection cannot be totally representative of the use of modern GC/MS, but shows the main areas into which the methodology has spread and will continue to do so. In addition, in special areas of application, such as the analysis of gases or aromas, isotope-specific measuring procedures and other powerful examples of the use of GC/MS are described.

Most of the applications described are compiled from the references cited and documented with various print-outs and lists. The analysis conditions are described in full to allow adaptation of the methods. If any of the methods have been published, the sources are given for each section. References to other directly related literature are also given.

4.1 Air Analysis According to EPA Method TO-14

Key words: air, SUMMA canister, thermodesorption, volatile halogenated hydrocarbons, cryofocusing, water removal, thick film column

The EPA (US Environmental Protection Agency) describes a process for sampling and analysis of volatile organic compounds (VOCs) in the atmosphere. It is based on the collection of air samples in passivated stainless steel canisters (SUMMA canisters). The organic components are separated by GC and determined using conventional GC detectors or by mass spectrometry (Fig. 4.1). The use of mass spectrometers allows the direct positive detection of individual components.

For an SIM analysis the mass spectrometer is programmed in such a way that a certain number of compounds in a defined retention time range are detected. These SIM ranges are switched periodically so that a list of target compounds can be worked through. In the full scan mode the mass spectrometer works as a universal detector during data acquisition.

A cryoconcentrator with a 3-way valve system is used for concentration. The sample can be let in by two routes without having to alter the screw joints on the tubing. Usually the inlet is via a mass flow regulator to a cryofocusing unit. The direct measurement of the sample volume provides very precise data. A Nafion drier is used to dry the air.

Handbook of GC/MS: Fundamentals and Applications, Second Edition. Hans-Joachim Hübschmann

ISBN: 978-3-527-31427-0

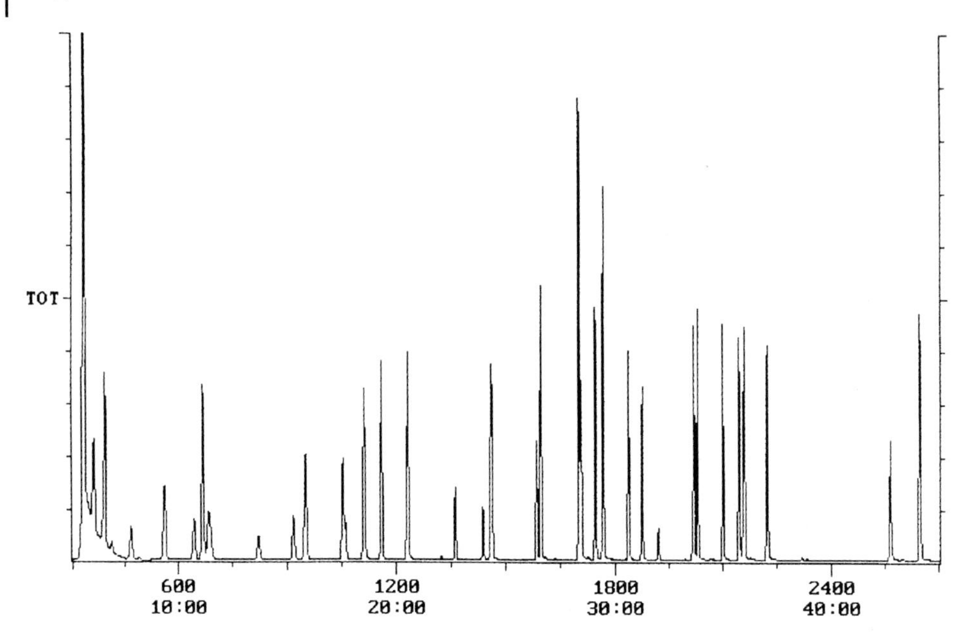

Fig. 4.1 A typical chromatogram for a 1000 mL sample of the 10 ppbv standard

SUMMA Canisters

The analysis of volatile organic compounds is frequently carried out by adsorption on to suitable materials. For this Tenax is mainly used (see Section 2.1.4). The limits of this adsorption method lie in the adsorption efficiency, which is dependent upon the compound, the breakthrough of the sample at higher air concentrations, the impossibility of multiple measurements on a sample and the possible formation of artefacts. Stainless steel canisters, whose inner surfaces have been passivated by the SUMMA process, do not exhibit these limitations. This passivation process involves polishing the inner surface and then applying a Cr/Ni oxide layer. Containers treated in this way have been used successfully for the collection and storage of air samples. Purification and handling of these canisters and the sampling apparatus must be carried out carefully, however, because of possible contamination problems.

Another means of sample injection is the loop injection. The sample is drawn through a 5 mL sample loop directly into the cryoconcentrator. This method is suitable for highly concentrated samples, as only a small quantity of sample is required. In this case the drier is avoided.

The Nafion drier is a system for removal of water from the air sample, which uses a semipermeable membrane. Nonpolar compounds pass through the membrane while polar ones, such as water, are held by the membrane and diffuse outwards. The outer side of the membrane is dried by a clean air stream and the water thus separated is removed from the system. The Nafion drier is recommended by the TO-14 method to prevent blockage of the cryofocusing unit by the formation of ice crystals.

In this application the GC is coupled to the mass spectrometer by an open split interface. A restrictor limits the carrier gas flow. Open coupling was chosen because the sensitivity of the ion trap GC/MS makes the concentration of large quantities of air superfluous. Open coupling dilutes the moisture, which may be contained in the sample, to an acceptable level so that cryofocusing can be used without additional drying.

Analysis Conditions

Open split interface:	SGE type GMCC/90 (SGE), mounted in the transfer line to the MS, restrictor 0.05 mm, adjusted to 2.5% transmission
GC/MS system:	Finnigan MAGNUM Mode: EI, 35–300 u Scan rate: 1 s/scan
GC separation:	J&W DB-5, 60 m × 0.25 mm × 1 µm. The thick film columns DB-1 or DB-5 guarantee chromatographic separation even at start temperatures just above room temperature if cryofocusing is used. Helium 1 ml/min.
Program:	Start: 35 °C, 6 min Ramp: 8 °C/min Final temp.: 200 °C
Concentrator:	Grasby Nutech model 3550A cryoconcentrator with 354A cryofocusing unit, Nafion drier
Autosampler:	Nutech 3600 16-position sampler
Sample injection:	The sample is drawn out of the SUMMA canister through the Nafion drier and reaches the cryoconcentrator cooled to –160 °C. It is then heated rapidly to transfer the sample to the cryofocusing unit of the GC (Fig. 4.2). At –190 °C the sample is focused in a fused silica column of 0.53 mm ID and then heated to 150 °C for injection.
Calibration:	A 6-point calibration based on 1 litre samples from a SUMMA canister was carried out. The concentration range for the calibration was between 0.1 ppbv and 20 ppbv (Fig. 4.3).

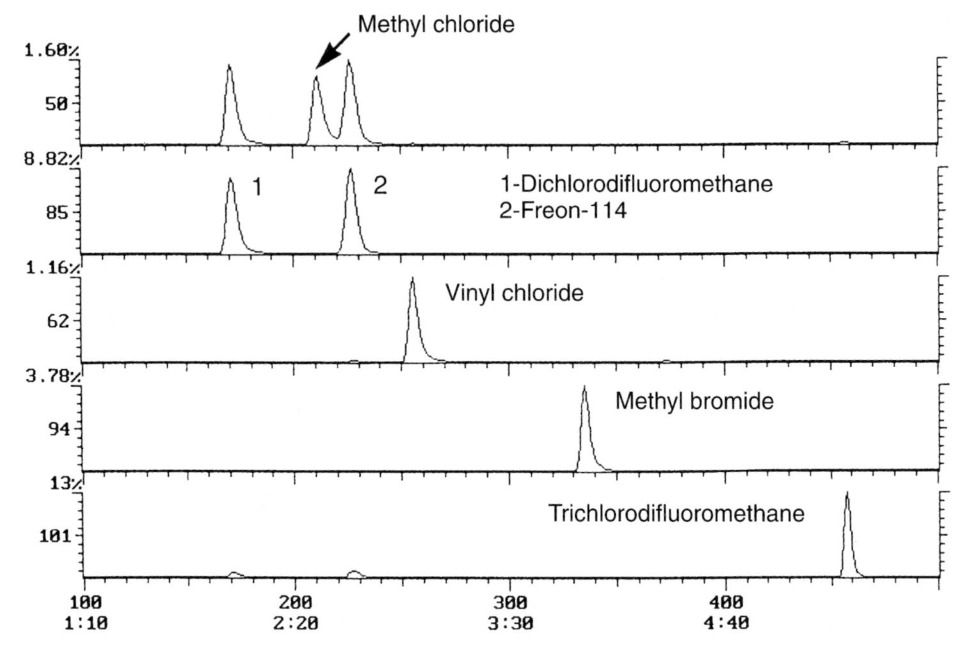

Fig. 4.2 Elution of gases following injection using cryofocusing

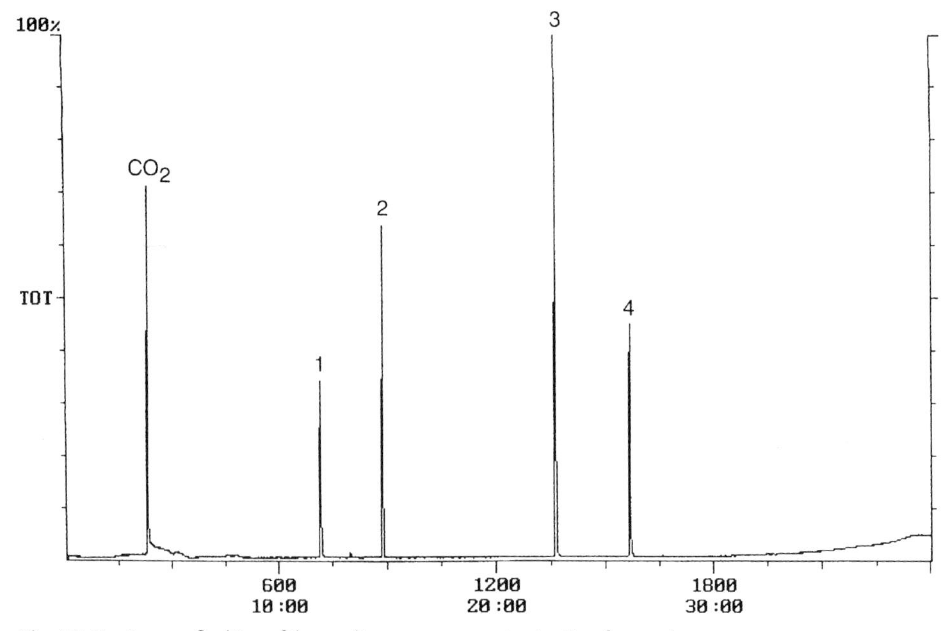

Fig. 4.3 Continuous flushing of the gas lines ensures contamination-free analyses Internal standards used here: (1) Bromochloromethane, (2) 1,4-Difluorobenzene, (3) Chlorobenzene- d_5, (4) Bromofluorobenzene

Limit of Detection

For the EPA method TO-14 (Figs. 4.4 and 4.5) a limit of detection of 0.1 ppbv is required. This requirement is achieved even at the high split rate of the interface.

Results

With this method data acquisition in the full scan mode allows mass spectra to be run with subsequent identification through library searching even at the required limit of detection of 0.1 ppbv. Figures 4.6 and 4.7 show examples of comparability of the spectra and the identification of dichlorobenzene at 20.0 ppbv and 0.1 ppbv. With the procedure described both the compounds required according to TO-14 and other unexpected components in the critical concentration range can be identified.

Dichlordifluormethan	0.01 ppbv	trans-1,3-Dichlorpropen	0.01 ppbv
Chlormethan	0.01 ppbv	Toluol	0.01 ppbv
Freon-114	0.02 ppbv	1,1,2-Trichlorethan	0.01 ppbv
Vinylchlorid	0.01 ppbv	1,2-Dibromethan	0.01 ppbv
Brommethan	0.01 ppbv	Tetrachlorethen	0.01 ppbv
Chlorethan	0.08 ppbv	Chlorbenzol	0.01 ppbv
Trichlorfluormethan	0.01 ppbv	Ethylbenzol	0.01 ppbv
1,1-Dichlorethen	0.02 ppbv	m/p-Xylole	0.01 ppbv
Methylenchlorid	0.01 ppbv	Styrol	0.02 ppbv
3-Chlorpropan	0.02 ppbv	o-Xylol	0.01 ppbv
Freon-113	0.01 ppbv	1,1,2,2-Tetrachlorethan	0.09 ppbv
1,1-Dichlorethan	0.01 ppbv	4-Ethyltoluol	0.02 ppbv
cis-1,2-Dichlorethen	0.01 ppbv	1,3,5-Trimethylbenzol	0.02 ppbv
Chloroform	0.01 ppbv	1,2,4-Trimethylbenzol	0.01 ppbv
1,1,1-Trichlorethan	0.01 ppbv	1,3-Dichlorbenzol	0.01 ppbv
1,2-Dichlorethan	0.01 ppbv	Benzylchlorid	0.08 ppbv
Benzol	0.01 ppbv	1,4-Dichlorbenzol	0.01 ppbv
Tetrachlorkohlenstoff	0.01 ppbv	1,2-Dichlorbenzol	0.01 ppbv
1,2-Dichlorpropan	0.02 ppbv	1,2,4-Trichlorbenzol	0.01 ppbv
Trichlorethen	0.02 ppbv	Hexachlorbutadien	0.02 ppbv
cis-1,3-Dichlorpropen	0.01 ppbv		

Fig. 4.4 Limits of detection for compounds used in the EPA method TO-14

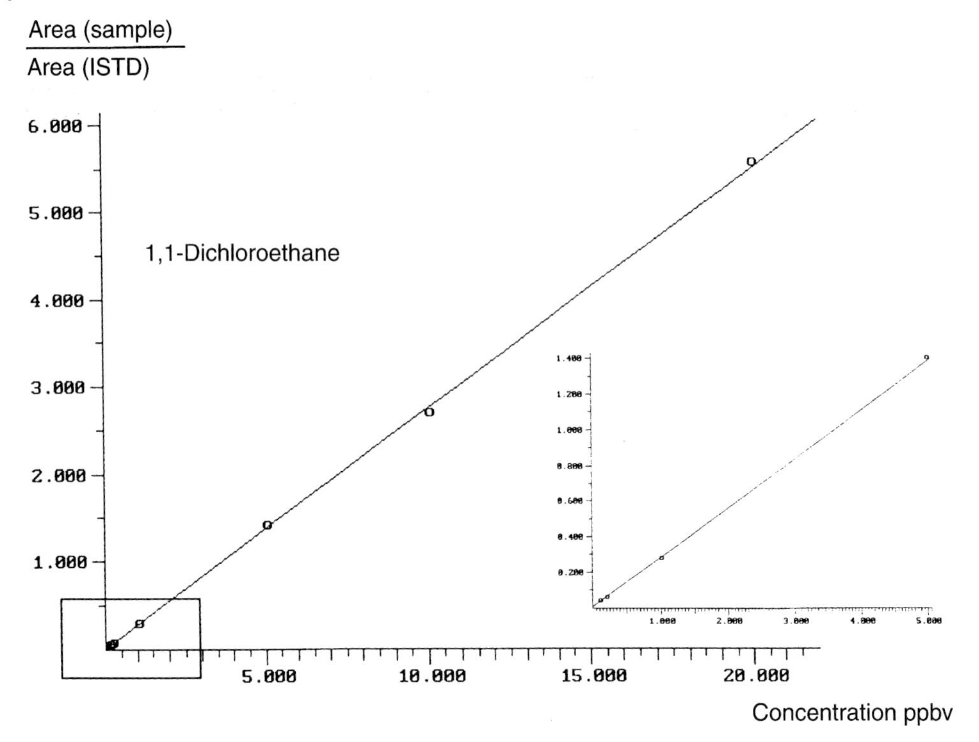

Fig. 4.5 Six point calibration from 0.1 to 20 ppbv. The lower region is shown magnified. ppbv = parts per billion in volume

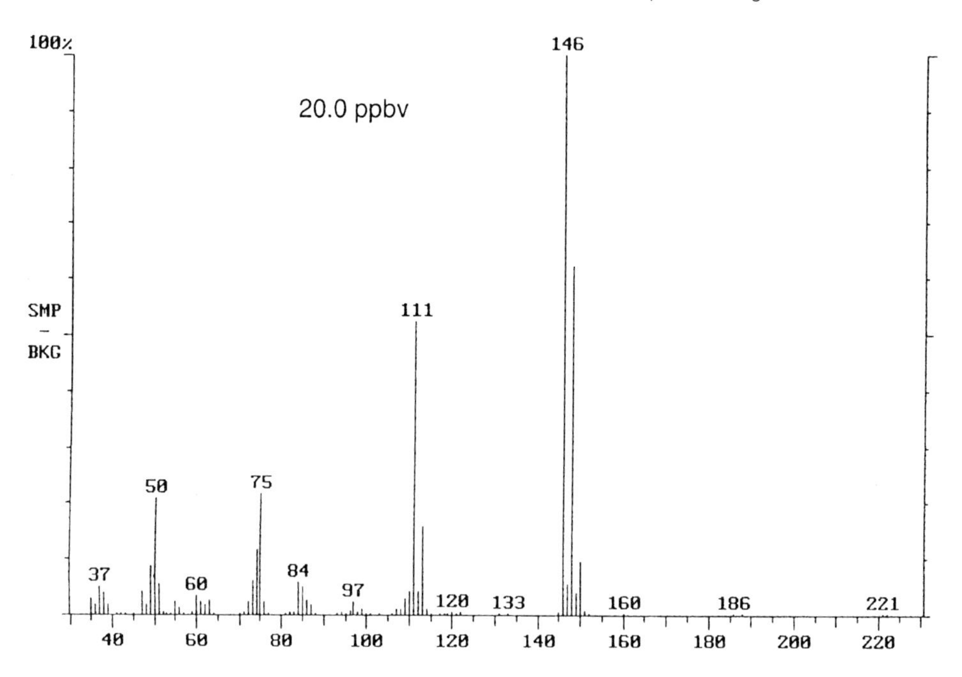

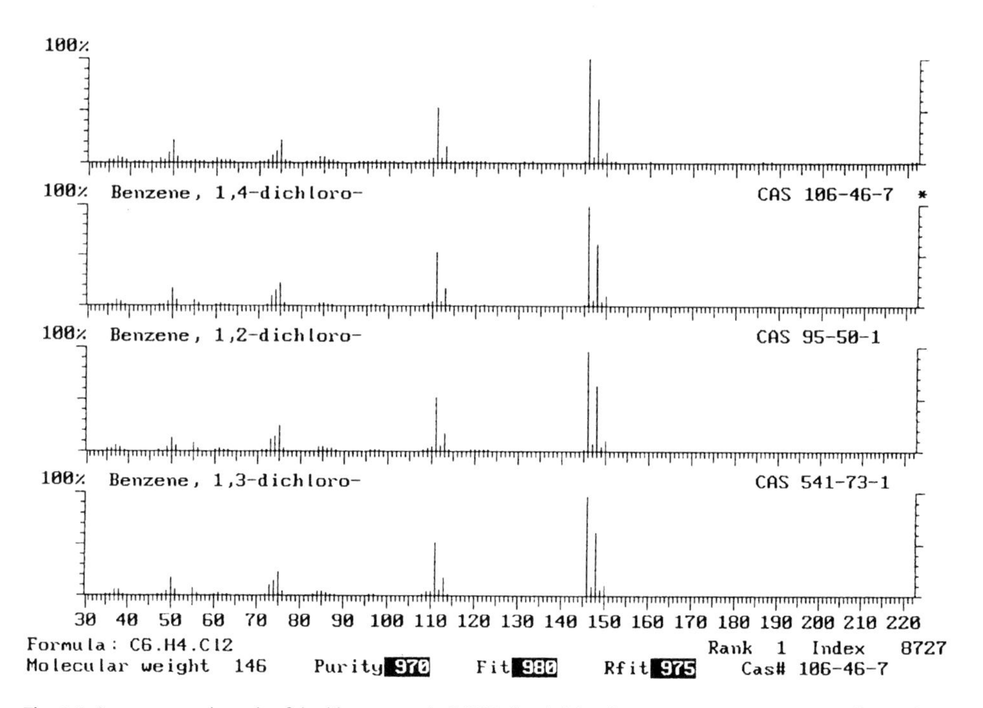

Fig. 4.6 Spectrum and result of the library search (NIST) for dichlorobenzene at a concentration of 20 ppbv

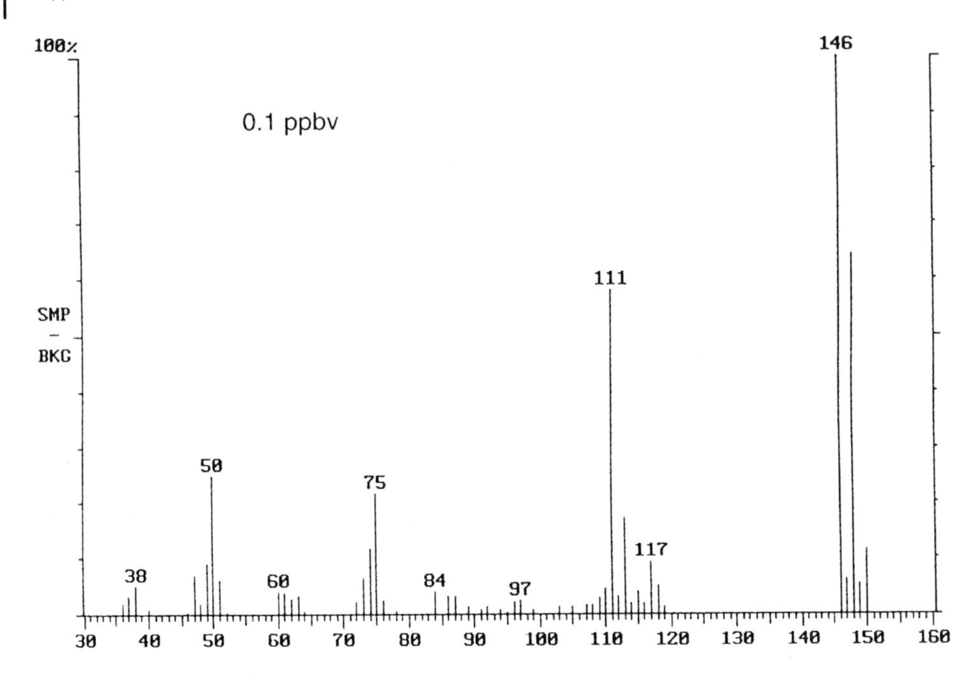

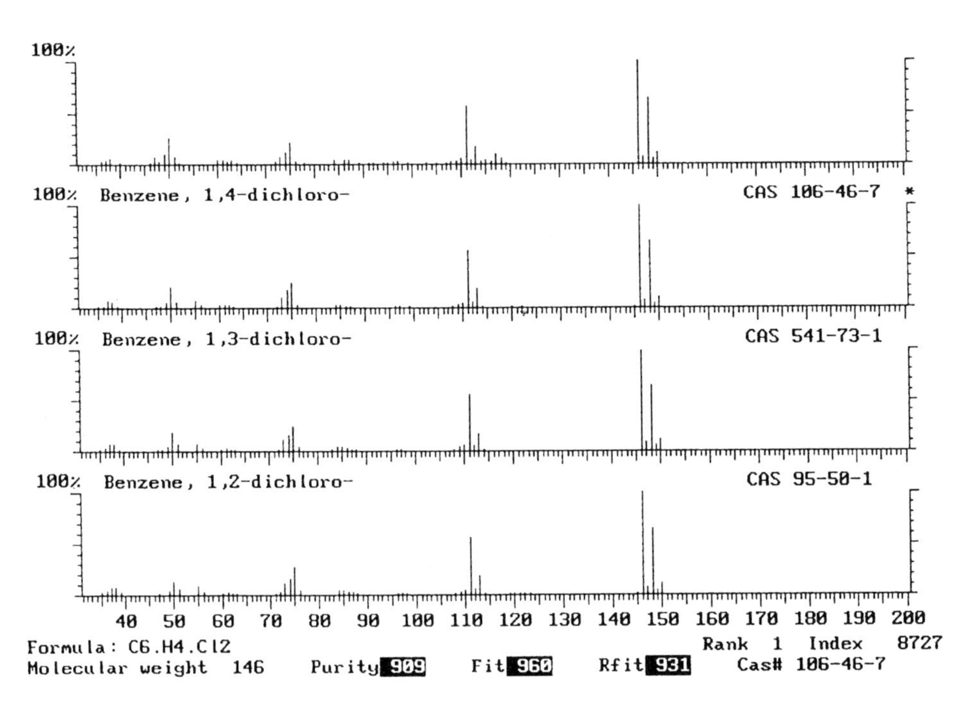

Fig. 4.7 Spectrum and result of the library search (NIST) for dichlorobenzene at a concentration of 0.1 ppbv

4.2
BTEX Using Headspace GC/MS

Key words: headspace, pressure balanced injection, BTEX, reproducibility, sensitivity, automatic evaluation

The determination of BTEX components using the static headspace technique, for example for soil and water samples, is a standard procedure designed for a high sample throughput. The reproducibility and stability of the procedure are of particular interest with regard to GC/MS coupling. The data given here demonstrate the capabilities of the system. Automatic evaluation is necessary for carrying out long series of analyses. By using GC/MS coupling the selectivity of the mass spectrometer can help to identify the analytes on the basis of the mass spectra in the evaluation of the analyses.

Analysis Conditions

Static headspace:	Perkin Elmer HS40 headspace sampler, 2 mL sample volume (blank value-free water fortified with 1 μg/L BTEX), incubation at 290 °C, 45 min (without shaker), sampling interval 0.05 min (sample injection)
GC/MS system:	Finnigan MAGNUM
Injector:	PTV/2 250 °C isotherm Split 10 mL/min
Column:	SGE HT8, 25 m × 0.25 mm × 0.25 μm
Program:	Start: 40 °C for 2 min Ramp: 10 °C/min Final temp.: 200 °C
Mass spectrometer:	Scan mode: EI, 33–240 u Scan rate: 0.75 s/scan

Results

Figure 4.8 shows the total ion chromatogram of the mass traces for benzene m/z 78, toluene m/z 91 and the isomers of xylene and ethylbenzene m/z 91, 106. A peak caused by CS_2 can be recognised from the mass trace m/z 78. Presumably CS_2 gets into the analysis because this is a product emitted from the butyl rubber septums of the headspace bottles. The isomers meta- and para-xylene elute at the same time from the column chosen. This concentration of 2 μg/L with an excellent signal to noise ratio of ca. 900 : 1 (Fig. 4.9) can be detected, from which a limit of detection of less than 0.01 μg/L can be expected.

A calibration file is used for detection and evaluation of the peaks. Each substance entry (Fig. 4.10) contains mass spectra, retention times and other parameters for identification of

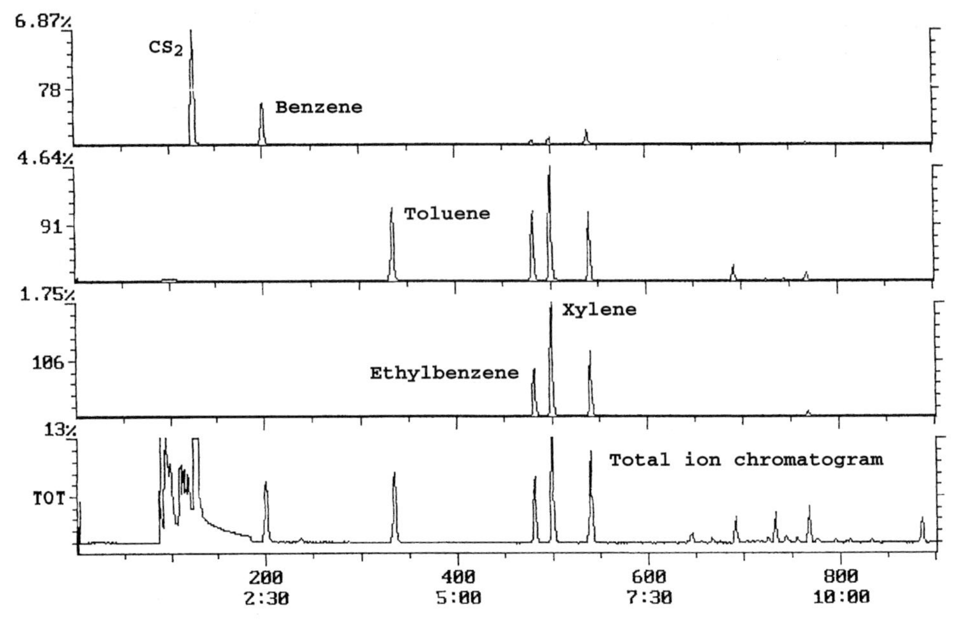

Fig. 4.8 Typical chromatogram from the headspace GC/MS analysis of BTEX components.
Above: Mass chromatogram for benzene (m/z 78), toluene (m/z 91) and xylene with ethylbenzene (m/z 91, 106)
Below: Total ion current chromatogram (33–240 u)

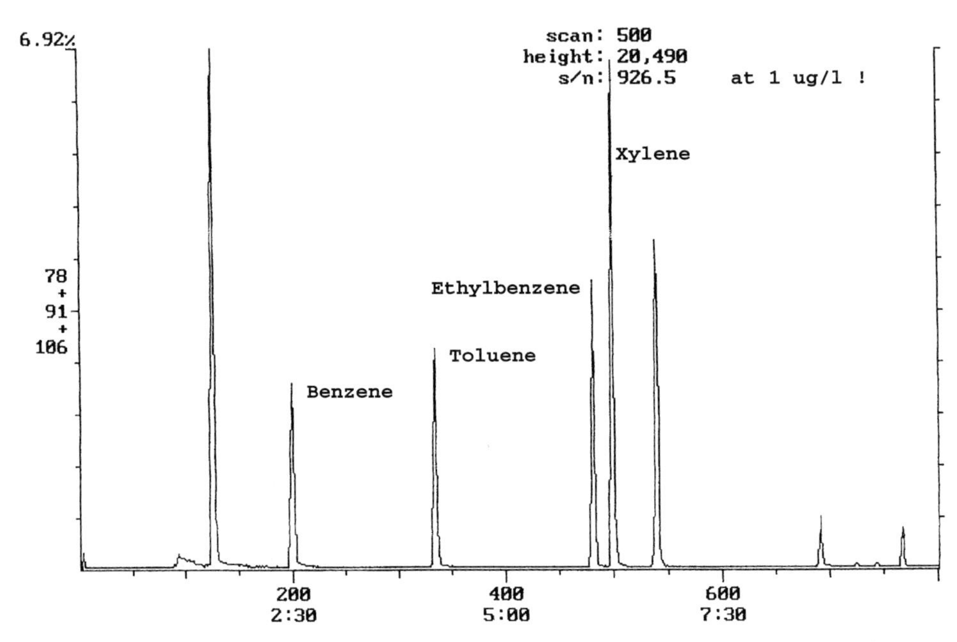

Fig. 4.9 Mass chromatogram for selective masses with the signal/noise ratio for the m,p-xylene peak

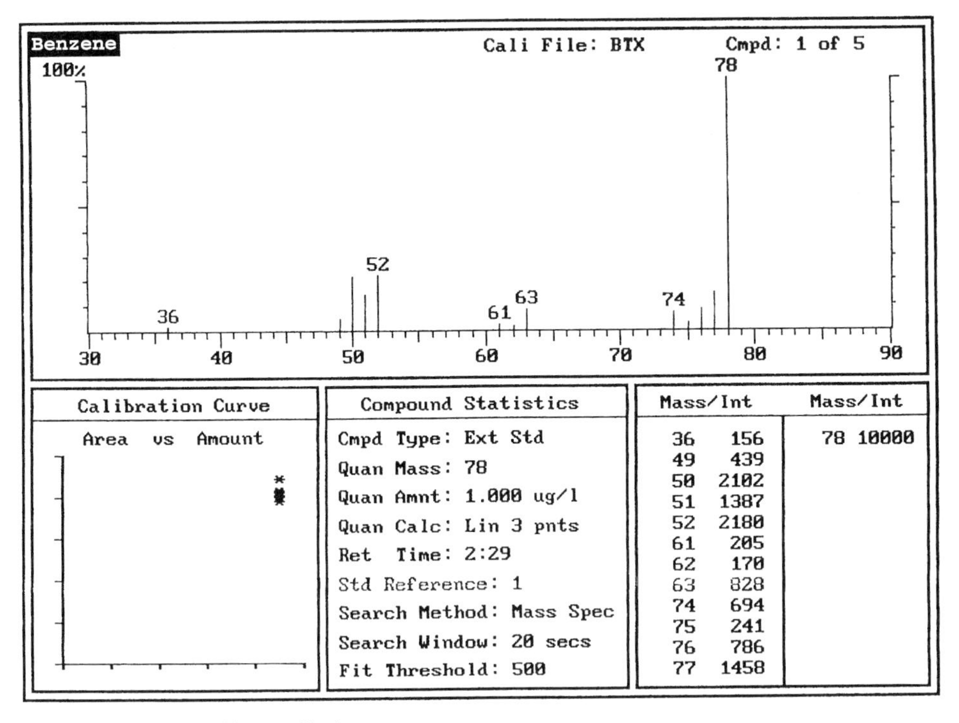

Fig. 4.10 Entry in the calibration file: benzene

Quantitation Report Quanfile: BTX-12 Quan Entries: 5
Comment: BTX 1ug/l HS40: 90 grdC, 45', 0.05' sampl.
Sorted via: Entry Number ↑ (S) = Standard

Cal	Name of Compound	S	Scan#	R Time	Me	Calc Amt(A)	Units
1	Benzene	E	201	2:31	BB	1.001	ug/l
2	Toluene	E	334	4:10	BB	1.001	ug/l
3	Xylene-1	E	481	6:01	BV	1.000	ug/l
4	Xylene-2	E	499	6:14	BB	1.000	ug/l
5	Ethylbenzene	E	540	6:45	BB	1.001	ug/l

External Standards Report Filename: BTX Quan Mass: 78
Benzene Compound: 1 of 5 Cali Pnts: 8
Sorted via: Entry Order ↑ Units: ug/l (S) = Standard

No	Ent Date	Time	Pk Area	Ret T.	Inj Amount	Ar/Amount
1	Aug-24-92	11:09	27,700	2:29	1.000	27700.000
2	Aug-24-92	11:10	27,771	2:29	1.000	27771.000
3	Aug-24-92	11:10	28,425	2:29	1.000	28425.000
4	Aug-24-92	11:11	28,630	2:30	1.000	28630.000
5	Aug-24-92	11:11	29,281	2:30	1.000	29281.000
6	Aug-24-92	11:11	31,354	2:30	1.000	31354.000
7	Aug-24-92	11:16	29,455	2:30	1.000	29455.000
8	Aug-24-92	11:20	28,425	2:29	1.000	28425.000

Fig. 4.11 Entry giving automatically determined peak areas in the quantitation list from eight analyses

Table 4.1 Limits of quantitation and average recoveries for the halogenated hydrocarbons and BTEX aromatics investigated

Compound	Limit of quantitation [μg/l]	Recovery [%]
Trifluoromethane	0.01	97.5
1,1-Dichloroethylene	0.01	98.2
Dichloromethane	0.01	101.2
1,2-Dichloroethane	0.01	94.6
1,1-Dichloroethane	0.01	93.8
Chloroform	0.01	102.7
1,1,1-Trichloroethane	0.01	97.6
Carbon tetrachloride	0.01	94.5
Trichloroethane	0.01	101.6
Bromodichloromethane	0.01	95.7
1,2-Dichloropropane	0.01	95.3
1,1,2-Trichloroethylene	0.01	94.8
Dibromochloroethane	0.01	91.8
Tetrachloroethylene	0.01	98.7
Chlorobenzene	0.05	87.4
Bromoform	0.01	89.9
1,1,2,2-Tetrachloroethane	0.02	90.8
Benzene	0.01	99.4
Toluene	0.01	96.7
Ethylbenzene	0.01	101.7
m/p-Xylene	0.01	95.3
o-Xylene	0.01	96.3

the BTEX components (see Section 3.3). The automatic evaluation compares the mass spectra at a given retention time in a search window and only carries out the peak integration after positive identification. The peak areas determined are entered into the file and are converted into concentration values if a calibration is present. In the present case the absolute peak areas from eight consecutive samples were determined and evaluated (Fig. 4.11). The values show very good agreement. The average deviation from the mean for all components is ca. 3% (e.g. benzene). Here the variances of the whole procedure from sample preparation to mass spectrometric detection are included because internal standardisation has not been carried out.

Using the same parameters the analysis of volatile halogenated hydrocarbons and BTEX can be carried out together.

4.3
Simultaneous Determination of Volatile Halogenated Hydrocarbons and BTEX

Key words: volatile halogenated hydrocarbons, BTEX, gas from landfill sites, seepage water, purge and trap (P&T), Tenax, internal standard

The analysis of seepage water (Fig. 4.12) and the condensate residue from the incineration of gas from landfill sites (Fig. 4.13) are two examples of the effective simultaneous control of the volatile halogenated hydrocarbon and BTEX concentrations. The purge and trap technique was chosen here so that as wide a spectrum of unknown components as possible could be determined together. The complex elution sequence can be worked out easily using GC/MS by means of selective mass signals. Toluene-d_8 is used as the internal standard for quantitation.

Analysis Conditions

Concentrator:	Tekmar purge and trap system LSC 2000	
	Trap:	Tenax
	Sample volume:	5 mL
	Sample temp.:	40 °C
	Standby:	30 °C
	Sample preheat:	2.50 min
	Purge duration:	10 min
	Purge flow:	40 mL/min
	Desorb preheat:	175 °C
	Desorb temp.:	180 °C
	Desorb time:	4 min
	MCM Desorb:	5 °C
	Bake:	10 min at 200 °C
	MCM bake:	90 °C
	BGB:	OFF
	Mount:	60 °C
	Valve:	220 °C
	Transfer line:	200 °C
	Connection to GC:	LSC 2000, linked into the carrier gas supply of the injector
GC/MS system:	Finnigan MAGNUM	
Injector:	PTV cold injection system	
Injector program:	Isotherm, 200 °C Split OPEN, ca. 20 mL/min	
Column:	Restek Rtx-624, 60 m × 0.32 mm × 1.8 μm	
Temp. program:	Start:	40 °C, 5 min
	Program:	15 °C/min
	Final temp.:	200 °C, 9.5 min
Mass spectrometer:	Scan mode:	EI, 45–220 u
	Scan rate:	1 s/scan

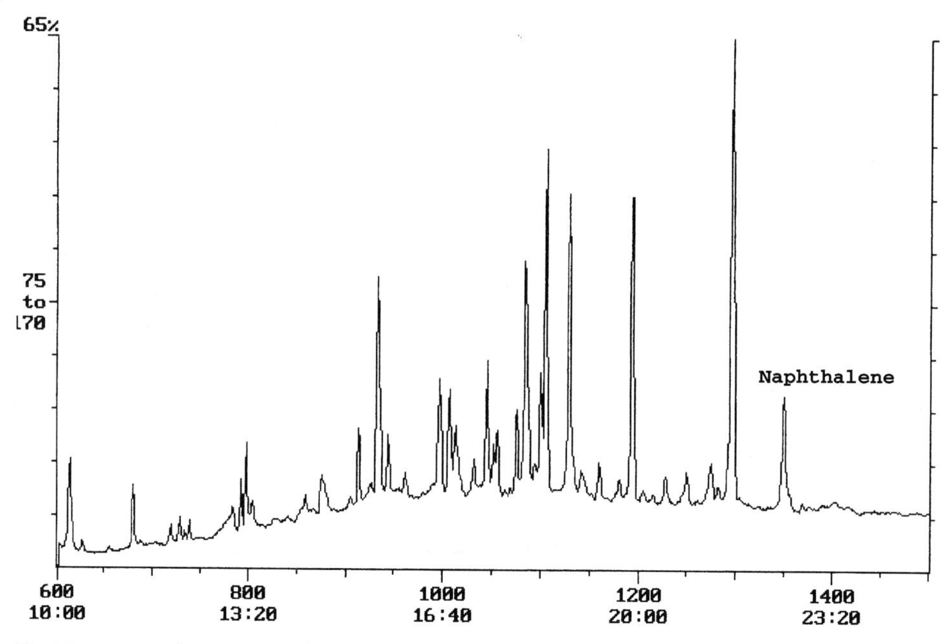

Fig. 4.12 Purge and trap GC/MS chromatogram for the analysis of a condensate from a block heating power station run on gas from landfill sites

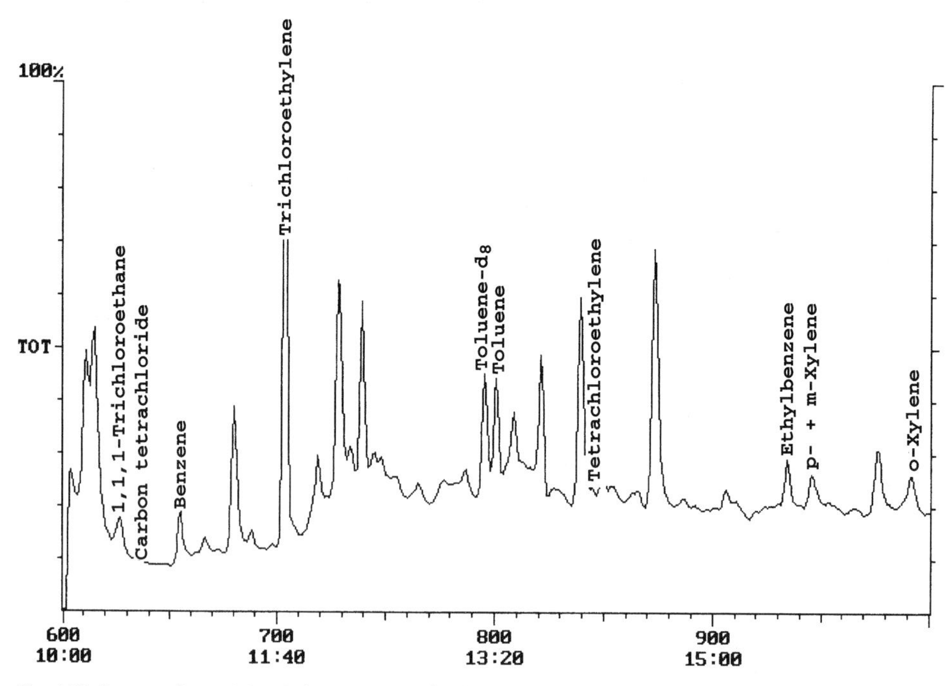

Fig. 4.13 Purge and trap GC/MS chromatogram for the analysis of a seepage water sample from a landfill site (section)

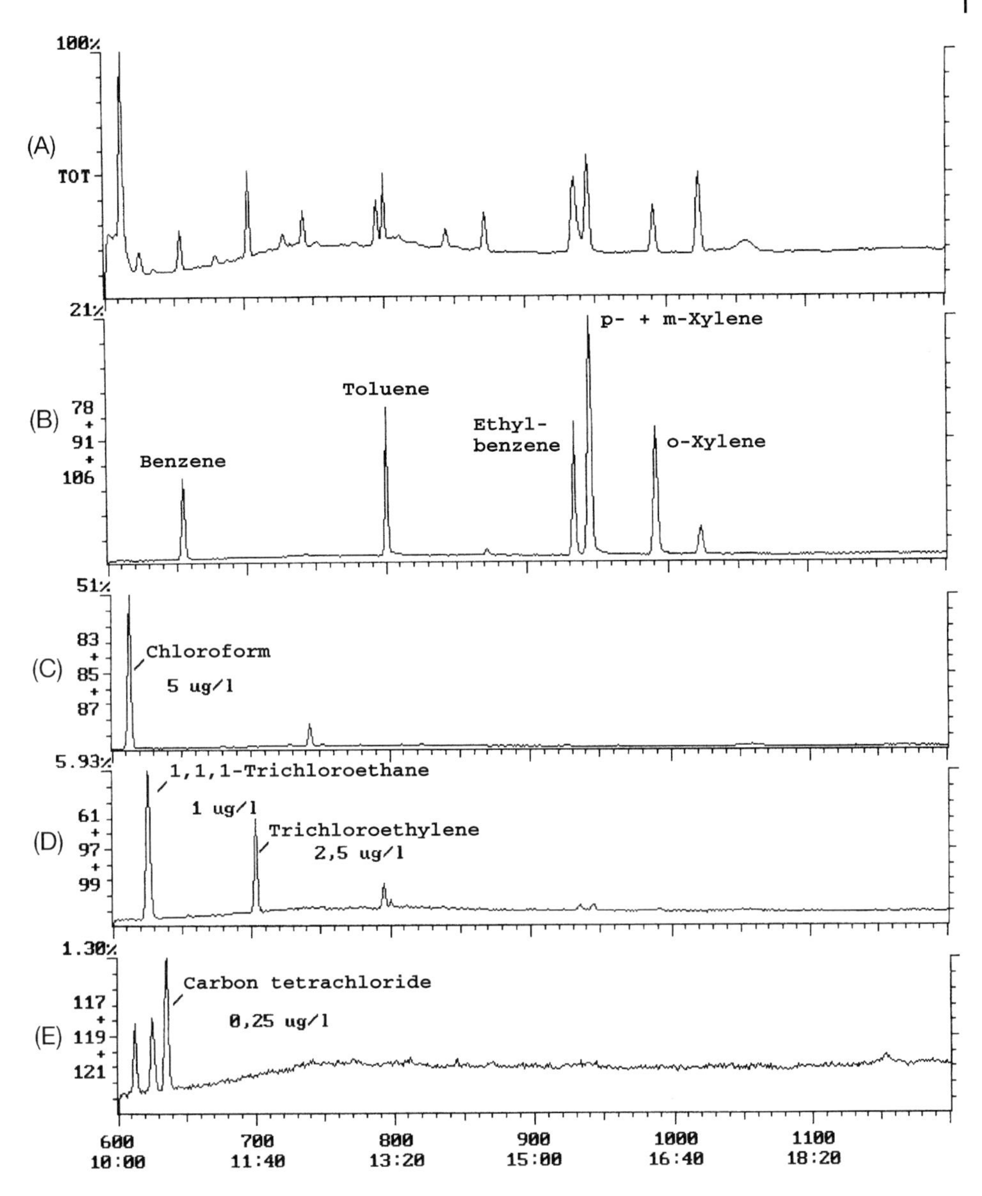

Fig. 4.14 Analysis of a standard volatile halogenated hydrocarbon/BTEX mixture
(A) Total ion current (for conditions see text)
(B) BTEX components (m/z 78, 91, 106)
(C) Chloroform (m/z 83, 85, 87)
(D) 1,1,1-Trichloroethane (m/z 61, 97, 99)
(E) Carbon tetrachloride (m/z 117, 119, 121)
(F) Trichloroethylene (m/z 95, 130, 132)
(G) Bromodichloromethane (m/z 83, 85, 129)
(H) Tetrachloroethylene (m/z 166)
(I) Dibromochloromethane (m/z 79, 127, 129)
(J) Bromoform (m/z 173)

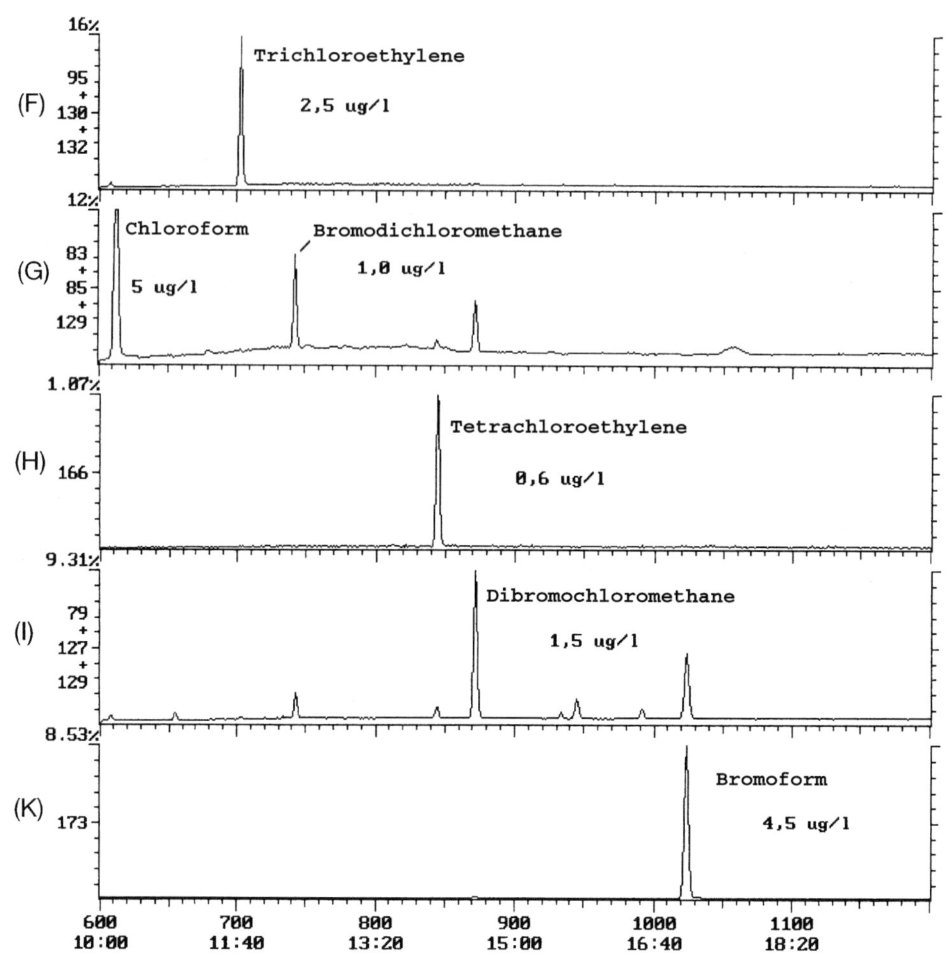

Fig. 4.14 (continued)

Results

The elution sequence of the volatile halogenated hydrocarbons compared to the BTEX aromatics is shown in Fig. 4.14 in the form of mass chromatograms using TIC in a standard analysis. The internal standard, toluene-d_8 (1 µg/L) elutes immediately before the toluene peak (see TIC). The column used has very high thermal stability until the end of the temperature program. This is shown by a completely flat base line. Benzene elutes here between tetrachloroethane and trichloroethylene, ortho-xylene, as the last BTEX component, between dibromochloromethane and bromoform. In the concentration range of around 1 µg/L volatile halogenated hydrocarbons and BTEX are determined together in an analysis with a good signal/noise ratio.

4.4
Static Headspace Analysis of Volatile Priority Pollutants

Key words: headspace, VOCs, water, soil, quantitation, internal standard

Many analytical techniques have been used for the quantitation of volatile organic compounds (VOC) in water and soil, including liquid-liquid microextraction (LLME), solid phase microextraction (SPME) and purge and trap (P&T). Automated static headspace analysis offers the advantages of simplicity and robustness especially when large sample throughput is required. A typical application of the static headspace method described here is the analysis of surface and waste water, see Fig. 4.15, other techniques such as purge & trap GC/MS are better designed for ultra-trace analysis of drinking water. The samples were transferred to 20 mL headspace vials, together with an internal standard (e.g. toluene-d_8). "Salting-out" was achieved by saturating the sample with sodium sulfate. Using the sample agitation feature, headspace equilibrium is reached very quickly, allowing all sampling operations to take place during the GC run time.

The GC was operated in accordance with the "whole column trapping technique" using a standard medium bore capillary column with a 1 µm stationary phase and liquid CO_2 as coolant.

The analytes were trapped in the column inlet at subambient temperature, without the need for a dedicated cold trapping device. Trapping of the VOCs helped to maintain optimum chromatographic efficiency by focusing the analytes at the column inlet. High sensitivity was achieved by injecting a relatively high volume of headspace using a low split flow. The injecting speed was controlled in order to prevent the injector from overflowing. The MS was operated in scanning electron impact (EI) mode, allowing acquisition of full mass spectra and thereby enabling both targeting analysis and identification of unknowns during a single analysis (Fig. 4.16). At 0.5 ug/L, all compounds were identified by automated library searching, using the NIST 98 library (Fig. 4.17).

The quantitation was based on an internal standard method, using an isotopically labelled analogue of toluene. A surrogate standard was also added to the samples in order to control the long-term spectral balance stability and to check the MS tuning criteria.

Analysis Conditions

GC/MS system:	Finnigan Trace MS system
Injector:	Split/splitless injector Injector temperature: 200 °C Split mode Constant flow at 3.0 mL/min
Autosampler:	CE Instruments HS2000 Headspace sample volume: 1 mL Incubation: 40 °C, 10 min
GC column:	J&W Scientific DB-1, 30 m × 0,32 mm × 1,0 µm

Temperature program:	−40 °C at start 10° C/min to 260 °C
Transfer line:	Direct coupling to MS ion source 260° C constant
Mass spectrometer:	EI mode Ion source at 200 °C
MS conditions:	Scan cycle time: 0.4 s Scan range: 45–270 u

Results

The technique of static headspace GC/MS offers significant benefits for laboratories tasked with running VOC analyses. The virtual elimination of sample carry-over, thanks to the programmable temperature cleaning cycle of the syringe and needle heaters in the headspace autosampler, obviates the need for running blank samples between specimen samples.

The wide linear dynamic range of the mass spectrometer, even running in full scan mode, permits target compounds to be accurately quantified over a concentration range of at least 3 decades. Full scan operation enables unknown peaks to be automatically detected and identified. Even after a period of several weeks of unattended operation, highly sensitive and reproducible results can be demonstrated.

The precision study, based on 72 replicate injections of the low standard (0.5 μg/L), represents analyses acquired over 3 weeks of continuous operation (see Table 4.2). Such high precision is normally associated with quadrupole mass spectrometers running in the selected ion monitoring (SIM) mode; yet these results were obtained in full scan mode. The R-square results show how linear the system is; the figures were derived from multiple calibration curves spanning 3 orders of magnitude (0.1 to 100 μg/L) and accumulated over 3 days (Fig. 4.18). The calculated limits of detection (LOD) for each target compound lie in the low ppb range, making this technique suitable for analyses of VOCs in water or other materials like soil or sediments (Fig. 4.19).

The system yields very high stability and sensitivity. The % RSDs and limits of detection were based on analyses performed at 0.5 μg/L ($n = 72$, over 3 weeks of continuous operation). The limits of detection computed according to LOD = 3SD. R-square values were obtained as described in the text.

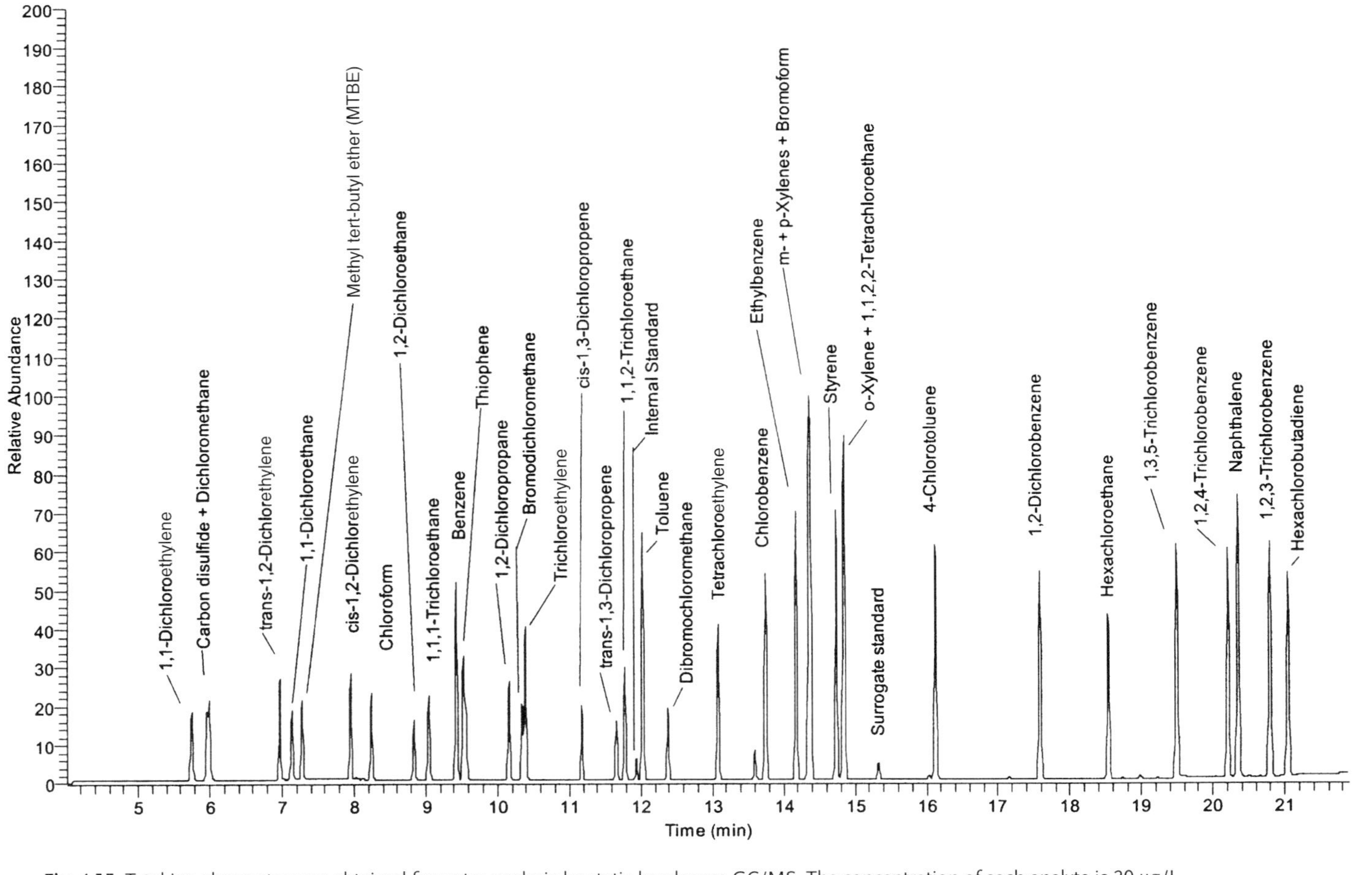

Fig. 4.15 Total ion chromatogram obtained for water analysis by static headspace GC/MS. The concentration of each analyte is 20 µg/L. The column (DB1, 30 m × 0.32 mm × 1 µm) is programmed from −40 °C to 260 °C at 10 °C/min. 1 mL of headspace is injected in the split mode.

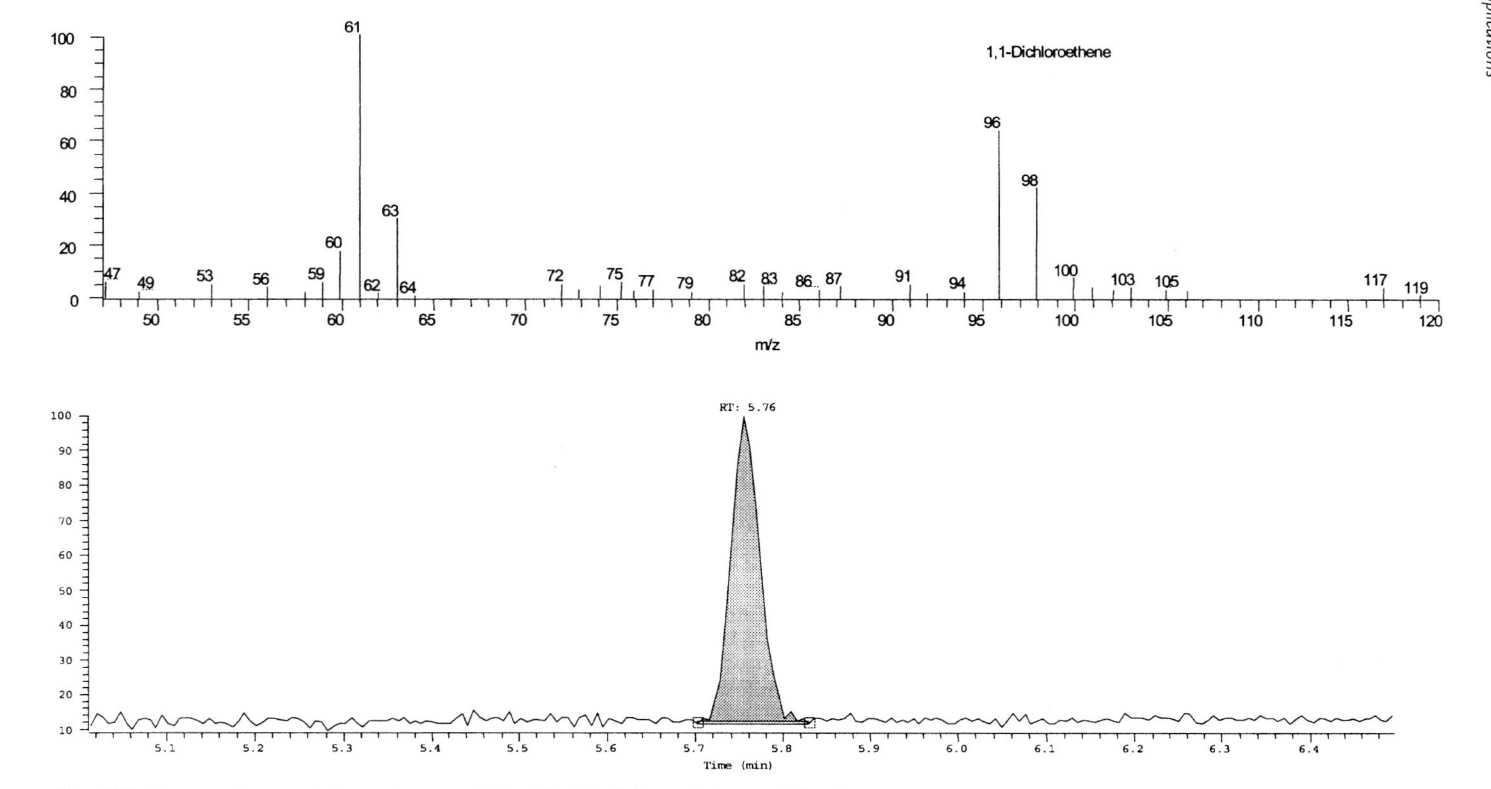

Fig. 4.16 Mass spectrum and chromatogram obtained for 1,1-dichloroethylene at 0.5 µg/L

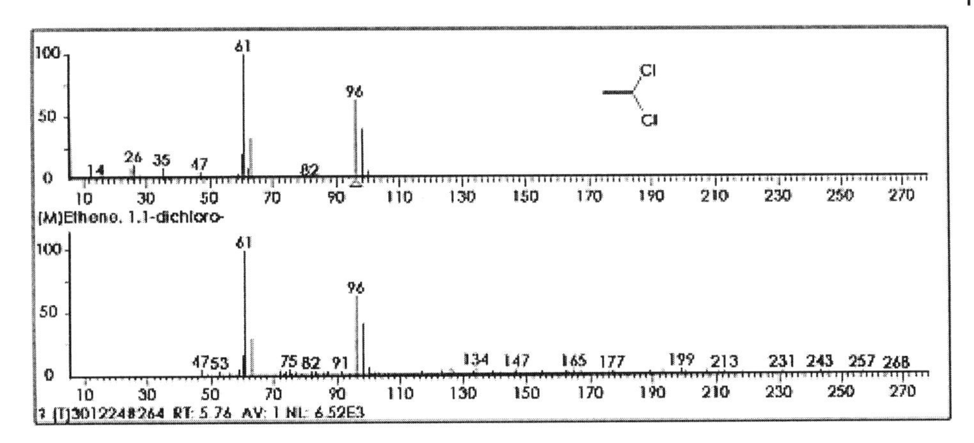

Fig. 4.17 Automated library searching using the NIST 98 library

Table 4.2 Substance list, LODs are given for static headspace method in water

Substance name	% RSD	R-square	LOD, µg/l
1,1-Dichloromethane	0.552	0.9998	0.010
Carbon disulfide	0.945	0.9990	0.013
Dichloromethane	1.150	0.9992	0.018
trans-1,2-Dichloroethane	0.698	0.9997	0.012
MTBE	0.697	0.9993	0.013
cis-1,2-Dichloroethane	1.318	0.9963	0.022
Chloroform	0.578	0.9996	0.013
1,2-Dichloroethane	0.362	0.9989	0.011
1,1,1-Trichloroethane	0.747	0.9998	0.014
Benzene	0.568	0.9994	0.016
Thiophene	0.793	0.9992	0.012
Carbon tetrachloride	0.385	0.9996	0.008
Bromodichloromethane	0.397	0.9995	0.010
Trichloroethane	0.585	0.9994	0.008
cis-1,3-Dichloropropene	0.909	0.9933	0.019
trans-1,3-Dichloropropene	1.017	0.9883	0.026
1,1,2-Trichloroethane	0.772	0.9989	0.015
Toluene	0.409	0.9997	0.016
Dibromochloromethane	0.225	0.9994	0.006
Tetrachloroethane	0.432	0.9997	0.006
Chlorobenzene	0.792	0.9990	0.007
Ethylbenzene	0.365	0.9996	0.011
Bromoform	0.397	0.9987	0.013
m- + p-Xylenes	0.219	0.9999	0.011
Styrene	0.833	0.9998	0.013
o-Xylene	0.193	0.9993	0.005
1,1,2,2-Tetrachloroethane	0.663	0.9983	0.013
4-Chlorotoluene	0.178	0.9996	0.007
1,2-Dichlorobenzene	0.605	0.9996	0.006
Hexachloroethane	0.496	0.9996	0.007
1,3,5-Trichlorobenzene	2.032	0.9936	0.021
1,2,4-Trichlorobenzene	2.550	0.9912	0.026
Naphthalene	1.216	0.9950	0.016
1,2,3-Trichlorobenzene	2.825	0.9933	0.027
Hexachlorobutadiene	1.569	0.9933	0.020

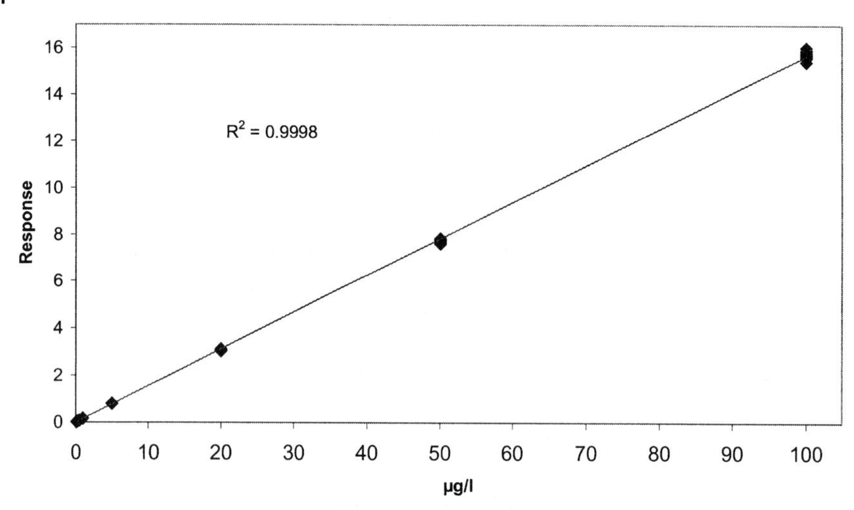

Fig. 4.18 Calibration graph obtained for 1,1-dichloroethylene over the 0.1 to 100 µg/L concentration range. The graph is based on nine calibrations (consecutive calibrations were repeated over three days).

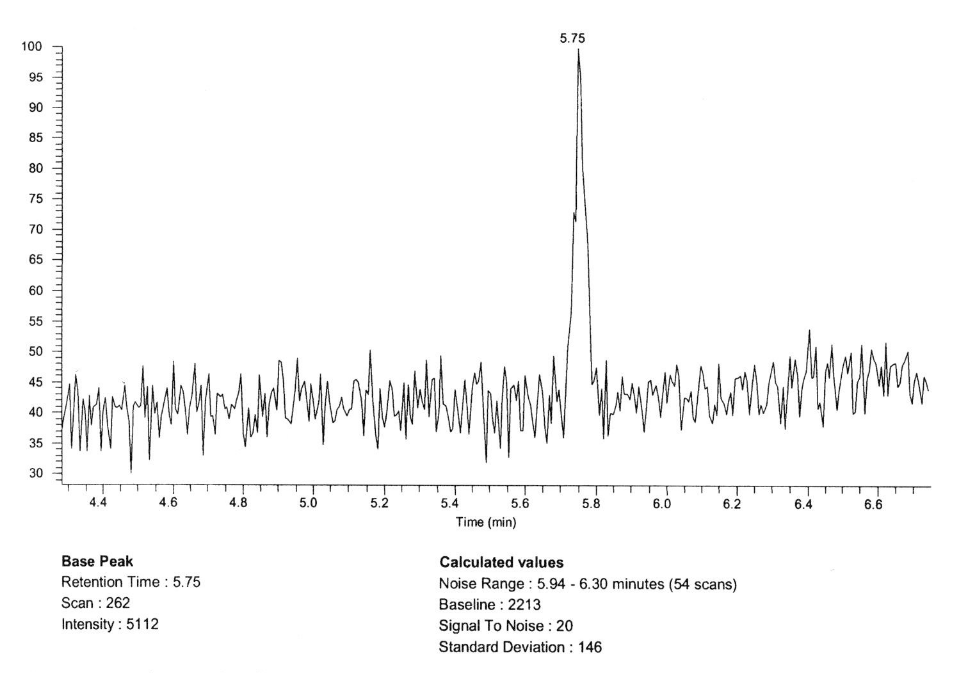

Fig. 4.19 Signal/noise data for 1,1-dichloroethylene at 0.1 µg/L

4.5 MAGIC 60 – Analysis of Volatile Organic Compounds

Key words: purge and trap, water, soil, volatile halogenated hydrocarbons, BTEX, drinking water, EPA, internal standards

The method described here for identification and determination of volatile organic compounds in water and soil is taken from the EPA methods 8260, 8240 and 524.2. These approximately 60 substances are referred to by the term MAGIC 60. The method also allows the determination of compounds in solid materials, such as soil samples, and is generally applicable for a broad spectrum of organic compounds which have sufficiently high volatility and low solubility in water for them to be determined effectively by the purge and trap procedure. Vinyl chloride can also be determined with certainty by this method (Figs. 4.20 and 4.21).

Gastight syringes (5 mL, 25 mL) with open/shut valves have been shown to be useful for the preparation of water samples. The syringe is carefully filled with the sample from the plunger, the plunger inserted, and the volume taken up adjusted. The standard solutions are added through the valve with a μL syringe. The syringe is then connected to the purge and trap apparatus via the open/shut valves and the prepared sample transferred to a needle sparger or frit sparger vessel.

The analysis of soil samples is similar to that of water, but, depending on the expected concentration of volatile halogenated hydrocarbons in the sample, a different preparation procedure is chosen. For samples where a concentration of less than 1mg/kg is expected, 5 g of the

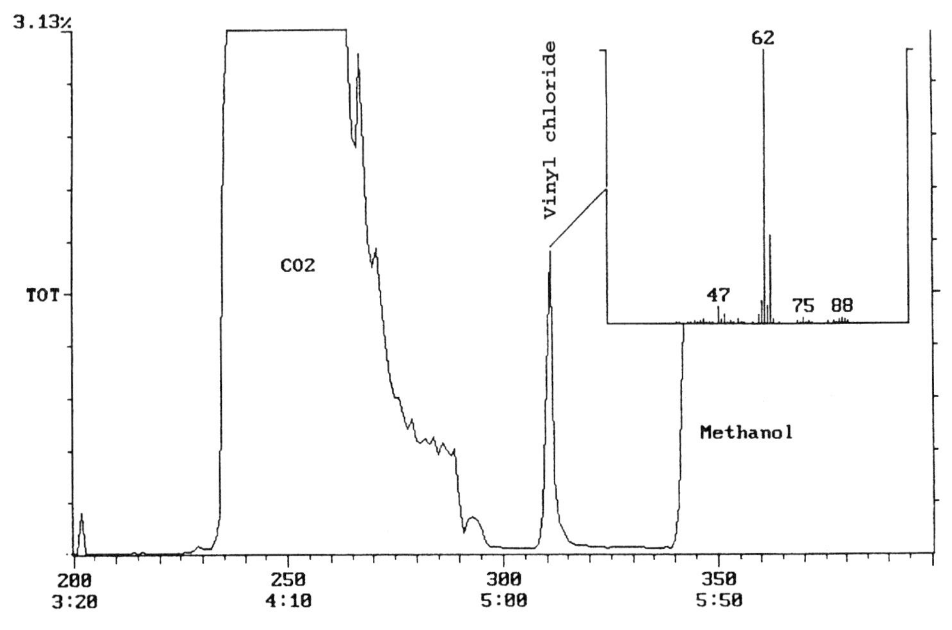

Fig. 4.20 Elution of vinyl chloride (50 μg/L) between CO_2 and methanol (DB-624, for analysis parameters see text). The methanol peak can be avoided by using poly(ethylene glycol) as the solvent for the preparation of standard solutions (Fig. 4.21)

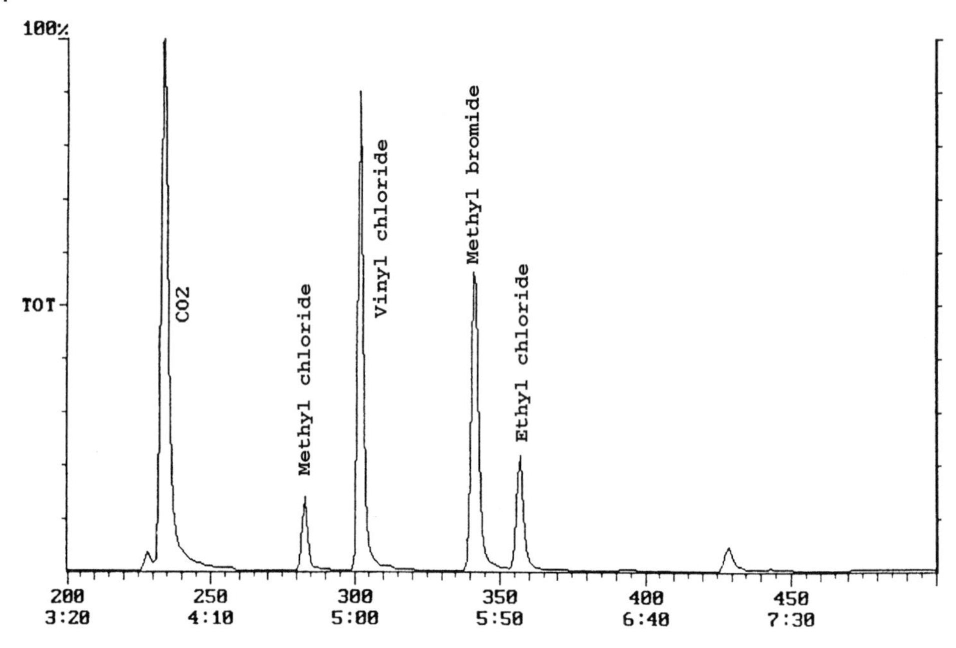

Fig. 4.21 Elution of vinyl chloride (50 μg/L) from a DB-VRX column (for analysis parameters see text) with poly(ethylene glycol) as the solvent

sample are placed directly in a 5 mL needle sparger and treated with 5 mL reagent water. Solutions of the internal standards are added with a μL syringe. At higher concentrations 4 g of the sample are weighed into a 10 mL vessel which can be closed with a Teflon-coated septum (e.g. headspace vial), and are treated with 9.9 mL methanol and 0.1 mL of the surrogate standard solution. After the solid phase has settled, 5–100 μL of the methanol phase are taken up, depending on the concentration, in a prepared 5 mL (25 mL) syringe together with reagent water. After addition of the internal standard, the sample is injected into the purge and trap apparatus (needle sparger).

Analysis Conditions

Sample material:	Water or soil	
Concentrator:	Tekmar purge and trap system LSC 2000	
Trap:	Supelco Vocarb 3000, room temperature	
Autosampler:	Tekmar ALS 2016 with 25 mL frit sparge glass vessels for water samples and needle sparge vessels for soil samples	
Purge and trap parameters:	Purge gas:	40 mL/min
	Pre-purge time:	0 min
	MCM:	0 °C/90 °C
	Sample temp.:	room temperature for water and soil

	Purge time:	12 min
	Dry purge:	0 min
	Desorb preheat:	255 °C for Vocarb 3000
	Desorption:	260 °C, 4 min
	Bake mode:	260 °C, 20 min
	Bake gas bypass:	ON after 120 s
	Valves:	110 °C
	Transfer line:	110 °C

GC/MS system: Finnigan MAGNUM

Injector: PTV-cold injection system, 200 °C isotherm, direct connection (insertion) of the purge and trap concentrator into the carrier gas supply line of the PTV

Column: J&W DB-VRX, 60 m × 0.32 mm × 1.8 μm
DB-624 or Rtx-624, 60 m × 0.32 mm × 1.8 μm

Program:	Start:	40 °C, 5 min
	Program 1:	7 °C/min to 180 °C
	Program 2:	15 °C/min to 220 °C
	Final temp.:	220 °C, 5 min
Mass spectrometer:	Scan mode:	EI, 33–260 u
	Scan rate:	1 s/scan

Calibration: Internal/surrogate standards: toluene-d_8, fluorobenzene, 4-bromofluorobenzene, 1,2-dichloroethane-d_4
Solvent for all standard solutions: methanol

Cleaning the purge and trap unit: In the case of contamination cleaning the concentrator can be necessary. Clean purge vessels filled with water are used and the following short program is put into operation:

Purge time:	5 min
Desorption:	1 min
Bake mode:	5 min
Valves, transfer line:	200 °C

All other parameters remain unchanged.

Results

The search and identification of individual analytes in the chromatograms (Fig. 4.22) are carried out automatically on the basis of reference data, such as the retention time and mass spectrum, which are stored in a calibration file (Fig. 4.23). For quantitation, a calibration with eight steps from 0.1 μg/L to 40 μg/L is constructed (Fig. 4.24). The purge and trap procedure gives very good linearity over this range. A standard analysis of the compounds listed in Table 4.3 in water is shown in Fig. 4.25.

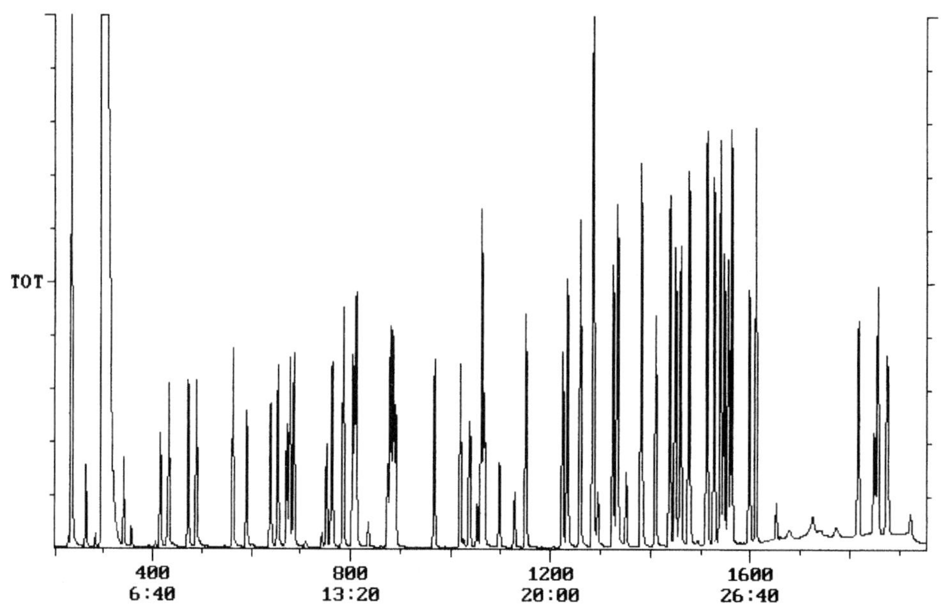

Fig. 4.22 MAGIC 60: standard chromatogram of a soil sample.
All analytes: 100 μg/kg, internal standard: 10 μg/kg (for analysis parameters see text)

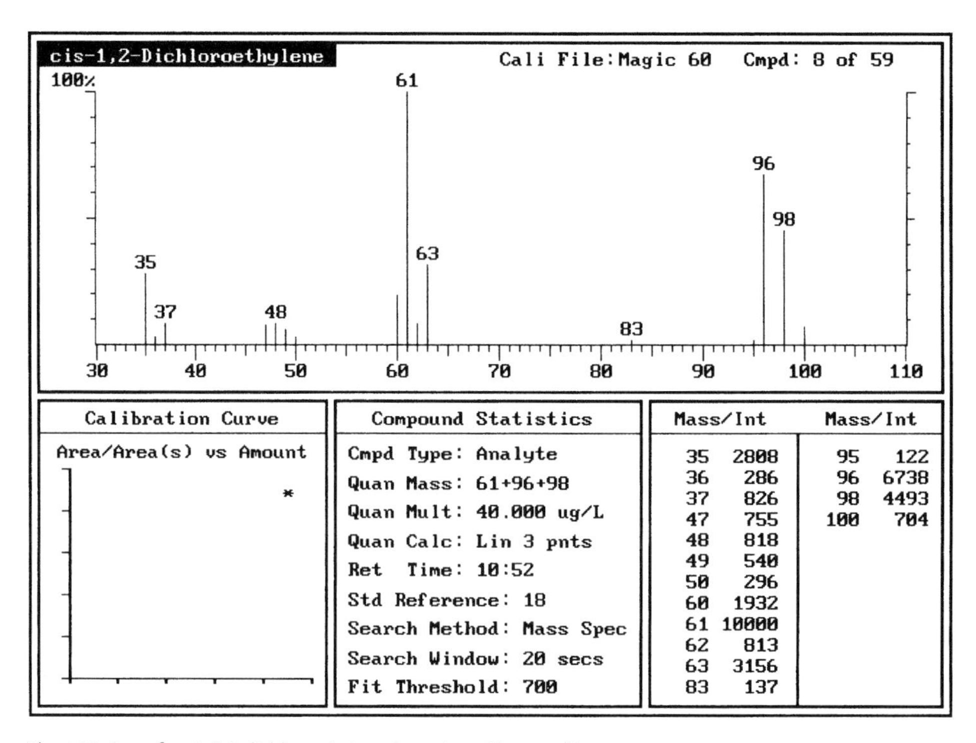

Fig. 4.23 Entry for cis-1,2-dichloroethylene from the calibration file

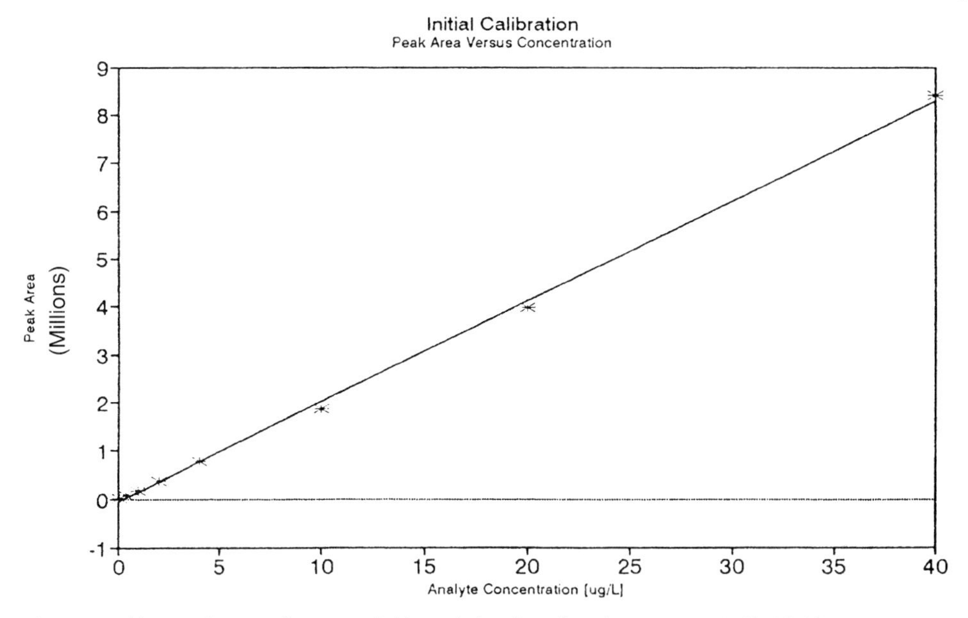

Fig. 4.24 Calibration function for cis-1,2-dichloroethylene based on the masses m/z 61, 96, 98. Correlation 0.9991, relative standard deviation: 4.96%

Table 4.3 MAGIC 60 substance list with details of quantitation masses, CAS numbers, limits of detection (MDL, minimum detection limit) and limits of quantitation (MQL, minimum quantitation limit), arranged alphabetically (for the method, see text)

Compound	m/z	CAS no.	MDL [μg/L]	MQL [μg/L]
Benzene	77, 78	71-43-2	0.05	0.1
Bromobenzene	77, 156, 158	108-86-1	0.1	0.2
Bromochloromethane	49, 128, 130	74-97-5	0.05	0.1
Bromodichloromethane	83, 85, 127	75-27-4	0.05	0.1
Methyl bromide	94, 96	74-96-4	n. b.	n. b.
Bromoform	173, 175, 252	95-25-2	0.05	0.1
n-Butylbenzene	91, 134	104-51-8	0.05	0.1
sec-Butylbenzene	105, 134	135-98-8	0.05	0.1
t-Butylbenzene	91, 119	98-06-6	0.05	0.1
Chlorobenzene	77, 112, 114	108-90-7	0.1	0.2
Ethyl chloride	64, 66	75-00-3	n. b.	n. b.
Methyl chloride	50, 52	74-87-3	n. b.	n. b.
Chloroform	83, 85	67-66-3	0.05	0.1
2-Chlorotoluene	91, 126	95-49-8	0.05	0.1
4-Chlorotoluene	91, 126	106-43-4	0.05	0.1
Dibromochloromethane	127, 129	124-48-1	0.05	0.1
1,2-Dibromo-3-chloropropane	75, 155, 157	96-12-8	0.25	0.5
Dibromomethane	93, 95, 174	74-95-3	0.1	0.2
1,2-Dibromoethane	107, 109	106-93-4	0.1	0.2
1,2-Dichlorobenzene	111, 146	95-50-1	0.15	0.2

Table 4.3 (continued)

Compound	m/z	CAS no.	MDL [µg/L]	MQL [µg/L]
1,3-Dichlorobenzene	111, 146	541-73-1	0.1	0.2
1,4-Dichlorobenzene	111, 146	106-46-7	0.1	0.2
1,1-Dichloroethane	63, 112	75-34-3	0.05	0.1
1,2-Dichloroethane	62, 98	107-06-2	0.1	0.2
1,1-Dichloroethylene	61, 63, 96	75-35-4	0.05	0.1
cis-1,2-Dichloroethylene	61, 96, 98	156-59-4	0.05	0.1
trans-1,2-Dichloroethylene	61, 96, 98	156-60-5	0.05	0.1
Dichlorodifluoromethane	85, 87	75-71-8	0.1	0.2
Dichloromethane	49, 84, 86	75-09-2	0.5	0.5
1,2-Dichloropropane	63, 112	78-87-5	0.1	0.2
1,3-Dichloropropane	76, 78	142-28-9	0.05	0.1
2,2-Dichloropropane	77, 97	590-20-7	0.05	0.1
1,1-Dichloropropylene	75, 110, 112	563-68-6	0.05	0.1
cis-1,3-Dichloropropylene	75, 110, 112	10061-01-5	0.05	0.1
trans-1,3-Dichloropropylene	75, 110, 112	10061-01-5	0.05	0.1
Ethylbenzene	91, 106	100-41-4	0.05	0.1
Hexachlorobutadiene	225, 260	87-68-3	0.15	0.2
Isopropylbenzene	105, 120	48-82-8	0.1	0.2
4-Isopropyltoluene	91, 119, 134	99-87-6	0.05	0.1
Naphthalene	128	91-20-3	0.5	0.5
Styrene	78, 104	100-42-5	0.1	0.2
1,1,1,2-Tetrachloroethane	131, 133	620-30-6	0.1	0.2
1,1,2,2-Tetrachloroethane	83, 85, 131	79-34-5	0.1	0.2
Tetrachloroethylene	129, 166, 168	127-18-4	0.1	0.2
Carbon tetrachloride	117, 119	56-23-5	0.05	0.1
Toluene	91, 92	108-88-3	*	*
1,2,4-Trichlorobenzene	180, 182	120-82-1	0.3	0.3
1,2,3-Trichlorobenzene	180, 182	87-61-6	0.4	0.4
1,1,1-Trichloroethane	61, 97, 99	71-55-6	0.05	0.1
1,1,2-Trichloroethane	83, 85, 97	79-020-5	0.1	0.2
Trichloroethylene	95, 130, 132	79-01-6	0.05	0.1
Trichlorofluoromethane	101, 103	75-69-4	0.05	0.1
1,2,3-Trichloropropane	75, 77	96-18-4	0.1	0.2
1,2,4-Trimethylbenzene	105, 120	95-63-6	0.1	0.2
1,3,5-Trimethylbenzene	105, 120	108-67-8	0.1	0.2
Vinyl chloride	62, 64	75-01-4	n. b.	n. b.
m-Xylene	91, 106	108-38-3	0.1	0.2
o-Xylene	91, 106	95-47-6	0.1	0.2
p-Xylene	91, 106	95-47-6	0.1	0.2
Internal standards:				
4-Bromofluorobenzene	95, 174			
1-Chloro-2-bromopropane	77, 79			
1,2-Dichlorobenzene-d_4	115, 150, 152			
1,2-Dichloroethane-d_4	65, 102			
Fluorobenzene	77, 96			
Toluene-d_8	70, 98, 100			

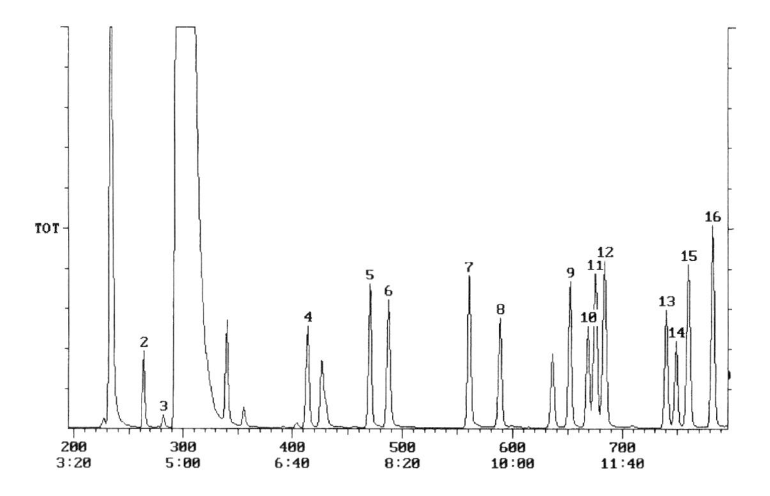

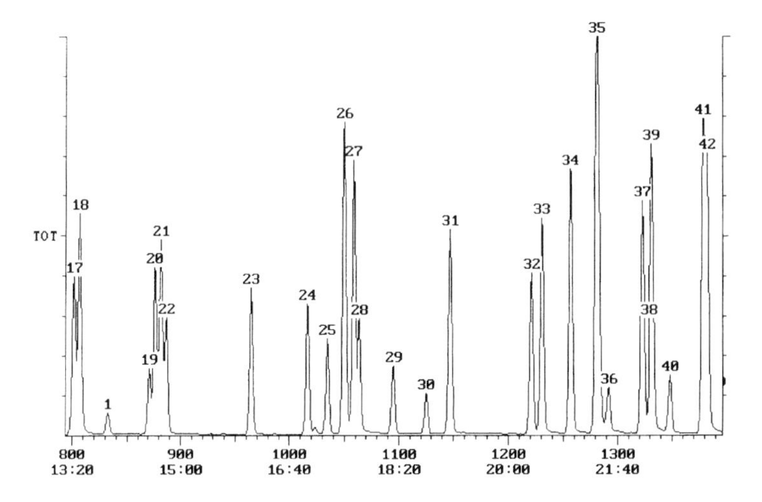

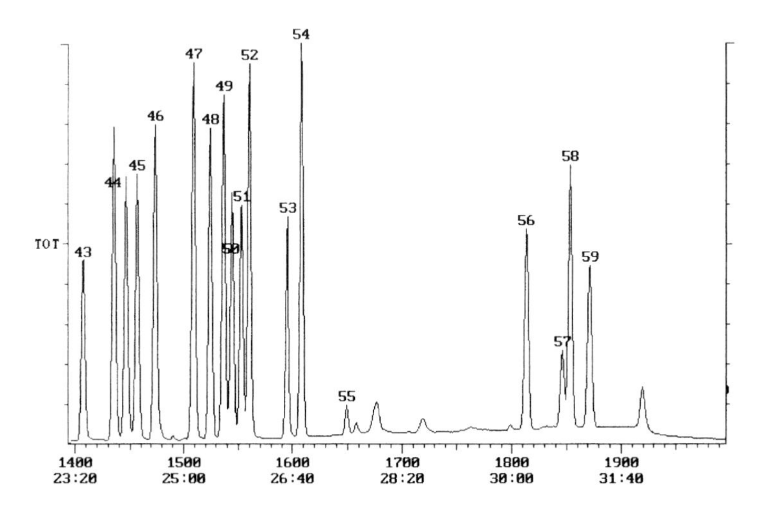

Fig. 4.25
(legend see p. 444)

1	Benzene	Active	I	13:52
2	Fluorobenzene	Active	A	4:24
3	Dichlorodifluoromethane	Active	A	4:42
4	Chloromethane	Active	A	6:54
5	Trichlorofluoromethane	Active	A	7:50
6	1,1-Dichloroethylene	Active	A	8:07
7	Methylene chloride	Active	A	9:20
8	trans-1,2-Dichloroethylene	Active	A	9:48
9	1,1-Dichloroethane	Active	A	10:52
10	cis-1,2-Dichloroethylene	Active	A	11:09
11	Bromochloromethane	Active	A	11:15
12	Chloroform	Active	A	11:24
13	2,2-Dichloropropane	Active	A	12:20
14	1,2-Dichloroethane-d4	Active	A	12:29
15	1,2-Dichloroethane	Active	A	12:40
16	1,1,1-Trichloroethane	Active	A	13:02
17	1,1-Dichloropropylene	Active	A	13:22
18	Carbon tetrachloride	Active	A	13:27
19	Dibromomethane	Active	A	14:31
20	1,2-Dichloropropane	Active	A	14:36
21	Trichloroethylene	Active	A	14:42
22	Bromodichloromethane	Active	A	14:47
23	cis-1,3-Dichloropropylene	Active	A	16:04
24	trans-1,3-Dichloropropyle	Active	A	16:56
25	1,1,2-Trichloroethane	Active	A	17:14
26	Toluene-d8	Active	A	17:30
27	Toluene	Active	A	17:38
28	1,3-Dichloropropane	Active	A	17:44
29	Dibromochloromethane	Active	A	18:15
30	1,2-Dibromoethane	Active	A	18:44
31	Tetrachloroethylene	Active	A	19:07
32	1,1,1,2-Tetrachloroethane	Active	A	20:21
33	Chlorobenzene	Active	A	20:30
34	Ethylbenzene	Active	A	20:56
35	m,p-Xylene	Active	A	21:21
36	Bromoform	Active	A	21:32
37	Styrene	Active	A	22:02
38	1,1,2,2-Tetrachloroethane	Active	A	22:09
39	o-Xylene	Active	A	22:11
40	1,2,3-Trichloropropane	Active	A	22:28
41	Isopropylbenzene	Active	A	22:58
42	4-Bromofluorobenzene	Active	A	23:02
43	Bromobenzene	Active	A	23:27
44	2-Chlorotoluene	Active	A	23:55
45	4-Chlorotoluene	Active	A	24:16
46	1,3,5-Trimethylbenzene	Active	A	24:34
47	tert.-Butylbenzene	Active	A	25:09
48	1,2,4-Trimethylbenzene	Active	A	25:24
49	sec.-Butylbenzene	Active	A	25:37
50	1,3-Dichlorobenzene	Active	A	25:44
51	1,4-Dichlorobenzene	Active	A	25:53
52	4-Isopropyltoluene	Active	A	26:00
53	1,2-Dichlorobenzene	Active	A	26:35
54	n-Butylbenzene	Active	A	26:48
55	1,2-Dibromo-3-Chloropropa	Active	A	27:29
56	1,2,4-Trichlorobenzene	Active	A	30:13
57	Naphthalene	Active	A	30:46
58	Hexachlorobutadiene	Active	A	30:53
59	1,2,3-Trichlorobenzene	Active	A	31:12

Fig. 4.25 MAGIC 60: Standard chromatogram for a water sample.
All analytes: 20 μg/L, internal standard: 2 μg/L (for components see above, for analysis parameters see text)

4.6
irm-GC/MS of Volatile Organic Compounds Using Purge and Trap Extraction

Key words: compound specific isotope analysis, VOC, purge and trap, sources of contamination, degradation pathways

The compound specific isotope analysis (CSIA) has already been successfully used in the assessment of *in situ* remediation of contaminated environments, identification of pollutant degradation pathways, or the verification of contaminant sources. In these types of studies, the sensitivity of isotope ratio monitoring GC/MS (irm-GC/MS) is often a limiting factor, since concentrations of organic contaminants in groundwater are very often in the low μg/L range. Hence, in order to be able to routinely use irm-GC/MS techniques in environmental studies, efficient extraction procedures are required.

While purge and trap (P&T), providing the lowest method detection limits for volatile organic compounds (VOCs), is a routinely used extraction method for the trace level quantification, the on-line coupling with irm-GC/MS has rarely been reported.

Since the P&T procedure includes various phase transition steps that may shift the isotopic signature of the analytes (evaporation, sorption, condensation), the P&T method parameters of purge time, desorption time, and injection temperature have been carefully evaluated for the determination of the $\delta^{13}C$-values. The compound specific isotope ratios of 10 different volatile organic compounds ranging from the unpolar benzene to the polar MTBE as listed in Table 4.4 were determined.

Analysis Conditions

Sample preparation:	Purge and trap concentrator Tekmar LSC3100 with liquid autosampler Tekmar AQUATek 70, (Tekmar-Dohrmann, Mason, OH, USA), coupled on-line to the pre-column of the irm-GC/MS system by a cryofocussing unit Aqueous samples were filled into 40 mL vials without headspace, 25 mL of the water samples were transferred by the autosampler into a fritted sparging glassware
Purge:	30 min with N_2 at 40 mL/min
Trap:	VOCARB 3000 (Supelco, Bellafonte, PA) at room temperature
Desorption:	250 °C for 1 min
Gas chromatograph:	Trace GC Ultra (Thermo Fisher, Milan, Italy)
Injector:	Cryofocussing unit at –120 °C using liquid N_2
Carrier gas:	Helium, constant pressure, 180 kPa
Column:	Deactivated pre-column, 0.5 m × 0.53 mm (BGB, Anwil, Switzerland) Rtx-VMS capillary column, 60 m × 0.32 mm i.d. × 1.8 μm film (Restek Corp., Bellefonte, PA, USA)

Oven temp. program: Start of temperature program with the flash-heating of the cryofocussing unit
40 °C, 2 min,
2 °C/min to 50 °C, 4 min,
8 °C/min to 100 °C, 2 min,
40 °C/min to 210 °C , 3.5 min.

Interface: Combustion interface GC-C III (Thermo Fisher, Bremen, Germany) at 940 °C.

Mass spectrometer: DELTA^{plus}XL Isotope Ratio Mass Spectrometer (Thermo Fisher, Bremen, Germany)

Ionisation: EI

Scan mode: simultaneous ion detection

Mass range: m/z 44, 45, 46

Aqueous solutions of the target analytes were obtained by spiking aliquots of methanolic stock solutions into tap water for method setup and parameter optimisation. The isotopic signatures of all the compounds relative to Vienna PeeDee Belemnite (VPDB) were obtained using CO_2 that was calibrated against referenced CO_2.

Table 4.4 Compounds in order of GC elution, method detection limits in water (MDL), accuracy and reproducibility of purge and trap extraction coupled to irm-GC/MS compared to elemental analyser technique (EA)

Compound	MDL µg/L	δ^{13}C EA-IRMS VPDB [‰]	δ^{13}C P&T irm-GC/MS VPDB [‰]
1,1-Dichloroethylene	3.6	−29.25 + 0.14	−29.07 + 0.08
trans-1,2-Dichloroethylene	1.5	−26.42 + 0.17	−25.61 + 0.22
Methyl-t-butylether	0.63	−28.13 + 0.15	−27.75 + 0.09
cis-1,2-Dichloroethylene	1.1	−26.61 + 0.06	−25.96 + 0.07c
Chloroform	2.3	−45.30 + 0.19	−46.22 + 0.14c
Tetrachloromethane	5.0	−38.62 + 0.01	−38.37 + 0.27
Benzene	0.3	−27.88 + 0.20	−27.27 + 0.20
Trichloroethylene	1.4	−26.59 + 0.08	−26.11 + 0.20
Toluene	0.25	−27.90 + 0.24a	−27.16 + 0.35
Tetrachloroethylene	2.2	−27.32 + 0.14	−26.76 + 0.19

Results

Purge and trap allowed MDLs ranging from 0.25 to 5.0 µg/L, depending on the analyte, which corresponds to the highest sensitivity of CSIA for volatile compounds, reported so far. These results were due both to the high sample volume (25 mL) as well as the high extraction efficiency (up to 80%) of the analytes. Thus, P&T-irm-GC/MS allows determining compound specific stable isotope signatures of contaminant concentrations frequently found in groundwater. As Fig. 4.27 and Table 4.4 show, P&T allowed highly reproducible CSIA measurements.

Expressed as the absolute amount of carbon injected on-column, the averaged MDLs for the analytes correspond to 0.4 + 0.1 (P&T) nmol C on-column. As can be seen from Fig. 4.26, the P&T method yields very clean chromatograms with very sharp peaks due to cryofocussing. The detection limit could even be slightly lowered.

Since the validated extraction technique shows little if any carbon isotopic fractionations due to the high extraction efficiency, it is also applicable for CSIA of D/H-ratios which require 10–20 times higher analyte concentrations than $\delta^{13}C$ analysis. P&T pre-concentration methods could also be used to lower analyte concentrations needed for $\delta^{15}N$, and is expected to work also for $\delta^{18}O$ analysis.

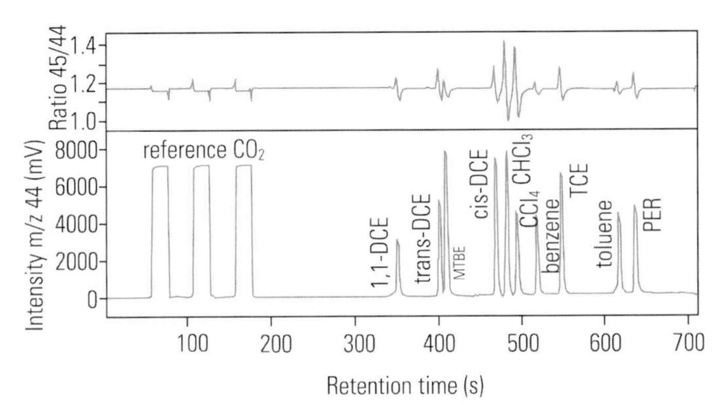

Fig. 4.26 Chromatogram of an on-line P&T-irm-GC/MS analysis. The concentrations of the different analytes were adjusted to achieve similar signal intensities (1.7 μg/L (Toluene) – 32 μg/L (CCl_4)). The three first peaks correspond to the reference CO_2 gas pulses.

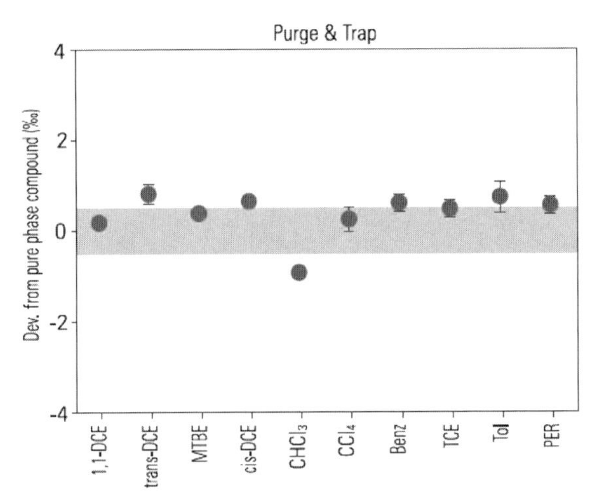

Fig. 4.27 Reproducibility and accuracy of P&T- irm-GC/MS. Plotted are the differences from the pure liquid standards measured by EA-IRMS and the horizontal bars correspond to a ^{13}C-measurement within a +0.5‰ interval of the EA-IRMS measurements allowing a direct comparison between the different compounds.

4.7
Vinyl Chloride in Drinking Water

Key words: vinyl chloride, purge and trap, Vocarb 4000, PLOT column, quantitation

For the sensitive and selective determination of vinyl chloride, chromatographic separation from other volatile components is necessary in special cases. Instead of a packed column, a fused silica column coated with Porapak was used. This type column can be used both with cryofocusing and also for direct coupling with the mass spectrometer.

Analysis Conditions

Sample material:	Drinking water	
Concentrator:	Tekmar purge and trap system LSC 2000 with AquaTek 50 sample dispenser 25 mL sample volume, 70 s sample transfer	
Trap:	Supelco Vocarb 4000 at room temperature	
Purge and trap parameters:	Purge gas:	40 mL/min
	Mount:	40 °C
	Pre-purge time:	1.5 min
	Sample temp.:	35 °C
	Purge time:	10 min
	Dry purge:	3 min
	MCM:	0 °C/90 °C
	Desorb preheat:	245 °C
	Desorption:	250 °C, 4 min
	Bake mode:	280 °C, 5 min
	Bake gas bypass:	OFF
	Valves:	110 °C
	Transfer line:	120 °C
GC/MS system:	Finnigan ITS 40	
Injector:	Tekmar cryofocusing –30 °C	
Column:	Chrompack PoraPlot Q, 25 m × 0.32 mm × 10 μm, cf. 2.5 m particle trap	
Program:	Start:	40 °C, 5 min
	Program 1:	10 °C/min to 100 °C
	Program 2:	20 °C/min to 250 °C
	Final temp.:	250 °C, 1.5 min
Mass spectrometer:	Scan mode:	EI, 55–249 u
	Scan rate:	0.5 s/scan
Calibration:	External standardisation	

Results

The PLOT column used (porous layer open tubular) exhibits considerable retention of chlorinated volatile components. For vinyl chloride the retention time is ca. 12:30 min with very good separation from the other volatile halogenated hydrocarbons. The mass spectrometric evaluation of vinyl chloride uses the masses m/z 62 and 64 (Fig. 4.28). The chromatogram of a standard with a concentration of 1 µg/L is shown in Fig. 4.29. The calibration for the determination of vinyl chloride in drinking water covers the concentration range 0.1 µg/L to 5.0 µg/L (Fig. 4.30). Figures 4.31 and 4.32 show chromatograms of a drinking water sample with and without addition of standard. The drinking water sample shows no signal for vinyl chloride (<0.1 µg/L).

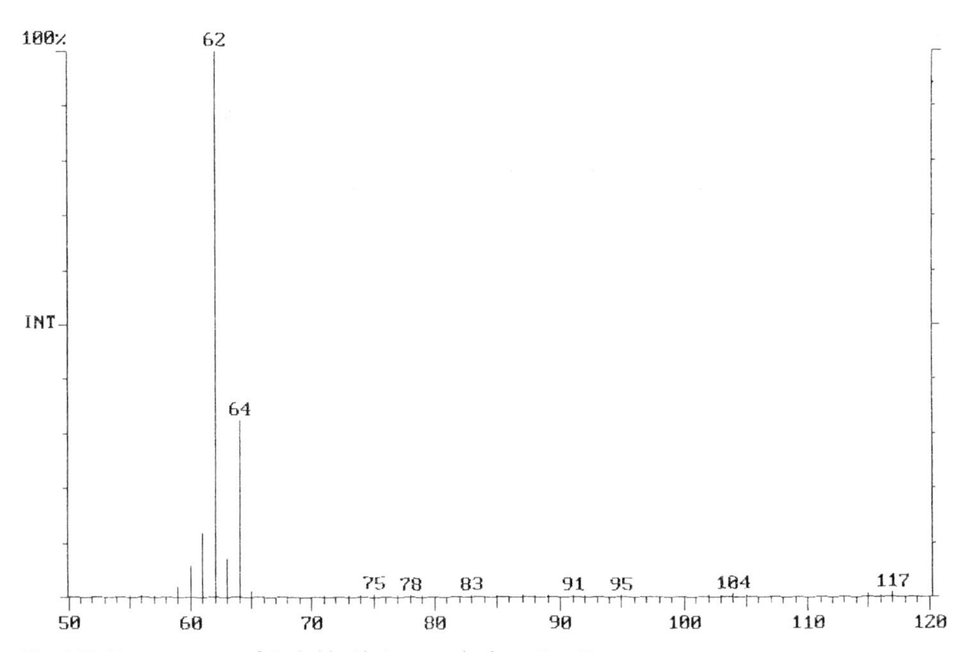

Fig. 4.28 Mass spectrum of vinyl chloride in a standard run, 1 µg/L

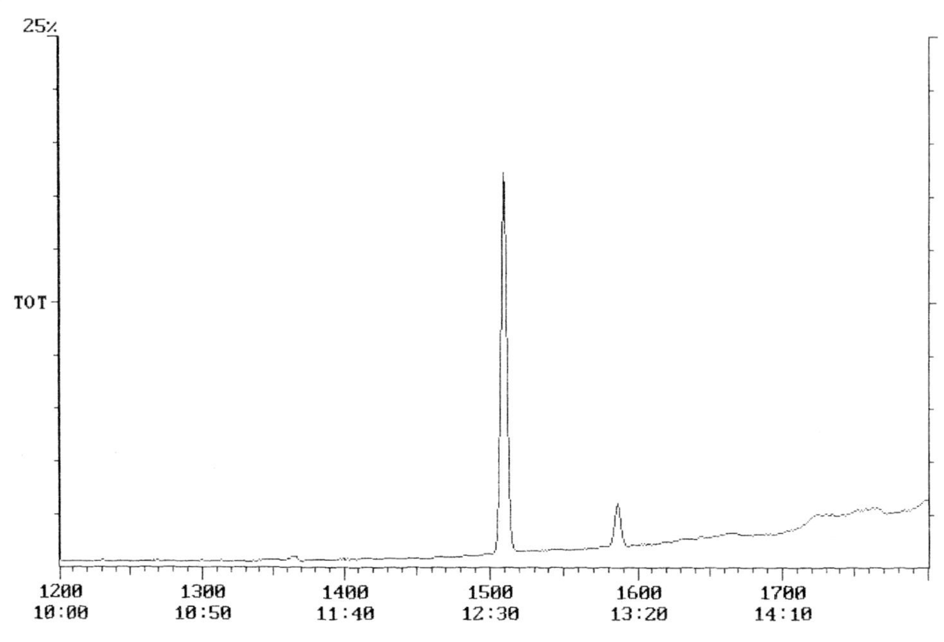

Fig. 4.29 Standard run for vinyl chloride on PoraPlot Q, 1 μg/L

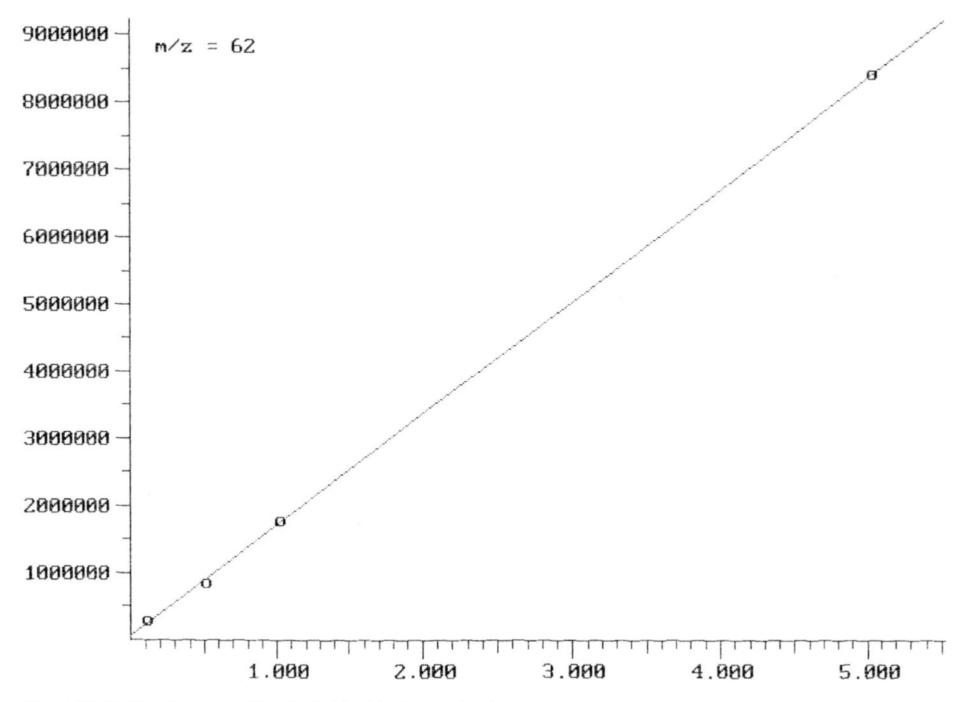

Fig. 4.30 Calibration curve for vinyl chloride determination

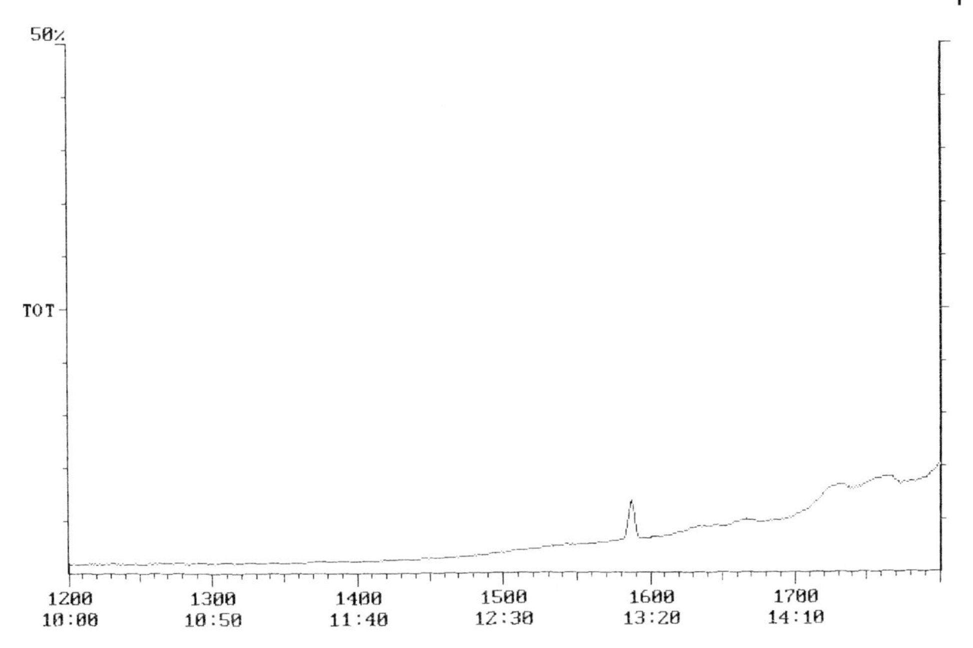

Fig. 4.31 Analysis of a drinking water sample for vinyl chloride

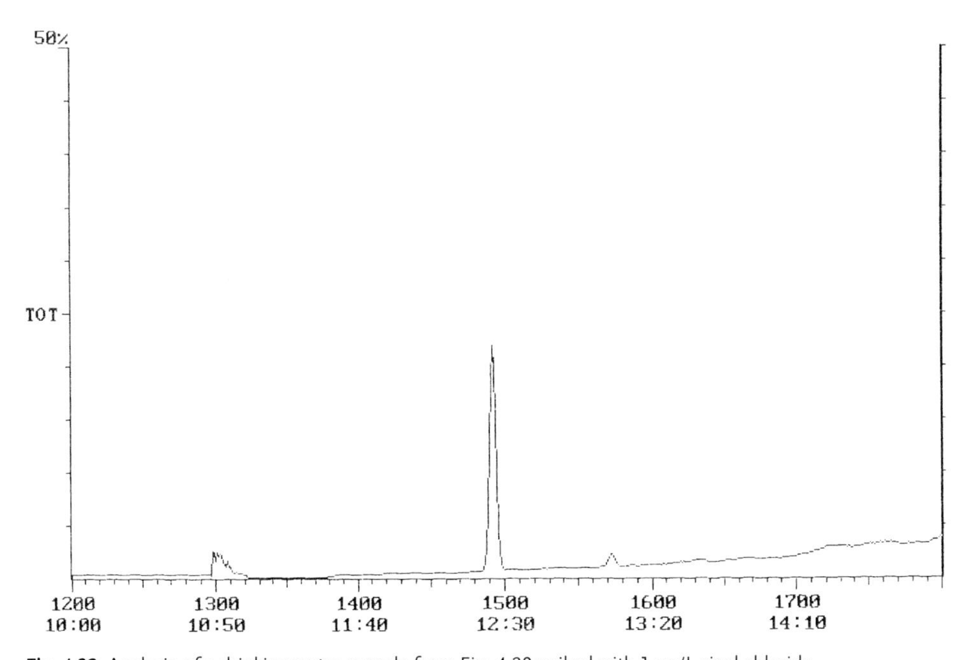

Fig. 4.32 Analysis of a drinking water sample from Fig. 4.29 spiked with 1 μg/L vinyl chloride

4.8
Chloral Hydrate in Surface Water

Key words: chloral hydrate, hydrolysis, chloroform, purge and trap, Vocarb 4000

Chloral hydrate can be determined analytically by quantitative conversion into chloroform in an alkaline medium:

$$CCl_3CH(OH)_2 + NaOH \rightarrow CHCl_3 + HCOONa + H_2O$$

For the determination of chloral hydrate in formulations and sprays, the samples are diluted appropriately and adjusted to pH 13.5 with 10 M NaOH. The reaction can take place within 5 min in the sample vessel of the purge and trap autosampler at 110 °C in the drying oven.

Analysis Conditions

Sample material:	Aqueous pesticide sprays, surface water	
Concentrator:	Tekmar purge and trap system LSC 2000 with AquaTek 50 sample dispenser 25 mL sample volume, 60 s sample transfer	
Trap:	Supelco Vocarb 4000, room temperature	
Purge and trap parameters:	Purge gas:	40 mL/min
	Mount:	40 °C
	Pre-purge time:	0 min
	Sample temp.:	35 °C
	Purge time:	10 min
	Dry purge:	3 min
	MCM:	0 °C/90 °C
	Desorb preheat:	245 °C
	Desorption:	250 °C, 4 min
	Bake mode:	280 °C, 5 min
	Bake gas bypass:	OFF
	Valves:	100 °C
	Transfer line:	200 °C
GC/MS system:	Finnigan ITS 40	
Injector:	Tekmar cryofocusing –30 °C	
Column:	Restek Rtx-1701, 30 m × 0.25 mm × 0.1 μm	
Program:	Start:	60 °C, 4 min
	Program 1:	5 °C/min to 150 °C
	Program 2:	20 °C/min to 280 °C
	Final temp.:	280 °C, 3.5 min

Mass spectrometer:	Scan mode:	EI, 60–99 u
	Scan rate:	0.5 s/scan
Calibration:	external standardisation	

Results

The chromatograms show a chloroform standard in Figs. 4.33 and 4.34 a standard run with 50 μg/L chloral hydrate, determined as chloroform. The chromatogram of a real sample with a concentration of ca. 50 μg/L is shown in Fig. 4.35. The sample also contains large quantities of toluene.

The standard solutions cannot necessarily be stored. A considerable decrease in concentration is observed even while standing in the autosampler.

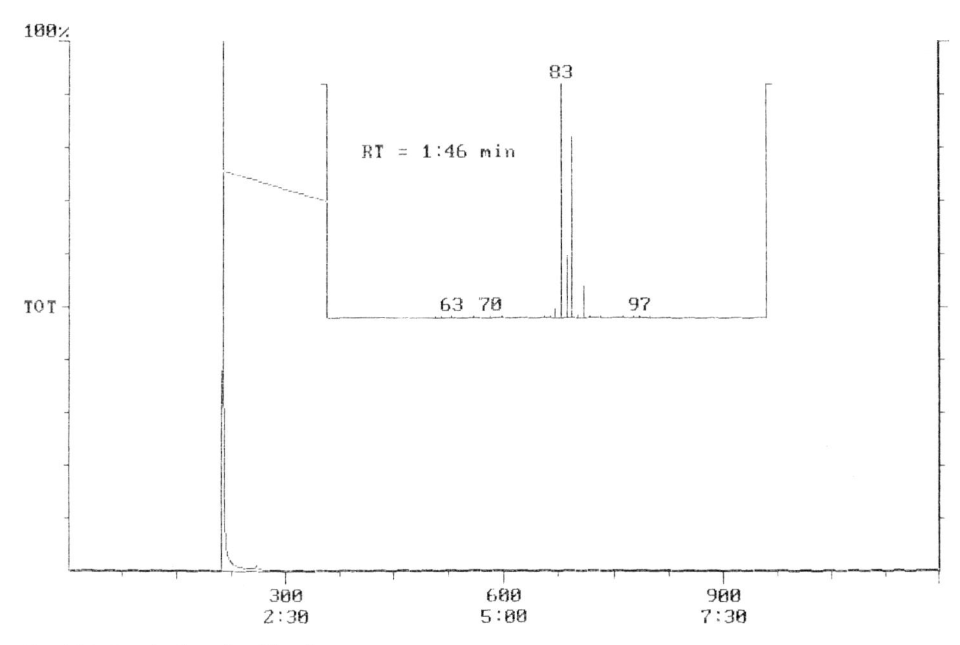

Fig. 4.33 Standard run for chloroform

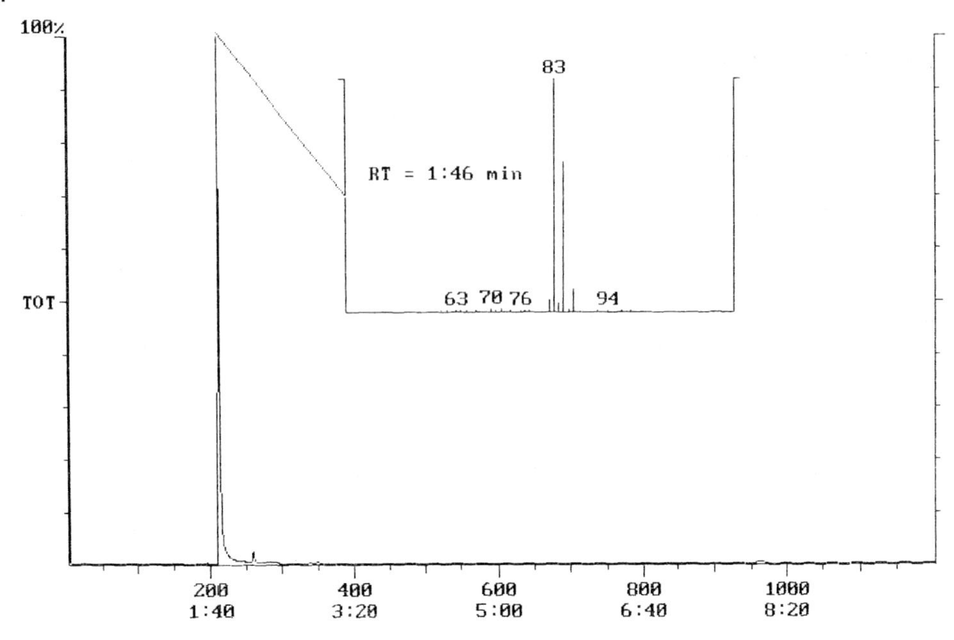

Fig. 4.34 Standard run for chloral hydrate, 50 µg/L

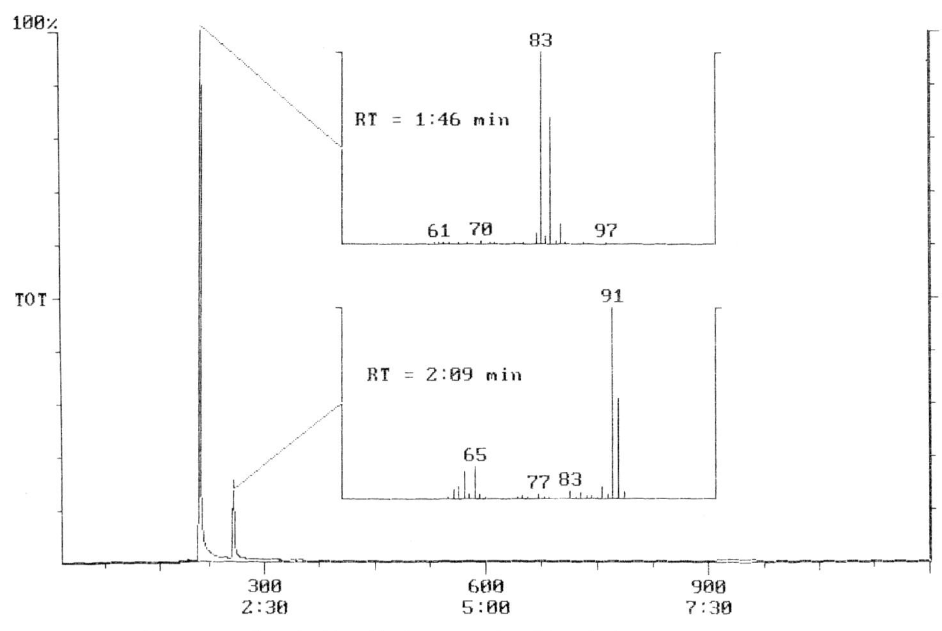

Fig. 4.35 Actual water sample with chloral hydrate (ca. 50 µg/L) and toluene

4.9
Field Analysis of Soil Air

Key words: volatile halogenated hydrocarbons, BTEX, adsorption, thermodesorption, soil air, mobile analysis, library search

Gaseous samples (soil air, gases from landfill sites, exhaust gases, air in enclosed spaces) are analysed for volatile halogenated and aromatic solvents by adsorption of the components in thermodesorption tubes. Because of the high affinity for volatile halogenated hydrocarbons and the low adsorption of water vapour, Tenax has proved particularly suitable for this purpose. Sampling can be actively effected using a pump, which, for example, can be connected to a purposely constructed bore column, and the volume flow can be controlled. Passive collection by diffusion can also be used.

A thermodesorption tube filled with Tenax, which should be fitted with a glass frit and which is closed with a silanised glass wool plug and conditioned, is used for sample preparation.

The mobile GC/MS system SpectraTrak 620 provides all the necessary equipment for active sampling, volume flow measurement and analysis, whereby the adsorption/desorption tube remains in the same piece of equipment. After sampling and analysis an evaluation is carried out automatically according to the requirements of the user.

Analysis Conditions

Sample collection:	0.5 L soil air	
GC/MS system:	VIKING SpectraTrak 620 with HP-MSD 5971 A	
Adsorbent:	ca. 100 mg Tenax	
Desorption:	200 °C, 1 min	
Column:	Restek Rtx-5, 20 m × 0.18 mm × 0.40 µm	
Carrier gas:	Helium, 1.5 bar (20 psi)	
Split:	10 mL/min	
Program:	Start:	50 °C
	Ramp:	15 °C/min
	Final temp.:	200 °C
Mass spectrometer:	Scan mode:	EI, full scan, 45–250 u
	Scan rate:	0.5 s/scan

Results

Figure 4.36 shows the chromatogram of a synthetic gas sample, which contains ca. 10 ppm of each of the six components. They are identified using a library search with the NIST library containing ca. 75 000 spectra (Fig. 4.37). The search results are printed out individually for visual examination of the matching of the spectra (Fig. 4.38).

In the field a rapid on-line analysis can be incorporated, which allows reliable identification of organic components in gases with an automatic recording device.

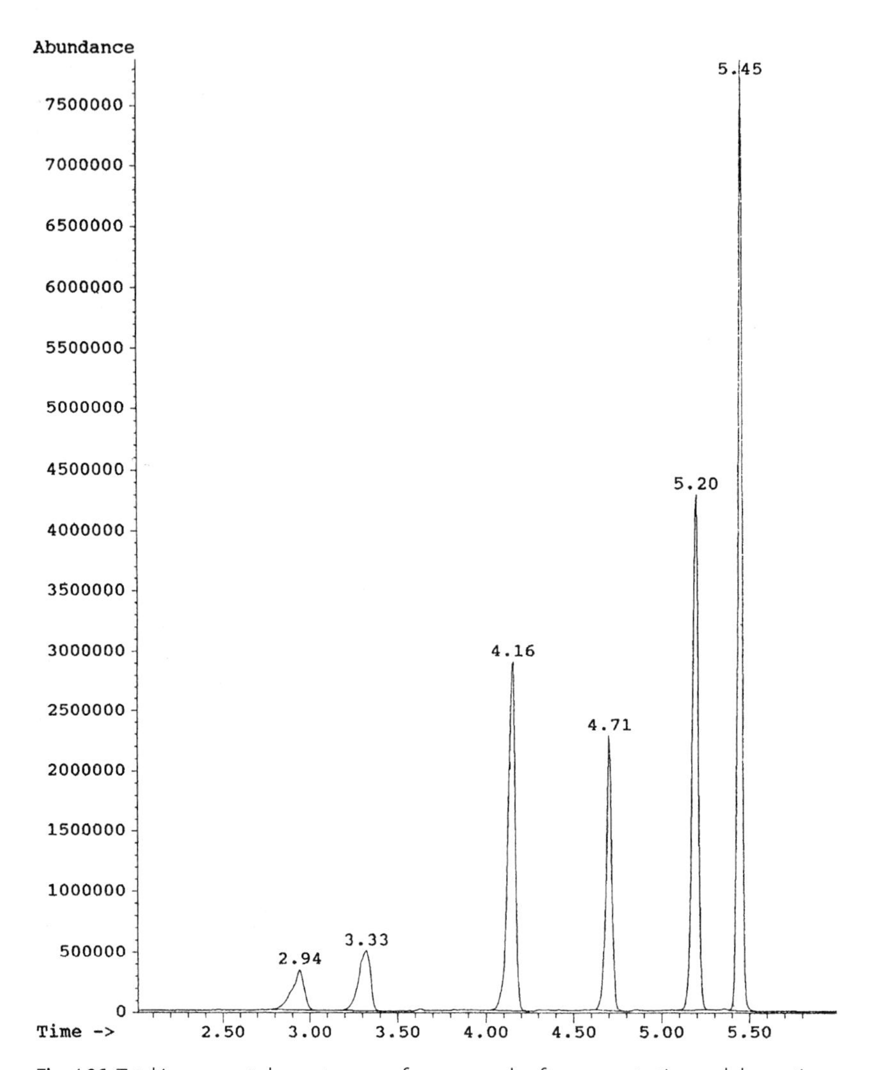

Fig. 4.36 Total ion current chromatogram of a gas sample after concentration and desorption

```
                    Viking SpectraTrak 600
                    Library Search Report

Disk File Name:       C:\CHEMPC\DATA\SCONGCMS
Operator:
Date Acquired:        22 Sep 93    1:56 pm
Cycle:                SCONGCMS
Run Type:
Data File:
Sample Name:          0.5 L Volumen auf tenax midtune @14psi
Misc Info:            Tenax only approx 100mg @15C/min

DA Method:            C:\CHEMPC\METHODS\DEFAULT.M
Integrator:           ChemStation
Integration Events:   AutoIntegrate
Search Libraries:     C:\DATABASE\NBS75K.L              Minimum Quality:   0
Unknown Spectrum:     Apex minus start of peak

 Pk#   RT  Area%          Library/ID                    Ref#     CAS#     Qual
_____________________________________________________________________________
  1   2.94   4.01 C:\DATABASE\NBS75K.L
                  Benzene                              62628 000071-43-2 91
                  Benzene                              62627 000071-43-2 91
                  Benzene                              62626 000071-43-2 91

  2   3.33   5.68 C:\DATABASE\NBS75K.L
                  Trichloroethylene                     5300 000079-01-6 97
                  Trichloroethylene                    65175 000079-01-6 97
                  Trichloroethylene                    65176 000079-01-6 94

  3   4.16  20.68 C:\DATABASE\NBS75K.L
                  Toluene                              63030 000108-88-3 94
                  Toluene                                965 000108-88-3 91
                  Toluene                              63029 000108-88-3 87

  4   4.71  12.65 C:\DATABASE\NBS75K.L
                  Tetrachloroethylene                  67757 000127-18-4 97
                  Tetrachloroethylene                  13222 000127-18-4 97
                  Tetrachloroethylene                  67759 000127-18-4 96

  5   5.20  24.74 C:\DATABASE\NBS75K.L
                  Benzene, chloro-                     63935 000108-90-7 94
                  Benzene, chloro-                     63934 000108-90-7 91
                  Benzene, chloro-                      2491 000108-90-7 87

  6   5.45  32.25 C:\DATABASE\NBS75K.L
                  p-Xylene                             63701 000106-42-3 91
                  p-Xylene                             63702 000106-42-3 87
                  Benzene, 1,3-dimethyl-                2027 000108-38-3 74
```

Fig. 4.37 Library search report after searching the NBS library

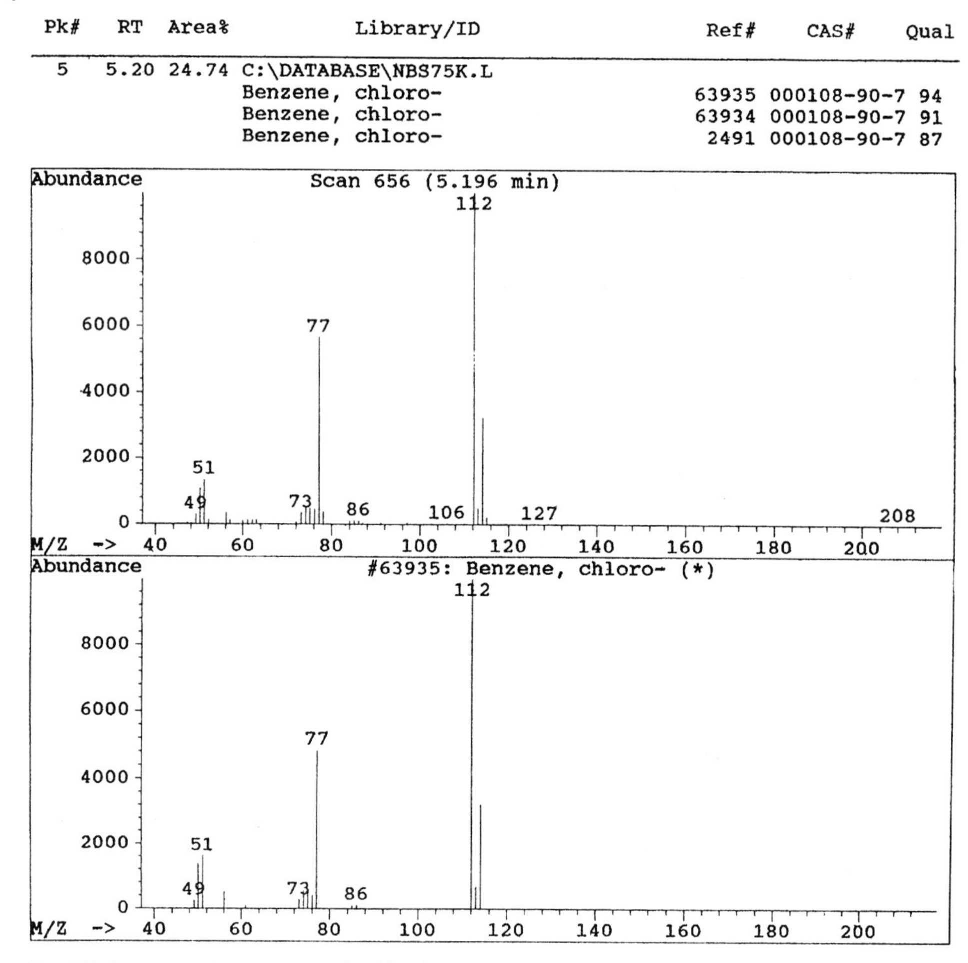

Pk#	RT	Area%	Library/ID	Ref#	CAS#	Qual
5	5.20	24.74	C:\DATABASE\NBS75K.L			
			Benzene, chloro-	63935	000108-90-7	94
			Benzene, chloro-	63934	000108-90-7	91
			Benzene, chloro-	2491	000108-90-7	87

Fig. 4.38 Spectroscopic comparison for chlorobenzene

4.10 Residual Monomers and Polymerisation Additives

Key words: purge and trap, polymers, residual monomers, additives

The purge and trap procedure is suitable for concentrating volatile components from polymers or materials in which plastics have been processed. The analysis of a group of these types of compounds is shown below. For the purge and trap analysis, the aqueous migration solutions of the corresponding incubation experiments are used.

Analysis Conditions

Concentrator:	Tekmar purge and trap system LSC 2000, needle sparge glass vessel, 5 mL	
Trap:	Supelco Vocarb 3000, cooled to –20 °C with TurboCool (liq. CO_2)	
Purge and trap parameters:	Purge gas:	40 mL/min
	Pre-purge time:	0 min
	MCM:	0 °C/90 °C
	Sample temp.:	room temperature
	Purge time:	12 min
	Dry purge:	0 min
	Desorb preheat:	255 °C with Vocarb 3000
	Desorption:	260 °C, 4 min
	Bake mode:	260 °C, 20 min
	Bake gas bypass:	ON after 120 s
	Valves:	110 °C
	Transfer line:	110 °C
GC/MS system:	Finnigan	
Injector:	PTV cold injection system 200 °C isotherm, split 30 mL/min direct connection (insertion) of the purge & trap concentrator into the carrier gas line of the PTV	
Column:	J&W DB-VRX, 60 m × 0.32 mm × 1.8 μm	
Program:	Start:	40 °C, 5 min
	Program 1:	7 °C/min to 180 °C
	Program 2:	15 °C/min to 220 °C
	Final temp.:	220 °C, 5 min
Mass spectrometer:	Scan mode:	EI, 33–260 u
	Scan rate:	1 s/scan

Results

Figure 4.39 shows the chromatogram of 26 selected monomers and additives. The components are assigned according to the entry number in the calibration file and from the retention time (Fig. 4.40). α-Bromostyrene is given as a typical entry. The standard mixture was injected in polyethylene glycol, which itself does not appear in the analysis. The first, not recorded, peak in the chromatogram is CO_2 from the sample vial. The concentrations of the substances (Fig. 4.41) are significantly higher than in the analysis of volatile halogenated hydrocarbons and BTEX in this application in the standard. To make the capacity suitable for the column and mass spectrometer used, the split ratio is varied according to the requirements. The analysis parameters are comparable with those of the volatile halogenated hydrocarbon/BTEX analysis (see Section 4.5).

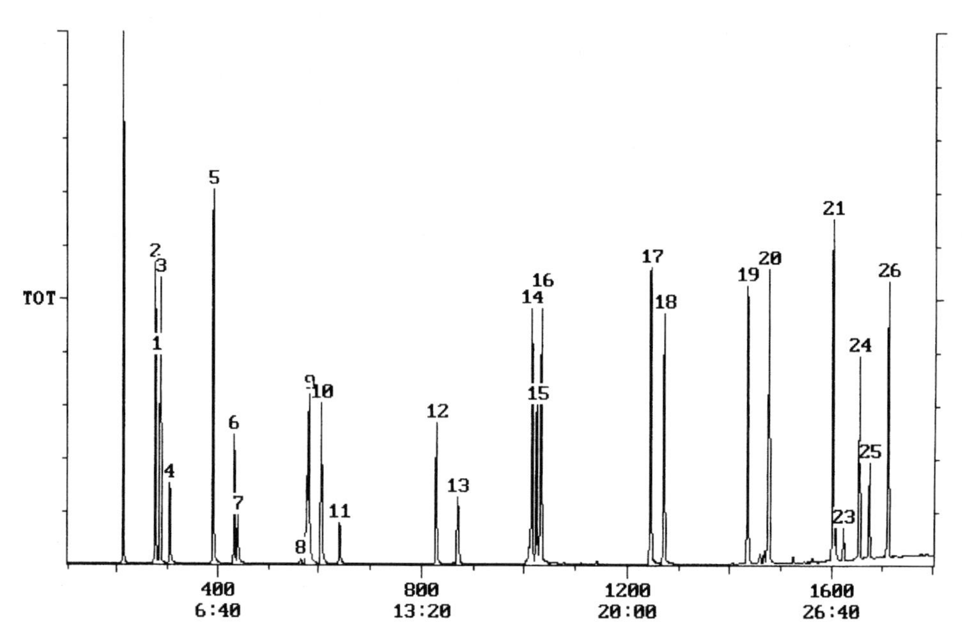

Fig. 4.39 Chromatogram of selected monomers and polymerisation additives (for analysis parameters see text; for peak list see Fig. 4.44)

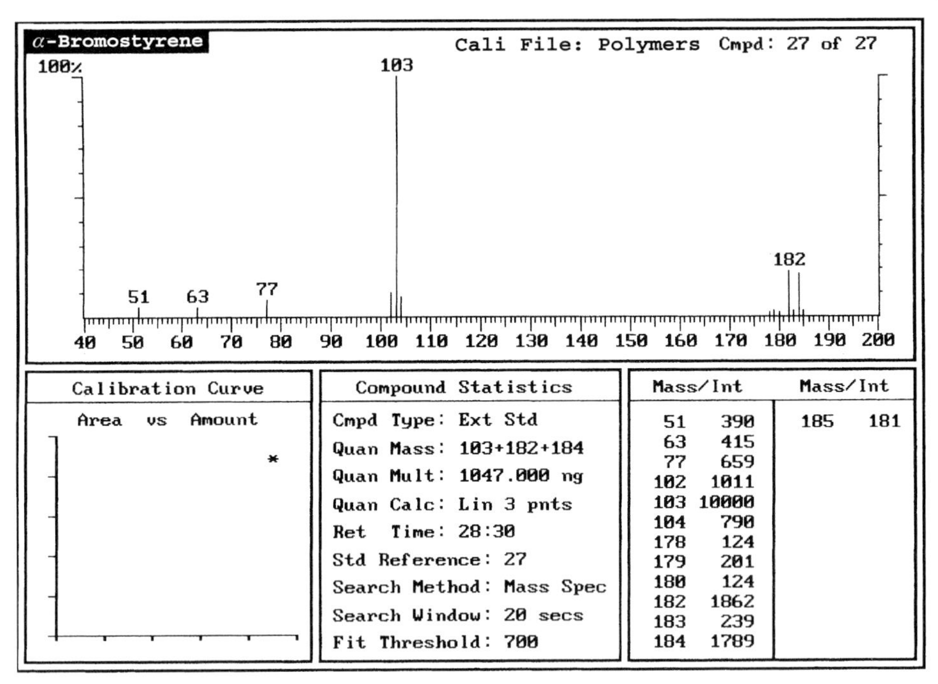

Fig. 4.40 Entry for α-bromostyrene in the calibration file

Cal	Name of Compound	S	Scan#	R Time	Me	Calc Amt(A)	Units
1	Vinyl chloride	E	276	4:36	BB	1811.001	ng
2	Acetaldehyde	E	275	4:35	BV	10010.001	ng
3	1,3-Butadiene	E	285	4:45	MM	1539.001	ng
4	Ethylene oxide	E	384	5:04	BB	5330.001	ng
5	Propylene oxide	E	391	6:31	BB	10129.001	ng
6	Vinylidene chloride	E	433	7:13	BB	1020.001	ng
7	Acrylonitrile	E	440	7:20	VB	1989.000	ng
8	Vinyl acetate	E	565	9:25	BB	1055.001	ng
9	Methyl vinyl ketone	E	576	9:36	MM	9558.001	ng
10	Butyraldehyde	E	581	9:41	BB	9830.001	ng
11	Methacrylonitrile	E	605	10:05	BB	1997.001	ng
12	Methyl acrylate	E	640	10:40	BB	1009.001	ng
13	Ethyl acrylate	E	828	13:48	BB	1911.001	ng
14	Epichlorohydrin	E	870	14:30	BB	9670.001	ng
15	1,7-Octadiene	E	1014	16:54	BV	981.001	ng
16	Ethyl methacrylate	E	1024	17:04	BB	1006.000	ng
17	1-Octene	E	1032	17:12	VB	1006.001	ng
18	Butyl acrylate	E	1245	20:45	BB	2023.000	ng
19	Styrene	E	1271	21:11	BB	1009.000	ng
20	α-Methylstyrene	E	1434	23:54	BB	1025.001	ng
21	p-Methylstyrene	E	1476	24:36	VB	987.001	ng
22	p-Chlorostyrene	E	1602	26:42	BB	986.001	ng
23	Ethylvinylbenzene isomer-1	E	1623	27:03	BB	221.000	ng
24	Ethylvinylbenzene isomer-2	E	1623	27:03	BB	221.000	ng
25	Divinylbenzene isomer-1	E	1655	27:35	BB	221.001	ng
26	Divinylbenzene isomer-2	E	1674	27:54	BB	221.001	ng
27	α-Bromostyrene	E	1711	28:31	BB	1047.001	ng

Fig. 4.41 Quantitation report with the components used in the standard analysis (the substance list refers to the peaks shown in Fig. 4.42)

4.11
Geosmin and Methylisoborneol in Drinking Water

Key words: drinking water, off odour contamination, SPME, headspace, automation, quantitation

Most complaints about the quality of drinking water found relate to odour and taste. Geosmin (1,2,7,7-tetramethyl-2-norborneol) and 2-MIB (2-methylisoborneol) are compounds mainly produced by blue-green algae (cyanobacteria) and actinomycete bacteria that cause musty, earthy odours in water supply reservoirs. Although these compounds have not been shown to be a health concern in public water supplies, the caused odors require removal and thus concentrations of geosmin and 2-MIB are monitored routinely in areas where they occur. The odour threshold for these compounds is very low and humans can typically detect them in drinking water at 30 and 10 ng/L (ppt) for geosmin and 2-MIB, respectively.

Past analytical techniques to determine concentrations of these off odour compounds used closed-loop stripping (CLS) and purge-and-trap sample preparation that provided the required sensitivity but have been either time consuming (CLS) or required a significant technical effort (P&T). Solid phase micro extraction (SPME) has shown in many applications to be a very versatile and sensitive sampling strategy with the potential of high sensitivity and full automation. This application describes the automated SPME technique from the sample headspace for the sensitive analysis of these compounds with high sample throughput.

Table 4.5 Geosmin and MIB compound characteristics

Compound name	Chemical structure	Molecular weight (amu)	Chemical formula	USGS parameter code	CAS registry number
Geosmin	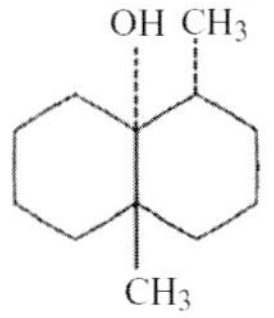	182.3	$C_{12}H_{22}O$	62719T	23333-91-7
2-methyisoborneol (MIB)	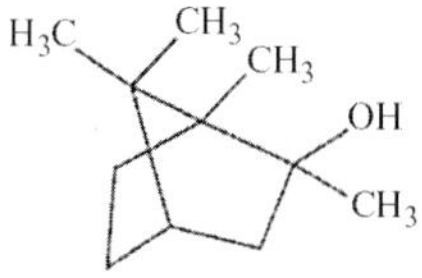	168.3	$C_{11}H_{20}O$	62749T	2371-42-8
2-isopropyl-3-methoxypyrazine (IPMP, surrogate compound	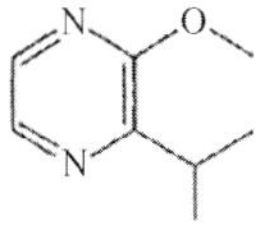	152.2	$C_3H_{12}N_2O$	–	25773-40-4

Analysis Conditions

GC/MS system	DSQ II with TRACE GC Ultra (Thermo Scientific, Austin TX, USA)
GC conditions:	
Oven program:	60 °C, 4.0 min 20 °C/min, 250 °C, 1.5 min (split flow: 15 mL/min)
Column:	Rxi-5MS 30 m × 0.25 mm × 0.25 µm
Carrier gas:	He 1.5 mL/min, constant flow
Injector:	PTV 60 °C, splitless time 1.0 min
Inject time:	0.1 min
Transfer rate:	14.5 °C/s
Transfer temp:	250 °C
Transfer time	4 min
SPME conditions:	
GC autosampler:	TriPlus Duo with SPME device
SPME fibre:	DVB/CAR/PDMS 50/30 cm (Supelco, Bellefonte, PA, USA)
Incubation:	6 °C, constant
Agitator on time:	10 sec
Agitator off time:	10 sec
Incubation time:	30 minutes
Injection depth:	20 mm
Pre- or post-injection:	no delay
Needle speed in vial:	20 mm/sec
SPME extraction time:	30 min
SPME desorption time:	4 min
Fiber conditioning station time:	10 min
Fiber conditioning port temp:	250 °C
Mass spectrometer conditions:	
Ionisation:	EI, 70 eV,
Source:	230 °C, closed-exit ion volume
SIM Mode	Segment 1 (7 min): m/z 137, m/z 152, 100 ms dwell time Segment 2 (8 min): m/z 95, m/z 107, 100 ms dwell time Segment 3 (10 min): m/z 112, m/z 149, 100 ms dwell time

SPME Method

The internal standard isopropylmethoxypyrazine was spiked to each sample at 300-ppt level. The addition of 30% (w/v) salt to the sample was used for salting out the analytes. The reproducibility of the SPME headspace analysis was depended on the same methanol content in the sample. The incubation temperatures and time were varied to maximize analyte concentrations. 60° C for 30 min was found optimal. Each calibration level (1, 10, 50, 100, 200, 500,1000 ppt) was prepared in 100 mL volumetric flask containing 300 ppt of internal standard. A 10 mL aliquot was transferred to a headspace vial with 3 g of salt. Each calibration standard was run in duplicate.

Results

SPME as a solventless sample preparation technique includes extraction, concentration and sample GC injection of the analytes in a single step. Automation of this sample preparation process offers a valuable tool to greatly extend sample throughput. SPME has gained wide spread acceptance as the technique of preference for the analysis of these odour compounds.

Excellent linearity was obtained for both Geosmin and 2-MIB with excellent linearity of the quantitative calibration over a range of 1 to 1000 ppt (see Figs. 4.42 and 4.43). Seven replicates were analyzed for method detection limits according to 40 CFR Part 136.

Average RSDs of the replicate injections were 0.6% for geosmin and 1.6% for 2-MIB.

These detection limits were run as 1:10 split injections, thus lower detection limits are possible with splitless injections. Excellent sensitivity and chromatographic performance was achieved across the calibration range for both 2-MIB and geosmin. Method detection limits of less than 0.3 ppb – less than the concentrations typically detected by humans – can be achieved by the described method.

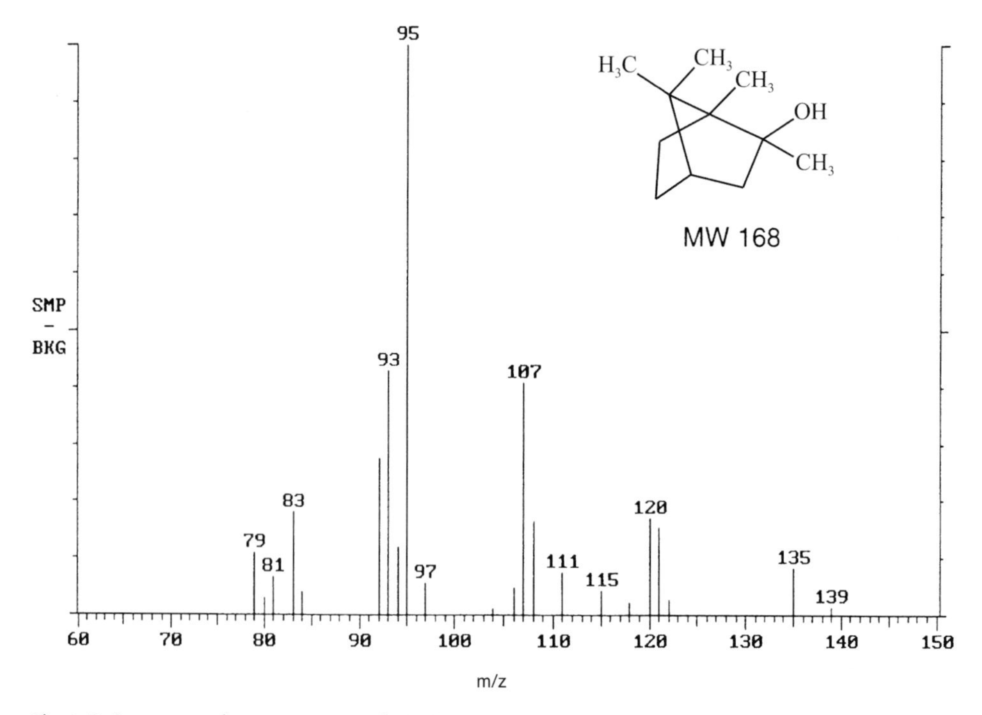

Fig. 4.42 Structure and mass spectrum of 2-MIB

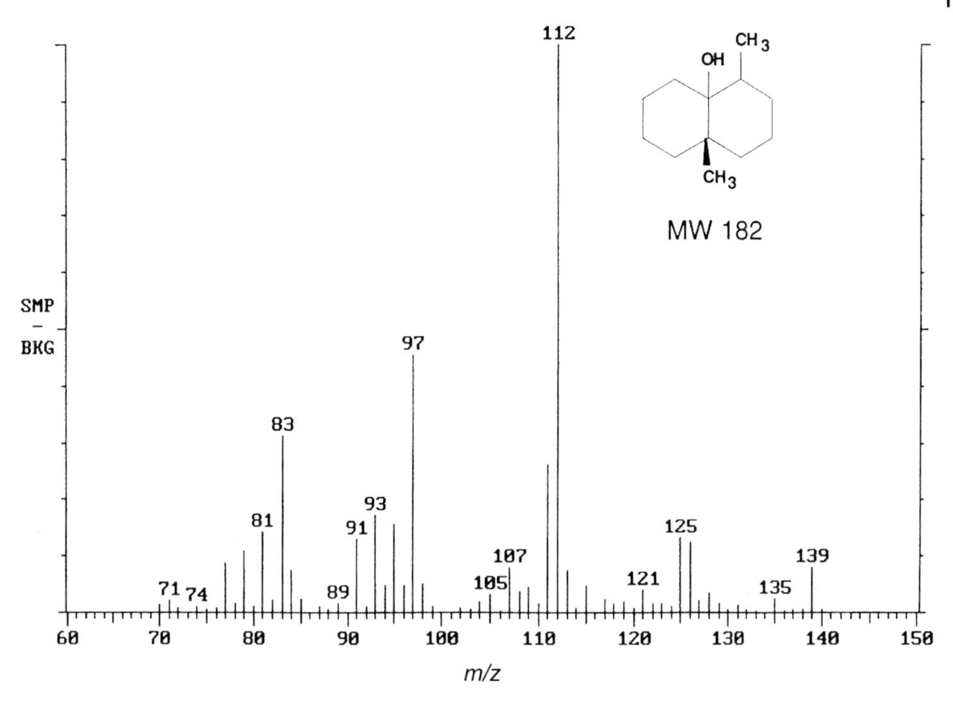

Fig. 4.43 Structure and mass spectrum of Geosmin

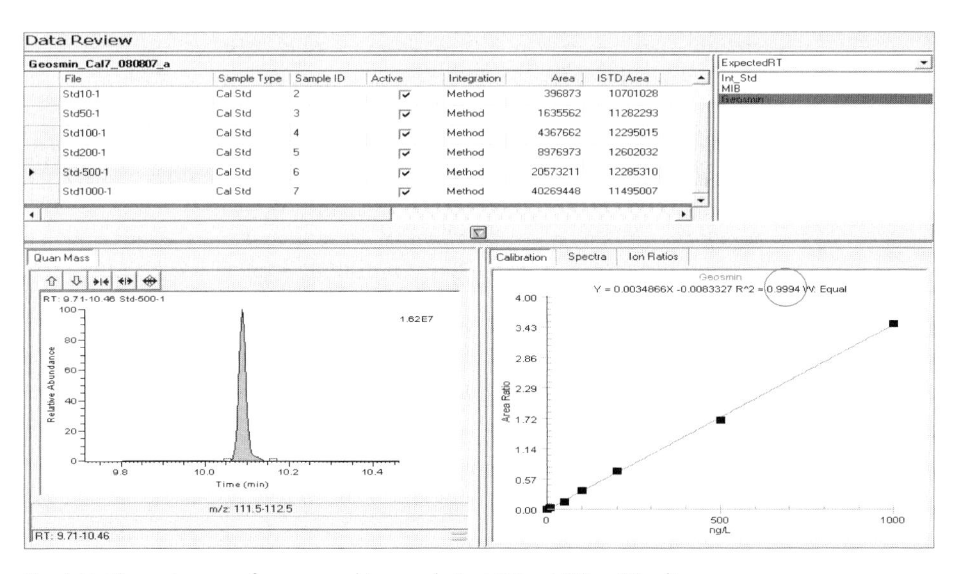

Fig. 4.44 Chromatogram of a purge and trap analysis at 2.5 ppt (50 pg/20 ml)

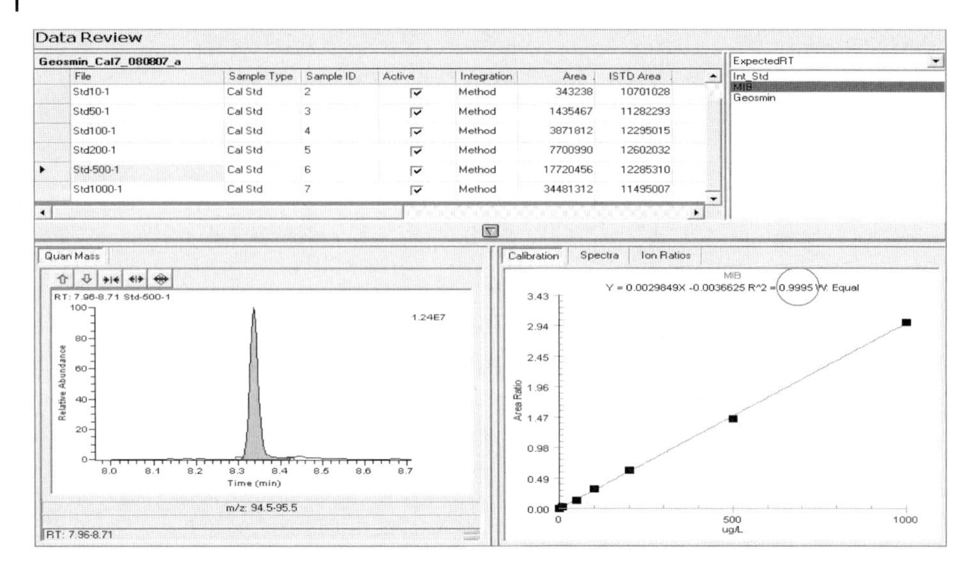

Fig. 4.45 Direct injection of 100 pg portions 2-MIB and Geosmin

4.12 Substituted Phenols in Drinking Water

Key words: odour contamination, drinking water, chlorophenols, bromophenols, alkylphenols, acetylation, chemical ionisation

Even at very low concentrations, substituted phenols lead to intense and unpleasant odours in drinking water. The various phenolic components can be formed from humic substances during chlorination of drinking water. The analysis procedures described up to now cannot guarantee reproducible and above all sensitive analyses. By combination of a special sample preparation procedure, known as extractive enrichment, with a sensitive mass spectrometer, phenols can be detected down to 1 ng/L in drinking, untreated and other water samples.

Analysis Conditions

Sample preparation:	Water sample 5000 mL (pH 11), addition of internal standard 2,4,6-trichlorophenol-$^{13}C_6$ + 20 mL acetic anhydride, 15 min + 100 mL methanol Then glass wool filtration
SPE extraction:	1 g C_{18}-material + 6 mL methanol + 6 mL water Sample flow 1000 mL/h Dry under N_2 (ca. 500 mL/min) Elution with 2 × 1 mL acetone Concentration to 0.5 mL
GC/MS system:	Finnigan ITS 40
Injector:	PTV cold injection system
Injector program:	Start: 60 °C, 1 min Program: 20 °C/s Final temp.: 250 °C
Column:	Rtx-1701, 30 m × 0.25 mm × 0.1 μm
Temp. program:	Start: 60 °C, 1 min Program 1: 5 °C/min to 150 °C Program 2: 20 °C/min to 280 °C Final temp.: 280 °C, 4.5 min
Mass spectrometer:	Scan mode: EI/CI, 50–350 u Scan rate: 1 s/scan

Results

The method described allows the determination of 29 substituted phenols (Table 4.6 and Fig. 4.46). To work out the analyses the individual components are isolated from the total ion current by selective mass chromatograms (Table 4.7, Figs. 4.47 and 4.48). By use of the ion trap mass spectrometer, the phenolic compounds contained in a drinking water sample can be identified from their mass spectra even at the very low limits of quantitation (Fig. 4.49). In the EI spectra the molecular ions do not appear or give very weak signals. Confirmation of molecular weights can be achieved using chemical ionisation with methane or methanol (Fig. 4.50).

Table 4.6 Composition of the standard from Fig. 4.56

No.	Phenol, analysed as the acetate	No.	Phenol, analysed as the acetate
1	2,6-Dimethylphenol	16	4-Chloro-3-methylphenol
2	2-Chlorophenol	17	2,4-Dichlorophenol
3	2-Ethylphenol	18	3,5-Dichlorophenol
4	3-Chlorophenol	19	2,3-Dichlorophenol
5	2,5-Dimethylphenol	20	3,4-Dichlorophenol
6	4-Chlorophenol	21	2,4,6-Trichlorophenol
7	2,4-Dimethylphenol	22	2,3,6-Trichlorophenol
8	3-Ethylphenol	23	2,3,5-Trichlorophenol
9	3,5-Dimethylphenol	24	2,4,5-Trichlorophenol
10	2,3-Dimethylphenol	25	2,6-Dibromophenol
11	3,4-Dimethylphenol	26	2,4-Dibromophenol
12	2-Chloro-5-methylphenol	27	2,3,4-Trichlorophenol
13	4-Chloro-5-methylphenol	28	2,4,6-Tribromophenol
14	2,6-Dichlorophenol	29	4,6-Dichlororesorcinol
15	4-Bromophenol	Internal Standard: 2,4,6-trichlorophenol-$^{13}C_6$	

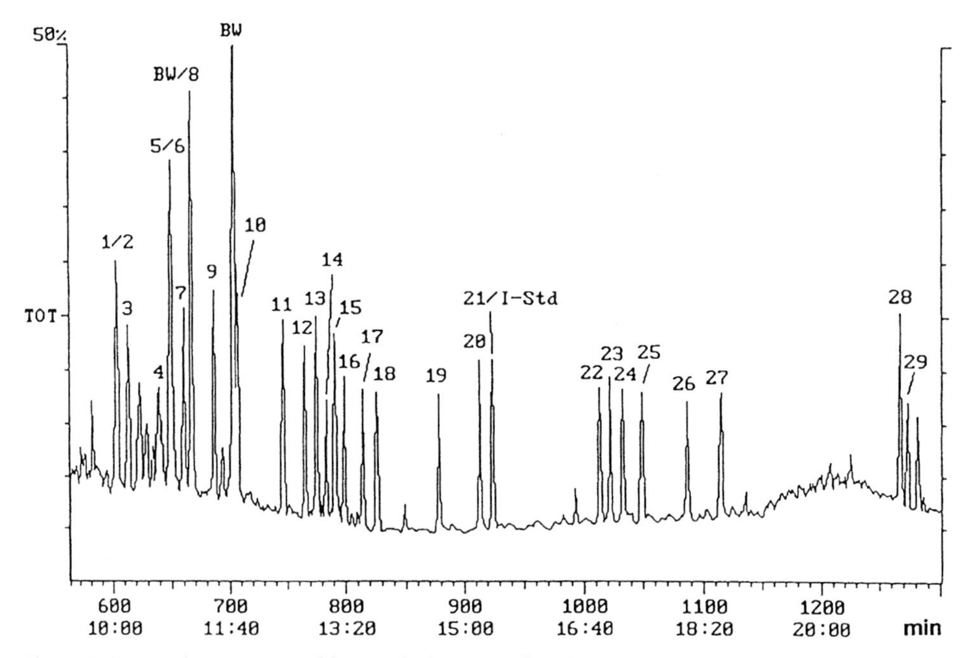

Fig. 4.46 GC/MS chromatogram of the standard mixture of 29 phenyl acetates (see Table 4.6)

Table 4.7 Specific masses of the phenyl acetates (EI)

Phenol, analysed as the acetate	m/z values
Ethylphenols	107, 122
Dimethylphenols	107, 122
Chloromethylphenols	107, 142
Chlorophenols	128, 130
Dichlorophenols	162, 164
Trichlorophenols	196, 198
Dichlororesorcinol	178, 180
Bromophenols	172, 174
Dibromophenols	250, 252
Tribromophenols	330, 332
Internal standard	
2,4,6-trichlorophenol-$^{13}C_6$	202

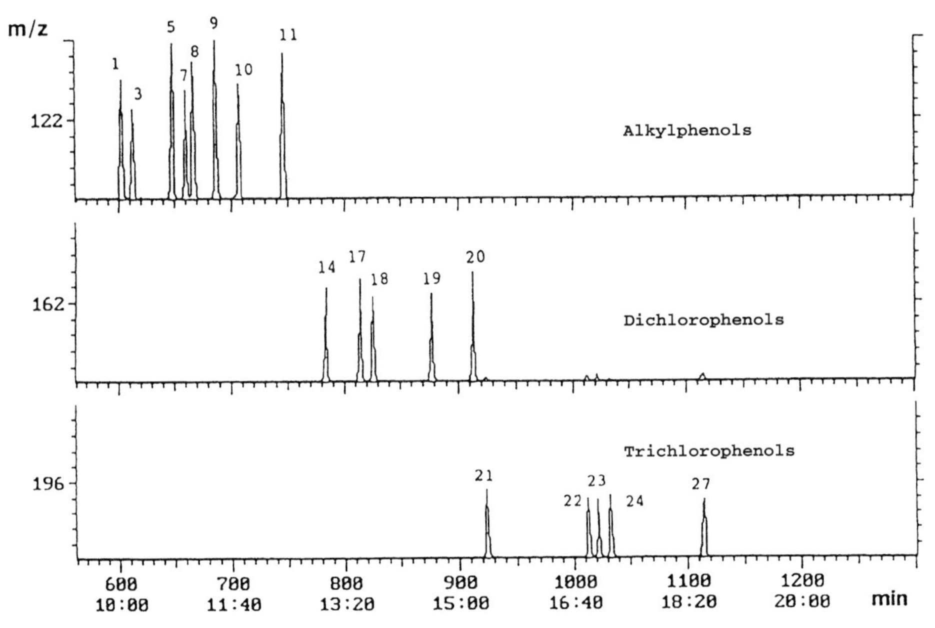

Fig. 4.47 Mass chromatogram of alkylphenols (m/z 122), dichlorophenols (m/z 162) and trichlorophenols (m/z 196)

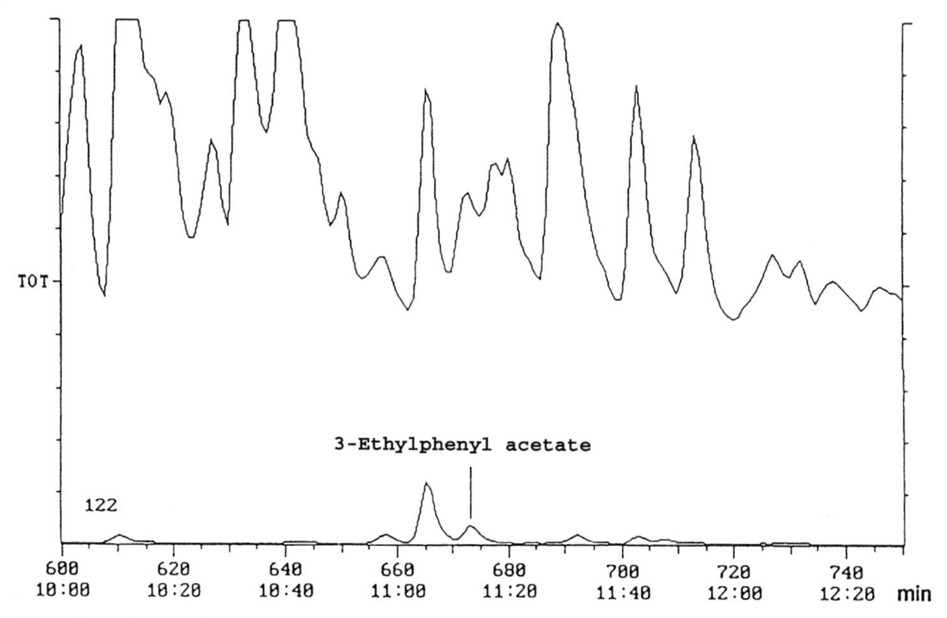

Fig. 4.48 Evaluation of a water sample with a total ion current and a selective mass trace m/z 122 for alkylphenyl acetates

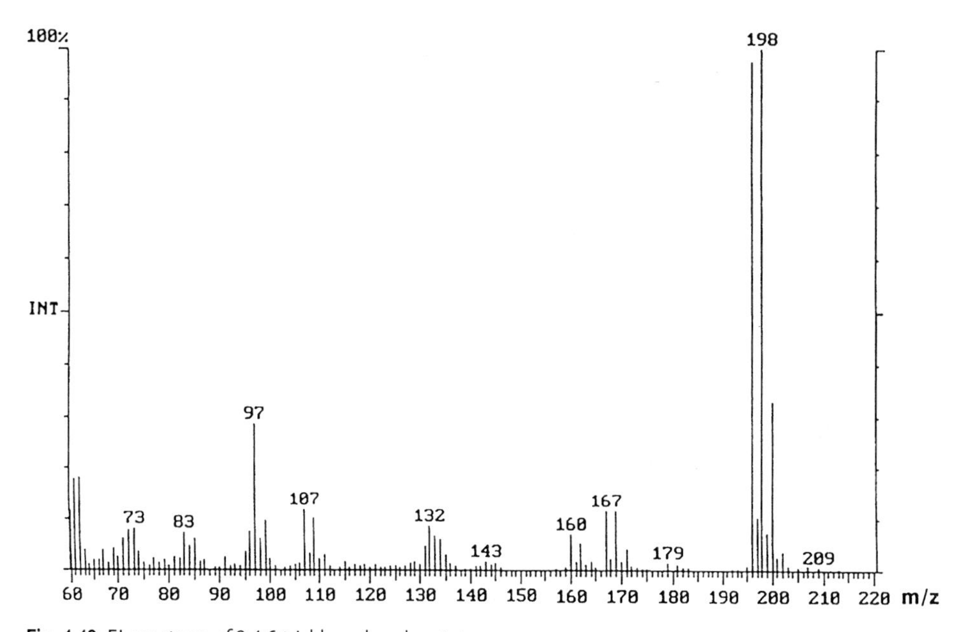

Fig. 4.49 EI spectrum of 2,4,6-trichlorophenyl acetate

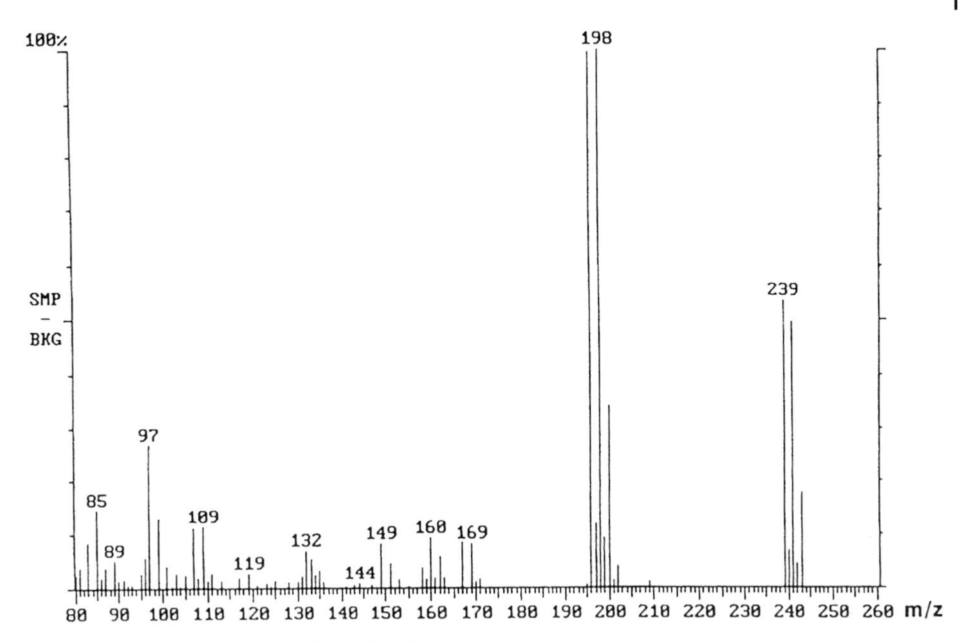

Fig. 4.50 CI spectrum of 2,4,6-trichlorophenyl acetate

4.13 GC/MS/MS Target Compound Analysis of Pesticide Residues in Difficult Matrices

Key words: pesticides, residue analysis, selectivity, MS/MS, confirmation, quantitation

The analysis of biological materials, such as plant and animal matter, for pesticide residue compounds using GC/MS has traditionally been difficult due to the very complex sample matrix. Approaches such as ECD and NPD detection, SIM, large volume injection or more sample concentration has been used for increasing the sensitivity of the method. But in matrices such as these, selectivity, another trace level requirement, is often the limiting factor. Simply injecting more sample necessarily means injecting more sample matrix and chemical background. The SIM approach, while enhancing the sensitivity at the expense of full scan mass spectral information, increases the monitored signal of ions common to both the background and the analyte within the retention time window. Real improvements in trace level complex matrix analyses depend on improving the detectivity (S/N) of the method by improving both the sensitivity and the selectivity.

The use of MS/MS in GC detection eliminates the matrix that causes the difficulty when using single stage MS. Thus, any background ions that may have been at the fragment ion m/z values are removed. The resultant product ion spectrum at the retention time is in presence and abundance due entirely to the m/z value of the selected pesticide precursor ion and not to the chemical background. Through this process the chemical background and matrix have been eliminated and low level detection is enhanced.

The samples used for the assay were fish tissue and river water from an industrial region. Aliquots of the homogenised fish tissue (typically 10 g) are treated with 100 g anhydrous sodium sulfate and extracted three times with 100 mL methylene chloride. The extract portions are combined, concentrated and purified using gel chromatography. Extracts are further cleaned using either a deactivated alumina/silica gel column or a Maxi Clean Florisil PR Cartridge (Alltech) and solvent elution. River water samples have been filtered and preconcentrated using C_{18}-SPE cartridges with standard methods described elsewhere.

For data acquisition multiple MS/MS scan events are time programmed as known from SIM acquisitions throughout the chromatographic run. Individual scan events are automatically set at the appropriate times so that as different target analytes elute, the conditions necessary for their optimal determination are activated. Precursor ions are usually chosen using the known SIM masses. MS/MS fragmentation is achieved using initial standard values with little individual response optimisation of the activation voltage.

Analysis Conditions

GC/MS system:	Finnigan Polaris Q
Injector:	Split/splitless injector 4 mm ID deactivated glass liner (for organochlorine pesticides with a 1 cm glass wool plug 3 cm from top) Splitless injection, split valve open after 1 min at 50 mL/min Injector temperature 300 °C
Autosampler:	AS 3000 Sample volume: 1 μL Air plug: 2 μL Hold time before injection: 2 s, after injection 0 s
GC column:	J&W Scientific DB-XLB, 30 m × 0,25 mm × 0,25 μm
Flow:	Constant flow at 1.5 mL/min (ca. 40 cm/s)
Temp. program:	60 °C for 1 min 40 °C/min to 150 °C, 1 min 5 °C/min to 300° C, 4.75 min
Transfer line:	Direct coupling to MS ion source 300 °C constant
Mass spectrometer:	EI mode Ion source at 200 °C
MS/MS conditions:	Organochlorine pesticides, see Table 4.8 NP-pesticides, see Table 4.9

Table 4.8 MS/MS acquisition parameters for organochlorine pesticides

Compound name	Acquisition mode	Precursor ion m/z	Excitation voltage V	q value	Product scan range m/z
4,4′-DDD	SRM MS/MS	235	0.80	0.225	40–440
4,4′-DDE	SRM MS/MS	318	0.80	0.225	40–440
4,4′-DDT	SRM MS/MS	235	0.80	0.225	40–440
Alachlor	SRM MS/MS	188	0.80	0.225	40–440
Aldrin	SRM MS/MS	293	0.80	0.225	40–440
α-BHG	SRM MS/MS	181	0.80	0.225	40–440
α-Chlordane	SRM MS/MS	373	0.80	0.225	40–440
β-BHC	SRM MS/MS	181	0.80	0.225	40–440
cis-Nonachlor	SRM MS/MS	407	0.80	0.225	40–440
Dieldrin	SRM MS/MS	279	0.80	0.225	40–440
Endosulfan I	SRM MS/MS	339	0.60	0.225	40–440
Endosulfan II	SRM MS/MS	339	0.60	0.225	40–440
Endosulfan sulfate	SRM MS/MS	387	0.65	0.225	40–440
Endrin	SRM MS/MS	281	0.80	0.225	40–440
Endrin aldehyde	SRM MS/MS	345	0.65	0.225	40–440
γ-BHC (Lindane)	SRM MS/MS	181	0.80	0.225	40–440
γ-Chlordane	SRM MS/MS	373	0.80	0.225	40–440
Heptachlor	SRM MS/MS	272	0.80	0.225	40–440
Heptachlor epoxide	SRM MS/MS	353	0.70	0.225	40–440
Methoxychlor	SRM MS/MS	227	0.90	0.225	40–440

Table 4.9 MS/MS acquisition parameters for nitrogen/phosphorus pesticides

Compound name	Retention time min:s	Acquisition mode	Precursor ion m/z	Excitation voltage V	q value	Product scan range m/z
Dichlorvos	9:10	SRM MS/MS	185.00	1.20	0.33	70–200
Mevinphos	11:18	SRM MS/MS	192.00	0.80	0.33	100–200
Heptenophos	13:31	SRM MS/MS	215.00	1.10	0.33	80–220
Propoxur	13:84	SRM MS/MS	152.00	1.00	0.33	60–155
Demeton-S-methyl	13:99	SRM MS/MS	142.00	0.90	0.33	50–150
Ethoprophos	14:17	SRM MS/MS	200.00	0.90	0.33	100–210
Desethylatrazin	14:65	SRM MS/MS	172.00	1.10	0.33	80–180
Bendiocarb	14:86	SRM MS/MS	166.00	0.80	0.33	80–170
Tebutan	14:95	SRM MS/MS	190.00	0.90	0.33	80–200
Dimethoate	15:73	SRM MS/MS	125.00	0.90	0.33	50–130
Simazine	15:96	SRM MS/MS	201.00	0.90	0.33	100–210
Atrazine	16:14	SRM MS/MS	200.00	1.30	0.33	100–210
Terbumeton	16:35	SRM MS/MS	169.00	1.00	0.33	70–175
Terbutylazine	16:62	SRM MS/MS	214.00	1.30	0.33	100–220
Dimpylate	17:03	SRM MS/MS	304.00	0.90	0.33	130–310

Table 4.9 (continued)

Compound name	Retention time min:s	Acquisition mode	Precursor ion m/z	Excitation voltage V	q value	Product scan range m/z
Terbazil	17:29	SRM MS/MS	161.00	1.20	0.33	80–170
Etrimfos	17:55	SRM MS/MS	292.00	0.90	0.33	150–300
Pirimicarb	17:95	SRM MS/MS	238.00	0.80	0.33	100–240
Desmethryn	18:23	SRM MS/MS	213.00	0.95	0.33	100–220
Ametryn	18:98	SRM MS/MS	227.00	0.90	0.33	100–230
Prometryn	19:10	SRM MS/MS	241.00	0.90	0.33	150–250
Terbutryn	19:56	SRM MS/MS	185.00	1.00	0.33	150–310
Pirimiphosmethyl	19:71	SRM MS/MS	305.00	0.75	0.33	60–190
Cyanazine	20:48	SRM MS/MS	225.00	1.10	0.33	60–230
Penconazole	21:62	SRM MS/MS	248.00	1.10	0.33	130–250
Triadimenol	22:05	SRM MS/MS	128.00	1.00	0.33	60–130
Methidathion	22:46	SRM MS/MS	145.00	0.80	0.33	60–150
Fluazifob-buthyl	24:63	SRM MS/MS	383.00	1.00	0.33	150–390
Oxadixyl	25:41	SRM MS/MS	163.00	0.80	0.33	60–170
Benalaxyl	26:19	SRM MS/MS	266.00	0.90	0.33	85–270
Hexazinon	26:95	SRM MS/MS	171.00	1.20	0.33	80–180
Azinphosmethyl	29:61	SRM MS/MS	132.00	1.10	0.33	60–140
Fenarimol	30:55	SRM MS/MS	139.00	1.00	0.33	60–145
Pyrazophos	30:83	SRM MS/MS	265.00	0.90	0.33	100–280

Results

Several commercially available mixes of pesticide compounds were used for calibration of the system by serial dilution and external standard techniques. A least squares regression was applied to curve fit the calibration data. Calibration curves were linear from below 0.01 ng/μL to 10 ng/μL (organochlorine pesticides) with correlation coefficients from 0.994 to 0.999.

The first example illustrates in Figs. 4.51 to 4.53 the determination of two Chlordane isomers. As can be seen in Fig. 4.51 the single stage technique, whether monitoring the total ion current (TIC) or three characteristic masses (m/z 373 + 375 + 266), can barely distinguish between the analyte signal and the complex matrix at this concentration level (550 ppb α-Chlordane in fish). MS/MS eliminates the matrix difficulties and allows much clearer detection of the analyte (Fig. 4.52) with an additional full product ion mass spectrum for confirmation (Fig. 4.53). Other compounds detected and confirmed are labelled. An example of a calibration curve is given in Fig. 4.54 for α-Chlordane showing a correlation factor of 0.998 over seven calibration points.

Another example shows the determination of Captan (Fig. 4.55) representative of nitrogen and phosphorus pesticides in difficult matrices. The single stage GC/MS analysis of a river water sample shows a susceptible peak at retention time 16.12 min indicating the occurrence of Captan deterioration (Fig. 4.56). The corresponding mass spectrum is dominated by chemical background ions (Fig. 4.57). The specific ions of Captan can be found within at

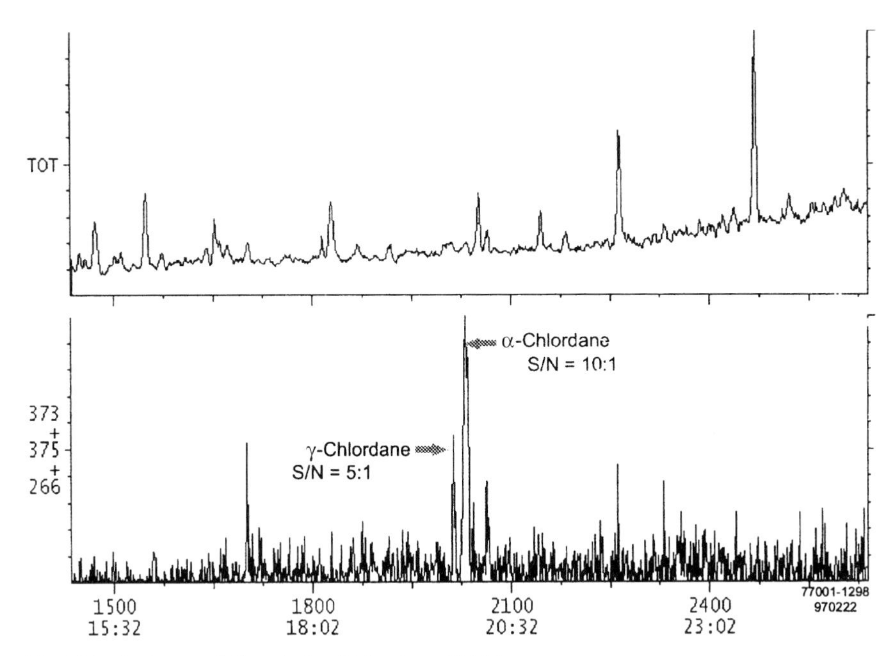

Fig. 4.51 Single stage MS analysis of fish extract 998 F2

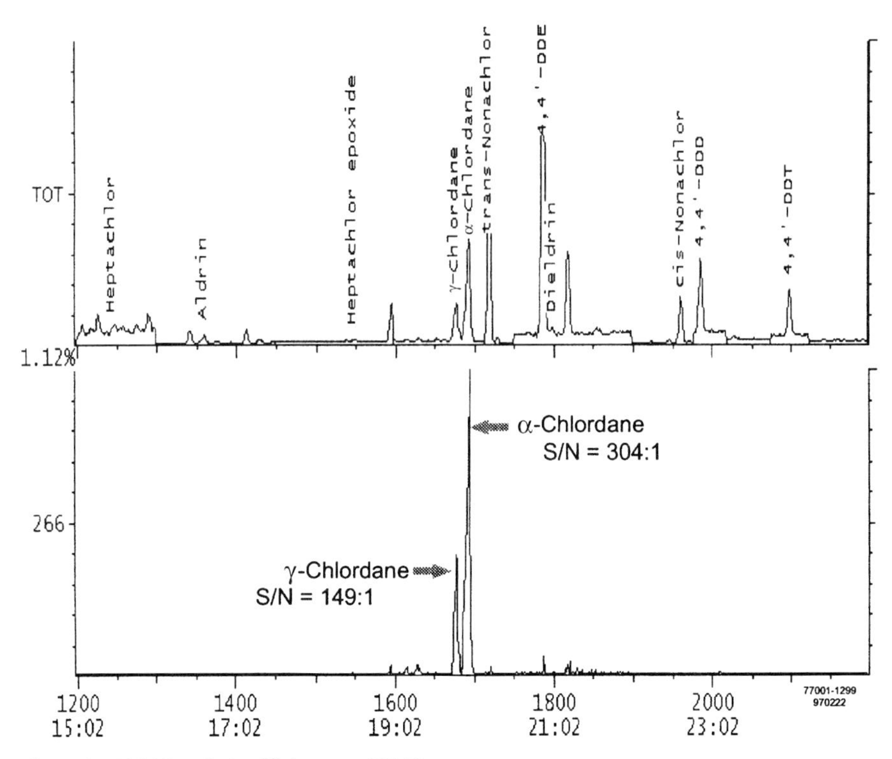

Fig. 4.52 MS/MS analysis of fish extract 998 F2

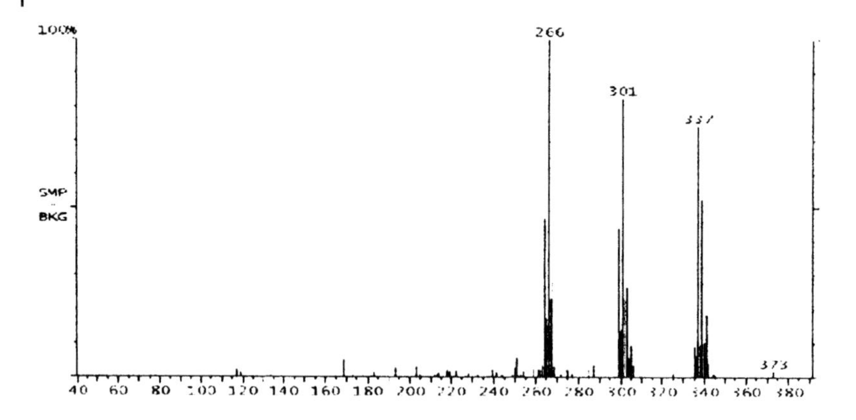

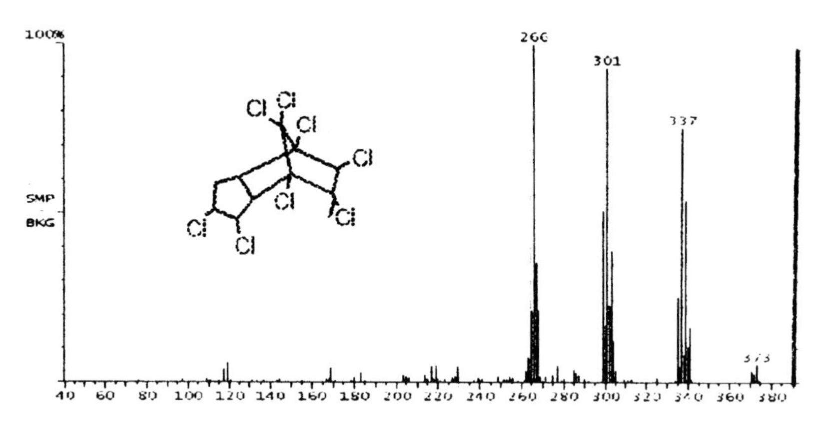

Fig. 4.53 Confirmation of α-Chlordane (the top spectrum represents the MS/MS library spectrum for reference)

lower levels. Mass m/z 149 is not diagnostic as there may be a contribution from ubiquitously occurring phthalate plasticisers. The high ion intensity of m/z 299 cannot be explained from the molecular ion of Captan in this case.

Using m/z 149 as the precursor ion for the MS/MS determination, the unambiguous confirmation of Captan in the sample at high S/N (signal/noise) levels is achieved (Fig. 4.58). The resulting product ion spectrum gives clear evidence of the structure-specific ions of Captan as proof (Fig. 4.59).

The retention and acquisition parameters of common nitrogen- and phosphorus-containing pesticides (phosphoric acid esters, atrazines, etc.) are listed in Table 4.7. It is recommended that the excitation parameters are used initially with equal standard values (0.9 V), which may be slightly adjusted in the case of response optimisation. The quality of the product ion spectra is not affected (Fig. 4.60). A typical standard run in MS/MS acquisition mode is shown in Figs. 4.61 and 4.62. Even at the low 10 pg/ul level the compounds are detected with high S/N values and full MS/MS product spectra for confirmation (Fig. 4.60).

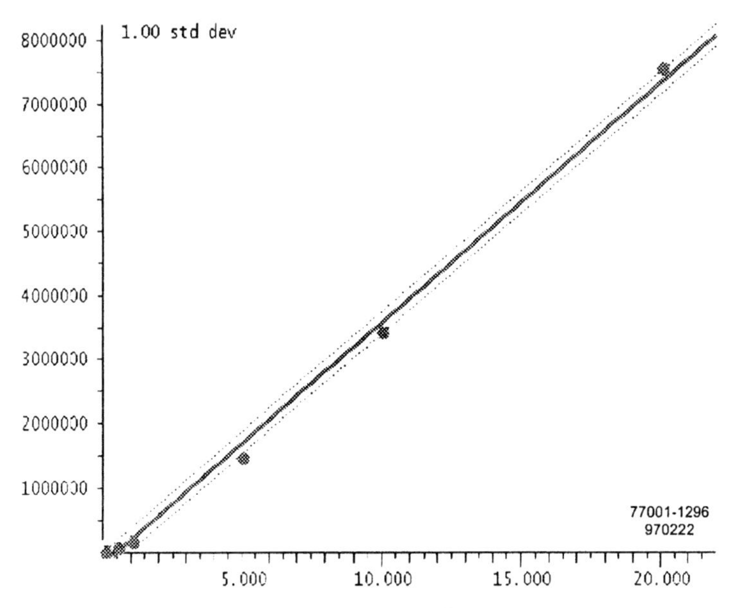

Fig. 4.54 α-Chlordane calibration curve (corr. coefficient 0.998)

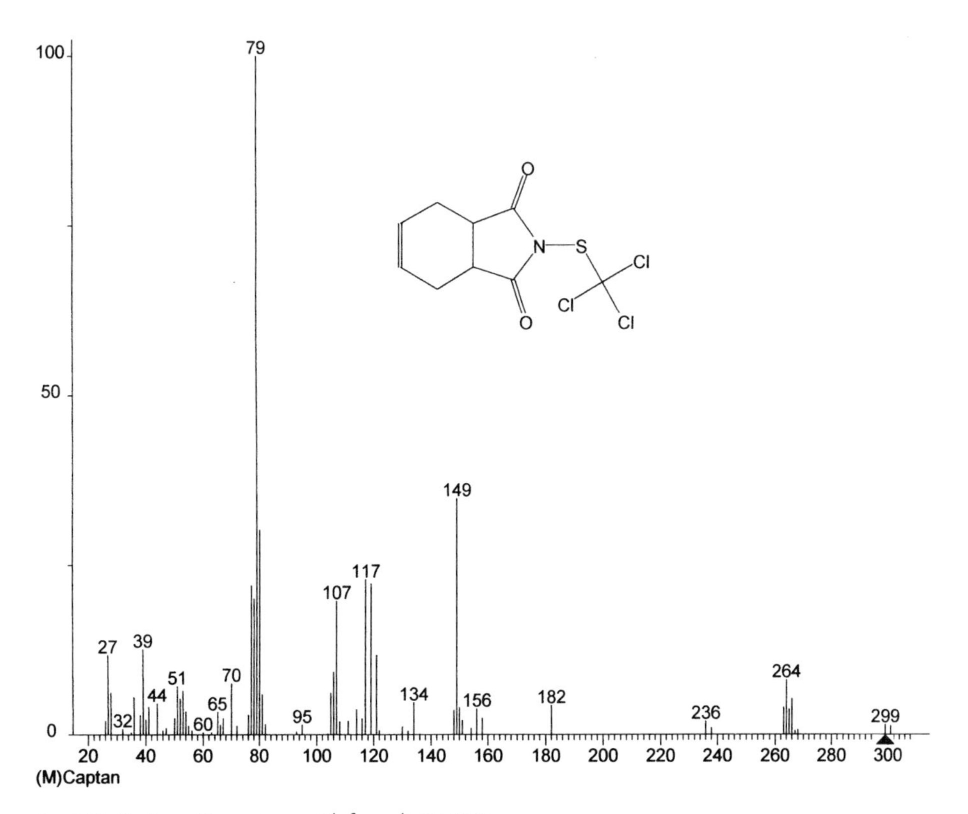

Fig. 4.55 Captan – EI spectrum with formula (M 299)

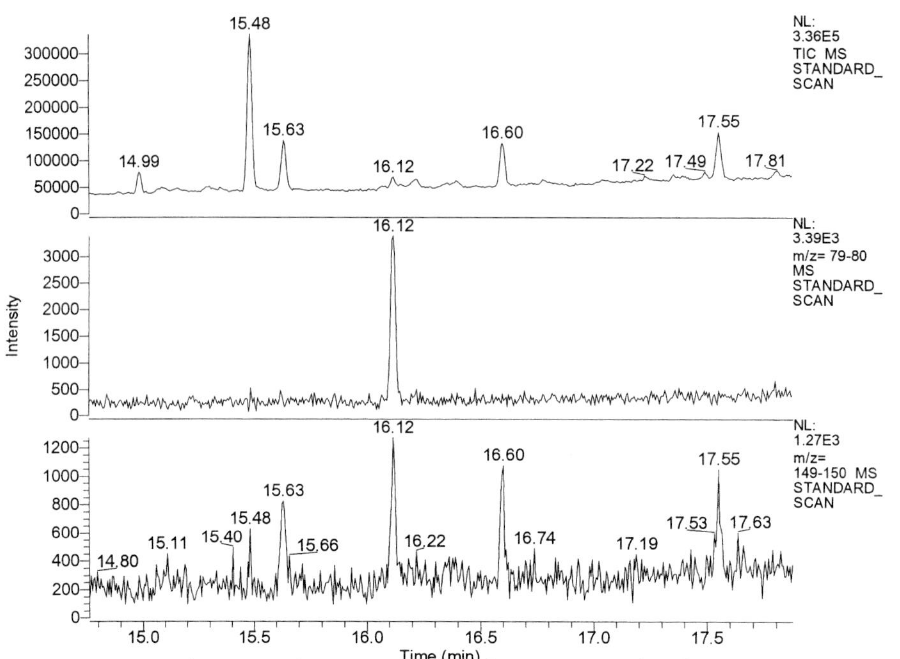

Fig. 4.56 Single stage MS analysis of river water sample with Captan mass chromatograms m/z 79, 149

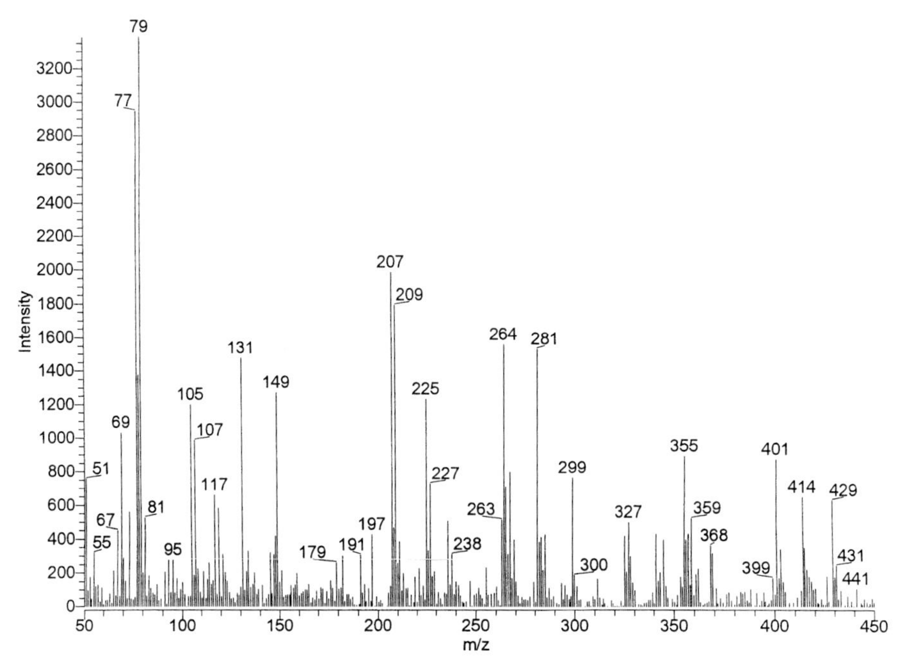

Fig. 4.57 Mass spectrum of GC peak eluting 16.12 min – Captan? Specific ions m/z 79, 107, 149 are within chemical background (m/z 149 may result from phthalate plasticiser)

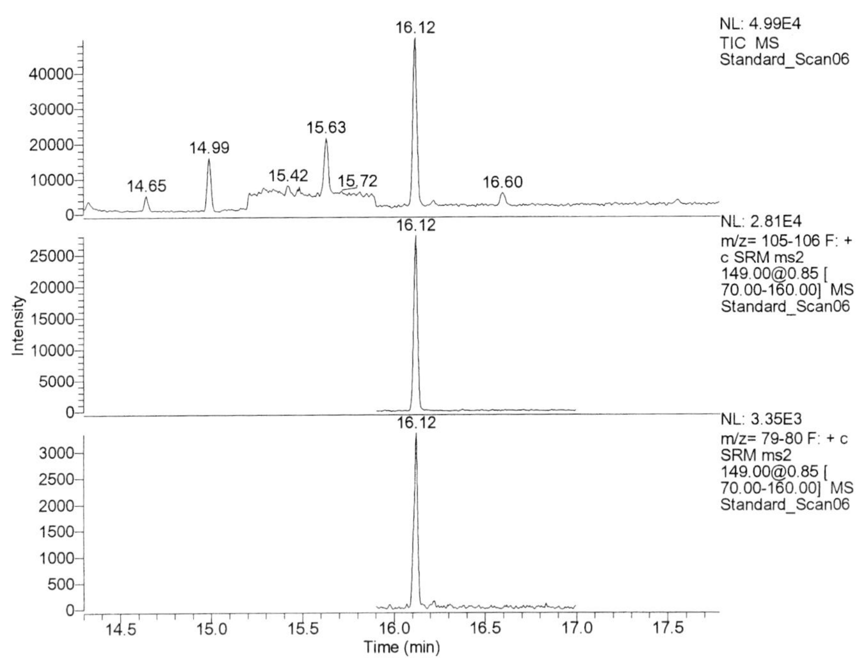

Fig. 4.58 GC/MS/MS analysis of a river water sample using m/z 149 as the precursor ion with Captan product ion chromatograms m/z 105, 79

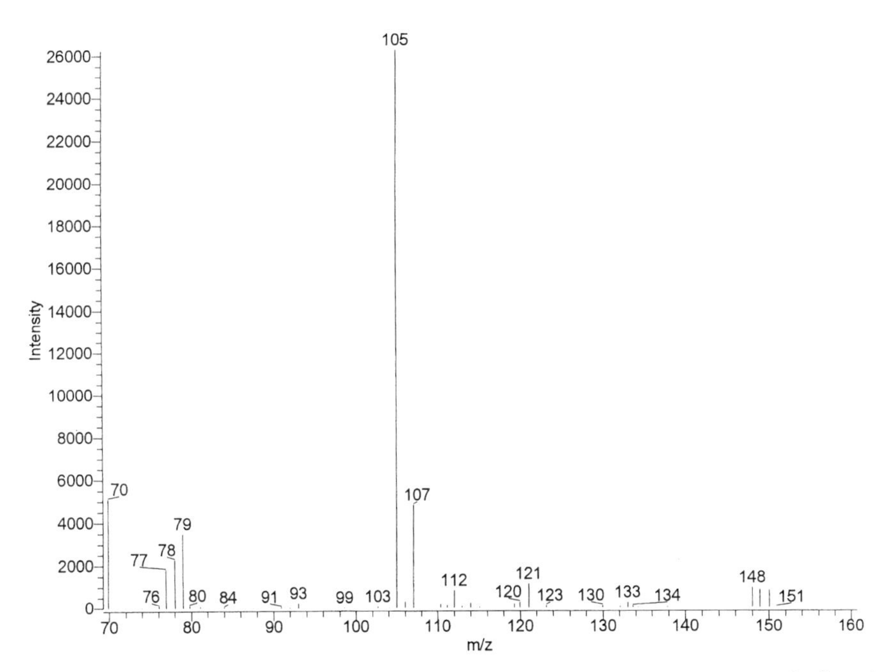

Fig. 4.59 MS/MS Spectrum of GC peak eluting at 16.12 min (m/z 149 as precursor ion) – Confirmation for Captan with structure specific ions m/z 79, 105, 107 free of chemical background

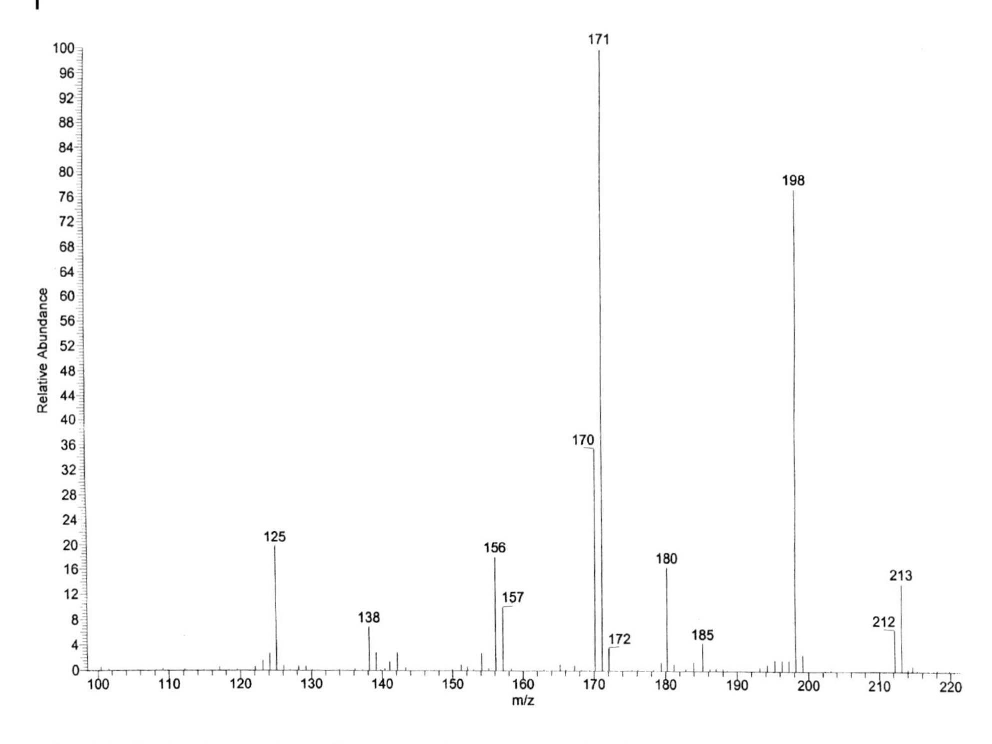

Fig. 4.60 Product ion spectrum of Desmetryn (precursor ion m/z 213)

The use of MS/MS techniques can provide a very reliable and indeed essential means of analysis in the determination of target compounds in complex matrices. The improved selectivity of the analysis from the MS/MS process, along with the inherent sensitivity of the ion trap mass spectrometer can often result in an order of magnitude or more of improvement in the signal to noise performance with related dramatic improvements in limits of detection, over single stage full scan or SIM methodologies.

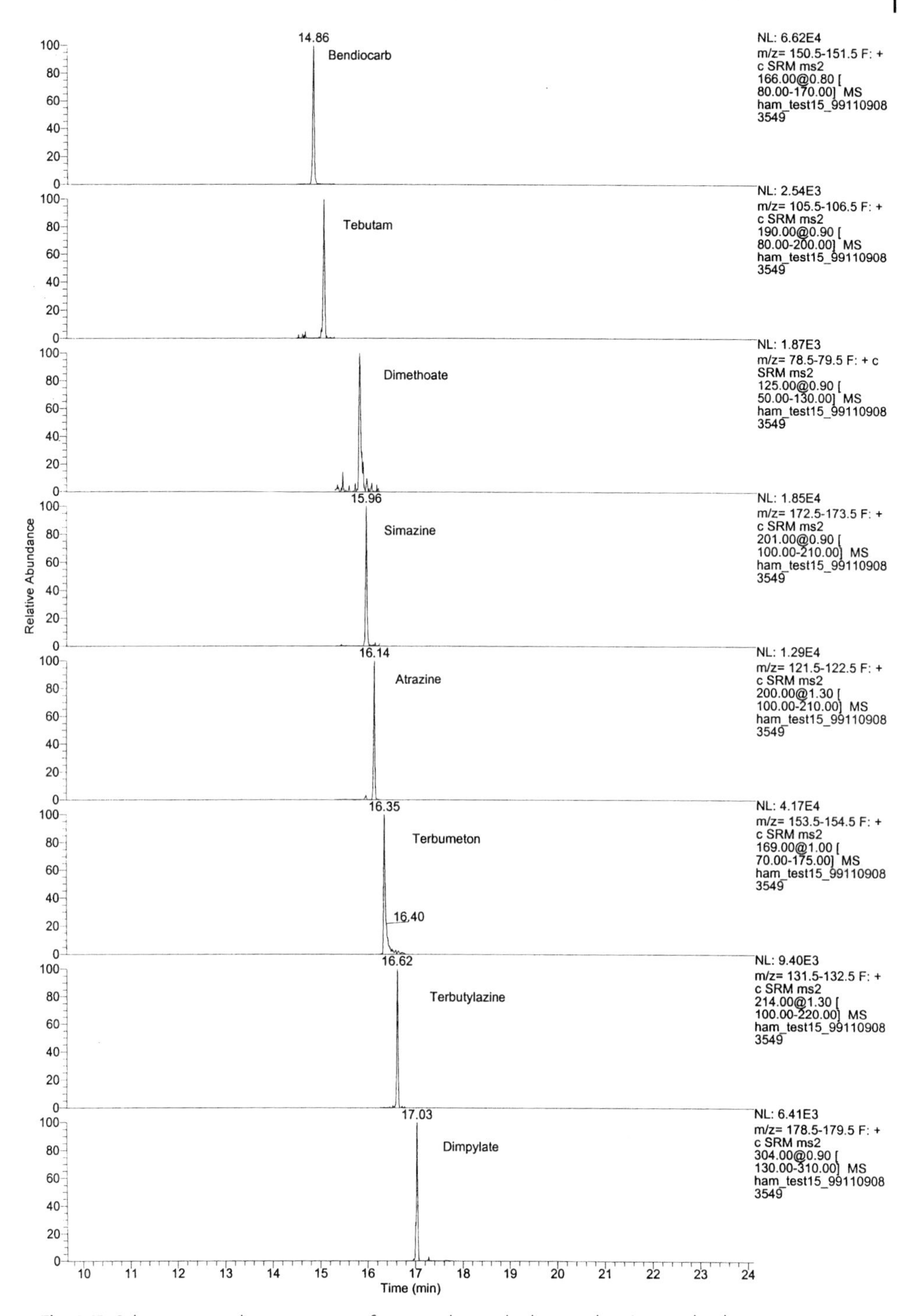

Fig. 4.61 Selective mass chromatograms of a pesticide standard run at the 10 pg/uL level (see Table 4.9)

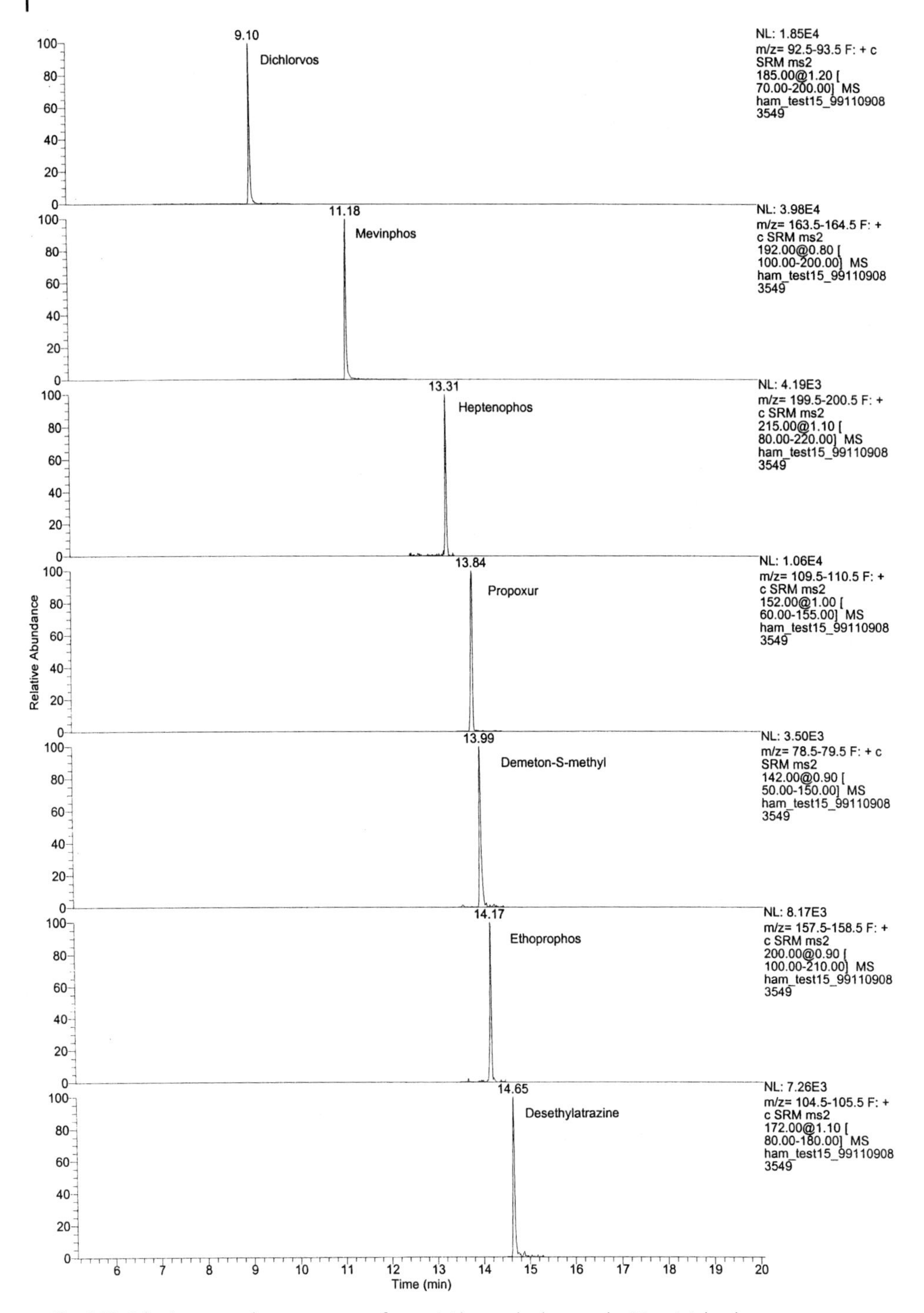

Fig. 4.62 Selective mass chromatograms of a pesticide standard run at the 10 pg/uL level (see Table 4.9)

4.14 Multi-component Pesticide Analysis by MS/MS

Key words: pesticides, food, multi-residue method, high throughput, QuEChERS, selected reaction monitoring SRM, H-SRM, MS/MS, confirmation, QED, data dependent acquisition

Food safety concerns are on the rise among consumers worldwide. In 2006, sweeping changes were made to the Food Hygiene Law in Japan regarding residual agricultural chemicals, including pesticides, in foods. As a result, residue standards were created for all pesticides, and standard residue values were established for more than 800 pesticides. Because each type has different physicochemical properties, there are limitations on simultaneous analysis. Among the pesticides for which standard values are currently set, GC/MS/MS can analyse more than 300 of them.
The superior selectivity and high speed acquisition rate of the GC/MS/MS technique allows interference-free quantification, even with peak co-elution, and provides positive confirmation in a single analytical run. To accurately monitor pesticide residues, a high throughput multi-residue method that can quantitate a large number of pesticide residues during a single analysis is described.

Sample Preparation

The QuEChERS sample preparation method is especially suited for low fat containing samples. Green pepper, carrot, grapefruit and banana samples were prepared for analysis using the QuEChERS method. A 10 g sample of food was homogenized in a food processor and placed in a polypropylene centrifuge tube. The sample was extracted with 20 mL of acetonitrile in a homogenizer. Then, 4 g of anhydrous magnesium sulfate and 1 g of sodium chloride were added and the resulting mixture was centrifuged. After centrifugation, the supernatant was loaded onto a graphite carbon/PSA dual layer solid phase extraction column and eluted with 50 mL of acetonitrile/toluene (3 : 1). After the eluate was concentrated under reduced pressure, it was dissolved (1 g/mL) in 10 mL of acetonitrile/n-hexane to give the test solution.

Analysis Conditions

Gas chromatograph:	TRACE GC Ultra with TriPlus liquid autosampler (Thermo Fisher Scientific, Milan, Italy)
Injector:	Split/splitless, 240 °C, surge pressure 200 kPa, 1 min splitless
Carrier gas:	Helium, 1.2 mL/min, constant flow
Column:	Rti-5MS, 30 m × 0.25 mm × 0.25 µm (Restek Corp., Bellefonte, PA, USA)

Oven temp. program:	80 °C, 1 min 20 °C/min to 180 °C 5 °C/min to 280 °C, 10 min
MS interface temp.:	280 °C
Mass spectrometer:	TSQ Quantum GC (Thermo Fisher Scientific, San Jose, CA, USA)
Ionisation:	EI
Scan mode:	Scan type: H-SRM Precursor peak width for H-SRM: Q1, 0.4 Da Product ion peak width: Q3, 0.7 Da Scan width: 0.002 m/z Scan time: 0.002 s, 0.005 s, 0.01 s Collision gas Ar at 1.2 mTorr
Mass range:	SRM transitions see Table 4.10
Cycle time:	0.5 s/scan

Results

Simultaneous analysis was carried out on multi-component pesticide residues in food products. Results obtained indicated excellent reproducibility (10% at 5 ppb) and linearity with R^2 better than 0.995 in the range of 0.1–100 ppb (0.1–100 pg/μL injected). Figure 4.63 shows as example the calibration curve for Propazine at 0.1–100 ppb with a corresponding chromatogram at 1 ppb, showing excellent reproducibility ($r^2 = 0.9998$).

Figure 4.64 shows examples of GC/MS/MS chromatograms of various pesticides in which 1 ppb of each pesticide was added to green pepper. Even at this extremely low concentration (1/10 of the uniform standard value for pesticides), it is possible to provide reliable quantitation measurements with remarkably high sensitivity using the highly resolved parent ions for MS/MS analysis. No cross talk was observed for the analysis of closely eluting multi-component mixtures. Using H-SRM, interferences from the sample matrix background were substantially reduced, leading to very lowlevel LOQs. Figure 4.65 illustrates the gain in selectivity when using the highly resolved Q1 precursor ion for SRM quantitation.

In addition, structural confirmation of the analytes during the quantification run have been provided using the data dependant acquisition of a MS/MS product ion spectrum (see Figs. 4.66 and 4.67).

In combination with the QuEChERS sample preparation GC/MS/MS using H-SRM turned out to provide the required sensitivity and certainty for a multicomponent quantitation of trace pesticides in food for a high throughput of samples.

Table 4.10 Retention time, SRM conditions, calibration range, linearity, and reproducibility of pesticide components investigated.

	RT	Precursor ion (m/z)	Product ion (m/z)	Collision energy	R^2	Calibration range	CV (%) n=5
Mevinphos	6.44	192	127	10	0.9999	0.1–100	4.03
XMC	7.52	122	107	10	0.9999	0.1–100	2.55
Tecnazene	8.03	261	203	15	0.9996	0.1–100	5.41
Ethoprpphos	8.22	200	114	10	0.9981	0.1–100	7.91
Ethalfluralin	8.42	316	276	10	0.9997	0.1–100	4.14
Benfluralin	8.62	292	264	10	0.9989	0.1–100	1.86
Monocrotophos	8.62	192	127	10	0.9754	5–100	19.47
α-BHC	9.03	219	183	15	0.9999	0.1–100	4.51
Dicloran	9.25	206	176	10	0.9994	0.1–100	2.30
Simazine	9.30	201	172	10	0.9999	0.1–100	4.33
Propazine	9.50	214	172	10	0.9998	0.1–100	1.99
β-BHC	9.57	219	183	15	1.0000	0.1–100	3.51
γ-BHC	9.73	219	183	15	0.9998	0.1–100	6.57
Cyanophos	9.78	243	109	10	0.9996	0.1–100	3.56
Pyroquilon	9.90	173	130	20	0.9996	0.1–100	2.95
Diazinon	4.94	304	179	15	0.9995	0.1–100	4.40
Phosphamidon-1	10.06	264	127	10	0.9989	0.1–100	10.31
Prohydrojasmon-1	10.12	184	83	20	0.9992	0.1–100	7.39
δ-BHC	10.26	219	183	15	0.9994	0.1–100	5.17
Prohydrojasmon-2	10.66	264	127	10	0.9972	0.1–100	17.11
Benoxacor	10.7	259	120	15	0.9999	0.1–100	3.30
Propanil	10.95	262	202	10	0.9993	0.1–100	3.65
Phosphamidon-2	10.97	264	127	10	0.9970	0.1–100	8.77
Dichlofenthion	10.99	279	223	15	0.9994	0.1–100	2.21
Dimethenamid	11.06	230	154	10	0.9996	0.1–100	2.51
Bromobutide	11.09	232	176	10	0.9990	0.1–100	5.91
Paration-methyl	11.24	263	109	10	0.9982	0.1–100	3.74
Tolclofos-methyl	11.38	265	250	15	0.9998	0.1–100	2.52
Ametryn	11.43	227	170	10	0.9999	0.1–100	0.90
Mefenoxam	11.57	249	190	10	0.9995	0.1–100	5.81
Bromacil	11.98	205	188	15	0.9988	0.1–100	3.87
Pirimiphos-methyl	12.00	305	276	10	0.9995	0.1–100	4.08
Quinoclamine	12.18	207	172	10	0.9989	0.1–100	4.24
Diethofencarb	12.34	225	125	15	0.9985	0.1–100	4.64
Cyanazine	12.52	225	189	10	0.9994	0.1–100	3.41
Chlorpyrifos	12.57	314	258	15	0.9991	0.1–100	3.37
Parathion	12.59	291	109	15	0.9962	0.1–100	9.76
Triadimefon	12.67	208	111	25	0.9986	0.1–100	6.10
Chlorthal-dimethyl	12.73	301	223	20	1.0000	0.1–100	1.23
Nitrothal-isopropyl	12.78	236	148	15	0.9974	0.1–100	5.53
Fthalide	13.04	272	243	10	0.9993	0.1–100	4.32
Fosthiazate	13.05 [a)]	195	103	10	0.9956	5–100	6.29
Diphenamid	13.10	239	167	10	0.9997	0.1–100	4.67
Pyrifenox-Z	13.64	262	200	15	0.9979	0.2–100	4.54
Fipronil	13.79	123	81	10	0.9991	0.1–100	3.49

Table 4.10 (continued)

	RT	Precursor ion (m/z)	Product ion (m/z)	Collision energy	R^2	Calibration range	CV (%) n=5
Allethrin	13.67	367	213	25	0.9991	5–100	3.79
Dimepiperate	13.87	145	112	10	0.9987	0.1–100	3.74
Quinalphos	13.87	274	121	10	0.9987	0.1–100	1.82
Phenthoate	13.88	146	118	10	0.9984	0.1–100	1.96
Paclobutrazol	14.45	236	125	15	0.9961	0.1–100	7.41
Endosulfan-á	14.67	241	206	15	0.9996	0.1–100	4.54
Butachlor	14.73	237	160	10	0.9998	0.1–100	5.26
Imazamethabenz-methyl	14.81	256	144	20	0.9932	2–100	12.09
Butamifos	15.00	286	202	15	0.9958	0.1–100	4.66
Flutlanil	15.06	173	145	15	0.9986	0.1–100	1.93
Hexaconazole	15.06	214	172	15	0.9924	0.1–100	8.98
Profenofos	15.28	337	267	15	0.9968	0.1–100	6.61
Uniconazole-P	15.38	234	137	15	0.9966	0.1–100	11.37
Pretilachlor	15.37	162	132	15	0.9982	0.1–100	6.72
Flamprop-methyl	15.66	276	105	10	0.9986	0.1–100	3.93
Oxyfluorfen	15.69	361	300	10	0.9980	0.5–100	6.07
Azaconazole	15.79	217	173	15	0.9981	0.1–100	7.07
Bupirimate	15.82	316	208	10	0.9982	0.1–100	4.65
Thifluzamide	15.84	449	429	10	0.9972	0.1–100	2.75
Fenoxanil	16.25	293	155	20	0.9989	0.1–100	3.73
Chlorbenzilate	16.43	251	139	15	0.9976	0.1–100	0.81
Pyriminobac-methyl-Z	16.76	302	256	15	0.9986	0.1–100	2.70
Oxadixyl	16.86	163	132	10	0.9998	0.1–100	3.72
Triazophos	17.30	257	162	10	0.9941	0.2–100	6.72
Fluacrypyrim	17.38	189	129	10	0.9988	0.1–100	2.15
Edifenphos	17.72	310	173	10	0.9927	0.1–100	7.95
Quinoxyfen	17.74	272	237	10	0.9993	0.1–100	4.50
Lenacil	17.78	153	136	15	0.9979	0.1–100	5.19
Trifloxystrobin	18.01	222	162	10	0.9966	0.1–100	8.47
Pyriminobac-methyl-E	18.19	302	256	15	0.9982	0.1–100	2.12
Tebuconazole	18.39	250	125	20	0.9907	0.2–100	13.03
Diclofop-methyl	18.51	253	162	15	0.9991	0.1–100	2.14
Mefenpyr-diethyl	19.15	253	189	20	0.9992	0.1–100	3.35
Pyributicarb	19.24	165	108	10	0.9973	0.1–100	2.00
Pyridafenthion	19.46	152	116	20	0.9940	0.2–100	4.71
Acetamiprid	19.39	340	199	10	1.0000	50–100	–
Bromopropylate	19.64	341	185	15	0.9956	0.1–100	3.72
Piperophos	19.84	320	122	10	0.9939	0.2–100	7.51
Fenpropathrin	19.98	265	210	10	0.9973	0.1–100	6.87
Etoxazole	20.06	300	270	20	0.9969	0.1–100	8.84
Tebufenpyrad	20.10	333	171	20	0.9978	0.5–100	13.35
Anilofos	20.31	226	157	15	0.9948	0.2–100	5.56
Phenothrin-1	20.49	183	165	10	0.9967	5–100	16.13

Table 4.10 (continued)

	RT	Precursor ion (m/z)	Product ion (m/z)	Collision energy	R^2	Calibration range	CV (%) n=5
Tetradifon	20.54	356	229	10	0.9998	0.2–100	4.17
Phenothrin-2	20.66	183	165	10	0.9968	0.1–100	3.79
Mefenacet	21.22	192	136	15	0.9955	0.1–100	4.90
Cyhalofop-buthyl	21.23	357	229	10	0.9967	0.1–100	5.52
Cyhalothrin-1	21.30	181	152	20	0.9975	0.2–100	3.21
Cyhalothrin-2	21.66	181	152	20	0.9984	0.2–100	6.67
Pyrazophos	22.06	373	232	10	0.9963	0.1–100	10.46
Bitertanol	22.80[b)]	170	141	20	0.9873	0.1–100	6.76
Pyridaben	23.18	147	117	20	0.9958	0.1–100	1.29
Cafenstrole	24.03	100	72	5	0.9958	0.1–100	9.77
Cypermethrin-1	24.72	181	152	20	0.9983	2–100	9.29
Halfenprox	24.79	263	235	15	0.9979	0.1–100	10.25
Cypermethrin-2	24.92	181	152	20	0.9982	2–100	6.91
Cypermethrin-3	25.06	181	152	20	0.9985	2–100	16.27
Cypermethrin-4	25.13	181	152	20	0.9948	2–100	13.79
Fenvalerate-1	26.47	167	125	10	0.9977	0.1–100	3.11
Flumioxazin	26.50	354	176	20	0.9937	0.1–100	9.66
Fenvarelate-2	26.91	167	125	10	0.9979	0.1–100	3.26
Deltamethrin + Tralomethrin	28.15	181	152	20	0.9967	0.2–100	8.20
Tolfenpyrad	29.11	383	171	20	0.9968	2–100	4.84
Imibenconazole	30.35	375	260	15	1.0000	50–100	–

a) and 13.12 min
b) and 22.97 min

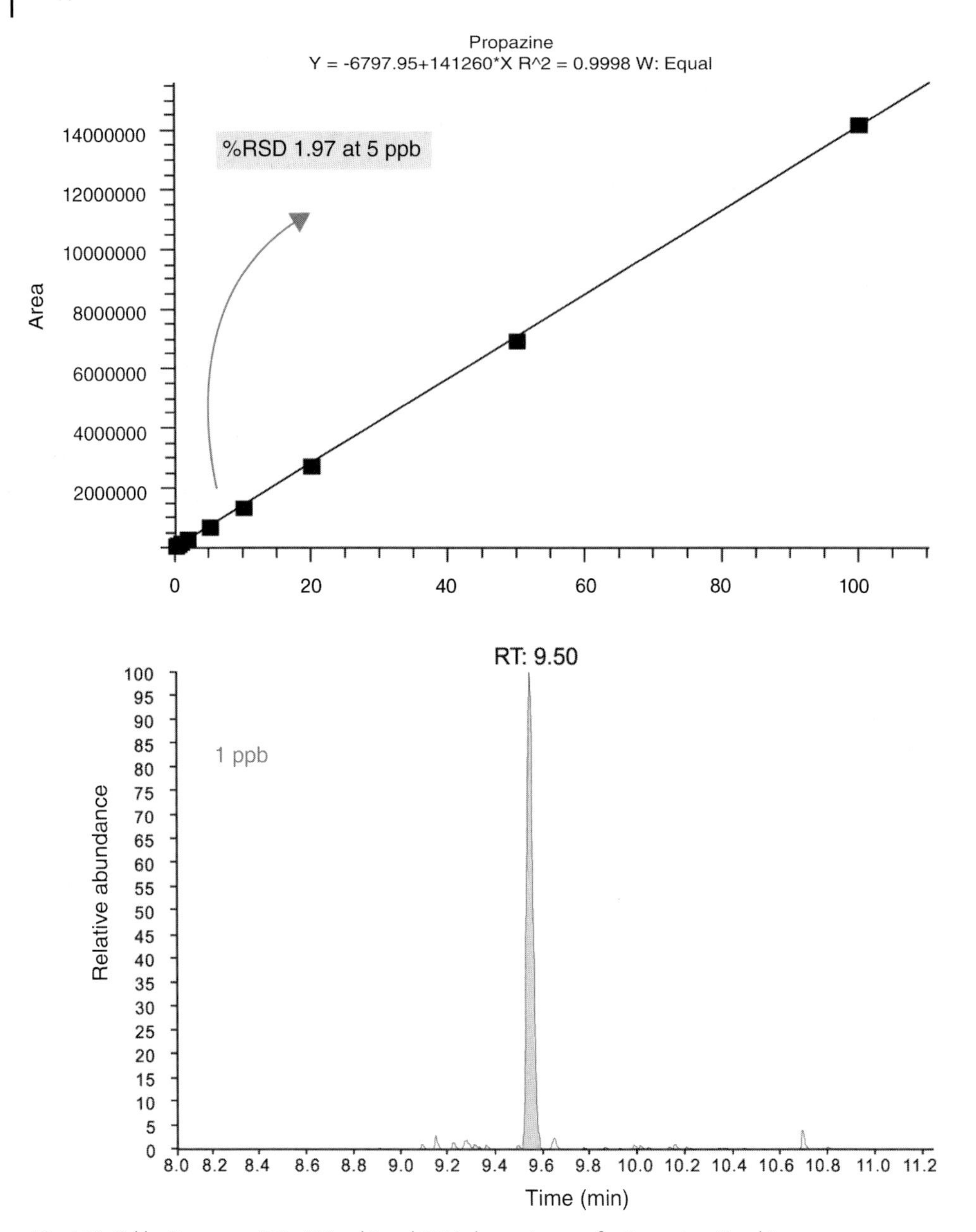

Fig. 4.63 Calibration curve (0.1–100 ppb) and SRM chromatogram for Propazine (1 ppb)

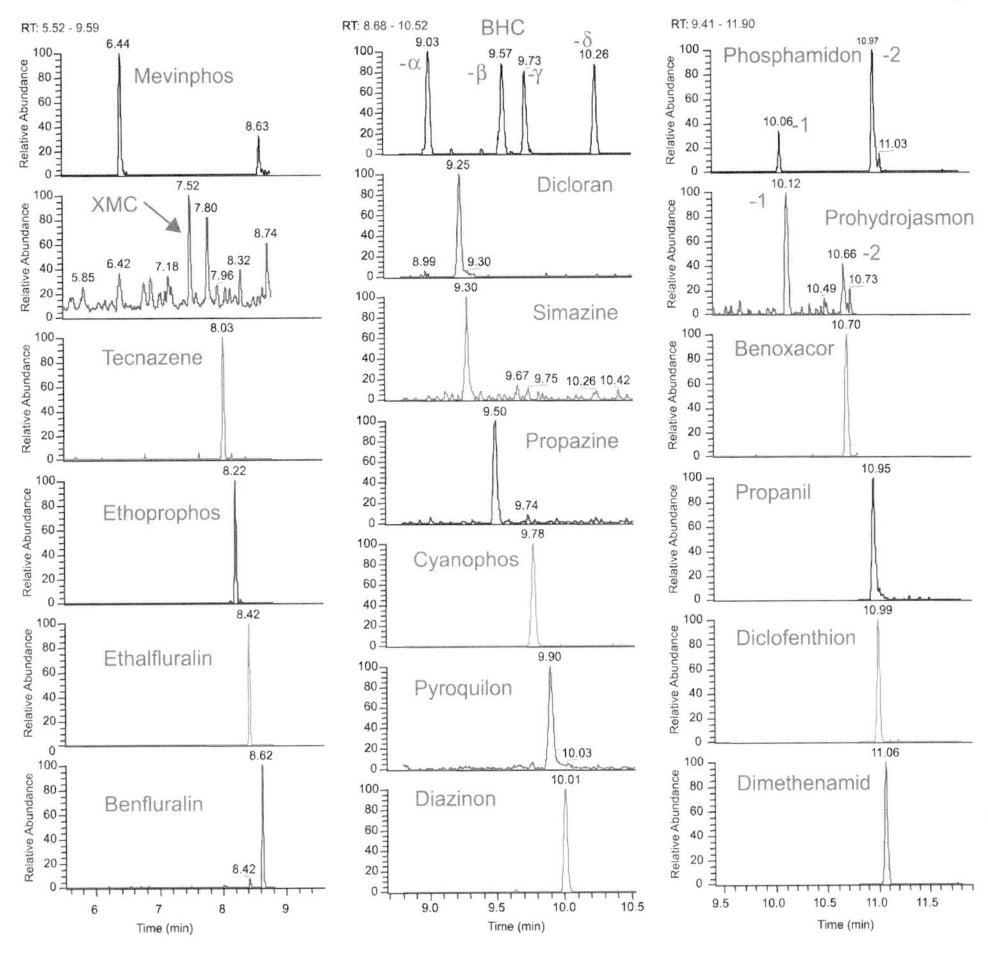

Fig. 4.64 GC/MS/MS chromatograms of various pesticides at 1 ppb in green pepper samples

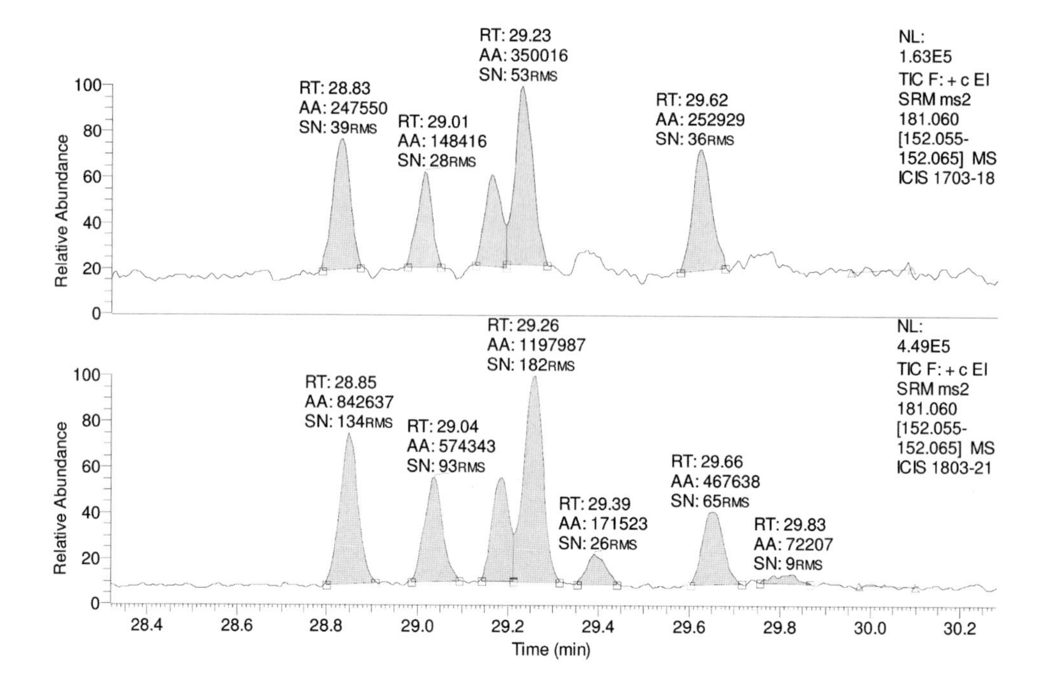

Fig. 4.65 SRM and H-SRM analysis of pyrethroids in cabbage matrix at a 5 ppb level.
Top: 2 μL injection, Q1 set to 0.7 Da peak width.
Bottom: 10 μL injection, Q1 set to 0.4 Da peak with shows increased selectivity with reduced background intensity even a 5-fold increased injection volume

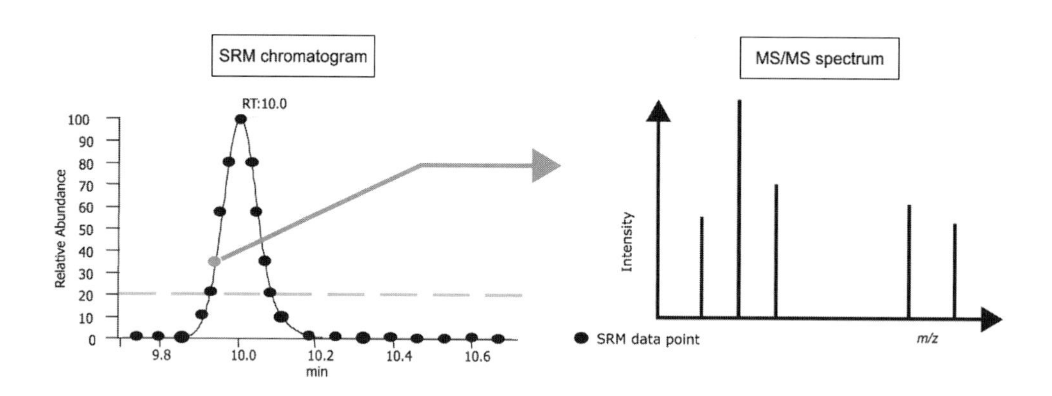

Fig. 4.66 Data dependant acquisition of the MS/MS product ion spectrum for confirmation of positive results during the H-SRM quantification run

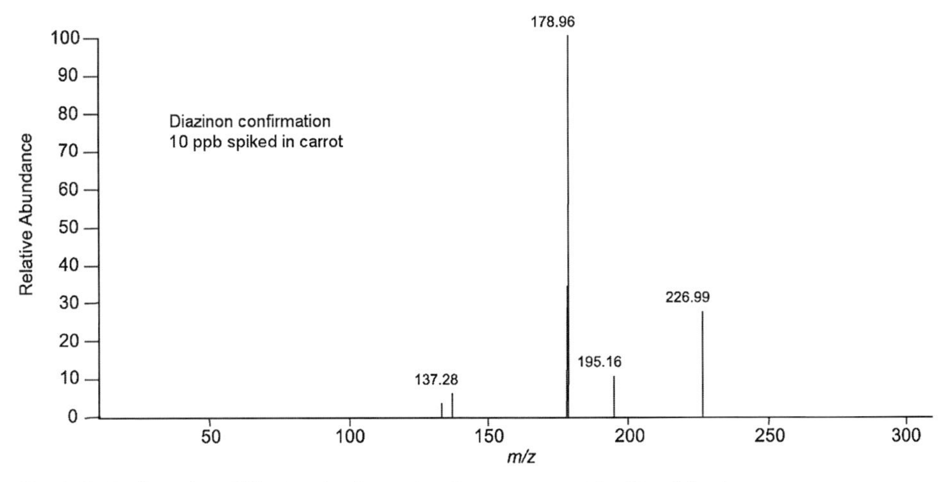

Fig. 4.67 Confirmation of Diazinon by the product ion spectrum at the 10 ppb level

4.15
Multi-method for the Determination of 239 Pesticides

Key words: multi-residue method, pesticides, fruits, vegetables, SPE cleanup

The effective monitoring of pesticides residues and its degradation products require fast and comprehensive multi-methods. The referenced method fulfills the requirement for monitoring a wide scope of pesticide compounds and degradation products in fruit and vegetable samples.

The described internal standard method is suitable for the quantification of more than 200 components, using a single quad GC/MS. The extraction process with acetonitrile is followed by a clean-up including a salting-out step and solid phase extraction. The compound detection is done by selected ion monitoring using the retention time and selected ion abundance ratios as qualifiers for identification. Two runs are required to determine the complete set of pesticides given in Table 4.11 according to the required SIM acquisition windows described as FV-1 and FV-2.

The quantitation using Aldrin as internal standard is based on blank matrix spikes to compensate for matrix effects.

In addition, the published method also includes the parallel determination of carbamates by HPLC with fluorescence detection (Fillion 2000).

Analysis Conditions

Sample preparation:	Extraction: Homogenize 50 g sample with 100 mL ACN for 5 min, add 10 g $NaSO_4$ and homogenize another 5 min. Clean-up: Condition with 2 mL ACN extract on a preconditioned C_{18} SPE tube and discard elute. Transfer 15 mL ACN extract onto cartridge an elute 13 mL, add $NaSO_4$ to 15 mL, shake and centrifuge. Evaporate 10 mL aliquot (equ. of 5 g sample) to 0.5 mL under N_2. Transfer extract to carbon to preconditioned SPE tube and elute with 20 mL ACN/toluen (3+1), evaporate and exchange solvent by adding 2×10 mL acetone, add 50 µL ISTD (Aldrin 1.0 ng/µL), add acetone to 2.5 mL. Transfer 0.5 mL for GC/MS analysis.
Gas chromatograph:	Hewlett-Packard 5890 Series with autosampler 7673A (Hewlett-Packard, Wilmington, DE, USA)
Injector:	Split/splitless 250 °C, 2 µL injection volume
Carrier gas:	Helium, constant flow mode
Column:	DB-1701, 30 m x 0.25 mm i.d. x 0.15 µm film (J&W Scientific, Folsom, CA, USA) 30 cm of same type is used as a retention gap pre-column
Oven temp. program:	70 °C, 2 min 25 °C/min to 130 °C 2 °C/min to 220 °C 10 °C/min to 280 °C, 6.6 min

Mass spectrometer: Mass selective detector 5972A (Hewlett-Packard, Wilmington, DE, USA)

Ionisation: EI

Scan mode: Selected ion monitoring, see Table 4.11

Results

For most of the components the achieved recovery has been between 70 and 120%. For mirex, EPTC, butylate, HCB, folpet, oxycarboxin, and chlorthiamid with recoveries below 50% the method is considered a screening procedure, for individual recoveries and discussion see the original reference (Fillion 2000). The published LODs range between 0.02 and 1.0 mg/kg with 80% of the compounds having LODs below 0.04 mg/kg.

Additional compounds have been tested but not included to the method. Benzoximate gave 4 peaks; chloroxuron, metoxuron, and oxydemeton-methy gave 3 peaks of degradation products; fluazinam and flualinate showed many degradation peaks and poor sensitivity; diclone showed poor sensitivity; ditalimfos was not recovered; and vamidothion degraded in solution.

Data analyses have been done by macro driven automation with speadsheet report and printouts of the selected ion chromatograms. It is reported that this method typically allows a sample throughput of 42 samples including blanks and spikes per week with results for priority samples within one day.

Table 4.11 List of compounds included in the multimethod with retention times, target and qualifier ions (Q) and ratios (Q/tgt); degr. = degradation product (Fillion 2000).

Compound	Method	Rt [min]	Ions monitored [m/z]				Abundance ratio of qualifier ion/target		
			Target	Q1	Q2	Q3	Q/tgt	Q2/tgt	Q3/tgt
Dichlorvos-naled	FV-1	7.24	185	109	220		4.61	0.19	
EPTC	FV-1	7.89	189	132			1.40		
Bendiocarb degr.	FV-2	7.99	166	151	126		1.86	1.82	
Allidochlor	FV-2	8.28	138	132	173		1.12	0.14	
Methamidophos	FV-2	8.88	141	94	95		2.31	1.60	
Butylate	FV-1	9.06	217	174			5.20		
Promecarb degr.	FV-1	9.09	135	150			0.35		
Chlorthiamid degr.	FV-2	9.36	171	173	136		0.64	0.18	
Dichlobenil	FV-1	9.37	171	173			0.62		
Dichlormid	FV-1	9.51	168	166			1.53		
Vernolate	FV-2	9.59	86	128			1.14		
Pebulate	FV-2	9.97	128	161	203		0.12	0.06	
Aminocarb degr.	FV-1	10.06	150	151	136		1.27	0.79	
Etridiazole	FV-2	10.12	211	183			0.97		
Chlormephos	FV-1	10.46	234	121			3.19		

Table 4.11 (continued)

Compound	Method	Rt [min]	Ions monitored [m/z]				Abundance ratio of qualifier ion/target		
			Target	Q1	Q2	Q3	Q/tgt	Q2/tgt	Q3/tgt
Nitrapyrin	FV-2	10.85	194	196	198		0.97	0.31	
Mexacarbate degr.	FV-2	11.18	165	150	134		0.81	0.34	
Bufencarb degr.	FV-1	11.24	121	122	107		0.37	0.22	
cis-Mevinphos	FV-1	11.82	127	192	164	109	0.20	0.06	0.23
Propham	FV-2	11.85	179	137			1.28		
trans-Mevinphos	FV-1	12.17	127	192	164	109	0.22	0.06	0.23
Chloroneb	FV-2	12.43	191	193	206		0.64	0.53	
o-Phenylphenol	FV-2	13.57	170	169	141	115	0.66	0.29	0.21
Tecnazene	FV-1	14.96	261	215			1.54		
Cycloate	FV-2	15.09	154	83	215		2.26	0.07	
Captafol degr.	FV-2	15.56	151	79	80		1.31	0.73	
Captan degr.	FV-1	15.60	151	79	80		1.26	0.71	
Heptenophos	FV-2	15.89	124	215	250		0.09	0.07	
Acephate	FV-1	15.89	136	94			0.54		
Demeton-S	FV-1	16.04	171	88	143		6.27	0.55	
Hexachlorobenzene	FV-1	16.51	284	286	282		0.77	0.56	
Ethoprophos	FV-2	16.98	158	242	139		0.08	0.47	
Diphenylamine	FV-1	17.09	169	167	168		0.34	0.63	
Di-allate 1	FV-1	17.46	234	236			0.36		
Chlordimeform	FV-2	17.67	196	181			0.75		
Propachlor	FV-1	17.68	120	176	211		0.23	0.05	
Demeton-S-methyl	FV-2	18.08	88	109	142		0.26	0.17	
Di-allate 2	FV-1	18.48	234	236			0.36		
Ethalfluralin	FV-1	18.61	276	316			0.50		
Phorate	FV-1	18.86	260	231			0.73		
Trifluralin	FV-1	19.27	306	264			1.18		
Sulfallate	FV-2	19.31	188	116	148		0.09	0.06	
Chlorpropham	FV-2	19.39	213	127			3.21		
Benfluralin	FV-1	19.45	292	264			0.23		
Sulfotep	FV-1	19.54	322	202			1.15		
alpha-BHC	FV-1	19.83	219	183			1.16		
Bendiocarb	FV-2	20.74	151	166	223		0.41	0.08	
Quintozene	FV-1	20.82	295	237			1.79		
Promecarb degr.	FV-1	20.93	135	150			0.62		
Omethoate	FV-1	21.15	156	110			1.10		
Terbufos	FV-2	21.71	231	153			0.38		
Demeton-O	FV-1	21.74	88	114	170		0.15	0.12	
Desethylatrazine	FV-1	21.86	172	174			0.31		
Clomazone	FV-2	21.92	204	125			2.14		
Prometon	FV-2	22.01	225	210	168		1.44	1.17	
Tri-allate	FV-1	22.32	268	270			0.67		
Fonofos	FV-1	22.40	246	109			4.50		
Diazinon	FV-1	22.67	304	179			3.58		
Terbumeton	FV-2	22.77	210	169	225		0.94	0.25	
Dicrotophos	FV-1	22.77	127	193			0.07		

Table 4.11 (continued)

Compound	Method	Rt [min]	Ions monitored [m/z]				Abundance ratio of qualifier ion/target		
			Target	Q1	Q2	Q3	Q/tgt	Q2/tgt	Q3/tgt
Lindane	FV-2	22.98	219	183			1.28		
Dioxathion	FV-1	23.18	125	153	270		0.29	0.35	
Disulfoton	FV-2	23.60	142	274			0.29		
Profluralin	FV-1	23.68	318	330			0.30		
Dicloran	FV-2	23.78	206	176			1.37		
Propazine	FV-1	24.04	229	214			1.85		
Atrazine	FV-2	24.12	215	200			1.88		
Etrimfos	FV-2	24.14	292	277			0.44		
Simazine	FV-1	24.21	201	186			0.70		
Heptachlor	FV-1	24.50	272	274			0.79		
Chlorbufam	FV-1	24.59	223	164			1.00		
Schradan	FV-2	24.59	199	135			1.74		
Aminocarb	FV-1	24.65	151	208	150		0.12	0.73	
Terbuthylazine	FV-1	24.78	214	173			0.55		
Monolinuron	FV-2	24.80	214	126			1.43		
Secbumeton	FV-2	25.33	196	210	225		0.20	0.14	
Dichlofenthion	FV-1	25.82	279	223			1.36		
Cyanophos	FV-1	26.09	243	125			1.34		
Isazofos	FV-2	26.10	161	257	313		0.28	0.04	
Mexacarbate	FV-2	26.28	165	222	150		0.21	0.86	
Propyzamide	FV-1	26.45	173	175			0.63		
Aldrin ISTD	FV-1	26.48	263	265			0.66		
Pirimicarb	FV-2	26.50	166	238			0.16		
Dimethoate	FV-1	26.60	87	229	143		0.04	0.09	
Monocrotophos	FV-2	26.68	127	192	109		0.09	0.11	
Chlorpyrifos-methyl	FV-1	26.93	286	288			0.65		
Fluchloralin	FV-2	27.56	306	326	264		0.82	0.83	
Fenchlorphos	FV-1	27.75	285	287			0.68		
Desmetryn	FV-1	27.91	198	213			1.49		
Dinitramine	FV-2	28.00	305	307	261		0.37	0.41	
Dimetachlor	FV-2	28.23	210	134	197		8.78	3.13	
Chlorothalonil	FV-1	28.75	266	264			0.78		
Alachlor	FV-1	28.80	188	160			1.27		
Prometryn	FV-1	29.22	241	226			0.68		
Metobromuron	FV-2	29.23	258	61			16.26		
Cyprazine	FV-2	29.33	212	227	229		0.89	0.12	
Ametryn	FV-2	29.34	227	212			1.12		
Simetryn	FV-1	29.45	170	155			0.93		
Pirimiphos-methyl	FV-1	29.53	290	305			0.52		
Vinclozolin	FV-1	29.55	285	287			0.62		
Thiobencarb	FV-2	29.79	100	257	125		0.09	0.25	
Metribuzin	FV-2	29.82	198	199			0.29		
beta-BHC	FV-1	29.90	219	183			1.17		
Terbutryn	FV-1	30.21	226	241			0.51		
Metalaxyl	FV-2	30.37	206	249			0.46		

Table 4.11 (continued)

Compound	Method	Rt [min]	Ions monitored [m/z]				Abundance ratio of qualifier ion/target		
			Target	Q1	Q2	Q3	Q/tgt	Q2/tgt	Q3/tgt
Panathion-methyl	FV-2	30.42	263	125			2.15		
Chlorpyrifos	FV-1	30.61	314	199			3.18		
Aspon	FV-1	30.97	211	253			0.27		
Dicofol	FV-1	31.14	250	139			5.88		
Oxychlordane	FV-2	31.21	115	185	149		0.62	0.51	
Malaoxon	FV-1	31.25	268	195			1.85		
Chlorthal-dimethyl	FV-2	31.26	301	299			0.81		
Phosphamidon	FV-1	31.31	264	193			0.28		
delta-HCH	FV-2	31.35	183	219	217		0.86	0.68	
Metolachlor	FV-1	31.63	238	162			2.14		
Terbacil	FV-2	31.91	160	161	216		1.33	0.02	
Fenthion	FV-1	32.01	278	169			0.33		
Bromophos	FV-1	32.15	331	329			0.77		
Dichlofluanid	FV-1	32.39	226	123			5.26		
Fenitrothion	FV-1	32.46	277	260			0.59		
Pirimiphos-ethyl	FV-2	32.67	333	304			1.26		
Malathion	FV-1	32.68	158	125			2.23		
Paraoxon	FV-1	32.71	275	109			11.06		
Heptachlor epoxide	FV-2	32.76	237	183	217		2.02	1.21	
Nitrothal-isopropyl	FV-2	33.21	236	212			0.75		
Butralin	FV-2	33.42	266	250			0.14		
Ethofumesate	FV-1	33.48	161	286	207		0.25	1.04	
Triadimefon	FV-1	33.99	208	210			0.30		
Parathion	FV-2	34.05	291	139			1.20		
Isopropalin	FV-1	34.15	280	238			0.60		
Pendimethalin	FV-2	34.34	252	281			0.08		
Fenson	FV-2	34.36	268	141			4.41		
Linuron	FV-2	34.40	248	160	250		1.44	0.62	
alpha-Endosulfan	FV-1	34.42	277	339	243		0.57	1.37	
Chlorthiamid	FV-2	34.45	205	170			2.36		
Chlorbenside	FV-1	34.54	125	268			0.07		
Allethrin	FV-2	34.82	123	136			0.28		
Chlorfenvinphos	FV-1	34.97	323	267			2.51		
trans-Chlordane	FV-1	35.22	373	375			0.93		
Bromophos-ethyl	FV-1	35.35	359	303			1.31		
Chlorthion	FV-2	35.56	297	125			4.38		
Quinalphos	FV-1	35.59	146	298			0.09		
Propanil	FV-2	35.69	161	163			0.66		
Diphenamid	FV-2	35.71	167	239			0.16		
Crufomate	FV-2	35.79	256	182			0.83		
cis-Chlordane	FV-1	35.91	373	375			0.93		
Isofenphos	FV-1	35.92	213	255			0.29		
Metazachlor	FV-2	36.11	209	133			1.82		
Phenthoate	FV-2	36.19	246	274			3.83		
Chlorfenvinphos	FV-1	36.26	323	267			2.51		

Table 4.11 (continued)

Compound	Method	Rt [min]	Ions monitored [m/z]				Abundance ratio of qualifier ion/target		
			Target	Q1	Q2	Q3	Q/tgt	Q2/tgt	Q3/tgt
Penconazole	FV-1	36.33	248	159			1.85		
Tolylfluanid	FV-1	36.61	238	137			3.56		
p,p′-DDE	FV-1	37.06	318	246			1'75		
Folpet	FV-2	37.21	260	262			0.69		
Prothiofos	FV-2	37.50	309	267			1.34		
Dieldrin	FV-1	37.55	277	263			1.47		
Butachlor	FV-1	37.75	176	160			0.84		
Chlorflurecol-methyl	FV-2	37.77	215	217	152		0.32	0.44	
Captan	FV-1	38.05	149	79			6.28		
Iodofenphos	FV-1	38.08	377	379			0.44		
Tribufos	FV-2	38.16	169	202			0.50		
Chlozolinate	FV-2	38.23	259	331			0.43		
Crotoxyphos	FV-2	38.56	193	127			3.65		
Methidathion	FV-1	38.70	145	85			0.75		
Tetrachlorvinphos	FV-1	38.83	329	331			0.95		
Chlorbromuron	FV-2	38-84	61	294			0.06		
Procymidone	FV-1	38.85	283	285			0.69		
Flumetralin	FV-2	38.98	143	157			0.14		
Endrin	FV-1	39.09	263	281			0.44		
Bromacil	FV-2	39.17	205	207			0.97		
Flurochloridone 1	FV-2	39.19	311	187			2.48		
Triadimenol	FV-2	39.22	112	168			0.40		
Profenofos	FV-1	39.28	339	337			1.06		
o,p′-DDD	FV-1	39.67	235	237			0.64		
Flurochloridone 2	FV-2	39.67	311	187			1.85		
Ethylan	FV-2	39.78	223	165			0.10		
Cyanazine	FV-1	40.04	225	240			0.49		
Chlorfenson	FV-1	40.09	302	175			6.83		
TCMTB	FV-2	40.40	238	180			6.23		
o,p′-DDT	FV-1	40.71	235	237			0.63		
Oxadiazon	FV-2	40.89	258	175			3.48		
Carbetamide	FV-2	41.06	119	120	236		0.22	0.04	
Tetrasul	FV-2	41.13	252	324			0.52		
Imazalil	FV-1	41.15	215	173	217		0.98	0.61	
Aramite 1	FV-1	41.17	185	319			0.18		
Fenamiphos	FV-2	41.48	303	217			0.59		
Erbon	FV-1	42.08	169	171			0.64		
Aramite 2	FV-1	42.13	185	319			0.08		
Methoprotryne	FV-2	42.18	256	213			0.37		
Chloropropylate	FV-1	42.24	251	139	253		1.14	0.63	
Methyl trithion	FV-2	42.40	157	314			0.10		
Nitrofen	FV-1	42.42	283	202			0.73		
Chlorobenzilate	FV-2	42.55	251	139			1.26		
Carboxin	FV-2	42.81	143	235			0.29		
Flamprop-methyl	FV-1	42.89	105	77	276		0.29	0.04	

Table 4.11 (continued)

Compound	Method	Rt [min]	Ions monitored [m/z]				Abundance ratio of qualifier ion/target		
			Target	Q1	Q2	Q3	Q/tgt	Q2/tgt	Q3/tgt
Bupirimate	FV-2	43.17	273	316	208		0.16	0.94	
beta-Endosulfan	FV-1	43.36	241	237			1.06		
p,p-DDD	FV-2	43.62	235	237			0.64		
Oxyfluorfen	FV-2	43.82	252	300			0.27		
Chlorthiophos	FV-1	43.92	325	360			0.46		
Ethion	FV-1	44.42	231	153	384		0.76	0.04	
Etaconazole 1	FV-1	44.42	245	173			1.51		
Sulprofos	FV-2	44.47	322	156	140		1.83	2.14	
Etaconazole 2	FV-1	44.60	245	173			1.70		
Flamprop-isopropyl	FV-2	44.85	105	276	363		0.07	0.01	
p,p′-DDT	FV-1	44.96	235	237			0.64		
Carbophenothion	FV-1	45.17	157	342	121		0.13	0.54	
Fluorodifen	FV-2	45.19	190	328	162			0.15	
Myclobutanil	FV-2	46.02	179	288			0.06		
Benalaxyl	FV-1	46.15	148	206	325		0.16	0.02	
Edifenphos	FV-1	46.84	173	310	201		0.27	0.27	
Propiconazole 1	FV-2	47.25	259	261			0.61		
Fensulfothion	FV-2	47.52	293	308			0.16		
Mirex	FV-1	47.61	272	274	237		0.80	0.62	
Propargite	FV-2	47.63	135	350	150	201	0.02	0.12	0.03
Propiconazole 2	FV-2	47.63	259	261			0.64		
Diclofop-methyl	FV-1	47.64	253	340	281		0.40	0.34	
Propetamphos	FV-2	48.22	124	208			2.24		
Triazophos	FV-2	48.40	162	161			1.54		
Benodanil	FV-1	48.92	231	323	203		0.17	0.21	
Nuarimol	FV-1	49.15	314	235	203		3.39	3.03	
Bifenthrin	FV-2	49.19	181	165	166		0.27	0.27	
Endosulfan sulfate	FV-1	49.43	272	387			0.27		
Bromopropylate	FV-2	50.26	341	183			1.11		
Oxadixyl	FV-2	50.29	163	132	278		0.79	0.06	
Methoxychlor	FV-1	50.41	227	228			0.16		
Benzoylprop-ethyl	FV-2	50.84	105	77	292		0.26	0.04	
Tetramethrin 1	FV-2	50.89	164	123			0.28		
Fenpropathrin	FV-1	51.21	181	265			0.33		
Tetramethrin 2	FV-2	51.25	164	123			0.29		
Leptophos	FV-2	51.51	171	377			0.29		
EPN	FV-1	51.70	169	157			2.46		
Norflurazon	FV-2	51.87	303	145			2.48		
Hexazinone	FV-1	52.08	171	128			0.13		
Phosmet	FV-1	52.16	160	161	317		0.12	0.02	
Iprodione	FV-2	52.30	314	316	187		0.64	1.31	
Tetradifon	FV-1	52.47	229	356			0.40		
Bifenox	FV-2	52.65	341	173			0.54		
Oxycarboxin	FV-2	53.16	175	267			0.26		
Phosalone	FV-1	53.43	182	367			0.08		

Table 4.11 (continued)

Compound	Method	Rt [min]	Ions monitored [m/z]				Abundance ratio of qualifier ion/target		
			Target	Q1	Q2	Q3	Q/tgt	Q2/tgt	Q3/tgt
Chloridazon	FV-2	53.48	221	220			0.51		
Azinphos-methyl	FV-2	53.49	160	132			0.79		
Nitralin	FV-2	53.59	274	316			0.95		
cis-Permethrin	FV-1	53.69	183	163	165		0.22	0.18	
Fenarimol	FV-2	53.79	219	139			2.18		
trans-Permethrin	FV-1	54.05	183	163	165		0.30	0.23	
Pyrazophos	FV-1	54.06	232	221	373		3.77	0.20	
Azinphos-ethyl	FV-1	54.37	160	132			1.18		
Dialifos	FV-2	54.40	210	208			3.13		
Cyfluthrin 1	FV-2	55.79	226	206			1.08		
Prochloraz	FV-2	55.80	180	308			0.33		
Coumaphos	FV-2	56.02	362	210			1.20		
Cypermethrin 1	FV-1	56.03	181	163			1.59		
Cyfluthrin 4	FV-2	56.33	226	206			1.24		
Cypermethrin 4	FV-1	56.58	181	163			1.91		
Fenvalerate 1	FV-2	57.67	167	225	419		0.44	0.06	
Fenvalerate 2	FV-2	58.16	167	225	419		0.44	0.06	
Deltamethrin	FV-2	59.50	181	251			0.51		

4.16 Nitrophenol Herbicides in Water

Key words: herbicides, nitrophenols, rotational perforator, methylation

The determination of nitrophenols increased in importance after this substance class was suspected of being one of the causes of the death of forests. Nitrophenols and their derivatives are determined using GC-NPD or GC/MS. The analytes are first extracted with a pentane/ether mixture and then methylated with diazomethane. The recoveries lie in the range 70–100%. The limit of detection of the procedure is 0.25 μg/L.

Analysis Conditions

Sample preparation:	Liquid/liquid extraction in Brodesser low density rotation perforator (Fig. 4.68), methylation with diazomethane.
Gas chromatograph:	Carlo Erba Mega (Rodano, Italy)
Injector:	On-column, 7.5 μL injection volume
Guard column:	Retention gap, length 5 m, ID 0.32 mm, deactivated

Column:	J&W DB-5, 30 m × 0.25 mm × 0.25 μm Helium 2.5 mL/min, 1.2 bar
Program:	Start: 70 °C, 1 min Program 1: 10 °C/min to 130 °C Program 2: 20 °C/min to 270 °C Final temp.: 270 °C, 10 min
Mass spectrometer:	Finnigan ITD 800 with direct coupling Scan mode: EI, full scan, 50–400 u Scan rate: 1 s/scan

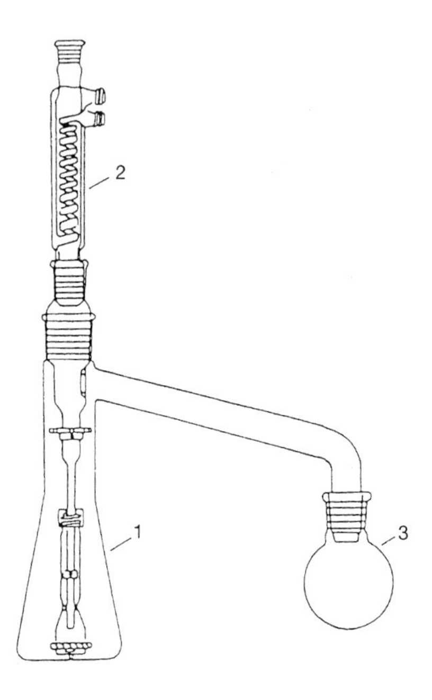

Fig. 4.68 Ludwig light phase rotation perforator for liquid/liquid extraction with specific light solvents.
1 Extraction vessel
2 Condenser
3 Solvent storage

Results

Bromoxynil, Ioxynil, 2,6-dimethyl-4.nitrophenol and DNOC can be assigned unambiguously from their characteristic molecular ions (Fig. 4.69). The evaluation of the chromatograms is carried out using mass chromatograms of selective ion signals (Fig. 4.70).

The procedure described is suitable for detecting and quantifying nitrophenols in aqueous media. Excess of the derivatisation agent leads to interference in the form of peak broadening arising from precipitation in the retention gap. While nitrophenol detection is unaffected by interference by background, contaminants formed by side reactions of the sample contents are detected in the GC/MS system when operating in the full scan mode.

Table 4.12 List of nitrophenols investigated (as methyl ethers)

No.	Nitrophenol derivative	Molecular mass [m/z]
1	4-Nitrophenol, methyl ether	153
2	2-Nitrophenol, methyl ether	153
3	3-Nitrophenol, methyl ether	153
4	2-Methyl-3-nitrophenol, methyl ether	167
5	3-Methyl-4-nitrophenol, methyl ether	167
6	4-Methyl-3-nitrophenol, methyl ether	167
7	5-Methyl-2-nitrophenol, methyl ether	167
8	4-Methyl-2-nitrophenol, methyl ether	167
9	Dinoseb, methyl ether	240
10	Dinoterb, methyl ether	240
11	Bromoxynil, methyl ether	291
12	Ioxynil, methyl ether	385
13	2,4-Dinitrophenol, methyl ether	198
14	3,4-Dinitrophenol, methyl ether	198
15	2,5-Dinitrophenol, methyl ether	198
16	2,6-Dinitrophenol, methyl ether	198
17	4,6- Dinitro-o-cresol, methyl ether	212
18	2,6- Dimethyl-4-nitrophenol, methyl ether	181

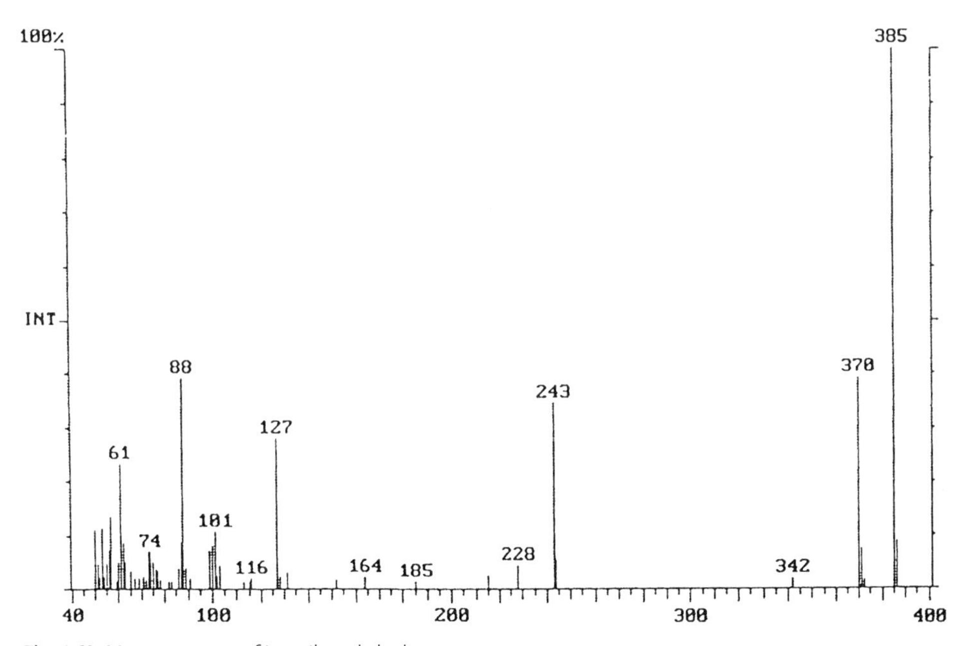

Fig. 4.69 Mass spectrum of Ioxynil methyl ether

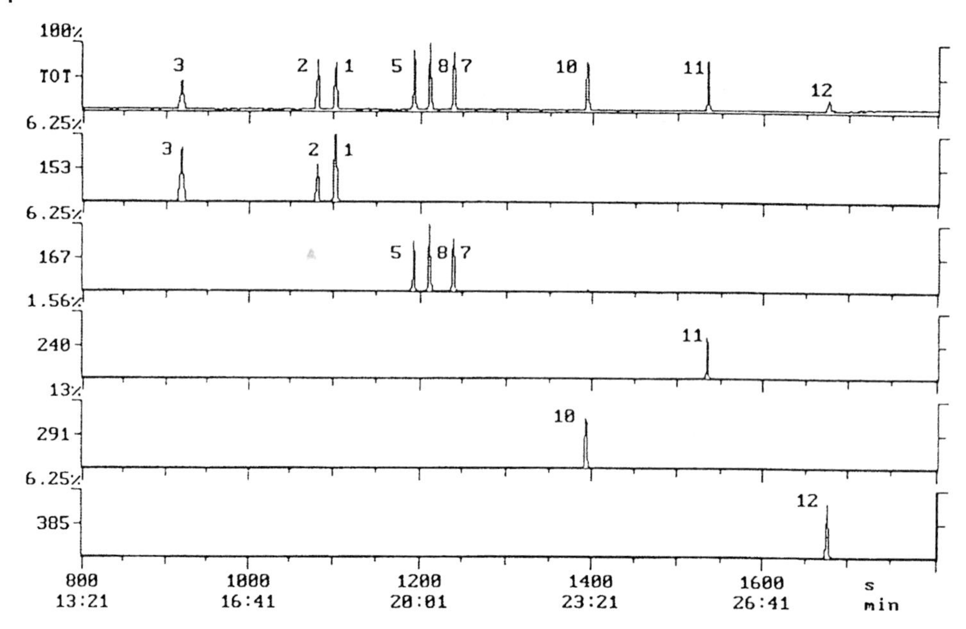

Fig. 4.70 Chromatogram of nitrophenols (as methyl ethers) in the total ion current and selective mass chromatogram (identification see Table 4.12)

4.17
Dinitrophenol Herbicides in Water

Key words: herbicides, dinitrophenols, drinking water, surface water, SPE, methylation, chemical ionisation

This method was drawn up for the determination of 2,4.dinitrophenol herbicides in drinking and raw water, taking into account the threshold values for drinking water control. Dinitrophenols and their derivatives are particularly important in drinking water treatment because of the toxicity of their metabolites (aminonitro- and diaminophenols). The following belong to this group of substances (Table 4.13):

- Type A (free phenol): Dinoseb, Dinoterb, DNOC
- Type B (esters): Dinoseb acetate, Dinoterb acetate, Binapacryl, Dinobuton

The active substances of type A are determined as the methyl ethers after treatment with diazomethane in the cartridge of the solid phase extraction before elution.

Analysis Conditions

Sample preparation:	Adsorption on glass cartridges with Carbopack B (120–400 μm, 250 mg, Supelco), conditioning with 10 mL methanol/acetic acid 1:1, 10 mL ethyl acetate, 2 mL methanol, 10 mL doubly distilled water Addition of 0.9 mL conc. HCl to 250 mL water sample, adsorption with ca. 4 mL/min flow rate, then drying with N_2 Derivatisation with 2 mL diazomethane in t-butyl methyl ether directly on the SPE cartridge (!), 30 min reaction time Elution with 5 mL ethyl acetate, concentration under N_2 to 100 μL final volume	
GC/MS system:	Finnigan MAGNUM with autosampler CTC A 200 S	
Injector:	PTV cold injection system	
Injector program:	Start:	80 °C
	Program:	300 °C/min to 250 °C
Column:	J&W DB-5, 30 m × 0.25 mm × 0.25 μm	
Program:	Start:	70 °C
	Program:	12 °C/min to 250 °C
Mass spectrometer:	Scan mode:	EI, full scan, 50–450 u CI, full scan, 50–600 u, reagent gas methanol
	Scan rate:	1 s/scan

Table 4.13 Structures of 2,4-dinitrophenol herbicides

Compound	R1	R2
DNOC	-OH	-methyl
Dinoterb	-OH	-t-butyl
Dinoseb	-OH	-s-butyl
Dinoseb acetate	-acetate	-s-butyl
Dinoterb acetate	-acetate	-t-butyl
Dinobuton	-carbonate	-s-butyl
Binapacryl	-acrylate	-s-butyl
2,4-Dinitrophenol	-OH	-H

R_1 NO_2 R_2 NO_2

Results

Because of the acidity of the compounds of type A, derivatisation with diazomethane to the methyl ethers is recommended for GC separation (Figs. 4.71 and 4.72). Dinoseb and Dinoterb are differentiated by means of their EI spectra (Figs. 4.73 and 4.74). If these compounds are analysed in the CI mode, the two substances must be differentiated using the retention time, as the CI mass spectra of both herbicides show the $(M+H)^+$ ion with m/z 255 (Figs. 4.75 and 4.76).

In chemical ionisation Dinoseb and Dinoterb acetates and Dinobuton show the protonated phenol fragments as the base peak. Dinobuton also shows the fragment m/z 177 (Table 4.14). Binapacryl does not form a quasimolecular ion with methanol as the reagent gas. In the analysis of compounds of type B Dinoseb and Dinoterb are formed as hydrolysis products and under certain circumstances can only be quantified as a total parameter (free phenols and esters). Hydrolysis can also occur during the analysis of compounds of type A (Heimlich 1994).

Table 4.14 Specific ions for EI/CI analyses of dinitrophenol herbicides (R_t = retention time, M = molecular mass) (Heimlich 1994)

Substance	R_t	M	EI [m/z (%)]	CI [m/z (%)]
DNOC, methyl ether	10:04	212	182(100), 165(70), 212(20)	213(100)
Dinoterb, methyl ether	11:25	254	239(100), 209(55), 254(20)	255(100)
Dinoseb, methyl ether	11:34	254	225(100), 195(70), 254(20)	255(100)
Dinoseb acetate	12:10	282	89(100), 211(55), 240(40)	241(100)
Dinoterb acetate	12:19	282	225(100), 177(50), 240(20)	241(100)
Dinobuton	13:33	326	211(100), 163(30), 205(60)	241(100), 177(20)
Binapacryl	15:07	322	83(100)	83(100)

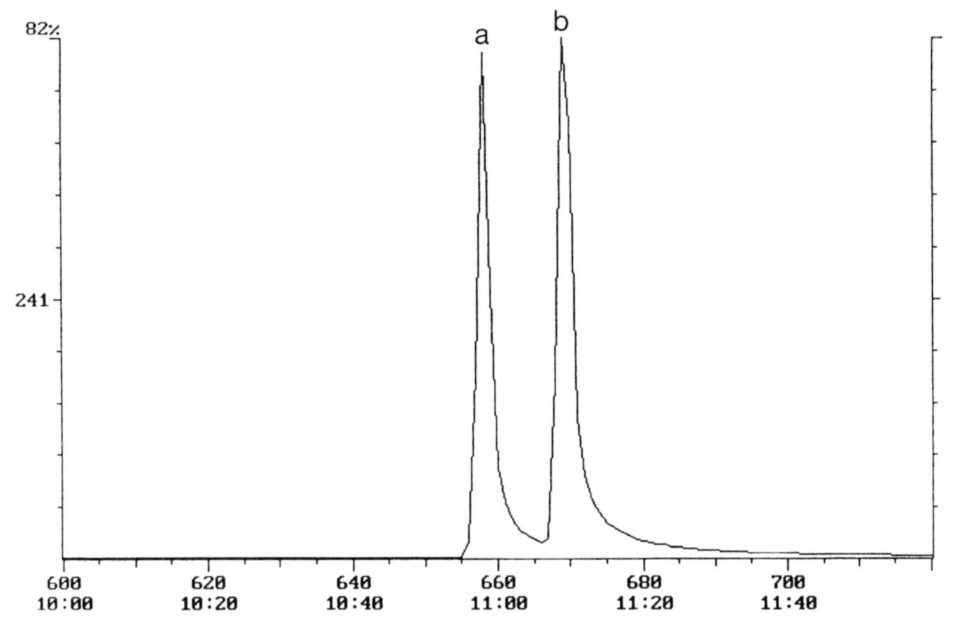

Fig. 4.71 The analysis of Dinoterb (a) and Dinoseb (b) shows severe peak tailing because of the acidity of the compounds

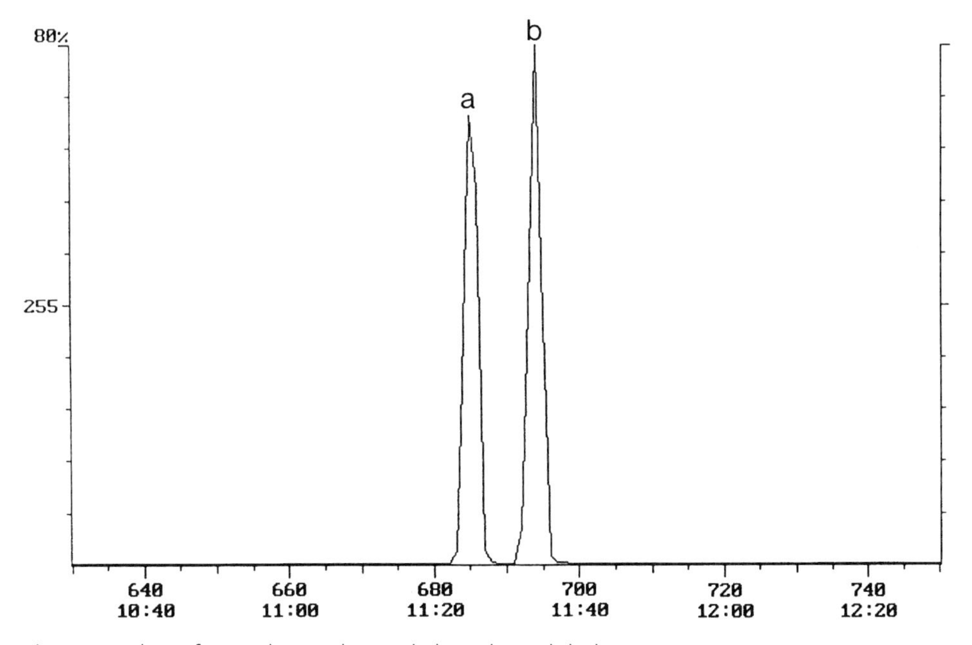

Fig. 4.72 Analysis of Dinoterb (a) and Dinoseb (b) as the methyl ethers

Fig. 4.73 EI spectrum of Dinoterb methyl ether

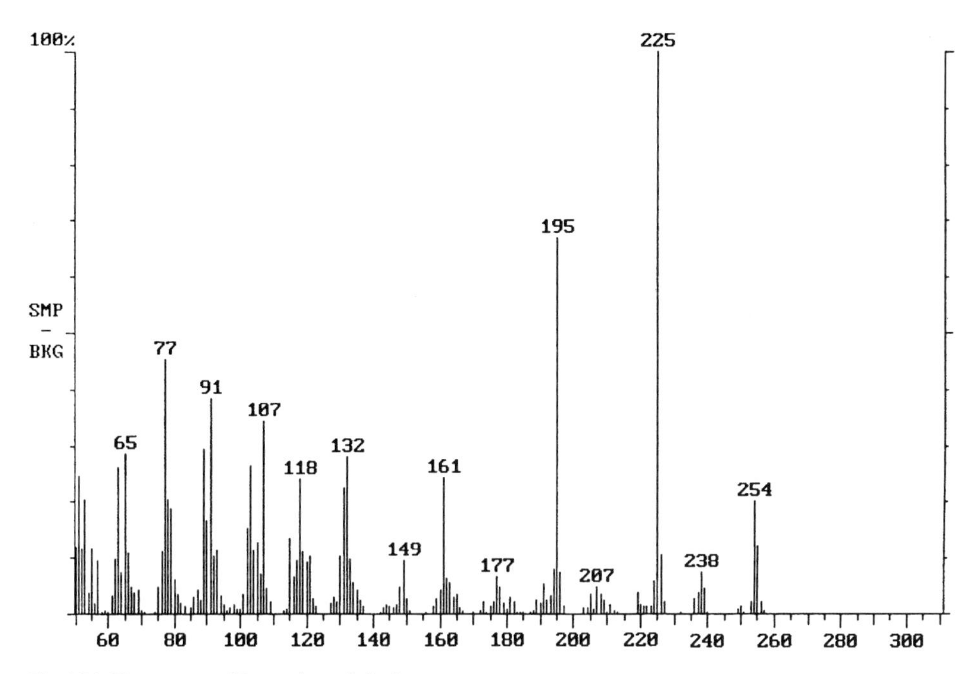

Fig. 4.74 EI spectrum of Dinoseb methyl ether

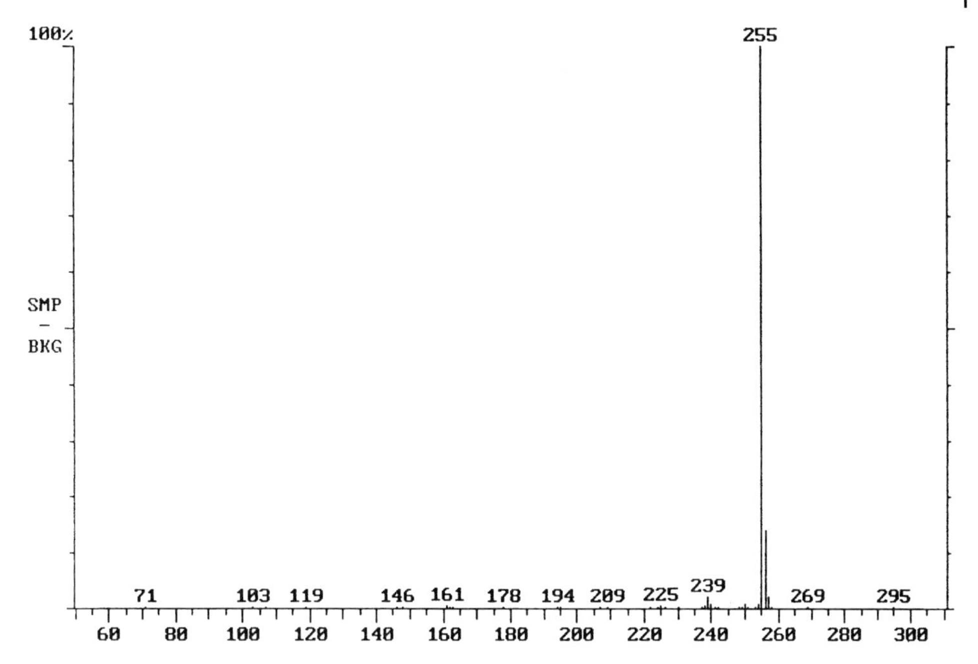

Fig. 4.75 CI spectrum of Dinoterb methyl ether, reagent gas methanol

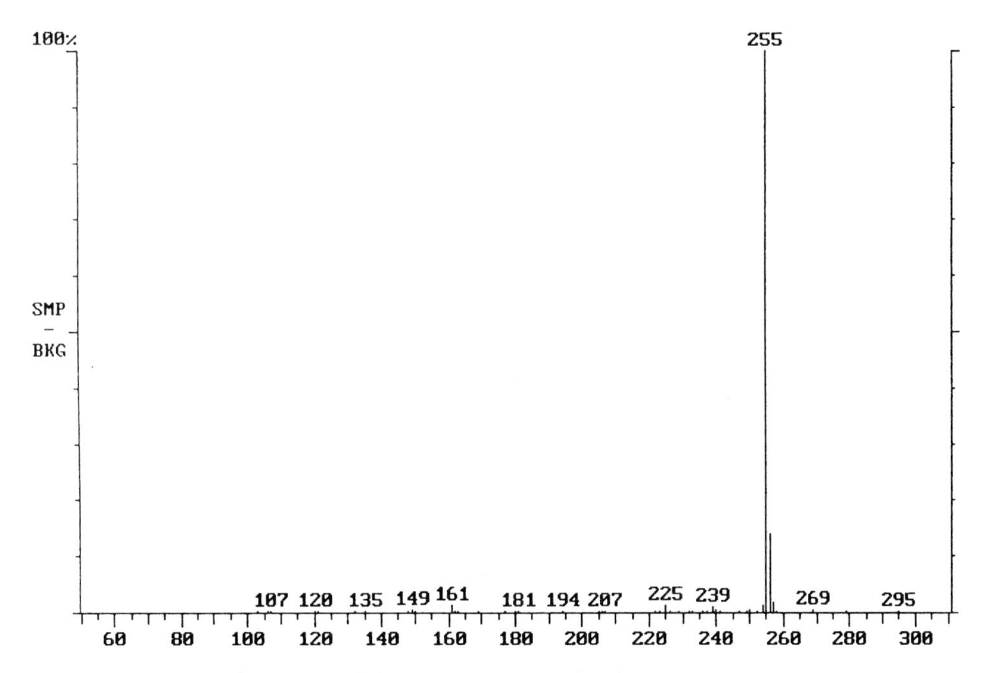

Fig. 4.76 CI spectrum of Dinoseb methyl ether, reagent gas methanol

4.18
Hydroxybenzonitrile Herbicides in Drinking Water

Key words: herbicides, dihalogenated hydroxybenzonitrile, drinking water, SPE, methylation, chemical ionisation

The method of determination of dihalogenated hydroxybenzonitrile herbicides was drawn up on the basis of the threshold values for the control of drinking water. During the development of a sensitive analysis procedure, the highly polar nature of these active substances in particular had to be taken into account. Quantitative desorption from the solid phase of the enrichment procedure (SPE) is made more difficult by their hydrophilic properties. Therefore in this procedure methylation is carried out on the SPE cartridge. The substances are determined as their methyl ethers after reaction with diazomethane (Table 4.15).

Table 4.15 Structures of the dihalogenated hydroxybenzonitriles and Mecoprop

Compound	Structure
Bromoxynil octyl ester Bromoxynil	NC–(ring, Br, Br)–O–C(=O)(CH$_2$)$_6$CH$_3$ → NC–(ring, Br, Br)–OH
Ioxynil	NC–(ring, I, I)–OH
Chloroxynil	NC–(ring, Cl, Cl)–OH
Mecoprop	Cl–(ring, CH$_3$)–O–C(H)(CH$_3$)–C(=O)OH

Analysis Conditions

Sample preparation:	Adsorption on glass cartridges with Carbopack B (120–400 μm, 250 mg, Supelco), conditioning with 10 mL ethanol/acetic acid 1:1, 10 mL ethyl acetate, 2 mL ethanol, 2 mL doubly distilled water Addition of 0.3 mL conc. HCl to 250 ml water sample, adsorption with ca. 4 mL/min flow rate, then drying with N_2 Derivatisation with 2 mL diazomethane in t-butyl methyl ether directly on the SPE cartridge (!), 30 min reaction time Elution with 5 mL ethyl acetate, concentration under N_2 to 100 μL final volume
GC/MS system:	Finnigan MAGNUM with autosampler CTC A200S
Injector:	PTV cold injection system
Injector program:	Start: 80 °C Program: 300 °C/min to 250 °C
Column:	J&W DB-5, 30 m × 0.25 mm × 0.25 μm
Program:	Start: 70 °C Program: 12 °C/min to 250 °C Transfer line: 220 °C
Mass spectrometer:	Scan mode: EI, full scan, 50–450 u CI, full scan, 100–600 u, reagent gas methanol Scan rate: 1 s/scan

Results

The herbicides are adsorbed well on the solid phase material. The methylation with diazomethane on the SPE cartridge is finished after ca. 30 min reaction time. The recoveries determined are 75% for Bromoxynil, 80% for Ioxynil, and 50% for Chloroxynil. For Mecoprop as the representative of the phenoxyalkylcarboxylic acids 95% recovery is achieved. Mass spectrometric detection using chemical ionisation with methanol as the reagent gas is better than EI detection (Fig. 4.77, Table 4.16). The limits of detection for Chloroxynil (Fig. 4.78) and Mecoprop (Fig. 4.79) are 10 ng/L and for Bromoxynil (Fig. 4.80) and Ioxynil (Fig. 4.81) 3 ng/L at a signal/noise ratio of 5. When using the method it should be noted that Bromoxynil can also appear as the degradation product of Bromoxynil alkylcarboxylic acid esters and of Bromfen oxime (Grass 1994).

Table 4.16 Specific ions for the EI/CI analysis of hydroxybenzonitrile herbicides and Mecoprop (R_t = retention time, M = molecular mass) (Grass 1994)

Substance	R_t	M	EI [m/z (%)]	CI [m/z (%)]
Chloroxynil methyl ether	7:23	201	201(100), 203(65)	202(100), 204(70)
Mecoprop methyl ester	8:33	228	228(100), 169(95), 107(90)	169(100), 229(35)
Bromoxynil methyl ether	9:22	291	291(100), 289(40), 293(45)	292(100), 290(50), 294(50)
Ioxynil methyl ether	11:52	385	385(100), 370(20)	386(100)

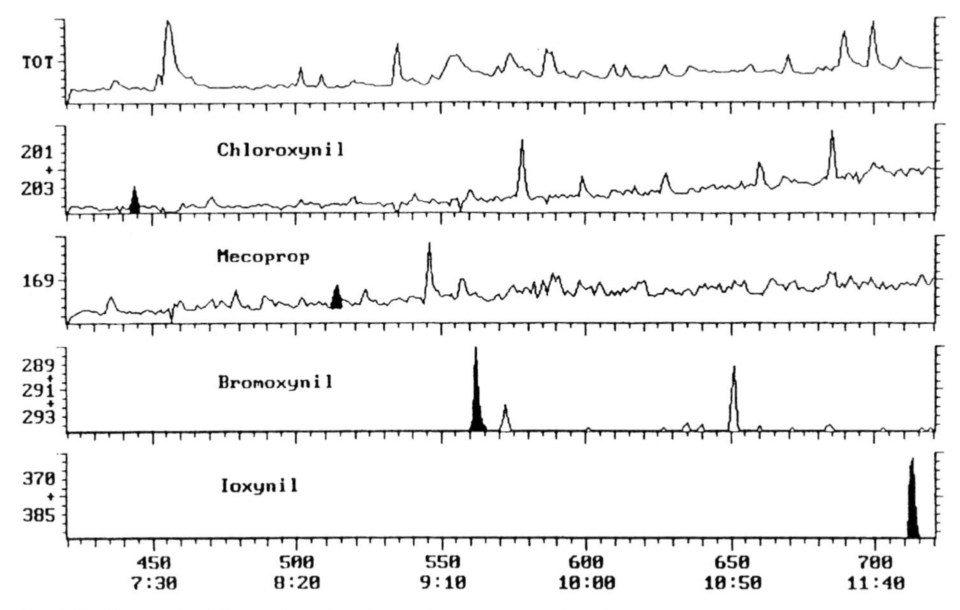

Fig. 4.77 EI analysis of the methyl ethers in a spiked water sample (10 ng/l)

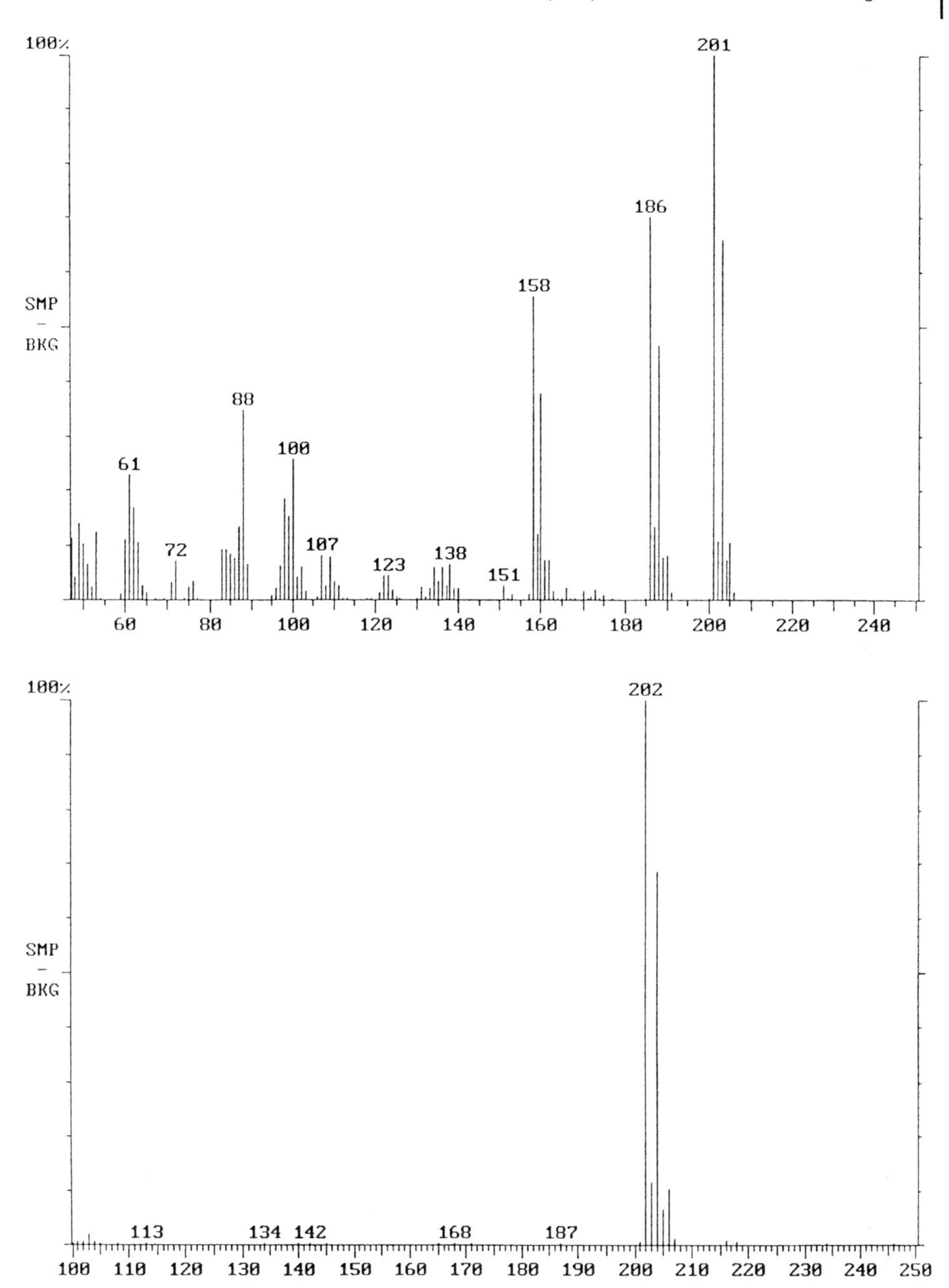

Fig. 4.78 Chloroxynil methyl ether (above: EI, below: methanol CI)

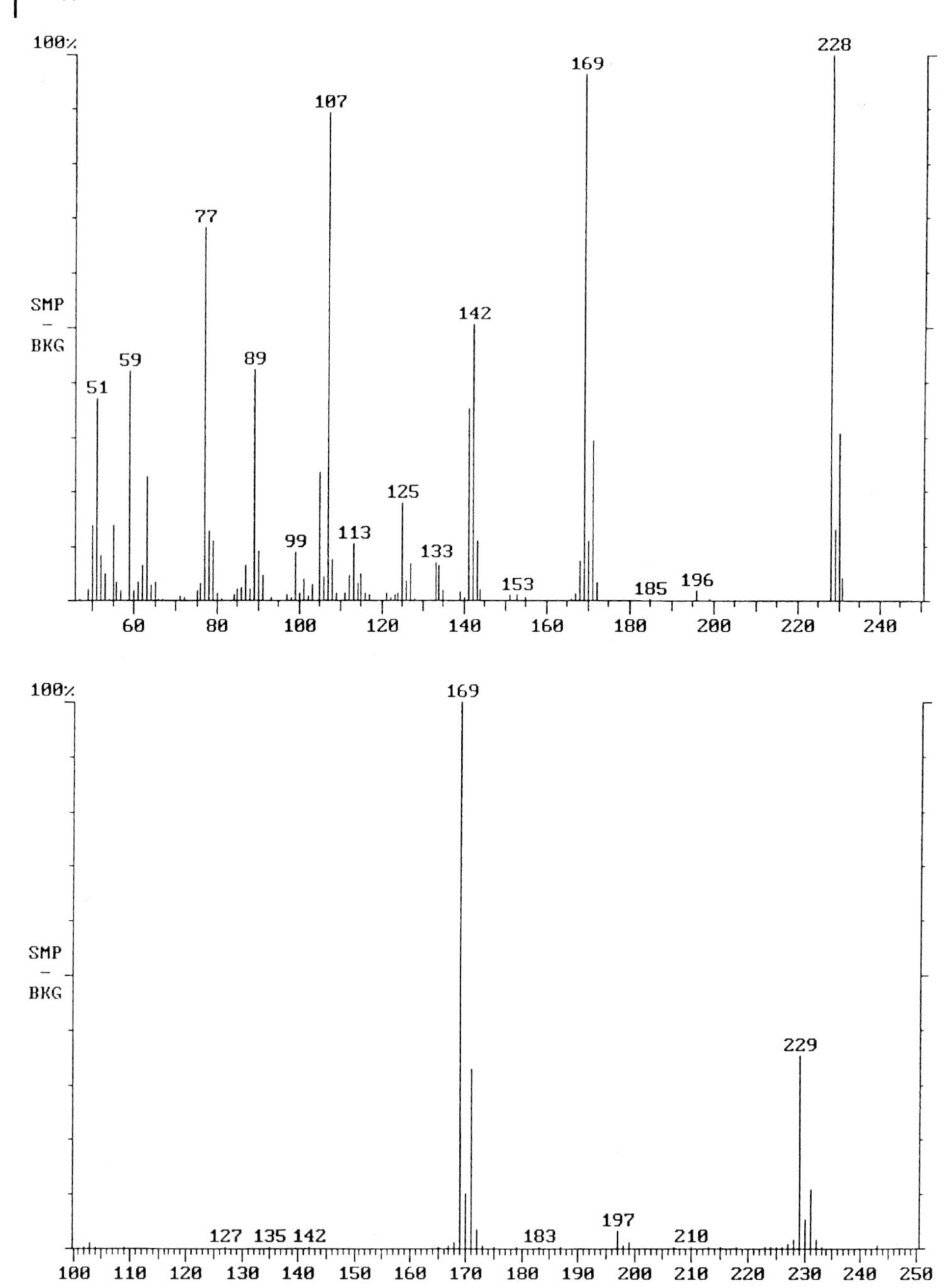

Fig. 4.79 Mecoprop methyl ether (above: EI, below: methanol CI)

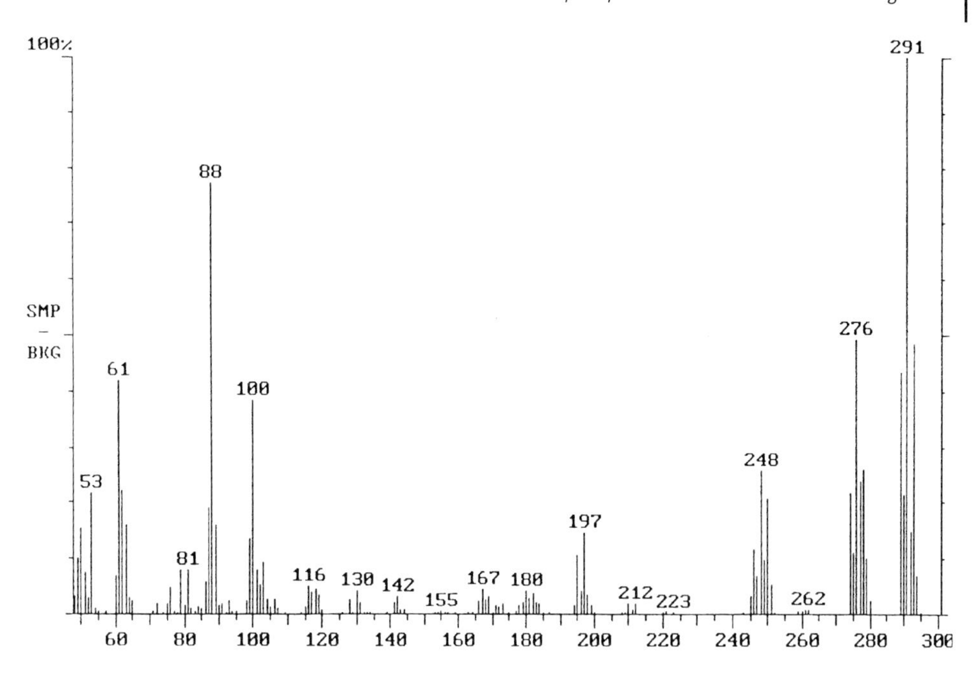

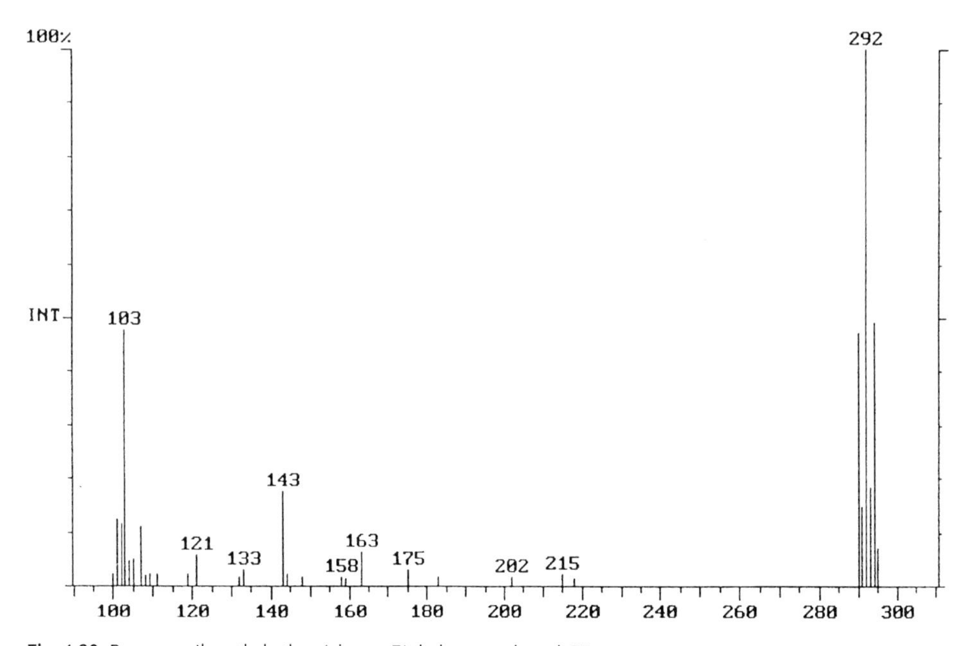

Fig. 4.80 Bromoxynil methyl ether (above: EI, below: methanol CI)

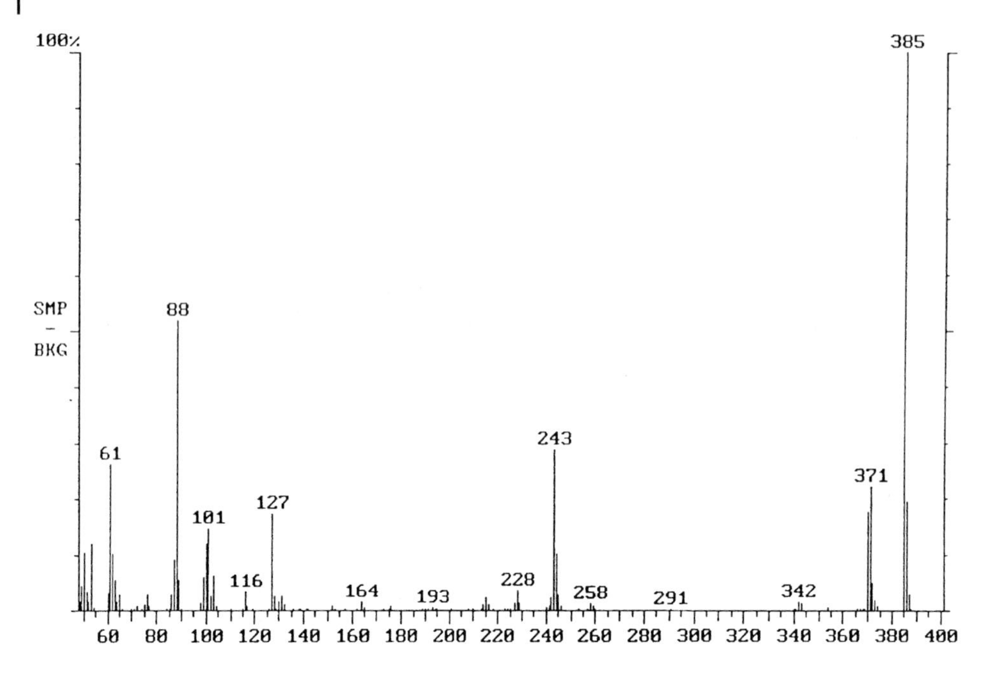

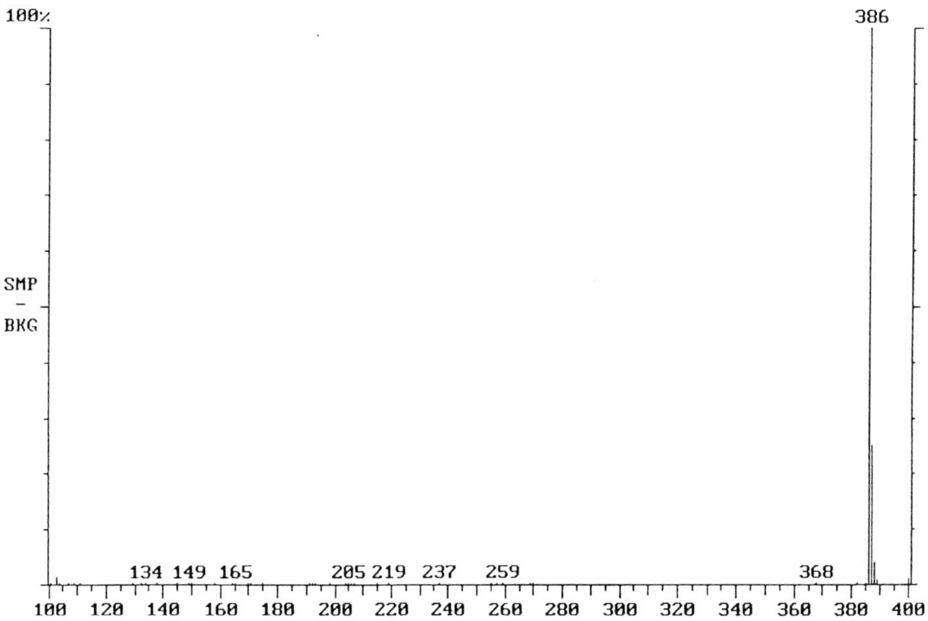

Fig. 4.81 Ioxynil methyl ether (above: EI, below: methanol CI)

4.19
Routine Analysis of 24 PAHs in Water and Soil

Key words: PAHs, SPE, Soxhlet, chemical ionisation, water, soil, dynamic region, deuterated standards

The process given here for the routine determination of PAHs is particularly suitable for samples which contain a strongly aliphatic hydrocarbon matrix, besides PAHs. Chemical ionisation with water is used as the ionisation procedure. This makes the undesired background selectively transparent and a higher response is achieved for the PAHs. Compared with EI analysis (Fig. 4.82) the response of compounds in the CI mode is about the same (Fig. 4.83). The signal/noise ratio with CI using water for compounds with a significant proton affinity exceeds that with EI.

Analysis Conditions

Sample preparation: Water: 500 mL sample, solid phase extraction (SPE) on C_{18}-cartridges; liquid/liquid extraction with cyclohexane; concentration to 0.2 mL
Soil: 10 g sample with 10 g sodium sulfate, Soxhlet extraction with cyclohexane, 6 h, concentration to 1 mL
The standard mixture of 6 deuterated compounds is added to the extraction mixture.GC/MS system:
Finnigan ITS 40,

Finnigan ITD 800 with Siemens GC Sichromat 1

Injector program:
Start: 40 °C
Program: rapidly to 320 °C
Final temp.: 320 °C, 15 min.

Column: Restek XTI-5, 30 m × 0.32 mm × 0.25 µm
Helium, 2 bar

Temp. program:
Start: 40 °C, 1 min
Program 1: 15 °C/min to 340 °C
Final temp.: 340 °C

Mass spectrometer:
Scan mode: CI, full scan, 100–300 u, reagent gas water
Scan rate: 1 s/scan

Calibration: Using deuterated internal standards (Table 4.17)

Results

The polycyclic aromatic hydrocarbons listed in Table 4.18 were investigated, of which 16 were analysed using the EPA method. A calibration file with the CI mass spectrum, retention time and calibration function are used for searching and identification of the compounds (Fig. 4.84). The specific mass trace $(M+H)^+$ is used to successfully identify the

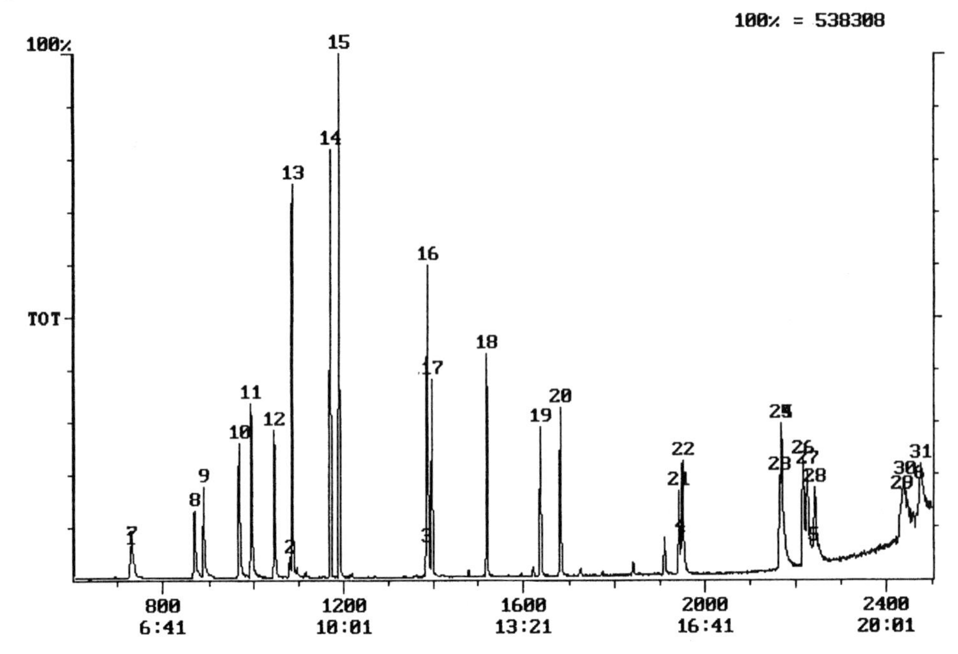

Fig. 4.82 PAH routine analysis: standard run in EI mode.
Internal standards: 0.5 ng/μL; PAHs: 5 ng/μL

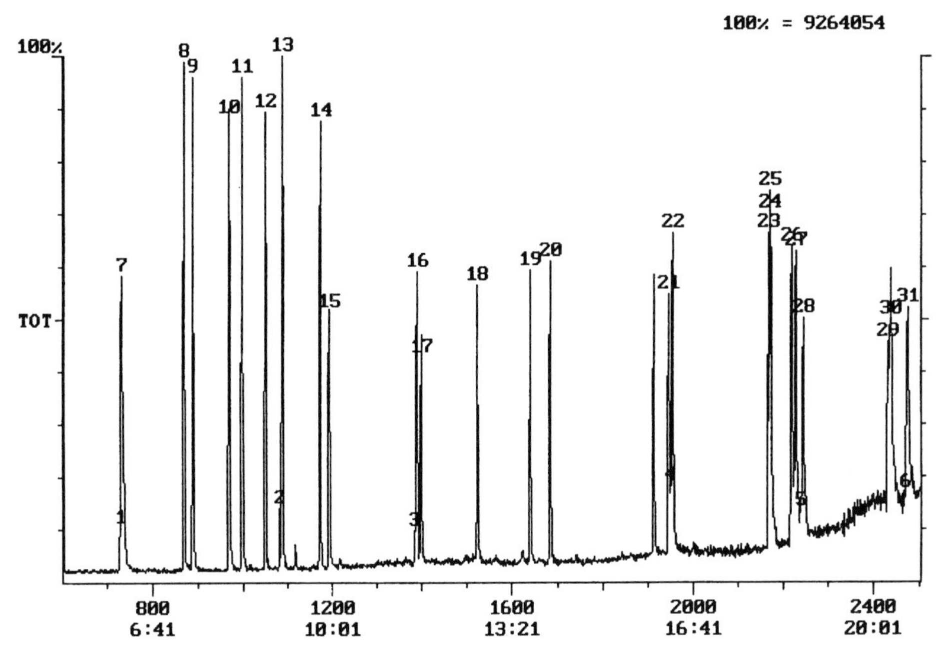

Fig. 4.83 PAH routine analysis: standard run in CI mode (reagent gas water).
Internal standards: 0.5 ng/μL; PAHs: 5 ng/μL

Table 4.17 Internal standards (dissolved in 80% dichloromethane-d_2 (99.6%) and 20% benzene-d_6 (99.6%), Promochem)

No.[1)]	Compound	M[2)]
0	1,4-Dichlorobenzene-d_4	150
1	Naphthalene-d_8	136
2	Acenaphthene-d_{10}	166
3	Phenanthrene-d_{10}	188
4	Chrysene-d_{12}	240
5	Perylene-d_{12}	264
6	Benzo(ghi)perylene-d_{12}	288

1) Entry number in the calibration file
2) The quasimolecular ions $(M+H)^+$ are formed in CI with water

Table 4.18 Polyaromatic hydrocarbon calibration substances (solution in toluene, certified by NIST, Promochem)

No.[1)]	Compound	ISTD No.[2)]	M[3)]
7	Naphthalene*	1	128
8	2-Methylnaphthalene	1	142
9	1-Methylnaphthalene	1	142
10	Biphenyl	2	154
11	2,6-Dimethylnaphthalene	2	156
12	Acenaphthylene*	2	152
13	Acenaphthene*	2	154
14	2,3,5-Trimethylnaphthalene	3	170
15	Fluorene*	3	166
16	Phenanthrene*	3	178
17	Anthracene*	3	178
18	1-Methylphenanthrene	4	192
19	Fluoranthene*	4	202
20	Pyrene*	4	202
21	Benz(a)anthracene*	4	228
22	Chrysene*	4	228
23	Benzo(b)fluoranthene*	5	252
24	Benzo(k)fluoranthene*	5	252
25	Benzo(e)pyrene	5	252
26	Benzo(a)pyrene*	5	252
27	Perylene	5	252
28	Indeno(123-cd)pyrene*	6	276
29	Dibenz(a,h)anthracene*	6	278
30	Benzo(ghi)perylene*	6	276

* PAHs according to EPA (16 components)
1) Entry number in the calibration file
2) Assignment of the PAHs to the internal standard from Table 4.16
3) The quasimolecular ions $(M+H)^+$ are formed through CI with water

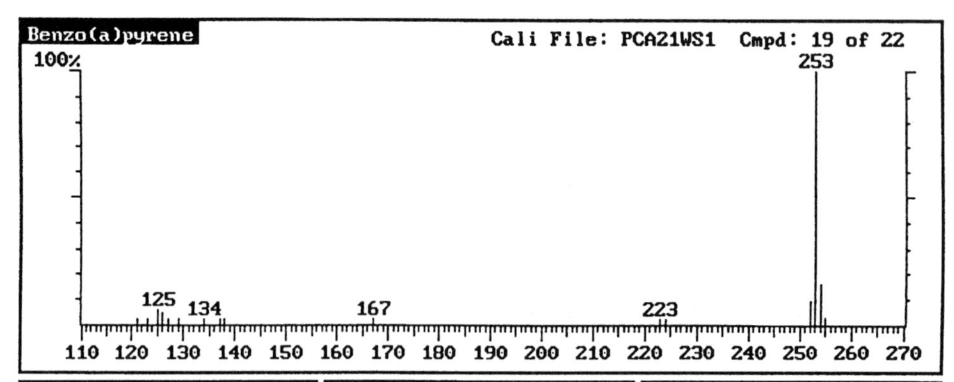

Fig. 4.84 Benzo(a)pyrene entry (CI spectrum) in the calibration file

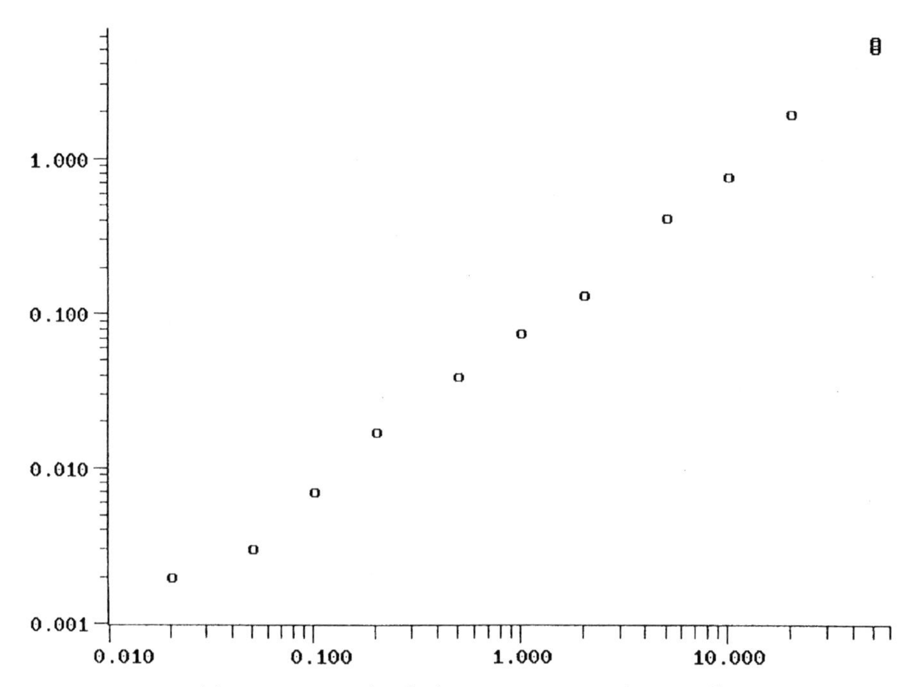

Fig. 4.85 Linearity of the water CI procedure for benzo(a)pyrene (correlation coefficient 0.998)

components, in order to check the minimum signal/noise ratio required and to determine the peak areas. Quantitation is then carried out with reference to the internal standard and the calibration function (Fig. 4.85).

The limits of detection (S/N = 3) of the procedure are between 2 ng/L for volatile PAHs and 10 ng/L for higher-boiling ones with CI using water. The absolute limit of detection of the CI method is between 2 and 10 pg/injection at a linear range greater than 10^4 (Landrock 1995).

4.20 Fast GC Quantification of 16 EC Priority PAH Components

Key words: PAH, benzo[a]pyrene, EU regulation, fast GC, high resolution, HRMS, food, meat, flavour, smoke, flavouring

The European Commission Regulation (EC) No 208/2005 of February 4, 2005, which came into force on April 1, 2005, regulates the maximum levels for benzo[a]pyrene in different groups of food of which have the strongest ruling foods for infants and young children with max. 1.0 µg/kg. The maximum level for smoked meats and smoked meat products is 5.0 µg/kg.

The best known carcinogenic compound – benzo[a]pyrene – is used as leading substance out of about 250 different compounds which belong to the PAH group. The German revision of the flavour directive (Aromenverordnung) of May 2, 2006 in § 2 (4) regulates the maximum level for benzo[a]pyrene at 0.03 µg/kg for all types of food with added smoke flavourings.

The Commission Recommendation of February 4, 2005 on the further investigation on the levels of polycyclic aromatic hydrocarbons in certain types of food is directed to analyse the levels of 15 PAH compounds which are classified as priority (Fig. 4.86) and to check the suitability of benzo[a]pyrene as a marker.

In addition, the Joint FAO/WHO Experts Committee on Food Additives (JECFA) identified the PAH compound benzo[c]fluorene as to be monitored as well (Fig. 4.87).

During GC/MS method setup it turned out that quadrupole desktop MS instruments could not provide the required selectivity and sensitivity at the low decision level for the list of 16 PAH for a reliable detection in real life samples. High resolution mass spectrometry with its particularly high selectivity provided the requested robustness and safety in the quantitative determination (Ziegenhals 2006).

The quantitation was done using isotope dilution technique by the addition of isotope-labelled and fluorinated standards before extraction, as well as for the determination of the response factors of all PAH. Recovery values have been determined by the addition of three deuterium labelled compounds.

Table 4.19 Exact masses of PAH and labeled internal standards

PAH native	Abbr.	Exact mass native [u]	PAH ISTD labeled	Exact mass labeled [u]
Benzo[c]fluorine	BcL	216.0934	5-F-BcL	234.0839
Benzo[a]anthracene	BaA	228.0934	$^{13}C_6$-BaA	234.1135
Chrysene	CHR	228.0934	$^{13}C_6$-CHR	234.1135
Cyclopenta[cd]pyrene	CPP	226.0777		
5-Methylchrysene	5MC	242.1090	d_3-5MC	245.1278
Benzo[b]fluoranthene	BbF	252.0934	$^{13}C_6$-BbF	258.1135
Benzo[j]fluoranthene	BjF	252.0934		
Benzo[k]fluoranthene	BkF	252.0934	$^{13}C_6$-BkF	258.1135
Benzo[a]pyrene	BaP	252.0934	$^{13}C_4$-BaP	256.1068
Indeno[123cd]pyrene	IcP	276.0934	d_{12}-IcP	288.1687
Dibenzo[ah]anthracene	DhA	278.1090	d_{14}-DhA	292.1969
Benzo[ghi]perylene	BgP	276.0934	$^{13}C_{12}$-BgP	288.1336
Dibenzo[al]pyrene	DlP	302.1090	13-F-DlP	320.0996
Dibenzo[ae]pyrene	DeP	302.1090	$^{13}C_6$-DeP	308.1291
Dibenzo[ai]pyrene	DiP	302.1090	$^{13}C_{12}$-DiP	314.1493
Dibenzo[ah]pyrene	DhP	302.1090		
PAH recovery standards deuterated			**PAH ISTD**	**Exact mass labeled [u]**
D_{12}-Benzo[a]anthracene			d_{12}-BaA	240.1687
D_{12}-Benzo[a]pyrene			d_{12}-BaP	264.1687
D_{12}-Benzo[ghi]perylene			d_{12}-BgP	288.1687

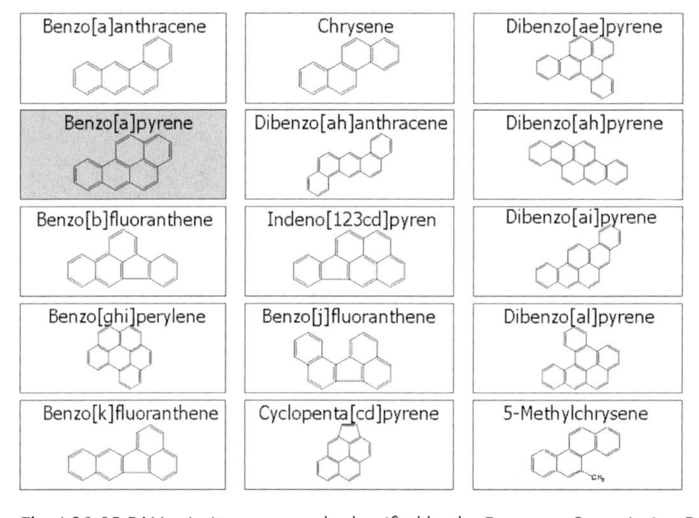

Fig. 4.86 15 PAH priority compounds classified by the European Commission Regulation

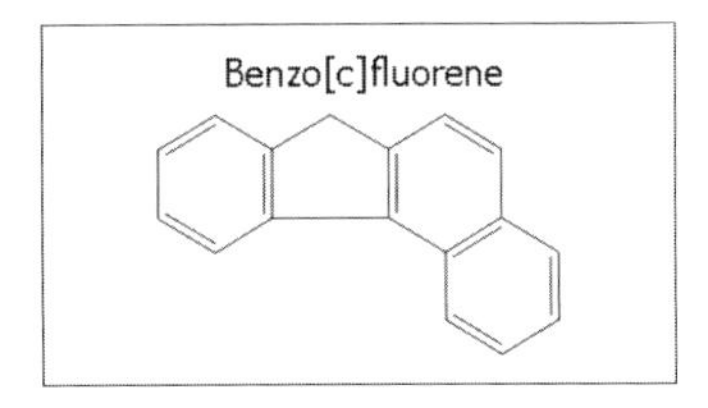

Fig. 4.87 Additional PAH priority compound to be monitored according to JECFA 2005

Analysis Conditions

Sample preparation:

1. Pressurized solvent extraction (ASE) for extraction of lipophilic substances:
Homogenised sample (4–6 g meat product, 1–2 g spice); agitate with same amount of the drying material poly(acrylic acid), partial sodium salt-graft-poly(ethylene oxide); add PAH standard mixture containing isotope labelled (^{13}C and ^{2}H) and fluorinated PAH compounds; carry out ASE extraction with n-hexane at 100 °C and 100 bar; flush volume 60% at purge time 120 s; two static cycles, each of 10 min; evaporate the solvent in a water bath (40 °C) using a nitrogen stream.

2. Size exclusion chromatography for the separation of higher molecular substances:
Dissolve dried ASE extract in 4.5 mL cyclohexane/ethylacetate (50:50 v/v) and filter through PTFE filter, pore size of 1 μm; GPC column (25 mm i.d.) with Bio-Beads S-X3 (height of filling 42 cm); elute samples at a flow rate of 5 mL/min with cyclohexane/ethylacetate (50:50 v/v); dump time 0–36 min, collect time 36–65 min; remove solvent with a rotary evaporator, and dried eluate in a nitrogen stream.

3. Solid phase extraction to remove polar substances:
Using a modified ASPEC Xli (Automatic. Sample Preparation with Extraction Columns) from Gilson (Middleton, USA) [6], modified with a fitting rack, Teflon® funnels and Teflon tubes silica, dried for 12 h at 550 °C, deactivated with 15% water; 1 g dried deactivated silica was filled into commercial 6-mL SPE columns (12 mm i.d.); after conditioning of columns apply samples with 3 mL cyclohexane; elute with 10 mL cyclohexane; add PAH recovery standard mixture; dissolve the dried eluate of SPE in 1 mL iso-octane and 50 μL of the PAH recovery standard mixture (benzo[a]anthracen-d_{12}, benzo[a]pyrene-d_{12} and benzo[ghi]perylene-d_{12}) and transfer to a 1 mL tapered vial; concentrate the sample carefully in a nitrogen stream to a volume of about 50 μL.

Gas chromatograph: Trace GC Ultra, Thermo Fisher Scientific, Milan, Italy

Injector: Split/splitless, 1 min, 320 °C, 1.5 μL injection volume

Carrier gas:	Helium, 0.6 mL/min, const. flow
Column:	TR-50MS, 10 m x 0.1 mm x 0.1 µm
Oven temp. program:	140 °C, 1 min 10 °C/min to 240 °C 5 °C/min to 270 °C 30 °C/min to 280 °C 4 °C/min to 290 °C 30 °C/min to 315 °C 3 °C/min to 330 °C
MS interface temperature:	Transfer line 300 °C Ion source 280 °C
Mass spectrometer:	DFS High Resolution GC/MS, Thermo Fisher Scientific, Bremen, Germany
Ionisation:	EI, 45 eV
Scan mode:	Multiple ion detection mode (MID), see Table 4.20
Resolution:	8,000
Cycle time:	0.7 s/scan

Results

Although an initial use of a 50 % phenyl capillary column of 60 m length (60 m × 0.25 mm × 0.25 µm, at constant pressure) provided the required chromatograph resolution of the various isomers, the retention time of more than 90 min turned out to be not appropriate for a control method with high productivity. The application of fast GC column technology reduced the required retention by more than ³/₄ to only 25 min maintaining the necessary chromatographic resolution (see Fig. 4.88). The critical separation components are shown in Figs. 4.89 and 4.90. For all components the fast GC method provides a robust peak separation and quantitative peak integration.

Applicability for different matrices has been shown for many critical matrices. Figure 4.91 shows the analysis of the extract from caraway seeds with a determined concentration of benzo[a]pyrene of 0.02 µg/kg. An LOD of 0.005 µg/kg and an LOQ of 0.015 µg/kg can be estimated for the analysis of spices, when the sample weight is 1 to 1.5 g. The recovery values achieved with the described sample preparation has been between 50 and 120 % [6].

Table 4.20 MID Descriptor for PAH Fast-GC/HRMS data acquisition

RT [min]	Exact mass [u]	Function	Dwell time [ms]
8:50	216.09375	native	82
	218.98508	lock	2
	226.07830	native	82
	228.09383	native	82
	234.08450	native	82
	234.11400	native	82
	240.16920	native	82
	263.98656	cali	6
13:00	218.98508	lock	2
	242.10960	native	82
	245.12840	native	82
	252.09390	native	82
	256.10730	native	82
	258.11400	native	82
	263.98656	cali	6
	264.16920	native	82
19:00	263.98656	lock	2
	276.09390	native	74
	278.10960	native	74
	288.13410	native	74
	292.19740	native	74
	313.98340	cali	6
22:00	263.98656	lock	2
	302.10960	native	120
	308.12970	native	120
	313.98340	cali	6
	314.13980	native	120
	320.10010	native	120

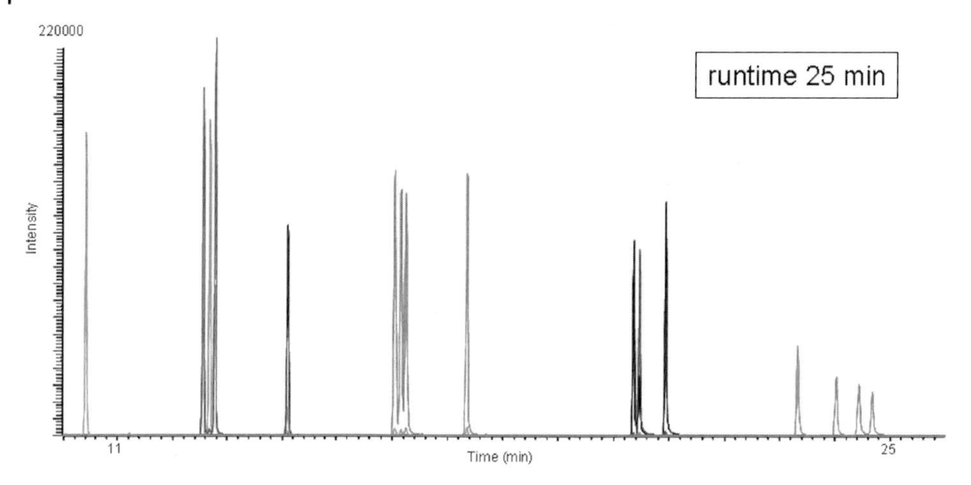

Fig. 4.88 Fast GC/HRMS separation of 16 EC priority PAHs in only 25 min

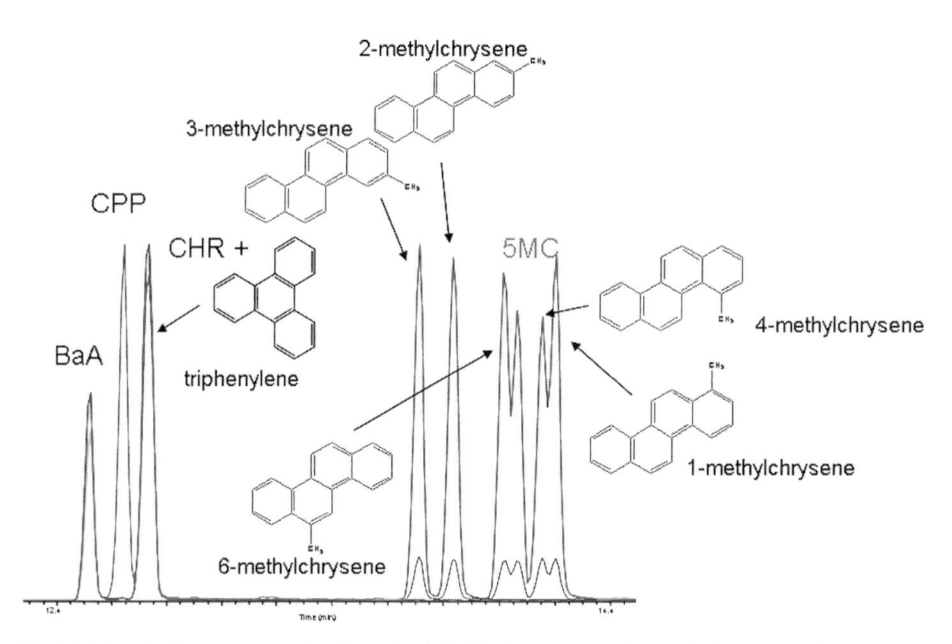

Fig. 4.89 Detail of isomer separation from Fig. 4.91. Elution sequence first peak cluster BaA, CPP, CHR, second peak cluster 3MC, 2MC, 6MC, 5MC, 4MC, 1MC

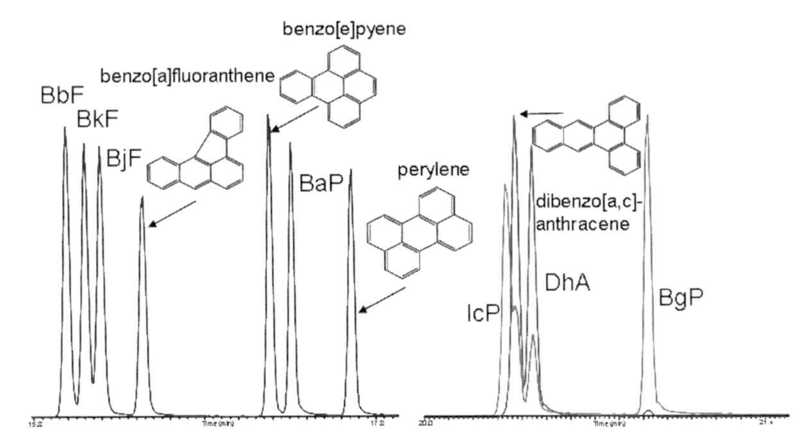

Fig. 4.90 Detail of isomer separation from Fig. 4.89. Elution sequence left peak cluster 15.8 to 17.8 min is BbF, BkF, BjF, BaF, BeP, BaP, PER, and right peak cluster 20.0 to 21.4 min is IcP, DcA, DhA, BgP

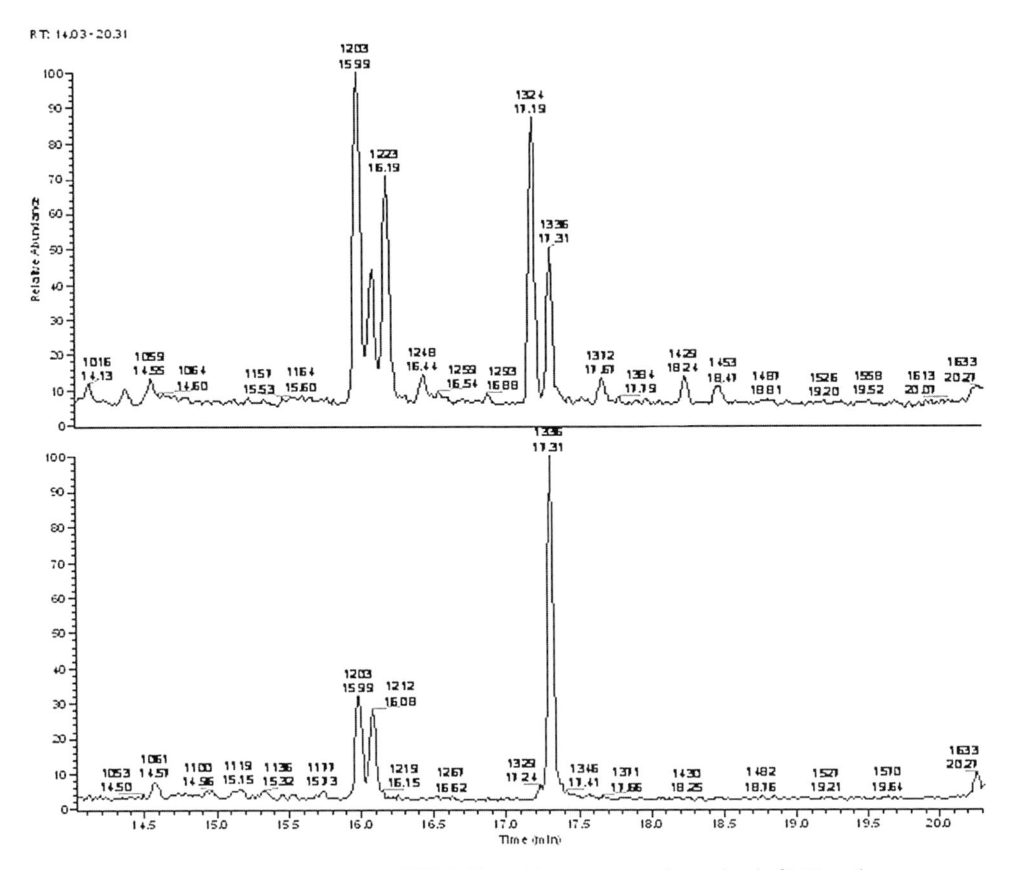

Fig. 4.91 Benzo[a]pyrene determination (RT 17:31 min) in caraway seeds at a level of 0.02 µg/kg (top native PAH, bottom ^{13}C-BaP, elution sequence BbF, BkF, BjF, BeP, BaP)

4.21
Analysis of Water Contaminants by On-line SPE-GC/MS

Key words: SPE-GC/MS, automation, water analysis, micro-contaminants, POPs, Lindane, HCH, BHC

The time consuming work load of a solid phase sample preparation with the extraction of waters in the size of liter volumes can be significantly reduced by automated on-line SPE extraction coupled to a GC/MS system. This technique offers by the large volume injection of a complete SPE extract the dramatic miniaturisation of the extraction step.

River water samples have been analysed for trace levels of benzene hexachloride/hexachlorocyclohexane (BHC/HCH) using SIM and MS/MS technologies.

Analysis Conditions

Sample preparation:	Filtration of river water On-line SPE: Autoloop SPE-GC-Interface (Interchro, Bad Kreuznach, Germany)
Gas chromatograph:	Trace GC (ThermoQuest, Milan, Italy)
Injector:	On-column injection with loop-type interface and direct retention gap coupling, transfer volume 100 µL, partial or fully concurrent solvent evaporation
Carrier gas:	Helium
Column:	DB-5MS, 30 m x 0.25 mm i.d. x 0.25 µm film
Oven temp. program:	
MS interface temperature:	
Mass spectrometer:	GCQ Ion Trap MS (ThermoQuest, Austin, TX, USA)
Ionisation:	EI
Scan mode:	SIM and MS/MS
Mass range:	SIM: m/z 219, 217, 183 MS/MS: m/z 219 transition to m/z 183
Cycle time:	0.5 s per scan

Results

During many years of continuous operation the validation and many applications have been reported (Jahr 1996, Louter 1996, Verma 1997). The on-line SPE-GC/MS system has been successfully applied to the trace analysis and quantification of more than 120 different target

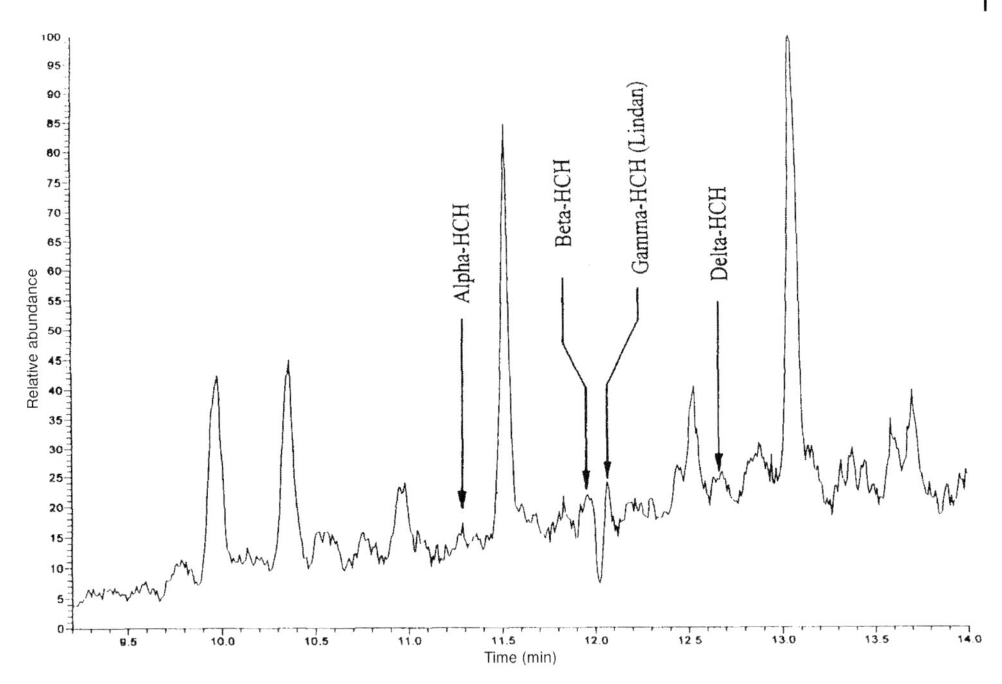

Fig. 4.92 Chromatogram of the HCH isomers in SIM mode m/z 183 (Courtesy D. Jahr 2000)

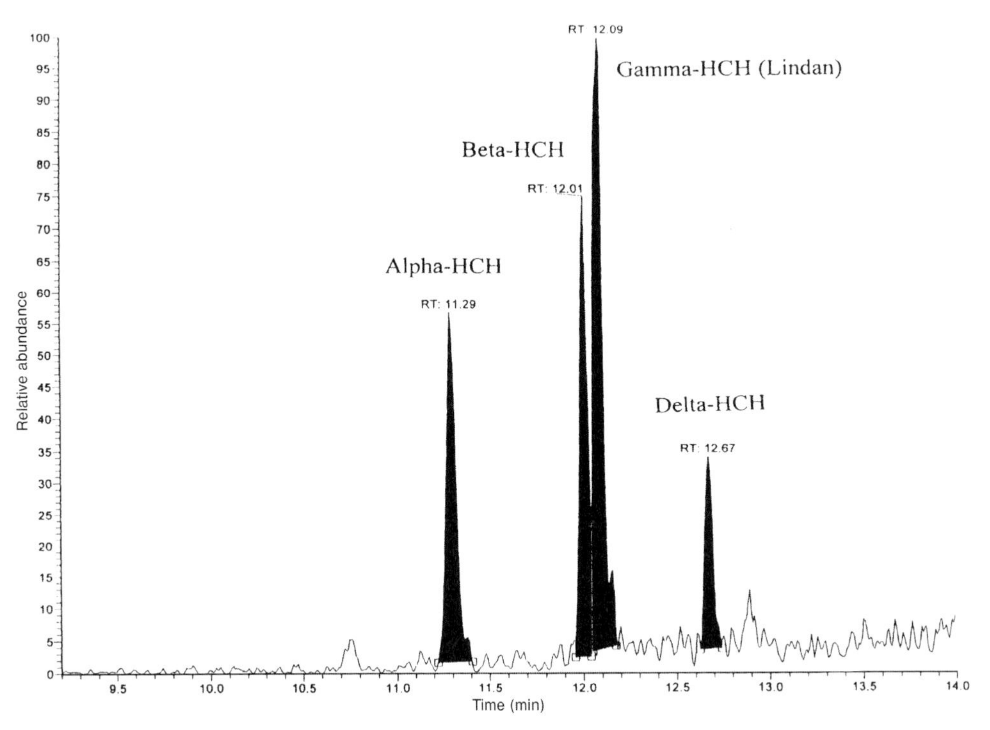

Fig.4-93 Chromatogram of the HCH isomers in MS/MS mode, transition m/z 219 to 183 (Courtesy D. Jahr 2000)

compounds e.g. pesticides, PAH, n-alkanes, nitroaromatic compounds, aromatic amines, phenols, nitro- and mosc-compounds, phthalates, and antioxidants) as well as the identification of unknown contaminants in ground and tap water. Identifications have been significantly facilitated by the high full scan sensitivity of ion trap MS/MS systems (Jahr 1998).

In the demonstrated example the SIM ion trace of a water sample analysis of the river Rhine is shown in Fig. 4.92 for the specific ion m/z 183. The retention time of α-, β-, γ- and δ-HCH is indicated showing a strong overlay of background ions for all HCH signals. In contrast to the selected ion monitoring acquisition, the MS/MS chromatogram monitoring the transition of m/z 219 to m/z 183 gives clear indication of all the four HCH isomers in the river water without interference from background, see Fig. 4.93.

The LOD for BHC/HCH was determined by using fortified river water samples as low as 2 ng/L in SIM detection. Using MS/MS technique the LOD could be lowered to 0.3 ng/L due to the increased specificity when detecting the product ions in real life samples (Jahr 2000).

4.22
Determination of Polar Aromatic Amines by SPME

Key words: nitroaromatic compounds, NAC, aromatic amines, *in situ* derivatisation, iodination, SPME, drinking water, EU regulation, full scan confirmation

Aromatic amines are known to form adducts with DNA and hence are suspected to cause a carcinogenic risk to humans. Six aromatic amines are classified as carcinogenic or probably carcinogenic by the International Agency for the Research of Cancer (IARC), and the source of these compounds is manifold from several industries. The major use is in the polyurethane production with possible emissions during production, use and disposal of such materials. A significant contribution to environmental distribution is the microbial reduction of nitroaromatic compounds (NAC) of which more than 70 compounds are mass produced with more than 1,000 mt per year. Up to today more than 30 aromatic amines have been identified as metabolites, also deriving from applied pesticides.

The described quantification method provides a highly sensitive and robust method for monitoring aromatic amines at the low ng/L level in water samples. The SPME method is particular useful for the quantification of aromatic amines in drinking water. In contrast to the time consuming SPE methodologies requiring high sample volumes, this SPME method utilizes a novel *in situ* derivatisation of the polar compounds in aqueous solution (Pan 1997) forming iodinated apolar derivatives. The automated SPME-GC/MS procedure allows the routine quantitation and full scan confirmation of a large number of water samples.

Analysis Conditions

Sample preparation:	*In situ* derivatisation: 10 mL water sample, acidified with 0.2 mL hydroiodic acid, shaken with 0.5 mL $NaNO_3$ solution (10g/L) for 20 min. Addition of 1 mL amidosulfonic acid solution (50 g/L) destroys the surplus of nitrite while shaking for 45 min, heating of the solution at 100 °C for 5 min, cool to RT. Addition of 0.25 mL $NaSO_3$ solution (sat.) destroys the surplus of iodine, pH is adjusted to approx. 8 with 0.25 mL K_2HPO_4 (0.25 mol/L) and 0.4 mL NaOH (5 mol/L). The solution is filled into 13 mL crimp top vials without any headspace, sealed with Al foil and can be stored at 4 °C until SPME analysis. Internal standard: aniline-d_5, 1 mg/L
SPME	65 µm PDMS/DVB, 30 min immersed into prepared sample vial using fibre vibration, "prep ahead" mode during GC runtime, automation with SPME Kit III (Varian)
Gas Chromatograph:	Varian 3800 (Varian, Darmstadt, Germany)
Injector:	Split/splitless, temperature programmable 1079 type (Varian), Siltek deactivated SPME liner, 0.8 mm i.d. (Restek, Bad Homburg, Germany)
SPME desorption	5 min total desorption time 50 °C initial temp. 250 °C final temp. after 1 min splitless Split 100:1 after 3 min
Carrier gas:	Helium 5.0, 2 mL/min constant flow mode
Column:	Stx-CLPesticides, 30 m × 0.25 mm × 0.25 µm (Restek) with 1.5 m retention gap and transfer capillary (Siltek deactivated fused silica, 0.25 mm, Restek)
Oven temp. program:	40 °C, 3 min 15 °C/min to 130 °C 30 °C/min to 160 °C 160 °C, 5 min 30 °C/min to 200 °C 20 °C/min to 250 °C
MS interface temp.:	280 °C
Mass Spectrometer:	Saturn 2000 ion trap mass spectrometer (Varian, Darmstadt, Germany)

Ionisation:	EI, 70 eV, emission current 40 µA, automatic gain control
Ion trap/manifold temp.:	200 °C/55 °C
Scan mode:	Full scan
Mass range:	60–450 u
Cycle time:	0.35 s/scan

Results

SPME has been successfully applied only for the extraction of non polar components from aqueous samples. The *in situ* derivatisation step lowers the polarity of the polar amines in an easy "one pot" reaction with extraction efficiencies of the aromatic iodine derivatives of better than 95%, with the only exception of 2A4,6DNT with only 77%. Spike solutions have been prepared with water and may not be diluted in alcohols, as the alcohol component will react as nucleophile.

SPME allows the complete transfer of the on the fibre accumulated analyte amount to the GC/MS analysis, also small sample sizes can be analysed with the given performance. The thermal stress to the fibre has been minimized by the temperature profile during injection resulting in the repeated use of one fibre bundle of up to 80 times.

For optimisation of the chromatographic conditions a standard sample diluted in ethylacetate was applied using identical conditions. Figure 4.95 shows the separation of a standard mixture with almost baseline separation of all components. The MS was operated in full scan mode to allow the additional detection of non target analytes.

An overview of the components included in the described method is given in Table 4.20. Listed with the names and used abbreviations of the compounds are the chromatographic retention times, molecular weight of the iodine derivative and its characteristic mass peaks. The base peaks used for compound selective quantitation has been typically the molecular ion. The observed $(M\text{-}17)^+$ base peaks are due to the ortho-effect fragmentation (see also Section 3.2.6.9). With the iodinated derivatives the quantitation mass has been shifted by 111 u per amino group to the higher mass range. This effect in particular allows the very sensitive detection of the derivatives with high S/N values as unspecific background typically appears in the lower mass range.

The calibration range has been selected in accordance with the European regulation for toxic organic pollutants in drinking water, with the regulated level of 0.1 µg/L for all target compounds. The calibration using aniline-d_5 as internal standard is linear from 0.05 µg/L over 2 orders of magnitude. A saturation of the SPME fibre has not been observed in this range. Limits of detection (LOD) have been calculated between 2 to 13 ng/L except for 2A4,6DNT, 4A2,6DNT and 2,4DA6NT with 27 to 38 ng/L, meeting excellently the EU regulatory levels.

Real water samples have been analysed from different sources and demonstrated the applicability of the method for waste water, groundwater, surface water and drinking water. Fig. 4.96 shows the analysis of a contaminated ground water sample from the area of a former ammunition plant. Repeated analyses gave RSDs in range of 3 to 10% (Zimmermann 2004).

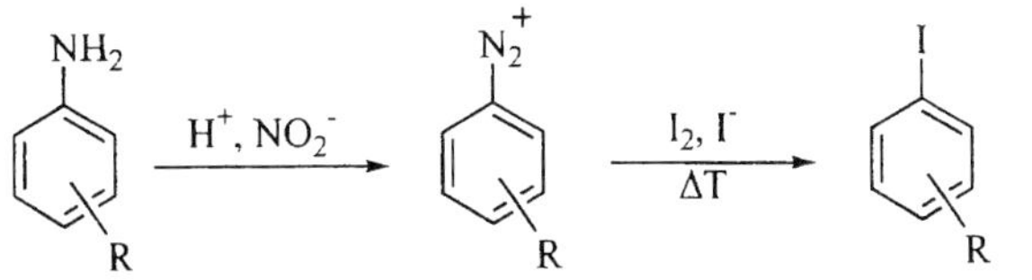

$$R\text{-}C_6H_4\text{-}NH_2 \xrightarrow{H^+,\ NO_2^-} R\text{-}C_6H_4\text{-}N_2^+ \xrightarrow[\Delta T]{I_2,\ I^-} R\text{-}C_6H_4\text{-}I$$

Fig. 4.94 Derivatisation reaction with diazotation followed by iodination

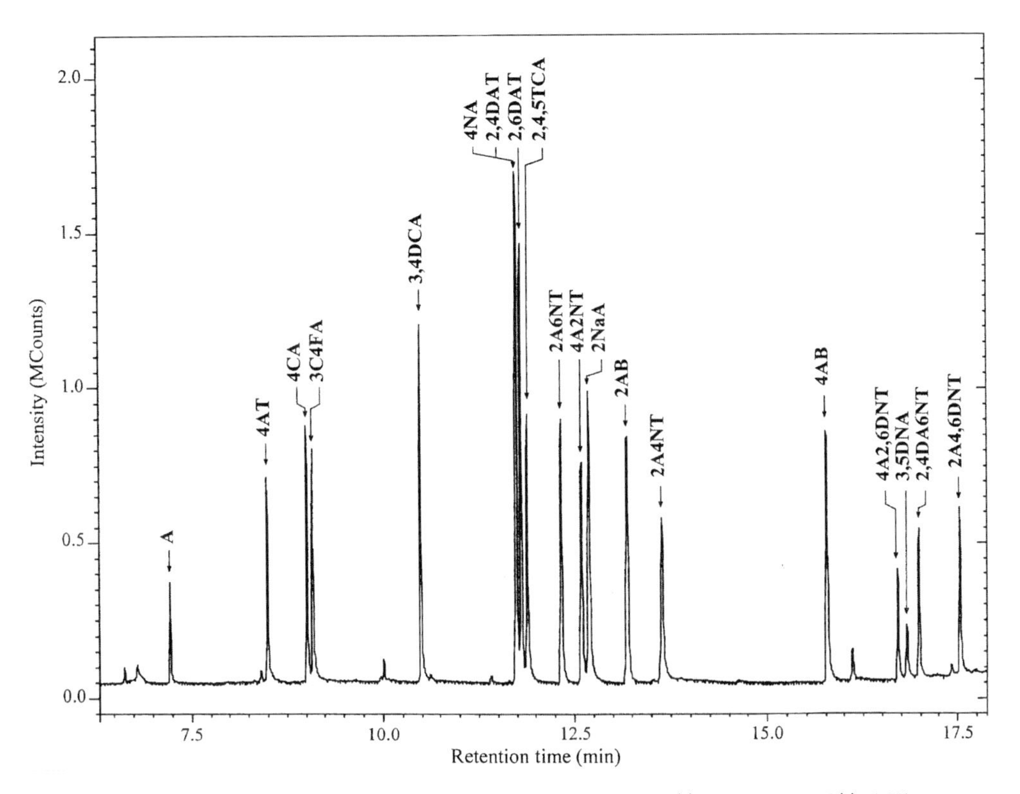

Fig. 4.95 Standard chromatogram of the derivatised aromatic amines (abbreviations see Table 4.21) (Zimmermann 2004; reprinted with permission of Analytical Chemistry, Copyright 2004 American Chemical Society)

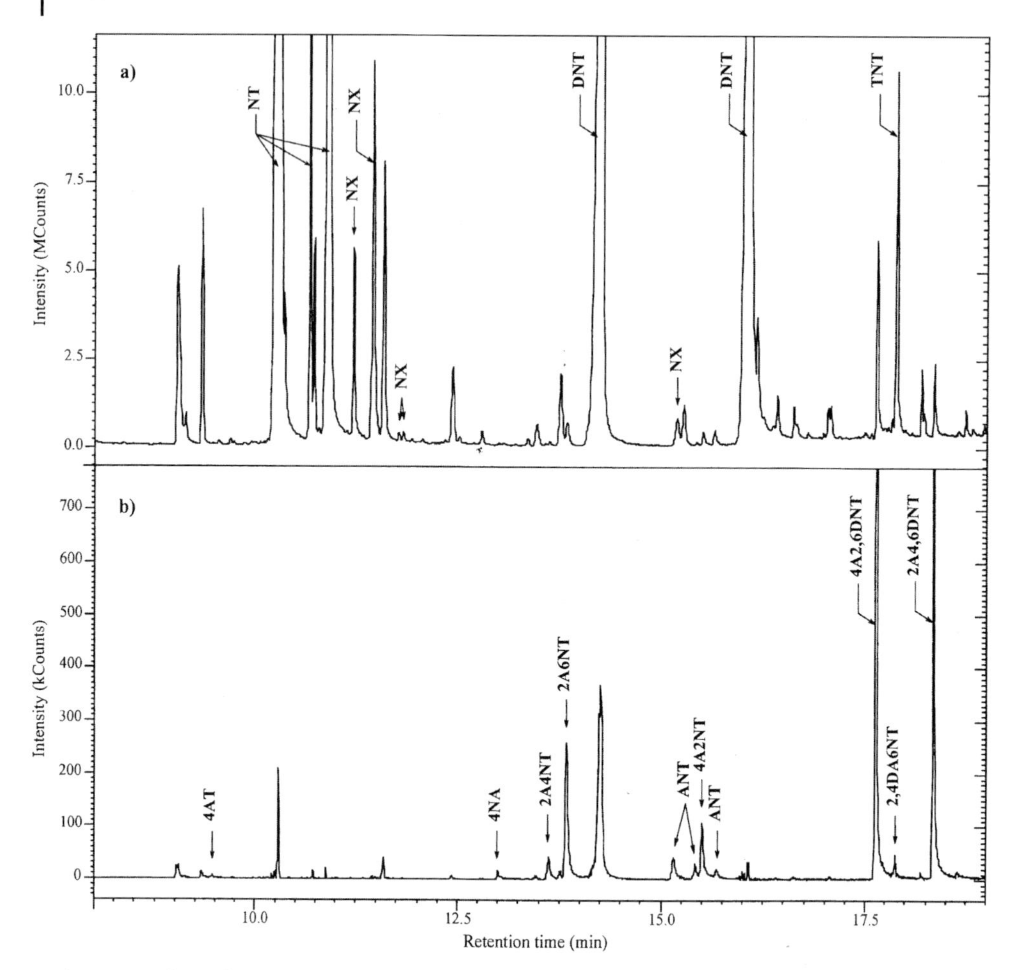

Fig. 4.96 Analysis of a contaminated ground water from the area of a former ammunition plant (Zimmermann 2004; reprinted with permission of Analytical Chemistry, Copyright 2004 American Chemical Society)

Table 4.21 List of aromatic amine compounds[a)] (Zimmermann 2004; reprinted with permission of Analytical Chemistry, Copyright 2004 American Chemical Society)

Compound	Abbrev.	Derivative	t_R (min)	M	Detected ions[a)] (m/z)	LOD (ng/L)	RSD % at 0.5 µg/L
aniline	A	Iodobenzene	8.36	204	**204**, 77, 127	4	6.0
4-aminotoluene	4AT	4-iodotoluene	9.54	218	**218**, 91, 65	12	4.3
4-chloroaniline	4CA	1-chloro-4-iodobenzene	9.98	238	**238**, 111, 75	2	3.8
3,4-dichloroaniline	3,4DCA	1,2-dichloro-4-iodobenzene	11.55	272	**272**, 145, 109	3	6.7
2,4,5-trichloroaniline	2,4,5TCA	1,2,4-trichloro-5-iodobenzene	13.41	306	**306**, 179, 143	6	11
3-chloro-4-fluoroaniline	3C4FA	2-chloro-4-iodo-1-fluorobenzene	10.04	256	**256**, 129, 109	3	4.3
2,4-diaminotoluene	2,4DAT	2,4-diiodotoluene	13.22	344	**344**, 217, 90	13	11
2,6-diaminotoluene	2,6DAT	2,6-diiodotoluene	13.31	344	**344**, 217, 90	7	9.2
2-naphthylamine	2NaA	2-iodonaphthaline	14.59	254	**254**, 127, 74	11	7.4
2-aminobiphenyl	2ABP	2-iodobiphenyl	15.20	280	**280**, 152, 127	5	14
4-aminobiphenyl	4ABP	4-iodobiphenyl	16.98	280	**280**, 152, 127	9	11
4-nitroaniline	4NA	1-iodo-4-nitrobenzene	13.22	249	**249**, 219, 203	5	20
2-amino-4-nitrotoluene	2A4NT	2-iodo-4-nitrotoluene	15.69	263	**263**, 90, 105	8	16
2-amino-6-nitrotoluene	2A6NT	2-iodo-6-nitrotoluene	14.05	263	**246**, 89, 119	2	14
4-amino-2-nitrotoluene	4A2NT	4-iodo-2-nitrotoluene	14.44	263	**246**, 89, 119	3	16
2-amino-4,6-dinitrotoluene	2A4,6DNT	2-iodo-4,6 dinitrotoluene	18.49	308	**291**, 164, 89	38	20
4-amino-2,6-dinitrotoluene	4A2,6DNT	4-iodo-2,6-dinitrotoluene	17.76	308	**291**, 89, 63	27	13
2,4-diamino-6-nitrotoluene	2,4DA6NT	2,4-diiodo-6-nitrotoluene	18.00	389	**372**, 344, 216	30	16
aniline-d_5 (ISTD)	A-d_5	iodobenzene-d_5	8.35	209	**209**, 82, 127	–	–

a) Quantitation mass in bold

4.23
Congener Specific Isotope Analysis of Technical PCB Mixtures

Key words: PCB, congener-specific δ^{13}C values, source, fate, environment, 2DGC, MCSS, IRMS, CSIA

Stable carbon isotopic analysis is increasingly applied for the understanding of the sources and fate of anthropogenic organic contaminants in the environment. In particular, compound-specific carbon isotope analysis (CSIA) of complex mixtures provides a powerful analytical tool to trace the origin and conversion of organic compounds in the environment. Although traditional approaches such as fingerprinting, which involves matching of the isomer profiles in samples with that in technical preparations, have been used to determine sources of anthropogenic chemicals e. g. PCBs, dioxins (Horii 2005, Horii 2008). GC-IRMS complements the existing methods to understand the sources and environmental destiny of anthropogenic chemicals.

In this application, 2DGC-C-IRMS technique is used for the isotopic analysis of carbon, and to determine congener-specific δ^{13}C values of PCB and PCN congeners in several technical mixtures produced in the United States, Japan, Germany, France, former USSR, Poland, and former Czechoslovakia (see Fig. 4.97).

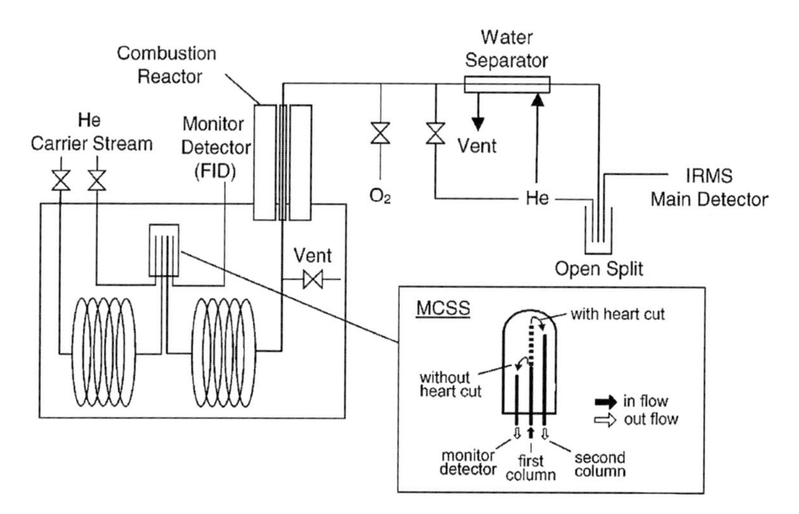

Fig. 4.100 Schematics of the 2DGC-C-IRMS instrument setup (reprinted with permission from Environ. Sci. Technology, Copyright 2005 American Chemical Society)

Analysis Conditions

Gas chromatograph:	TRACE GC™ 2000 with monitor detector FID (Thermo Fisher, Milan, Italy)
Injector:	Split/splitless, 260 °C, 1 min splitless, 1 μL injection volume

Carrier gas:	Helium
Column:	DB5-MS, 30 m × 0.25 mm × 0.25 μm (J&W Scientific, Foster City CA, USA), and as second dimension column Rtx-200, 15 m × 0.32 mm × 0.25 μm (Restek Corp., Bellefonte PA, USA)
Column switching:	MCSS, moving capillary switching system (CE Instruments, Mainz, Germany)
Oven temp. program:	70 °C, 1 min 15 °C/min to 180 °C 2 °C/min to 260 °C, 5 min
Oxidation reactor:	940 °C, reoxidation daily for 20 min by oxygen backflow
Mass Spectrometer:	MAT 252, Thermo Fisher, Bremen, Germany
Ionisation:	EI, 70 eV
Scan Mode:	Continuous monitoring of m/z 44, 45, 46

Results

Two-dimensional gas chromatography separation by using heart-cutting (2DGC-C-IRMS by MCSS) enabled significant improvement in the resolution and sensitivity of individual PCB isomers by at least an order of magnitude better than the traditional one-dimensional GC-IRMS method.

Only selected target compounds, which have been identified using the FID monitor detector, have been transferred onto a second column via heart cutting for online conversion with the oxidation furnace (Fig. 4.98). The background signal of m/z 44 was significantly reduced to 10 mV even at high temperature (260 °C) due to minimized column bleed. The estimated sensitivity of carbon using 2DGC-C-IRMS was less than 7 ng, which corresponds to 10–20 ng of individual PCB congeners injected. Thirty-one PCB congeners were selected in 18 technical PCB preparations for the determination of $\delta^{13}C$ values of individual PCB congeners (Table 4.22) (Horii 2005).

It has been observed that lower chlorinated PCB congeners showed higher $\delta^{13}C$ values in each technical PCB mixture, which might be influenced by isomer specific isotopic partitioning. Geographical differences in the $\delta^{13}C$ values among PCB preparations, particularly those of Delors, Sovol, Trichlorodiphenyl, and Chlorofen, indicate possible differences in the raw materials used during the production processes (Fig. 4.99).

The $\delta^{13}C$ values determined for PCB and also for separately investigated PCN congeners were similar to those from petroleum and terrestrial plants. However, these values are apparently different from those of carbonates and marine plants. It could be proven that GC-IRMS by accurate instrumental methods such as 2DGC-C-IRMS provides the necessary analytical tools to reveal isotopic partitioning by unknown environmental and geochemical processes.

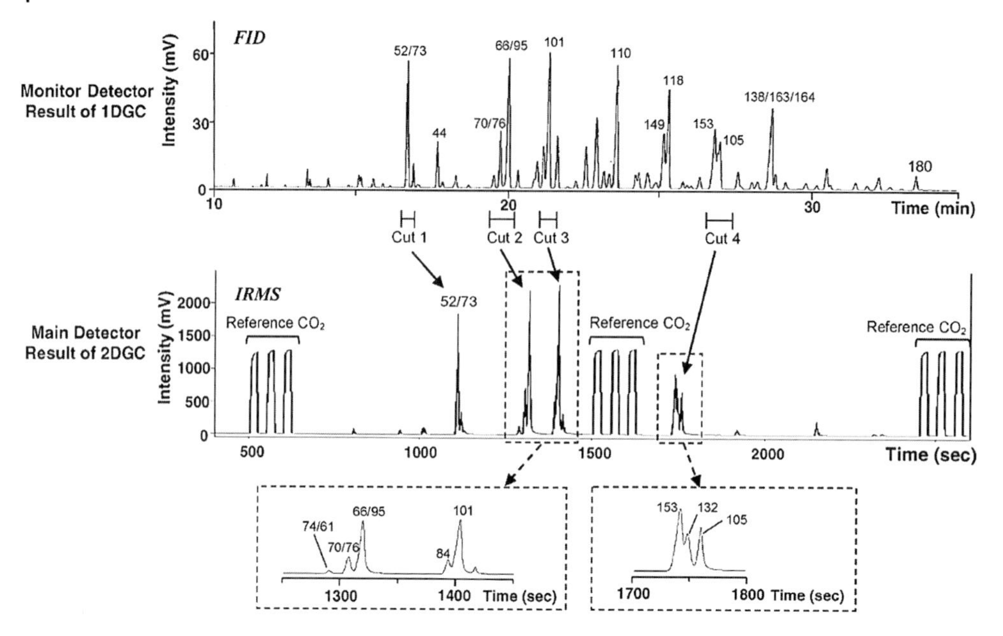

Fig. 4.98 Chromatograms of the monitor detector and IRMS for amixture of Kanechlor 500. The CO_2 reference gas injections occur at the beginning, between cuts and at the end of the separation (Horii 2005) (reprinted with permission from Environ. Sci. Technology, Copyright 2005 American Chemical Society)

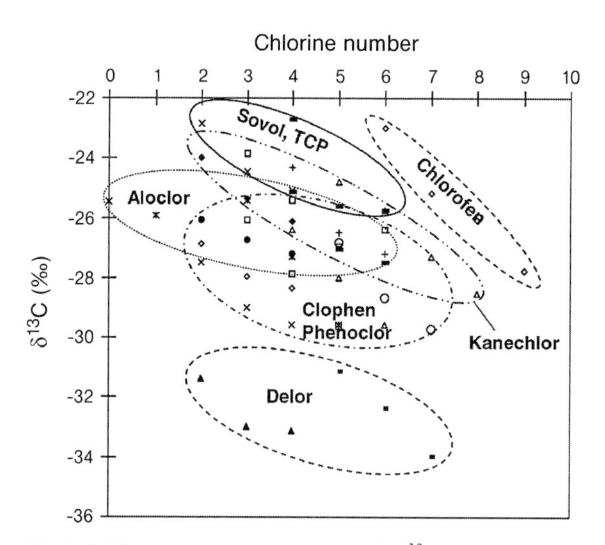

Fig. 4.99 PCB production characteristics by $\delta^{13}C$ value plot ag. chlorine number of the selected congeners from Table 4.22 (Horii 2005) (reprinted with permission from Environ. Sci. Technology, Copyright 2005 American Chemical Society)

Table 4.22 $\delta^{13}C$ values of selected PCB congeners from technical mixtures (extracted, for the full table refer to reference) (Horii 2005) (reprinted with permission from Environ. Sci. Technology, Copyright 2005 American Chemical Society)

	Country	United States			Germany			Czechoslovakia		Soviet Union	
	Year of production	1930–1975			1930–1982			1959–1984		1939–1993	
	Generic product name	Aroclor			Clophen			Delor			
	Cl content by weight [%]	21	40-42	52–54	40–42	48	52-54	NA		41	53
	Product	A 1221	A 1242	A 1254	A 30	A 40	A 50	D 103	D 106	Sovol	TCP
Congener	Structure Cl-PCB										
		−25.2									
1	2-	−27.2									
3	4-	−24.7									
10/4	2,6-/2,2′-		−25.9		−26.5			−31.3			−22.7
8/5	2,4′-/2,3-		−26.2		−27.3			−31.4			−23.1
18	2,2′,5-		−26.9		−28.2	−25.5		−33.2			−23.9
16/32	2,2′,3-/2,4′,6-		−26.7		−28.0	−25.8		−33.0			−24.7
31/28	2,4′,5-/2,4,4′-		−26.7		−27.7	−26.9		−32.8			−24.8
33/20/53	2,3′,4′-		−26.3		−28.1	−26.7		−33.1			−24.6
52/73	2,2′,5,5′-/2,3′,5′,6-		−27.7	−26.3	−28.8	−29.2	−27.1	−34.0		−23.4	−25.1
43/49	2,2′,3,5-/2,2′,4,5′-		−27.0	−25.0	−28.1	−28.1	−26.0	−32.8			−24.6
47/75/48	2,2′,4,4′-/2,4,4′,6-/2,2′,4,5-		−27.2			−26.6					
44	2,2′,3,5′-		−27.3	−25.3	−29.0	−28.6	−26.6	−33.9			−24.6
59/42	2,3,3′,6-/2,2′,3,4′-		−27.6		−27.2	−28.5		−31.7			−25.9
74/61	2,4,4′,5-/2,3,4,5-		−26.7			−26.5					
70/76	2,3′,4′,5-/2,3′,4′,5′-		−26.7	−24.0	−28.5	−27.6	−26.1	−33.1		−22.0	
66/95	2,3′,4,4′-/2,2′,3,5′,6-		−27.0	−26.5	−28.4	−28.4	−27.2	−33.1	−31.6	−23.6	−25.2

Table 4.22 (continued)

	Country	United States			Germany			Czechoslovakia		Soviet Union	
	Year of production	1930–1975			1930–1982			1959–1984		1939–1993	
	Generic product name	Aroclor			Clophen			Delor			
	Cl content by weight [%]	21	40-42	52–54	40–42	48	52-54	NA		41	53
	Product	A 1221	A 1242	A 1254	A 30	A 40	A 50	D 103	D 106	Sovol	TCP
89/101/90	2,2′,3,4,6′-/2,2′,4,5,5′-/2,2′,3,4′,5-			−27.4		−31.2	−28.3		−31.3	−27.1	
110	2,3,3′,4′,6-			−27.3		−29.9	−28.1		−31.2	−25.3	
149/139	2,2′,3,4′,5′,6-/2,2′,3,4,4′,6-			−27.8			−29.7		−32.6		
118/106	2,3′,4,4′,5-/2,3,3′,4,5-			−26.4		−27.8	−27.4		−30.9	−25.1	
153/132/168	2,2′,4,4′,5,5′-/2,2′,3,3′,4,6′-/2,3′,4,4′,5′,6-			−27.2			−28.4		−32.0	−25.1	
105	2,3,3′,4,4′-			−27.1			−28.2			−25.0	
164/163/138	2,3,3′,4′,5′,6-/2,3,3′,4′,5,6-/2,2′,3,4,4′,5′-			−27.5			−30.5		−32.5	−26.4	
182/187	2,2′,3,4,4′,5,6′-/2,2′,3,4′,5,5′,6-								−34.2		
174	2,2′,3,3′,4,5,6′-								−34.2		
180	2,2′,3,4,4′,5,5′-								−34.4		
170/190	2,2′,3,3′,4,4′,5-/2,3,3′,4,4′,5,6-								−33.2		
199	2,2′,3,3′,4,5,5′,6′-										
203/196	2,2′,3,4,4′,5,5′,6-/2,2′,3,3′,4,4′,5,6′-										
194	2,2′,3,3′,4,4′,5,5′-										
206	2,2′,3,3′,4,4′,5,5′,6-										
	Mean	−25.7	−26.9	−26.5	−28.0	−27.8	−27.8	−32.8	−32.5	−24.8	−24.5
	Maximum	−24.7	−25.9	−24.0	−26.5	−25.5	−26.0	−31.3	−30.9	−22.0	−22.7
	Minimum	−27.2	−27.7	−27.8	−29.0	−31.2	−30.5	−34.0	−34.4	−27.1	−25.9

4.24 Polychlorinated Biphenyls in Indoor Air

Key words: PCBs, sealants, air sampling, thermodesorption, Tenax, quantitation

Polychlorinated biphenyls (PCBs) were used in the period 1960 to 1980 as flameproofing plasticisers in sealants. The proportion used in polysulfide sealants was up to 30%.

Guidelines for the use of PCBs indoors were laid down by national authorities. For precautionary reasons these bodies suggested that PCB contamination below 300 ng/m^3 in the air could be tolerated. Within the range 300 to 3000 ng/m^3 it was recommended that the source of the contamination should be traced and removed. At concentrations above 3000 ng/m^3 measures should be introduced without question (e. g. a ban on the use of these rooms).

To sample PCBs from indoor air high volume samplers and polyurethane foam cartridges or porous polymers are generally used, which are later cleaned up in the laboratory according to the recommendations of national authorities. For the PCB determination described here a procedure was worked out which allows air sampling to be carried out on adsorption materials using an automatic thermodesorber without further sample preparation. An ion trap GC/MS was used as the detector.

The determination of all PCB congeners is not feasible. Because of this an overall determination according to DIN 51527 part 1 (polychlorinated biphenyls in waste oil) was recommended. According to DIN 51527 the calculation is based on the six congeners 28, 52, 101, 138, 153 and 180. The national authorities chose the congeners 28, 52, 101 and 153 for the determination of PCBs in air. To determine the overall concentration, the concentrations of these congeners are added and multiplied by the factor 6.

For air sampling the steel tubes of the thermodesorber are used packed with ca. 50 mg Tenax GR (35–16 mesh). The adsorbent filling is fixed with silanised glass wool and metal springs. In order to be able to determine the concentrations in the zero range of 50 ng/m^3, a sample volume of more than 200 L is necessary. Sampling thus takes ca. 3 h at a flow rate of 2 L/min. Du Pont P4000 personal air samplers were used.

Analysis Conditions

Sample collection:	Active over 3 h, 2 L/min	
	Adsorbent Tenax GR	
Thermodesorber:	Perkin Elmer ATD 400	
	Desorption temp.:	350 °C
	Desorption time:	30 min
	Cold trap filling:	20 mg Tenax GR (backflush)
	Cold trap temp.:	Low: 30 °C
		High: 350 °C, 5 min
	Desorption flow:	80 mL/min
	Input split:	none
	Output split:	30 mL/min
	Transfer line:	225 °C
	Valve temp.:	225 °C

Gas chromatograph:	Perkin Elmer model 8700	
Column:	SGE HT-5, 30 m × 0.23 mm × 0.25 μm	
Carrier gas:	Helium, 125 kPa	
Program:	Start:	60 °C, 1 min
	Program 1:	8 °C/min to 180 °C
	Program 2:	12 °C/min to 300 °C
	Final temp.:	300 °C, 10 min
Mass spectrometer:	Finnigan ITD 800, direct coupling	
	Scan mode:	EI, full scan, 100–400 u
	Scan rate:	0.5 s/scan
	Transfer line:	300 °C

Results

Figures 4.100 and 4.101 show a total ion current chromatogram and an individual mass chromatogram resulting from an air sample taken directly next to a sealant. The total calculated PCB concentration was ca. 7000 ng/m^3. An absolute calibration of 5 ng corresponds to a total PCB concentration of 600 ng/m^3 for 200 L of air. Figure 4.102 shows the calibration function for the PCB congener 52 from 1 to 50 ng per sample tube. The desorption of the PCB congeners from Tenax GR is readily reproducible with high transfer rates for the four congeners used for quantitation (Fig. 4.103). Using multiple desorption the transfer rates of the congeners 153, 138 and 180 can be increased further. However, hexa- and heptachloro-PCBs are insignificant because of their low vapour pressures. The thermodesorption system used has been shown to be extremely powerful and inert for PCB analysis so that even the use of PCB 209 (decachlorobiphenyl) as internal standard is possible (Tschickard 1998).

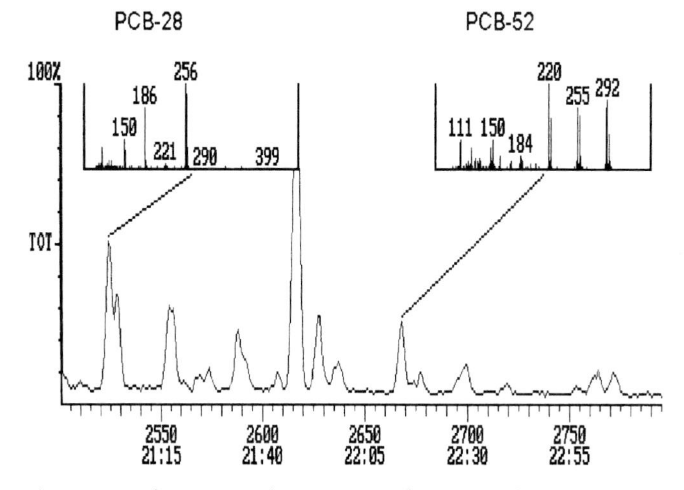

Fig. 4.100 Total ion current chromatogram of an air sample near the sealant

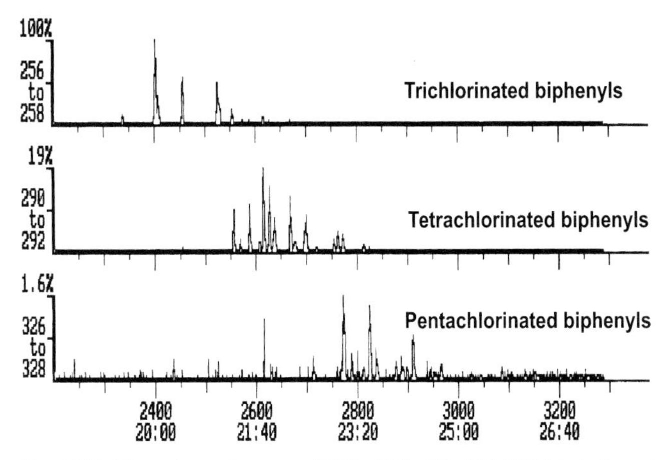

Fig. 4.101 Mass chromatograms of PCBs (3Cl-, 4Cl-, 5Cl-PCBs) in indoor air

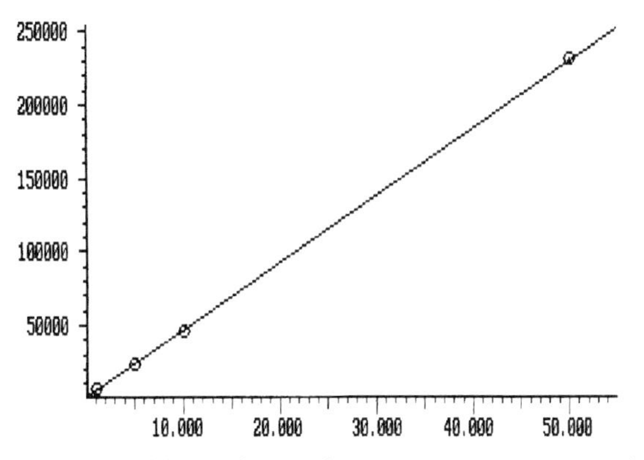

Fig. 4.102 Calibration function for PCB-52 (1 ng to 50 ng per tube)

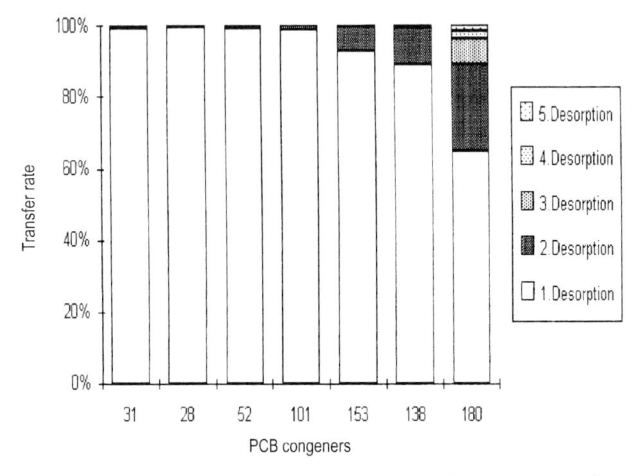

Fig. 4.103 Multiple desorption of 50 ng samples of PCB congeners from Tenax-GR

4.25
Confirmation Analysis of Dioxins and Dioxin-like PCBs

Key words: dioxin, PCDD, PCDF, dioxin-like PCB, TEF, TEQ, WHO, screening, confirmation, HRGC/HRMS, accurate mass, MID

Over the past 30 years dioxin TEQ levels and body burden levels in the general population have been on the decline and continue to decrease (Lorber 2002). But, more than 90% of human exposure to dioxins and dioxin-like substances is through food and hence lead to continuous control (USEPA 2000). With increasingly lower dioxin levels in food, feed, and tissues, more demanding limits of detection, selectivity, sensitivity and QC checks are required to trace their presence at these further decreasing levels.

Recent studies document the declines in exposure to and body burdens of dioxins in the U.S. The most current data available on body burdens in the general population have been provided by Patterson. The data and analyses highlight the importance of taking the age group into account. While the mean dioxin TEQ increases as age increases (Fig. 4.104) the sharpest increase is observed among the age group 60+ (Aylward 2002, Lorber 2002, Hays

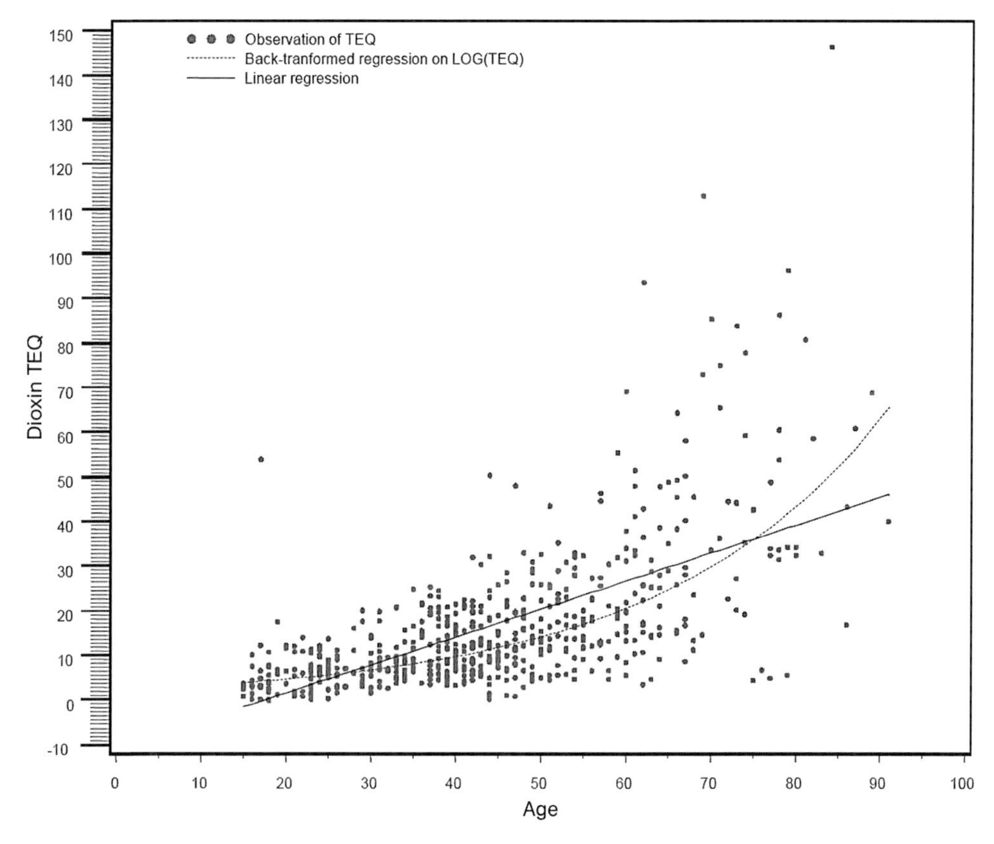

Fig. 4.104 Dioxin body burden in TEQ vs. age (Patterson 2004)

2003, Patterson 2004). As the dioxin and dl-PCB concentrations are related to the low fat level in blood, significantly improved sensitivities of the analytical instrumentation applied are required to further establish an efficient control at further decreasing levels, especially in the younger population. With children and toddlers the small available blood sample sizes becomes the limiting factor.

The described isotope dilution method for the quantification of polychlorinated dioxins, furans and dioxin-like PCBs (dl-PCBs) follows the EPA 1613 b method. The multiple ion detection scheme (MID) on the accurate target masses typically uses isotope ratio qualifiers besides the specific retention time for all native dioxin/furan congeners, as well as for their specific ^{13}C labelled internal standards, one quantification mass and one ratio mass. The analytical setup for the high resolution GC/MS is given with the MID descriptor, as shown in Table 4.23 below. A typical MID setup for the data acquisition of the individual groups of chlorinated compounds is given in Fig. 4.105. For a list of exact masses of PCDD/F and PCB including the ^{13}C labelled standards see Table 3.5. The mass spectrometer is operated at a mass resolution of equal or better than 10,000 as required by the EPA 1613 b method. The effective resolution achieved during the analysis of real-life samples has to be constantly monitored on the reference masses and documented in the data file for each MID window.

Analysis Conditions

Gas chromatograph:	TRACE GC Ultra (Thermo Fisher Scientific, Milan, Italy)
Injector:	Split/splitless, 260 °C, 2 µL injection volume, 1.5 min splitless, split flow 50 mL/min
Carrier gas:	Helium, 0.8 mL/min, constant flow
Column:	TRACE TR-5MS, 60 m × 0.25 mm i.d. × 0.1 µm film
Oven temp. program:	120 °C, 3 min 19 °C/min to 210 3 °C/min to 275 °C, 12 min 20 °C/min to 300 °C, 3 min
MS interface temp.:	280 °C
Mass spectrometer:	DFS High Resolution GC/MS (Thermo Fisher Scientific, Bremen, Germany)
Ionisation:	EI, 48 eV
Source temperature:	270 °C
Scan mode:	Multiple ion detection mode (MID) with lock-and-cali mass technique, FC 43 as reference compound
Resolution:	10,000 (10% valley definition)
Mass range:	MID descriptor see Table 4.23
Cycle time:	See Table 4.23

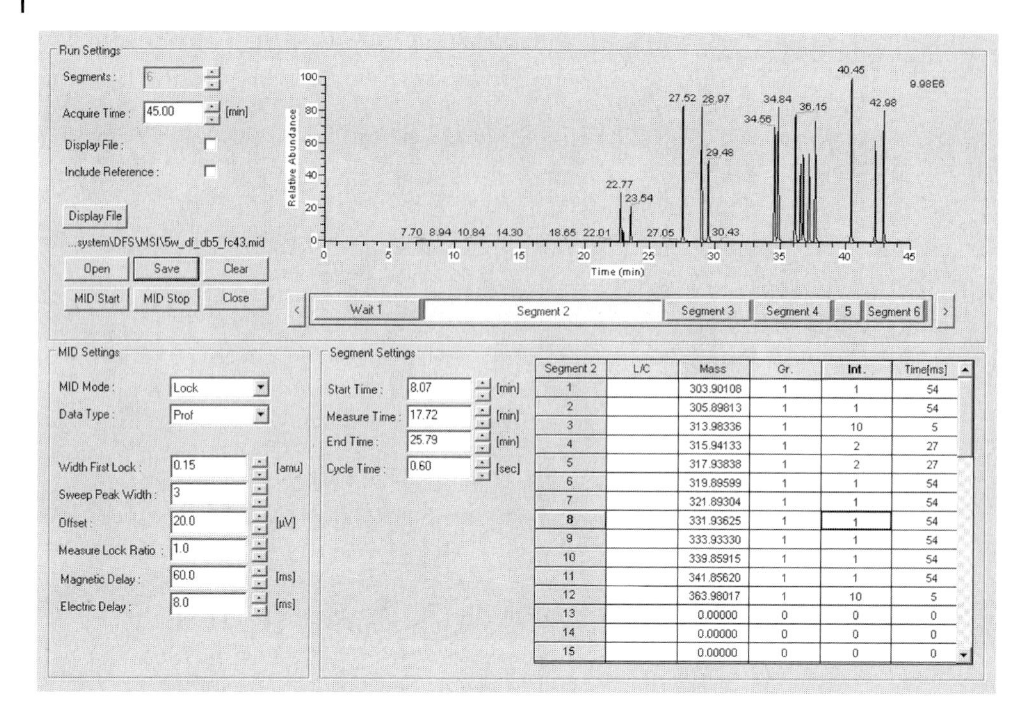

Fig. 4.105 Typical MID setup with the individual acquisition window (segment) for each eluting congener group of different chlorination degree

Results

A typical separation using the GC parameters from Table 4.23 of an EPA 1613 CS1 dioxin standard at 50 fg/µL of TCDD and TCDF is shown in Fig. 4.106. These GC parameters were also employed for the analysis of blood samples. For a quantitative analysis with the calculation of the TEQ values a summary of the WHO 1998 and the current WHO 2005 TEF values is given in Table 4.24.

The confirmation ratios (relative areas of quantification and ratio masses) for all dioxins/furans in repeated injections of a 17 fg/µL were evaluated for a blood pool sample (see Fig. 4.107). All 2,3,7,8-TCDD results of a sample measurement series over several days of the blood sample gave excellent results within the required ± 15% window at the lowest detection levels and provided the confirmation ion ratios in compliance with EPA 1613 requirements.

Table 4.23 Accurate mass MID set up for PCDD and PCDF analysis in MID lock-and-cali mode (width first lock: 0.3 u, voltage settling time delays: 10 ms)

MID window no. (time window)	Reference masses (FC43) m/z lock mass (L), cali mass (C)	Target masses m/z native (n) ^{13}C internal standard (is)	MID cycle time (intensity, dwell time ms)
1. Tetra-PCDD/F (9.00–19.93 min)	313.98336 (L), 363.98017 (C)	303.90088(n), 305.89813(n), 315.94133(is), 317.93838(is), 319.89651(n), 321.89371(n), 331.93680(is), 333.93381(is)	0.75 s (L/C: 30, 4 ms; n: 1, 137 ms; is: 7, 19 ms)
2. Penta-PCDD/F (19.93–23.52 min)	313.98336 (L), 363.98017 (C)	339.85889(n), 341.85620(n), 351.89941(is), 353.85702(n!), 353.89646(is!), 355.85400(n), 365.89728(is), 367.89433(is)	0.80 s (L/C: 30, 4 ms; n: 1, 147 ms; is: 7, 21 ms)
3. Hexa-PCDD/F (23.52 – 26.98 min)	375.97974 (L), 413.97698 (C)	371.82300(n), 373.82007(n), 385.86044(is), 387.85749(is), 389.81494(n), 391.81215(n), 401.85535(is), 403.85240(is)	0.80 s (L/C: 30, 4 ms; n: 1, 147 ms; is: 7, 21 ms)
4. Hepta-PCDD/F (26.98–32.06 min)	413.97698 (L), 463.97378 (C)	407.78101(n), 409.77826(n), 419.82147(is), 421.81852(is), 423.77588(n), 425.77317(n), 435.81638(is), 437.81343(is)	0.90 s (L/C: 35, 4 ms; n: 1, 169 ms; is: 7, 24 ms)
5. Octa-PCDD/F (32.06–36.00 min)	425.97681 (L), 463.97378 (C)	441.74219(n), 443.73929(n), (453.78250(is)), (455.77955(is)), 457.73706(n), 459.73420(n), 469.77741(is), 471.77446(is)	0.95 s (L/C: 40, 4 ms; n: 1, 183 ms; is: 7, 22 ms)

Table 4.24 Summary of WHO 1998 and WHO 2005 TEF values

Compound		WHO 1998 TEF	WHO 2005 TEF[a)]
Chlorinated dibenzo-p-dioxins			
2,3,7,8-TCDD		1	1
1,2,3,7,8-PeCDD		1	1
1,2,3,4,7,8-HxCDD		0.1	0.1
1,2,3,6,7,8-HxCDD		0.1	0.1
1,2,3,7,8,9-HxCDD		0.1	0.1
1,2,3,4,6,7,8-HpCDD		0.01	0.01
OCDD		0.0001	**0.0003**
Chlorinated dibenzofurans			
2,3,7,8-TCDF		0.1	0.1
1,2,3,7,8-PeCDF		0.05	**0.03**
2,3,4,7,8-PeCDF		0.5	**0.3**
1,2,3,4,7,8-HxCDF		0.1	0.1
1,2,3,6,7,8-HxCDF		0.1	0.1
1,2,3,7,8,9-HxCDF		0.1	0.1
2,3,4,6,7,8-HxCDF		0.1	0.1
1,2,3,4,6,7,8-HpCDF		0.01	0.01
1,2,3,6,7,8,9-HpCDF		0.01	0.01
OCDF		0.0001	**0.0003**
Non-ortho substituted PCBs			
3,3′,4,4′-tetraCB	(PCB 77)	0.0001	0.0001
3,4,4′,5-tetraCB	(PCB 81)	0.0001	**0.0003**
3,3′,4,4′,5-pentaCB	(PCB 126)	0.1	0.1
3,3′,4,4′,5,5′-hexaCB	(PCB 169)	0.01	**0.03**
Mono-*ortho substituted PCBs*			
2,3,3′,4,4′-pentaCB	(PCB 105)	0.0001	**0.00003**
2,3,4,4′,5-pentaCB	(PCB 114)	0.0005	**0.00003**
2,3′,4,4′,5-pentaCB	(PCB 118)	0.0001	**0.00003**
2′,3,4,4′,5-pentaCB	(PCB 123)	0.0001	**0.00003**
2,3,3′,4,4′,5-hexaCB	(PCB 156)	0.0005	**0.00003**
2,3,3′,4,4′,5′-hexaCB	(PCB 157)	0.0005	**0.00003**
2,3′,4,4′,5,5′-hexaCB	(PCB 167)	0.00001	**0.00003**
2,3,3′,4,4′,5,5′-heptaCB	(PCB 189)	0.0001	**0.00003**

a) Bold values indicate a change in TEF value.
Reference: Van den Berg, M., et al., The 2005 World Health Organization Re-evaluation of Human and Mammalian Toxic Equivalency Factors for Dioxins and Dioxin-like Compounds, 2005 WHO Re-evaluation of TEFs, Tox. Sci. 93(2) (2006), 223–241.

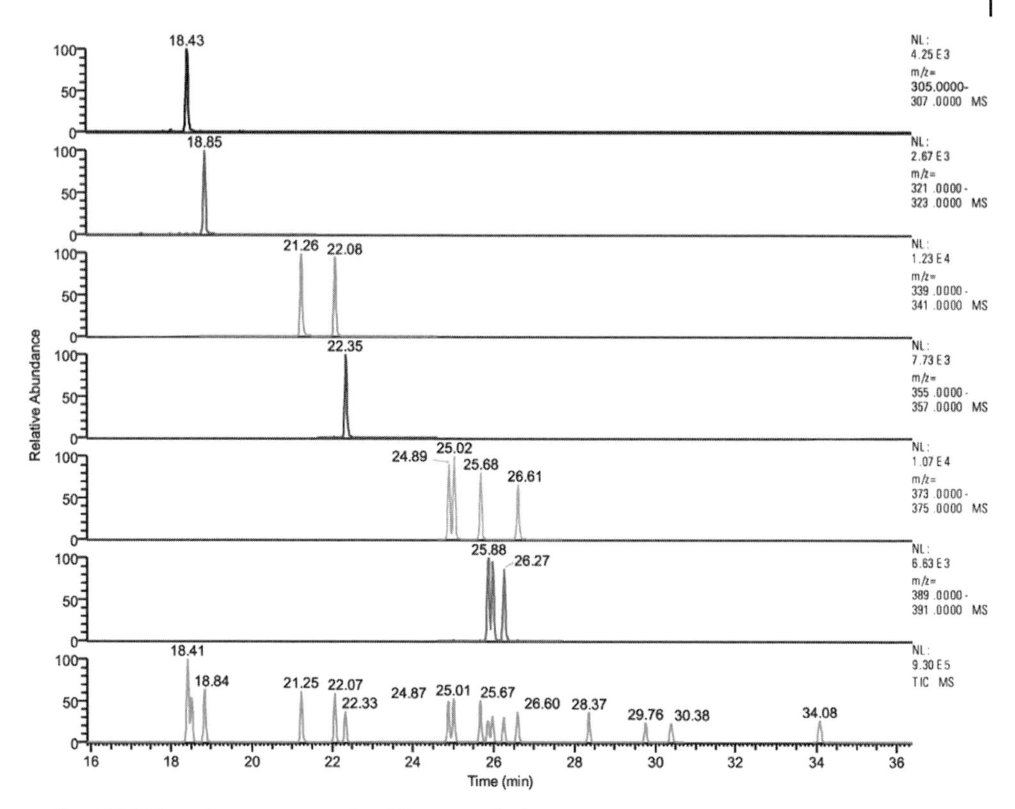

Fig. 4.106 Mass chromatograms of a 50 fg/μL standard

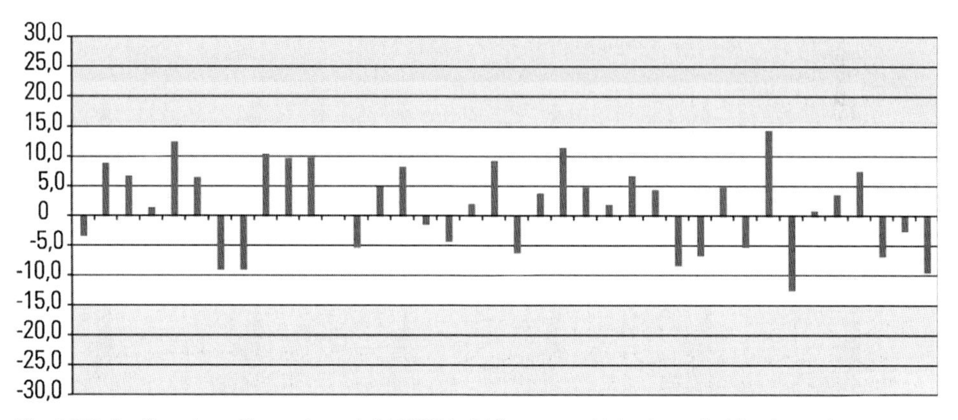

Fig. 4.107 Confirmation of ion ratios m/z 320/322 in % for repeated injections of a blood sample extract at 17 fg/μL, ±15% window complies with EPA 1613 method

4.26
Fast GC Analysis for PCBs

Key words: Fast GC, PCB, dl-PCBs, WHO-PCBs, sewage sludge, HT-8, separation power, carborane

Due to the widespread use of polychlorinated biphenyls (PCBs) as dielectric and heat transferring fluids in power transformers, hydraulic fluids, intense use as plasticizers and flame retardants and due to their stable molecular structure, PCBs with its 209 possible congeners, are still today subject to be monitored in environmental analysis. In particular the coplanar "dioxin like" dl-PCBs, or non-ortho-substituted PCBs, are of increasing analytical importance because of their toxicity similar to 2,3,7,8-TCDD with a significant contribution to the sample TEQ value.

The 12 so-called dl-PCBs or WHO-PCBs of toxicologic importance are the coplanar non-ortho substituted congeners of a total number of 68 coplanar congeners of which 20 in total are non-ortho substituted:

tetra Cl-PCB: 77, 81
penta Cl-PCB: 105, 114, 118, 123, 126
hexa Cl-PCB: 156, 157, 167, 169
hepta Cl-PCB: 189

The GC separation of these PCB congeners has been based for a long time on low and intermediate-polarity phases, such as SE-54 (typically DB-5MS, TR-5MS etc.). Problems with coelution, poor detection at low levels, and the need for better separation of the highly toxic coplanar PCBs demand columns with increased selectivity.

The HT8 carborane phase provides good selectivity for the above dl-PCBs as well as the seven indicator congeners (IUPAC 28, 52, 101, 118, 153, 138 and 180) and separates them from most of the potential co-eluting congeners. The analysis is usually performed with a 50 m and 0.25 mm ID capillary column and takes up to 60 minutes, depending on the used length of the column.

The analysis time for PCBs can be significantly reduced with Fast GC conditions by using a short 10 m column with 0.1 mm ID. The HT8 FAST PCB column has been successfully used as an excellent screening capillary column for MS detection.

Analysis Conditions

Gas chromatograph:	TRACE GC Ultra (Thermo Fisher, Milan, Italy)
Injector:	Split/splitless, 250 °C, 1 min splitless
Carrier gas:	Helium, 0.6 mL/min, constant flow
Column:	HT-8, 10 m × 0.1 mm × 0.1 μm (SGE Ringwood, Australia)
Oven temp. program:	90 °C, 1 min 30 °C/min to 220 °C 15 °C/min to 300 °C 300 °C, 5 min

MS interface temp.:	280 °C
Mass spectrometer:	DSQ II GC/MS (Thermo Fisher Scientific, Austin TX, USA)
Ionisation:	EI, 70 eV
Scan mode:	Full scan
Mass range:	100–500 u
Cycle time:	0.1 s/scan

Results

The Fast GC separation of PCBs still maintains excellent congener separation within a total analysis time of only 9 minutes including the internal standard decachlorobiphenyl (PCB209). Laboratories with a high sample throughput benefit in particular from the increased productivity by a higher sample throughput. The chromatogram of the test mix shows a very good separation of some of the critical pairs, for example congeners 31/28 (ca. 80%) and 163/138 (ca. 50%). This excellent separation can be achieved in a total analysis time of less than 10 minutes, see Fig. 4.108. Due to the very high scan speed of the employed MS detector a full scan mass spectrum of full integrity for substance confirmation is provided as given for HCB in Fig. 4.110. The achieved separation power (Fig. 4.109) shows that the FAST PCB capillary column in combination with a very fast scanning quadrupole mass spectrometer is an excellent choice for the screening and quantitation for PCBs.

Further potential for increased detection sensitivity could be achieved by running the MS in SIM detection mode. The sample analysis for contaminated sewage sludge acquired in full scan mode given in Fig. 4.111 shows the well known mass chromatogram patterns of the PCB chlorination degrees, but in a very short analysis runtime of less then 10 minutes including the elution of the internal quantification standard PCB209.

The unique selectivity of the HT8 column is attributed to the presence of the carborane unit having an affinity towards chlorinated biphenyls with the least number of ortho-substitutions. Fewer substitutions in the ortho position increase freedom of rotation, allowing the chlorinated biphenyl moiety to have greater interaction with the carborane unit (de Boer 1995). This phenomenon causes non-ortho substituted PCBs to have increased elution times compared to their ortho substituted congeners. This allows detection and quantitation of the important congeners used to monitor PCB occurrence and distribution in the environment.

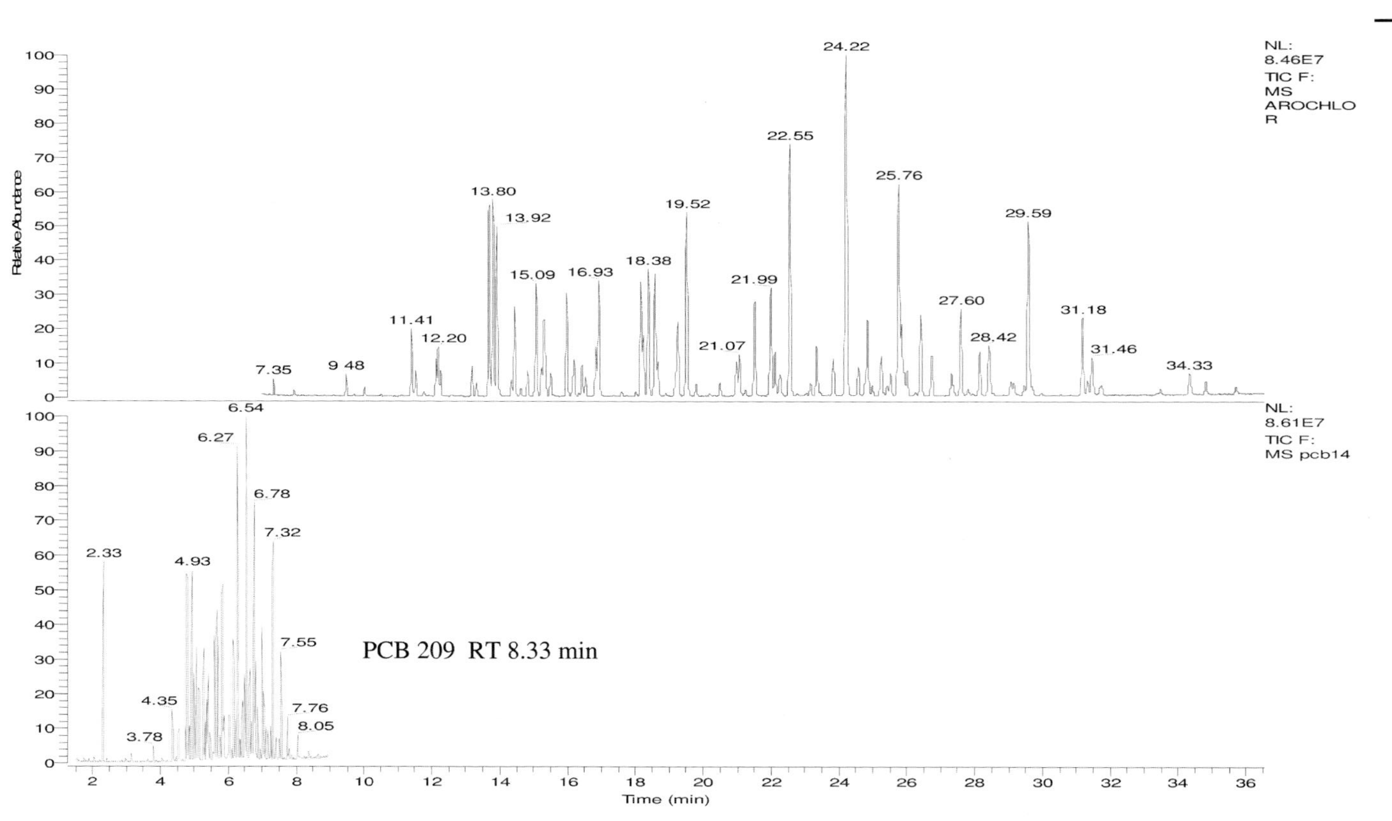

Fig. 4.108 Aroclor Mix A30, A40, A60: Comparison of Fast GC (10 m column length, 0.1 µm ID, bottom) to normal chromatography (30 m column length, 0.25 µm ID, top)

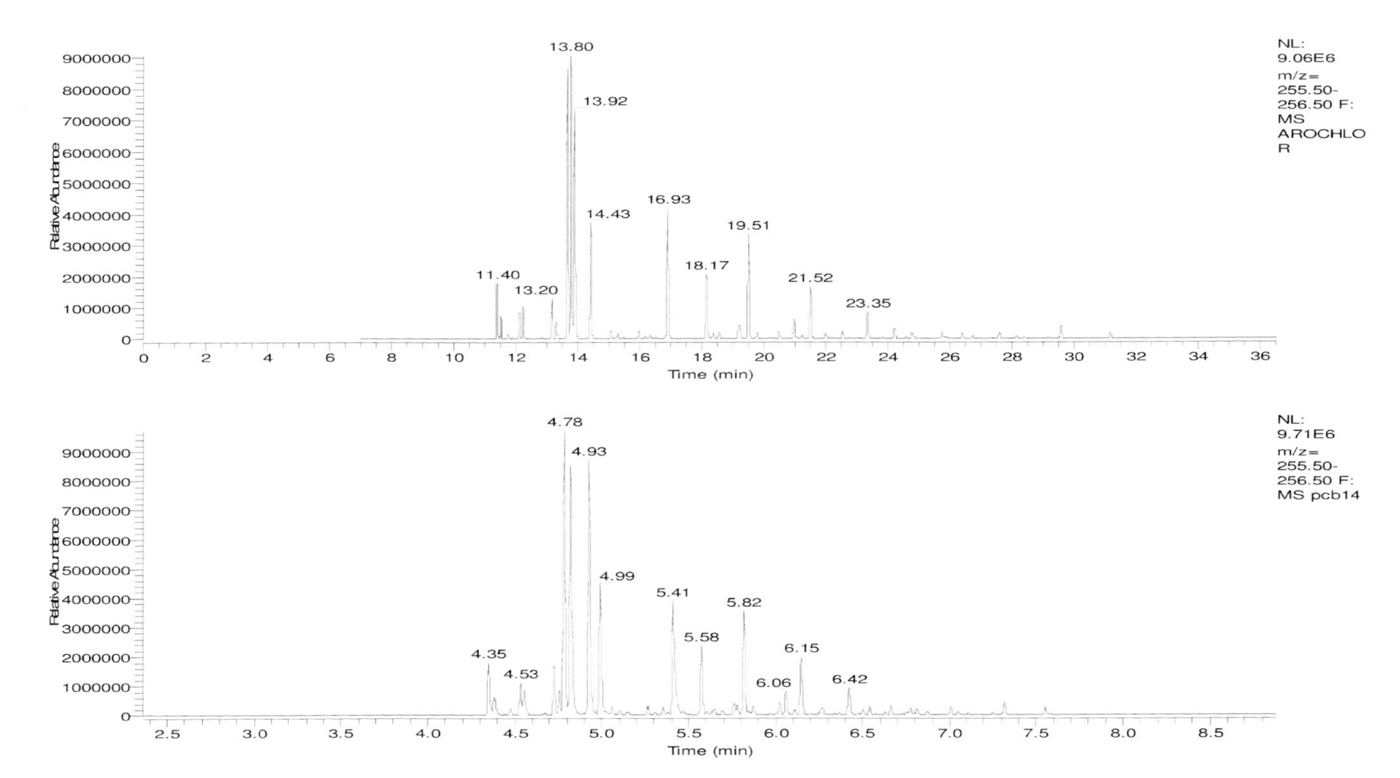

Fig. 4.109 Aroclor Mix A30, A40, A60: Separation power comparison of Fast GC (10 m column length, bottom) to normal chromatography (30 m column length, top)

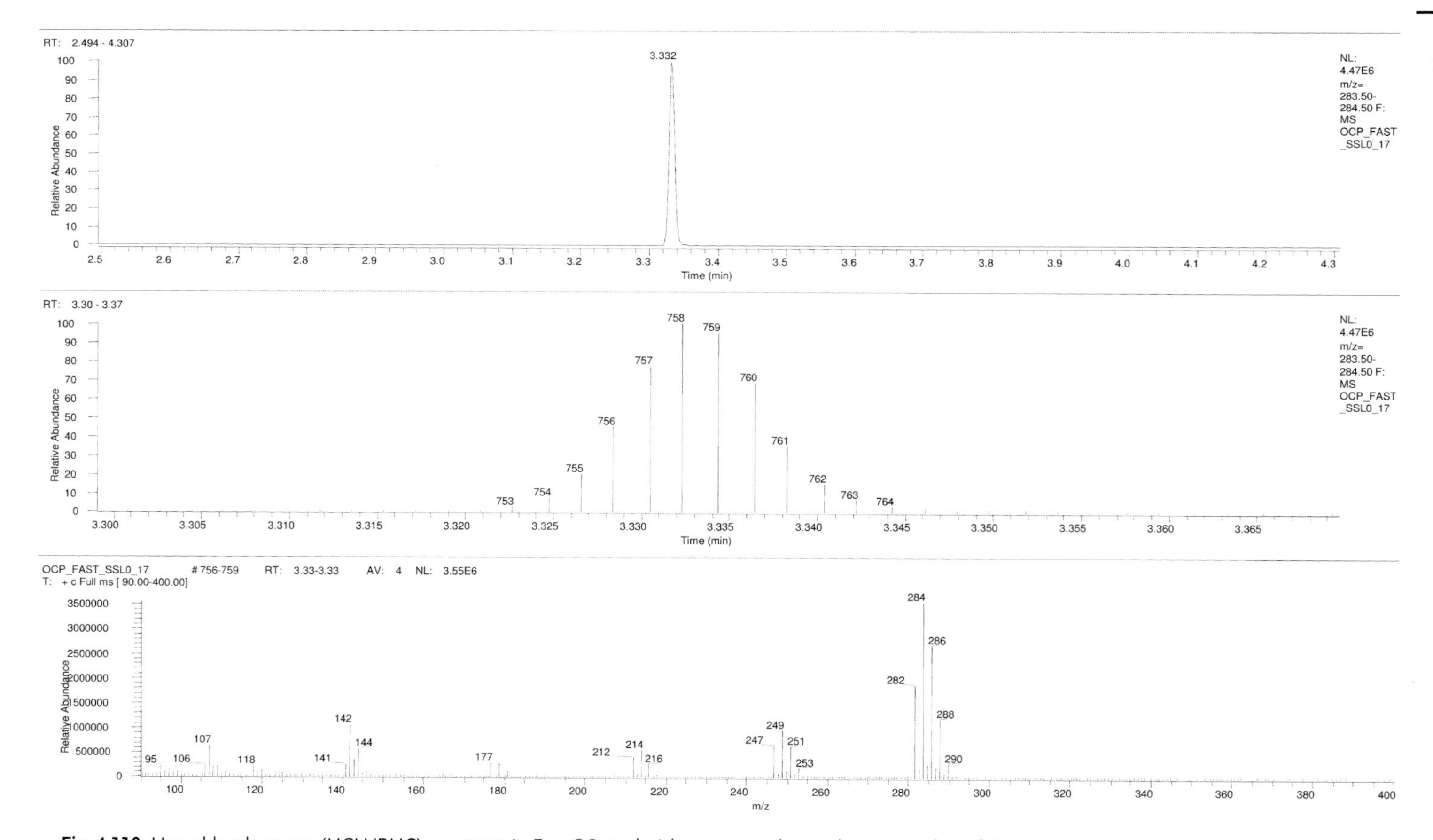

Fig. 4.110 Hexachlorobenzene (HCH/BHC) spectrum in Fast GC mode (chromatographic peak, top; number of data points over the Fast GC peak, middle; full scan spectrum, bottom)

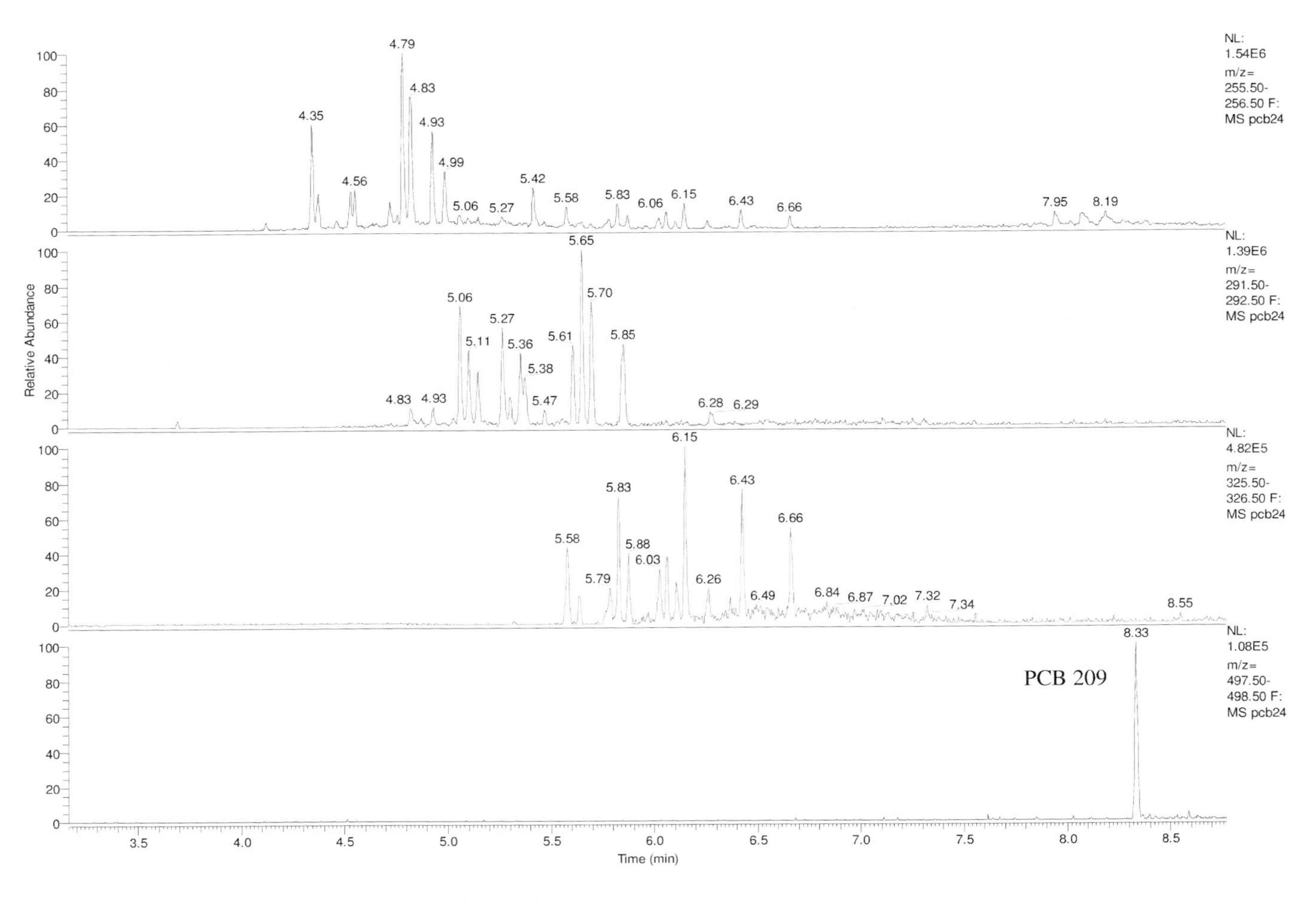

Fig. 4.111 Fast GC/MS analysis of a sewage sludge sample with the extracted mass chromatograms of tri-, tetra-, penta- and deca-chlorbiphenyl (ISTD)

4.27
Analysis of Brominated Flame Retardants PBDE

Key words: BFRs, PBDE, deca-BDE, MID, PTV, on-column, EPA 1614

Brominated diphenyl ethers (BDEs) are among the most important and most widely used flame retardants in a range of different industrial and consumer products. They are found worldwide in practically all types of matrices moving them into the focus of legislation resulting in a ban for certain BDE congeners. The EU directive 2003/11/EC prohibits the use of penta-BDE and octa-BDE for the member states of the European community.

As a result BDEs have received rising interest in recent years by the analytical community. The by far most efficient analysis technique is high resolution GC/MS using isotope dilution technique according to EPA 1614 for highest precision quantification with highest significance.

Analysis Conditions

Gas chromatograph:	TRACE GC Ultra™ (Thermo Fisher, Milan, Italy)
Injector:	Split/splitless, 280 °C, 1 min splitless, purge flow 50 mL/min
Carrier gas:	Helium, 1 mL/min
Column:	TR5-MS, 15 m x 0.25 mm x 0.1 μm (Thermo Fisher Scientific, Runcom, UK)
Oven temp. program:	120 °C, 2 min 15 °C/min to 230 °C 5 °C/min to 270 °C 10 °C/min to 330 °C, 5 min
MS interface temp.:	280 °C
Mass spectrometer:	DFS High Resolution GC/MS (Thermo Fisher, Bremen, Germany)
Ionisation:	EI, 40 eV
Ion source temp.:	270 °C
Scan mode:	MID, see Table 4.25
Resolution:	10,000 (10% valley definition)

Results

The full scan results proved that for all bromination degrees either the molecular ion or the fragment ion showing the loss of 2 Br atoms are the most abundant ions (see Fig. 4.112). The change of the most intense ion from M^+ to $[M\text{-}2Br]^+$ is typically observed with higher bromination degrees and starting to become significant already with penta- and hexa-BDEs, depending on the ion source conditions. The loss of Br is slightly temperature dependent and varies with GC elution and ion source temperatures. Therefore, with different instrument conditions the transition of the most abundant ion from M^+ to $[M\text{-}2Br]^+$ might be shifted to tetra/penta- or hexa/hepta-BDE. In general the relative intensity $[M\text{-}2Br]^+/M^+$ is increasing with the degree of bromination. For deca-BDE the intensity gain when using the $[M\text{-}2Br]^+$ mass peak for MID detection is at least a factor of 4 compared to M^+. For a list of exact masses including relative intensities see Tables 4.26 and 4.27.

Mass spectrometer tuning parameters similar to those typically used for dioxin/PCB analysis were found to give optimum sensitivity for BDE analysis as well. A slightly higher ion source temperature around 270 °C is recommended taking the high boiling characteristics of the BFRs into account (although highly brominated compounds appear at lower elution temperatures as same molecular weight hydrocarbons). The use of PFK as internal mass reference is mandatory, because lock and cali masses in the high mass range are needed (e. g. for deca-BDE). Autotuning for gaining highest sensitivity was carried out on PFK mass 480.9688.

All congeners in the employed Wellington BDE standard could be separated on the 15 m column (see Fig. 4.113). Similar to dioxins the BDE congeners are separated on unpolar columns group wise in the order of their bromination degree. The use of a short 10 to 15 m column with a thin film is recommended to analyse the thermolabile deca-BDE more efficiently.

The limits of quantitation (LOQs) achieved had been similar to those known for dioxin and PCB analysis. They can also be achieved for the analysis of the far higher boiling BDEs in the low femtogram range (see Fig. 4.114). Also the quantitation linearity proved to fulfil highest standards as shown in Fig. 4.115.

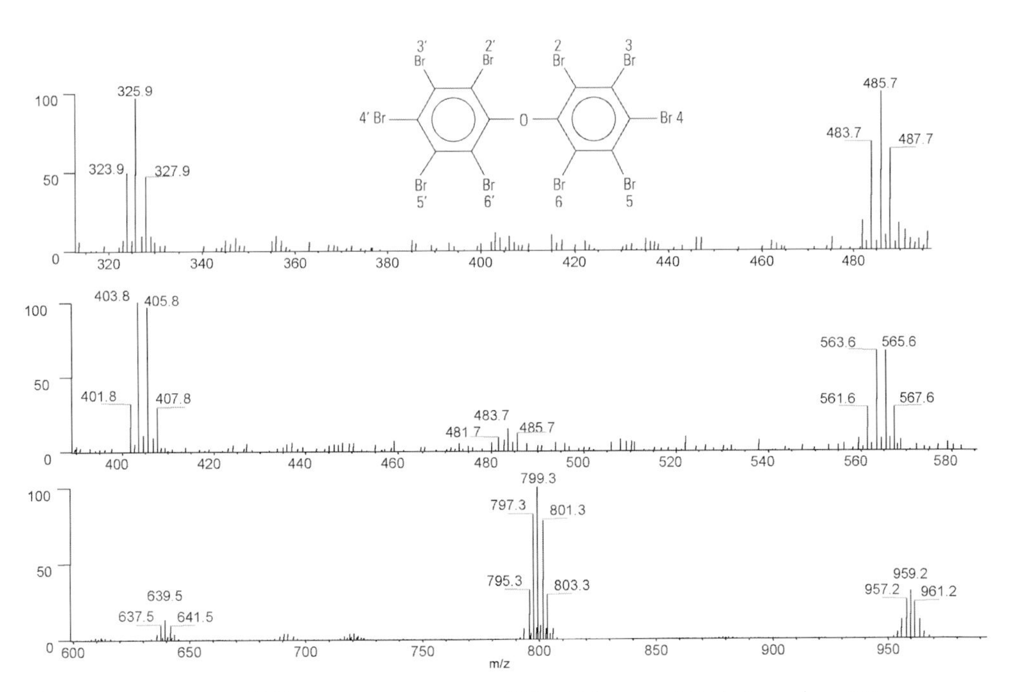

Fig. 4.112 Mass spectra of PBDE congeners; top: tetra-BDE, middle: penta-BDE, bottom: deca-BDE structure of deca-BDE

Table 4.25 Accurate mass MID setup: MID lock-and-cali mode (target masses in brackets: optional second ratio mass for native BDE)

MID window No.	Reference Masses (PFK) L = lock mass, C = cali mass	Target masses (second ratio mass native)	MID cycle time
1 tri-BDE	392.9753 (L), 430.9723 (C)	(403.8041), 405.8021, 407.8001, 417.8424, 419.8403	0.55 s
2 tetra-BDE	480.9688 (L), 492.9691 (C)	(483.7126), 485.7106, 487.7085, 495.7529, 497.7508	0.55 s
3 penta-BDE	554.9644 (L), 592.9627 (C)	(561.6231), 563.6211, 565.6190, 575.6613, 577.6593	0.60 s
4 hexa-BDE	480.9688 (L), 504.9691 (C)	481.6976, 483.6956, (485.6937), 493.7372, 495.7352	0.60 s
5 hepta-BDE	554.9644 (L), 592.9627 (C)	(559.6082), 561.6062, 563.6042, 573.6457, 575.6436	0.70 s
6 deca-BDE	754.9531 (L), 766.9531 (C)	797.3349, 799.3329, (801.3308), 809.3752, 811.3731	0.90 s

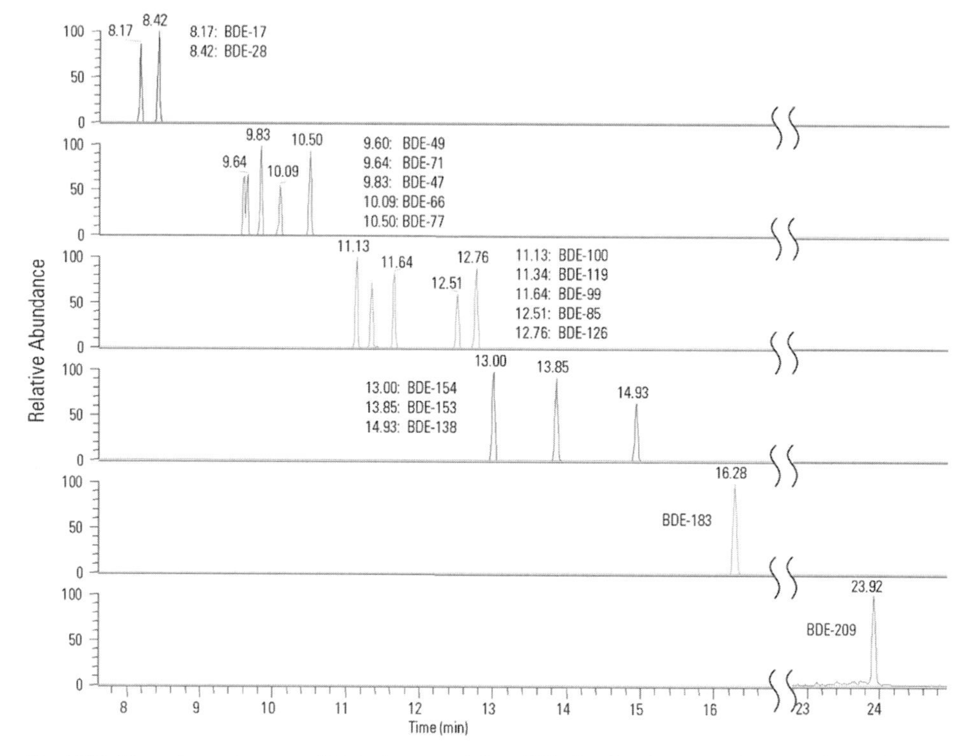

Fig. 4.113 Mass chromatograms of tri- to hepta- and deca-BDE showing the separation according to the bromination degree on a 5% phenyl phase column, length 15 m

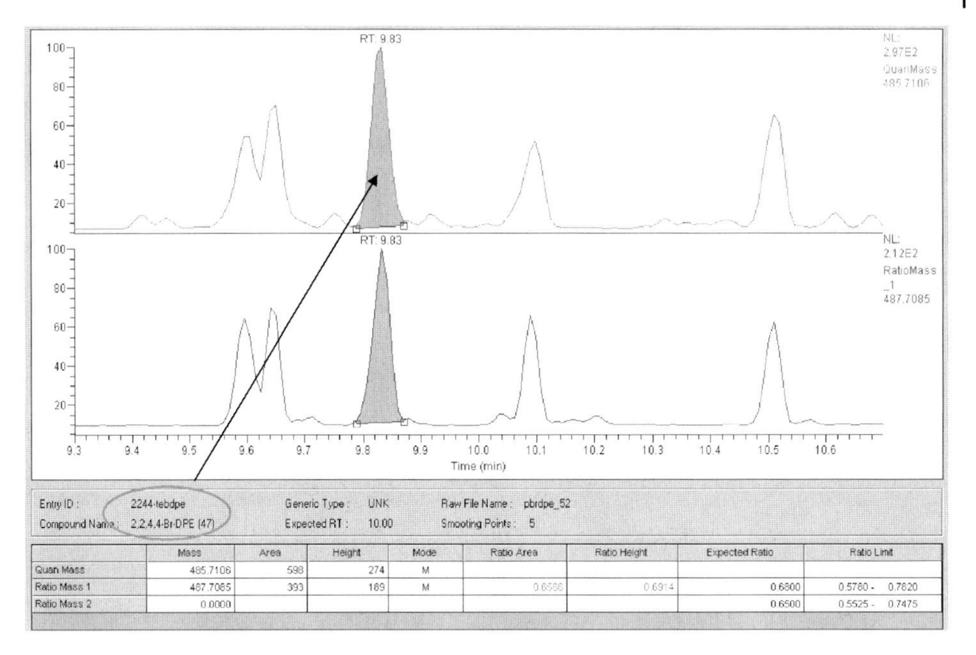

	Mass	Area	Height	Mode	Ratio Area	Ratio Height	Expected Ratio	Ratio Limit
Quan Mass	485.7106	598	274	M				
Ratio Mass 1	487.7085	393	189	M	0.6566	0.6914	0.6800	0.5780 - 0.7820
Ratio Mass 2	0.0000						0.6500	0.5525 - 0.7475

Fig. 4.114 25 fg BDE 47 (tetra-BDE); top: quan mass, bottom: ratio mass

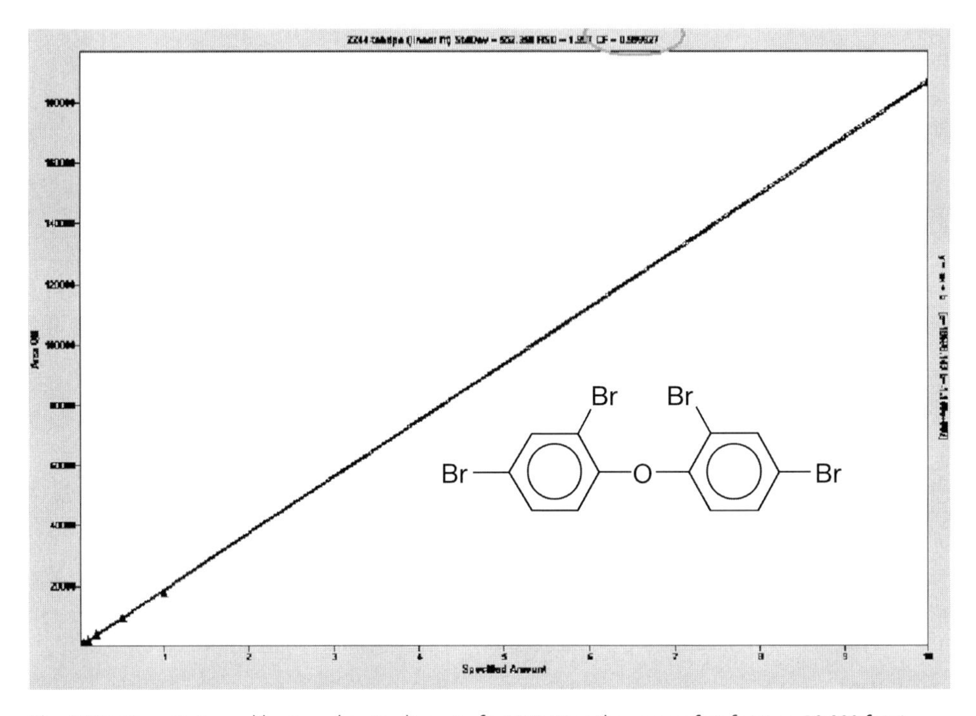

Fig. 4.115 Quantitative calibration showing linearity for BDE 47 in the range of 25 fg/μL to 10,000 fg/μL

Table 4.26 PBDE exact mass references (1) for natives and internal standards (molecular ions)

# Br	Native		$^{13}C_{12}$ Standard		$^{13}C_6$ Standard	
	Exact mass M^+ [u]	Relative intensity [%]	Exact mass M^+ [u]	Relative intensity [%]	Exact mass M^+ [u]	Relative intensity [%]
Br1	247,98313	100,0	260,02339	100,0	254,00326	100,0
	249,98108	97,3	262,02134	97,3	256,00121	97,3
Br2	325,89364	51,4	337,93390	51,4	331,91377	51,4
	327,89159	100,0	339,93185	100,0	333,91172	100,0
	329,88955	48,6	341,92981	48,6	335,90968	48,6
Br3	403,80415	34,3	415,84441	34,3	409,82428	34,3
	405,80211	100,0	417,84237	100,0	411,82224	100,0
	407,80006	97,3	419,84032	97,3	413,82019	97,3
	409,79801	31,5	421,83827	31,5	415,81814	31,5
Br4	481,71467	17,6	493,75492	17,6	487,73480	17,6
	483,71262	68,5	495,75288	68,5	489,73275	68,5
	485,71057	100,0	497,75083	100,0	491,73070	100,0
	487,70853	64,9	499,74878	64,9	493,72866	64,9
	489,70648	15,8	501,74674	15,8	495,72661	15,8
Br5	559,62518	10,6	571,66544	10,6	565,64531	10,6
	561,62313	51,4	573,66339	51,4	567,64326	51,4
	563,62109	100,0	575,66134	100,0	569,64122	100,0
	565,61904	97,3	577,65930	97,3	571,63917	97,3
	567,61699	47,3	579,65725	47,3	573,63712	47,3
	569,61495	9,2	581,65520	9,2	575,63508	9,2
Br6	637,53569	5,4	649,57595	5,4	643,55582	5,4
	639,53365	31,7	651,57390	31,7	645,55377	31,7
	641,53160	77,1	653,57186	77,1	647,55173	77,1
	643,52955	100,0	655,56981	100,0	649,54968	100,0
	645,52751	73,0	657,56776	73,0	651,54763	73,0
	647,52546	28,4	659,56572	28,4	653,54559	28,4
	649,52341	4,6	661,56367	4,6	655,54354	4,6
Br7	715,44620	3,1	727,48646	3,1	721,46633	3,1
	717,44416	21,1	729,48442	21,1	723,46429	21,1
	719,44211	61,7	731,48237	61,7	725,46224	61,7
	721,44006	100,0	733,48032	100,0	727,46019	100,0
	723,43802	97,3	735,47828	97,3	729,45815	97,3
	725,43597	56,8	737,47623	56,8	731,45610	56,8
	727,43392	18,4	739,47418	18,4	733,45405	18,4
	729,43188	2,6	741,47214	2,6	735,45201	2,6
Br8	793,35672	1,6	805,39697	1,6	799,37685	1,6
	795,35467	12,4	807,39493	12,4	801,37480	12,4
	797,35262	42,3	809,39288	42,3	803,37275	42,3
	799,35058	82,2	811,39084	82,2	805,37071	82,2
	801,34853	100,0	813,38879	100,0	807,36866	100,0
	803,34648	77,8	815,38674	77,8	809,36661	77,8
	805,34444	37,9	817,38470	37,8	811,36457	37,8
	807,34239	10,5	819,38265	10,5	813,36252	10,5
					815,36047	1,3

Table 4.26 (continued)

# Br	Native		$^{13}C_{12}$ Standard		$^{13}C_6$ Standard	
	Exact mass M^+ [u]	Relative intensity [%]	Exact mass M^+ [u]	Relative intensity [%]	Exact mass M^+ [u]	Relative intensity [%]
Br9	871,26723	0,9	883,30749	0,9	877,28736	0,9
	873,26518	7,8	885,30544	7,8	879,28531	7,8
	875,26314	30,2	887,30339	30,2	881,28327	30,1
	877,26109	68,5	889,30135	68,5	883,28122	68,5
	879,25904	100,0	891,29930	100,0	885,27917	100,0
	881,25700	97,3	893,29725	97,3	887,27713	97,3
	883,25495	63,1	895,29521	63,1	889,27508	63,1
	885,25290	26,3	897,29316	26,3	891,27303	26,3
	887,25086	6,4	899,29111	6,4	893,27099	6,4
Br10	949,17774	0,5	961,21800	0,5	955,19787	0,5
	951,17570	4,4	963,21595	4,4	957,19582	4,4
	953,17365	19,4	965,21391	19,4	959,19378	19,4
	955,17160	50,3	967,21186	50,3	961,19173	50,3
	957,16956	85,7	969,20981	85,7	963,18968	85,7
	959,16751	100,0	971,20777	100,0	965,18764	100,0
	961,16546	81,1	973,20572	81,1	967,18559	81,1
	963,16342	45,1	975,20367	45,1	969,18354	45,1
	965,16137	16,4	977,20163	16,4	971,18150	16,4
	967,15932	3,6	979,19958	3,6	973,17945	3,6

The calculated reference masses are based on the following values for isotopic masses: 1H 1.0078250321u, 12C 12.0000000000u, 13C 13.0033548378u, 16O 15.9949146221u, 79Br 78.9183376u and 81Br 80.9162910u. All listed masses refer to singly positively charged ions. Masses for isotope peaks have been calculated for a resolving power of 10,000 (10% valley definition). The mass of the electron (0.000548579911u) was taken into account for the calculation of the ionic masses.
Reference: Nuclear Phys. A 1995, 595, 409–480; J. Phys. Chem. Ref. Data 1999, 28 (6), 1713–1852 and references cited therein.

Table 4.27 PBDE exact mass references (2) for natives and internal standards (M-2Br ions)

# Br of M	Ratio [%] $M^+/(M\text{-}2Br)^+$	Native		$^{13}C_{12}$ Standard		$^{13}C_6$ Standard	
		Exact mass [u] $(M\text{-}2Br)^+$	Relative intensity [%]	Exact mass [u] $(M\text{-}2Br)^+$	Relative intensity [%]	Exact mass [u] $(M\text{-}2Br)^+$	Relative intensity [%]
Br1	100	na		na		na	
Br2	100	168.056966	100.0	180.097224	100.0	174.077095	100.0
Br3	100	245.967479	89.8	258.007737	100.0	251.987608	95.8
		247.965432	100.0	260.005690	97.7	253.985561	100.0
Br4	100	323.877991	47.5	335.918249	51.0	329.898120	49.4
		325.875945	100.0	337.916203	100.0	331.896074	100.0
		327.873898	54.4	339.914156	48.5	333.894027	51.4
Br5	85	401.788504	31.0	413.828762	33.9	407.808633	33.0
		403.786457	96.4	415.826715	100.0	409.806586	99.6
		405.784411	100.0	417.824669	97.4	411.804540	100.0
		407.782364	35.3	419.822622	31.4	413.802493	34.2
Br6	60	479.699017	16.0	491.739275	17.4	485.719146	16.7
		481.696970	65.7	493.737228	68.2	487.717099	66.9
		483.694923	100.0	495.735181	100.0	489.715052	100.0
		485.692877	69.5	497.733135	64.7	491.713006	66.9
		487.690830	17.9	499.731088	15.7	493.710959	17.0
Br7	55	557.609529	2.8	569.649787	3.0	563.629658	2.9
		559.607483	19.8	571.647741	21.0	565.627612	20.5
		561.605436	59.7	573.645694	61.7	567.625565	60.5
		563.603389	99.7	575.643647	100.0	569.623518	100.0
		565.601343	100.0	577.641601	97.4	571.621472	98.5
		567.599296	60.9	579.639554	56.9	573.619425	59.0
Br8	50	635.520042	4.8	647.560300	5.3	641.540171	5.1
		637.517995	29.5	649.558253	31.6	643.538124	30.6

Table 4.27 (continued)

# Br of M	Ratio [%] $M^+/(M\text{-}2Br)^+$	Native		$^{13}C_{12}$ Standard		$^{13}C_6$ Standard	
		Exact mass [u] $(M\text{-}2Br)^+$	Relative intensity [%]	Exact mass [u] $(M\text{-}2Br)^+$	Relative intensity [%]	Exact mass [u] $(M\text{-}2Br)^+$	Relative intensity [%]
Br8		639.515949	74.2	651.556207	77.2	645.536078	75.5
		641.513902	100.0	653.554160	100.0	647.534031	100.0
		643.511855	75.6	655.552113	73.1	649.531984	74.6
		645.509809	31.2	657.550067	28.4	651.529938	29.8
		647.507762	5.1	659.548020	4.6	653.527891	5.0
Br9	40	713.430554	2.8	725.470812	3.0	719.450683	2.9
		715.428508	19.8	727.468766	21.0	721.448637	20.5
		717.426461	59.7	729.466719	61.7	723.446590	60.5
		719.424414	99.7	731.464673	100.0	725.444544	100.0
		721.422368	100.0	733.462626	97.3	727.442497	98.5
		723.420321	60.9	735.460579	56.9	729.440450	58.9
		725.418275	20.7	737.458533	18.4	731.438404	19.6
		727.416228	2.8	739.456486	2.6	733.436357	2.8
Br10	25	791.341067	1.4	803.381325	1.6	797.361196	1.5
		793.339020	11.5	805.379278	12.3	799.359149	11.9
		795.336974	40.1	807.377232	42.2	801.357103	41.3
		797.334927	80.5	809.375185	82.1	803.355056	81.2
		799.332880	100.0	811.373138	100.0	805.353009	100.0
		801.330834	80.5	813.371092	77.8	807.350963	79.0
		803.328787	40.8	815.369045	37.8	809.348916	39.4
		805.326741	11.9	817.366999	10.5	811.346870	11.2
		807.324694	1.4	819.364952	1.3	813.344823	1.4

The calculated reference masses are based on the following values for isotopic masses: 1 H 1.0078250321 u, 12C 12.0000000000u, 13C 13.0033548378u, 16O 15.9949146221 u, 79Br 78.9183376u and 81 Br 80.9162910u. All listed masses refer to singly positively charged ions. Masses for isotope peaks have been calculated for a resolving power of 10,000 (10% valley definition). The mass of the electron (0.000548579911 u) was taken into account for the calculation of the ionic masses.
Reference: Nuclear Phys. A 1995, 595, 409-480; J. Phys. Chem. Ref. Data 1999, 28 (6), 1713-1852 and references cited therein.
The given ratio M+/(M-2Br)+ provides typical values, depends on actual ion source conditions

4.28
Trace Analysis of BFRs in Waste Water Using SPME-GC/MS/MS

Key words: PBBs, PBDEs, headspace, SPME, GC/MS/MS, quantitation

Polybrominated biphenyls (PBBs) and polybrominated diphenylethers (PBDEs) are among the most widely used flame retardants with a widely distributed contamination in the environment causing a high risk for subsequent accumulation in food and feedstuff. Although typically the highly brominated congeners are applied in flame retardants the low lower brominated species are found more often in environmental samples.

According to the focus of potential accumulation from environmental sources this application covers the congener range of up to the hexa brominated species of both PBBs and PBDEs. A fast and reliable automated online extraction method for brominated flame retardant from water samples is described. For the first time SPME is applied for the trace determination of BFRs. In contrast to conventional methods using the Soxhlet extraction with an extended extract cleanup this method uses the fast and straightforward extraction of the analytes from the sample headspace. This is possible due to the high water vapour volatility of the PBBs and PBDEs and the low partition coefficient of the these BFR compounds in aqueous media. The SPME method can be integrated in regular GC/MS autosamplers, is fast and prevents the chromatographic system from the typical background load of Soxhlet extracted samples and provides clean chromatograms. Tap water and urban waste water has been analysed (Polo 2004).

The mass spectrometric detection is achieved by MS/MS using an ion trap mass spectrometer providing excellent selectivity and sensitivity for trace level quantitation in environmental water samples. Highest selectivity and lowest determination levels has been achieved by using high resolution mass spectrometer (De Boer 2001, Krumwiede 2006).

Analysis Conditions

SPME load:	10 mL of a filtered aqueous sample are filled in 22 mL headspace vials, sealed with Teflon faced septum, equilibrated at 100 °C for 5 min before sampling, stir bar agitation during sampling, SPME PDMS fibres were immersed into the sample headspace for 30 min, after exposition period immediate transfer to GC injector for desorption
Gas chromatograph:	Varian 3800 (Varian, Walnut Creek, CA, USA)
Injector:	Split/splitless, 2 mL splitless, split flow 50 mL/min
SPME desorption:	300 °C for PDMS fibres
Carrier gas:	Helium, 1.2 mL/min constant flow
Column:	CP-Sil 8 CB, 25 m × 0.25 mm × 0.25 μm
Oven temp. program:	60 °C, 2 min 30 °C/min to 250 °C 5 °C/min to 280 °C 280 °C, 8 min

MS interface temp.:	280 °C
Mass spectrometer:	Saturn 2000 with MS/MS waveboard (Varian Walnut Creek, CA, USA)
Ionisation:	EI, 70 eV
Manifold temp.:	50 °C
Ion trap temp.:	250 °C
Scan mode:	Full scan
Mass Range:	40–650 u, 1 s/scan
MS/MS:	Resonant waveform

Results

Despite of the high molecular weight of the BFRs the investigated compounds have been extracted with higher recoveries from the headspace than from direct immersion into the aqueous sample. This is due to an obviously high water vapour volatility of the brominated aromatics which is known from PCBs as well. Using headspace SPME the transfer of high boiling matrix components to the chromatographic system is prevented usually causing high background levels during MS detection. At the same time the useful lifetime of the fibre is significantly extended. Extractions are supported by stir bar agitation in the headspace vial. The method parameters are based on a comprehensive and systematic optimization study (Polo 2004).

The described method shows excellent sensitivity, linearity and quantitative precision up to the hexabromo congeners. The detection limits are in the low pg/L range (7.5–190 pg/L) with a calibration range by spiking tap water between 120 fg/mL and 500 pg/mL, as given in Table 4.28. The achieved correlation values are excellent in the range of 0.9977 to 1.0. Method precision and LOD have been determined with triplicates measurements as given in Table 4.29. The recovery experiments have been performed with three different types of blank matrix samples including tap water, effluent and influent waste waters from an urban sewage plant. The results are given in Table 4.30.

Table 4.28 Linearity and LODs

Compounds	Concentration range (pg/mL)	Correlation factor (R^2)	LOD S/N 3, pg/L
BDE-3	1.00 – 498	0.9977	190
PBB-15	1.01 – 503	1.0000	9.0
PBB-49	0.95 – 476	1.0000	7.5
BDE-47	0.41 – 205	0.9998	20
BDE-100	0.12 – 60	1.0000	60
BDE-99	0.41 – 205	0.9999	47
BDE-154	0.34 – 17	1.0000	150
BDE-153	0.23 – 12	0.9995	100

Table 4.29 Repeatability at two different concentration levels (n=3)

Compounds	Concentration 1 (pg/mL)	Repeatability (RSD%)	Concentration 2 (pg/mL)	Repeatability (RSD%)
BDE-3	10.0	4.4	199	6.3
PBB-15	10.1	3.8	201	12
PBB-49	9.5	2.8	190	1.7
BDE-47	4.1	17	82	1.2
BDE-100	1.2	15	24	10
BDE-99	4.1	20	82	1.2
BDE-154	0.34	24	6.8	9.3
BDE-153	0.23	26	4.6	8.8

Table 4.30 Recovery in Real Water Matrix Samples (* Chromatogram see Fig. 4.116)

Compounds	Tap water spiked conc. (pg/mL)	Recovery ± RSD (%)	Effluent water spiked conc. (pg/mL)	Recovery ± RSD (%)	Influent water* spiked conc. (pg/mL)	Recovery ± RSD (%)
BDE-3	1.00	99 ± 1	10.0	100 ± 4	199	106 ± 10
PBB-15	1.01	97 ± 4	10.1	92 ± 2	201	90 ± 6
PBB-49	0.95	90 ± 21	9.5	94 ± 1	190	93 ± 5
BDE-47	0.41	91 ± 8	4.1	97 ± 7	82	87 ± 7
BDE-100	0.12	100 ± 4	1.2	83 ± 17	24	95 ± 4
BDE-99	0.41	87 ± 19	4.1	90 ± 12	82	92 ± 8
BDE-154	0.034	nd	0.34	100 ± 25	6.8	74 ± 11
BDE-153	0.023	nd	0.23	117 ± 13	4.6	82 ± 11

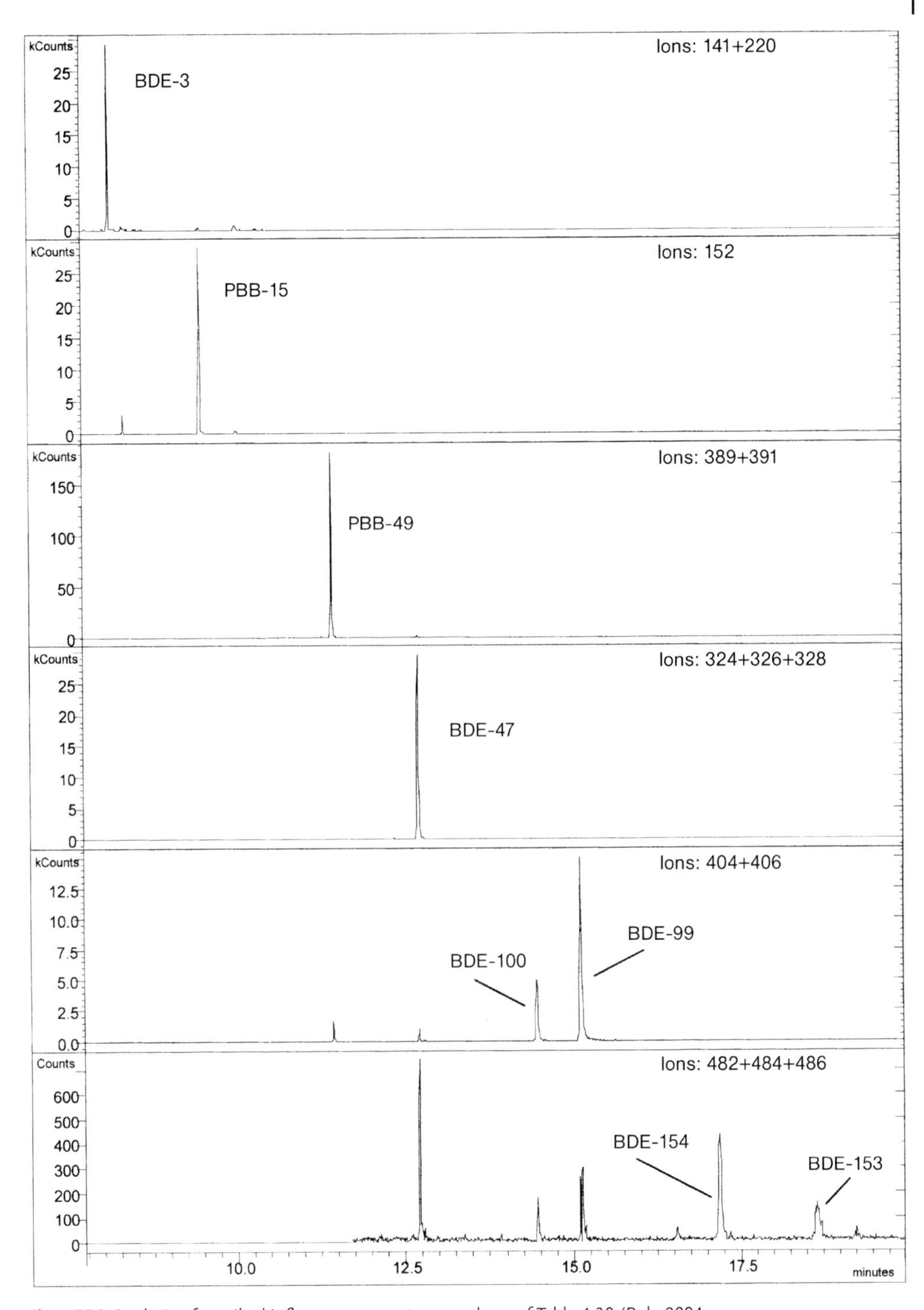

Fig. 4.116 Analysis of a spiked influent waste water sample as of Table 4.30 (Polo 2004; reprinted with permission of Analytical Chemistry, Copyright 2004, American Chemical Society)

4.29
Analysis of Military Waste

Key words: military waste, nitroaromatics, TNT, degradation products, carcinogenicity, Soxhlet, choice of column, ECD, PID, chemical ionisation, PAHs, water CI

In the past little attention was paid to the ecological consequences of rearmament and disarmament. Only frequently has the problem of military waste become a topic in residue analysis. The combined consequences of two world wars have not been considered for a long time and current political changes are leading to new types of contamination which will require analytical solutions within the framework of demilitarisation. This overall area will also therefore increase in importance in the future in the field of analysis.

There is no absolute definition of military waste. The term encompasses disused military sites where munitions or chemical weapons were manufactured, processed, stored or deposited. Soil and ground water are contaminated by the substances concerned and also by synthesis products, by-products, degradation products or by problematic fuels. Some of these compounds are liable to explode giving off toxic products. Explosives (nitroaromatics) and chemical warfare agents are of particular importance. The large quantity of water required for the manufacture of nitroaromatics led to the setting up of production plants in areas with large water supplies which are often still also used for public consumption. Many substances which are recovered today as metabolites or stabilisers are carcinogenic.

Besides HPLC for routine control, GC with ECD and PID has been the normal procedure for residue analysis up till now. GC/MS is increasing in importance as the number of substances in the trace region as well as in high concentrations which need to be determined is increasing. As the spectrum of expected contaminants is usually difficult to estimate, the mass spectrometer is indispensable as the universal and specific detector.

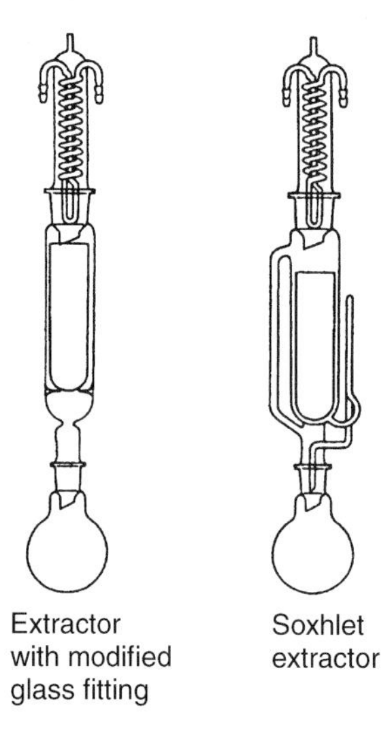

Compound/ extraction procedure	Methanol Soxhlet	Methanol flow extraction	Diethyl ether extraction
2-Nitrotoluene	85%	100%	91,7%
3-Nitrotoluene	82%	95%	90%
4-Nitrotoluene	73%	90%	85%
2,6-Dinitrotoluene	91%	95%	95%
2,4-Dinitrotoluene	72%	85%	90%
2,3-Dinitrotoluene	74%	90%	90%
3,4-Dinitrotoluene	34%	70%	80%
2,4,6-Trinitrotoluene	6%	70%	80%
1,3-Dinitrobenzene	66%	70%	85%
1,4-Dinitrobenzene	1%	70%	85%
4-Amino-2,6-dinitrotoluene	47%	45%	90%
2-Amino-4,6-dinitrotoluene	41%	45%	90%
3-Nitrobiphenyl	96%	95%	90%
2,2-Dinitrobiphenyl	92%	95%	95%
1,3-Dinitronaphthalene	56%	80%	95%
2,3-Diaminotoluene	–	35%	75%
2,4-Diaminotoluene	–	45%	80%
2,6-Diaminotoluene	–	50%	80%
2-Amino-4-nitrotoluene	–	65%	80%
2-Amino-6-nitrotoluene	–	65%	80%
2,6-Diamino-4-nitrotoluene	–	45%	80%

– Not determined

Fig. 4.117 Comparison of liquid extraction procedures for the analysis of nitroaromatics

The analysis of nitroaromatics and their metabolites is of particular importance in the assessment of military waste. 14 by-products and degradation products of 2,4,6-TNT (trinitrotoluene) alone are definitely classified or suspected as carcinogens (Table 4.31).

The analysis of TNT degradation products involves Soxhlet extraction or SFE for the clean-up of soil extracts, whereby the differing polarity of the substances to be extracted needs to be taken into consideration when choosing the extraction agent (Fig. 4.117). Nitroaromatics can be separated by HPLC or GC. GC/MS coupling is now the method of choice for determining them together with their metabolites. Classical GC detection using NPD, PID or ECD has limitations (Fig. 4.118 and Table 4.32). ECD is frequently used and exhibits high sensitivity for all compounds with two or more nitro groups. However, it can therefore only be used for the detection of certain metabolites (Fig. 4.119).

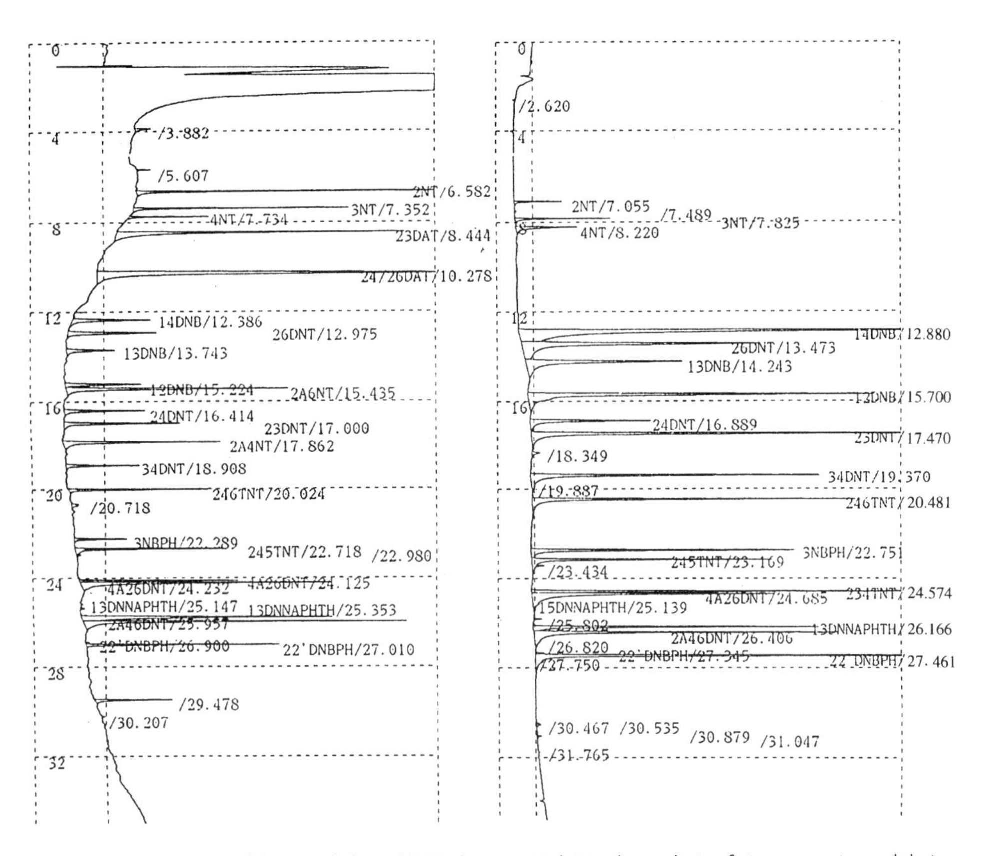

Fig. 4.118 Comparison of the PID (left) and ECD detector (right) in the analysis of nitroaromatics and their metabolites (column HT8, 400 pg/component, for conditions see text)

Table 4.31 By-products and degradation products of TNT with details of the carcinogenic potential

Compound	Assessment	Compound	Assessment
2-Nitrotoluene		2,6-Diaminotoluene	
3-Nitrotoluene		2,4,6-Triaminotoluene	
4-Nitrotoluene		2-Amino-6-nitrotoluene	
2,3-Dinitrotoluene	III A2	2-Amino-4-nitrotoluene	
2,4-Dinitrotoluene	III A2	6-Amino-2,4-dinitrotoluene	
2,6-Dinitrotoluene	III A2	4-Amino-2,6-dinitrotoluene	
3,4-Dinitrotoluene	III A2	2-Amino-4,6-dinitrotoluene	
2,4,6-Trinitrotoluene		3-Nitrobiphenyl	
2,4,5-Trinitrotoluene	III B	4-Nitrobiphenyl	III A2
2,3,4-Trinitrotoluene		3-Nitrobiphenyl	
1,2-Dinitrobenzene	III B	4-Nitrobiphenyl	III A2
1,3-Dinitrobenzene	III B	2,2′-Dinitrobiphenyl	
1,4-Dinitrobenzene	III B	2-Nitronaphthalene	III A2
2,3-Diaminotoluene		4-Aminobiphenyl	III A1
2,4-Diaminotoluene		1,3-Dinitronaphthalene	III B

The substances are assessed according to the ordinance on hazardous materials and the MAK value list:
III A1 = definitely carcinogenic in humans
III A2 = definitely carcinogenic in animals
III B = suspicion of potential carcinogenicity

Table 4.32 Detectors for the analysis of nitroaromatics and their metabolites

Detector	Advantages	Disadvantages
FID	High linearity	Low selectivity Low sensitivity
ECD	Good sensitivity	Low linearity Contamination when using highly concentrated samples, low sensitivity for substances with less than 2 nitro groups, correspondingly difficult detection of metabolites
NPD	Good selectivity for nitro compounds and metabolites	Low sensitivity
PID	The same response for nitro and amino compounds, field tests	Low sensitivity
ELD	N-specific, high linearity, simple calibration	Low sensitivity
GC/MS-EI	High sensitivity, good identification even of other accompanying substances	–
GC/MS-CI	High selectivity, additional confirmation from the molecular mass, PCI for nitroaromatics and metabolites, NCI only for nitro compounds	–

Analysis Conditions

Sample:	Soil samples
Sample preparation:	Continuous flow extraction with methanol or diethyl ether
GC/MS systems:	(1) HP MSD 5971A, EI analysis (2) Finnigan Magnum, EI/CI analyses
Injectors:	(1) HP split injector, isotherm 270 °C, split operation (2) PTV cold injection system, split operation Start: 40 °C, 1 min Program: 250 °C/min to 300 °C Final temp.: 300 °C, 15 min
Columns:	(1) Restek Rtx-200, 30 m × 0.25 mm × 0.25 µm (2) SGE HT8, 25 m × 0.22 mm × 0.25 µm Helium, 2 bar
Mass spectrometer:	Scan mode: (1) EI, full scan, 29–250 u (2) EI/CI, full scan, 100–400 u, reagent gas water Scan rate: (1) and (2) 1s/scan in each case

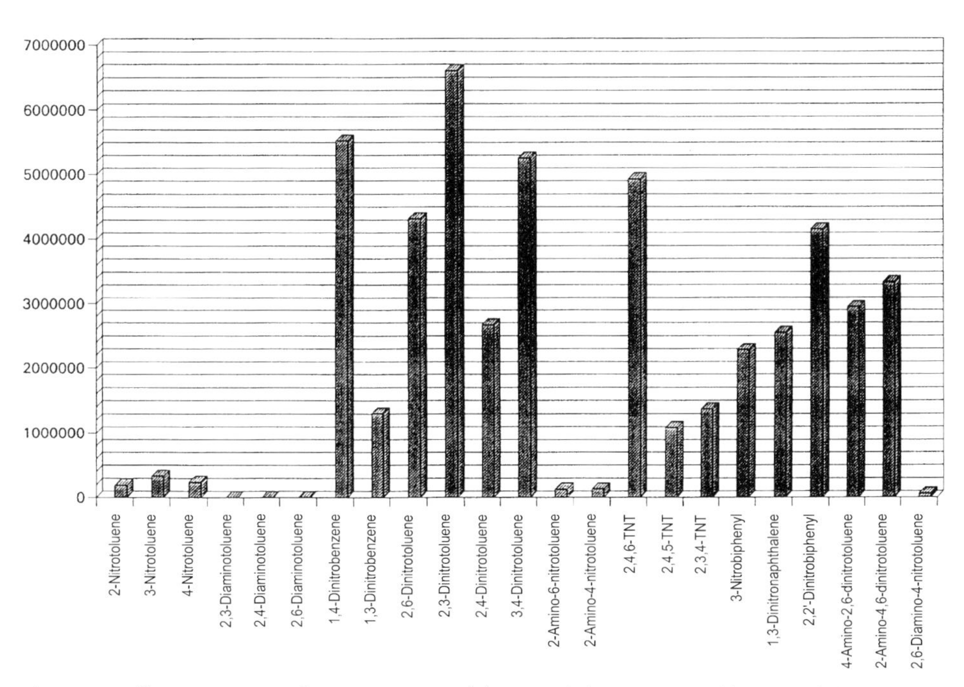

Fig. 4.119 Differing responses of nitroaromatics and their metabolites on ECD (400 pg samples)

Results

An analysis of nitroaromatics with ECD as the detector and hydrogen as the carrier gas is shown in Fig. 4.120. This separation is designed to be rapid and does not take metabolites into consideration. Rapid separation requires the appropriate choice of GC column. The critical pair 2,4. and 2,6-diaminotoluene can be separated well on the Rtx-200 column (trifluoropropylmethylsilicone phase) (Fig. 4.121), which is not possible with columns of the same length containing pure methylsilicone or carborane phases (HT8). As the isomers cannot be differentiated by mass spectrometry, gas chromatographic separation is a prerequisite for separate determination (Fig. 4.122). Other compounds from TNT analysis which can sometimes co-eluate with them can be differentiated by mass spectrometry.

The EI spectra of nitroaromatics frequently show an intense fragment at m/z 30 which results from rearrangement and fragmentation of the nitro group. The molecular ion signals are small and can therefore be completely undetectable in matrix-rich samples. With chemical ionisation all the substances involved form intense (quasi)molecular ions which can be detected with high S/N values because of their higher masses (Fig. 4.123). A series of typical CI spectra measured using water as the reagent gas are shown in Fig. 4.124 for various nitro- and amino-substitutions. The outstanding selectivity of chemical ionisation allows the determination of the equally important concentrations of polycondensed aromatics (PAHs) in the same analysis. As the PAHs are very accessible by CI, it is only necessary to assign the corresponding $(M+H)^+$ masses in order to be able to determine their concentrations (Fig. 4.125). The evaluation of the analysis is thus only directed to a further group of compounds. A second analysis run on the same sample is no longer necessary.

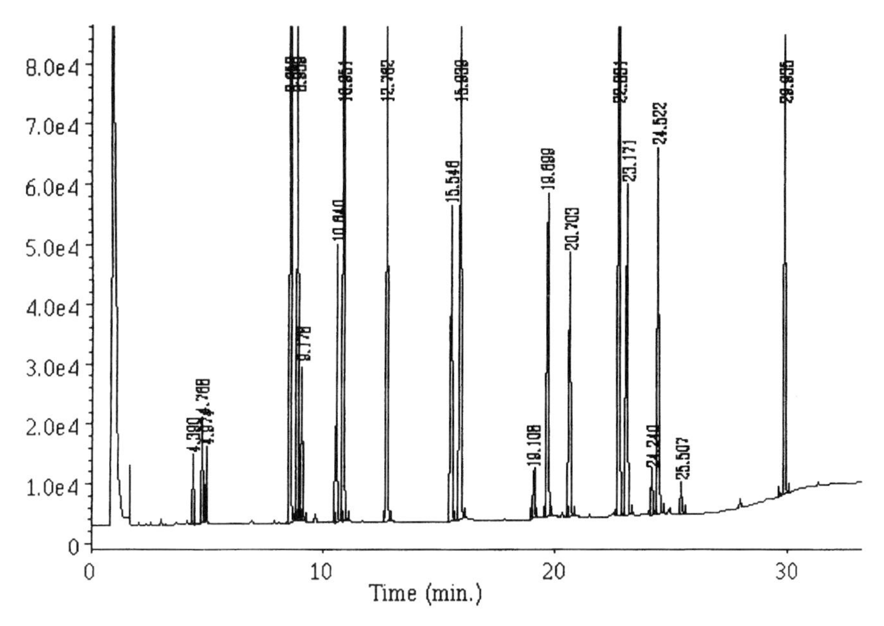

External Standard Report

Sample Name : 2 ppm Mixture Injection Number: 1
Instrument Method: NTH2_25.MTH

4.390	22076	BB	0.027	1	2.000	2MNT
4.768	30355	BB	0.025	1	2.000	3MNT
4.974	17798	BV	0.024	1	2.000	4MNT
8.658	265792	BB	0.042	1	2.000	14DNB
8.959	256766	BB	0.042	1	2.000	26DNT
9.178	78678	BB	0.046	1	2.000	13DNB
10.640	151300	BB	0.050	1	2.000	24DNT
10.951	374679	BB	0.047	1	2.000	23DNT
12.762	297095	BB	0.054	1	2.000	34DNT
15.546	213138	BB	0.063	1	2.000	3-NO2biphenyl
15.939	331069	BB	0.059	1	2.000	246TNT
19.108	30392	BB	0.057	1	2.000	245TNT
19.699	213808	BB	0.062	1	2.000	234TNT
20.703	188891	BB	0.067	1	2.000	1.3-DiNO2naphthalene
22.831	396779	BB	0.065	1	2.000	2.2'-DiNO2biphenyl
23.171	220906	BB	0.062	1	2.000	4NH2-2.6-DiNO2toluene
24.522	251604	BB	0.064	1	2.000	2NH2-4.6-DiNO2toluene
29.935	220999	BB	0.045	1	2.000	2.4-DiNO2Diphenylamine

Fig. 4.120 Separation of nitroaromatics on a Restek Rtx-1701 phase (60 m × 0.32 mm × 0.25 μm, H_2, HP 5890 GC/ECD, 80 °C, 2 min, 25 °C/min to 130 °C/min, 4 °C/min to 220 °C, 10 °C/min to 260 °C, 4.5 min)

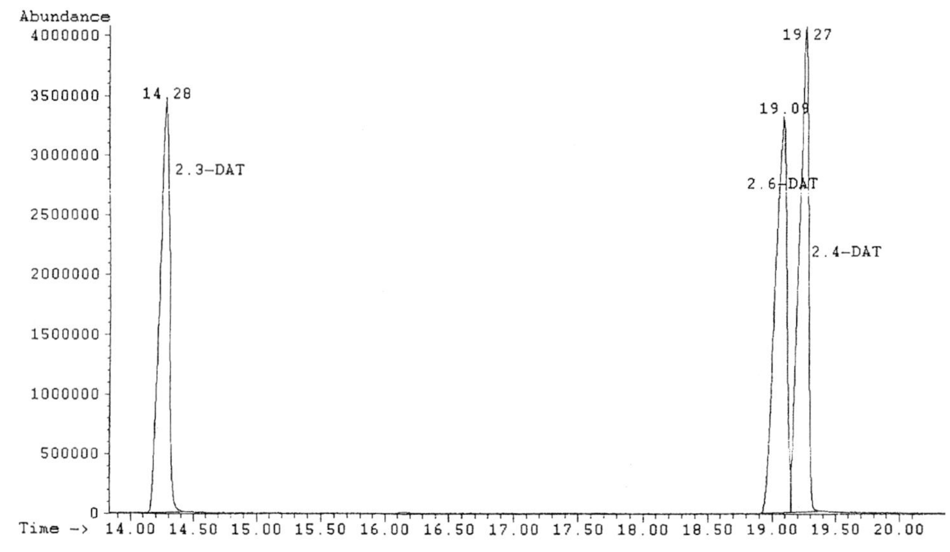

Fig. 4.121 Chromatographic separation of the critical pair 2,4- and 2,6-diaminotoluene on the Rtx-200 capillary column

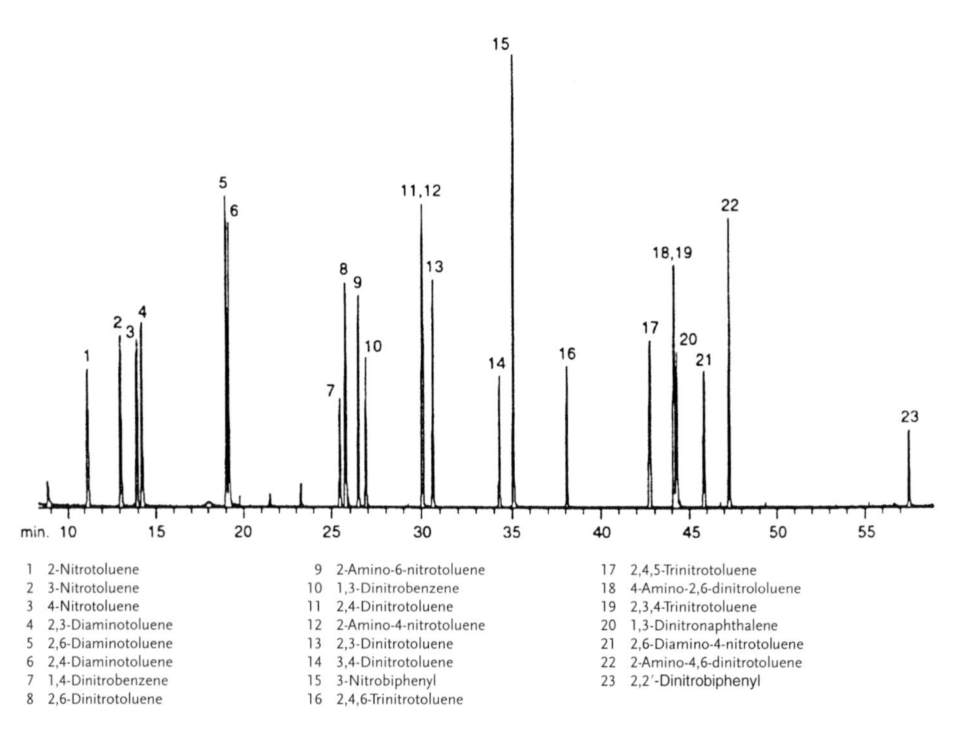

Fig. 4.122 Complete chromatographic separation of nitroromatics and metabolites on the Rtx-200 capillary column

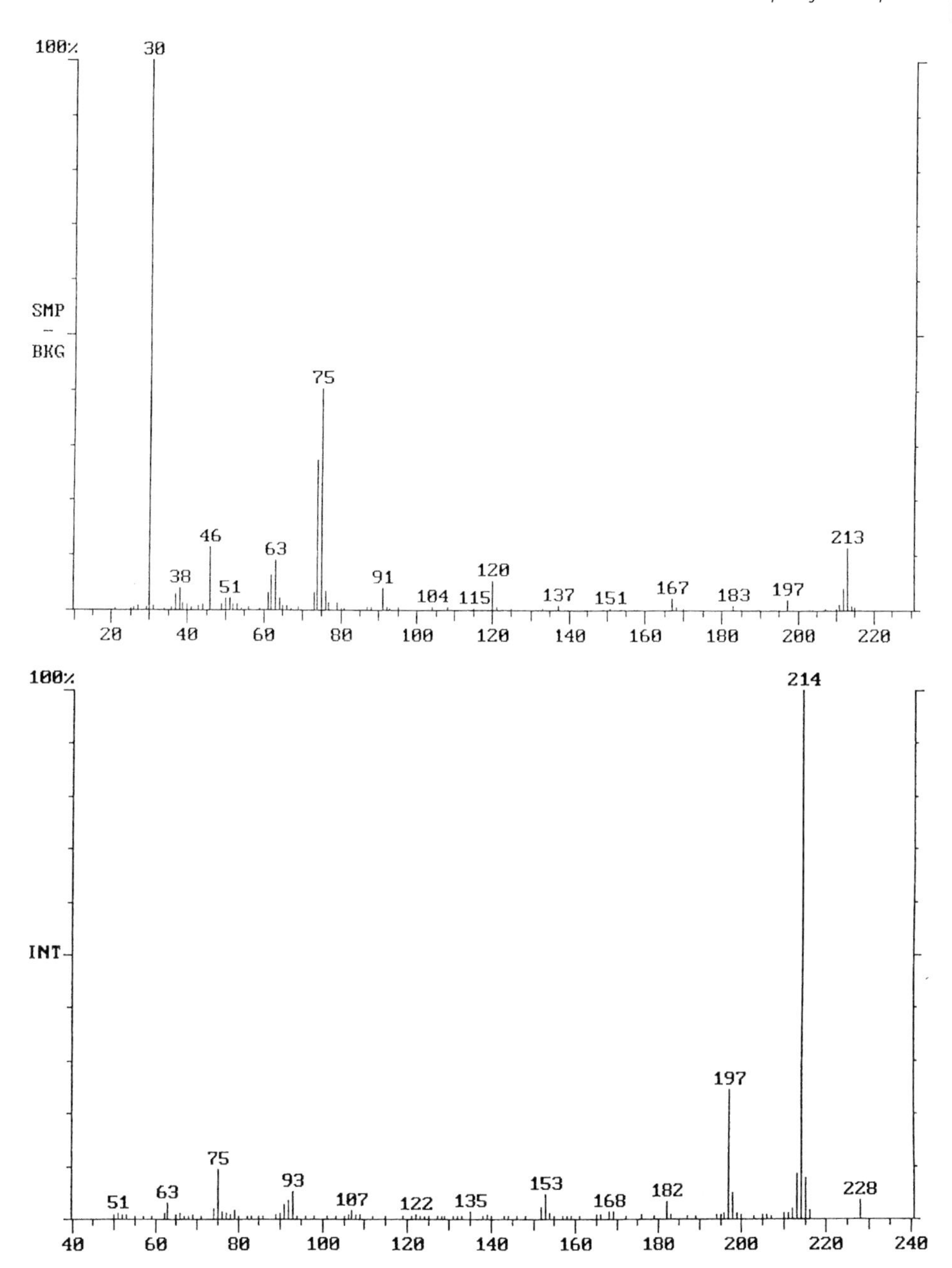

Fig. 4.123 Mass spectra of trinitrobenzenes, M 213
Above: EI spectrum with extensive fragmentation down to m/z 30 (NO^+) from the nitro group
Below: CI spectrum (methane) with $(M+H)^+$ as the base peak

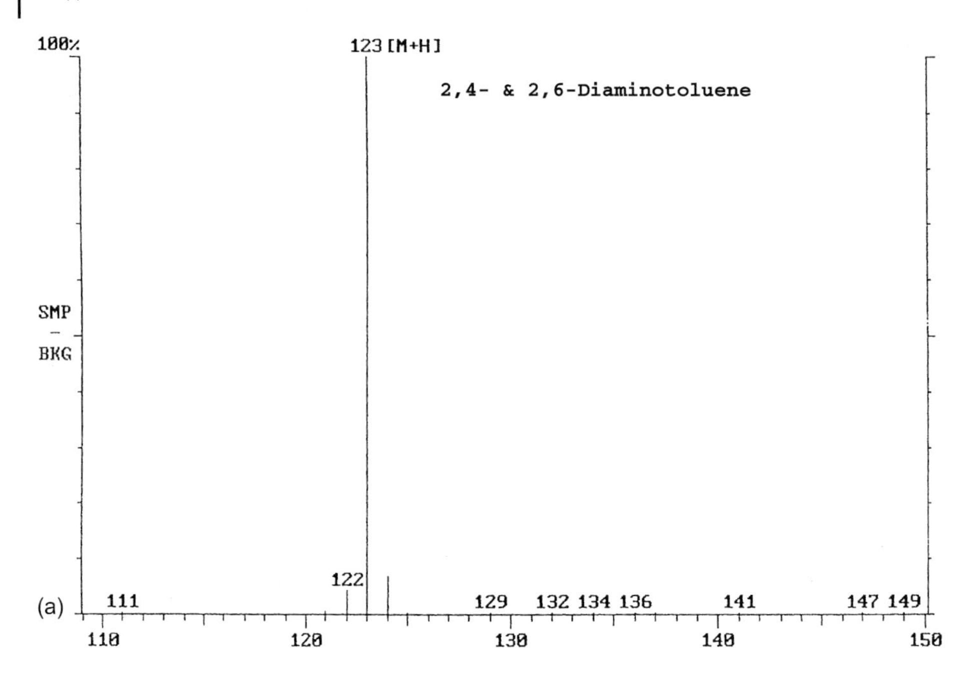

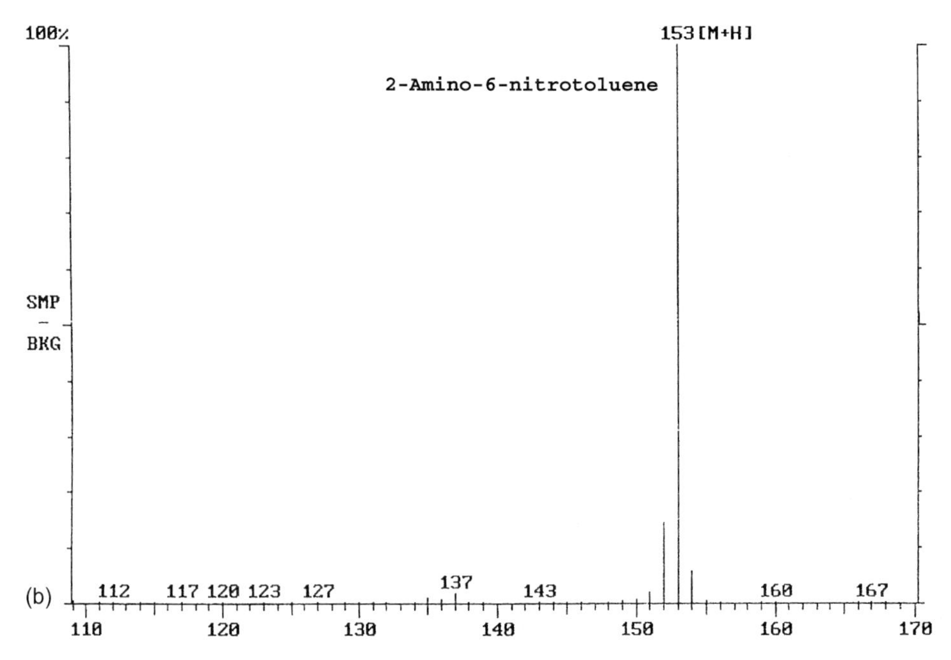

Fig. 4.124 Typical CI spectra with different amino and nitro substitutions
(a) Diaminotoluene
(b) Aminonitrotoluene

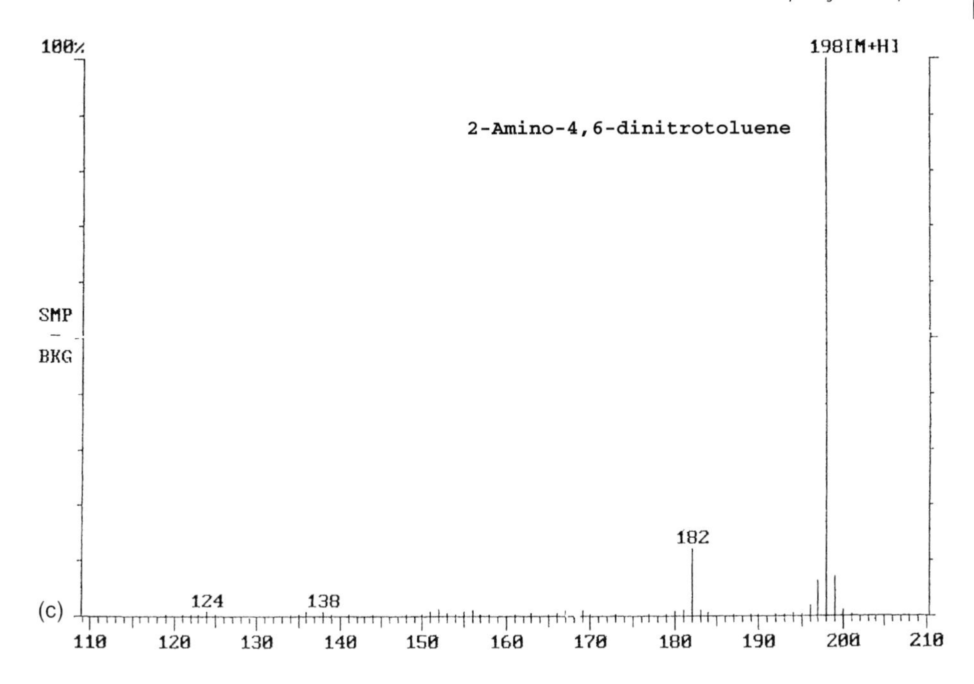

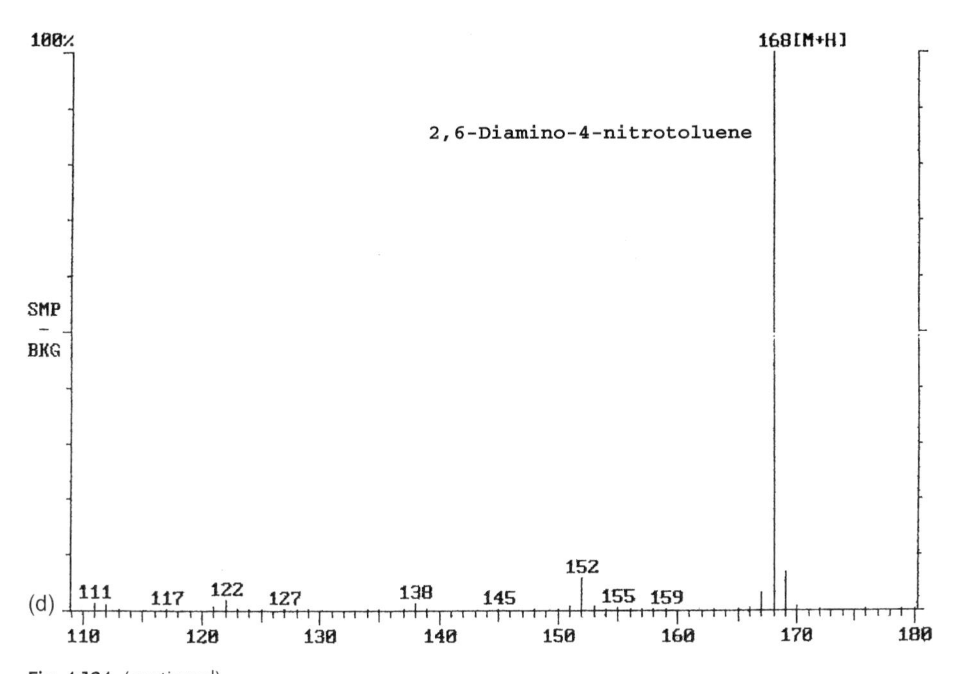

Fig. 4.124 (continued)
(c) Aminodinitrotoluene
(d) Diaminonitrotoluene

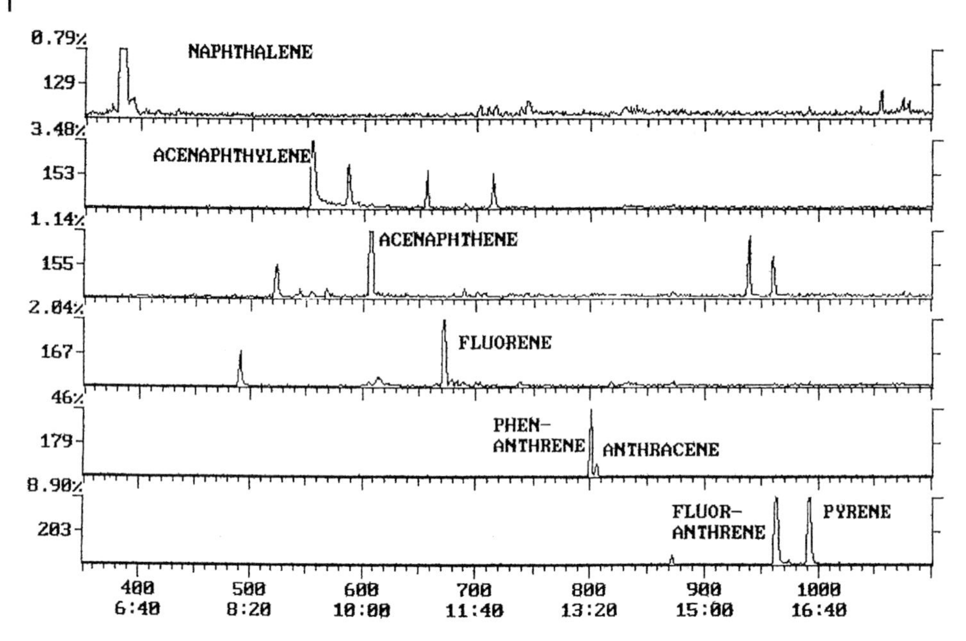

Fig. 4.125 Analysis of a TNT soil sample for PAH components (CI mode)

4.30 Detection of Drugs in Hair

Key words: SFE, opiates, pharmaceuticals, cocaine, methadone, fortifying, MS/MS

Analysis of hair allows information to be obtained concerning a period extending further back than that which can be examined using blood and urine samples. Testing of hair can be used to detect repeated and chronic drug misuse. Narcotics and other addictive drugs and, for example, medicines can be detected unambiguously using the method described and this satisfies the current legal requirements. With hair analysis the chronic misuse of these substances can be traced back over weeks and months (hair growth amounts to ca. 1 cm/month). Amphetamines, cannabis, cocaine, methadone and opiates, for example, can be detected. As this is a stepwise investigation, the development of drug use over a period can be visualised. Monitoring drug dependence in the methadone programme is one application of this type of analysis, as is doping detection in competitive sport. Hair analysis cannot be used to detect an acute drug problem, as a blood or urine test can. The quantity examined should be at least 50 mg. Sometimes a very small quantity such as the quantity of hair found daily in a shaver can be sufficient.

SFE is used for extraction in this application, whereby it is assumed that all the active substances which can be determined by GC can be extracted with supercritical CO_2. SFE has the advantage of a higher extraction rate, as, unlike other procedures, it takes only a few minutes. In addition the extraction unit can be coupled directly to the GC/MS allowing the method to be automated and ever smaller quantities to be analysed.

Analysis Conditions

Sample: A bundle of hair of pencil thickness is used, which is fixed in such a way that movement of the individual hairs over one another is inhibited. Sections of 1 cm length are ground and extracted using SFE.

Extraction: 20 mg hair powder are treated with 20 μL ethyl acetate in the extraction cartridge (volume 150 μL)

Oven temp.:	60 °C
Static extraction:	15 min, 200–300 bar
Dynamic extraction:	5 min, 200–300 bar
Extract isolation:	Transfer of the eluate to a sample bottle with ca. 1 mL ethyl acetate Derivatisation with pentafluoropropionic anhydride (PFPA)

GC/MS/MS system: Finnigan TSQ 700

Injector: Gerstel KAS cold injection system, splitless injection

Column: J&W DB-5, 30 m × 0.32 mm × 0.25 μm
Helium, 1 mL/min

Program:		
	Start:	60 °C, 2 min
	Program 1:	25 °C/min to 230 °C
	Isotherm:	230 °C, 5 min
	Program 2:	25 °C/min to 280 °C
	Final temp.:	280 °C, 5 min
Mass spectrometer:	Scan mode:	EI, MS/MS (product ion scan mode)
	CID gas:	Argon, 2 mbar; collision offset –25 eV

Results

In the method described the SFE was not coupled directly to the GC/MS (on-line SFE-GC/MS), but the extraction, derivatisation and analysis were carried out separately. The use of different pressures (100, 200, 300 bar) shows the different extraction possibilities. At 200 bar the extract produced showed the best signal/noise ratio for heroin (Fig. 4.126). Using derivatisation morphine is detected as morphine-2-PFP and monoacetylmorphine (MAM) as MAM-PFP together with underivatised heroin.

As it is known that the more lipophilic substances, such as heroin, cocaine and tetrahydrocannabinol carboxylic acid (THC), can be deposited directly in hair and do not need to be derivatised for chromatography, SFE can in fact be coupled directly with GC/MS/MS.

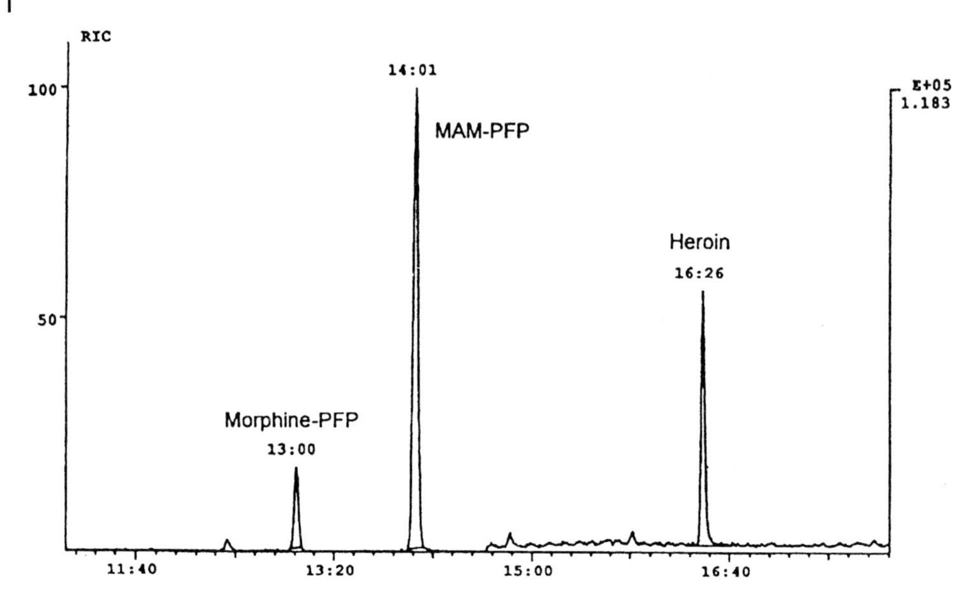

Fig. 4.126 Chromatogram of a hair sample extracted by SFE (200 bar), taken in product ion scan mode (MS/MS)

4.31 Detection of Morphine Derivatives

Key words: drug screening, heroin, codeine, morphine, hydrolysis, derivatisation

Detection of drug taking by the investigation of blood and urine samples is one of the tasks of a forensic toxicology laboratory in workplace drug testing. The routine analysis of drug screening can be carried out by TLC, HPLC or by immunological methods. Positive results require confirmation. Determination using GC/MS is recognised as a reference method. The full scan method is preferred over the SIM procedure for differentiating between different drugs in a comprehensive screening procedure because of the higher specificity and universality. The decision limits (Table 4.33) are laid down by the US Ministry of Health (HHS) and the US Ministry of Defense (DOD).

Heroin (3,6-diacetylmorphine) is usually not determined directly. The unambiguous detection of heroin consumption is carried out by determining 6-monoacetylmorphine (6-MAM) which is formed from heroin as a metabolite (Fig. 4.127). If morphine is difficult or impossible to detect when clean-up is carried out without hydrolysis, then the latter should be used. It should be noted that nonspecific hydrolases can also effect the degradation of 6-MAM to morphine. If the substitute drugs methadone and dihydrocodeine are present, these are also determined using the procedure described. As blood from corpses also has to be processed during routine operations to some extent, a comparatively time-consuming extraction including a re-extraction is carried out. For the gas chromatography of morphine derivatives derivatisation of the extract by silylation, acetylation, or pentafluoropropio-

Table 4.33 Decision limits in drug screening (data in ng/mL)

Active substance	Screening		Confirmation	
	HHS	DOD	HHS	DOD
THC	100 [a)]	50	15	15
BZE	300	150	150	100
Opiates	300	300	300	300
AMPs	1000	500	500	500
PCP	25	25	25	25
BARBs	k.A.	200	k.A.	200
LSD	k.A.	0.5	k.A.	0.2

a) HHS suggestion 50 ng/mL

THC 11-nor-Δ-9-tetrahydrocannabinol-9-carboxylic acid
PCP Phencyclidine
BZE benzoylecgonin
BARB barbiturates
AMP amphetamines
LSD lysergic acid diethylamide

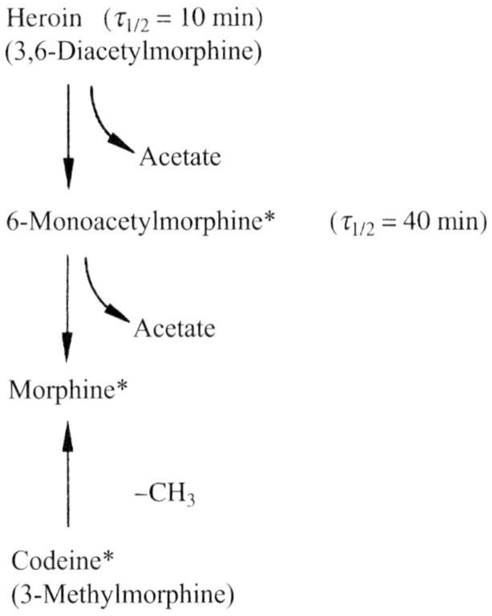

Heroin ($\tau_{1/2} = 10$ min)
(3,6-Diacetylmorphine)

↓ Acetate

6-Monoacetylmorphine* ($\tau_{1/2} = 40$ min)

↓ Acetate

Morphine*

↑ $-CH_3$

Codeine*
(3-Methylmorphine)

Fig. 4.127 Opiate metabolism (* determination by GC/MS)

nylation, for example, is basically necessary. Mass spectrometric detection using negative chemical ionisation is possible by using fluorinated derivatives. In the procedure described silylation with MSTFA (N-methyl-N-trimethylsilyl-trifluoroacetamide) and detection in the EI mode has been chosen.

Analysis Conditions

Clean-up for serum (blood):	1 mL sample, addition of the internal standard (e.g. 100 ng morphine-d_3 or codeine-d_3), dilution to 7 mL with phosphate buffer (pH 6) SPE extraction using Bond-Elut-Certify (130 mg, Varian No. 1211–3050) after conditioning with 2 mL methanol and 2 mL phosphate buffer (pH 6) Elution with 2 × 1 mL chloroform/isopropanol/25% ammonia (70:30:4). Derivatisation of the extract residue with 50 µL MSTFA (80 °C, 30 min)
Clean-up for urine:	Hydrolysis of 2 mL sample with 20 µL enzyme solution (β-glucuronidase) (60 min, 60 °C) Adjustment of sample to pH 8–9 with 0.1 N sodium carbonate solution Activation of C_{18}-cartridge with methanol and conditioning with 0.1 N sodium carbonate solution Application of sample and washing of cartridge with 0.1 N sodium carbonate solution Elution of C_{18}-cartridge with 1 mL acetone/chloroform (50:50) Evaporation of eluate to dryness, treatment of the residue with 50 µL MSTFA (80 °C, 30 min)
Gas chromatograph:	Finnigan GCQ GC (ThermoQuest, Austin, TX, USA) Split/splitless injector, 275 °C Splitless injection, split open at 0.1 min
Column:	J&W DB-1, 30 m × 0.25 mm × 0.25 µm Helium 40 cm/s
Program:	Start: 100 °C Ramp: 10 °C/min Final temp.: 310 °C
Mass spectrometer:	Finnigan GCQ (ThermoQuest, Austin, TX, USA) Scan mode: EI Scan range: 70–440 u Scan rate: 0.5 s/scan

Results

The chromatogram and mass spectra after clean-up of a typical serum sample are shown in Fig. 4.128. The mass traces for codeine, morphine-d_3 and morphine and the total ion current are shown. For morphine a concentration of 60 ng/L is calculated with reference to morphine-d_3 using a calibration curve. The resulting mass spectra allow identification, for example on comparison with relevant toxicologically orientated spectral libraries. The clean-up and analysis methods can be used for other basic drugs, such as amphetamine derivatives, methadone and cocaine and their metabolites.

A list of morphine derivatives and synthetic opiates which have been found by the procedure described after clean-up of serum samples is shown in Table 4.34. For the corresponding trimethylsilyl derivatives the specific search masses are given which allow identification in combination with the retention index.

Table 4.34 Selective masses of the opiates as TMS derivatives (after Weller and Wolf)

Opiate	Ret.-Index	m/z values
Levorphanol-TMS	2188	329, 314
Pentazocine-TMS	2262	357, 342
Levallorphan-TMS	2318	355, 328
Dihydrocodeine-TMS	2365	373, 315
Codeine-TMS	2445	371, 343
Hydrocodone-TMS	2447	371, 356
Dihydromorphine-TMS	2459	431, 416
Hydromorphone-TMS	2499	357
Oxycodone-TMS	2503	459, 444
Morphine-bis-TMS	2513	429, 414
Norcodeine-bis-TMS	2524	429
Monoacetylmorphine-TMS	2563	399, 340
Nalorphine-bis-TMS	2656	455, 440

Note: All TMS derivatives show an intense signal for the trimethylsilyl fragment at m/z 73, which does not appear when the scan is begun at m/z 100 and is unnecessary for substance confirmation. The retention indices were determined on DB-1, 30 m × 0.25 mm × 0.25 µm with an n-alkane mixture of up to C_{32}.

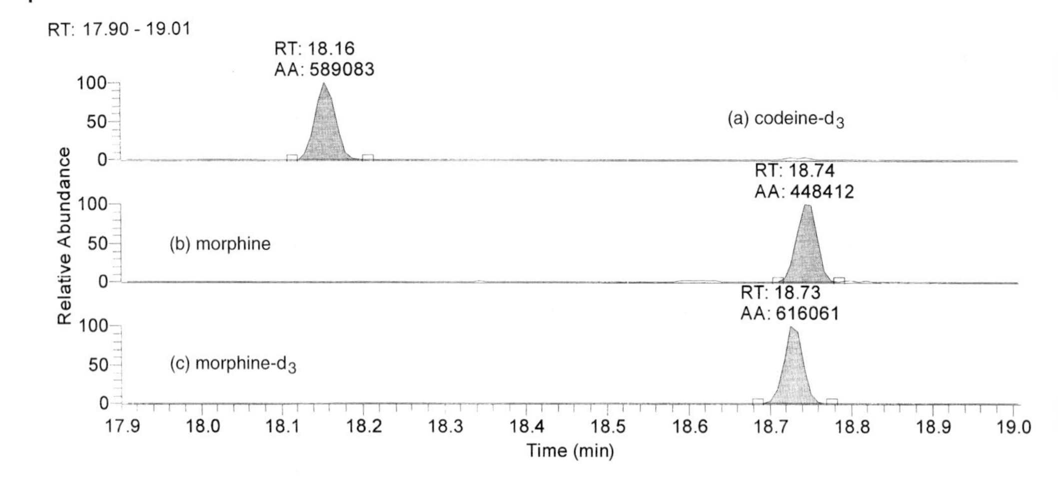

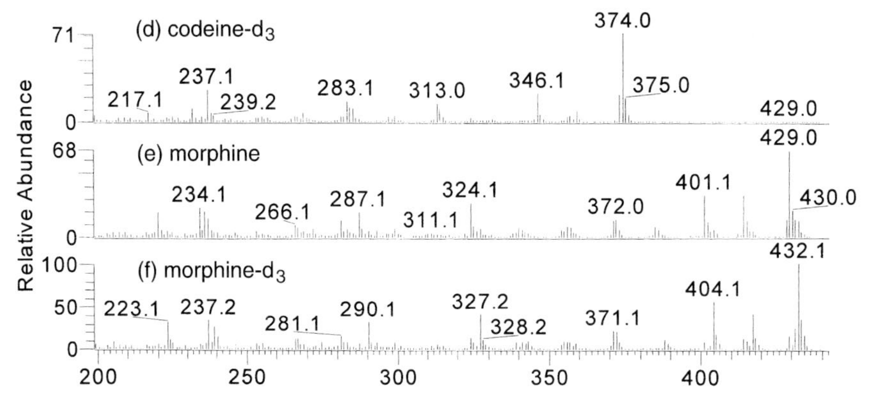

Fig. 4.128 Analysis of an authentic serum sample fortified with 100 ng morphine-d_3 and 100 ng codeine-d_3
Above: (a) Mass trace for codeine-d_3 (m/z 374), (b) Mass trace for morphine (m/z 429)
(c) Mass trace for morphine-d_3 (m/z 432)
Below: (d) Spectrum of codeine-d_3, (e) Spectrum of morphine, (f) Spectrum of morphine-d_3

4.32
Detection of Cannabis Consumption

Key words: THC, metabolite, SPE, derivatisation, quantitation, confirmation, MS/MS, internal standard

Because motorists and those in the workplace are now tested for alcohol and drugs, more samples are being produced. The use of a rapid and reliable analysis procedure is therefore necessary for routine operations. The clean-up of serum and urine samples by solid phase extraction, combined derivatisation and elution and GC/MS detection fulfils these requirements.

To detect cannabis consumption THC (tetrahydrocannabinol) and the main metabolite 11-nor-Δ-9-tetrahydrocannabinol-9-carboxylic acid (9-carboxy-THC), proving the body passage, are determined. The sample preparation method described here involves alkaline hydrolysis of THC bonded to glucuronide and clean-up by special solid phase extraction. Instead of the SPE cartridges usually used, a small SPE filter disc is employed, which is treated directly with the derivatisation agent after applying the sample, washing, and drying. This allows the derivatisation of the 9-carboxy-THC to take place at the same time as the elution and thus dispenses with several steps and the consumption of extra solvent. The 9-carboxy-THC is detected as the TMS derivative. Detection is confirmed from the mass spectrum. For quantitation the internal standard d_9-9-carboxy-THC is used and the mass chromatogram is evaluated. Alternatively MS/MS can be used for detection. Through the increased selectivity limits of detection in the lowest ng/mL range can be achieved for serum samples.

Analysis Conditions

Sample preparation:

Hydrolysis: 5 mL urine are treated with 200 μL 10 N KOH and incubated for 15 min at 60 °C

Clean-up: After cooling of the hydrolysate, 700 μL conc. acetic acid are added, the sample passed through a ToxiLab Spec extraction disc (SPE), the disc washed with 1 mL 20% acetic acid, and dried (15 min, vacuum). The SPE disc is placed in a sealable vessel, 75 μL BSTFA are added for derivatisation and elution and the disc left for 10 min at 90 °C. The sample is injected into the GC/MS system without further steps.

GC/MS/MS system: Finnigan Witness System

Column: SGE HT8, 25 m × 0.25 mm × 0.25 μm

Program:
Start: 60 °C, 1 min
Ramp: 25 °C/min
Final temp.: 330 °C

Mass spectrometer:
Scan mode: EI, full scan, 200–500 u
Scan rate: 0.6 s/scan

Calibration: For calibration d_9-9-carboxy-THC is added at a concentration of 50 ng/mL and 9-carboxy-THC in concentrations of 2.5, 5, 15, 50, 200 and 500 ng/mL as the internal standards to 5 mLTHC-free urine.

Results

The extracts from the disc clean-up (SPE) are very clean, and give a high recovery and good quantitative precision. 9-Carboxy-THC can be detected in urine samples down to the threshold value of 15 ng/mL recommended by the NIDA (US National Institute on Drug Abuse) by examining the whole mass spectrum (Fig. 4.129). In the calibration a linear response in the range of 5 ng/mL to 500 ng/mL is achieved. For this procedure involving urine the calibration graph is particularly precise.

In the analysis of serum samples for the nonmetabolised active substance THC there is a much stronger chemical background in the chromatogram, which does not allow the limits of quantitation given for urine to be achieved so easily. The effect of the matrix can be effectively counteracted by detection of THC in the MS/MS mode and effective limits of quantitation of below 1 ng/mL serum can be achieved. Fig. 4.130 shows the product ion spectrum of THC-TMS arising from the precursor ion m/z 386. THC-d_3 is used as the internal standard. The GC peak is successfully integrated on the mass chromatogram of the base peak m/z 371 and that of the internal standard m/z 374.

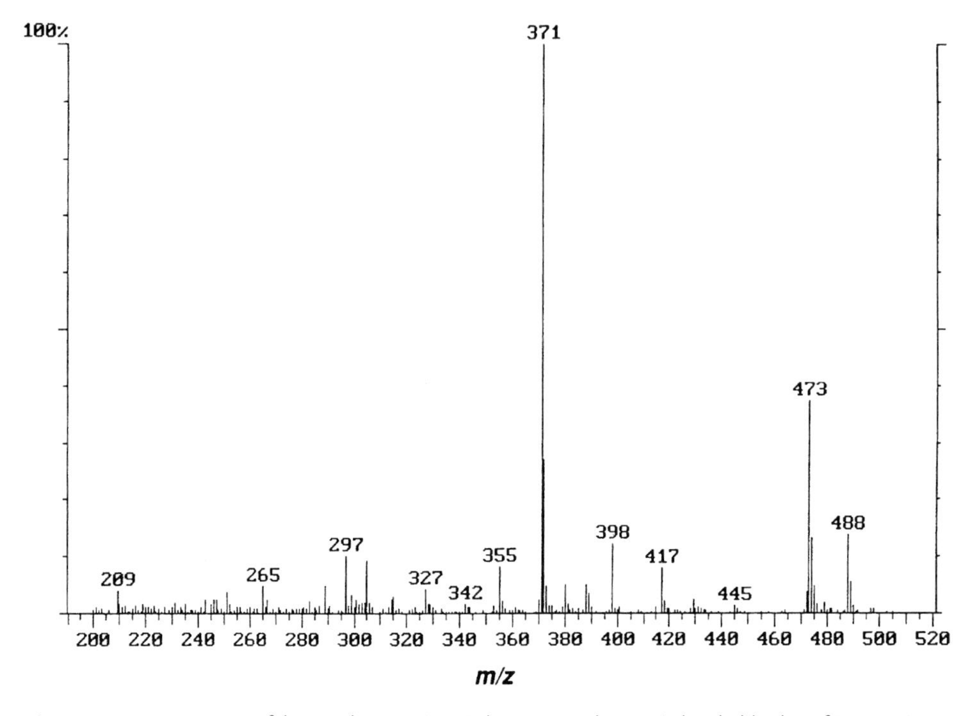

Fig. 4.129 Mass spectrum of the 9-carboxy-THC-TMS derivative at the NIDA threshold value of 15 ng/mL

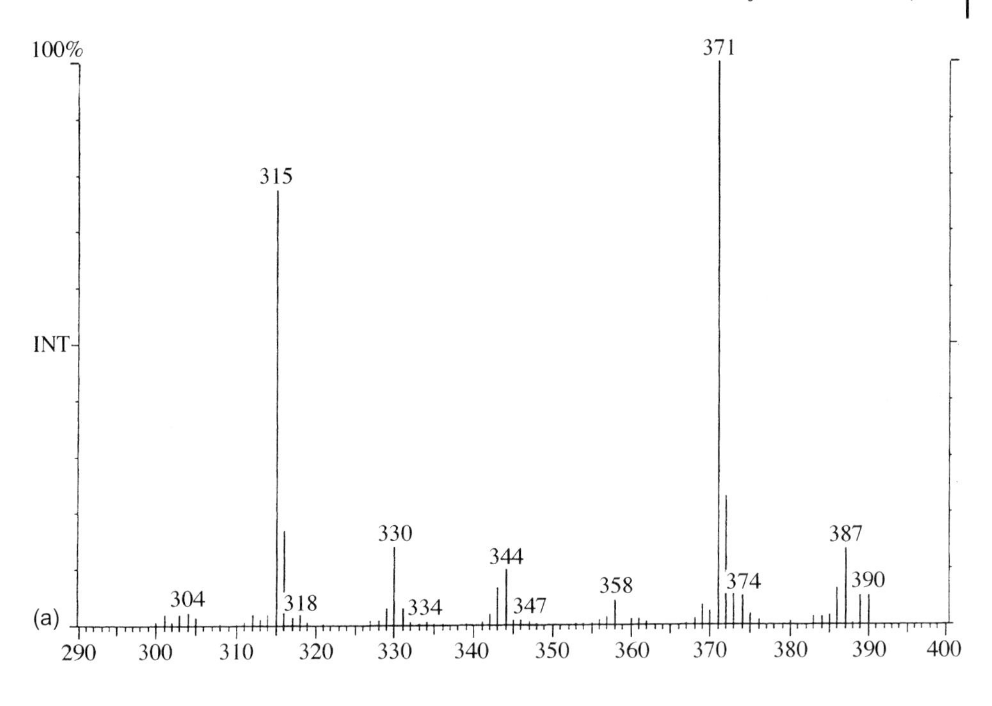

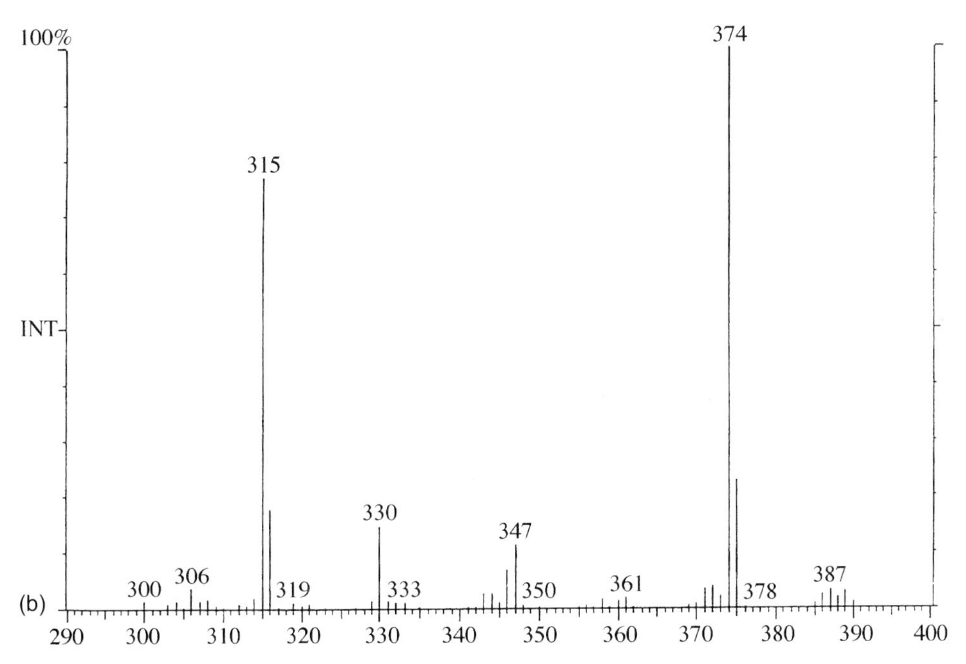

Fig. 4.130 MS/MS product ion spectrum of (a) THC-TMS, precursor ion m/z 386 and (b) THC-d_3-TMS, precursor ion m/z 389 (after Weller and Wolf 1995, Finnigan GCQ)

4.33
Analysis of Steroid Hormones Using MS/MS

Key words: anabolics, derivatisation, GC/MS, MS/MS, signal/noise ratio, selectivity

Besides the area of food safety, the analysis of steroid hormones is important in sports doping analysis. The steroids used as anabolics are derived from the natural androgenics. The use of the body's own steroids, such as testosterone, or synthetic compounds derived from them, such as methandienone tetrahydrogestrinon (THG), is forbidden. Details are regulated by the World Anti Doping Agency (WADA). They are detected by analysing a urine sample and detection must be unambiguous. One difficulty with the analysis arises from the usually very low concentrations of individual active substances together with a high concentration of matrix in the extracts. Many laboratories only recognise a finding as positive when a full scan mass spectrum of the substance can be produced.

Methandienone, a widely used anabolic, is metabolised in the body and excreted as 6-β-hydroxymethandienone. The method presented here demonstrates the effect of the GC/MS/MS technique on the selectivity of trace detection of the methandienone metabolite compared to GC/MS analysis in SIM mode. A standard solution of the metabolite at a dilution of 1 ng/µL and the extract of a fortified urine sample (10 ng/mL) with a concentration of 500 pg/µL are used for the measurement. For the gas chromatography the standard and the urine extract were converted into the tri-TMS derivatives.

Analysis Conditions

GC/MS system:	PolarisQ Ion Trap MS (Thermo Fisher, Austin, TX, USA)	
Column:	HP 1, 25 m × 0.2 mm × 0.1 µm	
Program:	Start:	90 °C,
	Ramp:	15 °C/min
	Final temp.:	320 °C
Mass spectrometer:	Scan mode:	EI, full scan, 50–600 u
	Scan rate:	0.7 s/scan
	Ion source:	175 °C

Results

The mass spectra of 6-β-hydroxymethandienone are shown in Fig. 4.131. The EI-GC/MS spectrum shows a dominant base ion at m/z 517 arising from the molecular ion at m/z 532 by loss of the methyl group. A further fragmentation is insignificant. The product ion spectrum of the EI fragment m/z 517 is shown in Fig. 4.131 b. The intensity of the precursor ion has almost completely disappeared in favour of the characteristic product ions. The product ion signals m/z 229 and 337 have high intensity.

The EI-GC/MS analysis of the urine extract at a concentration of 10 ng/mL is shown in Fig. 4.132 a together with the mass chromatograms of the molecular ion m/z 532 and the

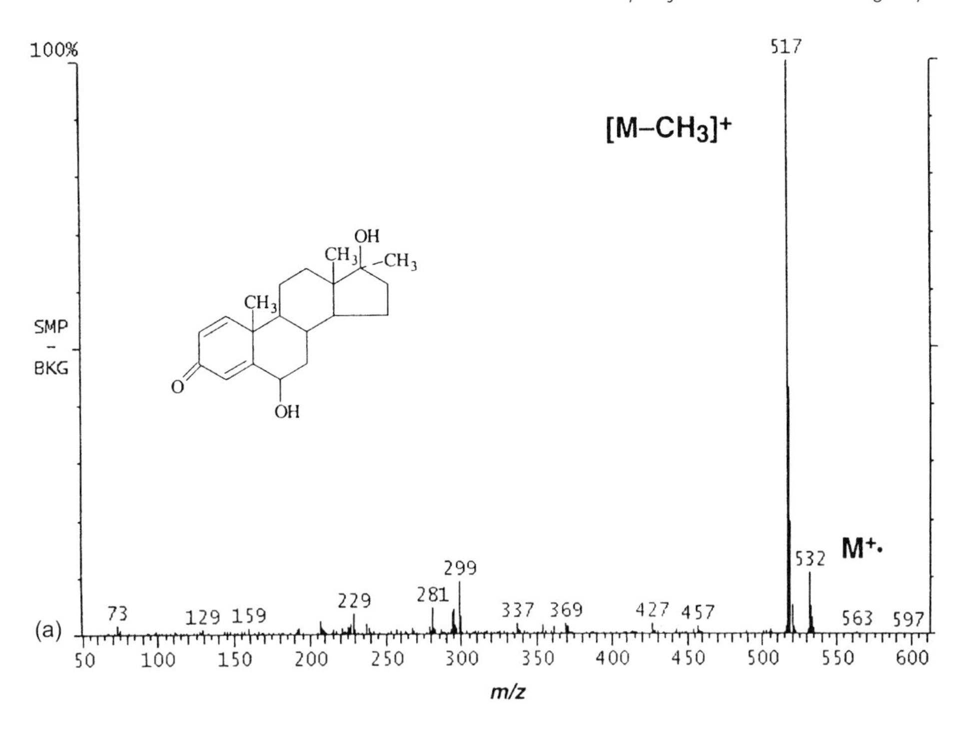

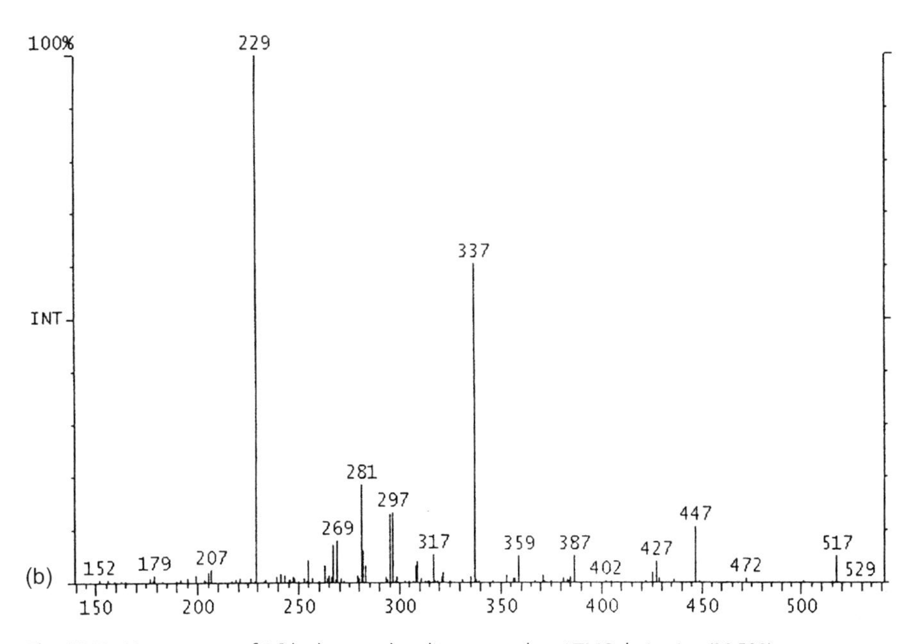

Fig. 4.131 Mass spectra of 6β-hydroxymethandienone as the tri-TMS derivative (M 532)
(a) EI-GC/MS
(b) EI-GC/MS/MS, precursor ion m/z 517

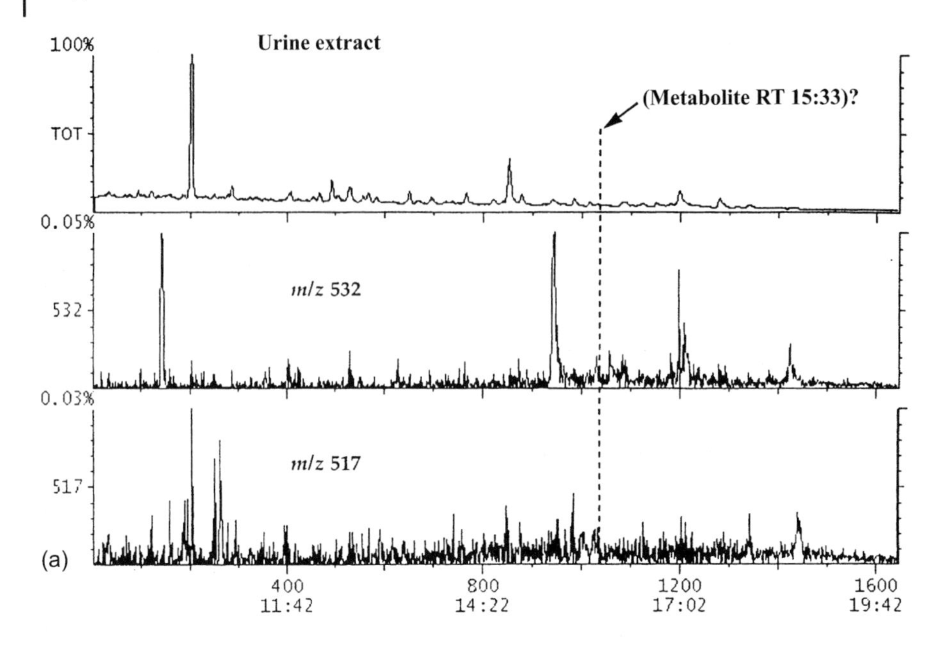

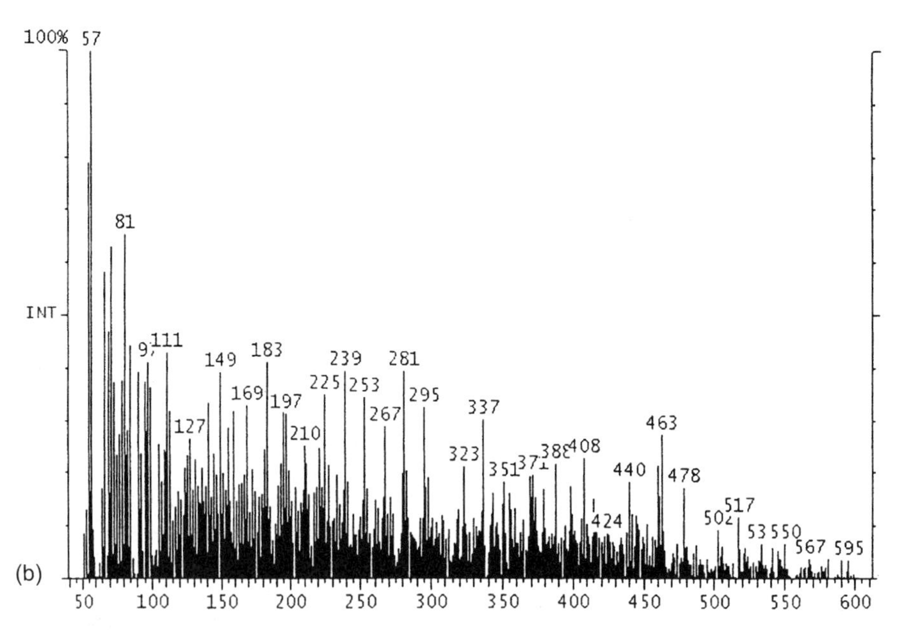

Fig. 4.132 EI-GC/MS analysis of a urine extract with a 6β-hydroxymethandienone concentration of 10 ng/mL
(a) Mass chromatograms m/z 532, 517
(b) Spectrum at the expected retention time 15:33 min

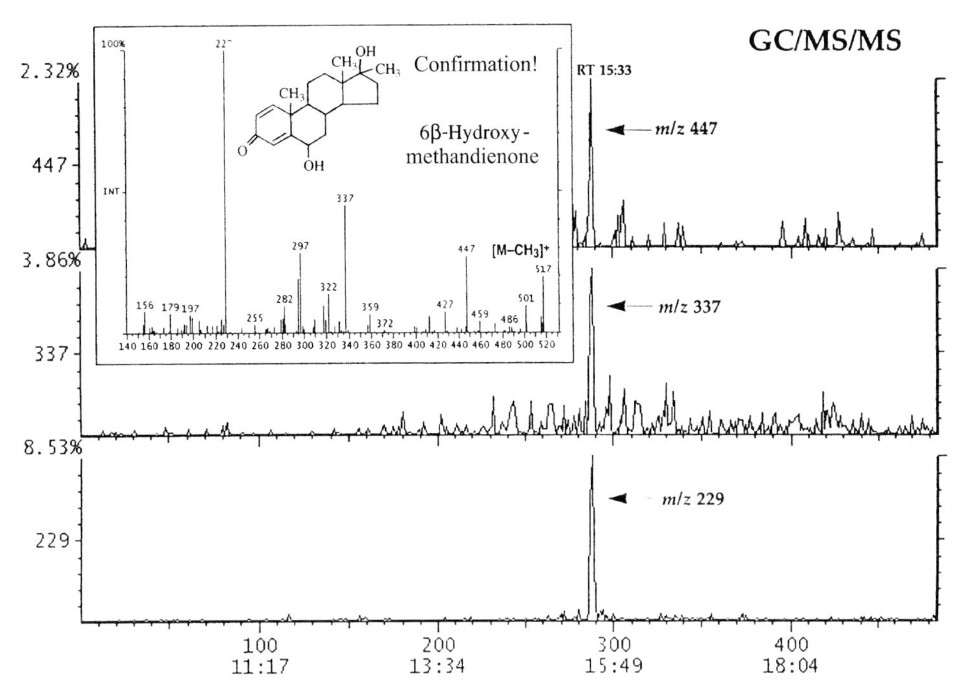

Fig. 4.133 EI-GC/MS/MS analysis of the urine extract from Fig. 4.132 with the product ion spectrum of 6 β-hydroxymethandienone using m/z 517 as the precursor ion

base peak m/z 517. At the expected retention time of the metabolites of 15:33 min no signal can be seen on any of the mass traces. Both masses are affected by chemical noise in the sample. It is not possible to obtain the mass spectrum of the metabolites at this point (Fig. 4.132 b). In the upper mass range also many unspecified matrix signals are present which completely mask the signal of the 6-β-hydroxymethandienone.

The effect of the co-eluting matrix can be eliminated using EI-GC/MS/MS analysis. The MS/MS analysis of the urine extract shows a clear picture (Fig. 4.133). Through the isolation of the precursor ion m/z 517 and the generation of the product ion spectrum, the appearance of the 6-β-hydroxymethandienone in the urine is unequivocal. The mass chromatograms of the product ions m/z 229, 337 and 447 show a common peak maximum at the expected retention time. For the intense product ion m/z 229 a peak with a high signal/noise ratio is detected, which is completely isolated from the matrix by the selectivity of the MS/MS technique. The product ion spectrum from the urine sample matches the standard very well and unquestionably confirms the detection of anabolics (Fig. 4.131 b).

4.34
Determination of Prostaglandins Using MS/MS

Key words: prostaglandins, urine, derivatisation, GC/MS, MS/MS, co-elution, quantitation

Prostaglandins are cyclic derivatives of arachidonic acid (5,8,11,14-eicosatetraenic acid). They occur widely in organisms in very low concentrations and exhibit many types of mediator activity similar to that of hormones. For example, blood pressure, stomach secretion, aching muscles, and in general pain and fevers are affected by prostaglandins. Many human prostaglandins are synthesised in the kidneys. Synthetic derivatives are also known. The analysis of the arachidonic acid metabolism and investigations of the pharmacological spectrum of activity of prostaglandins has been of great scientific interest for many years. While prostaglandins are degraded very rapidly in an organism, the concentration in urine can be used to check prostaglandin production and levels.

For the analysis of body fluids GC/MS and GC/MS/MS are used as a reference and addition to immunological procedures. The classical procedures are not specific enough so the values determined can be too high as a consequence of the accompanying matrix (Schweer 1986). The concentrations recorded for healthy children lie in the range of 100 pg/mL urine, and in the case of children with Zellweger syndrome the values can rise by more than 100 fold (8-iso-$PGF_{2\alpha}$).

For sample preparation (typically 5 mL urine) SPE is used as described here with subsequent PFB derivatisation and additional purification of the extracts by thin layer chromatography. For quantitative determinations reversed phase HPLC can be used for further purification of the extracts. This may however be omitted when MS/MS detection is used without loss of quality (Tsikas 1998). The keto and OH groups are then derivatised through formation of methoximes and TMS ethers (BSTFA).

In spite of the time-consuming sample preparation, with one stage GC/MS analysis co-elutions cannot be reliably excluded in spite of selectivity by single mass registration (SIM, MID). Only by the selective GC/MS/MS procedure described here (Schweer 1988, Tsikas 1998) can prostaglandins be determined reliably for diagnostic purposes. The possibilities for MS/MS are shown graphically. A direct comparison between triple quadrupole and ion trap instruments is shown.

Analysis Conditions

(1) (Tsikas 1998)

GC/MS system:	Finnigan TSQ 45 (triple stage quadrupole GC/MS/MS mass spectrometer) (Finnigan Corp., San Jose, CA, USA)
Gas chromatograph:	Finnigan GC 9611
Injector:	Splitless injection, 280 °C, 2 min splitless
Column:	SPB 1701: 30 m × 0.25 mm × 0.25 µm Helium, constant pressure 70 kPa

Program:	Start:	80 °C, 2 min
	Program 1:	250 °C, 25 °C/min
	Program 2:	280 °C, 2 °C/min
	Final temp:	280 °C, 5 min
MS interface:	280 °C	
Mass spectrometer:	Ionisation:	NCI, argon/methane (0.2/65 Pa)
	Ion source:	140 °C
	Scan mode:	MS/MS product ion scan
	Collision gas:	argon/methane (0.2/65 Pa)
	Collision energy:	18 eV

(2) (Gummersbach 1999)

GC/MS system:	Finnigan Polaris (ion trap GC/MS/MS mass spectrometer)	
Gas chromatograph:	CE Instruments Trace GC 2000	
Injector:	Splitless injection, 260 °C, 2 min splitless	
Column:	J&W 1301: 30 m × 0.32 mm × 0.25 μm Helium, constant flow ca. 30 cm/s	
Program:	Start:	80 °C, 2 min
	Program 1:	230 °C, 25 °C/min
	Program 2:	280 °C, 8 °C/min
	Final temp.:	280 °C, 22 min
MS interface:	280 °C	
Mass spectrometer:	Ionisation:	NCI, methane
	Ion source:	150 °C
	Scan mode:	MS/MS product ion scan
	Collision gas:	helium
	Collision energy:	1.4 V

Results

The GC-NCI-MS/MS procedure presented allows rapid and specific determination of prostaglandins in human urine. The retention times obtained with the most widely used GC columns are shown in Table 4.35. Table 4.36 shows the most important ions in the one-stage MS spectrum of the compounds arriving for analysis. In the case of $PGF_{2\alpha}$-PFB-TMS derivatives the fragment obtained after cleavage of the PFB group gives rise to the base peak. This ion is used as the precursor ion for the MS/MS analysis used in the selective determination. The resulting MS/MS spectra are summarised in Table 4.37.

To carry out the analyses only triple quadrupole systems are used in the literature cited. The use of the recently developed ion trap GC/MS/MS systems gives results of comparable selectivity at the required level of sensitivity (Figs. 4.134 and 4.135). The spectra obtained show the fragmentation pattern used for identification in the same way and with comparable intensities (Fig. 4.136). Similarly their use for analysis of other precipitated prostanoid components in the urine of sick children is also shown (Fig. 4.137) and thus makes the alternative use of ion trap systems worth considering, for example, for the diagnoses of Zellweger syndrome in children, as well as for research tasks.

Table 4.35 Gas chromatographic retention times of PFB-TMS derivatives of the prostanoids investigated using two different capillary columns (after Tsikas 1998)

F_2-prostaglandin derivative	Retention time (min)/rel. ret. time DB5-MS	Retention time (min)/rel. ret. time SPB-1701
9β,11α-PG$F_{2\alpha}$-PFB-TMS	22.88/1.0000	22.05/1.0000
$^2H_4$-8-iso-PG$F_{2\alpha}$-PFB-TMS	22.90/1.0009	22.19/1.0063
8-iso-PG$F_{2\alpha}$-PFB-TMS	22.97/1.0039	22.25/1.0091
9α-11β-PG$F_{2\alpha}$-PFB-TMS	23.48/1.0262	22.62/1.0258
$^2H_4$-PG$F_{2\alpha}$-PFB-TMS	23.70/1.0389	23.05/1.0454
PG$F_{2\alpha}$-PFB-TMS	23.77/1.0389	23.11/1.0481

Table 4.36 GC-NICI-MS mass spectra of PG$F_{2\alpha}$-PFB-TMS derivatives, I = PG$F_{2\alpha}$-PFB-TMS, II = 3,3′,4,4′-$^2H_4$- PG$F_{2\alpha}$-PFB-TMS (ions $[M\text{-}PFB]^-$ = $[P]^-$, after Schweer 1988)

Ion assignment	I	II	% Int.
$[P]^-$	569	573	100
$[P\text{-}(CH_3)_2Si{=}CH_2]^-$	481	485	8
$[P\text{-}TMSOH]^-$	479	483	10
$[P\text{-}TMSOH\text{-}(CH_3)_2Si{=}CH_2]^-$	407	411	1
$[P\text{-}2xTMSOH\text{-}(CH_3)_2Si{=}CH_2]^-$	317	321	1
$[P\text{-}3xTMSOH]^-$	299	303	1
$[C_6F_6CHO]^-$	196	196	1
$[C_6F_6CH_2]^-$	181	181	2
$[C_6F_4CH_2O]^-$	178	178	1

Table 4.37 Major mass fragments (intensity >5% is given in parentheses) in the GC-NCI-MS/MS mass spectra of the PFB-TMS derivatives of the prostaglandins (precursor ions $[M\text{-}PFB]^- = [P]^-$, after Tsikas 1998)

Ion assignment	8-iso-PGF$_{2\alpha}$	$^2H_4$-8-iso-PGF$_{2\alpha}$	PGF$_{2\alpha}$	$^2H_4$-PGF$_{2\alpha}$
$[P]^-$	569 (26)	573 (15)	569 (10)	573 (17)
$[P\text{-}TMSOH]^-$	479 (17)	483 (15)	479 (12)	483 (30)
$[P\text{-}2xTMSOH]^-$	389 (32)	393 (37)	389 (32)	393 (40)
$[P\text{-}2xTMSOH\text{-}(CH_3)_2Si{=}CH_2]^-$	317 (23)	321 (33)	317 (46)	321 (28)
$[P\text{-}3xTMSOH]^-$	299 (100)	303 (100)	299 (100)	303 (100)
$[P\text{-}2xTMSOH\text{-}(CH_3)_2Si{=}CH_2\text{-}CO_2]^-$	273 (33)	273 (60)	277 (50)	277 (37)
$[P\text{-}3xTMSOH\text{-}CO_2]^-$	255 (76)	255 (62)	259 (50)	259 (33)

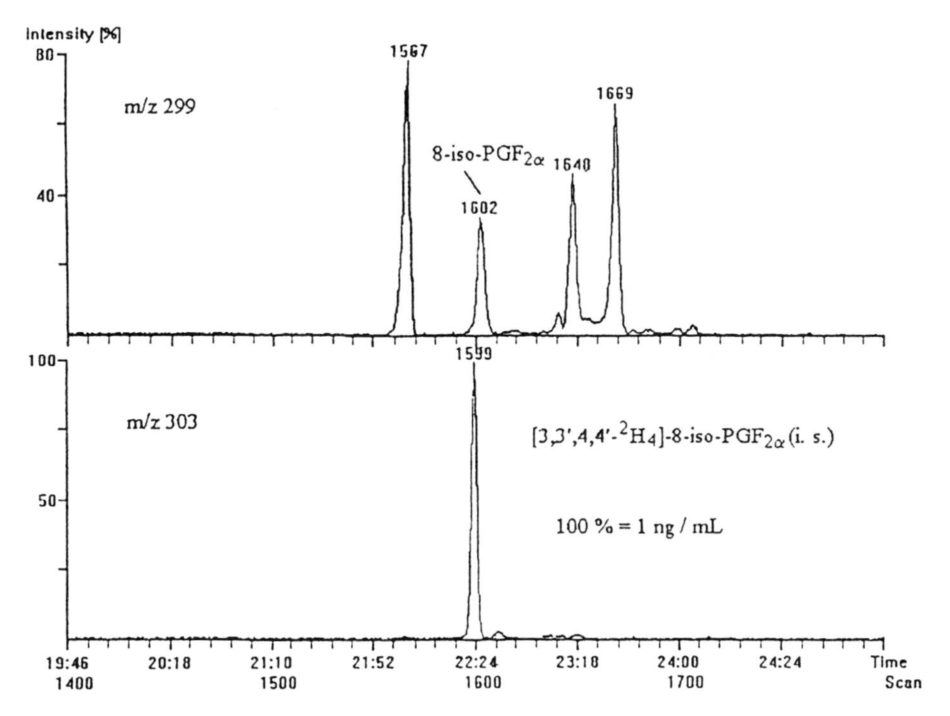

Fig. 4.134 Partial chromatogram from the TSQ GC-NCI-MS/MS analysis of urine samples from a healthy child (TLC fraction) containing the PFB derivatives of 8-iso-PGF$_{2\alpha}$ (m/z 299) and 3,3′,4,4′-$^2H_4$-PGF$_{2\alpha}$ (m/z 303), (cap. column SPB-1701, after Tsikas 1998)

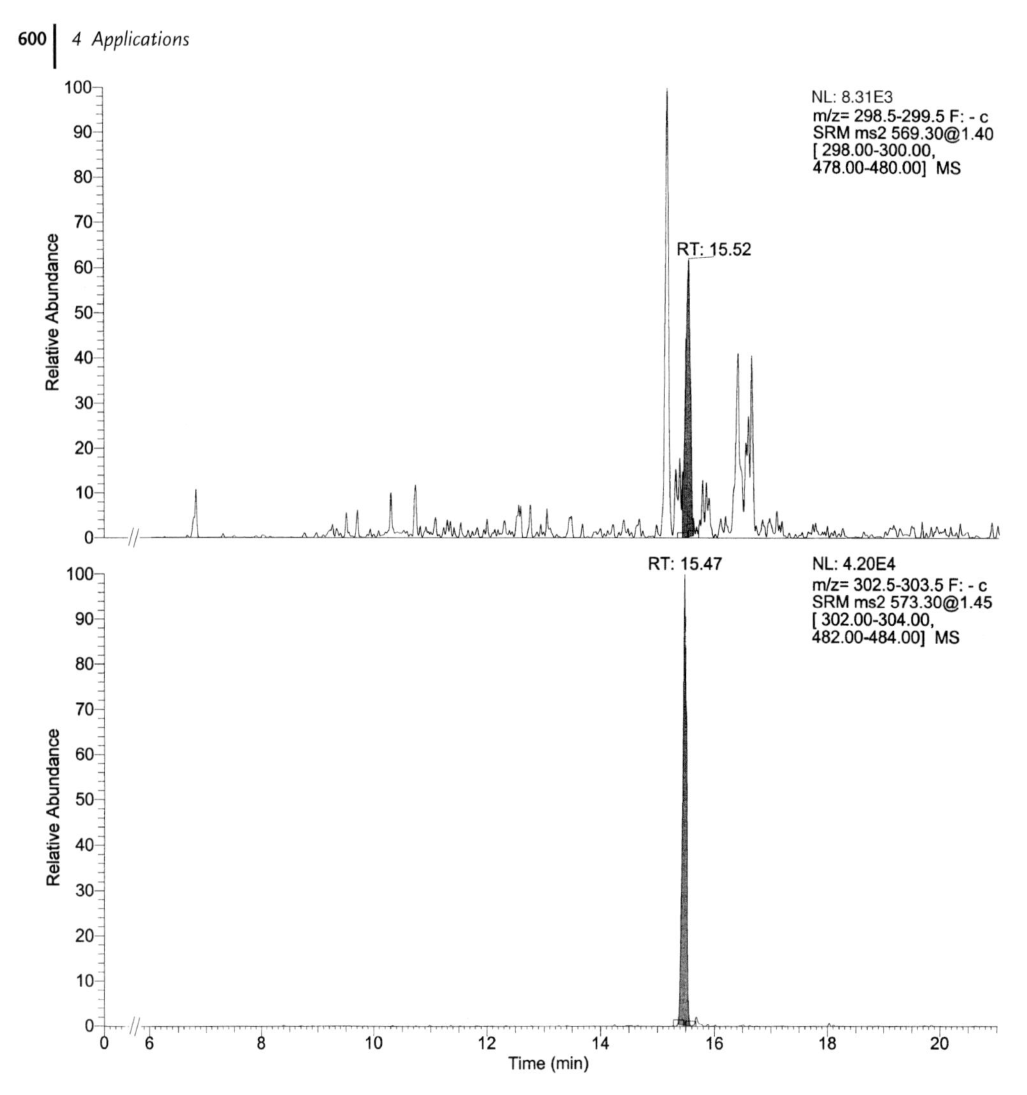

Fig. 4.135 Partial chromatogram from the polaris GC-NCI-MS/MS analysis of urine samples from a healthy child (TLC fraction) containing the PFB derivatives of 8-iso-$PGF_{2\alpha}$ (m/z 299, upper trace) and 3,3′,4,4′-$^{2}H_{4}$-$PGF_{2\alpha}$ (m/z 303, lower trace), (cap. column J&W 1301, Gummersbach 1999)

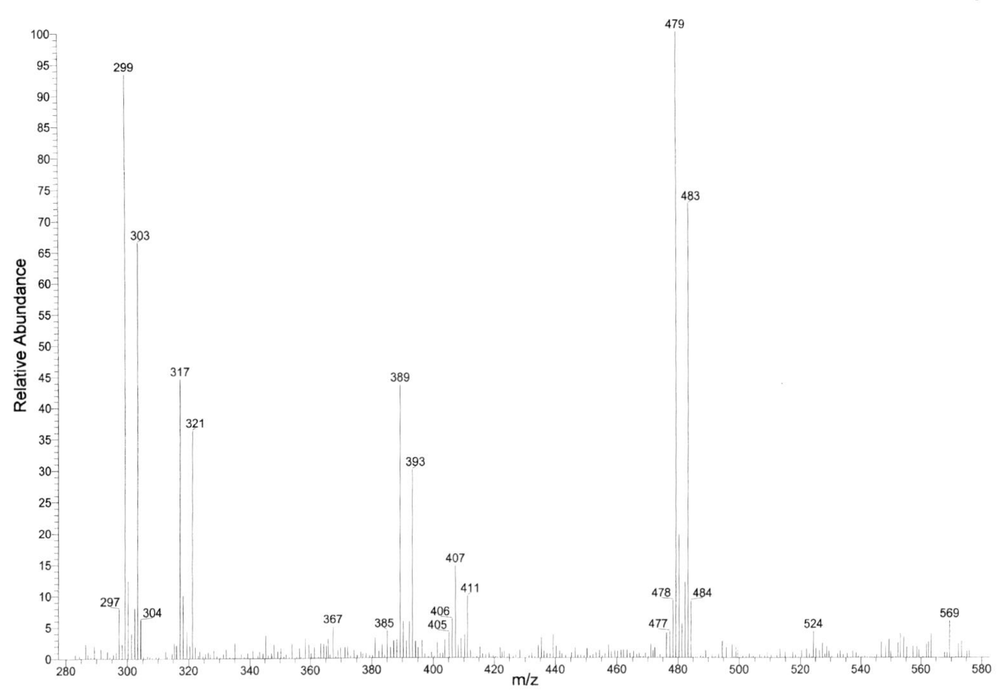

Fig. 4.136 Ion trap MS/MS mass spectrum of peaks RT: 15 : 52 min and 15 : 47 min of Fig. 4.135 (added 15 : 33 to 15 : 59) of $[P]^-$ precursor ions m/z 569 ($PGF_{2\alpha}$-PFB-TMS) and m/z 573 ($3,3',4,4'$-$^2H_4$-$PGF_{2\alpha}$-PFB-TMS)

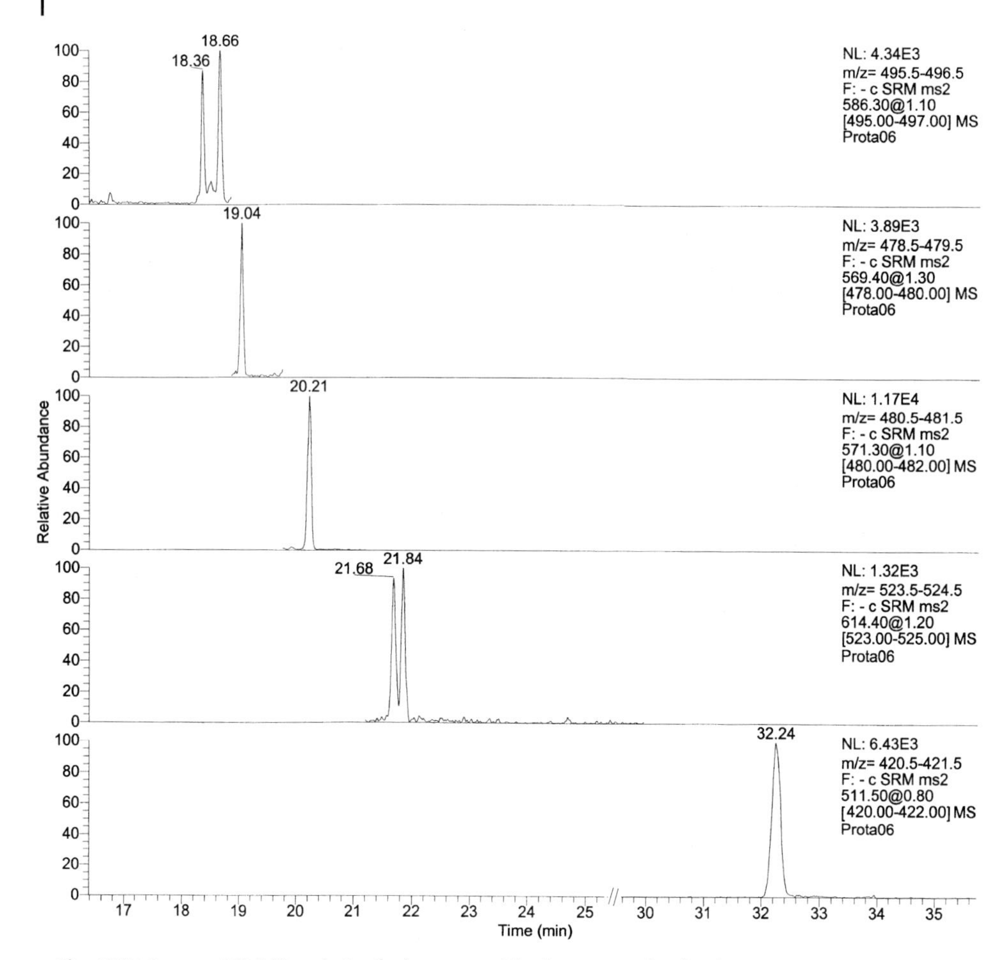

Fig. 4.137 Ion trap MS/MS analysis of other protanoid substances at low level concentration (50 pg/ul) 2,3-dinor-6-keto-$PGF_{1\alpha}$ (RT 18:36, 18:66 two isomers, precursor m/z 586.3), 8-iso-$PGF_{2\alpha}$ (RT 19:04, precursor m/z 569.4), $PGF_{1\alpha}$ (RT 20:21, precursor m/z 571.3), 6-keto- $PGF_{2\alpha}$ (RT 21:68, 21:84 two isomers, precursor m/z 614.4), 11-dehydro-TxB_2 (RT 32:24, precursor m/z 511.5)

4.35
Detection of Clenbuterol by CI

Key words: Clenbuterol, Salbutamol, chemical ionisation, internal standard quantification

β-Sympathomimetics are used both in human and animal medicine for treating diseases of the respiratory passages. In treatment of animals with higher concentrations an increase in muscle growth and an improvement in the meat/fat ratio is achieved. The use of these compounds for fattening purposes is forbidden in some countries. Clenbuterol has become known as a fattening and doping agent. Another active substance used from this group is Salbutamol.

HPLC, EIA and GC methods are applied for trace analysis. GC/MS procedures also allow the positive detection of the substance as a TMS derivative even in the trace region. On-column injection has proved to be particularly effective. However, because of the high matrix concentrations, the use of the hot split injector for routine analysis is preferred. Electron impact ionisation is unsuitable for residue analysis as intense fragments are formed exclusively in the lower mass region which is occupied by the matrix. Clenbuterol can be detected with greater sensitivity using GC/MS with positive chemical ionisation (protonation by methane) (Fig. 4.138).

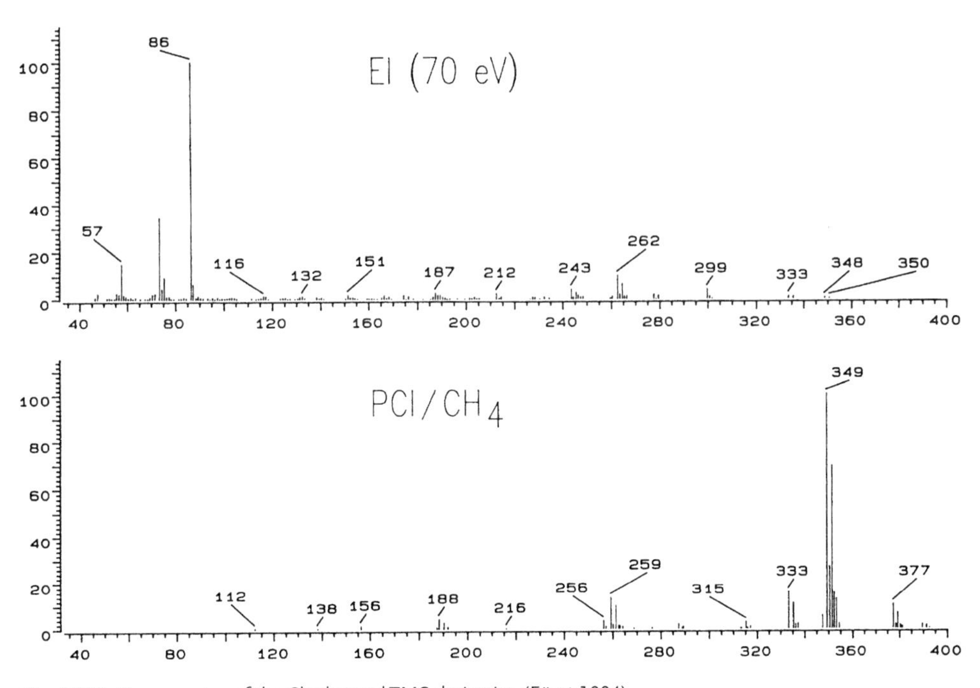

Fig. 4.138 Mass spectra of the Clenbuterol-TMS derivative (Fürst 1994)

Analysis Conditions

Derivatisation:	The extracts are dried in a stream of N_2 and treated with 200 µL HMDS (70 °C, 45 min). For injection, the excess HMDS can be blown out. The residue is taken up in acetone/5% HMDS or dichloromethane
GC/MS system:	Finnigan ITS 40
Split injector:	270 °C isotherm Splitless injection
Column:	J&W DB-5, 30 m × 0.25 mm × 0.25 µm
Program:	Start: 80 °C, 1 min Program 1: 20 °C/min to 160 °C Program 2: 10 °C/min to 250 °C Final temp.: 250 °C, 1 min
Mass spectrometer:	Scan mode: CI, full scan, 120–400 u, reagent gas methane Scan rate: 1 s/scan

Results

Figure 4.139 shows the analysis of an extract to which 2 ng/µL Clenbuterol has been added. With a total ion current only a few signals are present in the chromatogram, which can be detected using chemical ionisation. The selective masses of Clenbuterol-TMS m/z 349, 351 represent the only active substance peak in the mass chromatogram. The mass spectrum in the relevant trace region of the 100 pg injected shows the unambiguous fragments of Clenbuterol-TMS (Fig. 4.140). Chemical ionisation allows the selective detection of the active substance with an ion trap mass spectrometer down to the lower pg region (Fig. 4.141). The calibration shows good linearity over the region 25 pg to 2 ng shown (Fig. 4.142). Clenbuterol-d_9 can be used as the internal standard.

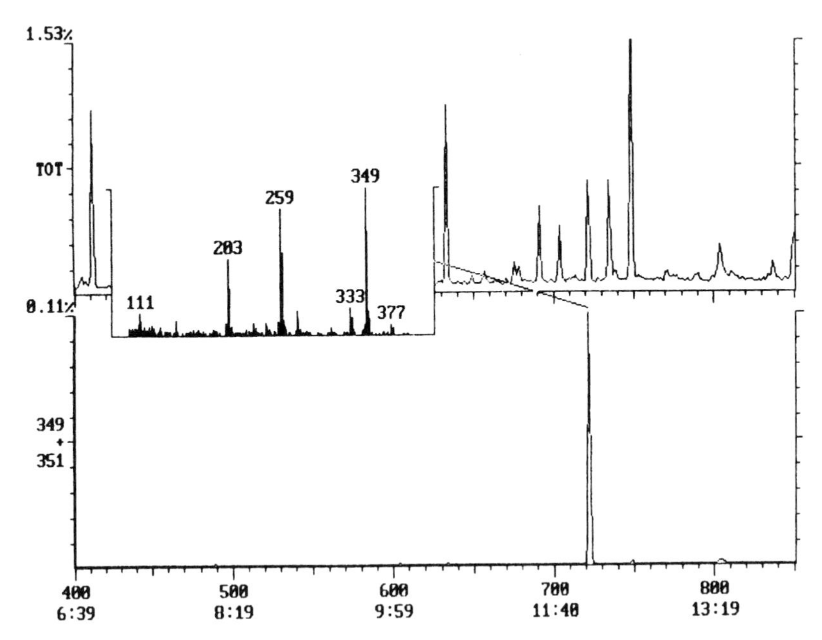

Fig. 4.139 PCI chromatogram of an extract fortified with 2 ng/µL Clenbuterol
Above: total ion current, below: selective mass chromatogram m/z 349, 351

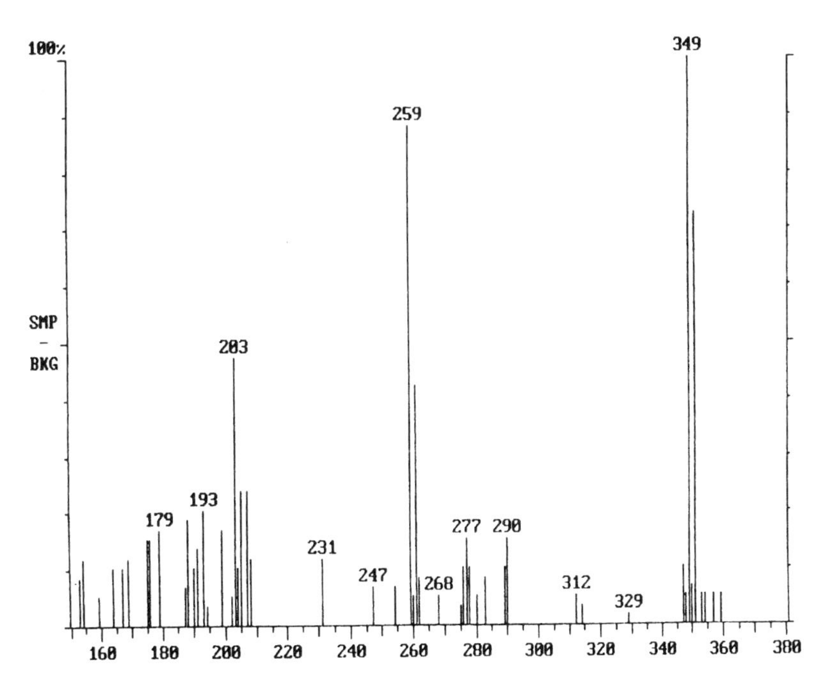

Fig. 4.140 CI (methane) mass spectrum of Clenbuterol, 100 pg

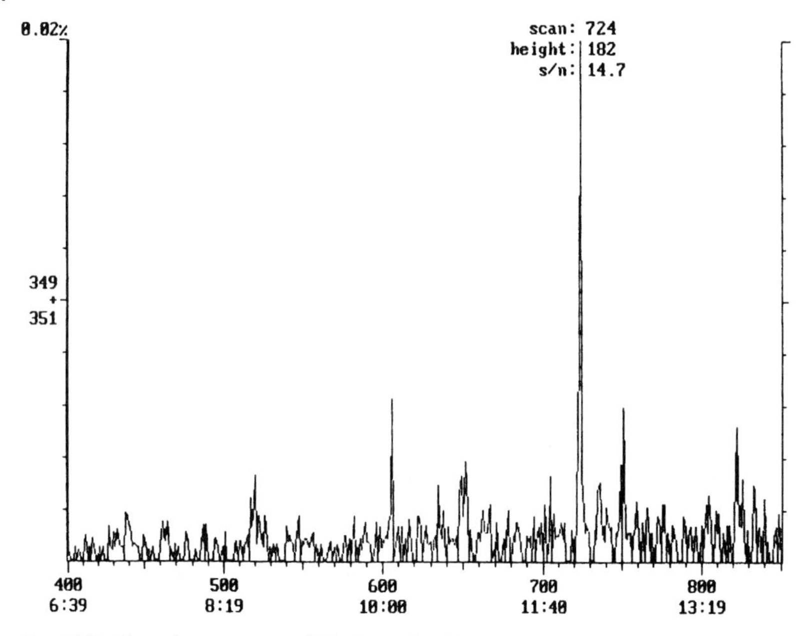

Fig. 4.141 Mass chromatogram of Clenbuterol at 50 pg/μL

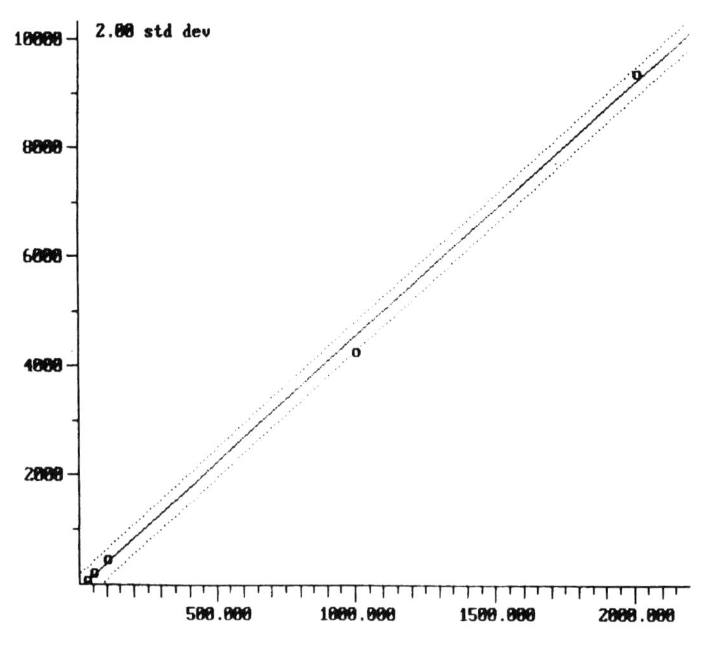

Fig. 4.142 Calibration graph for Clenbuterol determination between 25 pg and 2 ng

4.36
General Unknown Toxicological-chemical Analysis

Key words: diagnosis, emergency situations, general unknown analysis, detoxification, drug monitoring, quantitation, automatic evaluation, screening

Toxicological-chemical analysis can be very important for diagnosis and efficient treatment of acutely or chronically poisoned patients. In many cases, e.g. in emergency situations, poisoning is the result of the intake of mixtures. Therefore special requirements are made on an analysis system so that in one step as many active substances as possible can be identified and quantified by only one procedure. GC/MS is now the method of choice for the screening and subsequent unambiguous identification and quantitation of sufficiently volatile organic pharmaceutical substances and poisons (Figs. 4.143 and 4.144).

For the systematic toxicological analysis of general unknowns, an analysis procedure is described which allows the simultaneous detection and determination of a large number of pharmaceutical substances. Serum (or blood when no serum can be obtained, e.g. in the case of corpses) is analysed. Only the quantitative determination of serum or blood allows a good estimation of the degree of severity of poisoning, of the success of detoxification or therapy (drug monitoring) or the degree of damage to the body. In addition most active substances in serum are present in a nonmetabolised form, at least to some extent, so hydrolysis and derivatisation can be dispensed with. This cannot be circumvented for the determination of more hydrophilic metabolites from urine. Consequently sample preparation is quicker, as is the procedure as a whole. Besides isolation in sufficient yield, the separation of biogenic substances as far as possible is a requirement for the full utilisation of the dynamic range of the ion trap GC/MS system used.

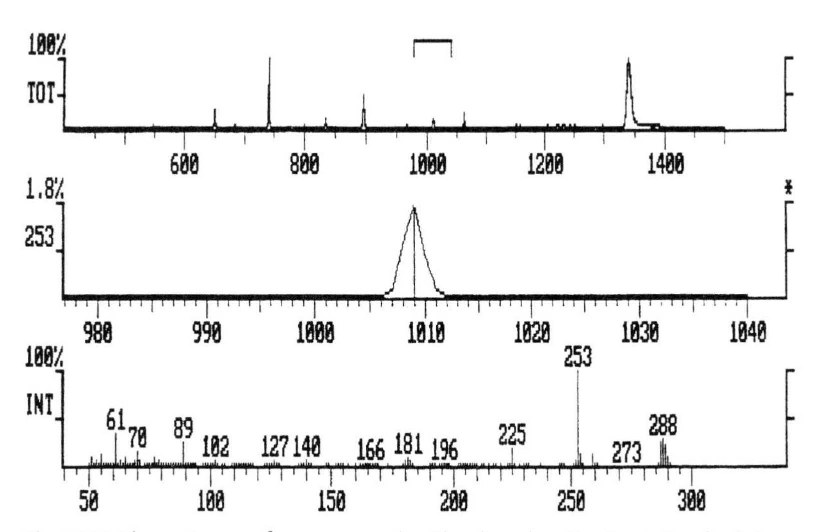

Fig. 4.143 Chromatogram of a serum sample with enlarged section (centre) and substance spectrum (below)

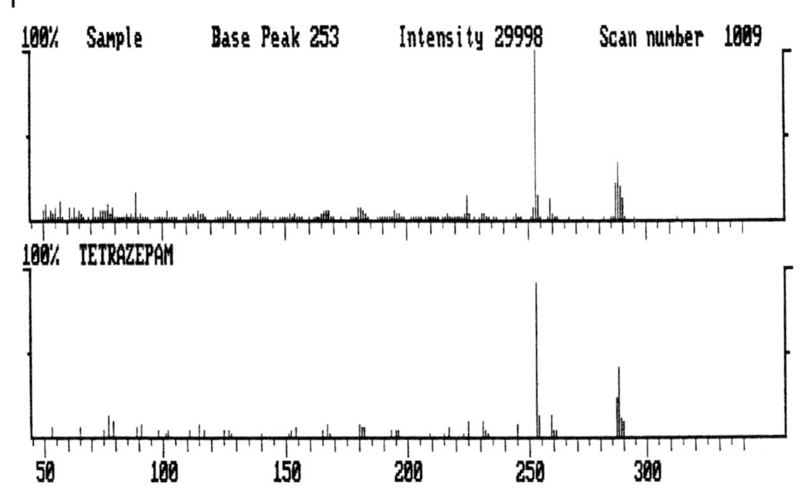

Fig. 4.144 Result of the library search through the toxicology library TX with the suggestion of Tetrazepam (FIT value 967, spectrum from Fig. 4.143)

Analysis Conditions

Sample preparation:	Adjust the pH of 1 mL serum to 9 with $NaHCO_3$ add the internal standard Etoloxamine (5 µL ≅ 100 ng Etoloxamine) Extract with 5 mL n-hexane/ethyl acetate 7 : 3 (v/v), 1% amyl alcohol, 2 min centrifuge Re-extract the organic phase with 2 mL 0.5 M H_2SO_4, centrifuge and discard the organic phase Adjust the pH of the aqueous phase to 9 with 0.2 ml 8 N NaOH and $NaHCO_3$ Extract with 2 mL $CHCl_3$, evaporate the organic phase to dryness and dissolve in 20 µL ethyl acetate, inject 1 µL Time required ca. 20–25 min
Gas chromatograph:	Varian 3400 with split injector and split injection
Column:	HP-1 15 m × 0.25 mm × 0.25 µm
Program:	Start: 80 °C, 1 min Ramp: 10 °C/min to 280 °C Final temp.: 280 °C, 4 min
Mass spectrometer:	Finnigan ion trap detector ITD 800 GC coupling: 270 °C Scan mode: EI, full scan, 50–450 u Scan rate: 1 s/scan

Calibration and automatic evaluation:	Search for active substances in the total ion current chromatogram using stored retention times and mass spectra	
	Retention time window:	20 s
	FIT threshold:	700
	S/N ratio:	min. 3
	Internal standard:	Etoloxamine
	Duration of the evaluation:	15 s

Results

At the current state of development of the procedure the 77 active substances listed in Table 4.37 are determined in an automatic evaluation program (AutoQuan). For most substances calibration is available which was drawn up at five concentrations using a blind serum (drug-free serum) (Fig. 4.145). Metabolites of active substances are not calibrated. The compounds marked with a * are only detected in cases of poisoning. The other active substances are also detected in therapeutic doses. Etoloxamine was chosen as the internal standard because it is very stable and readily chromatographed. It is no longer used as a pharmaceutical substance. The ease with which an active substance can be determined using the GC/MS procedure described depends on the recovery, the relative response and the concentration in the serum. For this reason the isolation procedure is set up to obtain basic substances, as many neutral and acidic pharmaceutical substances (barbiturates, Phenytoin) are present at such a high concentrations within the range of therapeutic activity that they are readily detectable by GC/MS in spite of the low recovery from sample clean-up. A more complete recovery in the case of these compounds often leads to overloading of the column and thus impairs quantitation. In Table 4.38 examples are given of the experimentally determined recovery and relative response together with the average therapeutic con-

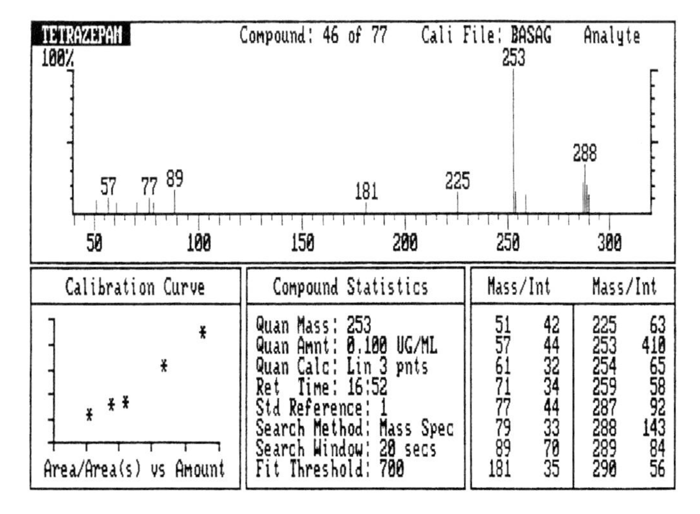

Fig. 4.145 Example of an entry from the calibration file with all data for the qualitative identification using retention time and mass spectrum as well as quantitative determination

Table 4.38 List of active substances which can be detected and quantified by the screening procedure (status 1994/95)

Amitriptyline	Doxepin	Phenacetin
Amitriptyline oxide*	Doxylamine	Phenazone
Biperiden	Etoloxamine – ISTD	Phenobarbital
Bornaprin**	Haloperidol*	Phenytoin
Bromazepam*	Hydrocodone	Primidone
Carbamazepine	Imipramine	Primidone-M**
Carbamazepine artefact 2**	Levomepromazine	Promazine
Carbamazepine-M**	Lidocaine	Promethazine
Quinidine	Medazepam	Promethazine-M(HO)**
Quinidine acetate**	Metamizol-Desalkyl**	Propranolol*
Quinine**	Methadone	Propranolol artefact**
Quinine acetate**	Methaqualone	Propyphenazone
Chlordiazepoxide	Methylphenobarbital	Propyphenazone-(HO-Phenyl)**
Chlormezanone	Midazolam**	Propyphenazone-(HO-Propyl)**
Chlorpromazine	Moclobemid	Propyphenazone-(Isopropenyl)**
Chlorprothixene	Naloxone*	Propyphenazone-(nor-di-HO)**
Clomipramine	Nicotine**	Propyphenazone-(nor-HO-Phenyl)**
Clozapine	Nitrazepam*	Remoxiprid*
Cocaine	Nordazepam	Tetrazepam
Codeine	Nortilidine**	Theophylline**
Caffeine	Nortrimipramine**	Tilidine**
Cotinine	Nortriptyline	Tramadol
Desmethylclomipramine**	Paracetamol*	Trimipramine
Desmethylmedazepam**	Pentobarbital	Trihexyphenidyl
Diazepam	Perazine	Zotepine
Diphenylhydramine	Pethidine	

* Active substances can only be determined in case of poisoning
** Only qualitative detection

Table 4.39 Determination of the product from recovery, response and concentration for ten selected active substances

Active substance	RI	Scan	m/z	a	c_{thp} [µg/mL]	RR	RR · a · c [µg/mL]
Diphenylhydramine	1870	701	58	0.91	0.05	1.58	0.072
Phenobarbital	1965	753	204	0.02	30	0.03	0.018
Methaqualone	2160	870	235	0.60	2	0.07	0.084
Amitriptyline	2205	904	58	0.91	0.1	1.82	0.166
Trimipramine	2225	910	58	0.81	0.12	0.46	0.045
Promethazine	2270	937	72	0.66	0.25	0.27	0.045
Biperiden	2280	960	98	0.91	0.07	0.59	0.038
Clomipramine	2455	1012	58	0.70	0.1	0.34	0.024
Chlorprothixene	2510	1060	58	0.66	0.16	0.48	0.051
Levomepromazine	2540	1076	58	0.85	0.09	0.98	0.075

centration for 10 active substances from different classes of pharmaceuticals. From this table the interaction of these three parameters for a successful determination using GC/MS is clear.

In principle all substances which can be extracted in sufficient quantity and have a retention index between 1200 and 3100 (cholesterol) can be determined using the procedure described. Besides basic pharmaceutical substances, neutral and acidic compounds, such as barbiturates, can be determined because of their high therapeutic dose, in spite of low recovery rates. From the times given for sample preparation, measurement and evaluation it can be concluded that the qualitative and quantitative analysis of a serum sample can be carried out in less than an hour. The development of the investigation programme by incorporating further substances is being carried out (Demme 1995).

4.37 Clofibric Acid in Aquatic Systems

Key words: Clofibric acid, pharmaceuticals, contamination, SPE, pentafluorobenzyl esters, SIM technique

In 1991 in the course of the investigation of a series of samples of Berlin (Germany) ground water for phenoxyalkylcarboxylic acid herbicides, an unknown chlorine-containing carboxylic acid was discovered, the structure of which was unambiguously found to be that of 2-(4-chlorophenoxy)-2-methylpropionic acid. This compound is used to lower lipid levels in human beings under the name of Clofibric acid. To find out how this substance got into the ground water the investigation procedure was changed to improve the sensitivity of detection. Using the methods described here tap water samples from different districts of Berlin were systematically investigated for residues of Clofibric acid. The isomeric compound 2-(4-chlorophenoxy)butyric acid (4-CPB) was added to a water sample as a surrogate standard. Clean-up involved solid phase extraction with a special RP-C_{18}-adsorbent. 2,4-Dichlorobenzoic acid (2,4-DCB) was added as the internal standard for quantitation. After derivatisation with pentafluorobenzyl bromide the Clofibric acid is detected as the pentafluorobenzyl ester by GC/MS in the SIM mode and determined quantitatively. It is identified from the retention time and the intensity ratio of three characteristic ions. Because two standards are added, the results are very reliable even at the lowest concentrations.

Analysis Conditions

Sample preparation: Adjustment of the pH of 1 L water to 1–2
Addition of 4-CPB standard (methanolic solution) and adjustment of the concentration to 200 ng/L; 1 g RP-C_{18}-SPE cartridge (washed with 10 mL acetone, 10 mL methanol, conditioned with 10 mL water pH <2), flow rate ca. 8 mL/min, drying with N_2, elution with 2.5 mL methanol, solvent removal with N_2 (Fig. 4.146).

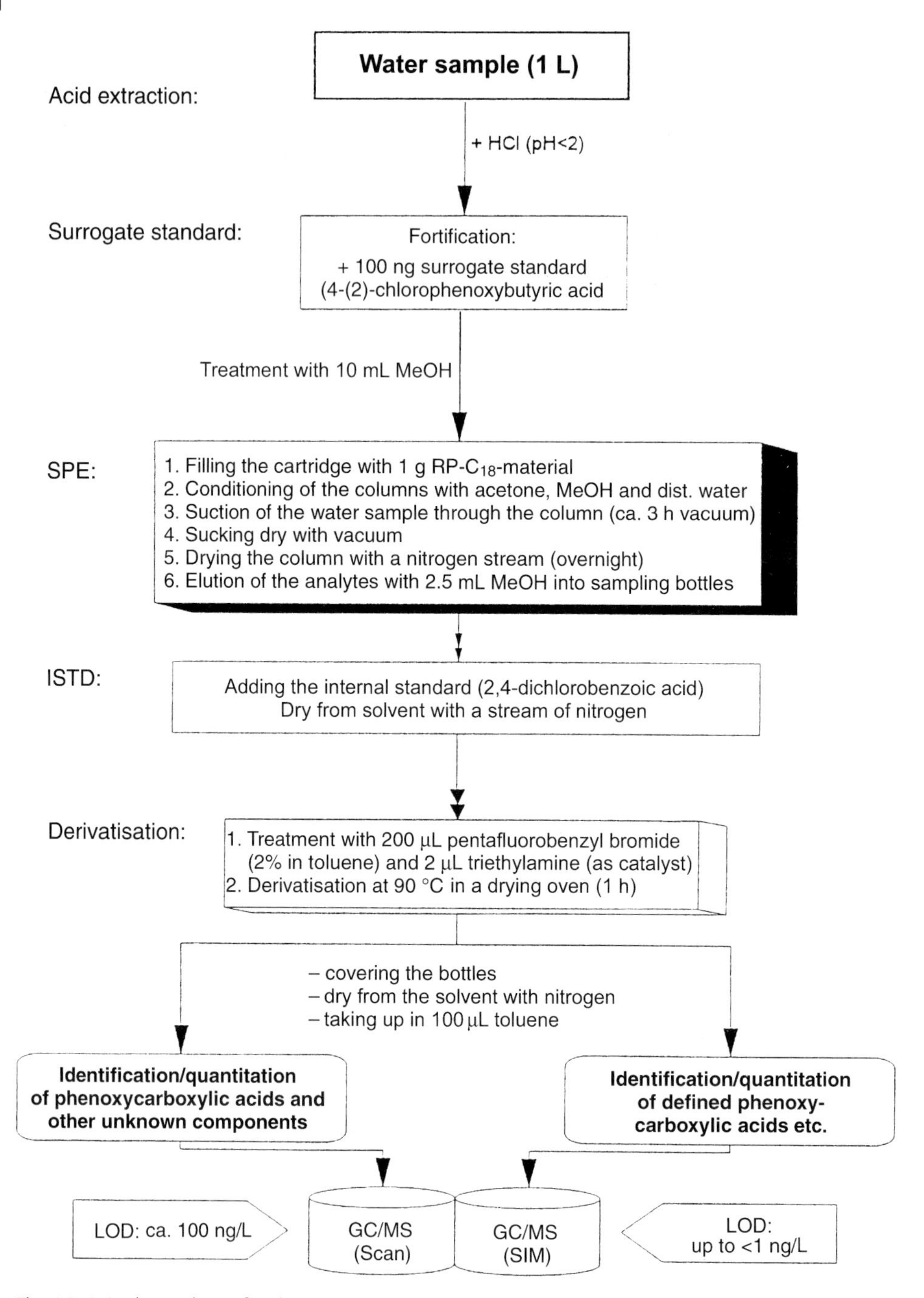

Fig. 4.146 Analysis scheme for phenoxycarboxylic acid analysis

Derivatisation: Addition of the internal standard 2,4-DCB (200 ng in methanol), transfer to autosampler vessels and evaporation to dryness with N_2, treatment with 200 μL pentafluorobenzyl bromide solution (100 μL pentafluorobenzyl bromide in 4.9 mL toluene) and 2 μL triethylamine, closure of the vessel, 1 h reaction time at 90 °C, excess reagent driven out with N_2, residue taken up in 100 μL toluene.

GC/MS system: HP 5890 with MSD HP 5970

Injection temp.: Split/splitless injector, 210 °C

Column: HP-5, 25 m × 0.2 mm × 0.33 μm

Program:
Start: 100 °C, 1 min
Program 1: 30 °C/min to 150 °C, 1 min
Program 2: 3 °C/min to 205 °C
Program 3: 10 °C/min to 260 °C, 23 min

Mass spectrometer:
Scan mode: EI, SIM mode (Table 4.32)
Scan rate: 1 s/scan (dwell times per ion, see Table 4.40)

Results

Recoveries of 90 to 100% were achieved for Clofibric acid, the surrogate standard 4-CPB, Mecoprop and Dichlorprop with the RP-C_{18}-material (Fig. 4.146). Interference in the analysis can be caused by humic substances. With such samples negative results do not necessarily mean that no Clofibric acid is present. The use of the additional internal standard demonstrates the reliability of the GC/MS analysis. The peak area ratio of the internal and the surrogate standards is a measure of the quality of the analysis.

The GC/MS detection procedure involves investigating five compounds in three time windows in a typical target compound analysis (Table 4.40). In the first time window Clofibric acid and 2,4-DCB are determined with six ions, in the second Mecoprop and 4-CPB with four ions and in the third window Dichlorprop with three ions. The results are plotted as an MID chromatogram to give a better overview (addition of SIM Signals). The absence of an ion here is a definite sign of the absence of the target substance (Fig. 4.147 a, b). For confirmation, besides the exact agreement of the retention times, the relative ion intensities are also checked. In the samples these may not differ by more than 20% from the calibration standards measured under the same conditions.

The limit of detection is 1 ng/L for Clofibric acid, and the limit of quantitation 10 ng/L. The method is also suitable for the analysis of highly contaminated surface water. Even with very contaminated samples other substances present do not appear to affect the results, for example the analysis of a run-off from a sewage works (Figs. 4.148 and 4.149). This selectivity is achieved by the derivatisation which gives derivatives with high molecular weights. Polar, nonderivatised components do not dissolve in the toluene. Possible impairment of the chromatography by highly contaminated samples can be controlled using the internal standard 2,4-DCB.

Table 4.40 SIM programming and retention times

No.	Substance	Retention time [min]	Time window [min]	SIM masses [m/z]	Dwell time[a)] [ms]
1	Clofibric acid	24.54	24.00–25.19	128, 130, 173	150
	2,4-DCB	24.78		370, 372, 394	
2	Mecoprop	25.64	25.20–26.39	128, 142	200
	4-CPB	25.84		169, 394	
3	Dichlorprop	27.07	26.40–27.50	162, 400, 402	300

a) Dwell time = duration of measurement per ion at 1 s-scan rate

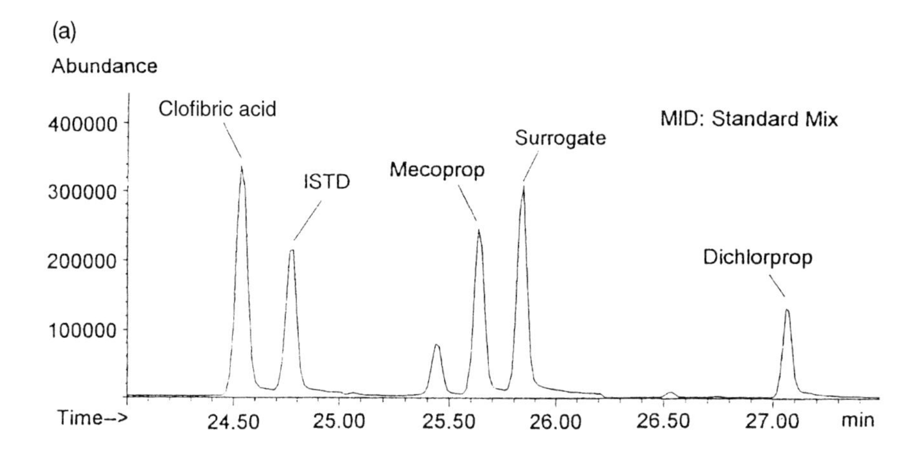

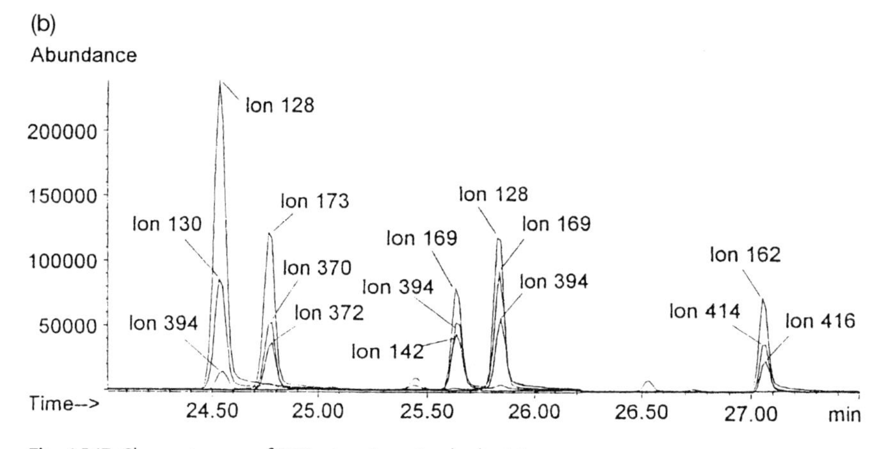

Fig. 4.147 Chromatogram of PFB esters in a standard mixture
(a) MID plot as a sum of SIM signals (2 ng/component)
(b) Individual ion traces of the PFB esters

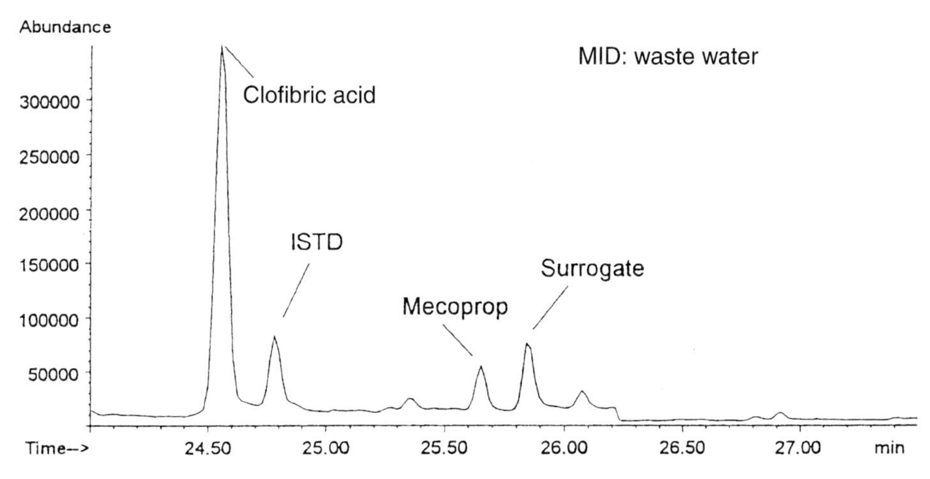

Fig. 4.148 Analysis of a waste water sample from a sewage works containing both Clofibric acid (1100 ng/L) and Mecoprop (175 ng/L)

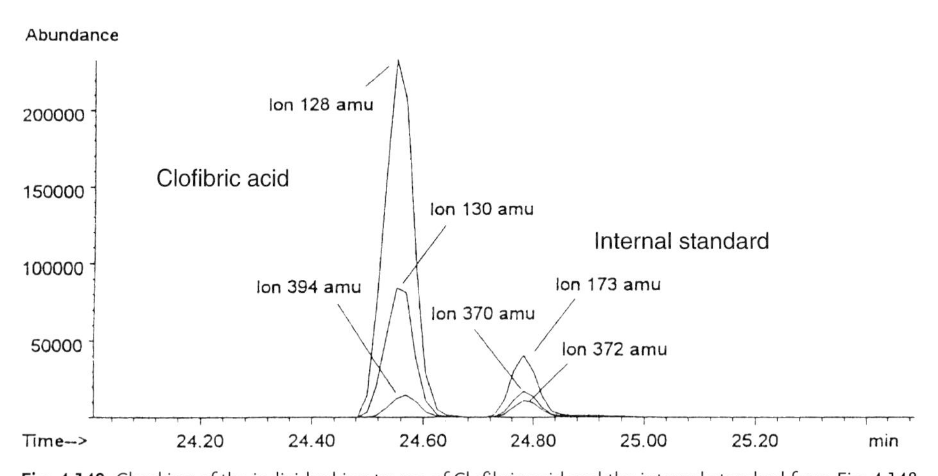

Fig. 4.149 Checking of the individual ion traces of Clofibric acid and the internal standard from Fig. 4.148

The presence of Clofibric acid in water has not only been found in Berlin. The substance has been found in random samples taken outside Berlin in rivers and drinking water. It has not been found in drinking water from areas without contact with the filtrate from the river banks. It is assumed that Clofibric acid from therapeutic use gets into surface water via municipal waste water. According to the working group this is the first known case of a medicine used for humans reaching aquatic systems in this way.

4.38
Polycyclic Musks in Waste Water

Key words: musk perfumes, surface water, waste water, solid phase extraction

Synthetic musk perfumes are added to washing and cleaning agents as well as fabric conditioners and many bodycare products to improve their odours. Because of the ecotoxicological problems caused by their poor degradability and lipophilic properties, there are reservations about their widespread use. In the examination of water, sewage sludge, sediments and fish for nitro-musk compounds (e. g. musk-xylene, 1-t-butyl-3,5-dimethyl-2,4,6-trinitrobenzene), three further compounds were found in the GC/MS chromatograms. These were the polycyclic nitro-musk compounds HHCB, AHTN and ADBI (Table 4.41). These substances are widely used in the cosmetics and perfumes industry and have already been detected in river water and various species of fish.

Table 4.41 Polycyclic musk perfumes

Specification	CAS-No.	Empirical formula	M	Structure
HHCB 1,3,4,6,7,8-Hexahydro-4,6,6,7,8,8-hexamethylcyclo-penta-(g)-2-benzopyran	1222-05-5	$C_{18}H_{26}O$	258	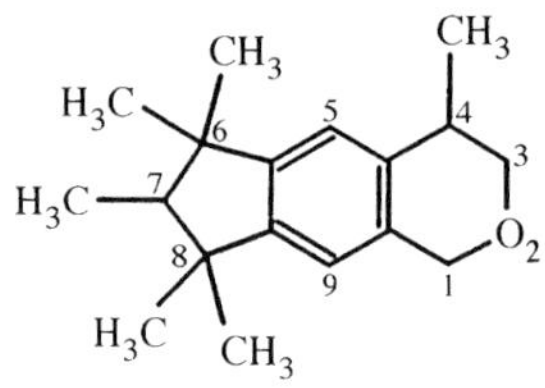 H_3C, CH_3, CH_3, H_3C, H_3C, CH_3, 1–9, O_2
AHTN 7-Acetyl-1,1,3,4,4,6-hexa-methyltetralin	1506-02-1	$C_{18}H_{26}O$	258	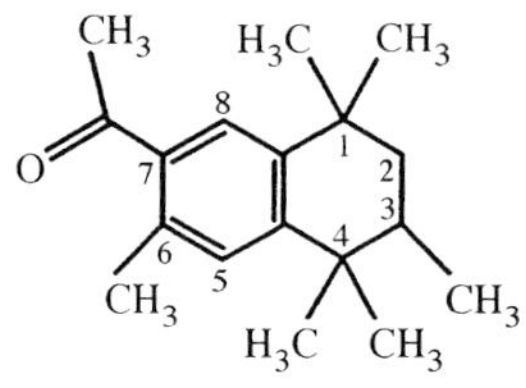 CH_3, H_3C, CH_3, O, CH_3, CH_3, H_3C, CH_3, 1–8
ADBI 4-Acetyl-1,1-dimethyl-6-tert.butylindane	13171-00-1	$C_{17}H_{24}O$	244	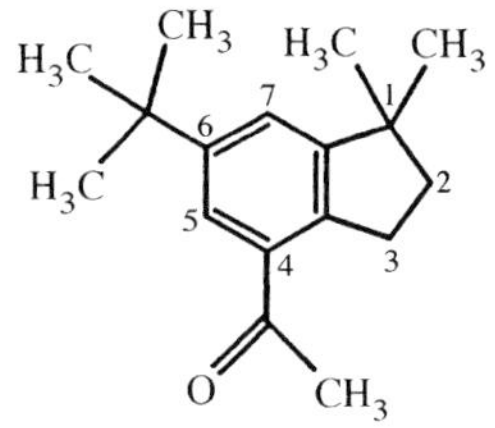 H_3C, CH_3, H_3C, CH_3, H_3C, O, CH_3, 1–7

Analysis Conditions

Sample preparation:	2 L surface water are enriched using SPE with 2 g Bakerbond PolarPlus C_{18}-cartridges and eluted with 10 mL acetone; nitrogen is blown through the eluate and the residue taken up in n-hexane. Waste water samples are extracted directly with n-hexane.
GC/MS system:	Finnigan ITS 40
Injector:	PTV cold injection system, splitless injection
Program:	Start: 60 °C Program : 300 °C/min Final temp.: 300 °C, 15 min
Column:	J&W DB-5MS, 30 m × 0.32 mm × 0.25 μm and SGE HT8, 25 m × 0.22 mm × 0.25 μm
Program:	Start: 60 °C, 1 min Program: 10 °C/min Final temp.: 300 °C, 6 min
Mass spectrometer:	Scan mode: EI, full scan, 40–400 u Scan rate: 1 s/scan

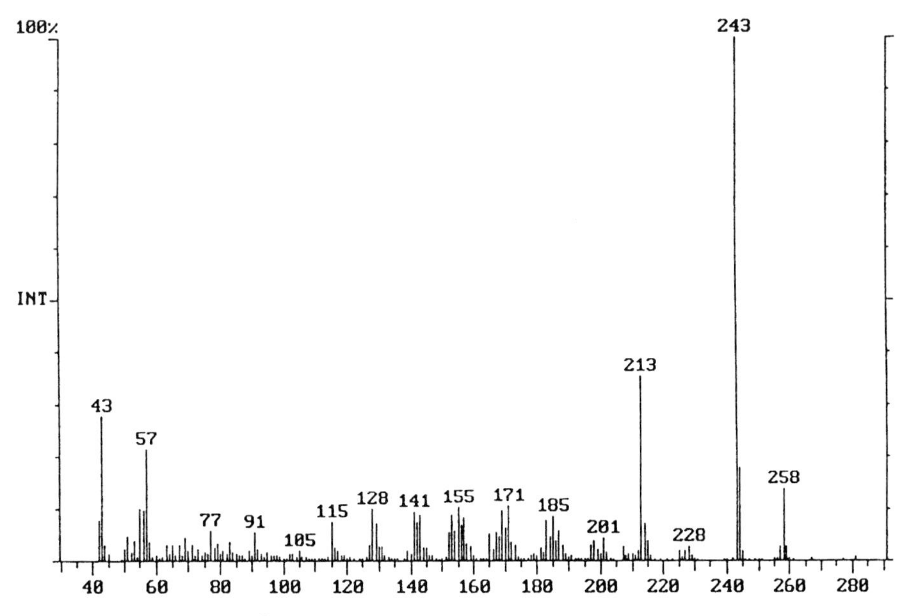

Fig. 4.150 Mass spectrum of HHCB, M 258

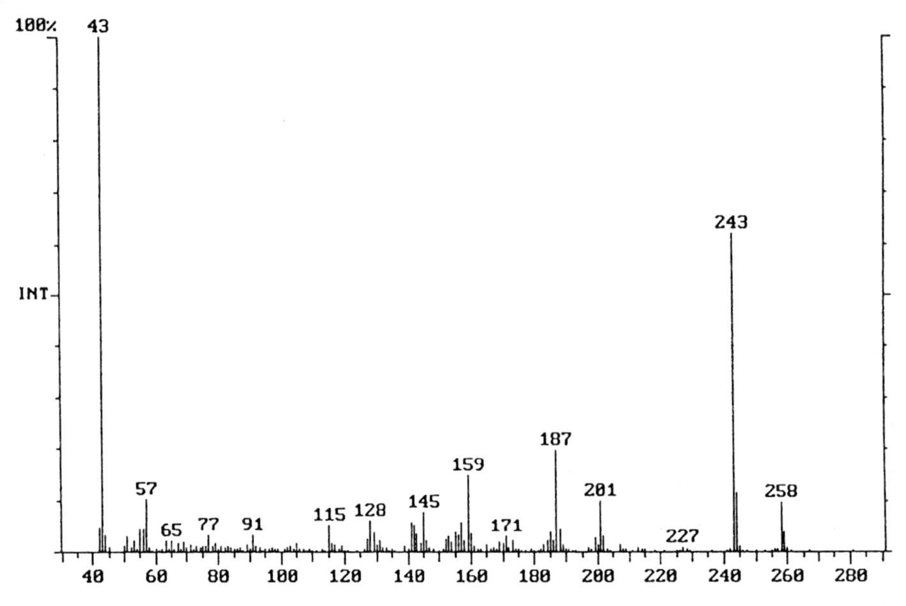

Fig. 4.151 Mass spectrum of AHTN, M 258

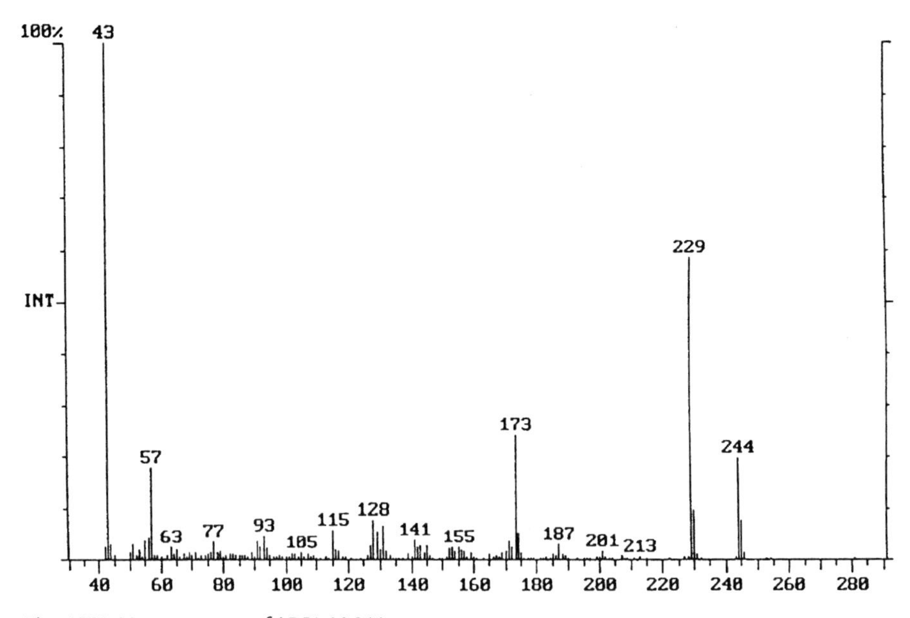

Fig. 4.152 Mass spectrum of ADBI, M 244

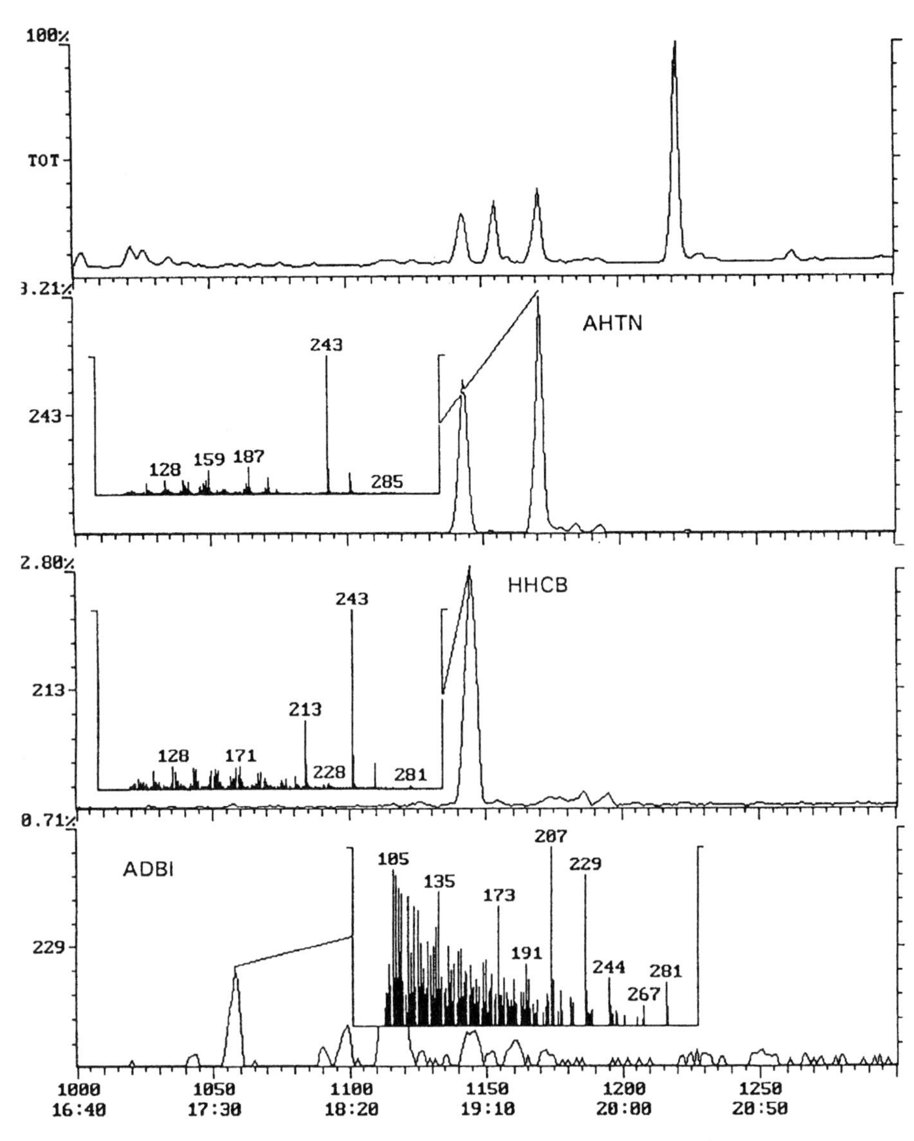

Fig. 4.153 Chromatogram of a waste water sample with mass chromatograms of HHCB (m/z 213/243), AHTN (m/z 243) and ADBI (m/z 229)

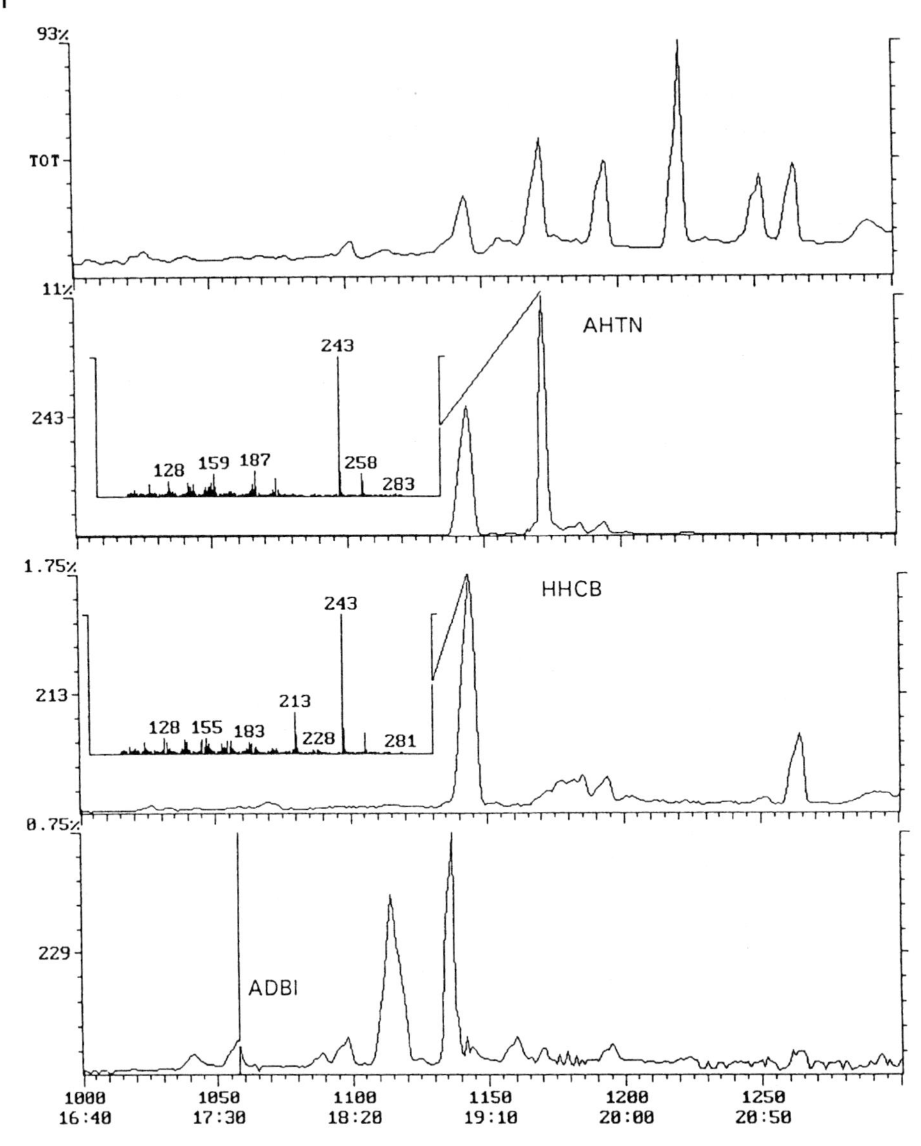

Fig. 4.154 Chromatogram of a fish sample (muscle tissue, eel) with the mass chromatograms of HHCB (m/z 213/243), AHTN (m/z 243) and ADBI (m/z 229)

Results

The EI mass spectra of the compounds taken in the full scan mode are suitable for unambiguous identification because of their typical fragmentation patterns. Intense fragment signals are seen at m/z 43, 258, 243 and 229. The mass spectrum of HHCB also contains the fragment m/z 213, unlike AHTN (Figs. 4.150 to 4.152). HHCB and AHTN are only partially separated on a 5%-phenyl column. All three substances are separated well on the slightly polar phase of the HT8 column. Figure 4.153 shows the chromatogram of a waste water sample with the mass chromatograms of the selective ions under the total ion current. In random samples significant measurable concentrations in the ppb range were found both in surface water and in the inflow and outflow of municipal sewage works. Very high concentrations were found in the muscle tissue of fish from these types of water (Fig. 4.154), indicating the low degradability and bioaccumulation of these compounds (Eschke 1994).

4.39
Identification and Quantification of Trichothecene Mycotoxins

Key words: mycotoxins, T2-toxin, HT-2 toxin, nivalenol, patulin, ricinine, bioterrorism, homeland security, food safety, MS/MS

Highly toxic mycotoxins can be found as contaminants from the metabolism of fungi in food, beverages and food preparations. Also, mycotoxins are on the list of biological weapons. Trichothecene mycotoxins and other toxins, in particular ricin, belong to those substances which are suspected to have a strong potential for terroristic attacks or biological warfare. Reliable and fast methods using the available equipment for the identification and quantification from complex matrices are essential. The described GC/MS method applies MS/MS technique in EI and chemical ionisation mode for a fast and reliable identification and quantification of mycotoxins and markers of toxin exposition.

Analysis Conditions

Sample preparation: Beverage samples for patulin and ricinine: 1 mL liquid sample, adjusted to pH 14 with 2 m KOH; extraction using Extrelut®-3 glass columns, eluted with 3 x 5 mL of $CHCl_3$ (drying with Na_2SO_4 was not necessary); evaporation to dryness and reconstitution in 100 µL MSTFA for derivatisation. Note: Patulin can be extracted completely without modifying the pH due to its neutral character.
Other sample materials, as published (Mateo 2001, Razzazi-Fazeli 2002, Lagana 2003).
Derivatisation with TSIM at room temperature for 1 h, with MSTFA at 60 °C for 20 min. For deoxynivalenol minor amounts of the partially substituted compound like the di-TMS-derivative can be observed when derivatised with MSTFA instead of TSIM.

Gas chromatograph:

Injector:	Split/splitless, 285 °C
Carrier gas:	Helium, 30 cm/s, constant velocity
Column:	Rtx-5MS with 5 m Integra-Guard, 30 m × 0.25 mm × 0.25 μm (Restek, Bellefonte, USA)
Oven temp. program:	120 °C, 1 min 40 °C/min to 295 °C 295 °C, 15 min
MS interface temp.:	298 °C
Mass spectrometer:	GCQ GC/MS/MS system (ThermoQuest, Austin TX, USA)
Ionisation:	EI, PCI and NCI using CH_4 as reagent gas, source temp. 170 °C
Scan mode:	MS/MS: excitation voltage: 1.00–10.00 V, q-value: 0.45 depending on the selected ions
Cycle time:	0.58 s/scan

Table 4.42 Mass spectrometric key fragments for different ionization modes

Substance	Key-fragments (m/e) and ionization mode		
	EI mode	**NCI mode**	**PCI mode**
Patulin	53, 55, 110, 136	136, 154, 108, 137	155, 81, 99, 71
Patulin-TMS	73, 170, 183, 75, 226	136, 108, 226, 123	227, 73, 255, 137
Ricinine	164, 121, 149, 82	149	nd
15-Acetoxyscirpenol-di-TMS	159, 73, 141, 91, 285, 131	121, 195, 306, 245, 246	319, 159, 73, 379
3-Acetyldeoxynivalenol-di-TMS	73, 287, 75, 117, 289, 105	230, 290, 482, 362, 229	289, 467, 377, 287
15-Acetyldeoxynivalenol-di-TMS	73, 75, 193, 197, 117, 77	135, 267, 215, 268, 216	407, 193, 393, 73
Deoxynivalenol-tri-TMS	73, 393, 333, 392, 259, 512	297, 298, 305, 215, 299	407, 497, 514, 269
Deoxynivalenol-di-TMS	73, 187, 321, 215, 261, 440	nd	nd
Diacetoxyscirpenol-TMS	91, 124, 73, 244, 105, 75	165, 59, 123, 245, 305	379, 183, 229, 319
Fusarenon X-tri-TMS	73, 141, 75, 185, 273, 275	297, 298, 321, 207, 135	555, 465, 273, 405
HT-2-toxin-di-TMS	73, 185, 157, 466, 143, 175	197, 89, 101, 212, 567	317, 185, 467, 303
Neosolaniol-di-TMS	73, 193, 75, 245, 244, 185	213, 179, 231, 226, 197	197, 317, 377,467
Nivalenol-tetra-TMS	73, 261, 289, 245, 259, 75	297, 298, 303, 207, 299	585, 495, 289, 273
T2-toxin-TMS	244, 73, 290, 245, 185, 75	101, 213, 197, 244, 89	197, 317, 377, 287
T-2-triol-tri-TMS	73,185, 157, 275, 143, 292	nd	nd
Verrucarol-di-TMS	73, 159, 91, 105, 143, 187	107, 122, 398, 168, 89	185, 231, 195, 169

nd: not determined.

Table 4.43 Parent and product ions used for confirmation and quantitative MS/MS determination

Substance	MS/MS mode, EI ionisation	
	Parent ion	Product ions
Patulin-TMS	136	108, 98, 104, 85
Ricinine	164	135, 134, 121, 149
Deoxynivalenol-tri-TMS	512	393, 333, 392, 496
Deoxynivalenol-tri-TMS	393	333, 259, 260, 305
Nivalenol-tetra-TMS	510	407, 317, 361, 289
Nivalenol-tetra-TMS	482	392, 362, 379, 377
T2-toxin-TMS	290	259, 274, 275, 257
T2-toxin-TMS	244	229, 214, 173
T2-toxin-TMS	185	142, 141, 157, 170
HT-2 toxin-di-TMS	466	287, 303, 284, 288
HT-2 toxin-di-TMS	185	142, 157, 141, 129

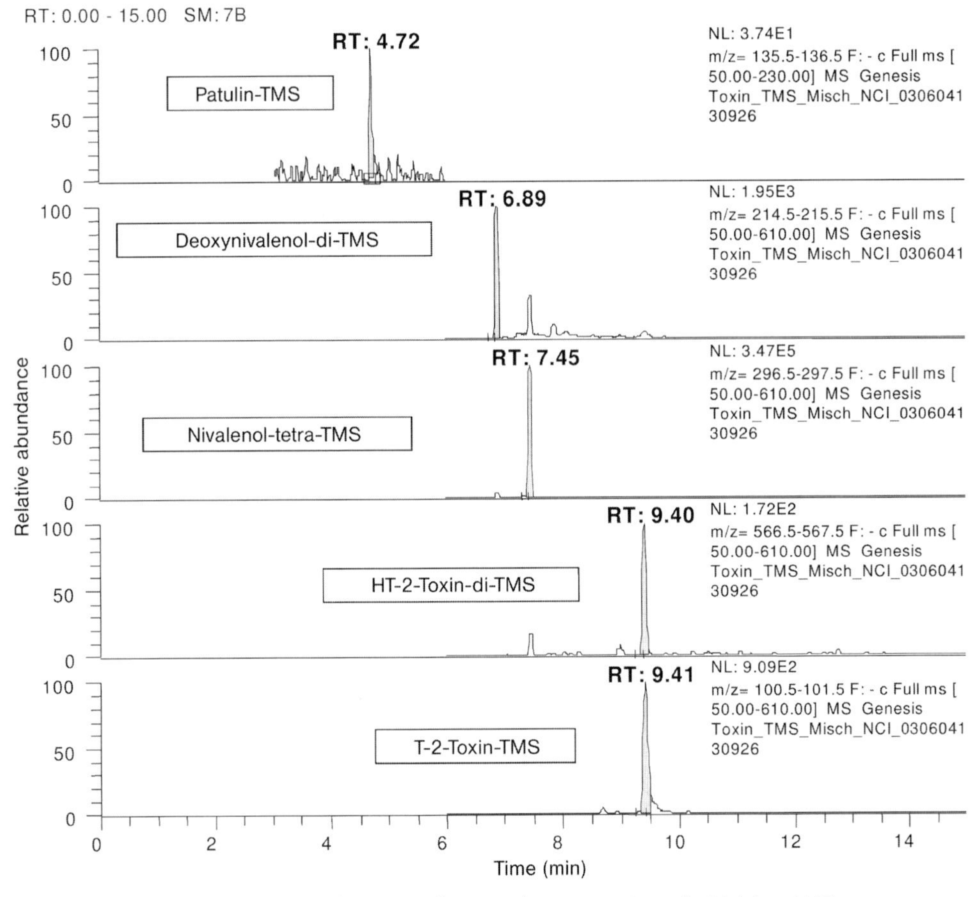

Fig. 4.155 Mass chromatograms of selected reference substances, NCI mode (Melchert 2007)

Results

Mycotoxins from the trichothecenes group as well as ricinine (as biological marker of ricin) and patulin can be analysed with high certainty by GC/MS/MS with short analysis times using the parent/product ion transitions given in Table 4.43. The mass spectra of the substances under investigation have been included by NIST into the last version of the spectrum library due to the importance of identification.

The substances are readily extracted using standardised methods for different matrices and derivatised using TSIM or MSTFA. A sample chromatogram and confirmation of ricinine is given in Fig. 4.156. The recovery of ricinine and patulin from spiked samples was 95 ± 5% and 85 ± 10% for spiked amounts of up to 200 ng/mL. The limits of quantitation of all mycotoxins in this study have been determined as 50 pg with a S/N of better 10 : 1 on the base peak. Linear calibration curves have been achieved up to 5,000 pg in EI mode. When analysing samples form complex matrices like food or soil samples, it is recommended for increased selectivity to monitor the MS/MS transition from PCI or NCI ionisation as given in Table 4.43.

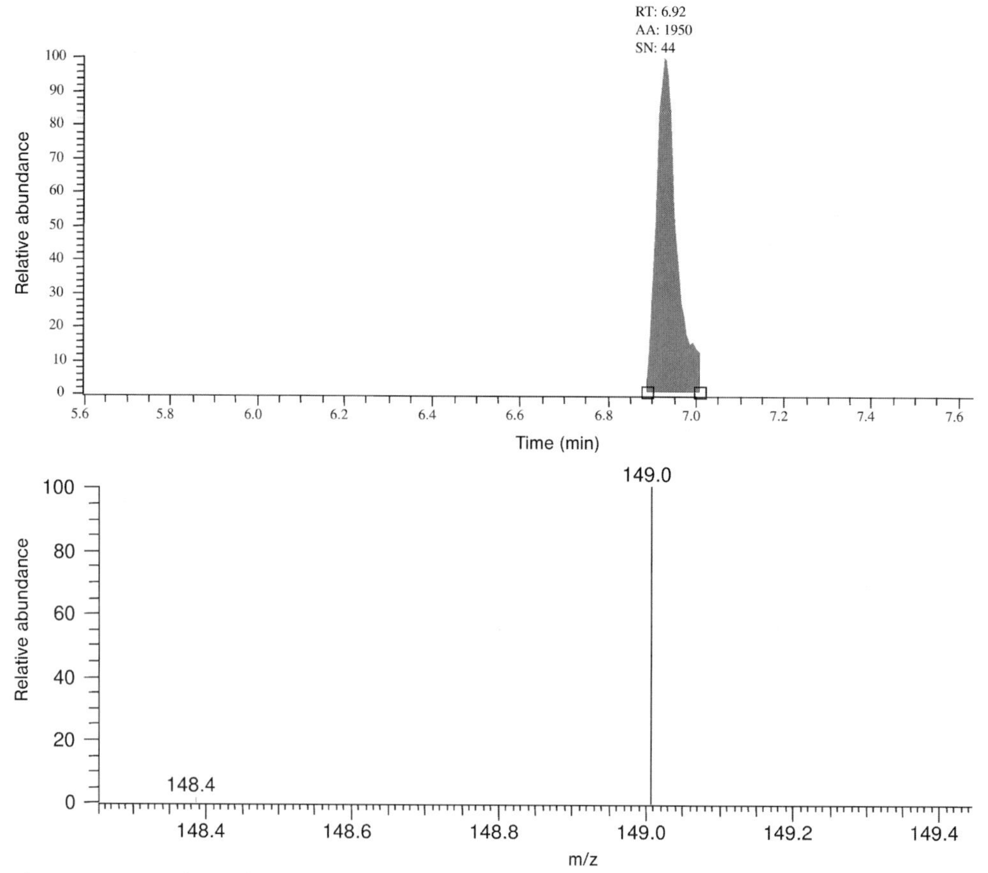

Fig. 4.156 Ricinine detected in a cola sample with MS/MS confirmation on the SRM transition m/z 164 > 149, quantified on the product ion m/z 149 (Melchert 2007)

Due to the good solubility of caffeine and quinine in $CHCl_3$ it is also possible to use this extraction procedure for the quantitative determination of these substances in soft drinks.

The described method is conceived to provide the reproducibility and specificity required for a safe and reliable detection and quantification in the context of homeland security analyses and is proposed also to be used for judicial proceedings.

Note: Ricin is a lectin derived from the castor bean plant, *Ricinus communis.* Ricinine is an alkaloid (3-cyano-4-methoxy-N-methyl-2-pyridone) that shares a common plant source with ricin, and its presence in urine infers ricin exposure. It is also used as biomarker for ricin exposure (Melchert 2007).

4.40
Highly Sensitive Screening and Quantification of Environmental Components Using Chemical Ionisation with Water

Key words: chemical ionisation, PCI, screening, quantification, ion trap

The trace analysis for screening and quantitation of environmental compounds can be significantly improved in terms of sensitivity and universal detection when using the liquids methanol and water as a chemical reagent gas for positive chemical ionisation in an ion trap mass spectrometer using internal ionisation (3D trap). Also, devices for providing an elevated pre-pressure from a liquid reservoir are available for use with conventional ion source mass spectrometers.

Electron impact ionisation, (EI), which is widely applied in routine GC/MS environmental analysis delivers strongly fragmented mass spectra with only poor molecular weight information. These information rich spectra are very useful for mass spectral library comparison. For quantitation, selective fragment ions need to be identified which may not be affected by matrix interferences. The ion population created by EI ionisation distributes along a wide mass range making highly fragmenting substances difficult to detect in matrix samples at low levels, even in selected ion monitoring mode.

In contrast to EI ionisation, the chemical ionisation, (CI), provides high quasi-molecular weight intensities with reduced or no fragmentation. When using methanol or water as the reagent gases, a highly effective and general CI ionisation with a up to 100 fold response gain for a wide variety of substance classes can be exploited. Applications for the trace analysis of polyaromatic hydrocarbons, trinitroaromatic compounds as well as glycols, which found application in routine analysis, are described.

Analysis Conditions

Gas chromatograph:	Sichromat 1 (Siemens, Karlsruhe, Germany)
Injector:	PTV 40, 40 °C ballistic to 250 °C, 10 min isothermal; Hot vaporizer, 200 °C, split 30 mL/min
Carrier gas:	Helium

Column:	XTI-5, 30 m x 0.32 mm x 0.25 µm (Restek, Bellafonte, USA); Rtx Volatiles, 30 m x 0.32 mm x 1.5 µm (Restek, Bellafonte, USA)	
Oven temp. program:	XTI-5:	40 °C, 1 min 15 °C/min to 340 °C
	Rtx Volatiles:	40 °C, 1 min 15 °C/min to 280 °C
Mass spectrometer:	ITD 800 Ion Trap Detector (Finnigan MAT, Bremen, Germany)	
Ionisation:	CI, reagent gases water and methanol from a liquid reservoir	
Scan mode:	Full scan	
Mass range:	30–500 u	

Results

Due to its medium proton affinity, water as Cl reagent gas is able to protonate almost all organic substances including apolar compounds of interest for trace analysis of environmental samples. The process is characterized by a high formation rate yielding high product ion signals. The fragmentation of the molecules is significantly reduced and the valuable substance signal is concetrated on just one ion species and shifted to the higher mass range of the molecular ion. The higher mass region is typically less affected by the unspecific low molecular weight matrix components. The analytical response is observed to be 10 to 50 times, for some substance classes up to 100 times, higher than known from EI ionisation. Even highly chlorinated substances like HCB can be protonated by water with a higher response compared to EI. The compound classes amenable to water CI cover the whole spectrum of industrial chemicals including the compound classes of environmental interest. A comprehensive overview is given in Table 4.46.

Methanol and water can be supplied easily to any mass spectrometer just by providing a liquid reservoir instead of a gas cylinder. For regular ion sources the reservoir need to be heated gently to provide the necessary heat for evaporation for a sufficient gas pressure. Commercial liquid-CI options are available.

The mass spectra achieved with water CI are easily interpretable and mainly consist of the quasi-moleceular ion $(M+H)^+$. If elements with isotopes e. g. Cl, Br are occurring the isotope pattern of the molecular ion reveals the number and identity of isotope carrying elements in the elemental formula clearly.

Table 4.44 Response comparison of trinitroaromatic compounds for methanol and water CI mode (Landrock 1995)

Compound	EI mode		Methanol CI^+ mode		Water CI^+ mode	
	S/N	Area	S/N	Area	S/N	Area
1,3,5-Trinitrobenzene	7.5	18,000	37	150,000	310	5,200,000
2,4,6-Trinitrotoluene	36	110,000	27	190,000	320	6,200,000

Table 4.45 a) Response gain of PAH compounds (a) and glycol derivatives (b) in water CI mode compared to EI ionisation (Landrock 1995).

a) PAH compounds

No.	Name of compound	EI^+ Peak area × 10^6	Water CI^+ Peak area × 10^6	Water CI^+ Gain in response
1	Naphthalene	0.25	23.47	92
2	2-Methylnaphthalene	0.32	17.24	54
3	1-Methylnaphthalene	0.35	18.03	51
4	Biphenyl	0.45	22.85	51
5	2,6-Dimethylnaphthalene	0.39	15.48	40
6	Acenaphthylene	0.4	23.76	60
7	Acenaphthene	0.92	23.49	26
8	2,3,5-Trimethylnaphthalene	0.69	17.74	26
9	Fluorene	1.12	13.45	12
10	Phenanthrene	0.57	14.99	26
11	Anthracene	0.41	12.47	30
12	1-Methylphenanthrene	0.43	12.56	30
13	Fluoranthene	0.35	15.57	45
14	Pyrene	0.37	16.09	44
15	Benzo(a)anthracene	0.24	15.31	64
16	Chrysene	0.35	20.21	58
17	Benzo(b)fluoranthene	0.22	17.29	80
18	Benzo(k)fluoranthene	0.64	26.46	42
19	Benzo(e)pyrene	0.32	20.05	63
20	Benzo(a)pyrene	0.40	21.63	54
21	Perylene	0.20	15.68	79
22	Indeno(1,2,3-cd)pyrene	0.33	21.61	66
23	Dibenzo(ah)anthracene	0.21	18.71	91
24	Benzo(ghi)perylene	0.29	19.45	68

b) Glycols and derivatives

No.	Name of compound	EI^+ Peak area × 10^6	Water CI^+ Peak area × 10^6	Water CI^+ Gain in response
1	Ethyleneglycol-monomethylether	0.63	33	52
2	Propyleneglycol-monomethylether	0.99	22	22
3	Ethyleneglycol-monomethylether	0.49	22	45
4	Propyleneglycol	0.78	19	24
5	Ethyleneglycol-monoisopropylether	1.28	21	16
6	Aceticacid-2-methoxyethylether	0.89	20	22
7	Propyleneglycol-monomethyletheracetate	2.29	24	10
8	Butylglycol	1.37	20	15
9	Ethylglycol-acetate	1.92	24	13
10	Diethyleneglycol-monomethylether	0.70	22	31
11	Diethyleneglycol-monoethylether	0.90	26	29
12	Diethyleneglycol	0.42	12	29
13	Diethyleneglycol-dimethvlether	1.32	19	14
14	Butyleneglycol-acetate	2.74	22	8
15	Diethyleneglycol-ethyletheracetate	2.98	24	8
16	Diethyleneglycol-monobutylether	1.55	18	12

Trinitroaromatics Even the electron rich trinitroaromatic substance class can be ionised using water CI to form the MH^+ ion without fragmentation and with high response. It also has been demonstrated that methanol as CI gas could not protonise these compounds sufficiently leading only to a weak response exhibiting some fragmentation. For a comparison of the analytical response with EI ionisation see Table 4.44.

Polycyclyc Aromatic Hydrocarbons (PAH) Due to the inherent stability of this compound class GC/MS with EI ionisation usually is the most favoured analytical technique. Using water CI the increase in response is approximately 30 to 90 fold compared to EI with an almost equal value for all the investigated PAH components, see Table 4.45 a. At the same time a high selectivity of the water CI ionisation is observed against hydrocarbon background which becomes even more transparent due to a lower ionisation efficiency.

Glycols and Its Derivatives These widely used industrial solvents are relevant to environmental analysis but typically not analysed by GC/MS. Due to the low stability in EI ionisation unspecific low molecular fragments are formed after elimination of H_2O. Using water CI the strong fragmentation is prevented and high intensity MH^+ ions are formed that can be used for identification and quantification. Table 4.45 b gives an overview of the relative response gain of various glycol derivatives compared to EI ionisation.

Table 4.46 Overview of compound classes identified and quantified by water CI GC/MS (Merten 1996)

Compound class	Typical compounds (selection)
Alcohols	Methanol, Ethanol, Propanol, Butanol
Aldehydes	Formaldehyde, Acetaldehyde, Benzaldehyde
Ketones	Acetone, Butanone, Ethylvinylketone, Lactone
Esters	Ethylacetate, Glycolester
Ethers	Benzofurane, Tetrahydrofurane
Organic acids	Propionic acid, Benzoic acid, Phthalic acid
Terpenes	Limonene, Terpineol, Campher
Cl-organic compounds	Vinylchloride, Chloroform, Chlorbenzene
Perchloro compounds	Hexachlorobutadien, Hexachlorobenzene, Perchlorostyrene
Alkanes, Cycloalkanes	All homologues
Aromatics, Alkylbenezenes	Benzene, Toluene, Styrene
Naphthylcompounds	Naphthalene, Tetraline, Methylnaphthalene
Glycols	Ethylglycol, Glycolether and ester
Nitrogencompounds	Dimethylformamide, Pyridine, Benzonitrile
Nitroaromatics	TNB, TNT, Nitroaniline, Chloronitroaniline
Sulfurcompounds	Sulfides, Disulfides, Thiophene, Benzothiophene
Phenols	Alkyl-, Chloro-, Nitrophenols
PAH	Anthracene, Phenanthrene, Pyrene
PCB	All congeners
Phalates	Dibuthylphthalate, Dioctylphthalate
Pesticides	DDT, Lindane, Triazines

4.41 Characterization of Natural Waxes by Pyrolysis-GC/MS

Key words: pyrolysis, Py-GC/MS, natural waxes, beeswax, lanolin, carnauba wax

Waxes can be of natural or synthetic origin and have a great importance as a raw material in many industries, e. g., pharmaceutical or cosmetic, as well as for applications in culture and art. Four groups of waxes are divided through their plant, animal, mineral and petrochemical origin. The composition of natural waxes is complex. Typically occurring mixtures are of waxy esters with fatty acids and alcohols as well as hydrocarbons. Due to the complexity of the composition, the analysis of waxes is difficult. Pattern recognition of pyrograms have been predominantly involved, but less effort has been spent on the structural identification of pyrolysis products from waxes. The coupling of pyrolysis with GC/MS offers extended possibilities for a characterization with detailed structural analysis of waxes.

Analysis Conditions

Pyrolyser:	Pyrola 9p (PyrolAB, Lund, Sweden)
Pyrolysis temp.:	800 °C, temp. rise time 8 ms, pyrolysis time 4 s
Ambient temp.:	190 °C
Gas chromatograph:	HP 5890 II (Agilent, Palo Alto, CA, USA)
Injector:	Split/splitless, 280 °C, split ca. 1:25
Carrier gas:	Helium, 80 kPa
Column:	HP5-MS, 30 m × 0.25 mm × 0.25 µm
Oven temp. program:	40 °C, 2 min 12 °C/min to 300 °C 300 °C, 10 min
MS interface temp.:	280 °C
Mass spectrometer:	HP 5972A (Agilent, Palo Alto, CA, USA)
Ionisation:	EI, 70 eV
Scan mode:	full scan
Mass range:	35–500 u
Cycle time:	1.6 scans/s

Results

The samples under investigation gave very characteristic fingerprints. With the positioning of the sample (about 30 µg each) in a small dip of the Pt foil, spreading of the sample could be avoided and replicate pyrolysis showed an excellent reproducibility of the pyrograms. The optimum pyrolysis temperature was found with 800 °C and a fast 8 ms temperature rise time. Increasing the temperature to 1000 °C resulted in the increase of lower molecular weight components being less suitable for the wax characterisation. The resulting pyrograms for beeswax, lanolin, carnauba wax are given in Figs. 4.157 to 4.159.

As model for comparative study a mix of high molecular weight esters, fatty acids and alcohols has been applied. The resulting pyrolysis products have mostly been identified by mass spectral library search using the Wiley registry. For the majority of the compounds a plausible explanation for the formation under pyrolysis conditions could be derived, e.g. cholesterol related compounds obtained from cholesterylstearate, see Fig. 4.160. For each of the waxes investigated individual substances or groups of products resulting from the original wax constituents could be identified for further characterization of the waxes or being suitable for serving as markers for discrimination purposes (Asperger 1999).

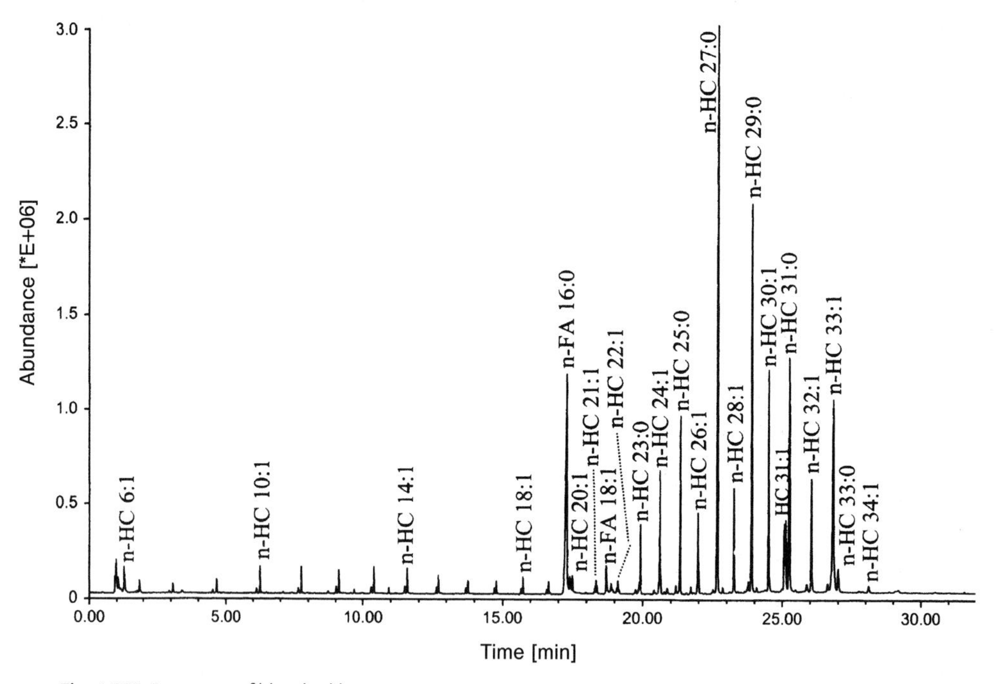

Fig. 4.157 Pyrogram of bleached beeswax
Abbreviations: FA fatty acid, HC hydrocarbon, X:Y chain length : double bonds

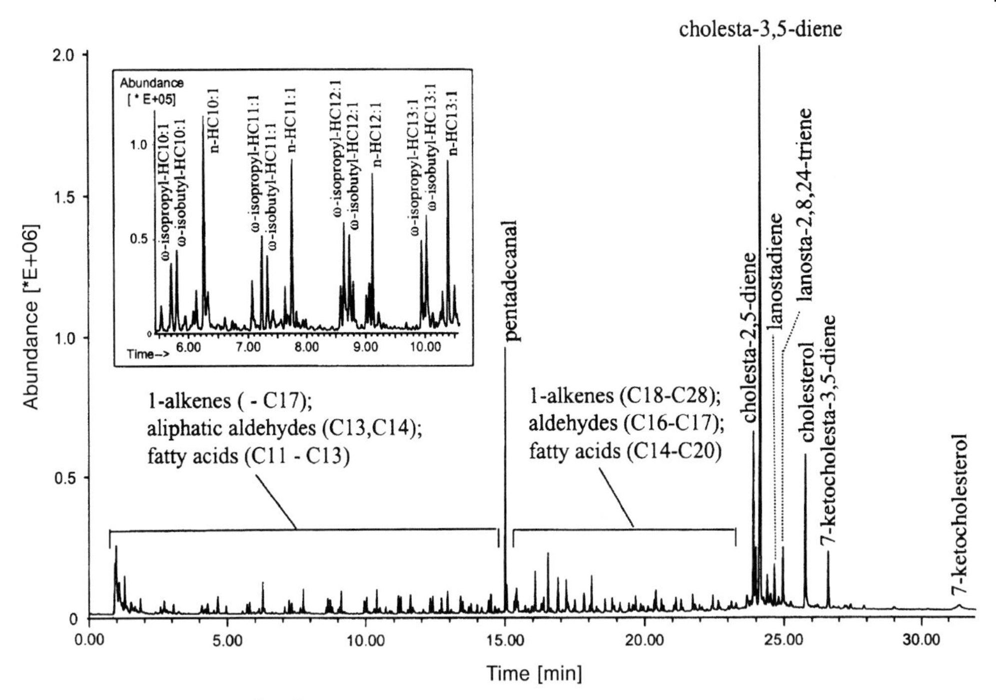

Fig. 4.158 Pyrogram of lanolin

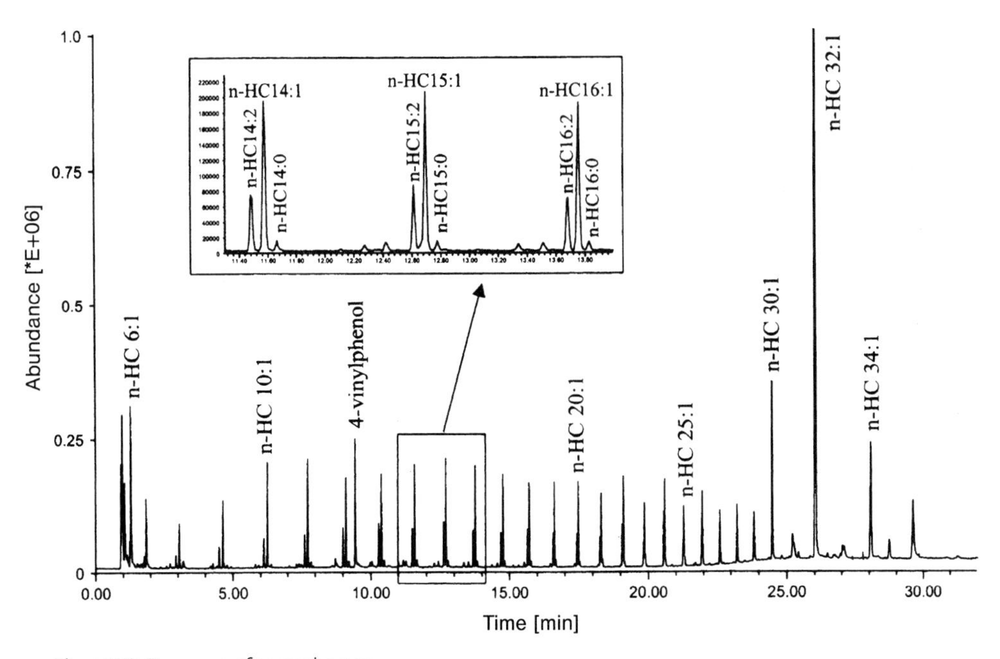

Fig. 4.159 Pyrogram of carnauba wax

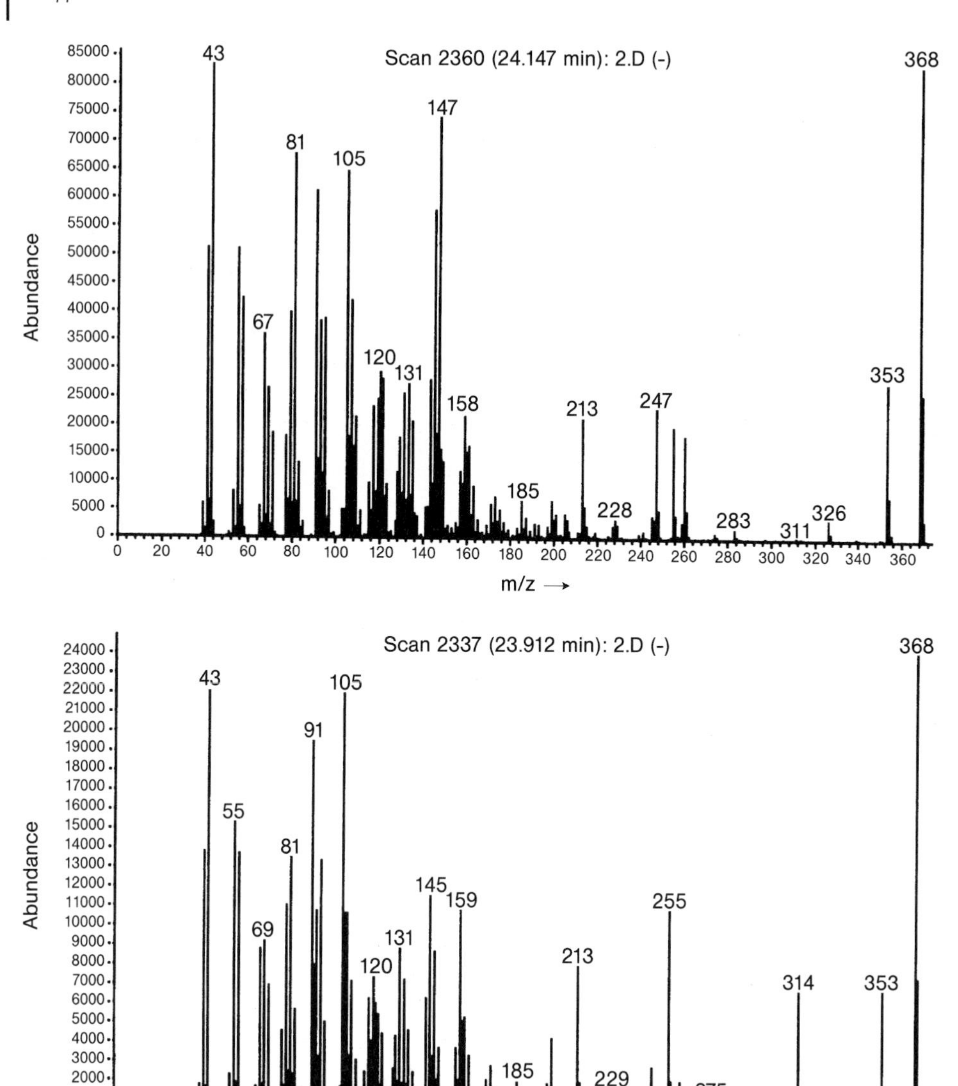

Fig. 4.160 Mass spectra of cholesterol related compounds (M 368) identified from the pyrolysis of the model substances cholesterylstearate (top) and cholesta-3,5-diene (bottom)

4.42
Quantitative Determination of Acrylate Copolymer Layers

Key words: pyrolysis, quantitation, acrylate, copolymer, coating, drug release

Acrylate copolymers are widely used as coating materials for retarding layers and provide a controlled release of the active agent after oral application. The copolymer layer is applied by spraying techniques to granules comprising drug and carrier material like cellulose. Correct thickness of the copolymer layer is crucial for the drug release kinetics. Weighting the granules before and after the coating treatment only gave results of insufficient precision.

A pyrolysis-GC/MS approach is described for quantification on the retarding copolymer layers. The given method offers the fast and reliable quantitative determination of poly(ethylacrylate-methylmethacrylate) layers (PEAMMA) on granules for pharmaceutical use and can serve as a template for similar applications.

Analysis Conditions

Sample preparation:	500 mg granulate sample 5 mL acetone, 1 h shaking on vibrator, centrifugation 2 µL aliquots are applied to the Pt filament of the pyrolyser
Pyrolyser:	PYROLA 9P filament pulse pyrolyser (Pyrolab, Lund, Sweden), with a dented Pt foil
Pyrolysis temperature:	700 °C
Pyrolysis time:	4 ms
Head temperature:	150 °C
Carrier gas:	He, 20 mL/min
Gas chromatograph:	HP 5890 II Series GC (Agilent, Palo Alto, USA)
Injector:	Split/splitless, 280 °C, split ratio approx. 1:15
Carrier gas:	Helium, 80 kPa
Column:	HP-5MS (Agilent, Palo Alto, USA) 30 m × 0,25 mm × 0,25 µm
Oven temp. program:	25 °C, 4 min, 20 °C/min 300 °C, 5 min
MS interface temp.:	280 °C

Mass spectrometer:	HP 5972A (Agilent, Palo Alto, USA)
Ionisation:	EI, 70 eV
Ion source temp.:	180 °C
Scan mode:	Full scan
Mass range:	35–500 u

Results

The individual identification of the copolymer components was attained by the characteristic mass spectral information (Fig. 4.161). For quantification the monomer pyrolysis products EA and MMA of both of the copolymer constituents are integrated on their specific mass traces m/z 55 resp. m/z 69.

As the intensity distribution of the pyrolysis products strongly depend on the pyrolysis temperature the optimum pyrolysis temperature was determined with 700 °C providing a low dimer and trimer production (Fig.4.162).

For satisfactory chromatographic separation of the volatile ethyl acrylate (EA) and methylmethacrylate (MMA) pyrolysis products a low start temperature of 25 °C is necessary for a reliable initial focussing with a fast heating ramp to 300 °C for the elution of the dimer and trimer products.

The analyses of real samples demonstrate the beneficial effect of the clean-up step with the application of the acetonic granulate extract (Fig.4.163). The pyrolyser filament was used with a dent to prevent the liquid to spread out on the filament. Quantitative reproducibility of the method was achieved with 4% RSD. A linear calibration between 4 and 25 mg PEAMMA lacquer mass was used (Fig. 4.164).

The obtained data showed good correlation with the much more time consuming dissolution tests (Asperger 1999).

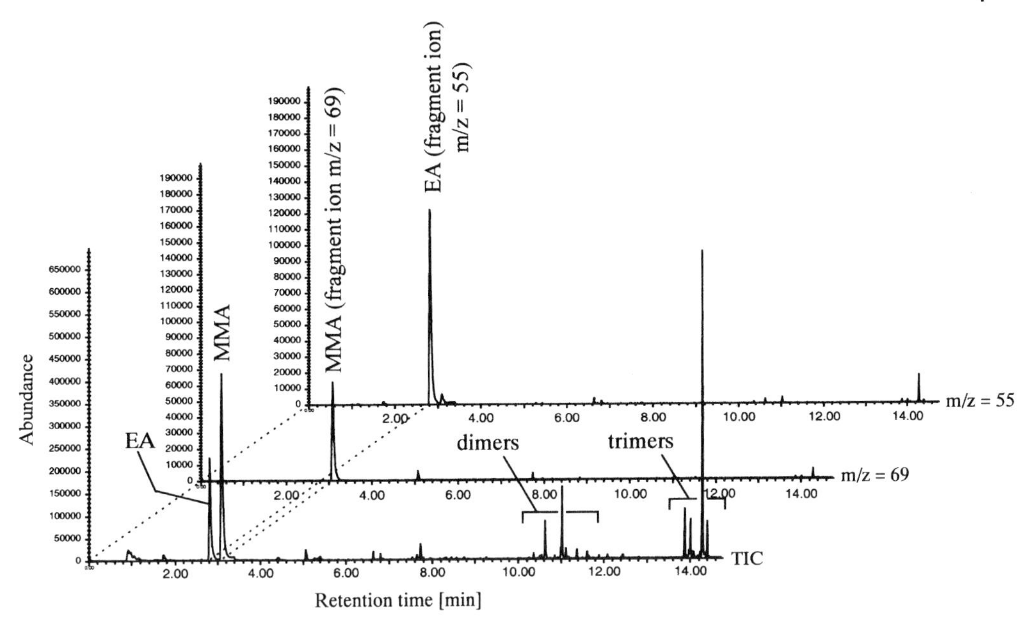

Fig. 4.161 Pyrogram of a PEAMMA lacquer material with extracted mass chromatograms for EA and MMA used for quantitation

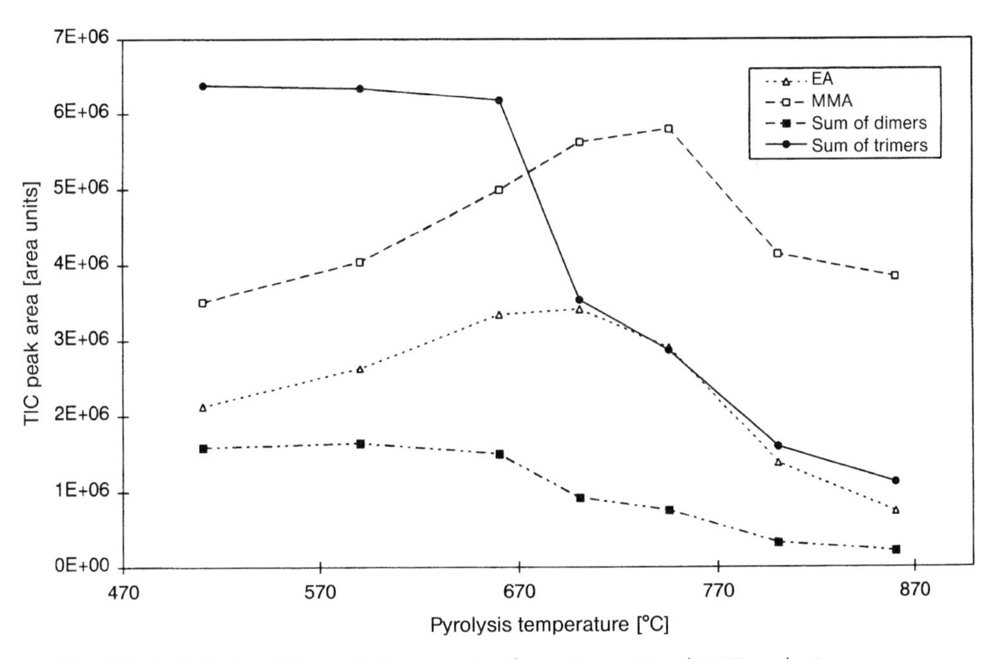

Fig. 4.162 Optimization of the pyrolysis temperature for maximum EA and MMA production

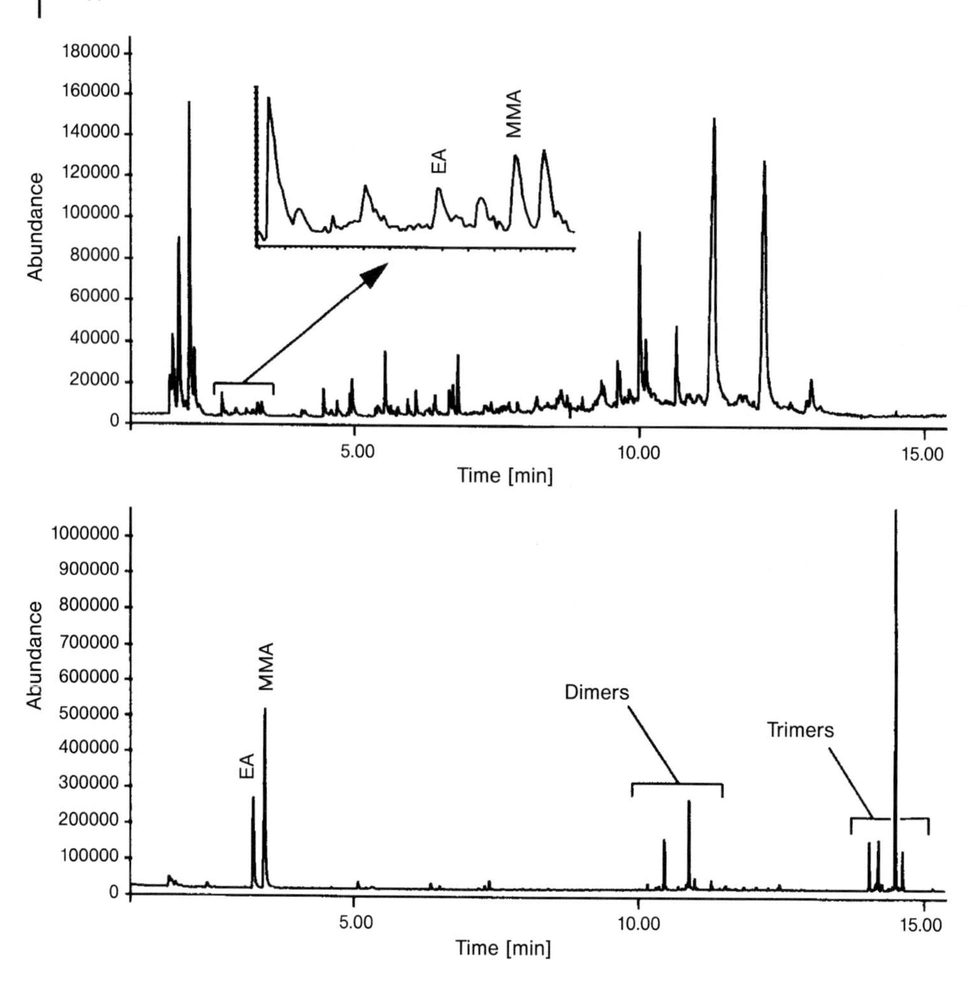

Fig. 4.163 Pyrograms of the granulate samples, top whole milled granulate w/o sample pre-treatment, bottom acetonic extract

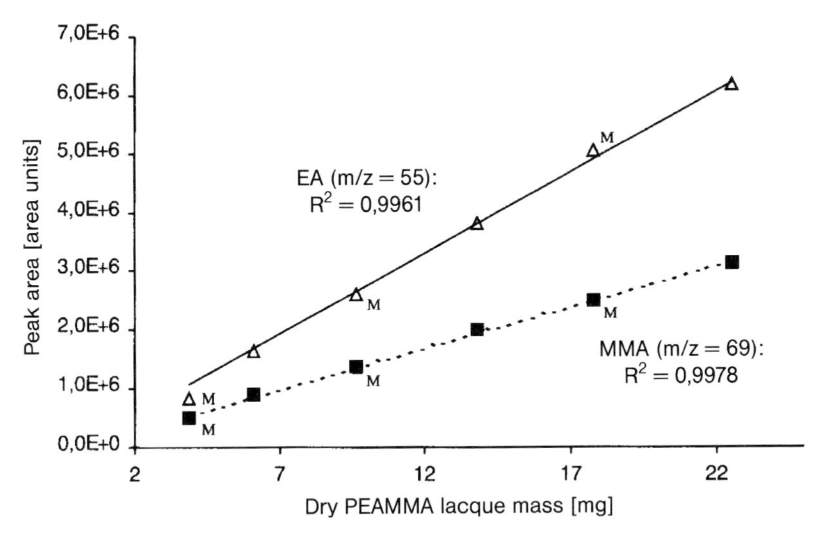

Fig. 4.164 Calibration plot for copolymer constituents EA and MMA

References for Chapter 4

Section 4.1

Madden, Al, Analysis of Air Samples for the Polar and Non-Polar VOCs Using a Modified Method TO-14, Tekmar Application Report Vol 5.3, Cincinatti 1994.

Schnute, B., McMillan, J., TO-14 Air Analysis Using the Finnigan MAT Magnum Air System, Application Report No. 230, 1993.

Section 4.4

Analysis of Volatile Priority Pollutants in Water by Static Headspace Gas Chromatography/Mass Spectrometry (GC/MS), Application Note AN 1000, ThermoQuest GC and GC/MS Division, Manchester, UK.

Belouschek, P., Brand. H., Lönz, P., Bestimmung von chlorierten Kohlenwasserstoffen mit kombinierter Headspace- und GC/MS-Technik, Vom Wasser 79 (1992), 3–8.

Rinne, D., Direkter Vergleich der Headspace-GC-Technik zum Extraktionsverfahren von leichtflüchtigen Halogen-Kohlenwasserstoffen, Gewässerschutz, Wasser, Abwasser 88 (1986), 291–325.

Section 4.6

Zwank, L., Berg, M., Enhanced Method Detection Limits for irm-GC/MS of Volatile Organic Compounds Using Purge and Trap Extraction, Application Note: 30053, Thermo Electron 2005.

Zwank, L., et al., Compound Specific Carbon Isotope Analysis of Volatile Organic Compounds in the Low µg/L-Range, Anal. Chem. 75 (2003), 5575–5583.

Section 4.7

Heimlich, F., Mayer, H., Nolte, J., Determination of Dinitrophenol Herbicides in Water, poster at the 8th IUPAC Congress of Pesticide Chemistry, Washington, DC, USA, Juli 9, 1994.

Schlett, C., Pfeifer, B., Gaschromatographische Bestimmung von Vinylchlorid nach Purge-and-Trap-Anreicherung und massenspektrometrischer Detektion, Vom Wasser 81 (1993), 1–6.

Section 4.11

Chang, J., Biniakewitz, R., Harkey, G., Determination of Geosmin and 2-MIB in Drinking Water by SPME-PTV-GC/MS, Application Note, Thermo Fisher Scientific, San Jose, CA, USA, AN10213, 2007.

McMillan, J., Analysis of 2-Methylisoborneol and Geosmin Using Purge and Trap on the Finnigan MAT MAGNUM GC/MS, Finnigan MAT Environmental Analysis Application Report No. 232, San Jose 1994.

Preti, G., Gittelman, T. S., et al., Letting the Nose Lead the Way – Malodorous Components in Drinking Water, Anal. Chem. 65 (1993), 699A–702A.

Section 4.12

Schlett, C., Pfeiffer, B., Bestimmung substituierter Phenole unterhalb des Geruchsschwellenwertes, Vom Wasser 79 (1992), 65–74.

Section 4.13

Edwards, J., Fannin, S. T., Klein, D., Steinmetz, G., The Analysis of Pesticide Residue Compounds in Biological Matrices by GC-MS, TR 9137 Technical Report, Finnigan Corp., Austin, TX, 7/1998.

Gummersbach, J., Thermo Fisher Scientific Applications Laboratory, Dreieich, Germany.

Matter, L., Bestimmung von Propham und Chlorpropham in Kartoffeln mittels hochauflösender Kapillargas-Chromatographie und Ion-Trap-Detection, GIT Fachz. Labor 11 (1989), 1116.

Sheridan, R.S., Meola, J.R., Analysis of Pesticide Residue in Fruits, Vegetables, and Milk by Gas Chromatography/Tandem Mass Spectrometry, AOAC Int. 82/4 (1999), 982–990.

Section 4.14

Anastassiades, M., Lehotay, S.J., Stajnbaher, D., Schenck, F.J. , A Fast and easy multiresidue method employing acetonitrile extraction/partitioning with dispersive solid-phase extraction for the determination of pesticide residues in produce, J. AOAC Int. 86 (2003), 412.

Okihashi, M., Kitagawa, Y., Akutsu, K., Obana, H., Tanaka, Y., Rapid method for the determination of 180 pesticide residues in foods by gas chromatography/mass spectrometry and flame photo-

metric detection", J. Pestic. Sci. 30(4) (2005), 368–77.
QuEChERS, see www.quechers.com.
Sugitate, K., Kanai, M., Okihashi, M., Ghosh, D., Multi-residue Analysis of Pesticides in Food using GC/MS/MS with the TSQ Quantum GC, Application Note 387, (2007), Thermo Fisher Scientific, San Jose, CA, USA.
The Japanese Ministry of Health, Labour, and Welfare: http://www.mhlw.go.jp/english/topics/foodsafety/positivelist060228/index.html (English), http://www.mhlw.go.jp/topics/bukyoku/iyaku/syoku-anzen/zanryu2/index.html (Japanese)

Section 4.15

Fillion, J., Sauvé, F., Selwyn, J., Multiresidue Method for the Determination of Residues of 251 Pesticides in Fruits and Vegetables by Gas CHormatography/Mass Spectrometry and Liquid Chromatography with Fluorescence Detection, J. AOAC Int. 83 (2000), 698–713.
Health and Welfare Canada, Analytical Methods for Pesticide Residues in Food, Health Canada, Ottawa, Canada, 2000.

Section 4.16

Brodesser, J., Schöler, H. F., Vom Wasser 1987, 69, 61–71.
Nick, K., Schöler, H. F., Bestimmung von Nitrophenolen mittels GC und GC/MS nach einer Derivatisierung mit Diazomethan, GIT Fachz. Labor, 5 (1993), 393–397.
Peldszus, S., Dissertation, Universität Bonn 1991.

Section 4.17

Heimlich, F., Mayer, H., Nolte, J., Determination of Dinitrophenol Herbicides in Water, Poster at the 8th IUPAC Congress of Pesticide Chemistry, Washington DC, USA, July 9, 1994.

Section 4.18

Grass, B., Nolte, J., Determination of Hydroxybenzonitrile Herbicides in Drinking Water, Poster at the 24th International Symposium on Environmental Analytical Chemistry, Ottawa, Canada, May 16–19, 1994.

Section 4.19

Hübschmann, H.-J., Niebel, J., Landrock, A., Richter, H., Merten, H., GC-MS Identification of Nitroaromatic Compounds with a New Selective and Highly Sensitive Detection Method for Ion Trap GC-MS Systems, In: Sandra, P., Devos, G. (Eds): 16th Int. Symp. Cap. Chrom., Riva del Garda, Sept. 27–30, 1994, Hüthig: Heidelberg 1994, 814–820.
Landrock, A., Richter, H., Merten, H., Water CI, a new selective and highly-sensitive detection method for ion trap mass spectrometers, Fresenius J. Anal. Chem. 351 (1995), 536–543.

Section 4.20

COMMISSION REGULATION (EC) No 208/2005 of 4 February 2005 amending Regulation (EC) No 466/2001 as regards polycyclic aromatic hydrocarbons.
COMMISSION RECOMMANDATION of 4 February 2005 on the further investigation into the level of polycyclic aromatic hydrocarbons in food.
Summary and Conclusion of the Joint FAO/WHO Expert Committee on Food Additives, Sixty-Fourth meeting, Rome, 8–17 February 2005, JCEFA/64/SC.
Ziegenhals, K., Jira, W., High sensitive PAH method to comply with the new EU directives, Presentation at the European High Resolution GC/MS Users Meeting, Venice, Italy, March 23–24, 2007.
Ziegenhals, K., Jira, W., Bestimmung der von der EU als prioritär eingestuften polyzyklischen aromatischen Kohlenwasserstoffe (PAK) in Lebensmitteln, Kulmbach Kolloquium, Sept. 2006.
Ziegenhals, K., Speer, K., Huebschmann, H.J., Jira, W., Fast GC/HRMS to quantify the EU priority PAH, J. Sep. Sci. 2008 (in print).
Kleinhenz, S., Jira, W., Schwind, K.-H.: Dioxin and polychlorinated biphenyl analysis: Automation and improvement of clean-up established by example of spices, Molecular Nutrition & Food Research 50 (4–5) (2006), 362–367.
COMMISSION DIRECTIVE 10/2005/EC of 4 February 2005 laying down the sampling methods and the methods of analysis for the official control of the levels of benzo(a)pyrene in foodstuffs.

Section 4.21

Jahr, D., Breier, D., Loeffelmann, C., Analyse von Wasserkontaminanten mit automatischer Online SPE/Ion Trap MS, GIT Fachz. Labor. 44 (2000), 654–657.

Jahr, D.,Vreuls, J.J., Louter, A.J.H., Loebel,W., Wasseranalytik mit on-line Festphasenextraktion-Gaschromatographie/Massenspektrometrie (Water analysis by on-line solid phase extraction Gas chromatography mass spectrometry), GIT Fachz. Labor. 40 (1996), 178–183.

Lerch, O., Automatisierte Spurenanalyse endokrin disruptiver Substanzen mittels GC/MS Methoden – Festphasenextraktion, Derivatisierung, Tandem-MS-Detektion, Dissertation, Ruhr University of Bochum, Germany, 2003.

Louter, A.J.H., Ramalho, S.,Vreuls, J.J., Jahr, D., Brinkman, U.A.Th., An improved approach for on-line solid phase extraction gas chromatography, J. Microcol. Sep. 8 (1996), 469– 477.

Verma, K.K., Louter, A.J.H., Jain, A., Pocurull, E., Vreuls, J.J., Brinkman, U.A.Th., On-line solid phase extraction gas chromatography ion trap tandem mass spectrometric detection for the nanogram per liter analysis of trace pollutants in aqueous samples, Chromatographia 44 (1997), 372–380.

Jahr, D., Determination of alkyl, chloro and mononitrophenols in water by sample acetylation and automatic on-line solid phase extraction gas chromatography mass spectrometry, Chromatographia 47 (1998), 49–56.

Section 4.22

IARC, see www.iarc.fr

Pan, L., Pawliszyn, J., J. Anal. Chem. 69 (1997), 196–205.

Zimmermann, T., Ensigner, W.J., Schmidt, T.C., In Situ Derivatisation/Solid-Phase Microextraction: Determination of Polar Aromatic Amines, Anal. Chem. 76 (2004), 1028–1038.

Section 4.23

Horii,Y., Kannan, K., Petrick, G., Gamo, T., Falandysz, J.,Yamashita, N., Congener-Specific Carbon Isotopic Analysis of Technical PCB and PCN Mixtures Using Two-Dimensional Gas Chromatography-Isotope Ratio Mass Spectrometry, Environ. Sci. Technol. 39 (2005), 4206–4212.

Section 4.24

M. Tschickardt, Landesamt für Umweltschutz und Gewerbeaufsicht, Mainz, personal communication.

Section 4.25

Aylward, L.L., Hays, S.M., Temporal trends in human TCDD body burden: Decreases over three decades and implications for exposure levels, J. Expo. Anal. Environ. Epidemiol. 12, (2002) 319–328.

CDC, National Report on Human Exposure to Environmental Chemicals, NCEH Pub. No. 02–0716, Atlanta GA, USA, see: www.cdc.goc/ exposurereport/

Council Directive 96/23/EC concerning the performance of analytical methods and the interpretation of results, Off. J. Europ. Communities L221/8, 17. 8. 2002.

Council Directive 2002/69/EC, July 26, 2002 laying down the sampling methods and the methods of analysis for the official control of dioxins and the determination of dioxin-like PCBs in foodstuffs, Off. J. Europ. Communities L209/5, 6. 8. 2002.

Council Directive 2002/70/EG, July 26, 2002 establishing requirements for the determination of levels of dioxins and dioxin-like PCBs in feeding stuffs, Off. J. Europ. Communities L209/15, 6. 8. 2002.

Council Directive 2006/13/EG, February 3, 2006 concerning the „Zur Änderung der Anhänge I und II der Richtlinie 2002/32/EG des Europ. Parlamentes und des Rates über unerwünschte Stoffe in Futtermitteln in Bezug auf Dioxine und dioxinähnliche PCB, Off. J. Europ. Communities L32/44, Feb 4, 2006.

EPA Method 1613 Rev.B, Tetra- through Octa-Chlorinated Dioxins and Furans by Isotope Dilution HRGC/HRMS, U.S. Environmental Protection Agency Office of Water Engineering and Analysis Division, Washington, Oct 1994.

Fishman, V.N., Martin, G.D., Lamparski, L.L., Comparison of a variety of gas chromatographic columns with different polarities for the separation of chlorinated dibenzo-p-dioxins and dibenzofurans by high-resolution mass spectrometry, J. Chrom. A, 1139 (2007), 285–300.

Hays, S.M., Aylward, L.L., Dioxin risks in perspective: past, present, and future, Regul. Toxicol. Pharmacol. 37(2) (2003), 202–217.

Krumwiede, D., Huebschmann, H.-J., Confirmation of Low Level Dioxins and Furans in Dirty

Matrix Samples using High Resolution GC/MS, Application Note, Thermo Fisher Scientific, Bremen, Germany, AN30112, 2006.
Lorber, M., A pharmacokinetic model for estimating exposure of Americans to dioxin-like compounds in the past, present, and future, Sci. Tot. Environ. 288 (2002), 81–95.
Patterson, D.G., Canady, R., Wong, L.-Y., Lee, R., Turner, W., Caudill, S., Needham, L., Henderson, A., Age specific dioxin TEQ reference range, Organohalogen Compounds 66 (2004), 2878–2883.
Patterson, D.G. Jr., Welch, S.M., Focant J.-F., Turner, W.E., The Use Of Various Gas Chromatography And Mass Spectrometry Techniques For Human Biomonitoring Studies, Lecture at the 26th Int. Symp. Halogenated Persistent Organic Pollutants, Oslo, Norway, 21–25. Aug. 2006.
Ragsdale, J., A Robust Screening Method for Dioxins and Furans by Ion Trap GC-MSMS in a variety of matrices, Thermo Electron Application Note AN10082, Dec 2004.
Toxicological profile for chlorinated dibenzodioxins and dibenzofurans, Agency for Toxic Substances and Disease Registry (ATSDR) (1998).
Turner, W.E., Welch, S.M., DiPietro, E.S., Whitfield, W.E., Cash, T.P., McClure, P.C., Needham, L.L., Patterson, D.G Jr., Instrumental approaches for improving the detection limit for selected PCDD congeners in samples from the general U.S. Population as background levels continue to decline, Poster at the 26th Int. Symp. Halogenated Persistent Organic Pollutants, Oslo, Norway, 21–25. Aug. 2006.
Turner, W., Welch, S., DiPietro, E., Cash, T., McClure, C., Needham, L., Patterson, D., The Phantom Menace – Determination of the True Method Detection Limit (MDL) for Background Levels of PCDDs, PCDFs, and cPCBs in Human Serum by High-resolution Mass Spectrometry, Organohalogen Comp. 66 (2004), 264–271.
US EPA, Exposure and Human Health Reassessment of 2,3,7,8-Tetrachlorodibenzo-p-Dioxin and Related Compounds, Washington D.C.: National Center for Environmental Assessment, US Environmental Protection Agency, EPA/600/P-00/001Be.
US EPA Method 8280B, Polychlorinated Diebnzo-p-dioxins and polychlorinated dibenzofurans by high resolution gas chromatography/low resolution mass spectrometry (HRGC/LRMS), Rev. 2, January 1998.
US EPA Method 8290B, Polychlorinated Diebnzo-p-dioxins and polychlorinated dibenzofurans by high resolution gas chromatography/high resolution mass spectrometry (HRGC/HRMS), Rev. 1, January 1998.
Van den Berg, M.L., et al., Toxic equivalent factors (TEFs) for PCBs, PCDDs, PCDFs for humans and wildlife, Environ. Health Persp. 106/12 (1998), 775–792.

Section 4.26

US EPA, PCB ID – BZ versus IUPAC, Comparison of PCB Nomenclature – Past and Present, see: http://www.epa.gov/toxteam/pcbid/bzviupac.htm
de Boer, J., Q. Dao, and R. van Dortmond, J. High Res. Chromatogr. 15:249–255 (1992).
Bøwadt, S., Larsen, B., Rapid screening of chlorobiphenyl congeners by GC-ECD on a carborane – polydimethylsiloxane copolymer, J. High Res. Chromatogr. 15/5 (2005), 350–351.
Gummersbach, J., Thermo Fisher Scientific, Application Laboratory Dreieich, Germany, personal communication.
SGE, HT8: The perfect PCB column, Publication No. AP-0040-C Rev:04 08/05.

Section 4.27

California Legislature Bill No. 302, Chaptered August 11, 2003.
Directive 2002/95/EC of the European Parliament and of the Council of the European Union, January 27, 2003
Krumwiede, D., Huebschmann, H.-J., DFS – Analysis of Brominated Flame Retardants with High Resolution Mass Spectrometry, Application Note 30098, Thermo Fisher Scientific, Bremen, Germany.
Krumwiede, D., Huebschmann, H.-J., Trace Analysis of Brominated Flame Retardants with High Resolution Mass Spectrometry, LCGC Europe Supplement, July 2006.
Krumwiede, D., Huebschmann, H.J., High Resolution GC/MS as a Viable Solution for Conducting Environmental Analyses, The Peak 11 (2007), 7–15.
Stapleton, H.M., Kelly, S.M., Allen, J.G., McClean, M.D., Webster, T.F., Measurement of Polybrominated Diphenyl Ethers on Hand Wipes: Estimating Exposure from Hand-to-Mouth Contact, Environ. Sci. Technol. 42 (9) (2008), 3329-3334.

Section 4.28

De Boer, J., Allchin, C., Zegers, B., Boon, J.P., Trends Anal. Chem. 10 (2001), 591–599.
Krumwiede, D., Huebschmann, H.J., Analysis of Brominated Flame Retardants by High Resolution GC/MS, Thermo Scientific Application Note 30098, Bremen 2006.
Polo, M., Gomez-Noya, G., Quintana, J.B., Llompart, M., Garcia-Jares, C., Cela, R., Development of a Solid-Phase Microextraction Gas Chromatography/Tandem Mass Spectrometry Method for Polybrominated Diphenyl Ethers and Polybrominated Biphenyls in Water Samples, Anal.Chem. 76 (2004), 1054–1062.

Section 4.29

Deutsche Forschungsgemeinschaft, Maximale Arbeitsplatzkonzentrationen und biologische Arbeitsstofftoleranzwerte, VCH: Weinheim 1990.
Kuitunen, M. L., Hartonen, K., Riekkola, M. L., Analysis of Chemical Warfare Agents from Soil Samples Using Off-Line Supercritical Fluid Extraction and Capillary Gas Chromatography. In: P. Sandra (Ed), 13th Int. Symp. Cap. Chrom., Riva del Garda May 1991, Huethig: Heidelberg 1991, 479–488.
Niebel, J., Cleanup und Detektion von Nitro- und Aminoaromaten mit dem GC-MSD/ECD/FID, Diplomarbeit, FH Technik Mannheim, 1992.
Schimmelpfeng, L. (Hrsg), Altlasten, Deponietechnik, Kompostierung, Teil 1: Erfassung, Untersuchung von Altlasten, Rüstungsaltlasten, Academia Verlag: Sankt Augustin 1993.
Wöhrle, D., Die neue Chemie-Waffen-Konvention, Nachr. Chem. Tech. Lab. 41 (1993), 291–296.
Yinon, J., Zitrin, S., Modern Methods and Applications of Explosives, John Wiley & Sons: Chichester 1993.

Section 4.30

Martz, R., Donelly, B., et al., The Use of Hair Analysis to Document a Cocaine Overdose following a Sustained Survival Period Before Death, J. Anal. Toxicol. 15 (1991), 279–281.
Möller, M. R., Fey, P., Wennig, R., Simultaneous Determination of Drugs of Abuse (Opiates, Cocaine and Amphetamine) in Human Hair by GC/MS and its Application to the Methadon Treatment Program, in: P. Saukko (Ed), Forensic Science International, Vol. 63, Special Issue: Hair Analysis as a Diagnostic Tool for Drugs of Abuse Investigation, Elsevier: Amsterdam 1993, 185–206.
Sachs, H., Institut für Rechtsmedizin, Munich (private communication).
Sachs, H., Raff, I., Comparison of Quantitative Results of Drugs in Human Hair by GC/MS, in: P. Saukko (Ed), Forensic Science International, Vol. 63, Special Issue: Hair Analysis as a Diagnostic Tool for Drugs of Abuse Investigation, Elsevier: Amsterdam 1993, 207–216.
Sachs, H., Uhl, M., Opiat-Nachweis in Haar-Extrakten mit Hilfe von GC/MS/MS und Supercritical Fluid Extraktion (SFE), T + K 59 (1992), 114–120.
Schwinn, W., Drogennachweis in Haaren, Der Kriminalist 11 (1992), 491–495.
Traldi, P., Favretto, D., Tagliaro, F., Ion Trap Mass Spectrometry, a New Tool in the Investigation of Drugs of Abuse in Hair, Forensic Science International 63 (1991), 239–252.

Section 4.31

Cowan, D. A., Woffendin, G., Drug Testing in Human Sports, Spectra 11 (1988), 4–9.
Donike, M., Derivatisierung für chromatographische Untersuchungen, GDCh-Kurs 306/92, Institut für Biochemie der Deutschen Sporthochschule Köln, 1992.
Jain, N. C., The HHS (formerly NIDA) Program for Drugs of Abuse Testing – The Views of a Certified Lab, Symposium Aktuelle Aspekte des Drogennachweises, Mosbach 15. April 1993.
Matthiesen, U., GC-MS in der klinischen Chemie, LaBo 1 (1988), 7–12.
Maurer, H. H., GC/MS contra Immunoassay?, Symposium Aktuelle Aspekte des Drogennachweises, Mosbach 15. Apr. 1993, Abstracts.
Maurer, H. H., Identifizierung unbekannter Giftstoffe und ihrer Metaboliten in biologischem Material, GIT Suppl. 1 (1990), 3–10.
Musshoff, F., Urinbefund nach Aufnahme von Heroin, GIT Labor Fachz. 1 (2007), 13–16.
Musshoff, F., Trafkowski, J., Madea, B., Validated assay for the determination of markers of illicit heroin in urine samples for the control of patients in a heroin prescription program, J. Chromatogr. B 811 (2004), 47–52.
Pfleger, K., Maurer, H. H., Weber, A., Mass Spectra and GC Data of Drugs, Poisons, Pesticides, Pollutants and Their Metabolites, Parts 1,2,3, 2. erw. Aufl., VCH: Weinheim 1992.
Stellungnahme der Arbeitsgruppe „Klinisch-Toxikologische Analytik“ der Deutschen Gesellschaft

für Klinische Chemie zu den notwendigen Drogensuchtests insbesondere bei der Methadon-Substitutionsbehandlung von i.v.-Heroinabhängigen, DG Klinische Chemie, Mitteilungen 24 (1993), 212–214.
Uhrich, M., Tillmanns, U., Identifizierung und Quantifizierung von Codein und Morphin, GIT Fachz. Lab. 10 (1990), 1265–1267.
Weller, J.P., Wolf, M., Nachweis von Morphinderivaten im Blut mittels Ion-Trap-Detektor, Finnigan MAT Application Report No. 76, 1990.
Weller, J. P., Wolf, M., Szidat, S., Enhanced selectivity in the determination of Δ^9-tetrahydrocannabinol and two major metabolites in serum using ion trap GC-MS-MS, J. Anal. Toxicol. 24 (2000), 1–6.

Section 4.32

Johnson, K., Uhrich, M., Tillmanns, U., Identifizierung und Quantifizierung von 11-Nor-9-Carboxy-9-Delta-Tetrahydrocannabinol, GIT Spezial-Chromatographie 1 (1991) 16–18.
Uhrich, M. D., On-Disc Derivatisation and Full Scan GC/MS Analysis of 9-Carboxy-THC Create Competitive Advantages for Laboratories, Finnigan MAT Application Data Sheet No. 53, Firmenschrift, 1992.
Weller, J. P., Wolf, M., Med. Hochschule Hannover, Institut für Rechtsmedizin 1995, unveröffentlichte Ergebnisse.

Section 4.33

Cowan, D. A., Woffendin, G., Drug Testing in Human Sports, Spectra 11 (1988), 4–9.
Schänzer, W., Thevis, M., Analytik von Steroidhormonen und Designer-Steroiden, GIT Lab. Fachz. 5 (2004), 440–443.
Schilling, G., Tetrahydrogestrinon – ein Dopingskandal mit neuen Dimensionen, GIT Labor Fachz. 11 (2003), 1118–1119.
Schweingruber, H., Wang, K., Determination of Estradiol in Plasma with Negative Chemical Ionization GC-MS/MS on TSQ Quantum GC, Application Note: 388, June 2007, Thermo Fisher Scientific, San Jose, CA, USA
WADA Laboratory Committee, Reporting and evaluation guidance for testosterone, epitestosterone, T/E ratio and other endogenous steroids, WADA Technical Document – TD2004EAAS, 13 August, 2004.
Woffendin, G., Analysis of Steroids by GC/MS and GC/MS/MS. Finnigan MAT Application Report 238, Firmenschrift, 1995.

Section 4.34

Gummersbach, J., ThermoQuest Applications Laboratory, Egelsbach, Germany
Schweer, H., Seybarth, H. W., Meese, C. O., Fürst, O., Negative Chemical Ionization Gas Chromatography/Mass Spectrometry and Gas Chromatography/Tandem Mass Spectrometry of Prostanoid Pentafluorobenzyl Ester/Methoxime/Trimethylsilyl Ether Derivatives, Biomed. Environm. Mass Spectrom. 15 (1988), 143–151.
Schweer, H., Meese, C. O., Watzer, B., Seybarth, H. W., Determination of Prostaglandin E_1 and its Main Plasma Metabolites 15-Keto-prostaglandin E_0 and Prostaglandin E_0 by Gas Chromatography/Negative Ion Chemical Ionization Triple-stage Quadrupole Mass Spectrometry, Biolog. Mass Spectrom. 23 (1994), 165–170.
Schweer, H., Seyberth, H. W., Schubert, R., Determination of Prostaglandin E_2 Prostaglandin $F_{2\alpha}$ and 6-Oxo-prostaglandin $F_{1\alpha}$ in Urine by Gas Chromatography/Mass Spectrometry and Gas Chromatography/Tandem Mass Spectrometry: A Comparison, Biomed. Environm. Mass Spectrom., 13 (1986), 611–619.
Strife, R. J., Kelley, P. E., Weber-Grabau, M., Tandem Mass Spectrometry of Prostaglandins: A Comparison of an Ion-Trap and a Reversed Geometry Sector Instrument, Rap. Comm. Mass Spectrom. 2/6 (1998), 105–109.
Tsikas, D., Schwedhelm, E., Fauler, J., Gutzki, F.-M., Mayatepek, E., Frölich, J. C., Specific and rapid quantification of 8-iso-prostaglandin $F_{2\alpha}$ in urine of healthy humans and patients with Zellweger syndrome by gas chromatography – tandem mass spectrometry, J. Chrom B, 716 (1998), 7–17.

Section 4.35

Courselle, P., Schelfaut, M., Sandra, P., et al., The Analysis of the β-Agonists Clenbuterol and Salbutamol by Capillary GC-MS. Considerations on Sample Introduction. In: P. Sandra (Ed), 13th Int. Symp. Cap. Chrom., Riva del Garda, May 1991, Huethig: Heidelberg 1991, 388–393.
Fürst P., GC/MS-Bestimmung von Rückständen und Kontaminanten, in: Matter L., Lebensmittel

und Umweltanalytik mit der Kapillar-GC, VCH: Weinheim 1994, 29–73.

Tillmanns, U., Nachweis von Clenbuterol mit dem ITS 40, Finnigan MAT Application Report No. 74, 1990.

Section 4.36

Demme, U., Thoben, M., Klinikum der Friedrich-Schiller-Universität Jena, Institut für Rechtsmedizin, Jena, 1995, private communication.

Pfleger, K., Maurer, H. H., Weber, A., Mass Spectral and GC Data, Part 1, VCH: Weinheim 1992.

Demme, U., Müller, U., Zur Ausbeutebestimmung toxikologisch-chemischer Analysenverfahren, Toxichem + Krimitech 57 (1990), 121.

Deutsche Forschungsgemeinschaft, Orientierende Angaben zu therapeutischen und toxikologischen Konzentrationen von Arzneimitteln und Giften in Blut, Serum oder Urin, VCH: Weinheim 1990.

Meyer, L. V., Hauck, G., Der Nachweis gebräuchlicher Antihistaminika nach therapeutischer Dosierung, Beiträge gerichtl. Medizin 34 (1976), 129.

Section 4.37

Butz, S., Heberer, Th., Stan, H.-J., Analysis of phenoxyalkanoic acids and other acidic herbicides at the low ppt-level in water applying solid phase extraction with RP-C18 material, J. Chromatogr. A 677 (1994), 63–74.

Heberer, Th., Butz, S., Stan, H.-J., Detection of 30 acidic herbicides and related compounds as their pentafluorobenzylic esters using GC/MSD, J. AOAC 77 (1994), 1587–1604.

Stan, H.-J., Institut für Lebensmittelchemie, Technical University, Berlin.

Stan, H.-J., Heberer, Th., Linkerhägner, M., Vorkommen von Clofibrinsäure im aquatischen System – Führt die therapeutische Anwendung zu einer Belastung von Oberflächen-, Grund- und Trinkwasser?, Vom Wasser 83 (1994), 57–68.

Section 4.38

Eschke, H.-D., Chemisches und Biologisches Laboratorium des Ruhrverbandes, Essen, personal communication.

Eschke, H.-D., Traud, J., Dibowski, H.-J., Analytik und Befunde künstlicher Nitromoschus-Substanzen in Oberflächen- und Abwässern sowie Fischen aus dem Einzugsgebiet der Ruhr, Vom Wasser 83 (1994), 373–383.

Eschke, H.-D., Traud, J., Dibowski, H.-J., Untersuchungen zum Vorkommen polycyclischer Moschus-Duftstoffe in verschiedenen Umweltkompartimenten, Z. Umweltchem. Ökotox. 6 (1994), 183–189.

Rimkus, G., Wolf, M., Analysis and Bioaccumulation of Nitro Musks in Aquatic and Marine Biota, in: P. Sandra, G. Devos (Eds): 16th Int. Symp. Cap. Chrom., Riva del Garda, Sept. 27th–30th 1994, Hüthig: Heidelberg 1994, 433–445.

Section 4.39

Lagana, A., et al., Rapid Comm. Mass Spectrom. 17 (2003), 1037.

Mateo, J.J., et al., J. Chromatogr. A 918 (2001), 99.

Melchert, H.-U., Pabel, E., J. Chromatogr. A 1056 (2004), 195–199.

Razzazi-Fazeli, E., et al., J. Chromatogr. A 968 (2002), 129.

Section 4.40

Hawthorne, S.B., Miller, D.J., Water Chemical Ionization Mass Spectrometry of Aldehydes, Ketones, Esters, and Carboxylic Acids, Appl. Spectroscopy, 40/8 (1986), 1200–1211.

Landrock, A., GC/MS-Analysenverfahren für Ion-Trap MS-Systeme mit vollautomatischer Identifizierung und Quantifizierung, GIT Spezial Separation I (2002), 29–31.

Landrock, A., Richter, H., Merten, H., Water CI, a new selective and highly sensitive method for the detection of environmental components using ion trap mass spectrometers, Fresenius J. Anal. Chem. 351 (1995), 536–543.

Lerch O, Zinn P., Derivatisation and gas chromatography-chemical ionisation mass spectrometry of selected synthetic and natural endocrine disruptive chemicals, J. Chromatogr. A. 991/1 (2003), 77–97.

Liquid CI option, see http://www.axelsemrau.de/Liquid_CI_Option.html

Merten, H., Richter, H., Landrock, A., Ion-Trap GC/MS: Chemische Ionisierung mit Wasser, eine zerstörungsarme Detektionsmethode, GIT Fachz. Lab 10 (1996), 1008–1017.

Section 4.41

Asperger, A., Engewald, W., Fabian, G., Analytical characterisation of natural waxes employing pyrolysis-gas chromatography-mass spectrometry, J. Anal. Appl. Pyrolysis 50 (1999), 103–115.

Ericsson, I., Influence of Pyrolysis Parameters on Results in Pyrolysis-Gas Chromatography, J. Anal. Appl. Pyrolysis 8 (1985), 73 (see also www.pyrolab.se)

Section 4.42

Asperger, A., Engewald, W., Wagner, T., Quantitative determination of acrylate-based copolymer retarsing layers on drug granules using pyrolysis-gas chromatography, J. Anal. Appl. Pyrolysis 49 (1999), 155–164.

5
Glossary

This glossary of chromatographic and mass spectrometric terms also contains selected terms of the third draft of recommendations for nomenclature, definitions of terms, and acronyms in mass spectrometry, currently undergoing review for publication in the IUPAC Journal Pure and Applied Chemistry.

A

α	Separation factor of two adjacent peaks; $\alpha = k_2/k_1$.
α-cleavage	Homolytic cleavage where the bond fission occurs between at the atom adjacent to the atom at the apparent charge site and an atom removed from the apparent charge site by two bonds.
A_p	Peak area.
a-ion	Fragment ion containing the N-terminus formed upon dissociation of a protonated peptide at a backbone C–C bond.
AC	Alternating current.
Accelerating voltage	Electrical potential used to impart translational energy to ions in a mass spectrometer.
Accelerator mass spectrometry (AMS)	Mass spectrometry technique in which atoms extracted from a sample are ionized, accelerated to MeV energies and separated according to their momentum, charge and energy.
Accuracy	The closeness of agreement between a test result and the accepted reference value, it is determined by determining trueness and precision.
Accurate mass	Experimentally determined mass of an ion that is used to determine an elemental formula. Note: accurate mass and exact mass are not synonymous. The former refers to a measured mass and the latter to a calculated mass.
ACN	Acetonitril.
Adiabatic ionization	Process whereby an electron is removed from an atom, ion, or molecule in its lowest energy state to produce an ion in its lowest energy state.
Adduct ion	Ion formed by the interaction of an ion with one or more atoms or molecules to form an ion containing all the constituent atoms of the precursor ion as well as the additional atoms from the associated atoms or molecules.

Handbook of GC/MS: Fundamentals and Applications, Second Edition. Hans-Joachim Hübschmann

ISBN: 978-3-527-31427-0

AFS	Amperes full scale.
AGC	Automatic gain control, controls the variable ionization time in ion trap mass spectrometers depending from the total ion current in the selected mass range by a quick pre-scan. This control provides the high inherent full scan sensitivity of ion trap mass spectrometers at low ion streams by storing ions until the full trap capacity is reached.
Alumina	A gas–solid adsorbent stationary phase.
AMDIS	Automated mass spectra deconvolution and identification system, analyses the individual ion signals and extracts and identifies the spectrum of each component in coeluting peaks analyzed by GC/MS.
Analog ion	Ions that have similar chemical valence, for example the acetyl cation CH_3-CO^+ and the thioacetyl cation CH_3-CS^+.
Analyte	The substance that has to be detected, identified and/or quantified and derivatives emerging during its analysis.
Analytical scan	The part of the ion trap MS scan function that produces the mass spectrum.
APE	Atom percent excess, commonly used expression in tracer experiment employing labeled substances for the degree of enrichment above the natural isotope content $APE = at.\% - at.\%_{nat}$
Appearance energy (AE)	Minimum energy that must be imparted to an atom or molecule to produce a detectable amount of a specified ion. In mass spectrometry it is the voltage which corresponds to the minimum electron energy necessary for the production of a given fragment ion. The term appearance potential (AP) is deprecated.
ARC	Automatic reaction control, variable ionization and reaction time control in internal ionisation ion trap mass spectrometers for chemical ionization. A built-up of CI ions in the ion trap is facilitated by the AGC control until the maximum capacity or preset maximum reaction time is reached. This results in the inherent high sensitivity of the ion trap analyzer for full scan data acquisition in full scan mode.
Array detector	Detector comprising several ion collection elements, arranged in a line or grid where each element is an individual detector.
Associative ion/ molecule reaction	Reaction of an ion with a neutral species in which the reactants combine to form a single ion.
Associative ionization	Ionization process in which two atoms or molecules, one or both of which is in an excited state, react to form a single positive ion and an electron.
Atmospheric pressure chemical ionization (APCI)	Chemical ionization that takes place using a nebulized liquid and atmospheric pressure corona discharge.
Atmospheric pressure ionization (API)	Ionization process in which ions are formed in the gas phase at atmospheric pressure.

Atmospheric pressure matrix-assisted laser desorption/ionization (AP MALDI) — Matrix-assisted laser desorption/ionization in which the sample target is at atmospheric pressure.

Atmospheric pressure photoionization (APPI) — Atmospheric pressure chemical ionization in which the reactant ions are generated by photoionization.

Atom % or at.% — Unit commonly used for the expression of isotope ratios, e.g. tracer experiments.

$$\text{at.\%} = n/\Sigma\, n_i \cdot 100 = 1/(1 + 1/R) \cdot 100\ [\%]$$

with n = number of isotope atoms, $n_i = R = {}^{13}C/{}^{12}C$

Autodetachment — Formation of a neutral species when a negative ion in a discrete state with an energy greater than the detachment threshold loses an electron spontaneously without further interaction with an energy source.

Autoionization — Formation of an ion when an atom or molecule in a discrete state with an internal energy greater than the ionization threshold loses an electron spontaneously without further interaction with an energy source.

Average mass — Mass of an ion or molecule calculated using the average mass of each element weighted for its natural isotopic abundance.

B

β — Phase ratio. The ratio of mobile- to stationary-phase volumes. Thicker stationary-phase films yield longer retention times and higher peak capacities.
For open-tubular columns, $\beta = V_G/V_L \sim d_c/4\,d_f$

β-cleavage — Homolytic cleavage where the bond fission occurs between at an atom removed from the apparent charge site atom by two bonds and an atom adjacent to that atom and removed from the apparent charge site by three bonds.

b-Ion — Fragment ion containing the N-terminus formed upon dissociation of a protonated peptide at a backbone C–N bond.

Backflush — Occurs when peaks at the end of a chromatogram are flushed from a column to vent or to another column by flow reversal.

Bake out — Generally a thermal cleaning step in various applications:
Gas chromatography – The process of removing contaminants from a column by operation at elevated temperatures, which should not exceed a column's maximum operating temperature (MAOT).
Purge and trap methodology – The purification step of the adsorption trap.
Mass spectrometry – The cleaning of the analyzer from dissolved gases and contaminants by heating the steel manifold for an extended time.

Balanced pressure injection — Headspace injection technique whereby the equilibrated sample vial is depressurised to its maximum against the column pre-pressure.

Band broadening	Several processes that cause solute profiles to broaden as they migrate through a column.
Base peak (BP)	The peak in a mass spectrum that has the greatest intensity. Note: This term may be applied to the spectra of pure substances or mixtures.
Beam instruments	Type of mass spectrometers in which beams of ions are continuously formed from an ion source, passed through ion optics and are resolved for mass analysis, typically magnetic sector and quadrupole instruments.
Benzyl cleavage	A fragmentation reaction of alkylaromatics forming the benzylic carbenium ion, which appears as the tropylium ion with m/z 91 in many spectra of aromatics.
Bias	The difference between the expectation of the test result and an accepted reference value.
Blank value	Many analysis methods require the determination of blank values in order to be able to compensate for non-specific analyte/matrix interaction. A differentiation is made between reagent blank samples and sample blank samples.
Bleed	The loss of material from a column or septum caused by high-temperature operation. Bleed can result in ghost peaks and increased detector baseline offset and noise.
Bonded phase	A stationary phase that has been chemically bonded to the inner column wall, → also Cross-linked phase.
BSIA	Bulk sample isotope analysis, the isotope ratio MS analysis of a sample after bulk conversion in a crucible of an elemental analyzer, contrary to CSIA.
BSTFA	Silylating agent for derivatisation reactions, Bis(trimethylsilyl)trifluoroacetamide
BTEX	Abbreviation for the analysis of the benzene, toluene, ethylbenzene and xylene isomers group of aromatics, mostly by headspace sample analysis.
BTV	Breakthrough volume, e. g. of an adsorption trap.
BTX	→ BTEX.
Buffer gas	Inert gas used for collisional deactivation of internally excited ions or of the translational energies of ions confined in an ion trap.

C

CAD	Collision activated decomposition in MS/MS experiments, → CID.
Capacity factor	The k' value of a column describes the molar ratio of a substance in the stationary phase to that in the mobile phase from the relationship of the net retention time to the dead volume.
Carbosieve	Carbon molecular sieve used as an adsorption material for air analysis, → also VOCARB (Supelco).

Carboxen	Carbon molecular sieve used as an adsorption material for air analysis, → also VOCARB (Supelco).
Carry over	Taking the analyte to the next analysis.
CAS No.	The unique registration number of a chemical compound or substance assigned by the Chemical Abstract Service, a division of the American Chemical Society. The intention is to make database searches more convenient, as chemicals often have many names. Almost all molecule databases today allow searching by CAS number. The CAS no. usually is given in the substance entries of many library mass spectra.
Cationized molecule	Ion formed by the association of a cation with a molecule M, for example $[M + Na]^+$ and $[M + K]^+$. The terms quasi-molecular ion and pseudo-molecular ion are deprecated.
CCM	A scan-by-scan calibration correction method used e.g. in high resolution selected reaction monitoring, → H-SRM.
CDEM	→ Continuous dynode electron multiplier.
CE	Coating efficiency. A metric for evaluating column quality. The minimum theoretical plate height divided by the observed plate height; $CE = H_{min}/H$.
Centroid	The calculated center of a mass peak acquired in scan mode. The centroid value can be calculated precisely independent of the resolution power of the mass spectrometer in use. Values displayed in the spectrum with three or more digits are often misleadingly associated with the resolving power of the instrument. In LRMS special care has to be taken as the centroid value gives the center of gravity of the mass peak composed of many compounds falling in the wide mass window. In HRMS centroid mass values are used to calculate a possible sum formula within a deviation of < 2 ppm.
Centroid acquisition	Procedure of recording mass spectra in which an automated system detects peaks, calculates their centroids, and assigns m/z values based on a mass calibration file. Only the centroid m/z and the peak intensity values are stored; → Continuum acquisition.
Charge exchange ionization	Type of CI reaction (PCI), interaction of an ion with an atom or molecule in which the charge on the ion is transferred to the neutral without the dissociation of either → charge transfer ionization.
Charge transfer reaction	Type of CI reaction (NCI), action of an ion with a neutral species in which some or all of the charge of the reactant ion is transferred to the neutral species.
Chemical ionization	Formation of a new ion in the gas phase by the reaction of a neutral analyte species with an ion. The process may involve transfer of an electron, a proton or other charged species between the reactants. Note 1: When a positive ion results from chemical ionization the term may be used without qualification. When a negative ion results the term negative ion chemical ionization should be used. Note 2: this term is not synonymous with chemi-ionization. Through chemical ionization usually a soft ionisation is achieved providing information on the (quasi-)molecular weight of a substance. The selectivity

of the reaction and extent of fragmentation are controlled by the choice of the reagent gas.

Chemi-ionization — Reaction of an atom or molecule with an internally excited atom or molecule to form an ion. Note that this term is not synonymous with chemical ionization.

CI — → Chemical ionization.

CID — Collision induces dissociation, leads in MS/MS to the formation of a product ion spectrum from a selected precursor ion.

C-ion — Fragment ion containing the N-terminus formed upon dissociation of a protonated peptide at a backbone N–C bond.

Clean-up — Generally the sample preparation procedure involving removal of the matrix and concentration of the analyte.

Cluster ion — Ion formed by a multi-component atomic or molecular assembly of one or more ions with atoms or molecules, such as $[(H_20)_nH]^+$, $[(NaCl)_nNa]^+$ and $[(H_3PO_3)_nHPO_3]^-$.

Co-chromatography — A procedure in which the extract prior to the chromatographic step(s) is divided into two parts. Part one is chromatographed as such. Part two is mixed with the standard analyte that is to be measured and chromatographed. The amount of added standard analyte has to be similar to the estimated amount of the analyte in the extract. This method is designed to improve the identification of an analyte, especially when no suitable internal standard can be used, → also Standard addition method.

Cold injection — An injection that occurs at temperatures lower than the final oven temperature, usually at or below the solvent boiling point.

Cold trapping — A chromatographic technique for focussing volatile compounds at the beginning of the column by cooling a section of the column e.g. with liquid CO_2 below the boiling point of the compounds or solvent, → also Cryofocussing.

Collision gas — Inert gas used for collisional excitation and ion/molecule reactions. The term target gas is deprecated.

Collision-induced dissociation (CID) — Dissociation of an ion after collisional excitation. The term collisionally activated dissociation (CAD) is deprecated.

Collision quadrupole — Transmission quadrupole to which an oscillating radio frequency potential is applied so as to focus a beam of ions through a collision gas with no m/z separation. Note: a collision quadrupole is often indicated by a lower case q as in QqQ.

Collisional activation (CA) — → Collisional excitation.

Collisional excitation — Reaction of an ion with a neutral species in which all or part of the translational energy of the collision is converted into internal energy of the ion, e.g. for the CID process in MS/MS.

Collision cell
: A transmission hexapole, octapole or square quadrupole collision cell to which an oscillating radio frequency potential is applied that is filled with a collision gas at low pressure and used to generate collision induced dissociation (CID) of ions to form a product ion spectrum. The collision cell has no mass separating capabilities.

Comprehensive GC (GC×GC)
: Two-dimensional GC technique in which all compounds experience the selectivity of two columns connected in series by a retention modulation device, thereby generating much higher resolution than that attainable with any single column.

Compressibility correction factor (j)
: This factor compensates for the expansion of a carrier gas as it moves along the column from the entrance, at the inlet pressure (p_i), to the column exit, at the outlet pressure (p_o).

Concurrent solvent recondensation (CSR)
: Large volume injection technique with splitless injectors, due to recondensation of solvent inside a retention gap the resulting pressure difference speeds up significantly the sample vapour transfer from the injector allowing larger samples volumes to be injected.

Confirmatory method
: A method that provides full or complementary information enabling the substance to be unequivocally identified and if necessary quantified at the level of interest.

Consecutive reaction monitoring (CRM)
: MS^n experiment with three or more stages of m/z separation and in which a particular multi-step reaction path is monitored.

Constant neutral loss scan
: A scan procedure for a tandem mass spectrometer designed to produce a constant neutral loss spectrum of different precursor ions by detection of the corresponding product ions produced by metastable ion fragmentation or collision-induced dissociation. Synonymous terms are constant neutral mass loss scan and fixed neutral fragment scan.

Constant neutral loss spectrum
: Spectrum of all precursor ions that have undergone an operator-selected m/z decrement, obtained using a constant neutral loss scan; → Constant neutral mass loss spectrum and fixed neutral mass loss spectrum.

Constant neutral mass gain scan
: Scan procedure for a tandem mass spectrometer designed to produce a constant neutral mass gain spectrum of different precursor ions by detection of the corresponding product ions of ion/molecule reactions with a gas in a collision cell.

Constant neutral mass gain spectrum
: Spectrum formed of all product ions produced by gain of a pre-selected neutral mass following ion/molecule reactions with the gas in a collision cell, obtained using a constant neutral mass gain scan.

Continuous dynode electron multiplier
: An ion-to-electron detector in which the ion strikes the inner continuous resistance surface of the device and induces the production of secondary electrons that in turn impinge on the inner surfaces to produce more secondary electrons. This avalanche effect produces an increase in signal in the final measured current pulse, → also SEM.

Continuous injection
: Process for the production of defined atmospheres for the calibration of thermodesorption tubes.

Conversion dynode — Surface that is held at high potential so that ions striking the surface produce electrons that are subsequently detected, → Post-acceleration detector.

CRM — Certified reference material, means a material that has had a specified analyte content assigned to it.

Cross-linked phase — A stationary phase that includes cross-linked polymer chains. Usually, it is bonded to the column inner wall, → also Bonded phase.

Cross-talk — MS/MS signal artifacts in a SRM transition from the previous SRM scan, can potentially occur when fragment ions from one SRM transition remain in the collision cell while a second SRM transition takes place, can be the source of false positives when different SRM events have the same product ions formed from different precursor ions.

Cryofocussing — Capillary GC injection technique for volatile compounds in headspace, purge and trap or thermodesorption systems. Instead of an evaporation injector, on-line coupling to the sample injection is set up in such a way that a defined region of the column is cooled by liquid CO_2 or liquid N_2 to focus a volatile sample. The chromatography starts with heating up of the focussing region.

CSIA — Compound specific isotope analysis, the isotope ratio MS analysis of individual compounds after chromatographic separation followed by the online conversion to simple gases, contrary to BSIA.

CSR — → Concurrent solvent recondensation.

Cyclotron motion — Circular motion of a particle of charge q moving at velocity v in a magnetic flux density B that results from the Lorentz force $qv \times B$.

D

2DGC — Two dimensional GC, abbreviation used for describing heart-cutting as well as comprehensive GCxGC methods. → Comprehensive GC.

3D trap — Three-dimensional ion trap mass spectrometer, usually reflected with this term is the typical internal ionisation capability; modern 3D traps also use external ion sources providing ion injection into the trap for storage and mass analysis.

δ notation — Notation in units per mil (‰), commonly used in isotope ratio mass spectrometry to express isotope abundance differences as a ‰ deviation between the sample against an international standard, e. g. for ^{13}C

$$\delta\,^{13}C = [(R_{Sample}/R_{Standard}) - 1] \cdot 10^3\ [‰]$$

with $R = {}^{13}C/{}^{12}C$ 0.01‰ equals 10^{-5} at.% (→ also atom %) The international standards are assigned a value of 0.0 ‰ on their respective δ scales.

d_c — Average column inner diameter.

d_f — Average stationary-phase film thickness.

D_G — Gaseous diffusion coefficient; approximately 0.05 for hydrocarbons in helium carrier gas and 0.1 for hydrogen carrier gas.

D_L — Liquid–liquid diffusion coefficient; approximately 1×10^{-5} for hydrocarbons in silicones.

d_p — Average particle diameter.

Dalton, Da — Non-SI unit of mass (symbol Da) that is identical to the unified atomic mass unit, based since 1961 on the mass definition of ^{12}C = 12.00000 g/mol

Daly detector — Detector consisting of a conversion dynode, scintillator and photomultiplier. A metal knob at high potential emits secondary electrons when ions impinge on the surface. The secondary electrons are accelerated onto the scintillator that produces light that is then detected by the photomultiplier detector.

DAT — Diaminotoluene isomers.

DC — Direct current.

DEA — The US Drug Enforcement Agency.

Dead time — The time taken for a substance which is not retarded e. g. air, methane for silicone phases, to pass through a chromatographic column. The carrier gas velocity is calculated from the dead time and the column length, → also HETP.

Dead volume — Extra volume experienced by analytes as they pass through a chromatographic system. Excessive dead volume causes additional peak broadening.

Decision limit — The limit at and above which it can be concluded with an error probability of α that a sample is non-compliant.

Deconvolution — The term is used in AMDIS in the broad sense of extracting one signal from a complex mixture. The treatment of noise, the correction for base line drift, and the extraction of closely co eluting peaks from one another are part of the deconvolution process.

DEGS — Diethylene glycol succinate; used as a stationary phase.

Delayed extraction (DE) — Application of the accelerating voltage pulse after a time delay in desorption/ionization from a surface. The extraction delay can produce energy focusing in a time-of-flight mass spectrometer.

Desorb preheat — Purge and trap technique step for heating up the adsorption trap. The effective desorption and transfer of analytes from the adsorption trap into the GC/MS system is effected by switching a 6-way valve first after a set preheat temperature of approx. 5 °C below of the desorption temperature has been reached.

Desorption ionization (DI) — Formation of ions from a solid or liquid material by the rapid vaporization of that sample in the ion source.

Detection capability — The smallest content of the substance that may be detected, identified and/or quantified in a sample with an error probability of β.

Detection limit — It describes the smallest detectable signal of the analyte that can be clearly differentiated from a blank sample (assessed in the signal domain, detection criterion). It is calculated as the upper limit of the distribution range of the blank value.

Diagnostic ion — Ion whose formation reveals structural or compositional information. For instance, the phenyl cation in an electron ionization mass spectrum is a diagnostic ion for benzene and derivatives.

Dimeric ion — An ion formed by ionization of a dimer or by the association of an ion with its neutral counterpart such as $[M_2]^{+\bullet}$ or $[M\text{-}H\text{-}M]^+$.

Direct injection — Occurs when sample enters an inlet and is swept into a column by carrier-gas flow. No sample splitting or venting occurs during or after the injection.

Direct insertion probe — Device for introducing a single sample of a solid or liquid, usually contained in a quartz or other non-reactive sample holder (e.g. crucible), into a mass spectrometer ion source.

Dissociative ionization — Reaction of a gas-phase molecule that results in its decomposition to form products, one of which is an ion.

Distonic ion — Radical cation or anion arising formally by ionization of diradicals or zwitterionic molecules (including ylides). In these ions the charge site and the unpaired electron spin cannot be both formally located in the same atom or group of atoms as it can be with a conventional ion.

dl-PCB — Dioxin-like PCBs, → PCB.

DMCS — Dimethylchlorosilane; used for silanizing glass GC parts.

DNB — Dinitrobenzene isomers.

DOD — The US Department of Defense.

Double-focusing mass spectrometer — Mass spectrometer that incorporates a magnetic sector and an electric sector connected in series in such a way that ions with the same m/z but with distributions in both the direction and the translational energy of their motion are brought to a focus at a point.

Dry purge — Purge & trap analysis step whereby moisture is removed from the trap by the carrier gas before desorption.

Duty cycle — Degree of the effective ion acquisition time of, for example, a mass spectrometer. Here the duty cycle is determined by the sum of the ion dwell times relative to the total scan cycle time including jump and stabilisation times of the analyser voltages.

Dwell time — Effective ion acquisition time typically in [ms] during a selected ion monitoring scan (→ SIM, MID, SRM).

E

ECD — Electron-capture detection, a detector ionizes analytes by collision with metastable carrier-gas molecules produced by β-emission from a radioactive source such as ^{63}Ni. The electron capture detector is one of the most sensitive detectors and it responds strongly to halogenated analytes and others with high electron-capture cross-sections. → also Electron capture dissociation.

Relative ECD-response to hydrocarbons:

10^1	Esters, ethers
10^2	Monochlorides, alcohols, ketones, amines
10^3	Dichlorides, monobromides
10^4	Trichlorides, anhydrides
10^5–10^6	Polyhalogenated, mono-, diiodides

Eddy diffusion — Multipath effect in chromatography, the cause of peak broadening through diffusion processes.

Effective plates — The number of effective theoretical plates in a column, taking the dead volume into consideration (→ also HETP).

Efficiency — The ability of a column to produce sharp, well-defined peaks. More-efficient columns have more theoretical plates (N) and smaller theoretical plate heights (H).

EI — Electron ionisation, with modern GC/MS instruments ionisation usually takes place at ionisation energy of 70 eV. Positive ions are formed predominantly.

Einzel lens — Three element ion lens in which the first and third elements are held at the same voltage. Such a lens produces focusing without changing the translational energy of the particle.

ELCD — Electrolytic-conductivity detector, also Hall detector; gives a mass-flow dependent signal. The detector catalytically reacts halogen-containing analytes with hydrogen (reductive mode) to produce strong acid by-products that are dissolved in a working fluid. The acids dissociate and the detector measures increased electrolytic conductivity. Other operating modes modify the chemistry for response to nitrogen- or sulphur-containing substances.

Electric sector — → Electrostatic energy analyzer.

Electron affinity, E_{EA} — Electron affinity of a species M is the minimum energy required for the process $M^{-\bullet} \rightarrow M + e^-$ where $M^{-\bullet}$ and M are in their ground rotational, vibrational and electronic states and the electron has zero translational energy.

Electron attachment ionization — Ionization of a gaseous atom or molecule by attachment of an electron to form $M^{-\bullet}$ ions.

Electron capture — The capture of thermal electrons by electronegative compounds. This process forms the basis of the ECD (electron capture detector) and is also made use of in negative chemical ionisation (NCI, ECD-MS).

Electron capture dissociation (ECD) — Process in which multiply protonated molecules interact with low energy electrons. Capture of the electron leads the liberation of energy and a reduction in charge state of the ion with the production of the $[M + nH]^{(n-1)+}$ odd electron ion, which readily fragments.

Electron energy	Magnitude of the electron charge multiplied by the potential difference through which electrons are accelerated, e.g. from a filament in order to effect electron ionization.
Electron impact ionisation	→ Electron ionization.
Electron ionization	Ionization of an atom or molecule by electrons that are typically accelerated to energies between 10 and 150 eV in order to remove one or more electrons from the molecule. The term electron impact is deprecated.
Electron volt, eV	Non-SI unit of energy (symbol eV) defined as the energy acquired by a particle containing one unit of charge through a potential difference of one volt. An electron volt is equal to $1.602\ 177\ 33(49) \times 10^{-19}$ J.
Electrospray ionization (ESI)	A process for LC/MS in which ionized species in the gas phase are produced from a solution via highly charged fine droplets, by means of spraying the solution from a narrow-bore needle tip at atmospheric pressure in the presence of a high electric field (1,000 to 10,000 V potential). Note: When a pressurized gas is used to aid in the formation of a stable spray, the term pneumatically-assisted electrospray ionization is used. The term ionspray is deprecated.
Electrostatic energy analyzer (ESA)	A device consisting of conducting parallel plates, concentric cylinders or concentric spheres that separates charged particles according to their ratio of translational energy to charge by means of a voltage difference applied between the pair.
Elution temperature	The temperature of the GC oven at which an analyte reaches the detector.
Energy	The SI unit is the Joule [J] 1 J = 1 Nm = 1 Ws In mass spectrometry the energy of ions is usually given in [eV] and is calculated from the elemental charge and the acceleration voltage in the ion source. Conversion factors: 1 eV = 23.0 kcal = 96.14 kJ (with 1 cal = 4.18 J)
E/2 mass spectrum	Mass spectrum obtained using a sector mass spectrometer in which the electric sector field E is set to half the value required to transmit the main ion-beam. This spectrum records the signal from doubly charged product ions of charge-stripping reactions.
EPA	The US Environmental Protection Agency.
ESI	Electrospray ionization, an LC/MS coupling technique, → Electrospray ionization.
ESTD	External standard, quantitation by using external standardization. The analyte itself is used for quantitative calibration as a clean standard or added to a blank standard matrix. The signal height for a known concentration of

the analyte is used for the calibration procedure. The calibration runs are carried out separately (externally) from the analysis of the sample.

Even-electron ion — An ion containing no unpaired electrons in its ground electronic state.

Exact mass — Calculated mass of an ion or molecule containing a single isotope of each atom, most frequently the lightest isotope of each element, calculated from the masses of these isotopes using an appropriate degree of accuracy.

External ionisation — Process for the production of ions for an ion storage mass spectrometer, e.g. ion trap MS. The ionisation does not take place inside of the ion trap analyser; instead, ions are formed in an attached ion source and transferred to the ion trap analyser. The special decoupling of ionisation and mass analysis in ion trap instruments allows the independent use of GC parameters, the application of negative chemical ionisation as well as several MS/MS scan techniques.

Extracted ion chromatogram — Chromatogram created by plotting the intensity of the signal observed at a chosen m/z value or series of values in a series of mass spectra recorded as a function of retention time, → Reconstructed ion chromatogram.

F

F_a — The column outlet flow-rate corrected to room temperature and pressure; for example, the flow-rate as measured by a flow meter. F_a can be calculated from the average carrier-gas linear velocity and the column dimensions.

F_s — The split-vent flow-rate, measured at room temperature and pressure.

FAME — Fatty acid methyl ester.

Faraday cup — A conducting cup or chamber that intercepts a charged particle beam and is electrically connected to a current measuring device.

Fast atom bombardment (FAB) — Ionization of any species by the interaction of a focused beam of neutral atoms having a translational energy of several thousand eV with a sample that is typically dissolved in a solvent matrix. Related term: secondary ionization.

Fast ion bombardment (FIB) — Ionization of any species by the interaction of a focused beam of ions having a translational energy of several thousand eV with a solid or liquid sample. For a liquid sample this is the same as liquid secondary ionization.

FC43 — Perfluorotributylamine (PFTBA), a widely used reference substance for calibration of the mass scale, M 671.

FFAP — Free fatty-acid phase.

FID — Flame ionization detector, providing a mass flow dependent signal, the detector ionizes most classes of organic compounds. FID is a universal detection technique. Little or no response have noble gases, CO, CO_2, O_2, N_2, H_2O, CS_2, NO_x, NH_3, per halogenated compounds, formic acid/aldehyde.

Field desorption (FD)	Formation of gas-phase ions in the presence of a high electric field from a material deposited on a solid surface. The term field desorption/ionization is deprecated.
Field-free region (FFR)	Section of a mass spectrometer in which there are no electric or magnetic fields.
Field ionization (FI)	Removal of electrons from any species, usually in the gas phase, by interaction with a high electric field.
Fixed product ion scan	In a sector instrument, either a high voltage scan or a linked scan at constant B^2/E. Both give a spectrum of all precursor ions that fragment to yield a preselected product ion. Note: The term "daughter ion" is deprecated.
Fortified sample	A sample enriched with a known amount of the analyte to be detected.
Forward library search	A procedure of comparing a mass spectrum of an unknown compound with a mass spectral library so that the unknown spectrum is compared in turn with the library spectra, considering only all the m/z peaks observed to have significant intensity in the unknown.
Fourier transform ion cyclotron resonance mass spectrometer (FT-ICR)	A mass spectrometer based on the principle of ion cyclotron resonance in which an ion in a magnetic field moves in a circular orbit at a frequency characteristic of its m/z value. Ions are coherently excited to a larger radius orbit using a pulse of radio frequency energy and their image charge is detected on receiver plates as a time domain signal. Fourier transformation of the time domain signal results in a frequency domain signal which is converted to a mass spectrum based in the inverse relationship between frequency and m/z.
FPD	Flame photometric detector, providing a mass flow dependent signal, the detector burns heteroatomic solutes in a hydrogen–air flame. The visible-range atomic emission spectrum is filtered through an interference filter and detected with a photomultiplier tube. Different interference filters can be selected for sulphur, tin or phosphorus emission lines. The flame photometric detector is sensitive and selective.
Fragment ion	A product ion that results from the dissociation of a precursor ion. Note: The term "daughter ion" is deprecated.
Fringe field	Electric or magnetic field that extends from the edge of a sector, lens or other ion optics element.
Frit sparger	U tube for the purge and trap analysis of water samples with built-in frit for fine dispersion of the purge gas.
Fritless sparger	U tube without a frit for the purge and trap analysis of moderately foaming water or solid samples.
FS	Fused silica.
FSOT	Fused-silica open-tubular column.
Full scan	Acquisition mode for the recording of complete mass spectra over a specified mass range.

FWHP — Full width at half peak, term used in the definition of peak resolution for the measurement of the peak width at half peak height.

G

GALP — Good automated laboratory practice.

Gas isotope ratio mass spectrometry — Common name for the area of isotope ratio mass spectrometry for the determination of the stable isotopes of H, N, C, O, S and Si. Compounds containing these elements can be quantitatively converted into simple gases for mass spectrometric analysis i. e. H_2, N_2, CO, CO_2, O_2, SO_2, SiF_4 fed by viscous flow or entrained into a continuous He flow into the ion source of a dedicated isotope ratio mass spectrometer.

GCB — Graphitised carbon black, → VORCARB (Supelco).

GC×GC — → Comprehensive GC.

Ghost peaks — Peaks not present in the original sample. Ghost peaks can be caused by septum bleed, analyte decomposition or carrier-gas contamination.

GIRMS — → Gas isotope ratio mass spectrometry.

Glass cap cross — → Werkhoff splitter.

GLC — Gas–liquid chromatography, this technique, analytes partition between a gaseous mobile phase and a liquid stationary phase. Selective interactions between the analytes and the liquid phase cause different retention times in the column.

GLP — Good laboratory practice.

GLPC — Gas–liquid phase chromatography, → Gas–liquid chromatography.

GPC — Gel permeation chromatography, gel chromatography used for sample preparation e. g. to remove lipids from fatty foodstuffs.

Gridless reflectron — A reflectron in which ions do not pass through grids in their deceleration and turn-around thereby avoiding ion loss due to collisions with the grid.

GSC — Gas–solid chromatography, this technique, analytes partition between a gaseous mobile phase and a solid stationary phase. Selective interactions between the analytes and the solid phase cause different retention times in the column.

H

H — Height equivalent to one theoretical plate. The distance along the column occupied by one theoretical plate; $H = L/N$.

H_{meas} — Height equivalent to one theoretical plate as measured from a chromatogram.

$$H_{meas} = \frac{L}{5.54\left(\frac{t_R}{w_h}\right)^2}$$

H_{min} — Minimum theoretical plate height at the optimum linear velocity, ignoring stationary-phase contributions to band broadening.
For open-tubular columns:

$$H_{min} = \left(\frac{d_c}{2}\right)\sqrt{\frac{1 + 6k + 11k^2}{3(1+k)^2}}$$

h_p — Peak amplitude.

H_{theor} — Theoretical plate height. For open-tubular columns (Golay equation):

$$H_{theor} = \left(\frac{2D_G}{\bar{u}}\right) + \bar{u}\left\{\left[\frac{(1 + 6k + 11k^2)}{96(1+k)^2}\right]\left(\frac{d_c^2}{D_G}\right) + \left[\frac{2k}{3(1+k)^2}\right]\left(\frac{d_f^2}{D_L}\right)\right\}$$

Hall detector — → ELCD.

Hard ionization — Formation of gas-phase ions accompanied by extensive fragmentation.

Headspace GC — Static headspace GC.

Headspace sweep — Technique in purge and trap analysis for treating foaming samples whereby the purge gas is passed only over the surface of the sample instead of through it.

Heartcut — GC/GC technique in which two or more partially resolved peaks that are eluted from one column are directed onto another column of different polarity or at a different temperature for improved resolution.

Heterolysis — → Heterolytic cleavage.

Heterolytic cleavage — Fragmentation of a molecule or ion in which both electrons forming the single bond that is broken remain on one of the atoms that were originally bonded, → Heterolysis.

HETP — Height equivalent to one theoretical plate; discontinued term for plate height (H). The dependence of the plate high value on the carrier gas velocity determined from the van Deemter curve is used to optimise chromatographic separation.

HFBA — Heptafluorobutyric anhydride, derivatisation agent for preparing volatile heptafluorobutyrates. It is frequently used for introducing halogens into compounds to increase the response in ECD or NCI.

High-energy collision-induced dissociation — Collision-induced dissociation process wherein the projectile ion has laboratory-frame translational energy higher than 1 keV.

High resolution MS — → HRMS.

Homolysis — → Homolytic cleavage.

Homolytic cleavage — Fragmentation of an ion or molecule in which the electrons forming the single bond that is broken are shared between the two atoms that were originally bonded. For an odd electron ion, fragmentation results from one of a pair of electrons that form a bond between two atoms moving to form a pair with the odd electron on the atom at the apparent charge site. Fragmentation results in the formation of an even electron ion and a radical. This reaction involves the movement of a single electron and is represented by a single-barbed arrow, → Homolysis.

HRGC	High resolution gas chromatography, the inherent meaning of this term is capillary gas chromatography in contrast to packed column chromatography.
HRMS	High resolution mass spectrometry, the separation of C, H, N, and O multiples in mass spectrometry. The empirical formula of an ion is usually obtained through accurate mass determination as part of the structure elucidation using HRMS. In GC/MS target compound analysis this term demands for a resolution power of better than 10,000 at 10% valley providing the accurate ion mass. Mass spectrometers providing resolution power of R > 10,000 (at 5% peak height) or R > 20,000 (at FWHP) are termed high resolution instruments in general.
HS	Headspace sampling: Gas-phase sampling technique in which the analyte is removed from an enclosed space above a solid or liquid sample.
HSGC	Headspace gas chromatography.
H-SRM	Highly resolved selected reaction monitoring, MS/MS target compound scan technique using an increased resolution at Cl1 for increased analyte selectivity.
HxCDD	Hexachlorodibenzodioxin isomers.
Hybrid mass spectrometer	A mass spectrometer that combines m/z analyzers of different types to perform tandem mass spectrometry.
Hydrogen/deuterium exchange	Exchange of hydrogen atoms with deuterium atoms in a molecule or preformed ion in solution prior to introduction into a mass spectrometer, or by reaction of an ion with a deuterated collision gas inside a mass spectrometer.

I

ICR-MS	Ion cyclotron resonance mass spectrometer, an ion storage mass spectrometer providing very high resolution power R > 100,000 and accurate mass measurements by measuring the cycle frequency of ions in a strong magnetic field followed by Fourier transformation.
INCOS	Integrated computer systems, the term refers to the search approach for mass spectra in former Finnigan GC/MS systems.
Indicator PCBs	→ PCB.
Inductive cleavage	A heterolytic cleavage of an ion. For an odd electron ion, inductive cleavage results from the pair of electrons that forms a bond to the atom at the apparent charge site moving to that atom while the charge site moves to the adjacent atom. The movement of the electron pair is represented by a double-barbed arrow.
In-house validation	→ Single laboratory study.

In-source collision-induced dissociation	The dissociation of an ion as a result of collisional excitation during ion transfer from an atmospheric pressure ion source and the mass spectrometer vacuum. This process is similar to ion desolvation but uses higher collision energy.
Interlaboratory study	The organisation, performance and evaluation of tests on the same sample by two or more laboratories in accordance with predetermined conditions to determine testing performance. According to the purpose the study can be classified as collaborative study or proficiency study.
Ion	An atomic, molecular or radical species with an unbalanced electrical charge. The corresponding neutral species need not be stable.
Ion desolvation	The removal of solvent molecules clustered around a gas-phase ion by means of heating and/or collisions with gas molecules.
Ionic dissociation	The dissociation of an ion into another ion of lower mass and one or more neutral species or ions with a lower charge.
Ion injection	The transfer of ions formed in an external ion source to the analyser of an ion storage mass spectrometer for subsequent mass analysis.
Ion/ion reaction	The reaction between two ions, typically of opposite polarity. The term ion-ion reaction is deprecated.
Ion mobility spectrometry (IMS)	Separation of ions according to their velocity through a buffer gas under the influence of an electric field.
Ion/molecule reaction	Reaction of an ion with a molecule. Note: the term ion-molecule reaction is deprecated because the hyphen suggests a single species that is both an ion and a molecule.
Ion/neutral complex	A particular type of transition state that lies between precursor and product ions on the reaction coordinate of some ion reactions.
Ion/neutral reaction	Reaction of an ion with an atom or molecule.
Ion/neutral exchange reaction	Reaction of an ion with a neutral species to produce a different neutral species as the product.
Ion-pair formation	Reaction of a molecule to form both positive ion and negative ion fragments among the products.
Ion source	Region in a mass spectrometer where ions are produced.
Ion storage MS	Mass spectrometer equipped with internal or external ion generation and ion collection. The collection and analysis of ions take place discontinuously e.g. ion traps (LRMS) and ICR MS (HRMS).
Ion-to-photon detector	Detector in which ions strike a conversion dynode to produce electrons that in turn strike a phosphor layer and the resulting photons are detected by a photomultiplier, → Daly detector.
Ion trap (IT)	Device for spatially confining ions using electric and magnetic fields alone or in combination.

Ionization cross section	A measure of the probability that a given ionization process will occur when an atom or molecule interacts with a photon, electron, atom or molecule.
Ionization efficiency	Ratio of the number of ions formed to the number of molecules consumed in the ion source.
Ionizing collision	Reaction of an ion with a neutral species in which one or more electrons are removed from either the ion or neutral.
IRMS	→ Isotope ratio mass spectrometry.
Isomers	Two substances with the same molecular sum formula, but different structural formula, or different spatial arrangement of the atoms (stereoisomers). Isomers differ chemically and physically. Frequently mass spectra of isomers cannot be differentiated, but because of different interactions with the stationary phase, they can be chromatographically separated.
Isotope	Although having the same nuclear charge (number of protons) most of the elements exist as atoms that have nuclides with varying numbers of neutrons, known as isotopes. They belong to the same chemical element but have different physical behaviour, e.g. masses. While in chemical synthesis the natural isotope composition is not taken into consideration (use of the average atomic weight), in mass spectrometry the distribution of the isotopes over the different masses is visible and the isotope pattern is assessed, e.g. dioxin analysis (molecular weight calculation based on the most common isotope). IRMS determines isotope ratios highly precise as a result of fractionation processes.
Isotope dilution mass spectrometry	A quantitative mass spectrometry technique based on the measurement of the isotopic abundance of a nuclide after isotope dilution with the test portion; an isotopically enriched compound is used as an internal standard.
Isotope pattern	Characteristic intensity pattern in a mass spectrum derived from the different abundance of the isotopes of an element. From the isotope pattern of an element in the mass spectrum conclusions can be drawn on the number of atoms of this element in a structure. Important isotope patterns in organic analysis are shown by Cl, Br, Si, C, and S. Organometallic compounds show characteristic patterns rich in lines. Molecular ions show the isotope pattern of all the elements in the formula.
Isotope ratio mass spectrometry	The measurement of the relative quantity of the different isotopes of an element in a material using a mass spectrometer.
Isotopologue ions	Ions that differ only in the isotopic composition of one or more of the constituent atoms. For example, $CH_4^{+\bullet}$ and $CH_3D^{+\bullet}$ or $^{10}BF_3$ and $^{11}BF_3$ or the ions forming an isotope cluster. The term isotopologue is a shortening of isotopic homologue.
Isotopomeric ions	Isomeric ions having the same numbers of each isotopic atom but differing in their positions. Isotopomeric ions can be either configurational isomers in which two atomic isotopes exchange positions or isotopic stereoisomers. The term isotopomer is a shortening of isotopic isomer.
ISTD	Internal standard.

ITD — Ion-trap detector: A mass spectrometric (MS) detector that uses an ion-trap device to generate mass spectra.

ITEX — In-tube extraction, automated dynamic headspace extraction technique using a packed syringe needle, injection by thermal desorption in a GC injector.

J

j — Mobile Phase Compressibility Correction Factor, a factor, applying to a homogeneously filled column of uniform diameter, that corrects for the compressibility of the mobile phase in the column, also called Compressibility Correction Factor. In liquid chromatography the compressibility of the mobile phase is negligible. In gas chromatography, the correction factor can be calculated as:

$$j = \frac{3}{2}\frac{p^2 - 1}{p^3 - 1} = \frac{3}{2}\frac{(p_i/p_o)^2 - 1}{(p_i/p_o)^3 - 1}$$

Jet separator — Interface construction for the coupling of mass spectrometers to wide bore and packed chromatography columns. The quantity of the lighter carrier gas is reduced by dispersion. Loss of analyte cannot be prevented.

K

k — Retention factor. A measurement of the retention of a peak; $k = (t_R - t_M)/t_M$

K — Partition coefficient. The relative concentration of an analyte in the mobile and stationary phases; $K = \beta k$.

Kovats index — The most used index system in gas chromatography for describing the retention behaviour of substances. The retention values are based on a standard mixture of alkanes: Alkane index = number of C-atoms × 100.

L

L — Column length.

Laboratory sample — A sample prepared for sending to a laboratory and intended for inspection or testing.

Laser ionization (LI) — Formation of ions through the interaction of photons from a laser with a material or with gas-phase ions or molecules.

Level of interest — The concentration of substance or analyte in a sample that is significant to determine its compliance with legislation.

Line spectrum — Representation of a mass spectrum as a series of vertical lines indicating the ion abundance intensities across the mass sale of m/z values.

Linear ion trap (LIT) — A two dimensional Paul ion trap in which ions are confined in the axial dimension by means of an electric field at the ends of the trap.

Linear range (LR) — Also called linear dynamic range. The range of analyte concentration or amount in which detector response per solute amount is constant within a specified percentage.

Linear velocity (u)	The speed at which the carrier gas moves through the column, usually expressed as the average carrier-gas linear velocity (u avg).
Linked scan	A scan in a tandem mass spectrometer with two or more m/z analyzers or in a sector mass spectrometer that incorporates at least one magnetic sector and one electric sector. Two or more of the analyzers are scanned simultaneously so as to preserve a predetermined relationship between scan parameters to produce a product ion, precursor ion or constant neutral loss or gain spectrum.
Liquid phase	In GC, a stationary liquid layer coated on the inner column wall (WCOT column) or on a support (packed, SCOT column) that selectively interacts with the analytes to produce different retention times.
LOD	Limit of detection, the lowest concentration of a substance that can still be detected unambiguously (assessed in the signal domain). The value is obtained from the decision limit (smallest detectable signal) using the calibration function or the distribution range of the blank value.
LOQ	Limit of quantitation, unlike the limit of detection, the limit of quantitation is confirmed by the calibration function. The value gives the lower limiting concentration which differs significantly from a blank value and can be determined unambiguously and quantitatively with a given precision. The LOQ value is therefore dependent on the largest statistical error which can be tolerated in the results.
Low-energy collision-induced dissociation	A collision-induced dissociation process wherein the precursor ion has translational energy lower than 1 keV. This process typically requires multiple collisions and the collisional excitation is cumulative.
LRMS	Low-resolution mass spectrometry, covers all MS analyzer technologies providing nominal mass resolution, in contrast to HRMS.
LVSI	Large volume splitless injection, using the concurrent solvent recondensation effect.

M

μ scan	The shortest scan unit in ion trap mass spectrometers, depending on the pre-selected scan rate of the chromatogram, several μ scans are accumulated to form the stored and displayed mass spectrum.
Magic 60	A term referring to the approx. 60 analytes of the combined volatile halogenated hydrocarbon/BTEX determination (VOC), which are analysed together in EPA methods e.g. by purge and trap GC/MS.
Magnetic sector	A device that produces a magnetic field perpendicular to a charged particle beam that deflects the beam to an extent that is proportional to the particle momentum per unit charge. For a monoenergetic beam, the deflection is proportional to m/z.
Magnetic sector MS	Single or double focussing mass spectrometer with magnetic (and electrostatic) analyser for the spatial separation of ions on individual flight paths and focussing on the exit slit at the detector, double focussing mass spectrometer are employed for high resolution MS, → also HRMS.

MAM	Monoacetylmorphine.
MAOT	Maximum allowable operating temperature, highest continuous column operating temperature that will not damage a column, if the carrier gas is free of oxygen and other contaminants. Slightly higher temperatures are permissible for short periods of time during column bakeouts.
Mass calibration	A means of determining m/z values in a scan from their times of detection relative to initiation of acquisition of a mass spectrum. Most commonly this is accomplished using a computer-based data system and a calibration file obtained from a mass spectrum of a compound that produces ions whose m/z values are known.
Mass defect	The difference between the mass number and the exact monoisotopic mass of a molecule or atom.
Mass excess	The negative of the mass defect.
Mass filter	A quadrupole analyser works as a mass filter. A mass spectrum is acquired by filtering out ions of individual m/z values from the large number of different ion species formed in the ions source. By cyclic changes in the control (scan) the whole mass range is covered (full scan). Switching the filter characteristics to pre-selected masses only ions of pre-selected m/z values are acquired (SIM, selected ion monitoring).
Mass gate	A set of plates or grid of wires in a time-of-flight mass spectrometer that is used to apply a pulsed electric field with the purpose of deflecting charged particles in a given m/z range.
Mass number	Sum of the number of protons and neutrons in an atom, molecule or ion, → Nucleon number.
Mass range	Range of m/z over which a mass spectrometer can detect ions or is operated to record a mass spectrum.
Mass resolution	Smallest mass difference Δm between two equal magnitude peaks so that the valley between them is a specified fraction of the peak height.
Mass resolving power	In a mass spectrum, the observed mass divided by the difference between two masses that can be separated: $m/\Delta m$. The procedure by which Δm was obtained and the mass at which the measurement was made should be reported.
Mass scale	A mass scale always implies the m/z scale (mass to charge value).
Mass selective axial ejection	Use of mass selective instability to eject ions of selected m/z values from an ion trap.
Mass selective instability	A phenomenon observed in a Paul ion trap whereby an appropriate combination of oscillating electric fields applied to the body and the end-caps of the trap leads to unstable trajectories for ions within a particular range of m/z values and thus to their ejection from the trap.
Mass spectral library	A collection of mass spectra of different compounds, usually represented as arrays of signal intensity vs. the m/z value rounded off to the integral

mass number. In some cases the library may consist of monoisotopic mass spectra.

Mass spectrometry/mass spectrometry (MS/MS) The acquisition and study of the spectra of the electrically charged products or precursors of m/z selected ion or ions, or of precursor ions of a selected neutral mass loss. MS/MS can be accomplished using beam instruments incorporating more than one analyzer (tandem mass spectrometry in space) or in trap instruments (tandem mass spectrometry in time). → Tandem mass spectrometry.

Mass spectrum A plot of the relative abundances of ions forming a beam or other collection as a function of the their m/z values.

Mass unit The SI unit for the atomic mass m is given in kg.
The additional unit used in chemistry is the atomic mass unit defined as

$1\ u = 1.660 \times 10^{-27}$ kg

In mass spectrometry the Dalton is also used as a mass unit, defined as 1/12 of the mass of the carbon isotope ^{12}C, → also Dalton.
In earlier literature mass units are given also in amu (atomic mass units), and the mmu (millimass unit) used for 1/1000 amu.

Mathieu stability diagram A graphical representation expressed in terms of reduced coordinates that describes the stability of charged particle motion in a quadrupole mass filter or quadrupole ion trap mass spectrometer, based on an appropriate form of the Mathieu differential equation.

Mattauch-Herzog geometry An arrangement for a double-focusing mass spectrometer in which a deflection of $n/(4\sqrt{(2)})$ radians in a radial electrostatic field is followed by a magnetic deflection of $n/2$ radians.

Maximising masses peak finder A routine method of data handling in GC/MS analysis. The change in each in individual ion intensity with time is analysed. If several ions have a common peak maximum at the same retention time, the elution of an individual substance is recognised and noted even in case of a co-elution. The peak finder is used for automatic analysis of complex chromatograms, → also AMDIS.

McLafferty rearrangement A dissociation reaction triggered by transfer of a hydrogen atom via a 6-member transition state to the formal radical/charge site from a carbon atom four atoms removed from the charge/radical site (the α-carbon); subsequent rearrangement of electron density leads to expulsion of an olefin molecule. This term was originally applied to ketone ions where the charge/radical site is the carbonyl oxygen, but it is now more widely applied.

MCP Microchannel plate detector, typically found as ion detector in TOF instruments, → Array detector.

MDL Method detection limit, the minimum amount of analyte that can be analysed within specified statistical limits of precision and accuracy, including sample preparation.

MDQ Minimum detectable quantity, the amount of analyte that produces a signal twofold that of the noise level.

MEPS	Micro-extraction by packed sorbent, miniaturized SPE sample preparation method using a packed syringe needle, GC injection via liquid desorption in the GC injector.
Merlin seal	Alternative injector septum solution "Merlin Microseal" using a septum less seal formed like a duck bill and sealed by the inside carrier pressure and spring load.
Membrane inlet (MI)	A semi-permeable membrane separator that permits the passage of analytes directly from solutions or ambient air to the mass spectrometer ion source.
Metastable ion	An ion that is formed with internal energy higher than the threshold for dissociation but with a lifetime great enough to allow it to exit the ion source and enter the mass analyzer where it dissociates before detection.
MHE	Multiple headspace extraction, a quantitation procedure used in static headspace involving multiple extraction and measurement from a single sample.
Microchannel plate (MCP)	A thin plate that contains a closely spaced array of channels that each act as a continuous dynode particle multiplier. A charged particle, fast neutral particle, or photon striking the plate causes a cascade of secondary electrons that ultimately exits the opposite side of the plate.
MID	Multiple ion detection, the recording of several individual ions (m/z values) for increased detection sensitivity in contrast to full scan, → also SIM.
Mixture search	Search mode of the PBM library search algorithm whereby a mixture is analysed by forming and searching difference spectra between sample and library.
MNT	Mononitrotoluene isomers.
Modifier	The addition of organic solvents in small amounts to the extraction agent in SFE. The modifier can be added directly to the sample (and is effective only in the static extraction step) or continuously by using a second pump. Typically up to 10% modifier is added to CO_2.
Molar mass	Mass of one mole of a compound: $6.022\,1415(10) \times 10^{23}$ atoms or molecules Note: The term molecular weight is deprecated because "weight" is the gravitational force on an object that varies with geographical location. Historically the term has been used to denote the molar mass calculated using isotope averaged atomic masses for the constituent elements.
Molecular ion	The nonfragmented ion formed by the removal of one or more electrons to form a positive ion or the addition of one or more electrons to form a negative ion.
Molecular sieve	A stationary phase that retains analytes by molecular size interactions.
Molecular weight	The sum of the atomic weights of all the atoms present in a molecule. The term molecular "weight" is commonly used although actually masses are involved. The average molecular weight is calculated taking the natural isotopic distribution of the elements into account (stoichiometry), → Molar mass.

Molecular weights in MS: The calculation is carried out exclusively using the atomic masses of the most frequently occurring isotope, → Nominal mass.

Monoisotopic mass	Exact mass of an ion or molecule calculated using the mass of the most abundant isotope of each element.
MQL	Minimum quantitation limit, → also LOQ (limit of quantitation).
MRM	Multiple reaction monitoring, MS/MS scan to monitor selected product ions only, essentially the same experiment as selected reaction monitoring, → SRM.
MRPL	Minimum required performance limit, means minimum concentration of an analyte in a sample, which at least has to be detected and confirmed. It is used to harmonise the analytical performance of methods for substances for which no permitted limited has been established.
MS/MS^n, MS^n	This symbol refers to multi-stage MS/MS experiments designed to record product ion spectra where n is the number of product ion stages (progeny ions). For ion traps, sequential MS/MS experiments can be undertaken where $n > 2$ whereas for a simple triple quadrupole system $n = 2$, → Multiple-stage mass spectrometry.
M-series	n-Alkyl-bis(trifluoromethyl)phosphine sulfides, a homologous series used to construct a retention time index system. The compounds can be used in gas chromatography with FID, ECD, ELCD, NPD, FPD, PID and MS detectors.
MSTFA	N-Methyl-N-trimethylsilyltrifluoroacetamide, a derivatisation agent for silylation, typically used e.g. in anabolic steroid derivatisation.
Multidimensional	Separations performed with two or more columns in which peaks are selectively directed onto or removed from at least one of the columns by a timed valve system, → Backflush, → Heartcut, → Pre-cut.
Multiphoton ionization (MPI)	Photoionization of an atom or molecule in which in two or more photons are absorbed.
Multiple-stage mass spectrometry	Multiple stages of precursor ion m/z selection followed by product ion detection for successive progeny ions.
m/z	Symbol m/z is used to denote the dimensionless quantity formed by dividing the mass of an ion in unified atomic mass units by its charge number (regardless of sign). The symbol is written in italicized lower case letters with no spaces. Note 1: The term mass-to-charge-ratio is deprecated. Mass-to-charge-ratio has been used for the abscissa of a mass spectrum, although the quantity measured is not the quotient of the ion's mass to its electric charge. The symbol m/z is recommended for the dimensionless quantity that is the independent variable in a mass spectrum. Note 2: The proposed unit Thomson (Th) is deprecated.

N

η	Viscosity. Carrier-gas viscosity increases with increasing temperature.
N	Number of theoretical plates: $N = 5.54\ (t_R/w_h)^2 \sim 16\ (t_R/w_b)^2$
N_{eff}	The number of effective plates. This term is an alternate measurement of theoretical plate height that compensates for the non-partitioning nature of an unretained peak. $N = 16\ (t'_R/w_b)^2$
N_{req}	The number of theoretical plates required to yield a particular resolution (R) at a specific peak separation (α) and retention factor (k): $N_{req} = 16R^2 \left(\frac{\alpha}{\alpha - 1}\right)^2 \left(\frac{k+1}{k}\right)^2$
Nafion dryer	Device for drying analytical gas streams using capillary membrane tubes of polar Nafion material with outer counter flow of dry carrier gas.
NB	Nitrobenzene.
NBS	The US National Bureau of Standards, now NIST.
ndl-PCB	Non dioxin-like PCBs, → PCB.
Needle sparger	Glass vessel used in purge and trap analysis of solids or foaming samples. The purge gas is passed into the sample via a needle perforated on the side or by means of a headspace sweep.
Negative ion	An atomic or molecular species having a net negative electric charge.
Negative ion chemical ionization (NCI, NICI)	Chemical ionization that results in the formation of negative ions.
Neutral loss	Loss of an uncharged species from an ion during either a rearrangement process or direct dissociation.
Neutral loss scan	MS/MS experiment for analysing precursor ions which undergo a common loss of neutral particles of the same mass. Analytes with functional groups in common are detected.
Nier-Johnson geometry	Arrangement for a double-focusing mass spectrometer in which a deflection of $n/2$ radians in a radial electrostatic field analyzer is followed by a magnetic deflection of $n/3$ radians.
NIH	US National Institute of Health.
NIST	US National Institute of Standards and Technology, part of the US Department of Commerce.
Nitrogen rule	An organic molecule containing the elements C, H, O, S, P, or a halogen has an odd nominal mass if it contains an odd number of nitrogen atoms.

Nominal mass — Mass of an ion or molecule calculated using the mass of the most abundant isotope of each element rounded to the nearest integer value and equivalent to the sum of the mass numbers of all constituent atoms.
According to this convention CH_4 and CD_4 have the same nominal mass!

Nominal mass resolution — The mass spectrometric resolution for the separation of mass signals with a uniform peak width of 1000 mu (1 mass unit) in the quadrupole or ion trap analyser, → LRMS. The resulting mass numbers are generally given as nominal mass numbers or to one decimal place (unlike high resolution data).

NPD — Nitrogen–phosphorus detection, the nitrogen–phosphorus detector, providing a mass flow dependent signal, catalytically ionizes nitrogen- or phosphorus-containing solutes on a heated rubidium or cesium surface in a reductive atmosphere. The nitrogen–phosphorus detector is highly selective and provides sensitivity that is better than that of a flame ionization detector.

NT — Nitrotoluene isomers.

Number of theoretical plates — Describes the separating capacity of a column.

O

OCDD — Octachlorodibenzodioxin.

OCI — On-column injection, sample enters the column directly from the syringe and does not contact other surfaces. On-column injection usually signifies cold injection for capillary columns.

Odd-electron ion — → Radical ion.

Odd-electron rule — Odd-electron ions may dissociate to form either odd or even-electron ions, whereas even-electron ions generally form even-electron fragment ions.

On-column — Sample injection technique in GC whereby a diluted liquid extract is injected directly (without evaporation) on to a pre-column or on to the separation column itself. The temperature of the injection site at the beginning of the column is controlled by the oven temperature.

Onium ion — A positively charged hypervalent ion of the nonmetallic elements. Examples are the methonium ion CH_5^+, the hydrogenonium ion H_3^+ and the hydronium ion H_3O^+. Other examples are the oxonium, sulfonium, nitronium, diazonium, phosphonium, and halonium ions.

Orbitrap — An ion trapping device that consists of an outer barrel-like electrode and a coaxial inner spindle-like electrode that form an electrostatic field with quadro-logarithmic potential distribution. The frequency of harmonic oscillations of the orbitally trapped ions along the axis of the electrostatic field is independent of the ion velocity and is inversely proportional to the square root of m/z so that the trap can be operated as a mass analyzer using image current detection and Fourier transformation of the time domain signal.

Orthogonal extraction — Pulsed acceleration of ions perpendicular to their direction of travel into a time-of-flight mass spectrometer. Ions may be extracted from a directional ion source, drift tube or m/z separation stage.

P

P — Relative pressure across the column; $P = p_i/p_o$.

Δp — Pressure drop across the column; $\Delta p = p_i - p_o$.

p_i — Absolute inlet pressure.

p_o — Absolute outlet pressure.

PAH — Polyaromatic hydrocarbons, the group of polycyclic hydrocarbons from naphthalene to and beyond coronene.

Partition coefficient — The partition coefficient k used in the headspace analysis technique is given by the partition of a compound between the liquid and the gaseous phases $c_{liqu.}/c_{gas}$.

PAT — Purge and trap technique, → P&T.

Paul ion trap — Ion trapping device that permits the ejection of ions with an m/z lower than a prescribed value and retention of those with higher mass. It depends on the application of radio frequency voltages between a ring electrode and two endcap electrodes to confine the ion motion to a cyclic path described by an appropriate form of the Mathieu equation. The choice of these voltages determines the m/z below which ions are ejected. The term cylindrical ion trap is deprecated.
The name was given to the device developed by Prof. Paul, University of Bonn, Germany, and known as the ion trap analyser. Prof. Paul received the Nobel price in 1989 for his work on the QUISTOR development at the beginning of the 1950s.

PBB — Polybrominated biphenyl.

PBDE — Polybrominated diphenylether.

PBM — Probability based match, library search comparison procedure for mass spectra developed by Prof. McLafferty.

PCB — Polychlorinated biphenyls with a total number of 209 congeners.
7 indicator PCBs are routinely monitored:
tetra Cl-PCB: 28, 52
penta Cl-PCB: 101, 118
hexa Cl-PCB: 138, 153
hepta Cl-PCB: 180
12 so-called dioxin-like PCBs or WHO-PCBs are the coplanar non-ortho substituted congeners of a total number of 68 coplanar congeners of which 20 in total are non-ortho substituted:
tetra Cl-PCB: 77, 81
penta Cl-PCB: 105, 114, 118, 123, 126
hexa Cl-PCB: 156, 157, 167, 169
hepta Cl-PCB: 189

PCI	Positive chemical ionisation.
PCN	Polychlorinated naphthalene.
PCXE	Polychlorinated xanthene.
PCXO	Polychlorinated xanthone.
Peak capacity	In quantitative GC: The amount of solute that can be injected without a significant loss of column efficiency. In multidimensional GC: The maximum number of peaks that can be resolved.
Peak (in mass spectrometry)	Localized region of relatively large ion signal in a mass spectrum. Although peaks are often associated with particular ions, the terms peak and ion should not be used interchangeably.
Peak intensity	Height or area of a peak in a mass spectrum.
Peak matching	Procedure for measuring the accurate mass of an ion using scanning mass spectrometers, in which the peak corresponding to the unknown ion and that for a reference ion of known m/z are displayed alternately on a display screen and caused to overlap by adjusting the acceleration voltage.
Peak overload	If too much of analyte is injected, its peak can be distorted into a triangular shape exceeding the column capacity.
PEEK	Polyether ether ketone, hard plasticizer-free polymer with the general structure (with $x = 2, Y = 1$): 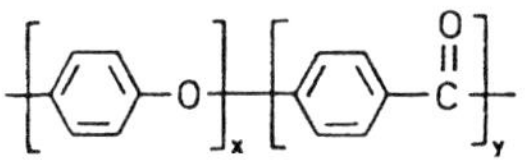 PEEK is used as a sealing and tubing material and for screw joints in HPLC, SFE, SFC and in high vacuum areas of MS, thermally stable up to 340 °C.
PEG	Polyethylene glycol.
Permitted limit	The maximum residue limit, maximum level or other maximum tolerance for substances established in legislation.
PFBA	Pentafluoropropionic anhydride, derivatisation agent, it is also used for introducing halogens to increase the response in ECD and NCI.
PFK	Perfluorokerosene, calibrant in mass spectrometry, widely used with magnetic sector MS instruments.
PFPD	Pulsed flame photometric detector, provides two simultaneous signals for S and P by measurement of a fluorescence/time profile in the range of 2–25 ms with ca. 5 Hz cycle time (after Prof. Aviv Amirav, University Tel Aviv, Israel).
PFTBA	Perfluorotributylamine, calibrant in mass spectrometry, widely used with quadrupole, ion trap and magnetic sector MS instruments, → FC43.
Phase ratio	In the headspace analysis technique the phase ratio of $V_{gas}/V_{liqu.}$ gives the degree of filling the headspace bottle.

In capillary GC the phase ratio describing the ratio of the internal volume of a column (volume of the mobile phase) to the volume of the stationary phase. High performance columns are characterised by high phase ratios. A table showing different combinations of internal diameters and film thicknesses can be used to optimise the choice of column with regard to analysis time, resolution and capacity.

Photodissociation — Process wherein the reactant ion is dissociated as a result of absorption of one or more photons.

Photoionization (PI) — Ionization of an atom or molecule by a photon, $M + h\nu \rightarrow M^{+\bullet} + e^-$. The term photon impact is deprecated.

PID — Photoionization detection, the photoionization detector ionizes analyte molecules with photons in the UV energy range, provides a concentration dependent signal. The photoionization detector is a selective detector that responds to aromatic compounds and olefins when operated in the 10.2 eV photon range, and it can respond to other materials with a more energetic light source.

PIONA — Paraffins, isoparaffins, olefins, napthenes and aromatic compounds.

PLOT — Porous-layer open-tubular column, a capillary column with a modified inner wall that has been etched or otherwise treated to increase the inner surface area or to provide gas–solid chromatographic retention behaviour.

PONA — Paraffins, olefins, napthenes and aromatic compounds.

Porous polymer — A stationary-phase material that retains analytes by selective adsorption or by molecular size interaction.

Positive ion — Atomic or molecular species having a net positive electric charge.

Post-acceleration detector — Detector in which a high voltage is applied after m/z separation to accelerate the ions and produce an improved signal, → Conversion dynode.

Post-source decay (PSD) — Technique specific to reflectron time-of-flight mass spectrometers where product ions of metastable transitions or collision-induced dissociations generated in the flight tube prior to entering the reflectron are m/z separated to yield product ion spectra.

PPINICI — Pulsed positive ion negative ion chemical ionisation, the alternating data acquisition of positive and negative ions formed during chemical ionisation, patented by former Finnigan Corp., San Jose, CA, USA.

Precision — The closeness of agreement between independent test results obtained under stipulated (predetermined) conditions. The measure of precision usually is expressed in terms of imprecision and computed as standard deviation of the test result. Less precision is determined by a larger standard deviation.

Precursor ion — Ion that reacts to form particular product ions. The reaction can be unimolecular dissociation, ion/molecule reaction, isomerization, or change in charge state. The term parent ion is deprecated.

Precursor ion scan
MS/MS scan function or process that records a precursor ion spectrum. It is used to detect substances with related structures which give common fragments. The term parent ion scan is deprecated.

Precursor ion spectrum
Mass spectrum recorded from any spectrometer in which the appropriate m/z separation function can be set to record the precursor ion or ions of selected product ions. The term parent ion spectrum is deprecated.

Pre-cut
Peaks at the beginning of a chromatogram are removed to vent or directed onto another column of different polarity or at a different temperature for improved resolution.

Pre-purge
A preliminary step in purge and trap analysis, the atmospheric oxygen is removed by the purge gas before the sample is heated to avoid side reactions.

Pre-scan
The step in the ion trap scan function before the analytical scan. During the pre-scan the variable ionisation time or ion collection time is adjusted to fill the trap to its optimum capacity with ions.

Pre-search
Part of library searching of mass spectra in which a small group of candidates is selected from the whole number for detailed comparison and ranking.

Press fit
Glass tube connectors for fused silica capillaries. The cross cut column end is simply pushed into the conical opening and seal is achieved with the external polyimide coating. Caution is necessary when applying high temperatures during oven ramping for weakening the polymer at the sealing site.

Pressure units
The SI unit is given in Pascal:

$1\ \text{Pa} = 1\ \text{N/m}^2$
$10^5\ \text{Pa} = 10^5\ \text{N/m}^2 = 1\ \text{bar}$

Pressure values are often given in traditional units. In MS vacuum technologies pressures are given in [Torr] or [mTorr] and gas pressures of GC supplies frequently in [kPa], [bar] or [psi].

Conversion table:

	Pa	bar	Torr	psi	at	atm
Pa	1	$1 \cdot 10^{-5}$	$7.5 \cdot 10^{-3}$	$1.45 \cdot 10^{-4}$	$1.02 \cdot 10^{-5}$	$9.87 \cdot 10^{-6}$
bar	**$1 \cdot 10^5$**	**1**	**750**	**14.514**	**1.02**	**0.987**
Torr	133	$1.33 \cdot 10^{-3}$	1	$1.94 \cdot 10^{-2}$	$1.36 \cdot 10^{-3}$	$1.32 \cdot 10^{-3}$
psi	$6.89 \cdot 10^3$	$6.89 \cdot 10^{-2}$	51.67	1	$7.03 \cdot 10^{-2}$	$6.80 \cdot 10^{-2}$
at	$9.81 \cdot 10^4$	0.981	736	14.224	1	0.968
atm	$1.0133 \cdot 10^5$	1.0133	760	14.706	1.033	1

at technical atmosphere 1 kp/cm^2
atm physical atmosphere 1.033 kp/cm^2
psi pound per square inch

Primary reaction
Conversion of the reagent gas used for CI into the reagent ions by electron ionisation, → also CI.

Principal ion — Most abundant ion of an isotope cluster, such as the $^{11}B^{79}Br_2{}^{81}Br^{+\bullet}$ ion of m/z 250 of the cluster of isotopologue molecular ions of BBr_3. The term principal ion has also been used to describe ions that have been artificially isotopically enriched in one or more positions such as $CH_3\text{-}^{13}CH_3^{+\bullet}$ or $CH_2D_2^{+\bullet}$, but those are best defined as isotopologue ions.

Product ion — An ion formed as the product of a reaction involving a particular precursor ion. The reaction can be unimolecular dissociation to form fragment ions, an ion/molecule reaction, or simply involve a change in the number of charges. The term fragment ion is deprecated. The term daughter ion is deprecated.

Product ion scan — Specific scan function or process that records a product ion spectrum. The terms fragment ion scan and daughter ion scan are deprecated.

Product ion spectrum — Mass spectrum recorded from any spectrometer in which the appropriate m/z separation scan function is set to record the product ion or ions of selected precursor ions. The terms fragment ion scan and daughter ion scan are deprecated. Note: The term MS/MS spectrum is deprecated; a scan-specific term, e.g. precursor ion spectrum or second-generation product ion spectrum should be used.

Proficiency study — Analysing the same sample, allowing laboratories to choose their own methods, provided these methods are used under routine conditions. The study has to be performed according to ISO guide 43-1 and 43-2 and can be used to assess the reproducibility of methods.

Proton abstraction — Type of NCI reaction, ionisation is effected by transfer of a proton from the analyte (abstraction) to the reagent ion, e.g. from analytes with phenolic OH groups.

Proton affinity — Proton affinity of a species M is the negative of the enthalpy change for the reaction $M + H^+ \rightarrow [M + H]^+$ at 298 K.

Protonated molecule — An ion formed by interaction of a molecule with a proton, and represented by the symbolism $[M + H]^+$. Note 1: The term protonated molecular ion is deprecated; this would correspond to a species carrying two charges Note 2: The terms pseudo-molecular ion and quasi-molecular ion are deprecated; a specific term such as protonated molecule, or a chemical description such as $[M+Na]^+$, $[M\text{-}H]^-$, etc. should be used.

Protonation — Type of PCI reaction, ionisation is effected by transfer of one ore more protons to the substance molecule. Protonating reagent gases include methane, methanol, water, isobutene and ammonia.

PTGC — Programmed-temperature GC, the column temperature changes in a controlled manner as peaks are eluted.

PTI — Programmed-temperature injection, a cold injection technique in which the inlet temperature is specifically programmed from the gas chromatograph.

PTV — Programmed-temperature vaporizer, an inlet system designed to perform programmed-temperature injection, a cold injection system for direct liquid injection for split or splitless injection, solvent split technique and cryo-enrichment.

P&T — Purge-and-trap sampling, dynamic headspace procedure, a concentration technique for volatile solutes. Sample is purged with an inert gas that entrains volatile components onto an adsorptive trap. The trap is then heated to desorb trapped components into a GC column.

Pure search — A mode of the PBM search procedure, which only uses the forward search capability of the library search, → also Forward library search.

PyGC — Pyrolysis GC, the sample is pyrolysed (decomposed) in the inlet before GC analysis.

PyMS — → Pyrolysis mass spectrometry.

Pyrogram — Chromatogram received from the separation of pyrolysis products.

Pyrolysis mass spectrometry (PyMS) — A mass spectrometry technique in which the sample is heated to the point of decomposition and the gaseous decomposition products are introduced into the ion source.

Q

QED — Quantification enhanced by data dependant MS/MS, a QED scan on a triple quadrupole instrument delivers an information rich product ion mass spectrum that can be used to confirm the existence of compounds while they are being quantified by an in-built MS/MS library.

QIT — Quadrupole ion trap, → Paul ion trap.

QMS — Quadrupole mass spectrometer, → Transmission quadrupole mass spectrometer.

QqQ — Triple quadrupole mass spectrometer. Note: The lower case q denotes the collision cell.

Quasimolecular ion — An ion to which the molecular mass is assigned, which is formed by, for example, chemical ionization as $(M+H)^+$, $(M+NH_4)^+$, $(M-H)^+$, $(M-H)^-$. The term "quasimolecular ion" is deprecated by the IUPAC to use "cationized molecule" instead.

QuEChERS — Quick Easy Cheap Effective Rugged Safe, acronym for a "Fast and easy multi-residue method employing acetonitrile extraction/partitioning" and "dispersive solid-phase extraction for the determination of pesticide residues in produce", → www.quechers.com, M. Anastassiades, S.J. Lehotay, D. Stajnbaher and F.J. Schenck, J AOAC Int 86, (2003) 412.

QUISTOR — Quadrupole ion storage trap, → Paul ion trap.

R

r — Relative retention. For peak *i* relative to standard peak *s*; $r = k_i/k_s$.

R — Resolution, the quality of separation of two peaks. In GC for two closely eluted peaks – $R = (t_{R,2} - t_{R,1})/w_{b,2}$ – where the subscripts 1 and 2 refer to the first and second peaks.

From N, k_2 and α

$$R = \left(\frac{\sqrt{N}}{4}\right)\left(\frac{\alpha - 1}{\alpha}\right)\left(\frac{k_2}{k_2 + 1}\right)$$

where k_2 is the retention factor of the second peak. A resolution of 1.5 is said to be baseline resolution. R incorporates both efficiency and separation.

Radical ion
: An ion, either a cation or anion, containing unpaired electrons in its ground state. The unpaired electron is denoted by a superscript dot alongside the superscript symbol for charge, such as for the molecular ion of a molecule M, that is, $M^{+\bullet}$. Radical ions with more than one charge and/or more than one unpaired electron are denoted such as $M^{(2+)(2\bullet)}$. Unless the positions of the unpaired electron and charge can be associated with specific atoms, superscript charge designation should be placed before the superscript dot designation.

Reagent gas cluster
: The spectrum of ions formed from the reagent gas through chemical ionisation, monitored to adjust the correct reagent gas pressure in the ion source.

Reagent ion
: An ion produced in large excess in a chemical ionization source that reacts with neutral sample molecules to produce an ionized form of the molecule through an ion/molecule reaction.

Reagent ion capture
: Type of CI reaction in PCI/NCI. The ionization of the analyte is achieved by an addition reaction of the reagent ion.

Recombination energy
: Energy released when an electron is added to an ionized molecule or atom, that is, the energy involved in the reverse process to that referred to in the definition of vertical ionization energy.

Recovery
: The percentage of the true concentration of a substance recovered during the analytical procedure. It is determined during validation, if no certified reference material is available.

Reference ion
: Stable ion whose structure is known with certainty. These ions are usually formed by direct ionization of a molecule of known structure, and are used to verify by comparison the structure of an unknown ion.

Reference material
: A material of which one or several properties have been confirmed by a validated method, so that it can be used to calibrate an apparatus or to verify a method of measurement.

Reflectron
: Constituent of a time-of-flight mass spectrometer that uses a static electric field to reverse the direction of travel of the ions entering it. A reflectron improves mass resolution by assuring that ions of the same m/z but different translational kinetic energy arrive at the detector at the same time.

Repeatability
: Precision under repeatability conditions, where independent test results are obtained with the same method on identical test items in the same laboratory by the same operator using the same equipment.

Reproducibility	Precision under reproducibility conditions, where test results are obtained with the same method on identical test items in different laboratories with different operators using different equipment.
RER	Reduced energy ramp, collision energy regime during the generation of a MS/MS product spectrum in which the applied collision energy is ramped down to produce a richer product ion spectrum for structure elucidation and confirmation, especially in the higher mass range part of the spectrum.
Residual gas analyzer (RGA)	Mass spectrometer used to measure the composition and pressure of gases in an evacuated chamber.
Resonance-enhanced multiphoton ionization (REMPI)	Multiphoton ionization in which the ionization cross section is significantly enhanced because the energy of the incident photons is resonant with an intermediate excited state of the neutral species.
Resonance ion ejection	Mode of ion ejection in a Paul ion trap that relies on an auxiliary radio frequency voltage that is applied to the endcap electrodes. The voltage is tuned to the secular frequency of a particular ion to eject it.
Resonance ionization (RI)	→ Resonance-enhanced multiphoton ionization.
Response	The specific height of the detector signal, usually calculated as the ratio of the peak area or height to the quantity of the analyte.
Retention gap	A short piece of deactivated but uncoated column as a pre-column or guard column placed between the inlet and the analytical column. There is no retention of the analytes. A retention gap often helps relieve solvent flooding. It also contains non-volatile sample contaminants from on-column injection. After evaporation of the solvent the analytes are focused at the beginning of the analytical column.
Retention index	A uniform system of retention classification according to a solute's relative location between a pair of homologous reference compounds on a specific column under specific conditions. It compares the time a compound is eluting from the column to the times of a set of standard compounds. The most common set of compounds used for retention indices are hydrocarbons typically from C5 to C30, separated in the order of boiling points.
Retention time	The time that the compound is held up on the GC column. Usually a retention index is used as compound specific rather than a retention time which is method dependant.
Retention volume	The carrier gas volume required to elute a component.
Reverse library search	A procedure of comparing a mass spectrum of an unknown compound with a mass spectral library so that the unknown spectrum is compared in turn with the library spectra, considering only the m/z peaks observed to have significant intensity in the current library spectrum. This comparison procedure serves to expose mixed spectra or to search for a substance in a GC/MS chromatogram.
RF	Response factor; Radio frequency.

RI — Retention Index, e. g. Kovats Index.

RIC — Reconstructed ion chromatogram. The total ion current (TIC) is calculated from the sum of the intensities of all of the acquired mass signals in a mass spectrum. This value is generally stored with the mass spectrum and shown as the RIC or TIC value in the mass spectrum.
Plotting the RIC or TIC value along the scan axis (retention time) gives the conventional chromatogram diagram (older magnetic sector instruments used a dedicated total ion current detector for this purpose). For this reason the shape of the chromatogram in GC/MS is dependent from the mass range scanned.

RSD — Relative standard deviation.

RT — Retention time.

Ruggedness — The susceptibility of an analytical method to changes in experimental conditions which can be expressed as a list of the sample materials, analytes, storage conditions, environmental and/or sample preparation conditions under which the method can be applied as presented or with specified minor modifications.

S

s — Split ratio. The ratio of the sample amount that is vented to the sample amount that enters the column during split injection. Higher split ratios place less sample on the column. s is usually measured as the ratio of total inlet flow to column flow; $s = (F_s + F_c)/F_c$.

Sandwich technique — An injection technique in which a sample plug is placed between two solvent plugs in the syringe to wash the syringe needle with solvent and obtain better sample transfer into the inlet, → also Solvent flush.

SBSE — Stir Bar Sorptive Extraction, a glass coated magnetic stir bar is coated with PDMS which serves as high capacity sorption phase, extraction by thermal desorption in a dedicated injector system.

Scan function — The control of ion trap or quadrupole mass analysers by changing the applied voltages in time, represented by a diagram of voltage $[U]$ against time t [ms].

SCD — Sulphur chemiluminescence detection, a sulphur chemiluminescence detector responds to sulphur-containing compounds by generating and measuring the light from chemiluminescence.

SCOT — Support-coated open-tubular column, a capillary column in which stationary phase is coated onto a support material that is distributed over the column inner wall. A SCOT column generally has a higher peak capacity than a WCOT column with the same average film thickness.

Screening method — Methods that are used to detect the presence of a substance or class of substances at the level of interest. These methods have the capability for a high sample throughput and are used to sift large numbers of samples for potential non-compliant results. They are specifically designed to avoid false compliant results.

Secondary reaction	The chemical reaction of analyte molecules with the reagent ions in a CI ion source to form stable product ions, → also CI.
Sector mass spectrometer	Mass spectrometer consisting of one or more magnetic sectors for m/z selection in a beam of ions. Such instruments may also have one or more electric sectors for energy dispersion.
Selected ion monitoring (SIM)	Operation of a mass spectrometer for target compound analysis in which the abundances of several ions of specific m/z values are recorded rather than the entire mass spectrum.
Selected reaction monitoring (SRM)	Data acquired from specific product ions corresponding to m/z selected precursor ions recorded via two or more stages of mass spectrometry. Selected reaction monitoring for target compound analysis can be preformed as tandem mass spectrometry in time or tandem mass spectrometry in space. The term multiple reaction monitoring is deprecated.
Selectivity	The fundamental ability of a stationary phase to retain substances selectively based upon their chemical characteristics, including vapour pressure and polarity. In mass spectrometry the ability to distinguish small mass differences, e.g. between target ions and matrix.
Selectivity tuning	Several techniques for adjusting the selectivity of separations that involve more than one column or stationary-phase type. Serially coupled columns and mixed-phase columns can be selectivity-tuned.
SEM	Secondary electron multiplier for high amplification factors, usually built from discrete dynodes which are electrically connected by resistors to provide a voltage ramp across the number of dynodes (contrary to a CDEM).
Sensitivity	The degree of detector response to a specified solute amount per unit time or per unit volume.
Separation α	The degree of separation of two peaks in time, → α and R.
Septum	Silicone or other elastomeric material that isolates inlet carrier flow from the atmosphere and permits syringe penetration for injection (alternative septum solution → Merlin seal)
Septum purge	The carrier gas swept across the septum face to a separate vent so that material emitted from the septum does not enter the column.
SFC	Supercritical fluid chromatography, uses a supercritical fluid as the mobile phase, can be coupled directly to mass spectrometry by using capillary columns.
SFE	Supercritical fluid extraction, as extraction medium mostly CO_2 is used, modifiers e.g. methanol or ethyl acetate can be added to optimize matrix specific extraction efficiencies.
Silica gel	A gas–solid adsorbent.
SIM	Selected ion monitoring, recording of individual pre-selected ion masses, as opposed to full scan, → also MID.

SIM descriptor — Data acquisition control file for SIM analyses, contains the selected specific masses of the analytes (m/z values), the individual dwell times and the retention times for timely switching the detection to other analytes in the same run.

SIMDIS — Simulated distillation, a boiling-point separation technique that simulates physical distillation of petroleum products.

Single laboratory study — An analytical study involving a single laboratory using one method to analyse the same or different test materials under different conditions over justified long time intervals (in-house validation).

SIS — Selected ion storage, describes the SIM measuring technique in ion trap MS, → SIM, → also Waveform ion isolation.

SISCOM — Search for identical and similar compounds, search procedure for mass spectra in libraries developed by Henneberg, Max Planck Institute for Coal Research, Mühlheim, Germany.

Skewing — Reversal of the relative intensities in a mass spectrum caused by changing the substance concentration during a scan. Skewing occurs with beam instruments during slow mass scans in the steep rising or falling slopes of GC peaks.

SN — Separation number or Trennzahl (TZ). A measurement of the number of peaks that could be placed with baseline resolution between two sequential peaks, z and $z + 1$, in a homologous series such as two hydrocarbons:

$$SN = \frac{t_{R(z+1)} - t_{R(z)}}{w_{h(z+1)} + w_{h(z)}} - 1$$

S/N — Signal-to-noise ratio, the ratio of the peak height to the noise level. Peak height is measured from the average noise level to peak top. Noise is measured as the width of the noise band (typically 4σ), excluding known signals.

Soft ionization — Formation of gas-phase ions without extensive fragmentation.

Solutes — Chemical substances that can be separated by chromatography.

Solvent effect — Focussing the analyte to a narrow band at the beginning of a capillary column by means of the condensation of the solvent and directed evaporation.

Solvent flooding — A source of peak-shape distortion caused by excessive solvent condensation inside the column during and after splitless or on-column injection.

Solvent flush — GC injection technique, also known as sandwich technique, whereby a liquid (solvent, derivatisation agent) is first drawn up into the syringe, followed by some air to act as a barrier, and finally the sample.

Solvent flushing — A column rinsing technique that can remove non-volatile sample residue and partially restore column performance.

Solvent split — GC injection technique using the PTV injector so that larger quantities of diluted extracts can be applied from more than 2 µL up to a LC-GC coupling.

Space charge effect	Result of mutual repulsion of particles of like charge that limits the current in a charged-particle beam or packet and causes some ion motion in addition to that caused by external fields.
SPE	Solid-phase extraction, a sample clean-up technique.
Specificity	The ability of a method to distinguish between a particular analyte being measured and other substances. This characteristic is predominantly a function of the measuring technique, but can vary according to class of compound or matrix.
SPI	Septum equipped programmable injector, special design of a cold injection system which can be used exclusively for total sample transfer.
Split injection	The sample size is adjusted to suit capillary column requirements by splitting off a major fraction of sample vapours in the inlet so that as little as 0.1% enters the column. The rest is vented.
Splitless injection	A derivative of split injection. During the first 0.5–4 min of sampling, the sample is not split and enters only the column. Splitting is restored afterward to purge the sample remaining in the inlet. As much as 99% of the sample enters the column.
SPME	Solid-phase micro extraction, a sample clean-up technique that uses a removable sorptive micro extraction device.
Spray and trap	Extraction procedure for foaming aqueous liquids, the liquid stream is sprayed into a purge gas stream.
SRM	Selected reaction monitoring, MS/MS scan technique, whereby a product ion spectrum is produced by collision induced decomposition of a selected precursor ion, in which only one or more selected fragment ions are detected. In MS/MS instruments the SRM technique provides the highest possible selectivity together with the highest possible sensitivity for target compound quantitation.
SSL	Split-/splitless injector.
Stable ion	Ion with internal energy sufficiently low that it does not rearrange or dissociate prior to detection in a mass spectrometer.
Standard addition method	A procedure in which the test sample is divided in two (or more) test portions. One portion is analysed as such and known amounts of the standard analyte are added to the other test portions before analysis. The amount of the standard analyte added has to be between two and five times the estimated amount of the analyte in the sample. This procedure is designed to determine the content of an analyte in a sample, taking account of the recovery of the analytical procedure. Often used quantitative calibration method in headspace analysis for matrix independent calibrations.
Static field	Electric or magnetic field that does not change in time.
Stationary phase	Liquid or solid materials coated inside a column that selectively retain analytes.

SUMMA canister — Passivated stainless steel canisters for air analysis (e.g. EPA method TO 14/15) for collection of samples and standardisation. The inner surface is deactivated by a patented procedure involving a Cr/Ni oxide layer.

Surrogate standard — The internal standard for quantification, which is added to a sample extract before analysis. The ratio of the surrogate standard to other internal standards added during the course of the clean-up is used to calculate the recovery.

SWIFT — Stored waveform inverse Fourier transformation, technique to create excitation waveforms for ions in FT-ICR mass spectrometer or Paul ion traps. An excitation waveform in the time-domain is generated by taking the inverse Fourier transform of an appropriate frequency domain programmed excitation spectrum, in which the resonance frequencies of ions to be excited are included. This procedure may be used for selection of precursor ions in MS/MS experiments.

T

T_o — Room temperature.

T_c — Column temperature.

t_M — Unretained peak retention time. The time required for one column volume (V_G) of carrier gas to pass through a column.

t_R — Retention time. The time required for a peak to pass through a column.

t'_R — Adjusted retention time; $t'_R = t_R - t_M$.

Tandem-in-space — MS/MS analysis with beam mass spectrometers. Ion formation, selection, collision induced decomposition, and acquisition of the product ion spectrum take place continuously in separate sections of the mass spectrometer, e.g. triple quadrupole MS.

Tandem-in-time — MS/MS analysis with ion storage mass spectrometers. Ion formation, selection, collision induced decomposition, and acquisition of the product ion spectrum take place in the same location of the mass spectrometer but sequentially in time, e.g. ion trap MS, ICR MS.

Tandem mass spectrometry — → Mass spectrometry/mass spectrometry (MS/MS).

Target compound — Analyte to be quantitatively determined, usually taken from a list of compounds taken from regulations or directives.

TCA — Target compound analysis, multicomponent analysis, multimethod. The analytical strategy with the setup and data evaluation for the quantitative monitoring of a selected group of target compounds. For data evaluation the analytes are grouped together in a target compound list for analysis and are searched for using automated routines by spectrum comparison in a retention time window, identified and quantified when found.

TCD — Thermal-conductivity detection, a thermal conductivity detector measures the differential thermal conductivity of carrier- and reference-gas flows. Solutes emerging from a column change the carrier-gas thermal conductivity and produce a response. TCD is a universal detection technique with moderate sensitivity depending on the thermal conductivity of the analytes.

Analytes	Thermal conductivity [10^5 cal/cm · s · °C]
H_2	49.9
He	39.9
N_2	7.2
Ethane	7.7
H_2O	5.5
Benzene	4.1
Acetone	4.0
CH_3Cl	2.3

TCDD
Tetrachlorodibenzo-p-dioxin isomers.

TCDF
Tetrachlorodibenzofuran isomers.

TCEP
Tris(cyanoethoxy)propane.

Tenax
Nonpolar synthetic polymer based on 2,6-diphenyl-p-phenyleneoxide used as an adsorption material for the concentration of air samples or in purge and trap analysis.

TE-GC/MS
Thermal extraction GC/MS using a thermal extraction unit for solid sample material as inlet system of the gas chromatograph.

Theoretical plate
A hypothetical entity inside a column that exists by analogy to a multiple-plate distillation column. As solutes migrate through a column, they partition between the stationary phase and the carrier gas. Although this process is continuous, chromatographers often visualize a step-wise model. One step corresponds approximately to a theoretical plate.

Thermal ionization (TI)
Ionization of a neutral species through contact with a high temperature surface.

TIC
Total ion current, sum of all acquired intensities in a mass spectrum, → also RIC.

Time lag focusing
Energy focusing in a time-of-flight mass spectrometer that is accomplished by introducing a time delay between the formation of the ions and the application of the accelerating voltage pulse. Ion formation may be in the gas phase or at a sample surface. Related term: delayed extraction.

Time-of-flight mass spectrometer (TOF-MS)
Instrument that separates ions by m/z in a field-free region after acceleration to a fixed acceleration energy.

TMCS
Trimethychlorosilane, trimethylsilyl chloride, silylation agent (catalyst).

TMS derivative
Trimethylsilyl derivative, a chemical derivative to increase substance volatility to facilitate chromatographic application, typical diagnostic mass m/z 73.

TMSH	Trimethylsulfonium hydroxide, derivatisation agent for esterification and methylation, e.g. free fatty acids, phenylureas, etc. TMSH was used successfully to save on extra steps for the derivatisation in the insert of the PTV cold injection system.
TNT	Trinitrotoluene.
Total ion current (TIC)	Sum of all the separate ion currents carried by the ions of different m/z contributing to a complete mass spectrum or in a specified m/z range of a mass spectrum.
Total ion current chromatogram	Chromatogram obtained by plotting the total ion current detected in each of a series of mass spectra covering the acquired mass range recorded as a function of retention time.
Total sample transfer	GC injection technique whereby the entire injection volume reaches the column with or without evaporation, splitless injection.
TPH	Total petroleum hydrocarbons.
Transmission	The ratio of the number of ions leaving a region of a mass spectrometer to the number entering that region.
Transmission quadrupole mass spectrometer	A mass spectrometer that consists of four parallel rods whose centers form the corners of a square and whose opposing poles are connected. The voltage applied to the rods is a superposition of a static potential and a sinusoidal radio frequency potential. The motion of an ion in the x and y dimensions is described by the Mathieu equation whose solutions show that ions in a particular m/z range can be transmitted by oscillation along the z-axis.
Triple quadrupole mass spectrometer	A tandem mass spectrometer comprising two transmission quadrupole mass spectrometers in series, with a nonresolving (RF-only) quadrupole between them to act as a collision cell.
Tropylium ion	The characteristic ion in the spectrum of alkylaromatics, $C_7H_7^+$ m/z 91, is formed by benzyl cleavage of the longest alky chain. The structure of the seven-membered ring (after internal rearrangement) is responsible for the high stability of the ion.
TRT	Temperature rise time, heating up period in pyrolysis until the set pyrolysis temperature is reached. A fast TRT is a quality feature of the pyrolyser.
TSD	Thermionic-specific detection, → NPD (nitrogen–phosphorus detection).
TSIM	N-Trimethysilyl-imidazole, a potent derivatisation agent for silylation.
TSP	Thermospray, LC/MS coupling device.
TZ	Trennzahl, → Separation number.

U

$\bar{u}_{avrg}$ — Average linear carrier-gas velocity; $\bar{u}_{avrg} = L/t_M$.

u_o — Carrier-gas velocity at the column outlet; $u_o = \bar{u}_{avrg}/j$.

u_{opt} — Optimum linear gas velocity. The carrier-gas velocity corresponding to the $\bar{u}$ minimum theoretical plate height, ignoring stationary-phase contributions to band broadening:

$$u_{opt} = 8\left(\frac{D_G}{d_c}\right)\sqrt{\frac{3\,(1+k)^2}{1+6\,k+11\,k^2}}$$

u — The unified atomic mass unit u, a non-SI unit of mass defined as one twelfth of the mass of one atom of ^{12}C in its ground state and equal to $1.6605402(10) \times 10^{-27}$ kg. The term atomic mass unit is deprecated. Note: The term atomic mass unit (amu) is ambiguous as it has been used to denote atomic masses measured relative to a single atom of ^{16}O, or to the isotope-averaged mass of an oxygen atom, or to a single atom of ^{12}C.

UAR — Unknown analytical response, this term refers to GC peaks in a multicomponent target compound analysis (TCA), which are not included in the list of target compounds.

Unimolecular dissociation — Fragmentation reaction in which the molecularity is treated as one, irrespective of whether the dissociative state is that of a metastable ion produced in the ion source or results from collisional excitation of a stable ion.

Unstable ion — Ion with sufficient energy to dissociate within the ion source.

UTE, UTE% — Utilization of theoretical efficiency, → CE (coating efficiency).

V

V_G — The volume of carrier gas contained in a column. For open tubular columns and ignoring the stationary-phase film thickness: $(d_f), V_G = L(\,d_c^2/4)$.

V_L — The volume of (liquid) stationary phase contained in a column.

Validation — The confirmation by examination and the provision of effective evidence that the particular requirements of a specific intended use are fulfilled.

van Deempter plot — → HETP.

Viton — The brand name of synthetic rubber and fluoropolymer elastomer commonly used in O-rings and other moulded or extruded goods. The name is trademarked by *DuPont Performance Elastomers L.L.C., → http://en.wikipedia.org/wiki/ Viton.

VOC — Volatile organic compounds, a group of analytes which are determined together using an EPA method and which contain both volatile halogenated hydrocarbons and BTEX.

VOCARB	Nonpolar adsorbent filling for concentration of air samples, and in purge and trap analysis, multilayer filling based on graphitized carbon black (Carboxen) and carbon molecular sieves (Carbosieve), Supelco.
VOCOL	Volatile organic compounds column, Sigma-Aldrich.

W

w_b	The peak width at its base, measured in seconds. For a Gaussian peak, $w_b = 1.596\,(A_p/h_p)$.
w_h	The peak width at half height, measured in seconds. For a Gaussian peak, $w_h = 0.940\,(A_p/h_p)$.
Waveform ion isolation	Ion storage technique for the ion trap analyser. By using resonance frequencies the ion trap can exclude ions of several m/z values from storage (e.g. from the matrix) and collect selectively determined pre-selected analyte ions, also SIS.
WBOT	Wide-bore open-tubular column, open-tubular (capillary) column with a nominal inner diameter of 530 µm.
WCOT	Wall-coated open-tubular column, a capillary column in which the stationary phase is coated directly on the column wall.
Werkhoff splitter	Adjustable flow splitter for sample injection into two capillary columns, or a split located after the analytical column, also know as the "glass cap cross divider".
Wiswesser line notation	Code for chemical structures using an alphanumerical system, e.g. for Lindane, L6TJ AG BG CG DG EG FG *GAMMA, contained in the Wiley library of mass spectra.
Within-laboratory reproducibility	Precision obtained in the same laboratory under stipulated (predetermined) conditions (concerning e.g. method, test materials, operators, environment) over justified long time intervals.

X

XAD	Synthetic polymer (resin) used as an adsorption material in air or purge and trap analysis as well as in the clean-up for chlorinated compounds e.g. dioxins, furans, PCBs.
x-ion	Fragment ion containing the C-terminus formed upon dissociation of a protonated peptide at a backbone C–C bond.

Y

y-ion	Fragment ion containing the C-terminus formed upon dissociation of a protonated peptide at a backbone C–N bond.

Z

z-ion	Fragment ion containing the C-terminus formed upon dissociation of a protonated peptide at a backbone N–C bond.

References

1. Hinshaw, J.V., A Compendium of GC Terms and Techniques, LC•GC, 10(7) (1992) 516–522.
2. Chromatography Topics, Library4Science, http://www.chromatography-online.org/topics/
3. IUPAC Nomenclature for Chromatography, Pure Appl. Chem., 65 (1993) 819–872.
4. IUPAC Task Group MS Terms, see at http://www.msterms.com: Murray, K.K., Boyd, R.K., Eberlin, M.N., Langeley, G.J., Li, L., Naito, Y., Standard Definitions of Terms Relating to Mass Spectrometry, International Union of Pure and Applied Chemistry Analytical Chemistry Division, 3rd Draft Document, August 2006.
5. Mass Spectrometry Wiki, Articles on Mass Spectrometry Based on the Wikipedia Mass Spectrometry Category, see: http://mass-spec.lsu.edu/mswiki/index.php/Main_Page

Subject Index

Handbook of GC/MS: Fundamentals and Applications, Second Edition. Hans-Joachim Hübschmann

ISBN: 978-3-527-31427-0

e

f